Springer Proceedings in Mathematics & Statistics

Volume 73

For further volumes:
http://www.springer.com/series/10533

Springer Proceedings in Mathematics & Statistics

This book series features volumes composed of selected contributions from workshops and conferences in all areas of current research in mathematics and statistics, including OR and optimization. In addition to an overall evaluation of the interest, scientific quality, and timeliness of each proposal at the hands of the publisher, individual contributions are all refereed to the high quality standards of leading journals in the feld. Thus, this series provides the research community with well-edited, authoritative reports on developments in the most exciting areas of mathematical and statistical research today.

Alberto Adrego Pinto • David Zilberman
Editors

Modeling, Dynamics, Optimization and Bioeconomics I

Contributions from ICMOD 2010
and the 5th Bioeconomy Conference 2012

 Springer

Editors

Alberto Adrego Pinto
University of Porto
Department of Mathematics
Porto
Portugal

David Zilberman
University of California
Department of Agricultural and Resource
 Economics
Berkeley, CA
USA

ISSN 2194-1009   ISSN 2194-1017 (electronic)
ISBN 978-3-319-04848-2   ISBN 978-3-319-04849-9 (eBook)
DOI 10.1007/978-3-319-04849-9
Springer Cham Heidelberg New York Dordrecht London

Library of Congress Control Number: 2014941096

Mathematics Subject Classification (2010): 37-XX, 49-XX, 91-XX, 58-XX, 60-XX, 62-XX, 97M10, 97M40

Printed on acid-free paper

Springer is part of Springer Science+Business Media (www.springer.com)

To our families

Alberto Adrego Pinto

I first met Alberto Pinto when he was taking a master's course at the University of Warwick. He was very enthusiastic about the subject of my lectures that were about the boundary between order and chaos in dynamical systems, particularly those, like period doubling, that could be analysed using renormalisation. He worked with me on his master's thesis that studied the work of Feigenbaum and Sullivan on scaling functions, and he went on to a PhD on the universality features of other classes of maps that form the boundary between order and chaos.

During this time he met a number of leaders in dynamical systems, notably Dennis Sullivan and Mauricio Peixoto, and this had a great impact on his career. As a result he and his collaborators have made many important contributions to the study of the fine-scale structure of dynamical systems, and this has appeared in leading journals and in his book *Fine Structures of Hyperbolic Diffeomorphisms* co-authored with Flávio Ferreira and myself.

I would like to pick out his important work with Welington de Melo. While doing postdoc with Dennis Sullivan at the Graduate Center at City University of New York he met Edson de Faria and through Mauricio Peixoto he got in contact with Welington de Melo. With de Melo he proved the rigidity of smooth unimodal maps in the boundary between chaos and order extending the work of McMullen. Furthermore, de Faria, de Melo and he proved the conjecture raised in 1978 in the work of Feigenbaum and Coullet-Tresser which characterises the period-doubling boundary between chaos and order for unimodal maps. This appeared in the *Annals of Mathematics* and was based in particular on the previous works of Sandy Davie, Dennis Sullivan, Curtis McMullen and Mikhail Lyubich.

Since then Alberto has branched out into more applied areas. He has contributed across a remarkably broad area of science including optics, game theory and mathematical economics, finance, immunology, epidemiology, and climate and energy. In these applied areas he has published widely and edited two books.

With Michel Benaim he founded the *Journal of Dynamics and Games* of the *American Institute of Mathematical Sciences* (AIMS) and they are the editors in chief. He has also increasingly taken on important administrative tasks. For example, he is currently the President of International Center for Mathematics

(CIM), Portugal and has started the CIM Mathematical Sciences Series to be published by Springer-Verlag.

I was very lucky to have had Alberto as a student and I have greatly enjoyed collaborating with him. He is a deep thinker and extremely focused and determined and working with him has been always fun.

Coventry, United Kingdom   David Rand
December 2013

David Zilberman

David Zilberman is a Professor and holds the Robinson Chair in the Department of Agricultural and Resource Economics at U.C. Berkeley. David has made major contributions to several major areas of research in agricultural and environmental economics. In water economics, he introduced a framework to evaluate and estimate adoption of advanced irrigation technology and analyze the performance of alternative water allocations, including water rights vs. water trading. He also introduced the damage control model to assess pesticides productivity and a framework to consider existing policies in designed new regulations.

His work provided empirical evidence of the gains associated with adoption of Genetic modified varieties, and he introduced a clearinghouse for intellectual property to enable development of technologies for the poor. He was among the first to identify the trade-offs between food and the fuel associated with introduction of biofuels and developed quantitative methods to assess the economic and environmental impacts of biofuel. David also made a major contribution to the economics of payment for ecosystem services, adoption, and risk.

David is a Fellow of the American Agricultural Economics Association (AAEA) and the Association of Environmental and Resource economics (AERE) and is the recipient of the 2000 Cannes Water and the Economy Award. He won the AAEA 2002 and 2007 Quality of Research Discovery Award and the 2005 and 2009 AAEA Publication of Enduring quality award. He edited 17 books and coauthored 280 papers in refereed journals. David received his B. A. in Economics and Statistics at Tel Aviv University, Israel, and his Ph.D. at the University of California, Berkeley, in 1979. He has served as a consultant to the World Bank, FAO, USDA, EPA, and CDFA. He served as Department Chair from 1994 to 1999 and was on the boards of the AAEA and C-FARE and on three NRC panels.

Preface

The aim of the project of this book "Modeling, Dynamics, Optimization and Bioeconomics I" is the exploration of emerging and current cutting-edge theories and methods of modeling, optimization, dynamics and bioeconomy. The theories and techniques presented here originated from dynamics, statistics, control theory, computer science and informatics and are applied to novel and innovative real-world applications. During the past decades, the use of dynamic systems, control theory, computing, data mining, machine learning and simulation has gained the attention of numerous researchers from all over the world. Smart or intelligent algorithms are often called heuristics and model-free. They are usually less firm in mathematical rigor but liberated from the strictness of calculus in order to integrate nature-inspired and, especially, bio-inspired approaches to solve hard problems efficiently. Herewith, these intelligent algorithms have evolved in parallel with the development of model-based, mathematical models.

Admirable scientific projects using both model-free and model-based methods coevolved today at research centers and are introduced in conferences around the world, yielding new scientific advances and contributing to the solution of important real-world problems. One important area of progress is the bioeconomy—where advances in life sciences are used to produce new products in a sustainable and clean manner. In this book, scientists from all over the world share their latest insights and important results in the field.

We are very thankful to the editors, Alberto Adrego Pinto and David Zilberman, for having given these experts the opportunity and honor of publishing their contributions. We express our gratitude to them for having prepared a premium work of a remarkable scientific and social value!

Ankara, Turkey   Gerhard-Wilhelm Weber
December 2013

Acknowledgements

We thank the authors of the chapters for having shared their vision with us in this book and we thank the anonymous referees.

We are grateful to Gerhard-Wilhelm Weber for contributing the preface of the book.

We thank David Rand for contributing the foreword to the book.

We thank the Executive Editor for Mathematics, Computational Science and Engineering at Springer-Verlag, Martin Peters, for his invaluable suggestions and advice throughout this project.

We thank João Paulo Almeida, João Passos Coelho, Ricardo Cruz, Helena Ferreira, Miguel Ferreira, Alan John Guimarães, Susan Jenkins, José Tenreiro Machado, Filipe Martins, José Martins, Abdelrahim Mousa, Bruno Oliveira, Telmo Parreira, Diogo Pinheiro and Renato Soeiro for their invaluable help in assembling this volume and for their editorial assistance.

The majority of the contributed papers for this volume come from the participants of the International Conference on Modeling, Optimization and Dynamics— ICMOD 2010,[1] a satellite conference of EURO XXIV Lisbon 2010, that took place at Faculty of Sciences of University of Porto, Portugal, and from the Berkeley Bioeconomy Conference 2012[2] at University of California, Berkeley, USA. We thank all authors for their contribution to this volume.

---

[1] https://sites.google.com/site/workshopeuro2010/.

[2] http://www.berkeleybioeconomy.com/conference/2012-conference/.

Contents

Chapter 1
Dynamic Management of Fossil Fuel, Biofuel, and Solar Energy

Scott Kaplan, Charles Séguin, Karl W. Steininger, and David Zilberman

1.1 Introduction

Cheap energy has been key to the modern economy. The use of electricity and the internal combustion engine have been crucial for current patterns of civilization, and reduction in availability or increases in the cost of energy have serious consequences for current activities of society. Concerns about climate change, depletion of fossil fuel, and exchange rates are causing societies to transition from fossil to renewable fuel. The transition is challenging both in terms of modeling and policy design. There is an emerging portfolio of alternative technologies, but the extent and order in which they will be introduced is uncertain and presents a major challenge. While there are many renewable technologies that present alternatives to fossil fuel, we will concentrate on two here, biofuel and solar energy, and try to develop basic economic understanding of the forces that lead to a transition to them. Our approach will be to move from a simple modeling effort to a more complex one in order to understand some of the considerations that affect the process of technological transition in energy.

S. Kaplan • D. Zilberman (✉)
Department of Agricultural and Resource Economics at the University of California, Berkeley, CA, USA
e-mail: scottkaplan@berkeley.edu; zilber11@berkeley.edu

C. Séguin
Department of Economics, University of Quebec in Montreal (UQAM), Montreal, QC, Canada
e-mail: seguin.charles@uqam.ca

K.W. Steininger
Department of Economics and Wegener Center for Climate and Global Change, University of Graz, Graz, Austria
e-mail: karl.steininger@uni-graz.at

A.A. Pinto and D. Zilberman (eds.), *Modeling, Dynamics, Optimization and Bioeconomics I*, Springer Proceedings in Mathematics & Statistics 73, DOI 10.1007/978-3-319-04849-9_1,
© Springer International Publishing Switzerland 2014

1.2 The Model

We consider a model where there is demand for energy, derived from the benefit from energy consumption denoted by $B(t, Y_t)$ where t represents the time period and Y_t the energy consumption in period t. We assume that $\frac{\partial B}{\partial Y_t} > 0$ and $\frac{\partial^2 B}{\partial Y_t^2} < 0$ (i.e. the function is increasing in and has decreasing marginal utility of energy use). We will assume that benefit is increasing over time because of increases in income and population growth, and declines over time because of technological change that includes increases in energy use efficiency. If the population growth and income effects are greater than the energy use efficiency effect, then $\frac{\partial B}{\partial t} > 0$, but if the energy use efficiency effect is dominant, then $\frac{\partial B}{\partial t} < 0$. Energy is generated from many sources, including fossil fuels and several renewable alternatives. Let x_{it} denote the output of fuel i at time t where $i = 0$ denotes the fossil fuel source and $i = 1, \ldots I$ renewable fuel sources.

Let β_i be the energy conversion efficiency of fuel i. Let $Y_t = \sum_{i=1}^{I} \beta_i x_{it}$ be the total energy consumed at time t. Because the fossil fuel is a nonrenewable resource, let S_t be the stock of the fossil fuel energy source at time t. The equation of motion of S_t is

$$\dot{S}_t = \frac{\partial S}{\partial t} = -x_{0t}. \tag{1.1}$$

Namely, the stock of the fossil fuel is decreasing with the level of extraction in each period. We can add complexity to the analysis by considering the effect of new discoveries, which we have not done here.

Each energy source has short-term production costs. In the case of fossil fuel, the variable production costs depend on reserves, and are likely to increase as reserves decline. In the case of renewables, the variable production cost declines with investment in the technology. The variable cost of x_{0t} at time t is $c_0(x_{0t}, S_t)$, where $\frac{\partial c_0}{\partial x_{0t}} > 0$ and $\frac{\partial^2 c_0}{\partial x_{0t}^2} > 0$ and $\frac{\partial c_0}{\partial S_t} < 0$. Namely, the marginal cost of fossil fuel is increasing with use but declining with stock size.

The variable cost of the renewable fuels declines with a stock of knowledge, which for the ith renewable fuel we denote by R_{it}. It tends to increase with investment in the technology, denoted by I_{it} in period t, and declines through depreciation. The equation of motion of R_{it} is:

$$\dot{R}_{it} = \frac{\partial R}{\partial t} = h(I_{it}) - \delta R_{it}, \tag{1.2}$$

where $h(I_{it})$ denotes the new knowledge associated with investment in R&D, where $\frac{\partial h}{\partial I_{it}} > 0$ and $\frac{\partial^2 h}{\partial I_{it}^2} < 0$, and δ is the depreciation of knowledge. The variable cost of the ith renewable fuel is $c_i(x_{it}, R_{it})$ with $\frac{\partial c_i}{\partial x_{it}} > 0$, $\frac{\partial^2 c_i}{\partial x_{it}^2} > 0$ and $\frac{\partial c_i}{\partial R_{it}} < 0$.

All fuels contribute to a stock pollution (greenhouse gas) level, which at time t is denoted by Z_t. The equation of motion of this stock pollution is:

$$\dot{Z}_t = \frac{\partial Z}{\partial t} = \sum_{i=0}^{I} \varphi_i x_{it} - \omega Z_t, \tag{1.3}$$

where φ_i is the level of emission per unit of fuel i, and assuming that $\varphi_0 = 1$ and $\varphi_i > \varphi_{i+1}$, namely a renewable fuel with a higher indicator i generates less pollution per unit fuel. For example, $i = 1$ may be biofuel and $i = 2$ solar energy. ω represents the depreciation of the greenhouse gas stock in each period.

The environmental cost associated with energy use in general depends on the stock of pollution, and can be denoted by $E(Z_t)$ where $\frac{\partial E}{\partial Z_t} > 0$, $\frac{\partial^2 E}{\partial Z_t^2} > 0$. The social optimization problem can thus be built in the following form:

$$\int_{t=1}^{\infty} e^{-rt} \left[B(Y_t, t) - c_0(x_{0t}, S_t) - \sum_{i=1}^{I} c_i(x_{it}, R_{it}) - \sum_{i=1}^{I} I_{it} - E(Z_t) \right] \partial t. \tag{1.4}$$

Subject to:

(1) $\dot{S}_t = \frac{\partial S}{\partial t} = -x_{0t}$,
(2) $\dot{R}_t = \frac{\partial R}{\partial t} = h(I_{it}) - \delta R_{it}$,
(3) $\dot{Z}_t = \frac{\partial Z}{\partial t} = \sum_{(} i = 0)^I \varphi_i x_{it} - \omega Z_t$,

given S_{0t}, R_{0t}, and Z_0 when r is the discount rate. This problem can be solved using the maximum principle.

Let H_t be the temporal Hamiltonian at time t, and for simplicity, inserting $Y_t = \sum_{i=1}^{I} \beta_i x_{it}$ it becomes:

$$H_t = \begin{matrix} max \\ x_{it}, I_{it}, S_t, R_{it}, Z_t, \lambda_t, \mu_{it}, \rho_t \end{matrix} \left\{ \left[B\left(\sum_{i=1}^{I} \beta_i x_{it}, t \right) - c_0(x_{0t}, S_t) \right. \right.$$

$$- \sum_{i=1}^{I} c_i(x_{it}, R_{it}) - \sum_{i=1}^{I} I_{it} - E(Z_t) \right] - \lambda_t x_{0t} + \sum_{i=1}^{I} \mu_{it}(h_i(I_{it}) - \delta_i R_{it})$$

$$\left. - \rho_t \left[\sum_{i=0}^{I} \varphi_i x_{it} - \omega Z_t \right] \right\} \tag{1.5}$$

Where λ_t is the dynamic shadow price of pollution, i.e. the extra cost of reducing the stock of the nonrenewable fuel at period t, μ_{it} is the shadow price of knowledge to improve productivity of renewable fuel i, and ρ_t is the dynamic shadow price of the stock of pollution. The maximum principle will be used to identify the first order conditions for this problem. The first order conditions are derived and analyzed from condition (1.6) to condition (1.14):

$$\frac{\partial H_t}{\partial x_{0t}} = \frac{\partial B(Y_t, t)}{\partial Y_t} \beta_0 - \frac{\partial c_0(x_{0t}, S_t)}{\partial x_{0t}} - \lambda_t - \rho_t \varphi_0 = 0. \tag{1.6}$$

Condition (1.6) states that the socially optimal level of fossil fuel use at time t is where the marginal benefit of fossil fuel (which corresponds to the marginal benefit of energy $\frac{\partial B(Y_t, t)}{\partial Y_t}$, which is also the price under competition), times the energy efficiency of fossil fuel β_0, is equal to the sum of the marginal cost of producing fossil fuel, denoted by $MC_{X0t} = \frac{\partial c_0(x_{0t}, S_t)}{\partial x_{0t}}$, plus the shadow price of the energy stock λ_t plus the marginal social cost of the pollution generated by the fossil fuel $\rho_t \varphi_0$ (which is the product of its pollution intensity φ_0 times the shadow price of the pollution stock).

$$\frac{\partial H_t}{\partial x_{it}} = \frac{\partial B(Y_t, t)}{\partial Y_t} \beta_i - \frac{\partial c_i(x_{it}, S_t)}{\partial x_{it}} - \lambda_t - \rho_t \varphi_i = 0. \tag{1.7}$$

Condition (1.7) states that the socially optimal use level of the ith renewable fuel at time t is where the marginal benefit of this fuel $\frac{\partial B(Y_t, t)}{\partial Y_t}$ is equal to the sum of the marginal cost producing of producing the fuel denoted by $MC_{Xit} = \frac{\partial c_i(x_{it}, S_t)}{\partial x_{it}}$ plus the marginal social cost of the pollution generated by the fossil fuel $\rho_t \varphi_i$.

$$\frac{\partial H_t}{\partial I_{it}} = -1 + \mu_{it} \frac{\partial (h_i(I_{it}))}{\partial I_{it}} = 0. \tag{1.8}$$

Condition (1.8) states that the socially optimal investment in the stock of knowledge to enhance the productivity of the ith renewable fuel at time t is where the marginal benefit of this investment, $\mu_{it} \frac{\partial (h_i(I_{it}))}{\partial I_{it}}$ (the marginal value of stock of knowledge times the marginal productivity of investment creating knowledge), is equal to the marginal cost of investment, which is one.

In addition to the above FOCs with respect to the control variables, the dynamic optimality conditions consist of the equations of motions of the stocks given by Eqs. (1.1)–(1.3) above. The equations of motion of the costate variables are presented below. They are based on the maximum principle, and the rate of change in the shadow price of the fossil fuel is derived from:

$$\frac{\partial H_t}{\partial S_t} = -\dot{\lambda}_t + r\lambda_t. \tag{1.9}$$

Since $\frac{\partial H_t}{\partial S_t} = -\frac{\partial c_0(x_{0t}, S_t)}{\partial S_t}$, condition (1.9) implies:

$$\dot{\lambda}_t = r\lambda_t + \frac{\partial c_0(x_{0t}, S_t)}{\partial S_t}. \tag{1.10}$$

Condition (1.10) suggests that under socially optimal pricing of the fossil fuel stock, the increase in value associated with holding a resource stock an extra period of time at time t, denoted $\dot{\lambda}_t$, is equal to the interest cost of holding the resource an extra period, $r\lambda_t$, minus the marginal cost reduction of mining because of increased stock $\left(-\frac{\partial c_0(x_{0t},S_t)}{\partial S_t}\right)$. The change in the dynamic shadow price of the knowledge for producing the ith renewable is derived from:

$$\frac{\partial H_t}{\partial R_{it}} = -\dot{\mu}_{it} + r\mu_{it}\delta_i. \tag{1.11}$$

Since $\frac{\partial H_t}{\partial R_{it}} = -\frac{\partial c_i(x_{it},R_{it})}{\partial R_{it}} - \mu_{it}$, condition (1.11) implies:

$$\dot{\mu}_{it} = (r - \delta_i)\mu_{it} + \frac{\partial c_i(x_{it}, R_{it})}{\partial R_{it}}. \tag{1.12}$$

Condition (1.12) suggests that the socially optimal marginal change in value of the stock of knowledge for producing the ith renewable is equal to the interest cost minus depreciation of holding the asset $(r - \delta_i)\mu_i t$ an extra period minus the marginal cost reduction of producing the renewable fuel because of increased stock $\left(-\frac{\partial c_i(x_{it},R_{it})}{\partial R_{it}}\right)$. The rate of change of the dynamic shadow price of the pollution is derived from:

$$\frac{\partial H_t}{\partial Z_t} = \dot{\rho}_t - r\rho_t. \tag{1.13}$$

Since $\frac{\partial H_t}{\partial Z_t} = -\frac{\partial E(Z_t)}{\partial Z_t} - \rho_t\omega$, condition (1.13) implies:

$$\dot{\rho}_t = \rho_t(r - \omega) - \frac{\partial E(Z_t)}{\partial Z_t}. \tag{1.14}$$

Condition (1.14) suggests that under the optimal allocation, the marginal value of a one period delay in introducing a unit of pollution stock is equal to the interest cost minus depreciation of the delay, $\rho_t(r - \omega)$, minus the marginal pollution cost (saved by the delay), $\frac{E(\partial Z_t)}{\partial Z_t}$.

We assume that the functions are well behaved[1] so that a solution to the social optimization exists. We will not analyze the dynamic properties of the solution but instead its implications.

1.2.1 Policy Implications

Let us start by assuming that production is undertaken by competitive industries with many small energy producers that are producing different fuel. The price of

---

[1] Benefit function is concave in energy and the cost functions are convex.

energy to consumers at time t is denoted by P_t where:

$$P_t = \frac{\partial B(Y_t, t)}{\partial Y_t}. \tag{1.15}$$

The consumer price for the ith fuel is $P_t \beta_i$. However the producer's calculus is affected by various policy interventions suggested by the optimality conditions. Since a competitive economy will not yield the social optimum, there is an externality problem (pollution) and a public good problem (both lack of investment in the knowledge generation and because R&D requires some basic knowledge that has public good properties).

The tax rate on fuels will evolve over time. Since different fuels have different levels of pollution intensity, the pollution tax per unit of energy at time t is $\varphi_i \rho_t$. Thus fuel with lower energy intensity will pay a lower tax. For example, in the case of solar energy, when the concern is GHG emissions the tax may be close to zero; the highest tax is for fossil fuel, and biofuels will have different tax rates based on their GHG emissions. For example, corn ethanol in the US generated about 80 % and sugarcane about 30 % of the GHG emissions of gasoline, so the tax rate should be designed accordingly.[2]

The model did not consider the decline in the GHG emissions of various fuels, but this has been the case for biofuels where learning by doing and research has led to reduced costs and emissions per unit of energy [19], and has also been the case to an even larger degree for photovoltaics (PV) [14]. Thus, even if the GHG emission price (ρ_t) increases over time, the tax on renewables that go through improvement resulting in lower pollution intensity may decline. On the other hand, a higher pollution tax and resorting to the use of dirtier fossil fuels (tar sands) may lead to increases in the tax per unit of fossil fuel over time. The growing gap between the pollution tax on fossil fuels and biofuel over time is likely to drive the adoption of the latter and lead to more intensive use.

The increase in the use of fossil fuel will also depend on technological change, and the model suggests policies to enhance investment that will provide a foundation for this change. The optimal investment levels are given by the model at each period for each fossil fuel (the optimal I_{it}'s). The reality is that R&D is a joint public and private venture,[3] and design of a technology strategy is a complex challenge beyond the scope of this paper. The analysis suggests that we have the ability to quantify knowledge generated by research and production as well as understand the relationship between investment and knowledge generation. The optimal outcome would have been obtained by paying μ_{it} per unit of knowledge about producing renewable i at time t. Indeed, start-ups are valued for their intellectual property,

---

[2]Exact numbers are subject to debate—not to mention the controversy around indirect land use [34]. The numbers here are used for illustrative purposes.

[3]See the analysis on technology transfer from public to private institutions in Heiman et al. [15].

and researchers in both the private and public sector are compensated based on their discoveries.

In the case of the fossil fuel, there is a need for another possible intervention. If the fossil fuel is extracted by competitive producers from a common aquifer, no one producer will have the incentive to consider the future, so a central planner may be needed to impose an extra tax, λ_t, per unit of fossil fuel. Thus, for the fossil fuel sector the equilibrium condition occurs where $P_t \beta_0 = MC_{X0t} + \lambda_t + \rho_t \varphi_0$, and for the renewable fuel sectors the equilibrium conditions occur where $P_t \beta_i = MC_{Xit} + \rho_t \varphi_i$.

Baumol and Oates [2] emphasized that optimal environmental policies under certainty can be achieved using both prices and quantities. The main difference between policies that attain the optimal resource allocation is the distributional implications. In principle, instead of using taxes, the government can tell all producers of fossil and renewable fuel how much they can produce. It can also determine the total quantity of each fuel and allocate tradable permits, and profit-maximizing firms will reach the optimal outcome of how much to produce. Indeed there is a growing tendency to rely on tradable permits, perhaps because they are preferred by producers when the permits are given to the industry and because the government can designate a maximum level of pollution.

The assumption of a common pool resource used to derive the shadow price of the fossil fuel energy stock is also unrealistic. Mining and oil companies (public or private) own their resource stocks (or have a long term lease) and thus have their own dynamic calculus based on their reserves, time preferences, etc. Thus, they may have their own (implicit) shadow prices of the resource stocks that affect their level of extraction, and as a result regulation of the stock by a shadow price may not be needed or the formulas for intervention may need to take into account that we do not have a perfect common pool problem in the extraction of fossil fuels. However, the reality of the energy sector is much more complex, and requires nuanced interventions, taking into account political feasibility considerations that can be guided qualitatively by the model we present in the previous section. The energy sector is also not monolithic, and it can be divided into two important sub-sectors, power and transportation, which will be discussed in detail in the next section.

The two main generic problems that challenge the implementation of the policy recommended by the model are:

1. *The lack of a global authority that has the power to impose policies addressing global externalities.* The main externality from the power sector is GHG emissions, and policies tax fuels based on carbon content per unit of energy over time in order to incentivize the transition to renewable fuel that emits less GHGs. However, the formulas presented here suggest principles to establish a carbon tax or to determine quotas of GHG generation for different renewable fuels. The question remains: who can impose these policies globally? International agreements like the Kyoto Protocol may provide a base for cooperation among nations to meet certain standards, but agreements like this are voluntary and have

limited impact. Many of the actions are taken by national governments in an uncoordinated fashion, and as the perils of climate change become clearer over time, coordination and cooperation are more likely to increase. As mentioned earlier, one of the challenges in environmental policy is not the aggregate action taken, but the distributional effect of policy decisions, which is a major problem in enacting new policies. Nevertheless, there are policies aiming to curtail fossil fuel use and move towards subsidizing the production of renewables in a way that is related to their respective GHG emissions per unit of energy use. For example, policies include the Renewable Fuel Standard (RFS), which provides preferential treatment to cleaner ethanol produced from sugarcane or second generation crops, or the Low Carbon Fuel Standard (LCFS) that sets an upper bound on the carbon content of fuels, implicitly taxing fossil fuel and subsidizing clean renewable fuels. These policies are implemented on a piecemeal basis that may reduce their effectiveness and provide perverse incentives to shuffle more polluting fuels to regions that have lighter regulatory standards. There are a significant number of subsidization programs for solar energy, and one way to assess the program is the implicit price of carbon that they imply. Frondel et al. [12] use this approach to critically review the subsidization of solar energy in Germany. In addition to subsidies for clean energy sources, there is a significant amount of investment in R&D to develop alternative fuels [30].

2. *Noncompetitive behavior in different energy sectors.* The oil sector is dominated by OPEC, which is referred to by Hochman et al. [17] as a "cartel of nations" rather than of firms. They document the large wedge between the high export price compared to the low domestic price among the various OPEC nations, and suggest that this difference is designed to maximize revenue from the export market while sharing the welfare gain from oil consumers domestically between consumers and producers. Hochman et al. [17] further assume OPEC takes other producers of fossil fuel energy as well as producers of renewable energy sources as given.[4] While the monopolistic behavior of OPEC reduces consumption in importing countries compared to outcomes under competitive behavior, the reluctance of OPEC countries to impose incentives to reduce GHG emissions also leads to increased use of fossil fuel. There are deviations from the competitive model both in the transportation and power sectors that may reflect political economic situations and the recognition that energy is a necessity good. Thus, power and transportation for the poor are being subsidized [13]. In richer countries or in countries that may be concerned about balance of trade issues, energy may be taxed to reduce foreign exchange losses (which may occur due to an increase in imports relative to exports) or to provide a source of revenue for governments. Another factor not considered in the analysis thus far is conservation. Governments may provide incentives for energy conservation that will reduce demand, and indeed incentives are a crucial element in any strategy attempting to address the GHG challenge, but they are not included in this

---

[4]The non-OPEC suppliers are treated by OPEC as a competitive fringe in a leading firm model.

analysis for simplicity. While our use of the competitive model is a benchmark that deviates from reality, it is useful in that forces that lead to changes in energy use under these models will also lead to significant changes under more complex institutional arrangements. However, to better understand the dynamics of alternative forms of energy, the power and transportation sectors are looked at separately in more detail in the following sections.

1.2.1.1 Analyzing the Different Energy Applications

While the model we have presented is generic, the energy sector can produce several distinct products. It is useful to distinguish between two types of output: energy for electric power and energy for other uses. Energy for power is generated from many sources that are converted by a power plant to electricity and is sent through the electric grid to final users. Energy for other uses consists of wood, liquid fuels, dung, or gas and has to be transported to the location of use. One important sub-category in energy for other uses is energy for transportation. In both cases, energy is the intermediate input in a production process whether it be a household production process [3] or industrial production process, and the final use of the energy depends on the location and time of consumption. Thus, one of the main challenges of the energy system is transporting energy from the source of origin to its final use. The transportation of fuel throughout the production process can be quite complex. For example, in the case of electric power from coal, coal is moving from a mine to a power plant, but after converted to electricity it moves through a grid. With solar energy, large-scale photovoltaics (PV) provide an intermediate source to harness the sun's power and feed the energy straight into the grid. In the case of liquid fuel, there may be multiple stages of transportation: from the well to the refinery and the refinery to the gas station. When the fuel is used for transportation, it will move with the vehicle while being consumed.

There is a large literature on spatial modeling that addresses issues of transportation strategies and location [21, 29], water conveyance and application systems [6], and a recent literature on optimal design of PV [14]. These bodies of literature develop frameworks establishing that the efficient price of a product, like power, should reflect the cost of generation, the cost of conveyance, the cost of storage, and the externality cost. The same should apply to different sources of energy for different applications. Thus, one can expand the generic model presented in the first section of this paper to add a spatial and storage element, which is beyond the scope of this paper, but we discuss how spatial and storage considerations affect the relative performance of fossil fuel, renewable biofuel, and solar power. This is especially important because there is some substitution between these energy sources [5].

Table 1.1 Fuel types and their applications

Type of fuel	Power sector	Transportation sector
Fossil fuel	Coal and natural gas	Oil and natural gas
Biofuel	Biomass	Ethanol and biodiesel
Solar	Photovoltaic	Batteries

1.2.1.2 Assigning Different Fuels for Different Energy Applications

It is useful to distinguish between types of energy and their applications, and for simplicity we will distinguish between two applications: power (electricity) and transportation. Because of space limitations, we do not address other energy applications, such as heating. There is some overlap between power and transportation since electric power is a source of energy for a growing amount of transportation services. However, the principles we establish in our discussion of power and transportation also apply to other uses of energy. For simplicity, we only consider three types of fuels: fossil fuel, biofuel, and solar. Table 1.1 categorizes the major types of fuels used for power and transportation.

The table is not exclusive; we mention two renewable fuels, biofuel and solar, but detailed analysis should include wind, hydroelectric, geothermal, and wave power as additional renewable fuels as well as nuclear power as an additional nonrenewable fuel. As we mentioned in the introduction to the conceptual model, an important option is conservation, which is not discussed here. As the model in the second section suggests, the selection of specific fuels for different applications depends on the infrastructure cost, operational cost, storage cost, and environmental cost of a fuel system. The extent to which a technology will be used depends on investment in R&D and the results of this research. The share of different fuels in meeting energy demand will change over time, as increased knowledge about fuels will change their relative value.[5] We consider below the factors that will affect the use of the specific fuels mentioned in Table 1.1.

1.2.1.3 The Power Sector

Coal is a very abundant fossil fuel resource, and without considering environmental cost, it dominates other fuels for production of power in many cases. However, even without considerations of climate change, air quality considerations as well as landscape integrity considerations may restrict its use. Overall, the optimal quantity of coal is at a level where net benefits are equal to the sum of production and processing costs of the coal, environmental costs, and scarcity costs. The environmental quality impact of coal includes the health effect because of air pollution, the cost from environmental damage due to mining, and the climate change effect. We expect the

---

[5]This value is measured by net benefit that includes both economic and environmental gains, as the previous section suggests.

climate change effect to increase over time, which may lead to a phase-out of coal. However, technologies that sequester the carbon emissions of coal into the ground [8] will expand its use. The increases in availability of natural gas and its superior environmental performance makes it useful for power generation, especially in the United States, and it will be a dominant fuel source for a long period of time as long as the combination of increased scarcity and environmental costs are not binding. Carbon dioxide (CO_2) emissions of natural gas are about half those of coal [23] and about 60 % of those generated by oil [16]. While there is concern about the environmental consequences of hydraulic fracturing (also known as fracking), ongoing research will continue to improve the technology, and its use is likely to be expanded to other regions [1]. Currently, there are significant differences between the price of natural gas between the U.S. and other countries. However, increased utilization of fracking globally and improved export facilities of liquefied natural gas (LNG) from the U.S. to the rest of the world will lead to increased supply of natural gas in the future [31]. While presently the price of natural gas in the U.S. is below $5 per million BTU and the price in Europe and Asia may be double or even triple the price in the U.S., increases in global use of LNG will push prices closer [9]. Introduction of GHG policies that will penalize the use of coal relatively more so than natural gas will be another factor in speeding up the transition from coal to natural gas. However, natural gas is a nonrenewable resource, and at a certain point in the future, scarcity will drive its price up, which will increase the attractiveness of renewable resources for power.

There is a limited use of biomass for power, primarily through wood products, agricultural waste, and municipal garbage.[6] The marginal cost of producing, harvesting, shipping, and processing biomass may be higher than alternatives, but its low GHG effect and its renewability are significant advantages. Biofuel is especially attractive when burning of waste materials solves disposal problems, which generates additional benefit.[7] The European Union introduced wood pellets as part of their power generation portfolio as a result of a renewable energy directive, which generated significant exports of these wood pellets [27]. Ehrig and Behrendt [11] identify conditions where wood pellets are an efficient strategy to meet the European Renewable Energy Directive (RED). However, the efficiency of the directive is questioned based on the high implied shadow price of GHGs [24].

Solar energy has been an expensive source of power, and has been used primarily in remote locations or on a small scale in situations where the cost of generation is smaller than cost of connection to the grid. However, due to research and learning by doing, the cost of PV has been declining rapidly, and it is projected to decline even further, thus making it competitive with other sources of power in many regions of the world [14]. This reduction in manufacturing cost combined with the expected

---

[6]Biomass supplies about 52 gigawatts of global power generation capacity [33] while the world capacity for power generation is 4.3 terawatts [10].

[7]There is increased solid waste in China, which is providing the benefit of renewable energy and waste disposal [7]

increase in the shadow price of GHGs is likely to make PV a more competitive source of power in years to come. There are two major constraints to the application of solar power. The first is that while it is relatively cheap at the source when it is generated in regions with the appropriate climate and large amounts of sunlight, its transportation across regions is costly [14]. Second, there are issues of variability of cost over space because of expensive conveyance facilities and constraints on availability over time resulting from both seasonality but mostly variable exposure to the sun throughout the day (including both daytime and nighttime). Building excess capacity and storage have been the traditional mechanisms to address variability of energy supply and demand over time, but the high cost of energy storage may affect the overall cost of solar power systems and reduce its competitiveness. Grossmann et al. [14] suggest developing a distributed system of generation embedded in an energy transmission system that will take advantage of reduced cost of energy transmission and reduced overall costs through savings in storage requirement by interlinking production sites that have sunlight at complementary times. Through such a network, each solar energy source will be connected to several intermediate points that will supply power to final destinations in order to counter variability in demand and supply. Generally speaking, there will be longitudinal movement of energy between the equator and northern and southern hemispheres of the globe to overcome seasonal differences and an east-west movement of energy to overcome the day and night cycle. Grossmann et al. [14] developed an algorithm that will reduce the cost of such systems.

1.2.1.4 The Transportation Sector

While there is still a large portion of transportation that relies on draft animals, oil is currently the major source of fuel for transportation. The major forms of fuels for transportation derived from oil are gasoline and diesel; approximately one third of the oil is used to produce gasoline and diesel and the rest is used to produce jet fuel, kerosene, and other fuels. Some diesel and much of the kerosene and other oil derivatives are used for heating, cooking, and other energy applications [25]. While oil emits fewer GHGs and other forms of air pollution than coal, oil is not a homogeneous product. Oil produced from tar sands may be 50 % more polluting than conventional oil, which also varies in emissions [4]. Improvement of refining technology is likely to reduce GHG emissions, especially for nonconventional fossil fuel sources.

Natural gas can be a major source of transportation fuel in different forms, either compressed or liquified, and it can even be converted to gasoline and diesel [28]. Converted natural gas has advantages of fewer GHG emissions than gasoline and diesel by about 40 % [16], and is competitive in terms of price. However, use of natural gas with small, private cars requires extra storage space in the car as well as a new distributional infrastructure that is generally not in place (e.g. natural gas fuel pumps). Yet, there is significant growth in the use of natural gas in commercial vehicles, which is likely to continue in the future [32].

Biofuels have been toted as the transportation alternative to fossil fuel since production of the feedstock sequesters carbon. However, while production of biofuel may benefit consumers by reducing fuel prices, it also diverts land from the production of food and as a result may increase the price of food, and only advancements in agricultural productivity can counter price increases, which can be substantial [26]. Furthermore, the production of the feedstock and its processing require energy, and thus the net effect in terms of GHG production is not obvious. The expansion of agricultural production because of the introduction of biofuel will result in indirect GHG emissions through land use expansion. As our model in the second section suggests, the impact of the introduction of biofuel on prices and GHG emissions is not obvious and depends on policies as well as technological innovation. Understanding this complex relationship has been a subject of major research and policy debate [20]. Some biofuels aim to replace gasoline and others to replace diesel. Ethanol produced from sugarcane, corn, cassava, and other products is a major biofuel used commercially. It replaces more than 50 % of the gasoline in Brazil, 10 % of the gasoline used in the U.S., and a lesser percentage in other countries. Biodiesel is produced from animal fats, soybean, rapeseed, oil residues, and palm oil, and is mostly used in Europe but also in the U.S. and Brazil as well. Overall, ethanol substitutes for gasoline to a larger extent than biodiesel substitutes for diesel because of economic and environmental factors. Biofuel produced from food crops are referred to as first generation biofuels, second generation biofuels are produced mostly from cellulosic ethanols, and third generation biofuels are produced primarily from algae and other exotic sources. The introduction of biofuels was enhanced by subsidies, mandates, and the subsidization of research [25].

The economic viability of sugarcane and corn ethanol as well as their GHG emissions reductions have been enhanced by adoption of technological innovation. In the case of corn ethanol, the use of residue for animal feed and the higher octane level of ethanol compared to gasoline increases its profitability. This combined with the increased reliance on natural gas for production and processing as well as improvement in the refining process have reduced direct and indirect GHG emissions. Still, sugarcane ethanol is preferable from a GHG perspective to corn ethanol,[8] which in turn is preferable to other forms of ethanol and biodiesel [19,22]. The application of second-generation biofuel thus far has been limited, and the extent of their adoption depends on policies, but they are likely to reduce GHG emissions significantly relative to gasoline [18].

There are several commercial models of electric vehicles in the market. Battery technology is the major constraint, as it restricts range of operation and increases in range capacity are expensive. Future research is likely to improve batteries, reduce the price of the car, and result in increased supply of batteries that will make the electric car more competitive. Electric vehicles have already been competitive, as was seen in the very beginning of the car's history in the early 1900s, and

---

[8]Sugarcane ethanol may reduce up to 80 % of the GHG emissions of conventional gasoline while corn ethanol may reduce up to 40 % of the GHG emissions of conventional gasoline.

have the potential to become the dominant vehicle technology. They fell behind the now dominant internal combustion engines due to the construction of the extended road network in the 1920s that allowed driving for longer distances for the first time, which the electric vehicles (EVs) were not well equipped for. Environmental concerns in 1990 led to the enactment of the Low Emission Vehicle (LEV I) Program of the California Air Resources Board, which boosted EVs. Again, it was the very heavy battery with limited range of up to 140 miles and high production costs that prevented market penetration. However, with the recent battery developments in the early 2010s, (most notably by Tesla), we observe signs that a breakthrough in battery technology has emerged or is emerging.

1.3 Conclusion

The challenge of climate change and availability of cheap energy requires government intervention. This intervention can be derived from a dynamic model that recognizes properties of production, research, consumption, and the environment, and derives optimal pricing of fuel commodities and environmental amenities. This paper uses optimal control to determine guiding mechanisms in designing these policies, suggesting that optimal intervention requires restriction on GHG emissions of different forms of energy as well as investment in different lines of research. However, these policies can take the form of taxes, subsidies, or tradable permits, and these policies must vary across fuels, production processes, and location over time. Our discussion of current energy situations suggests there is a slow transition away from fossil fuel to biofuel and solar energy, but the Stern Review and most recent IPCC report suggest that more aggressive policies are needed to achieve reasonable GHG levels that meet global targets.

The analysis in this paper suggests that both solar energy and biofuels are likely successors to fossil fuel as major sources of energy. However, the two renewable sources are complementary. While solar energy is likely to be a dominant source of power, biofuel will play a major role in providing energy for transportation. Research and development as well as sound policies are key to this transition.

References

1. Alonso, A., Mingo, M.: The expansion of "non conventional" production of natural gas (tight gas, gas shale and coal bed methane). A silent revolution. In: Energy Market (EEM), 2010 7th International Conference on the European, pp. 1–8. IEEE (2010)
2. Baumol, W.J., Oates, W.E.: Economics, Environmental Policy, and the Quality of Life. Gregg Revivals, Aldershot (1993)
3. Becker, G.S. Human capital: A Theoretical and Empirical Analysis, with Special Reference to Education. University of Chicago Press, Chicago (2009)

4. Brandt, A.R., Farrell, A.E.: Scraping the bottom of the barrel: greenhouse gas emission consequences of a transition to low-quality and synthetic petroleum resources. Climatic Change **84**(3–4), 241–263 (2007)
5. Chakravorty, U., Roumasset, J., Tse, K.: Endogenous substitution among energy resources and global warming. J. Polit. Econ. **105**, 1201–1234 (2009)
6. Chakravorty, U., Hochman, E., Umetsu, C., Zilberman, D.: Water allocation under distribution losses: comparing alternative institutions. J. Econ. Dyn. Control **33**(2), 463–476 (2009)
7. Cheng, H., Hu, Y.: Municipal solid waste (MSW) as a renewable source of energy: current and future practices in China. Bioresourc. Technol. **101**(11), 3816–3824 (2010)
8. Chung, T.S., Patino-Echeverri, D., Johnson, T.L.: Expert assessments of retrofitting coal-fired power plants with carbon dioxide capture technologies. Energy Pol. **39**(9), 5609–5620 (2011)
9. Economides, M.J., Oligney, R.E., Lewis, P.E.: US natural gas in 2011 and beyond. J. Nat. Gas Sci. Eng. **8**, 2–8 (2012)
10. Energy Information Administration (EIA): International energy statistics: total electricity installed capacity (million kilowatts). Available from: http://www.eia.gov/cfapps/ipdbproject/IEDIndex3.cfm?tid=2&pid=2&aid=7 (2013)
11. Ehrig, R., Behrendt, F.: Co-firing of imported wood pellets: an option to efficiently save CO_2 emissions in Europe? Energy Pol. **59**, 283–300 (2013)
12. Frondel, M., Ritter, N., Schmidt, C.M., Vance, C.: Economic impacts from the promotion of renewable energy technologies: the German experience. Energy Pol. **38**(8), 4048–4056 (2010)
13. Gangopadhyay, S., Ramaswami, B., Wadhwa, W.: Reducing subsidies on household fuels in India: how will it affect the poor? Energy Pol. **33**(18), 2326–2336 (2005)
14. Grossmann, W.D., Grossmann, I., Steininger, K.W.: Distributed solar electricity generation across large geographic areas, part I: a method to optimize site selection, generation and storage. Renew. Sustain. Energy Rev. **25**, 831–843 (2013)
15. Heiman, A., Zilberman, D., Graff, G.: University research and offices of technology transfer. Calif. Manag. Rev. **45**(1), 88–115 (2002)
16. Hekkert, M.P., Hendriks, F., Faaij, A., Neelis, M.L.: Natural gas as an alternative to crude oil in automotive fuel chains well-to-wheel analysis and transition strategy development. Energy Pol. **33**, 579–594 (2005)
17. Hochman, G., Rajagopal, D., Zilberman, D.: Are biofuels the culprit? OPEC, food, and fuel. Am. Econ. Rev. **100**(2), 183–187 (2010)
18. Huang, H., Khanna, M., Önal, H., Chen, X.: Stacking low carbon policies on the renewable fuels standard: economic and greenhouse gas implications. Energy Pol. **56**, 5–15 (2012)
19. Khanna, M., Crago, C.L.: Measuring indirect land use change with biofuels: implications for policy. Annu. Rev. Resour. Econ. **4**(1), 161–184 (2012)
20. Khanna, M., Scheffran, J., Zilberman, D.: Handbook of Bioenergy Economics and Policy, vol. 33. Springer, New York (2010)
21. Kulmer, V., Koland, O., Steininger, K.W., Fürst, B., Käfer, A.: The interaction of spatial planning and transport policy: a regional perspective on sprawl. J. Transport Land Use **7**(1), 57–77 (2014)
22. Laborde, D., Valin, H.: Modeling land-use changes in a global CGE: assessing the EU biofuel mandates with the MIRAGE-BioF model. Climate Change Econ. **3**(03) (2012)
23. Logan, J., Heath, G., Macknick, J., Paranhos, E., Boyd, W., Carlson, K.: Natural Gas and the Transformation of the US Energy Sector: Electricity. National Renewable Energy Laboratory (NREL), Golden (2012)
24. López-Pena, À., Pérez-Arriaga, I., Linares, P.: Renewables vs. energy efficiency: the cost of carbon emissions reduction in Spain. Energy Pol. **50**(C), 659–668 (2012)
25. Rajagopal, D., Zilberman, D.: Environmental, economic and policy aspects of biofuels. Found. Trends® Microeconomics **4**(5), 353–468 (2008)
26. Rajagopal, D., Sexton, S.E., Roland-Holst, D., Zilberman, D.: Challenge of biofuel: filling the tank without emptying the stomach? Environ. Res. Lett. **2**(4), 044004 (2007)
27. Sikkema, R., Steiner, M., Junginger, M., Hiegl, W., Hansen, M.T., Faaij, A.: The European wood pellet markets: current status and prospects for 2020. Biofuels Bioproducts Biorefining **5**(3), 250–278 (2011)

28. Sudiro, M., Bertucco, A.: Production of synthetic gasoline and diesel fuel by alternative processes using natural gas and coal: process simulation and optimization. Energy **34**(12), 2206–2214 (2009)
29. Takayama, T., Judge, G.G.: Spatial and temporal price and allocation models. Contrib. Econ. Anal. **73** (1971)
30. Veugelers, R.: Which policy instruments to induce clean innovating? Res. Pol. **41**(10), 1770–1778 (2012)
31. Weizhong, L., Hongtao, Z., Meng, M.: LNG exports from the United States and their impact on the global LNG market. Nat. Gas Ind. **6**, 027 (2012)
32. White, B.: The Long Road for Natural Gas Vehicles. Office of the Federal Coordinator: Alaska Natural Gas Transportation Projects, Washington, DC/Anchorage, AL (2011)
33. World Watch Institute.: Study: biofuels more efficient as electricity source. Available from http://www.worldwatch.org/node/6109 (2013)
34. Zilberman, D., Barrows, G., Hochman, G., Rajagopal, D.: On the indirect effect of biofuel. Am. J. Agric. Econ. **95**(5), 1332–1337 (2013)

Chapter 2
Optimal Localization of Firms in Hotelling Networks

Alberto Adrego Pinto and Telmo Parreira

2.1 Introduction

Hotelling [1] introduce a model of spatial competition in a city represented by a line segment, where a uniformly distributed continuum of consumers have to buy a homogeneous good. Consumers have to support the transportation costs when baying the good in one of the two firms of the city. In this framework, firms simultaneously choose their location and afterwards set their prices in order to maximize their profits. In [2], the Hotelling's line model is extended to networks, where the firm's location is taken exogenously, that is, the focus consists in studying the second stage of the two-stages game where the firms compete in prices. In this model, the edges are inhabited by consumers uniformly distributed, in the spirit of Hotelling's line model and Salop's [4] circular model.

The *Hotelling town* model consists of a network of *consumers* and *firms*. The consumers (*buyers*) are located along the *edges* (*roads*) of the network and the firms (*shops*) are located at the *vertices* (*nodes*) of the network. Every road has two vertices and at every vertex is located a single firm. The *degree k* of the vertex is given by number of incident edges. If the degree $k > 2$ then the vertex is a crossroad of k roads; if the degree $k = 2$ then the vertex is a junction between two roads; and if $k = 1$ the vertex is in the end of a no-exit road. Every consumer will buy one unit

A.A. Pinto
LIAAD-INESC TEC, Porto, Portugal

Department of Mathematics, Faculty of Sciences, University of Porto, Rua do Campo Alegre, 687, 4169-007, Porto, Portugal,
e-mail: aapinto@fc.up.pt

T. Parreira (✉)
Department of Mathematics, University of Minho, Campus de Gualtar, 4710-057 Braga, Portugal

LIAAD-INESC TEC, Porto, Portugal
e-mail: telmoparreira@hotmail.com

A.A. Pinto and D. Zilberman (eds.), *Modeling, Dynamics, Optimization and Bioeconomics I*, Springer Proceedings in Mathematics & Statistics 73, DOI 10.1007/978-3-319-04849-9__2,
© Springer International Publishing Switzerland 2014

of the commodity to only one of the firms of the network and every firm will charge a price for the commodity that will be the same for all its customers.

A Hotelling town *price strategy* **P** consists of a vector whose coordinates are the prices p_i of each firm F_i. A consumer located at a point x of the network that decides to buy at firm F_i *spends* $E(x; i, \mathbf{P}) = p_i + t\, d(i, x)$, where p_i is the price charged by the firm F_i and $t\, d(i, x)$ is the *transportation cost* that is proportional t to the minimal distance $d(i, x)$ between the position i of the firm F_i and the position x of the consumer measured in the network. The *local firms* of a consumer located at a point x in a road $R_{i,j}$ with vertices i and j are F_i and F_j. In every road $R_{i,j}$, there is at most one consumer located at a point $\mathbf{x}_{i,j} \in R_{i,j}$ that is *indifferent* to in which local firm is going to buy his commodity, i.e. $E(x; i, \mathbf{P}) = E(x; j, \mathbf{P})$. A price strategy **P** determines a *local market structure* if every road $R_{i,j}$ has an indifferent consumer $\mathbf{x}_{i,j} \in R_{i,j}$.

We introduce the BLC condition that gives a bound for the different production costs of firms and for the different road lengths of the network in terms of the transportation cost, the minimal road length of the network and the maximum node degree (see Sect. 2.2). For the Hotelling town network satisfying the BLC condition, we exhibit the unique optimum Nash price equilibrium strategy **P*** (see the proof in [3]).

We say that a price strategy has the *profit degree growth* property if the profits of the firms increase with the degree of the nodes. We introduce the DB condition that gives a bound for the different production costs of firms and for the different road lengths of the network in terms of the transportation cost, the minimal road length of the network and the maximum node degree (see Sect. 2.2). We show that for a Hotelling town network satisfying the DB and BLC conditions the Nash price equilibrium strategy has the profit degree growth property.

If all firms have the same production costs and all roads have the same length then we say that the Hotelling town is uniform. A uniform Hotelling town satisfies the BLC and DB conditions and so has a unique Nash price equilibrium with the profit degree growth property.

We show that a Hotelling town network satisfying the BLC condition and with all degrees of the nodes above 2 has the property that for small perturbations on the localization of the firms the Nash price equilibrium is unique and it is a small perturbation of **P*** (see Sect. 2.3). Furthermore, the profit of a firm is optimal at the node for small perturbations on its own localization. Hence, the node localization of the firms is locally optimal. All the results presented in this chapter are proved in [3].

2.2 Nash Equilibrium Price Strategy

Given a price strategy **P**, the consumer will choose to buy in the firm $F_{v(x,\mathbf{P})}$ that minimizes his expenditure in the Hotelling town

$$v(x, \mathbf{P}) = argmin_{i \in V} E(x; i, \mathbf{P}),$$

where V is the set of all vertices of the Hotelling town. Hence, for every firm F_i, the *market*

$$M(i, \mathbf{P}) = \{x : v(x, \mathbf{P}) = i\}$$

consists of all consumers that minimize their expenditure by opting to buy in firm F_i.

The *road market size* $l_{i,j}$ of a road $R_{i,j}$ is the Lebesgue measure (or length) of the road $R_{i,j}$, because the buyers are uniformly distributed along the roads, and the *market size* $S(i, \mathbf{P})$ of the firm F_i is the Lebesgue measure of $M(i, \mathbf{P})$. The Hotelling town *production cost* $\mathbf{C}$ is the vector whose coordinates are the production costs c_i of the firms F_i. The Hotelling town *profit* $\Pi(\mathbf{P}, \mathbf{C})$ is the vector whose coordinates

$$\pi_i(\mathbf{P}, \mathbf{C}) = (p_i - c_i) S(i, \mathbf{P})$$

are the *profits* of the firms F_i. A price strategy $\mathbf{P}$ determines a *local market structure* if every consumer buys in a local firm

$$M(i, \mathbf{P}) \subset \bigcup_{j \in N_i} R_{i,j}.$$

Hence, for every road $R_{i,j}$ there is an *indifferent buyer* located at a distance

$$0 < x_{i,j} = (2t)^{-1}(p_j - p_i + t\, l_{i,j}) < l_{i,j} \qquad (2.1)$$

of firm F_i. Thus, a price strategy $\mathbf{P}$ determines a local market structure if, and only if, $|p_i - p_j| < t\, l_{i,j}$ for every road $R_{i,j}$.

The firms F_i and F_j (or vertices i and j) are *neighbors* if there is a road $R_{i,j}$ with end nodes i and j. Let N_i the set of all vertices that are neighbors of the vertex i and, so, k_i is the cardinality of the set N_i that is equal to the degree of the vertex i. If the price strategy determines a local market structure then $S(i, \mathbf{P}) = \sum_{j \in N_i} x_{i,j}$ and

$$\pi_i(\mathbf{P}, \mathbf{C}) = (p_i - c_i) \sum_{j \in N_i} x_{i,j} = (2t)^{-1}(p_i - c_i) \sum_{j \in N_i} (p_j - p_i + t\, l_{i,j}).$$

Given a pair of price strategies $\mathbf{P}$ and $\mathbf{P}^*$ and a firm F_i, we define the price vector $\tilde{\mathbf{P}}(i, \mathbf{P}, \mathbf{P}^*)$ whose coordinates are $\tilde{p}_i = p_i^*$ and $\tilde{p}_j = p_j$, for every $j \in V \setminus \{i\}$.

The price strategy $\mathbf{P}^*$ is a *best response* to the price strategy $\mathbf{P}$, if

$$(\tilde{p}_i - c_i) S(i, \tilde{\mathbf{P}}(i, \mathbf{P}, \mathbf{P}^*)) \geq (p_i' - c_i) S(i, \mathbf{P}_i'),$$

for all $i \in V$ and for all price strategies $\mathbf{P}_i'$ whose coordinates satisfy $p_i' \geq c_i$ and $p_j' = p_j$ for all $j \in V \setminus \{i\}$. A price strategy $\mathbf{P}^*$ is a Hotelling town *Nash equilibrium* if $\mathbf{P}^*$ is the best response to $\mathbf{P}^*$.

We denote by c_M (resp. c_m) the maximum (resp. minimum) production cost of the Hotelling town

$$c_M = \max\{c_i : i \in V\} \quad \text{and} \quad c_m = \min\{c_i : i \in V\}.$$

We denote by l_M (resp. l_m) the maximum (resp. minimum) road length of the Hotelling town

$$l_M = \max\{l_e : e \in E\} \quad \text{and} \quad l_m = \min\{l_e : e \in E\},$$

where E is the set of all edges of the Hotelling town. Let

$$\Delta(c) = c_M - c_m \quad \text{and} \quad \Delta(l) = l_M - l_m.$$

We denote by k_M be the maximum node degree of the Hotelling town

$$k_M = \max\{k_i : i \in V\}.$$

Definition 2.1. A Hotelling town satisfies the *bounded length and costs (BLC)* condition, if

$$\Delta(c) + t\Delta(l) \leq \frac{(t\, l_m - \Delta(c)/2)^2}{2\, t\, k_M\, l_M}. \tag{2.2}$$

The Hotelling town *admissible market size* **L** is the vector whose coordinates are the *admissible local firm market sizes*

$$L_i = \frac{t}{k_i} \sum_{j \in N_i} l_{i,j}.$$

The Hotelling town *neighboring market structure* **K** is the matrix whose coordinates are (1) $k_{i,j} = k_i^{-1}$, if there is a road $R_{i,j}$ between the firms F_i and F_j; and (2) $k_{i,j} = 0$, if there is not a road $R_{i,j}$ between the firms F_i and F_j. Let **1** denote the identity matrix.

Theorem 2.1. *If a Hotelling town satisfies the BLC condition then there is a unique Hotelling town Nash equilibrium price strategy*

$$P^* = \frac{1}{2}\left(1 - \frac{1}{2}K\right)^{-1}(C + L) = \sum_{m=0}^{\infty} 2^{-(m+1)}K^m (C + L). \tag{2.3}$$

We note that the Nash equilibrium price strategy for the Hotelling town satisfying the BLC condition determines a local market structure, i.e. every consumer located at $x \in R_{i,j}$ spends less by shopping at his local firms F_i or F_j than in any other firm in the town and so the consumer at x will buy either at his local firm F_i or at his local firm F_j.

We say that a price strategy **P** has the *profit degree growth* property if

$$k_i > k_j \Rightarrow \pi_i(\mathbf{P}, \mathbf{C}) > \pi_j(\mathbf{P}, \mathbf{C})$$

for every $i, j \in V$.

Lemma 2.1. *Let F_i be a firm located in a node of degree k_i and F_j a firm located in a node of degree k_j. Let $\bar{p}_i = p_i^* - c_i$ and $\bar{p}_j = p_j^* - c_j$ represent the unit profit of firm F_i and F_j, respectively. Then, $\pi_i^* > \pi_j^*$ if and only if*

$$\frac{k_i - k_j}{k_j} > \frac{\bar{p}_j^2 - \bar{p}_i^2}{\bar{p}_i^2}.$$

Definition 2.2. A Hotelling town network satisfies the *degree bounded lengths and costs (DB)* condition if

$$\Delta(c) + t\,\Delta(l) < \left(\sqrt{1 + 1/k_M} - 1\right)(t\,l_m - \Delta(c)/2). \tag{2.4}$$

Theorem 2.2. *A Hotelling town network satisfying the BLC and DB conditions has the profit degree growth property.*

A *cost uniform* Hotelling town is a Hotelling town with $\Delta(c) = 0$. A *length uniform* Hotelling town is a Hotelling town with $\Delta(l) = 0$. A *uniform* Hotelling town is a Hotelling town with $\Delta(c) = \Delta(c) = 0$. We note that the degrees of the nodes can be different.

Remark 2.1. In a cost uniform Hotelling town, if $2\,k_M\,l_M\,\Delta(l) \leqslant l_m^2$ there is a unique network Nash price strategy; in a length uniform Hotelling town, if $2\,t\,k_M\,l_M\,\Delta(c) \leqslant (t\,l_M - \Delta(c)/2)^2$ there is a unique network Nash price strategy; and in a uniform Hotelling town there is a unique network Nash price strategy that satisfies the profit degree growth property

2.3 Local Stability

Consider that a firm F_i located a node i changes its location to a point y_i in a road $R_{i,j}$ at distance x for the node i. Let $\mathbf{P}(x; i, j)$ denote the Nash equilibrium price strategy taking in account the new localization of the firm F_i and let $\pi_i(x; i, j)$ denote the profit of firm F_i with respect to the price strategy $\mathbf{P}(x; i, j)$.

Definition 2.3. We say that a firm F_i is *node local stable* if there is $\epsilon_i > 0$ such that $\pi_i(0; i, j) > \pi_i(x; i, j)$ for every $0 < x < \epsilon_i$, with respect to the local optimal equilibrium price strategy. A Hotelling network is *firm position local stable* if every firm in the network is node stable.

We denote by k_M be the maximum node degree of the Hotelling town

$$k_m = \min\{k_i : i \in V\}.$$

Theorem 2.3. *A Hotelling town satisfying the BLC condition and with $k_m \geq 3$ is firm position local stable.*

Firms F_i with $k_i = 2$ are node local unstable, except for networks satisfying special symmetric properties. Firms F_i with $k_i = 3$ whose neighboring firms have nodes degree greater or equal to 3 are node local stable. Furthermore, firms F_i with $k_i \geq 4$ whose neighboring firms have nodes degree greater or equal to 2 are node local stable.

2.4   Conclusion

We presented a model of price competition in a network, extending the linear city presented by Hotelling to a network where firms are located at the nodes and consumers distributed along the edges. Under a condition on the production costs, road lengths and maximum node degree we found the unique pure Nash price strategy for which the Hotelling tow has a local market structure, i.e. the consumers prefer to buy at the local firms. Finally, we show that for small perturbations of its own localization, a Hotelling network where all nodes have degree greater that two has an optimal localization strategy consisting in all firms located at the nodes.

Acknowledgements We acknowledge the financial support of LIAAD-INESC TEC through 'Strategic Project—LA 14—2013–2014' with reference PEst-C/EEI/LA0014/2013, USP-UP project, IJUP, Faculty of Sciences, University of Porto, Calouste Gulbenkian Foundation, FEDER, POCI 2010 and COMPETE Programmes and Fundação para a Ciência e a Tecnologia (FCT) through Project "Dynamics and Applications", with reference PTDC/MAT/121107/2010.

References

1. Hotelling, H.: Stability in competition. Econ. J. **39**, 41–57 (1929)
2. Pinto, A.A., Parreira, T.: A hotelling-type network. In: Peixoto, M., Pinto, A.A., Rand, D. (eds.) Dynamics, Games and Science. I. Springer Proceedings in Mathematics, vol. 1, pp. 709–720. Springer, Berlin (2011)
3. Parreira, T., Pinto, A.A.: Local market structure in a hotelling town. Submited Paper, 2013
4. Salop, S.: Monopolistic competition with outside goods. Bell J. Econ. **10**, 141–156 (1979)

Chapter 3
On the Dynamics and Effects of Corruption on Environmental Protection

Elvio Accinelli, Laura Policardo, and Edgar J. Sánchez Carrera

3.1 Introduction

Few studies have been devoted to model the dynamics of firms' bribing behavior and corruption driven by imitation, and in fact most studies are static instead. Little effort has been made to model the equilibrium level of bribery in an economy by taking into consideration both the macro environment and the micro-bribing behavior. Corrupt behavior is defined as bribes paid by firms to public officials (auditors).

Related references on corrupt behavior began with Tirole's seminal paper [18] as one of the first attempts to model group reputation and the persistence of corruption as an aggregate of individual reputations. He studies the joint dynamics of individual and collective reputations and derives conditions to rebuild group reputations. In his work, group reputation is modelled as an aggregate of individual reputations, and new members joining a group "inherit" the good or bad reputation of the coalition. Stereotypes about the expected quality of a group are history dependent since collective reputation is a long term, path dependent and long-lasting process because new members inherit the reputation of the elders. Despite the model by Tirole, few studies have been devoted to model firms' bribing behavior (see [16], for a literature review).

E. Accinelli (✉)
Facultad de Economía, Universidad Autónoma de San Luis Potosí, San Luis Potosí, México
e-mail: elvio.accinelli@eco.uaslp.mx

L. Policardo
Department of Economics, University of Siena, Siena, Italy
e-mail: policardo@unisi.it

E.J. Sánchez Carrera
Facultad de Economia, UASLP Mexico, Department of Economics and Statistics, University of Siena, Siena, Italy
e-mail: edgar.carrera@uaslp.mx

A.A. Pinto and D. Zilberman (eds.), *Modeling, Dynamics, Optimization and Bioeconomics I*, Springer Proceedings in Mathematics & Statistics 73, DOI 10.1007/978-3-319-04849-9__3,
© Springer International Publishing Switzerland 2014

Mishra [11] considers a group of firms facing a certain pollution standard to illustrate how pervasive corruption can become a social norm. He shows that corruption or non-compliant behavior can be the equilibrium outcome in some cases and in such situations, corruption is the norm rather than deviant behavior.

Carilllo [6] develops a dynamic model of corruption in which agents are aware of their "propensity for corruption" and their clients choose an optimal level of bribe to offer. Such a framework provides an explanation for different implicit prices for illegal services (bribes or kick-backs) for similar countries (or organizations within similar countries), based on an analysis of clients' reaction.

Fredriksson and Svensson [8] develop a theory of environmental policy formation, taking into consideration the degrees of corruptibility and political turbulence. They find an empirical interaction between corruption and political instability, i.e. political instability has a negative effect on the stringency of environmental regulations if the level of corruption is low, but a positive effect when the degree of corruption is high.

Wydick [20] argues that in a free market, firms with well-defined property rights have no incentives to bribe public officials. However, if the government uses monopoly power to interfere with and restrict the market, then firms may be forced to bribe public officials. Firms face the "prisoner's dilemma" in the sense that if all the firms refuse to bribe, they will all be better off, but since a single deviation will make the deviant firm better off when the other players are playing honestly, every firm realises that the others will cheat and must therefore bribe to remain competitive and they will be collectively worse off as a result (see, e.g., [12, 17]). Hardin [9], in his seminal essay, calls it the "tragedy of the commons". Fortunately, in reality, we observe that the tragedy of the commons does not occur everywhere. In some societies, firms paying bribes to the government are very rare (e.g., the Scandinavian countries), although other negative examples of widespread corruption exist.

Shleifer and Vishny point out two different types of corruption: (1) "Corruption without theft" where the corrupt official accepts a bribe to provide whatever service, but then turns over the legal price of the service to the government and (2) "Corruption with theft" in which the corrupt officer accepts the bribe, but then doesn't turn over anything to the government at all. According to Whydick [20], this latter type of corruption is rampant and hard to stop, and is the type of corruption we consider in this paper.

Our aim is to study the joint dynamics of corruption and polluting choices undertaken by firms, with the novel assumption in this context that the learning process is simply imitation, in a context of environmental protection regulation. Our approach comes from evolutionary game theory and dynamic optimisation.[1] The hypothesis of evolutionary dynamics driven by imitation helps us to understand the strategic foundations of the stable corrupt behaviour equilibrium. In the real world indeed, pervasive corruption sometimes is a "social norm", although some other opposite examples of almost absence of corruption exist (according to Transparency

---

[1]Evolutionary analysis is well-documented in the game theory literature (see [19]).

International on the Global Corruption Barometer[2] New Zealand, Singapore or Finland are very little corrupted). In Mexico, for instance, corruption is widespread at all levels in public offices, and this behaviour is sustained by imitation, because people get corrupted because the others are. Corruption comes at all levels in Mexico as a kind of "cultural behavior", since the word for bribe, mordida, literally means bite, and getting bitten in Mexico is regrettably common (see [20]). In Mexico the mordida permeates every level of society and institutions where individuals act because it is a norm and they just do what the others are doing (to be corrupt or not). Bribes in Mexico are common and indeed are often necessary for obtaining business licenses and other types of permits. A popular Mexican saying: "*el que no transa no avanza, who does not corrupt does not move on*", highlights how corruption is fundamental for personal attainment.

The point of departure of our evolutionary model is that people's beliefs are not always rational. In general, in evolutionary games strategies emerge from a trial-and-error learning process according to which players find that some strategies perform better than others, and afterwards, they decide to adopt—or simply imitate—them.[3]

We assume that firms face a given pollution standard, decided by the government. The government exercises a control over those firms through public officials, who have to check the "quality" of the firm by writing down a report stating whether the level of pollution produced by the firm itself is above or lower the standard. A negative report (a report stating a level of pollution above the standard) implies a fine to be paid by the firm to the government. This fact may induce a corrupt firm which does not respect the standard to offer a bribe to the officer, who accepts it if he is corrupt as well or refuse it if he is not. If a firm instead respects the standard, this does not mean that the fine is avoided, because if the officer is corrupt, he may thread to write down a negative report and ask a bribe. It is assumed that even though the firm can appeal to the court against the unfair report, this requires a long bureaucratic process so the firm strictly prefers to pay the requested bribe.

It will be clear later that one possible "bad" outcome characterised by a whole society of polluting firms and corrupt officers can be sustained by rational agents who learn by imitation, despite the existence of multiplicity of equilibria of a perfectly honest population and a more realistic simultaneous presence of honest and dishonest agents.

The remainder of this paper is organized as follows. Section 3.2 presents a one-shot 2 × 2 game to model firms' pollution decisions and officers' decisions about to be corrupt. Section 3.3 develops a model of evolutionary dynamics of officers' imitative behavior about to accept (or ask for) a bribe or not. We consider that official behavior is driven by imitation of the most successful. Section 3.4 develops a model for the dynamics of payoff-maximizing firms. Subsequently, we

---

[2]Available at: http://www.transparency.org/policy_research/surveys_indices/gcb/2010.

[3]See [13] for a definition of imitative behaviour.

consider a dynamic decision problem for a firm facing intertemporal externalities from pollution. Section 3.5 concludes the paper.

3.2   The One-Shot Game

This section introduces a 2×2 game between an inspection official and a firm. As previously mentioned, the game assumes two types of individuals, firms and public officials, who can be either corrupt or not. We assume that there exists a regulatory institution, namely a court or environmental authority, which decides a given pollution standard (assumed to be zero for convenience) that has to be respected by the firms. This authority checks the compliance to this standard by means of public officials who have the duty to inspect and measure the level of emissions produced by each firm, and write down a report declaring whether the firm respects the standard or not.

Each player decides to be corrupt or honest: a corrupt firm does not respect the pollution standard decided by the government and, when inspected by a public official, offers a bribe to avoid the fine. A corrupt officer, instead, accepts the bribe when offered by a corrupt firm, or asks for one when he inspects an honest firm (meaning a firm which respects the standard), by threading her to write down an unfair report and let her pay the fine. Even though the firm can appeal to the court against the unfair report, this requires a long bureaucratic process so we assume that the firm strictly prefers to pay the requested bribe.

An honest officer refuses the bribe offered by a corrupt firm, and writes down fair reports, irrespective of the fact he is meeting an honest firm or not.

Moreover, consider that:

- There are strategic complementarities, i.e. a polluting p-firm prefers matching a dishonest d-official and a non-polluting n-firm prefers matching an honest h-official.
- With some probability the d-official is caught and fined by court. If the official is detected by the court, he is charged a fine $M > 0$. The probability to detect a dishonest officer is denoted by $P \in [0, 1]$.
- For each firm inspected, the officer receives a monetary reward $W > 0$, and bears an effort $e \leq W$ which does not change according to the level of emissions produced or whether the firm is corrupt or not.
- Nonpolluting firms pay a fixed environmental cost of production, denoted by $C > 0$, which represents the additional cost of buying newer an environmentally friendly machinery, while polluting firms do not consider this cost.
- A polluting firm found guilty by the court is charged a fine denoted by $\mathscr{F} > 0$. A nonpolluting firm inspected by an honest official does not pay the fine.
- A polluting firm inspected by a dishonest official pays a bribe R and avoids the fine. Notice that the bribe R is a monetary quantity, $R = \theta \mathscr{F} > 0$, with $\theta \in (0, 1)$.

- A non-polluting firm inspected by a dishonest official must pay a bribe, $r_d = \theta W > 0$ with $\theta \in (0, 1)$, for having a fair report. That is non-polluting firms must pay a bribe when inspection is done by a dishonest official.

The 2×2 game between the inspection officer, O, and a firm, F, is introduced in the following definition:

Definition 3.1. The normal-form representation of this game is the following payoff matrix:

O/F	p	n
d	$W + R - e - PM, \ \pi_p - R$	$W + r_d - e - PM, \ \pi_n - r_d - C$
h	$W - e, \ \pi_p - \mathscr{F}$	$W - e, \ \pi_n - C$

where $\pi_p > 0$ is the gross-payoff (revenues) of the polluting firm and $\pi_n > 0$ is the revenues of the non-polluting firm. The natural choice of parameters $R - PM \geq 0$ and $\pi_p - R \geq \pi_n - r_d - C$, makes it a coordination game.

The game can be represented as a two-population normal form game denoted by the list:

$$\Gamma = \langle(\{F, O\}); \ i \in \{p, n; \ h, d\}; \ E(i)\rangle.$$

for each population of officers and firms $\{O, F\}$ with their respective vector of strategies for each i-strategic player, and respective expected payoffs $E(i)$.

Let us denote by $O = (O_h, \ O_d) \in \Delta^O$ the profile distribution of officers' type in a given period of time t_0, where O_h is the share of honest officers, and O_d is the share of dishonest officers. At the same period of time the profile distribution of firms' type is given by $F = (f_p, \ f_n) \in \Delta^F$ where f_p is the share of polluting firms, and f_n is the share of non-polluting firms. Hence the strategy distribution of each population is given by:

$$\Delta^O = \{O \in R_+^2 : O_h + O_d = 1\}.$$

$$\Delta^F = \{F \in R_+^2 : f_p + f_n = 1\}.$$

Note that the expected payoff of a non-polluting firm is given by:

$$E(f_n|O) = [\pi_{np} - C]O_h + [\pi_n - C - r_d]O_d \tag{3.1}$$

and the expected payoff of a polluting firm is given by:

$$E(f_p|O) = [\pi_p - \mathscr{F}]O_h + [\pi_p - R]O. \tag{3.2}$$

where π_p is the profit of a polluting firm. Firms prefer polluting if $E(f_p|O) > E(f_n|O)$ and this happens if the share of honest officials is not large enough, i.e.

$$O_h < \bar{O}_h = \frac{\pi_p - \pi_n + C + r_d - R}{\mathscr{F} + r_d - R}.$$

Moreover, if $\pi_p - \mathscr{F} > \pi - C$ firms prefer to be polluting.

Similarly, the expected payoffs of the honest, h-official, and dishonest, d-official, is given by:

$$E(O_h|F) = W - e.$$

$$E(O_d|F) = W + R + r_d - e - PM + f_n(r_d - R).$$

Note that $E(O_h|F) \geq 0$ since $W \geq e$ and $E(O_d|F) \geq 0$ if $f_n \geq \frac{e + PM - (W + R)}{r_d - R}$, since we are considering the case of non-negative expected payoffs. Therefore, officials prefer to be h-official if $E(O_h|F) > E(O_d|F)$, and it happens if the share of non-polluting firms is large enough, i.e.:

$$f_n > \bar{f}_n = \frac{R + r_d - PM}{R - r_d}.$$

And this happens if either the fine for a dishonest official or the probability that the court monitoring a dishonest behavior increase.

Remark 3.1. For the above one-shot game (Definition 3.1), in order to eradicate polluting firms and dishonest officials, the punishments (fines $\mathscr{F}$ and PM) must be greater than the environmental costs of production and the bribes (C, R and r_d).

Of course, under the quantitative relationships between payoffs described above, we can state that:

Remark 3.2. The game Γ has two pure Nash equilibria. One is a high-compliance equilibrium with no firms choosing to pollute and officials remaining honest, $(n, h) = (f_n, f_p; O_h, O_d) = (1, 0; 1, 0)$. The other is the low-compliance equilibrium where all the officers are dishonest and all the firms choose to pollute, $(p, d) = (f_n, f_p; O_h, O_d) = (0, 1; 0, 1)$. There is also a mixed strategy Nash equilibrium given by:

$$(\bar{O}_h, (1 - \bar{O}_h); \bar{f}_n, (1 - \bar{f}_n)), \tag{3.3}$$

where firms and officials are indifferent between to be corrupt or not.

Hence a firm's decision regarding whether to be polluting or not will depend on the probability to encounter an honest officer. Hence the "evolution" of honesty amongst the officers over time will affect the level of pollution.

3.3   The Officers' Imitative Behavior

This section presents the key innovative feature of the paper. We study an analysis of corruption amongst officers by using imitative dynamics, since we argue that "imitation" of corrupt and successful strategy has a lot to do with the spread and persistence of corruption. From this perspective, we present the evolutionary dynamics of corruption driven by imitative behavior.

To explain why individuals imitate we should think of it as a kind of rational behavior (see [2]). Imitation results in agents performing a spectrum of tasks "as others do". We assume that occasionally each individual in a finite population gets an impulse to revise her (pure) strategy choice (be corrupted or non-corrupted). There are two basic elements in imitation theory. The first is a specification of the time rate at which individuals in the population review their current strategy choice. This rate may depend on the current performance of the agent's pure strategy and on other aspects of the current population state. The second element is a specification of the choice probabilities of a reviewing individual. The probability an i-strategist will switch to some pure strategy j may depend on the current performance of these strategies and other aspects of the current population's state. If these impulses arrive according to i.i.d. Poisson processes, then the probability of simultaneous impulses is zero, and the aggregate process is also a Poisson process. Moreover, the intensity of the aggregate process is just the sum of the intensities of the individual processes. If the population is large, then one may approximate the aggregate process by deterministic flows given by the expected payoffs from corruptive and non-corruptive behaviors.

Björnerstedt and Weibull [5] study a number of such models, where individuals who revise may imitate other agents in their population of players, and show that a number of payoff-positive selection dynamics, including the replicator dynamics, may be so derived. In particular, if an individual's revision rate is linearly decreasing in the expected payoff of her strategy (or of the individual's latest payoff realization), then the intensity of each pure strategy's Poisson process will be proportional to its population share, and the proportionality factor will be linearly decreasing in its expected payoff. If every revising agent selects her future strategy by imitating a randomly drawn agent in their own player population,[4] then the resulting flow approximation is the replicator dynamics.

In the sequel, we consider that officials follow an imitative behavior of the best performed strategy given a fixed distribution of the share of non-polluting and polluting firms. A reviewer official i is willing to review her current strategy $i = \{h, d\}$, sometimes resulting in a change on it, with probability $r_i(O) \in [0, 1]$. $r_i(O)$ is then the time rate at which officials review their strategy choice. This

---

[4]Evolutionary game theory considers populations of decision makers, while analysing the player profiles within these populations, instead of single players. We can therefore identify a population game, where N large populations strategically interact, as an N-player form game, where each player has a large population behind him (see [10]).

probability depends on the actual distribution of the honest and dishonest officers and in the benefits associated with her current behavior.[5] It is natural to assume that the likelihood that an official will be willing to change her current strategy depends inversely with the performance of her current behavior. Having opted for a change, the official will adopt a strategy followed by the first successful person met from the population (her neighbour), i.e. there is a probability $p_{ij}(O) \in [0, 1]$, that a reviewing i-official really switch to some pure strategy $j = \{h, d\}$, $j \neq i$. Assuming a continuum of officers, independence of switches across officials' same type, and the process of switches from type i to type j as a Poisson process with arrival rate $o_i r_i p_{ij}$, by the law of large numbers we model these aggregate stochastic process as a deterministic flow, and this means that the probability that an i-officer, $i = \{h, d\}$ will review his own strategy will be denoted by $r_i(O)$.

From these considerations it follows that:

- The outflow from the i-types' officials is:

$$\sum_{j \neq i} O_i r_i(\cdot) p_{ij}(\cdot).$$

- While the inflow is:

$$\sum_{j \neq i} O_j r_j(\cdot) p_{ji}(\cdot).$$

Being $O = (O_h, O_d)$ the profile distribution of officials' behavior, we apply the behavioral rule according to which a reviewing official who decides to change her current strategy takes into consideration imitating a strategy which performs better than her own current strategy. With the use of behavioral rules we can define an evolutionary dynamic as an inflow-outflow model. That is, rearranging terms, we get the system of differential equations characterising the dynamic flow of officials:

$$\dot{O}_h = r_d p_{dh} O_d - r_h p_{hd} O_h$$

$$\dot{O}_d = -\dot{O}_h,$$

(3.4)

meaning that the inflow of honest officers is given by the difference between the number of dishonest officers who decide to become honest ($r_d p_{dh} O_d$) and the number of honest officers who decide to stay honest ($r_h p_{hd} O_h$)

Assume that $r_i, i = d, h$ is population specific and is linear and decreasing in the level of the expected utility, i.e.

---

[5]This is the "behavioral rule with inertia" (see [5, 14, 15, 19]) that allows an agent to reconsider her action with probability $r \in (0, 1)$ each round.

$$r_i = \alpha - \beta E(\cdot), \tag{3.5}$$

where $\alpha > 0$, $\beta \geq 0$ and $\frac{\alpha}{\beta} \geq E(\cdot)$ assure that $r_i(\cdot) \in [0, 1]$. The parameter α is interpreted as the degree of dissatisfaction for following a behavior $i = \{h, d\}$ and β measures the weight of the payoff on the probability to be a reviewer. As long as the expected payoff level of the i-official, $E(\cdot)$ increases, her average reviewing rate r_i will decrease.

Reviewing officers evaluate their current strategy and decide to imitate only the successful one. Therefore, by the above considerations, the system (3.4) can be written as:

$$\dot{O}_h = -O_h \left[(\alpha - \beta E(O_h)) \, p_{hd} - (\alpha - \beta E(O_d)) \, p_{dh}\right] + (\alpha - \beta E(O_d)) \, p_{dh}$$

$$\dot{O}_{dh} = -\dot{O}_h,$$

$$\tag{3.6}$$

The share f_n of non-polluting firms is a constant number at any period of time t, and gross-payoffs π_i, salaries and effort (W, e) are given. Then $E(O_h)$ and $E(O_d)$ are constant too. Defining by $A = [(\alpha - \beta E(O_h)) \, p_{hd} - (\alpha - \beta E(O_d)) \, p_{dh}]$ and $B = (\alpha - \beta E(O_d)) \, p_{dh}$, the solution of the differential equation (3.6) is

$$O_h(t) = \left(O_h(0) - \frac{B}{A}\right) \exp(-At) + \frac{B}{A}. \tag{3.7}$$

where $O_h(0)$ is the share of h-type officials at time $t = 0$ and,

$$\frac{B}{A} = \frac{(\alpha - \beta E(O_d)) \, p_{dh}}{(\alpha - \beta (W - e)) \, p_{hd} - (\alpha - \beta E(O_d)) \, p_{dh}}. \tag{3.8}$$

According to the solution of the differential equation it follows that

$$\lim_{t \to \infty} O_h(t) = \frac{B}{A}.$$

This share depends on the expected value of the possible strategies and on the probabilities that the reviewers change their behaviour.

- this indicates that corrupt behaviour can persist over time.
- However, this decreases as the probability that a non corrupt official become corrupt, and $O_d(t) \to 0$ as $p_{hd} \to 0$.

Consider that reviewer officials copy successful behaviours according to a payoff-monotonic updating.

An evaluation rule that seems fairly natural in a context of simple imitation is the "positive differences rule", whereby a strategy is evaluated according to the differences in payoffs observed in the reference group (see [4]).

That is, each i-reviewer changes her strategy if and only if $E(O_i) < E(O_j)$, $\forall i \neq j = \{h, d\}$,

$$p_{ij}(x) \equiv \begin{cases} 1 \text{ if } E(O_j) - E(O_i) > 0. \\ \\ 0 \text{ otherwise.} \end{cases}$$

According to this rule, that supposes that every reviewer knows the true value of $E(O_h)$ and $E(O_d)$, it happens that:

(i) the share of honest officials converges to one, $\frac{B}{A} = 1$ if $f_n > \bar{f}_n$ (no-corruption),

(ii) the share of honest officials converges to zero if $f_n < \bar{f}_n$ because in this case $\frac{B}{A} = 0$ (all corrupt),

where

$$\bar{f}_n = \frac{R + r_d - PM}{R - r_d}.$$

In the other case there is a mixed strategy equilibrium where the share of honest officials is given by:

$$\frac{B}{A} = \frac{(\alpha - \beta E(O_d)) p_{dh}}{(\alpha - \beta (W - e)) p_{hd} + (\alpha - \beta E(O_d)) p_{dh}} \in (0, 1).$$

Therefore, as the share of non-polluting firms becomes larger, the share of honest officials increases at a rate depending on the reviewing rate r_i. This means it is possible the coexistence of corrupt and non corrupt officials if the central authority has not an appropriate policy to fight corruption.

How to make corruptive behaviour disappear? Suppose that in time $t = t_0$ the share of non-polluting firms is f_n and consider an high probability to detect corrupt behaviour, i.e. $P = 1$ and the particular case of $R = \theta \mathscr{F}$, $r_d = \theta W$ then the threshold value is

$$\hat{f}_n^* = \frac{\theta(\mathscr{F} + W) - M}{\theta(\mathscr{F} - W)}.$$

Now, if M verifies the inequality

$$M(f_n) > \theta(\mathscr{F} + W) - f_n \theta(\mathscr{F} - W),$$

then the actual share of non polluting firms overpasses the threshold and then $E(O_j) - E(O_i) > 0$.

3.4   The Payoff-Maximizing Actions of the Firms

Recall that we have a mixed model where firms are profit maximizers and the officers are imitators. The chosen behavior by firms depends on the expected payoff associated with each of the possible strategies, to be polluting or not. Of course a polluting firm can switch to being non-polluting, and in this case the accumulated waste is assumed to be transported quickly and efficiently to the nearby garbage collector, so we do not study this fact. However potentially switches are allowed because the proportion of honest officers might be different at different points in time.

Consider that firms maximize intertemporal profits in a market on which they face the next assumptions:

1. A demand structure that gives to each firm sales revenues $g(x(t))$ where $x(t)$ represents the capacity[6] of production at time t, and $x(t) > 0$. We consider that g is a differentiable $C^2(0, \infty)$ function such that: $g(0) = 0$, $g \geq 0$, $g' > 0$ and $g'' < 0$.
2. The capacity of production is a differentiable function $x : R_+ \to R$ for each time t. Capacity is finite and bounded, so at every time $t \geq 0$, $x(t) \in [0, x']$.
3. A fraction $(1 - u(t)) g(x(t))$ of the revenues is consumed, and what is left, $u(t)g(x(t))$, is invested in new production capacity at a price $1/a > 0$.
4. An official inspects the firm at the end of the planning period T.
5. Each firm can buy an initial production capacity at the unitary price c and can sell it at price w at the end of the planning period T.
6. The production process generates pollution. Let $z(t)$ be the accumulated waste at time t. The instantaneous variation of waste $\dot{z}(t)$ is proportional to the used production capacity (level of production), so $\dot{z}(t) = bx(t)$, with $b > 0$.
7. Nonpolluting firms have an environmental cost of production (or cost to keep clean the environment) that is proportional to the waste accumulated until time t, $C(t) = \int_0^t bz(t)e^{-rt}d\tau$, $b \in (0, 1)$. So, at the end of the period T, $C(c) = C(T) = \int_0^T bz(t)e^{-rt}dt$. The intertemporal discount rate is constant and equal to $r > 0$.

Therefore the profit of a non-polluting firm, $\pi_n(T)$, in period T, is given by the following maximization program:

$$
\begin{cases}
\pi_n(T) - C(T) = \\
\max_{u(t) \in [0,1]} \int_0^T [(1 - u(t))g(x(t)) - bz(t)]e^{-rt}dt - cx(0) + wx(T)e^{-rT} \\
\\
\dot{x}(t) = au(t)g(x(t)), \quad x(0) \text{ free}, \quad x(T) \text{ free} \\
\dot{z}(t) = bx(t), \quad z(0) = z_0, \quad z(T) \text{ free}.
\end{cases}
$$

$$(3.9)$$

---

[6]For simplicity, we assume that the production's capacity is always fully exploited, so $x(t)$ can be interpreted as both capacity and level of production.

Since a polluting firm does not pay for the environmental cost of production, its maximization program is given by:

$$
\begin{cases}
\pi_p(T) = \max_{u(t)\in[0,1]} \int_0^T [(1-u(t))g(x(t))]e^{-rt}\,dt - cx(0) + wx(T)e^{-rT}. \\
\dot{x}(t) = au(t)g(x(t)), \quad x(0) \text{ free}, \quad x(T) \text{ free}.
\end{cases}
$$

$$(3.10)$$

Notice that the fraction of the sales, $(1-u(t))g(x(t))$, is consumed and the rest, $u(t)g(x(t))$, is invested in making a new production bought at a price $1/a$. Without loss of generality, let us consider for the cases (3.9) and (3.10) the following inequality:

$$
c > 1/a > w. \tag{3.11}
$$

Because the initial capacity is financed by a loan having an interest rate higher than r while the gradual increase in capacity $ug(x(t))$ is paid for immediately. To ensure that it pays to invest, i.e. $x(0) > 0$, we consider that:

$$
r^{-1}(1-e^{-rT})(g'(x(0))) > c - we^{rT}. \tag{3.12}
$$

Hence, the Hamiltonian for the problem (3.9), with $p_0 = 1$,[7] is:

$$
H(x, z, u, p_1, p_2, t) = [(1-u)g(x) - bz]e^{-rt} + p_1 aug(x) + p_2 bx. \tag{3.13}
$$

and the candidate for optimality $(x^*(t), z^*(t), u^*(t))$ satisfies:

$$
u^*(t) \text{ maximizes } u(t)g(x)\left[p_1 a - e^{-rt}\right].
$$

So

$$
u^*(t) = \begin{cases}
1 & \text{if } p_1(t) > (1/a)e^{-rt} \\
0 & \text{if } p_1(t) < (1/a)e^{-rt} \\
\in (0,1) & \text{if } p_1(t) = (1/a)e^{-rt}
\end{cases} \tag{3.14}
$$

and,

$$
\dot{p}_1(t) = -\frac{\partial H^*}{\partial x} = -\left[(1-u^*(t))e^{-rt} + p_1(t)au^*(t)g'^*(x^*(t))\right]. \tag{3.15}
$$

---

[7]Notice that, if $p_0 = 0$ then: $\dot{p}_2 = -\frac{\partial H}{\partial z} = 0 \Rightarrow p_2(t)$ is a constant. From the transversality condition it follows that $p_2(t) \equiv 0$. From the maximum principle is necessary that: $(p_0, p_1(t), p_2(t)) \neq (0,0,0)$ ∀ t, then $p_1(T) \neq 0$, but this contradicts the transversality condition.

Taking into account the value of $u^*(t)$ it follows that:

$$\dot{p}_1 = -\left[\max\left\{e^{-rt},\, ap_1(t)\right\}\right]g'^*(x^*(t)), \tag{3.16}$$

and then $\dot{p}_1(t) < 0$. So $p_1(t)$ is a decreasing function.

From the necessary conditions it follows that:

$$p_1(0) = c \quad \text{and} \quad p_1(T) = 0. \tag{3.17}$$

Proposition 3.1. *There is a single moment* T^* *verifying* $p_1(T^*) = (1/a)e^{-rT^*}$.

Proof. Denote by $\phi(t) = \frac{1}{a}e^{-rt}$. Since $c > 1/a > w$ it follows that $p_1(0) > \phi(0)$ and $p_1(T) = 0 < \frac{1}{a}e^{-rT}$, being $p_1(t)$ strictly decreasing, there exists at least one moment T^* such that $p_1(t) = \frac{1}{a}e^{-rt}$.   □

Proposition 3.2. *There is a single solution for the maximization problem* (3.9).

Proof. To see this recall that $u(t) = 1$, $\forall t \in [0, T^*]$ and $u(t) = 0$, $\forall t \in [T^*, T]$. Then, according to Eq. ((B)3.9) it follows that $x^*(t) = x^*(T)$, $\forall t \in [t^*, t]$. In turns $x^*(t)$ is uniquely determined by the equation $\dot{x} = ag(x)$ and $x^*(T^*) = x^*(T)$, $\forall t \in [0, T^*]$ and $p_1(t) < \frac{1}{a}e^{-rt}$, $\forall t \in [T^*, T]$. Thus, in time $t = T^*$ the equality $p_1(T^*) = (1/a)e^{-rT^*}$ follows.

Hence:

$$x^*(t) = \begin{cases} \text{a solution of } \dot{x}(t) = ag\,(x(t)) \ \forall \, t \, \in [0, T^*] \\ x(T^*) \qquad\qquad\qquad\qquad\quad \forall \, t \in [T^*, T] \end{cases} \tag{3.18}$$

$$\dot{p}_2(t) = -\frac{\partial H^*}{\partial z} = be^{-rt} \rightarrow p_2(t) = -\frac{b}{r}e^{-rt} + \bar{p}_2 \text{ where } \bar{p}_2 \text{ is a constant.}$$

□

Note that the differences in profits between the polluting and nonpolluting firms is given by:

$$\pi_p(T) - \pi_n(T) = b\left[\int_0^{T^*} x(t)e^{-rt}dt + x(T^*)\frac{1}{r}\left(e^{-rT} - e^{-rT^*}\right)\right]. \tag{3.19}$$

Since the expression that appears in the integrand is always positive, the higher the value of T^* the greater the difference between these two benefits.

The same result holds for the expected values,

$$E(\pi_p) - E(\mathscr{I}_n) =$$

$$= b \left[\int_0^{T^*} x(t)e^{-rt}dt + \frac{x(T^*)}{r} \left(e^{-rT} - e^{-rT^*} \right) \right] \qquad (3.20)$$

$$- R - r_d + [-F + R + r_d]O_h,$$

according with Eqs. (3.1) and (3.2).

To get an analytical solution we consider the usual case $g(x) = x^{\frac{1}{2}}$. From Eq. (3.16) it follows that:

$$\dot{x}(t) = a \, (x(t))^{\frac{1}{2}}, \forall t \leq T^*.$$

So,

$$x^*(t) = \begin{cases} \frac{1}{4}a^2t^2 + x(0) & \forall t \in [0, T^*] \\ \\ \frac{1}{4}a^2(T^*)^2 + x(0) & \forall t \in [T^*, T] \end{cases} \qquad (3.21)$$

From Eq. (3.16), $\forall t \leq T^*$ we get that:

$$\dot{p}_1 = ap_1(t)\frac{1}{2} \left(x^*(t) \right)^{-\frac{1}{2}}, \qquad (3.22)$$

Substituting (3.21) in the above equation we obtain:

$$\dot{p}_1 = -ap_1\frac{1}{2} \left[\frac{1}{4}a^2t^2 + x(0) \right]^{-\frac{1}{2}}, \qquad (3.23)$$

and integrating, we obtain that:

$$\ln p_1 = -\arctan \frac{a}{2\sqrt{x_0}}t + C, \qquad (3.24)$$

where C is a constant on integration. Taking exponential on both sides of (3.24) and $p_1(0) = c$, hence the expression:

$$p_1(t) = ce^{-\arctan \frac{a}{2\sqrt{x_0}}t} \qquad (3.25)$$

holds.

The difference in benefits between a polluting and nonpolluting firm is given by:
$$\pi_p(T) - \pi_n(T) =$$

$$\begin{cases} \int_0^{T^*} [\frac{1}{4}a^2t^2 + x(0)]e^{-rt}\,dt & \text{if } T < T^* \\[2mm] \int_0^{T^*} [\frac{1}{4}a^2t^2 + x(0)]e^{-rt}\,dt - [\frac{1}{4}a^2(T^*)^2 + x(0)]\frac{1}{r}[e^{-rT} - e^{-rT^*}] & \text{if } T > T^*. \end{cases}$$
$$(3.26)$$

Notice that such a difference between the benefits of the polluting and nonpolluting firm increases with T^*.

3.4.1   The Rate of Pollution and the Discount Factor

According with the equation $\dot{z} = bx^*(t)$ the instantaneous velocity of pollution accumulation, (or the rate at which pollution accumulates) increases with time $t \leq T^*$ and after this moment it is a constant: $\dot{z} = bx(T^*)$, $\forall\, t > T^*$. Then, the period during which the instantaneous velocity of contamination grows, it is an increasing function of the discount rate. These facts are summarized in the following proposition:

Proposition 3.3. *If there exists a solution T^* for the equation $\chi(T, r)) = p_1(T) - \frac{1}{a}e^{-rT} = 0$, then there exists a neighborhood V_{r^*} of r^* such that for each $r \in V_{r^*}$ there exists only one optimal time $T^*(r)$ such that the instantaneous velocity at which the pollution is created increases until $t = T^*$, after that time the rate of pollution does not increase. This optimal time increases with the discount factor r.*

Proof. Consider the function,

$$\chi(T, r)) = p_1(T) - \frac{1}{a}e^{-rT}$$

Since, for a given $r^* > 0$ there exists T^* such that $\chi(T^*, r^*) = 0$. From the implicit function theorem there exists a continuous function $T^*(r)$ such that $\chi(T(r), r) = 0$. Now using the derivative of the implicit function, it follows that:

$$\frac{dT}{dr} = -\frac{\partial\chi/\partial r}{\partial\chi/\partial T} = -\frac{\frac{1}{a}Te^{-rT}}{\dot{p}_1 + \frac{1}{a}re^{-rt}}$$

Since $\dot{p}_1 < -\frac{1}{a}re^{-rt}$ then $\frac{dT}{dr} > 0$. This means that as low is the discount rate, lower is the optimal time until the rate of pollution increase (see Fig. 3.1). $\square$

Finally, for the second case (program (3.10)) we have a similar situation except for $b = 0$. The Hamiltonian to this problem (3.10) with $p_0 = 1$, is given by:

$$H(x, u, p_1, t) = (1 - u)g(x)e^{-rt} + p_1 au f(x) \qquad (3.27)$$

Fig. 3.1 Optimal time as a function of r

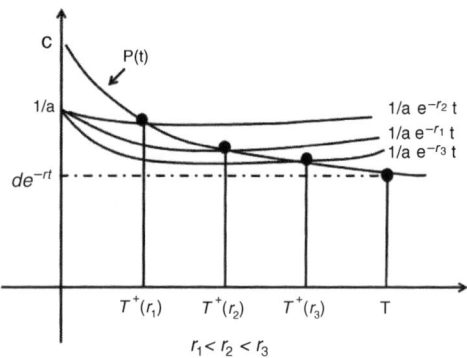

and the candidate for optimality $(x^*(t), u^*(t))$ verifies the similar conditions of the previous one. The maximized Hamiltonian $H^* = \max \{e^{-rt}, ap_1(t)\} g(x)$ is strictly concave on x if $g(x)$ is a strictly concave function, and so (x^*, u^*) is a solution to this problem.

For the particular case where $g(x) = x^{\frac{1}{2}}$, we can get the equation:

$$ce^{-\arctan \frac{a}{2\sqrt{x_0}} T} = \frac{1}{a} e^{-rT} \tag{3.28}$$

and it follows that T^* is a solution of the equation:

$$-\arctan \frac{a}{2\sqrt{x_0}} T + rT + \ln ac = 0 \tag{3.29}$$

Choosing the parameters of the model verifying the conditions (3.12) and (3.11), then a solution T^* for this Eq. (3.29) exists.

3.4.2 A Threshold Value for the Dynamics of Firms

In this section we show that there exists a threshold value such that once the share of honest officials exceeds this value, then a process in which polluting firms prefer to become nonpolluting begins, and the current nonpolluting firms remain non-polluting.

Recall that firms maximize their expected profits, so they prefer to be polluting if:

$$E(f_p) - E(f_n) > 0. \tag{3.30}$$

and this happens if the share of honest officials is smaller, i.e.

$$O_h < \frac{b\left[\int_0^{T^*} x(t)e^{-rt}dt + \frac{x(T^*)}{r}\left(e^{-rT^*} - e^{-rT}\right)\right] - R - r_d}{\mathcal{F} - R - r_d}. \tag{3.31}$$

Recall that the profile distribution of firms' type is given by $(f_p(t), f_n(t)) \in \Delta^F$ in time t. Consider that at the end of each period, firms choose their behavior for the next one. Assume that at each time t, firms know the officials' distribution $O(t)$, (i.e. they know the probability to be inspected by an honest or a dishonest official) then the dynamics of the share of firms is given by the next law of motion:

$$\begin{cases} \dot{f}_n = \left[E(f_n) - E(f_p)\right] f_p \\ \dot{f}_p = -\dot{f}_n. \end{cases} \tag{3.32}$$

If $E(f_n) - E(f_p) > 0$ then the share of non-polluting firms increases.

Let us introduce the function: $O_h^T : R_+ \to R$ defined by:

$$O_h^T(r) = \frac{b\left[\int_0^{T^*} x(t)e^{-rt}dt + \frac{x(T^*)}{r}\left(e^{-rT^*} - e^{-rT}\right)\right] - R - r_d}{\mathcal{F} - R - r_d}. \tag{3.33}$$

This function defines (for the discount rate r) a threshold value of honest officials $O_h^T(r)$ such that if $O_h^T(r) < O_h$, then the share of non-polluting firms increases. The next proposition summarizes the above consideration.

The following proposition shows that the threshold value is an increasing function of r. It is straightforward the intuition behind this proposition: if firms care less about the future, more should be the society's efforts to prevent pollution.

Proposition 3.4. *The threshold value, $O_h^T(r)$, is increasing function with r.*

Proof. To prove this theorem let us consider the auxiliary function:

$$\phi(r) = -\int_0^{T^*(r)} tx(t)e^{-rt}dt + x(T^*(r))\int_{T^*(r)}^T e^{-rt}dt.$$

From Proposition 3.3 we know that $T^*(r)$ is an increasing function of r, that $(T^*)'(r) > 0$ and that $0 < T^*(r) < T$. It follows that $\phi'(r) > 0$, so this function is increasing on r, i.e.

$$\phi'(r) = -\int_0^{T^*(r)} tx(t)e^{-rt}dt - x(T^*(r))e^{-rT^*(r)}T^{*\prime}(r)$$
$$+\dot{x}(T^*(r))T^{*\prime}(r)\int_{T^*}^T e^{-rt}dt + x(T^*(r))e^{-rT^*(r)}T^{*\prime}(r) - x(T^*(r))\int_{T^*(r)}^T te^{-rt}dt$$
$$= -\left[\int_0^{T^*(r)} tx(t)e^{-rt}dt + x(T^*(r))\int_{T^*(r)}^T te^{-rt}dt\right] > 0$$

The second equality is a consequence of the fact that $x(t) = x(T^*), \forall t \in [T^*, T]$ hence $\dot{x}(t) = 0$. $\square$

Two important insights are:

1. The intuition of Proposition 3.4 is given by the fact that when future does not matter at all, the discount rate is high and the current environmental cost of production is low, since we do not care about environment. When we care about future, then we may clean the current environment such that the environmental cost of production is higher. Then if the current cost of cleaning is lower and honest officials are few, $E(f_p) < E(f_n)$.
2. Note that if:

$$\int_0^{T^*} x(t)e^{-rt}dt + \frac{x(T^*)}{r}\left(e^{-rT} - e^{-rT^*}\right) > \frac{\mathscr{F}}{b}, \tag{3.34}$$

then independently of the officials' distribution, firms prefer to be polluting. That is contrary to intuition, because if the value $\int_0^{T^*} x(t)e^{-rt}dt + \frac{x(T^*)}{r}\left(e^{-rT} - e^{-rT^*}\right)$ is higher than the fine, a policy of raising the fine F may not be efficient. This may explain why increasing the fine makes the option of offering a bribe more attractive, thus inducing more corruption. The right value of $\mathscr{F}$ should be fixed (exogenously to our model) to the correct value that the society worries about the future, so is reducing the size of the bribes and implement a policy for increasing the probability of the time when there is compliance and no corruption.

Therefore it is important to develop a policy aimed at creating awareness about the future, so as to diminish the value of the discount rate.

3.5 Conclusion

This paper develops a model of corruption based on imitation, in an environmental policy context where (potentially corrupt) officers report pollution produced by firms. Officers might be honest or dishonest while firms may be polluting or not.

We identify several equilibria in the static game, which are confirmed by extending such a game in an evolutionary setting where officials' imitate the others' strategy and firms maximise profits. Equilibria range from stable corruption to honesty depending on the parameters of the model (i.e. fines, bribes and environmental damages as well as the firms' discount rate).

When firms care about future (i.e. a low discount rate) and officials are honest then we get the good outcome implying an economy without corruption, but the worst scenarios occurs when all the firms are briber and officials are dishonest.

To encourage an honest behavior, that is to say, a situation where firms prefer to be clean and officers prefer to be honest, bribes' size must be reduced, fines (M) must be increased, and P, the probability to detect dishonest behaviours performed by an officer must be greater, that is to say, the government must invest in increasing its effectiveness in detecting corrupt officers.

When this effectiveness is increased by means of the firm which receives an unfair report could be the object of future research, since the hypothesis that a firm prefers not to appeal to a court because of the cost and the long bureaucratic process may seem not completely realistic.

Acknowledgements We thank comments and feedbacks from the seminar participants of DGS II 2013—International Conference and Advanced School Planet Earth, Dynamics, Games and Science II. The financial support of CONACYT MEXICO is acknowledged.

References

1. Accinelli, E., Sanchez Carrera, E.: Strategic complementarities between innovative firms and skilled workers: the poverty trap and the policymaker's intervention. Struct. Change Econ. Dyn. **22**(1), 30–40 (2011)
2. Accinelli, E., Brida, J.G., Carrera, E.S.: Imitative behavior in a two-population model. In: Breton, M., Szajowski, K. (eds.) Advances in Dynamic Games, Annals of the International Society of Dynamic Games 11. Springer, New York (2010)
3. Aglietta, M., Reberioux, A.: Corporate Governance Adrift. Edward Elgar Publishing Limited, Cheltenham (2005)
4. Apesteguia, J., Huck, S., Oechssler, J.: Imitation-theory and experimental evidence. J. Econ. Theory **136**, 217–235 (2007)
5. Björnerstedt, J., Weibull, J.: Nash equilibrium and evolution by imitation. In: Arrow, K., et al. (eds.) The Rational Foundations of Economic Behavior. Macmillan, London (1996)
6. Carilllo, J.: Graft, bribes, and the practice of corruption. J. Econ. Manag. Strat. **9**, 257–286 (2000)
7. Coffee, J.: A theory of corporate scandals: why the usa and eu'rope differ. Oxf. Rev. Econ. Pol. **21**, 198–211 (2005)
8. Fredrikssona, P., Svenssonb, J.: Political instability, corruption and policy formation: the case of environmental policy. J. Public Econ. **87**, 1383–1405 (2003)
9. Hardin, G.: The tragedy of the commons. Science **162**, 1243–48 (1968)
10. Hofbauer, J., Sigmund, K.: Evolutionary Games and Population Dynamics. Cambridge University Press, Cambridge, MA (2002)
11. Mishra, A.: Persistence of corruption: some theoretical perspectives. World Dev. **34**(2), 349–358 (2006)
12. Rose-Ackerman, S.: The political economy of corruption. In: Elliott, K.A. (ed.) Corruption and the Global Economy. Institute for International Economics, Washington, DC (1997)
13. Sanditov, B.: Essays on social learning and imitation. Ph.D. thesis, Maastricht University (2006)
14. Schlag, K.H.: Why imitate, and if so, how? a boundedly rational approach to multi-armed bandits. J. Econ. Theory **78**(1), 130–156 (1998)
15. Schlag, K.H.: Which one should i imitate. J. Math. Econ. **31**(4), 493–522 (1999)
16. Svensson, J.: Eight questions about corruption. J. Econ. Perspect. **19**(3), 19–42 (2005)
17. Shleifer, A., Vishny, R.: Politicians and firms. Q. J. Econ. **1094**, 995–1025 (1994)
18. Tirole, J.: A theory of collective reputations (with applications to the persistence of corruption and to firm quality). Rev. Econ. Stud. **63**(1), 1–22 (1996)
19. Weibull, J.W.: Evolutionary Game Theory. MIT Press, Cambridge, MA (1995)
20. Wydick, B.: Games in Economic Development. Cambridge University Press, Cambridge, MA (2008)

Chapter 4
A Bayesian Pricing Model for CAT Bonds

Frieder Ahrens, Roland Füss, and A. Sevtap Selcuk-Kestel

4.1 Introduction

Over the last two decades, a number of major catastrophes have struck the (re)insurance industry and challenged its mitigation capacity causing a remarkable peak in overall insured losses. Hurricane Katrina was one of those which resulted in an insured loss of more than USD 68 billion [19]. As most of the catastrophe-prone areas are more densely populated, and thus, more prone to high property losses and more frequent catastrophes leading to a higher level of risk perception with more insureds and higher exposures to insurers, CAT bonds has become a useful instrument for capacity expansion.

Insurance and reinsurance companies finance catastrophe claims, either by building internal reserves, or by accumulating risk capital from external sources such as the capital markets. The securitization of catastrophe risk has created a broad range of insurance-linked securities (ILS), such as sidecars, industry loss warrants, catastrophe risk swaps, and catastrophic equity puts [6]. Besides its advantages, accurate pricing of CAT bonds is still an open question as it requires a careful analysis of many diverse factors such as the catastrophic risks, market and the

F. Ahrens
IBB Beteiligungsgesellschaft GmbH, Bundesallee 171, 10715 Berlin, Germany
e-mail: fahrens@ibb-bet.de

R. Füss
Swiss Institute of Banking and Finance (s/bf), University of St. Gallen, Rosenbergstrasse 52,
9000 St. Gallen, Switzerland
e-mail: roland.fuess@unisg.ch

A.S. Selcuk-Kestel (✉)
Institute of Applied Mathematics, Actuarial Sciences Program, Middle East Technical University,
06800 Ankara, Turkey
e-mail: skestel@metu.edu.tr

A.A. Pinto and D. Zilberman (eds.), *Modeling, Dynamics, Optimization and Bioeconomics I*, Springer Proceedings in Mathematics & Statistics 73, DOI 10.1007/978-3-319-04849-9_4, © Springer International Publishing Switzerland 2014

parties involved. In literature, many studies done on CAT bonds pricing take into account the expected value principles [11, 13, 18]. The basic principle is mostly the evaluation of the defined probability or equivalently, bond default, will occur based on the catastrophic risk according to [22]. Cox and Pederson [5] and Egami and Young [7] determine the price through representative agent utility optimization. In some studies, a specific martingale measure is employed in order to cope with the incomplete CAT bonds market [12]. Additionally, behavioral and purely risk theoretical approaches are available in literature [1, 23].

This study aims to determine the pricing of CAT bonds under the influence of hurricane triggered catastrophes, especially, the impact of the Hurricane Katrina (2005) whose financial impacts are underestimated. On pricing of CAT bonds, it is questioned that, whether highly rated CAT bonds demonstrate a different relationship than subinvestment bonds between objective risk measures and the spread. In particular, we consider two factors: (1) the impact of a severe event, specifically, the 2005 hurricane season, with three major Hurricanes Katrina, Wilma and Rita, and (2) the potential difference in the relationship between risk measures and the spread of highly rated CAT bonds compared to subinvestment-grade bonds. The theoretical framework for this relationship is based on the Lance Financial (LFC) model, introduced by Lane [13]. Treed Bayesian estimation method is implemented to estimate the severity component of the spread and to obtain detailed information about the pricing mechanisms of CAT bonds. Additionally, the influence of conditional expected loss is also investigated to determine the investment-grade ratings. An econometric model is fitted based on data from Willis for the period from March 2003 to July 2008.

The LFC model which captures the interrelationships between spread, expected loss, risk load, and conditional expected loss is presented in Sect. 4.2. Section 4.3 explains the econometric model of the treed Gaussian process as an extension of the Bayesian classification and regression tree (CART) model. Section 4.4 describes the data, the analyses and the estimation process in which an empirical reference model to cross-validate the hypotheses on the impact of Katrina and investment-grade ratings is performed. The final section summarizes the study and provides concluding remarks.

4.2   The CAT Bond Pricing Method

From a risk theoretical point of view, CAT bond prices should reflect the expected loss associated with the trigger event. In a world with risk-neutral individuals and no systematic risk, this would equal exactly the price of the bond. However, as these conditions clearly do not exist, an important price determinant is how investors gauge the attractiveness of CAT bonds within a broader investment environment implying behavioral analysis of CAT bonds prices [1].

Assuming that catastrophe risk is diversifiable in equilibrium, independent of total wealth, and that catastrophe models are either unbiased or there is no better estimation technique available, the premium being equal to the expected loss would be fair. Based on this approach and with the assumption of the bond is issued at par value, the spread of CAT bonds can be taken as a measure of its market price [18]. Suppose P_t is the price of a corporate bond at time t, and C is its coupon. The current yield $Y_t = \frac{C}{P_t}$ is then a convenient way to express bond prices. The secondary market prices of CAT bonds are quoted similarly. The difference is that the LIBOR is subtracted from the total yield. Hence, they are quoted as yield spreads over the 3-month LIBOR [16].

However, one drawback of this approach is that CAT bond spreads depend strongly on the underlying expected loss, which can be derived directly from the issue prospectus. This hinders the comparison of bonds with differing expected losses. From an insurance viewpoint, the multiple of spread over expected loss is a better measure of price, because it normalizes the spread with respect to expected loss and is closer to what (re)insurance buyers would consider the cost of (re)insurance. On the other hand, from an investor perspective, the multiple is similar to an expected excess return or a primitive form of risk-adjusted expected profit [16].

4.2.1 CAT Bond Spread and Its Components

Based on Capital Asset Pricing Model (CAPM), CAT bonds constitute "zero-beta" assets, whose correlations between catastrophe risk and other asset classes are nearly zero [8]. CAT bonds were thus, regarded as excellent instruments for portfolio diversification because of the absence of systematic risk and investors should theoretically be compensated only for the expected loss and the risk-free rate. However, the existence of such an equilibrium price is not empirically observable in CAT bond markets [13–15].

Consider a coupon paying CAT bond, where the spread or the premium, respectively, is defined as the difference between the coupon and 6-month LIBOR as follows:

$$S = \text{Coupon} - \text{LIBOR}.$$

The spread of CAT bonds is composed of two components, the expected loss (**el**), $E[L]$, and the risk load, R.

$$S = E[L] + R \quad \Rightarrow \quad R = S - E[L]. \tag{4.1}$$

Dividing both sides of Eq. (4.1) by the expected loss, $E[L]$, yields the multiple, M, which measures the risk load proportional to the expected loss:

$$\frac{S}{E[L]} - 1 = \frac{R}{E[L]} \quad \Rightarrow \quad M = \frac{S - E[L]}{E[L]}. \tag{4.2}$$

Here, the expected loss is the probability-weighted sum of the possible losses the bond can incur. If investors are indifferent between two bonds with the same coupon and the same expected loss, the expected loss would be an adequate risk measure. Intuitively, we would expect other aspects of the distribution of outcomes (such as standard deviation or behavioral determinants like loss or ambiguity aversion) to also play a role [13].

The risk load component of the CAT bond spread captures all the other unexpected loss factors such as deviations from the expected loss [13]. This measure is influenced directly by behavioral risks such as the appeal of the bond for investors and sponsors. Two sources of variation in the risk load can be distinguished: (1) the characteristics of the loss distribution and (2) the demand and supply factors.

We can state the calculation of (re)insurance premiums with respect to the standard deviation principle as

$$S = E[L] + \psi.(\text{Var}[L])^{\frac{1}{2}}, \tag{4.3}$$

where S is the spread over LIBOR, $\text{Var}[L]$ and ψ denote the variance of the losses and the loading factor, respectively [24].

The standard deviation is a good approximation of the unexpected loss with symmetric distributions. However, CAT bonds loss distributions tend to be very asymmetric. Hence, allocating precise probability measures to each CAT bond allows for a more appropriate assessment of the unexpected loss [13].

Another component in pricing is the likelihood of loss which is the probability of a first dollar loss (**pfl**) or the probability of exhaustion. The pfl represents the inverse of the "frequency". Therefore, by the total probability principle, pfl is expressed as [13]:

$$\text{pfl} = \sum_i p_i \qquad \forall i > 0 \tag{4.4}$$

where p_i denotes the probability of the ith loss when L_i occurs.

Another important measure associated to pfl is the severity of the loss if the bond is triggered. An ex-ante measure of the severity from the conditional expected loss (**cel**) states that having the assumption of independence between the frequency and severity of a catastrophe, the expected loss given a catastrophe is:

$$
\begin{aligned}
\text{cel} &= \sum_{i>0} \left[\frac{p_i}{\sum_i p_i} \right] \cdot L_i \\
&= \sum_i \frac{p_i \cdot L_i}{\sum_i p_i} \\
&= \frac{E[L]}{\text{pfl}}.
\end{aligned}
\tag{4.5}
$$

Because the pfl and the cel are appropriate measures for the unexpected loss in a CAT bonds setup, we use them to construct the LFC model.

4.2.2 The Pricing Model of Lane Financial (LFC)

The model proposed by Lane [13] is of general form and provides initial insights
into CAT bond pricing. It also gives a more general reasoning for the application
of the severity (or magnitude) and the frequency in CAT bonds pricing. For these
reasons, LFC model is taken as the underlying pricing methodology in the context
of this study. The Lane model states that the risk load expressed in Cobb-Douglas
power relation should take into account the shape of the distribution which is
approximated by the frequency and the severity of loss:

$$R = \gamma \cdot (\text{pfl})^{\alpha} \cdot (\text{cel})^{\beta}, \tag{4.6}$$

where γ, α, and β are the parameters. In this form, the risk load states that given
the likelihood of a loss, a lower severity leads to a lower risk load and vice versa.
Additionally, it is approximately related to the standard deviation principle and
estimates the total spread as a multiple of expected loss as follows:

$$S = E[L] + \gamma.R. \tag{4.7}$$

This empirical model, however, does not provide a full and rational description of
investor's behavior, and hence does not result in arbitrage-free prices [13].

4.3 The Treed Gaussian Process Model

The estimation model in CAT bonds pricing proposed in this study is an extension of
the Bayesian Classification and Regression Tree (CART) model [3, 4]. This variant
of the model independently fits stationary Gaussian Processes (GP) that incorporate
linear trends into their Limiting Linear Models (LLM) within $\mathbf{R} = \{r_v\}_{v=1}^{R}$ regions,
where the regions are the leaves of a tree T. In this Bayesian approach, the tree is
not built by deterministic rules, but by a tree-generating prior process. Essentially,
the input space is partitioned by recursively making binary splits on single vectors
of the predictor matrix $\mathbf{X}$. Each subtree is thus a subset of its parent. During the
estimation process, each variable may be revisited, so that the binary form does
not restrict the model's generality. The regressions are thus performed by a "fully
Bayesian nonstationary, semiparametric nonlinear model" implemented through
treed Gaussian processes with jumps to the limiting linear model [9]. If stationary
GP models are not flexible enough, a partition model which divides the parameter
space by independently estimating a separate model on each data partition increases
the flexibility. The estimation model used for regression within each partition is a
"mixture" of the GP and LLM models. However, the method used to generate the
treed partitions is a Bayesian CART model.

The stationary GP with an LLM is a hierarchical Bayesian model for the data $D_v = \{\mathbf{X}_v, \mathbf{Z}_v\}$. For each region r_v, D_v is composed of n_v pairs of m_X covariates and one response $\{(x_{i1}, \ldots, x_{m_X}), z_i\}_{i=1}^{n_v}$. The model defines a multivariate normal likelihood as follows [9]:

$$\mathbf{Z}_v | \beta_v, \sigma_v^2, \mathbf{K}_v \sim N_{n_v}(\mathbf{F}_v \beta_v, \sigma_v^2 \mathbf{K}_v) \tag{4.8}$$

$$\sigma_v^2 \sim IG(\alpha_\sigma/2, q_\sigma/2)$$

$$\beta_v | \sigma_v^2, \tau_v^2, \mathbf{W}, \beta_0 \sim N_m(\beta_0, \sigma_v^2 \tau_v^2 \mathbf{W})$$

$$\tau_v^2 \sim IG(\alpha_\tau/2, q_\tau/2)$$

$$\beta_0 \sim N_m(\boldsymbol{\mu}, \mathbf{B})$$

$$\mathbf{W}^{-1} \sim W((\rho \mathbf{V})^{-1}, \rho)$$

where, β_v are linear trend coefficients, having a common unknown mean β_0 and variances $\sigma_v^2 \tau_v^2$. The $\mathbf{Z}_v$ variances are regional specifications given by σ_v^2, as well as the $n_v \times n_v$ correlation matrices $\mathbf{K}_v$. The $n \times m_X$ dimensional design matrix $\mathbf{X}_v$ is extended by an intercept term to the $n_v \times m$ matrix $\mathbf{F}_v = (1, \mathbf{X}_v)$, where $m \equiv m_X + 1$ and $\mathbf{W}$ is an $m \times m$ matrix governing the correlation structure between the β_v coefficients. Here, N, IG and W denote the (multivariate) normal, the inverse-gamma and the Wishart distributions, respectively. The constant hyperparameters $\boldsymbol{\mu}, \mathbf{B}, \mathbf{V}, \rho, \alpha_\sigma, q_\sigma, \alpha_\tau, q_\tau$ are treated as known. Note that the model might not be continuous near the borders of nearby regions.

The LLM can be obtained simply by changing the "likelihood line" presented in Eq. (4.9)

$$\mathbf{Z}_v | \beta_v, \sigma_v^2 \sim N_{n_v}(\mathbf{F}_v \beta_v, \sigma_v^2 \mathbf{I}_v),$$

where $\mathbf{I}_v$ is the $n_v \times n_v$ identity matrix. The parameters in Eq. (4.9) are sampled by using the Gibbs algorithm.

The correlation structure, $\mathbf{K}_v$, has the form $K(\mathbf{x}_j, \mathbf{x}_k) = K^*(\mathbf{x}_j, \mathbf{x}_k) + g\delta_{j,k}$, where g denotes the "nugget" or "jitter", $\delta_{\cdot \cdot}$ is the Kronecker delta function, and $K^*(\cdot, \cdot)$ represents the "true" correlation structure parameterized by an exponential power function. The jitter term ($g > 0$) introduces the potential to account for measurement errors in the stochastic process, which may increase the robustness in the estimations and assures $\mathbf{K}$ to be numerically non-singular. The parameterization of $K^*(\cdot, \cdot)$ may be from an isotropic power family, or from a separable power family. Both families are stationary, meaning that the correlations are measured in the same way throughout the entire input domain. The first parameterization is isotropic, and the correlations are measured as a function of the Euclidean distance between $\mathbf{x}_j$ and $\mathbf{x}_k$, $\| \mathbf{x}_j - \mathbf{x}_k \|$:

$$K^*(\mathbf{x}_j, \mathbf{x}_k | d) = \exp\left\{ -\frac{\| \mathbf{x}_j - \mathbf{x}_k \|^{h_0}}{d} \right\}, \tag{4.9}$$

where $d > 0$ is the range parameter, and $0 < h_0 \leq 2$ represents the smoothness of the underlying process. If $h_0 = 2$ the process is Gaussian and indefinitely differentiable, whereas if $0 < h_0 < 2$ the process is indifferentiable.

The separable process is an enhancement to the isotropic process, because it allows for a unique range parameter d_i in each dimension of the design matrix $(i = 1, \ldots, m_X)$. We consider the isotropic power process as a special case when $d_i = d$, for all $i = 1, \ldots, m_X$.

$$K^*(\mathbf{x}_j, \mathbf{x}_k | \mathbf{d}) = \exp \left\{ - \sum_{i=1}^{m_X} \frac{|x_{ij} - x_{ik}|^{h_0}}{d_i} \right\}. \tag{4.10}$$

Thus, the correlations of some input variables may be weighted more strongly with the separable power family than with others. However, this may result in overfitting.

Note that the GP regression is definitively more flexible than the standard linear regression. However, if the coherence is linear, then the Gaussian process regression must deal with parsimony, overfitting, and numerical stability. For the Bayesian case, it is possible to combine both approaches by constructing a prior for a "mixture" of both models [10]. This can be done by extending the parameter space with a vector of Boolean indicator variables $\mathbf{b} = \{b_i\}_{i=1}^{m_X} \in \{0, 1\}^{m_X}$. This indicator variable is multiplied with the range parameter of the correlation structure $K^*(\cdot, \cdot | \mathbf{b}' \mathbf{d})$, which indicates whether the ith dimension is estimated by the GP model ($b_i = 1$), or by the LLM model. The probability mass function, P, of the prior on $\mathbf{b}$ is given as:

$$P_{\gamma, \theta_1, \theta_2}(b_i = 0 | d_i) = \theta_1 + \frac{\theta_2 - \theta_1}{1 + \exp\{-\gamma(d_i - 0.5)\}}, \tag{4.11}$$

where $0 \leq \theta_1 \leq \theta_2 < 1$. The minimum ($\theta_1$) and the maximum ($\theta_2$) are the probabilities of switching from the GP to the LLM model, while $\gamma > 0$ determines how fast $P(b_i = 0 | d_i)$ approaches θ_2 when d_i increases. This prior clearly represents the preference that, with increasing range parameter d_i, we expect the GP model to be more likely to jump to the LLM.

Note that the process resulting from the mixture is still a GP unless all $b_i = 0$. As stated in Gramacy [9], the numerical stability increases and a unique transitory model lying between the GP and the LLM is obtained by intermediate results. Therefore, a full linear model (LM) is only implemented with a prior probability of

$$P(\text{LM}) = \prod_{i=1}^{m_X} P(b_i = 0 | d_i) = \prod_{i=1}^{m_X} \left[\theta_1 + \frac{\theta_2 - \theta_1}{1 + \exp\{-\gamma(d_i - 0.5)\}} \right]. \tag{4.12}$$

The prediction of mean and variance under the covariance structure is straightforward. Conditioned on the covariance structure, normally distributed $Z(\mathbf{x})$ with mean and variance follows:

$$\hat{z}(\mathbf{x}) = E(\mathbf{Z}(\mathbf{x})|\text{data}, \mathbf{x}) = \mathbf{f}'(\mathbf{x})\tilde{\beta} + \mathbf{k}(\mathbf{x})'\mathbf{K}^{-1}(\mathbf{Z} - \mathbf{F}\tilde{\beta}), \quad (4.13)$$

$$\hat{\sigma}(\mathbf{x})^2 = \text{Var}(\mathbf{Z}(\mathbf{x})|\text{data}, \mathbf{x}) = \sigma^2[\kappa(\mathbf{x}, \mathbf{x}) - \mathbf{q}'(\mathbf{x})\mathbf{C}^{-1}\mathbf{q}(\mathbf{x})], \quad (4.14)$$

where $\tilde{\beta}$ is the posterior mean estimate of β and

$$\mathbf{C}^{-1} = (\mathbf{K} + \mathbf{F}\mathbf{W}\mathbf{F}'/\tau^2)^{-1}$$

$$\mathbf{q}(\mathbf{x}) = \mathbf{k}(\mathbf{x}) + \tau^2\mathbf{F}\mathbf{W}\mathbf{f}(\mathbf{x})$$

$$\kappa(\mathbf{x}, \mathbf{y}) = K(\mathbf{x}, \mathbf{y}) + \tau^2\mathbf{f}'(\mathbf{x})\mathbf{W}\mathbf{f}(\mathbf{y})$$

where $\mathbf{f}'(\mathbf{x}) = (1, \mathbf{x}')$, and $\mathbf{k}(\mathbf{x})$ is an n-dimensional vector with $k_j(\mathbf{x}) = K(\mathbf{x}, \mathbf{x_j})$ for all $\mathbf{x_j} \in \mathbf{X}$. From Eq. (4.14) we see that $\hat{\sigma}(\mathbf{x})^2$ does not directly depend on the predicted responses $\mathbf{Z}$.

Parameter prediction then becomes simpler for the LLM case. If $\mathbf{K} = (1 + g)\mathbf{I}$, the predicted value of $z(\mathbf{x})$ is normally distributed with mean $\hat{z}(\mathbf{x} = \mathbf{f}'(x)\tilde{\beta}$ and variance $\hat{\sigma}(\mathbf{x})^2 = \sigma^2[1 + \mathbf{f}'(\mathbf{x})\mathbf{V}_{\tilde{\beta}}\mathbf{f}(\mathbf{x})]$, where $\mathbf{V}_{\tilde{\beta}} = (\tau^{-2} + \mathbf{F}'\mathbf{F}(1 + g))^{-1}$. And the $m \times m$ inversion of this case is faster than the $n \times n$ when taking the form $\mathbf{K} = \mathbf{I}(1 + g)$, as in Eq. (4.14) [10].

The Bayesian tree-generating process is governed by two priors: (1) the size of the tree and (2) the location of each split. The size prior is specified in order to form more bushy trees [3], because the probability of a split (p_{SPLIT}) in one predictor variable decreases as the number of parent nodes increases: $p_{SPLIT}(\eta, T) = a(1 + q_\eta)^{-b}$, where q_η is the number of parent nodes of $\eta \in T$, and a and b are parameters to determine the size and spread of the tree. A restriction may also be imposed on the minimum number of observations each leaf contains. Eventually, the tree dependence is integrated out using Markov Chain Monte Carlo (MCMC) simulation with the Metropolis-Hastings algorithm [3,4].

In the content of this study, the application of treed Gaussian approach on pricing CAT bonds data captures the impact of conditional events without setting rigid parametric assumptions. For this purpose, reversible jump MCMC to integrate out the tree structure and to sample from the joint posterior (Θ, T) by alternately drawing $\Theta|T$ and $T|\Theta$ is used [10].

4.4 Empirical Results

4.4.1 Data

The data on CAT bond prices has been received from Willis which is collected from vendors such as Goldman Sachs, SwissRe Capital Markets, Aon, USAA. The 199 observations between March 2003 and July 2008 include information about sponsors, issue date, risk period, maturity, size in 1,000s, trigger, ratings, risk premium, attachment probability, peril type, and expected loss. CAT bond-specific

characteristics, such as type of trigger, covered perils, or rating, are provided by
S&P, Moody's, Fitch, or A.M. Best and are used to make analyses based on these
clusterings.

The annual yield spread (risk premium), S is calculated over the 6-month LIBOR.
S is converted to a 365-days per year convention to render the risk premium
comparable to the expected losses, and then, is adjusted for the reinsurance price
cycle. The annual expected loss of a CAT bond (**el**), probability of first loss (**pfl**),
conditional expected loss (**cel**), and finally, the risk load (**R**) are determined as
presented previously. Additionally, based on the analyses on the data, a significant
structural break after Hurricane Katrina is observed. Therefore, subsequent to
Katrina, in 2006, catastrophe models were revised, which led to a 58 % increase
in the expected losses of outstanding CAT bonds from March to June 2006 [14]. To
include this effect, we introduce a dummy variable "**katrina**", which we set equal
to 1 for all bonds issued from November 2005 onward. Finally, we use a dummy
variable "**rating**" to indicate whether a CAT bond is rated BBB or above being set
to one and zero otherwise.

4.4.2 Cyclical and Seasonal Components

Reinsurance prices as well as CAT bond premiums exhibit strong long-term
cyclical behavior [17]. Because of the shortage of capital in the aftermath of major
catastrophes, such mega-catastrophes are typically followed by a hard market,
leading to substantial premium increases. This can be seen from Fig. 4.1 which
shows the strong cyclical behavior of reinsurance prices and a significant increase
in CAT bond premiums due to the shortage of capital in the aftermath of major
catastrophes.

In the CAT bond market, the impact of non-price adjustments in terms and
conditions might be approximated by the capital-weighted relationship between
single- and multi-peril bonds. Multi-peril bonds are triggered by each trigger event,
while single-peril bonds are subject to only one trigger event which differs in terms
and conditions of the bond protection. In hard markets, we might expect to see
fewer multi-peril CAT bonds issues, because of their relatively high prices. This
might cause sponsors to retain some minor risks, and insure only specific or peak
risks [17].

Seasonality is not an issue for all types of CAT bonds however, and depends on
the peril the bond covers. CAT bonds covering the U.S. hurricanes and European
wind are exposed to seasonality, whereas earthquake triggered CAT bonds do not
show any seasonality [14]. In contrast to single-peril bonds, where seasonality is
relatively easy to predict, multi-peril bonds pose a certain amount of unpredictability
with respect to seasonality analysis. The seasonal treatment for hurricane CAT
bonds should include at least four dummies (D) based on its behaviour within a
year: ($D1$) the beginning of the American hurricane season in June, ($D2$) the end of
the American hurricane season in October, ($D3$) the beginning of the European

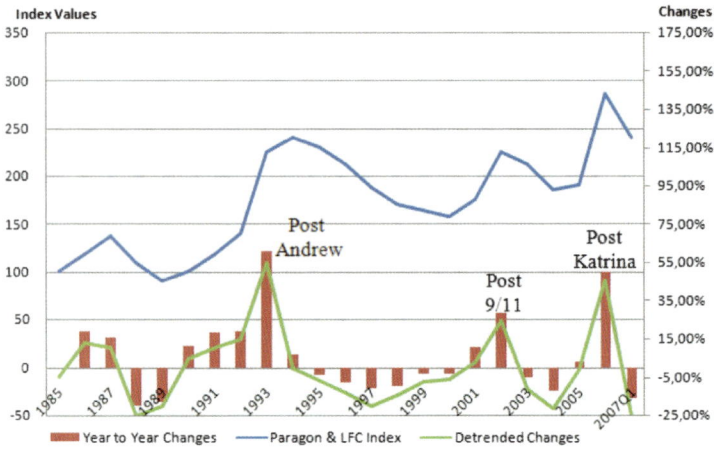

Fig. 4.1 U.S. catastrophe reinsurance price index [18]

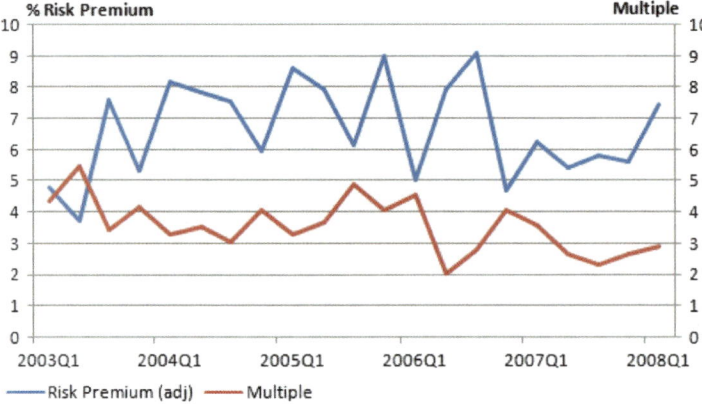

Fig. 4.2 Quarterly means of adjusted risk premium and its multiple

wind season in November, and ($D4$) the end of the European wind season in January. The interaction term between ($D2$) and the cyclical variable to capture the emergence of a hard market is ignored, because the spread is already adjusted for cyclical behavior. However, because of the scarcity of data, only one dummy variable referring the beginning of the hurricane season is defined.

The Lane Financial LLC index was originally available only on a quarterly basis, but it was interpolated to match each CAT bond's issuance date. We perform the spread adjustment by simply dividing it by the index. Figure 4.2 shows significant seasonal fluctuations (i.e., between quarters) in the adjusted spreads and multiples based on our dataset. The spread is adjusted by dividing the original spread series by the Paragon & LFC index.

4.4.3   Estimation

Ordinary least square (OLS) technique can be used to estimate the parameters of the log-transformed S given in Eq. (4.7) as:

$$S = \text{el} + \gamma((\text{pfl})^\alpha (\text{cel})^\beta)$$

$$\Rightarrow \ln(S - \text{el}) = \ln(\gamma) + \alpha \ln(\text{pfl}) + \beta \ln(\text{cel}), \tag{4.15}$$

However, the estimation of Eq. (4.15) with the Bayesian version of a treed regression results in a treed posterior distribution which yields a richer structure than OLS, and is more flexible than parametric estimation techniques. It also handles non-linear relationships with fewer trees than a conventional treed regression model [4, 10].

The graphical presentation of the results obtained by the Bayesian approach constitutes a surface plot of the marginal predictive mean over the different values of the respective variables. The measure of uncertainty within each region of the surface is given by plotting the 90 % quantile difference (95 % quantile $-$ 5 % quantile) of the distribution around its mean.

A sensitivity analysis based on the "Sobol" sensitivity indices allows computing the contribution of each input variable through the main effects and all interaction effects to the output's variance. The main effects by using the first-order sensitivity index, SI, is determined as follows:

$$SI(i) = \frac{\text{Var}(E[f|x_i])}{\text{Var}(f)}, \tag{4.16}$$

where x_i is the ith input variable, and $f(\cdot)$ is the posterior distribution function. A measure for the contribution through interaction effects is given by the total effects index, SI^T:

$$SI^T(i) = \frac{\text{Var}(E[f|x_{-i}])}{\text{Var}(f)}. \tag{4.17}$$

Here, x_{-i} indicates the exclusion of the ith variables from the condition. A significant difference between $SI(i)$ and $SI^T(i)$ for a given variable x_i implies the existence of relevant interaction terms for that variable. Thus, in purely additive models, it would follow that $\sum_{i=1}^{m_X} SI(i) = 1$ [20].

These indices are estimated by drawing random Latin hypercube samples at each MCMC iteration, and then illustrating them by the plots of three measures: (1) the estimated main effects over the range of each predictor variable (x_1) on the dependent variable ($E[\mathbf{Z}|\mathbf{x_i}]$) (2) the first-order sensitivity index, and (3) the total effects index. Note, however, that the total effects index may exhibit negative values, which are due to errors resulted in Monte Carlo simulations. The posterior intervals represent the uncertainty of the function response and the integration estimates.

Fig. 4.3 Surface and error plots for the base and linearized models

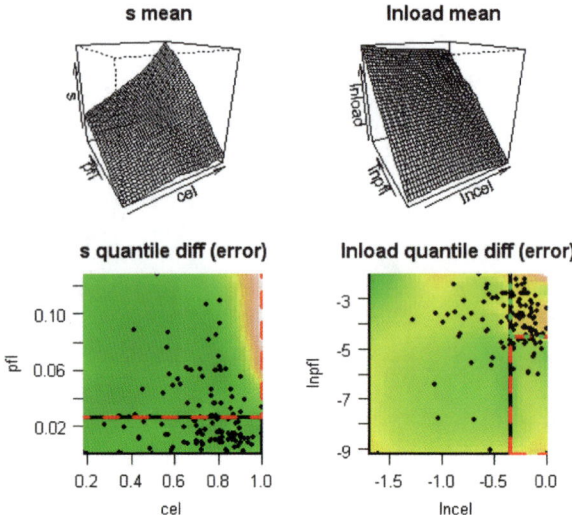

To perform the final step in the evaluation of our results, we sample nearly thirty input locations from the posterior, and compare the root mean squared errors (RMSE) and the mean absolute percentage errors (MAPE) for efficiency reasons.

4.4.4 Impact of Hurricane Katrina

The main focus of the LFC model estimation is set in a non-linear context. However, this study evaluates the model under several hypotheses. The "Cobb-Douglas" and its linearized form are two base specifications of the LFC model. The linearized version is difficult to interpret. Nevertheless, we employ it mainly to cross-validate the original form, which provides a good description of the relation among variables. If the Cobb-Douglas form describes the data well, the relation among the variables in the linearized version should have a plain surface [13]. We see from Fig. 4.3 that the choice of the functional form is justified. It shows the LFC base model (left hand side) and the LLFC model (right hand side); the upper graphs serves to indicate if the functional form of the LFC model is well specified; the plain surface on the right top indicates that the LLFC model describes the data well; *cel*, *pfl*, and *S* denote the conditional expected loss, the attachment probability, and the risk premium, respectively. Note also that the input matrices for the base model are taken as $Z=S$, $X=(el, pfl, cel)$. In the aftermath of the severe 2005 hurricane season, it was necessary for investors to update their catastrophe models. If the relationship between different probabilities of loss measures and the spread changes, as it actually did, then two claims arise due to either a change in supply (**Supply Hypothesis**) or a change in perception (**Perception Hypothesis**).

A drop in the spread at the given *el*, *pfl*, and *cel* might indicate an overall price reduction due to the increased supply of risk capital on the CAT bonds. On the other

Table 4.1 Specifications for the analysis of hypotheses

Spec. No.	Specification	Hypothesis	No. of obs.	RMSE
1	The base models augmented by the Katrina dummy	Supply	131	0.0422
2	Katrina = 0, estimated with the base models	Perception	17	0.0116
3	Katrina = 1, estimated with the base models	Perception	85	0.0118
4	Hurricane Peril = 1, tested with the base models, augmented by the Katrina dummy	perception	66	0.0189
5	Hurricane Peril = 0, tested with the base models and the Katrina dummy	Perception	67	0.0195

hand, spreads of hurricane-exposed CAT bonds may react differently than spreads of bonds with other perils because of price drivers such as cost of risk capital, erosion of deductibles, ongoing loss estimations, and the credit ratings of bond and swap partners. We expect other perils to experience spread **reductions** due to (1) a shift in the perception of hurricane risk and (2) diversification credit for other perils. To analyze (1) and (2), we infer the shape of the posterior separately based on different subsamples in the dataset and the two base model specifications.

Table 4.1 summarizes the hypotheses, specifications (Spec. No.1–5), number of observations for each specifications, and their RMSE values. Specification 1 includes a **katrina** dummy that captures the main effects of risk measures with respect to the spread. Specifications 2 and 3 are estimated to determine the changes before and after the 2005 hurricane season with respect to the dummy variable. The last two specifications, 4 and 5, aim to determine whether Katrina affected the relative risk perception among hurricane perils. It can be observed that bonds with a rating above BBB yield a significant increase on the multiple.

We next analyze the **Supply** Hypothesis that corresponds to Specification 1. The surfaces depicted in Panel (a) of Fig. 4.4 indicate that Katrina impacted relationships between the expected loss, the attachment probability, and the conditional expected loss. Specification 1 refers to the base models augmented by the Katrina dummy, as can also be seen in Table 4.1. The difference in scaling should be noted. At a given expected loss and attachment probability, it is found that the spreads declined after Katrina (see Panel (b) of Fig. 4.4). In contrast, the reaction of the spread at a given conditional expected loss was the opposite. Based on Fig. 4.4, we can conclude that after the 2005 hurricane season, the spreads decreased relative to *el* and *pfl*. As spreads are cyclically adjusted in the dataset, we find that the **Supply** Hypothesis is strongly supported.

The **"Perception"** Hypothesis is analyzed in two groups. The first group Specifications 2 and 3, represent the impact of Hurricane Katrina in the base model, when we control for this event by setting the dummy 0 before and 1 after Katrina occurs, separately. From Fig. 4.5 we see that the spread increases slightly in *cel* over

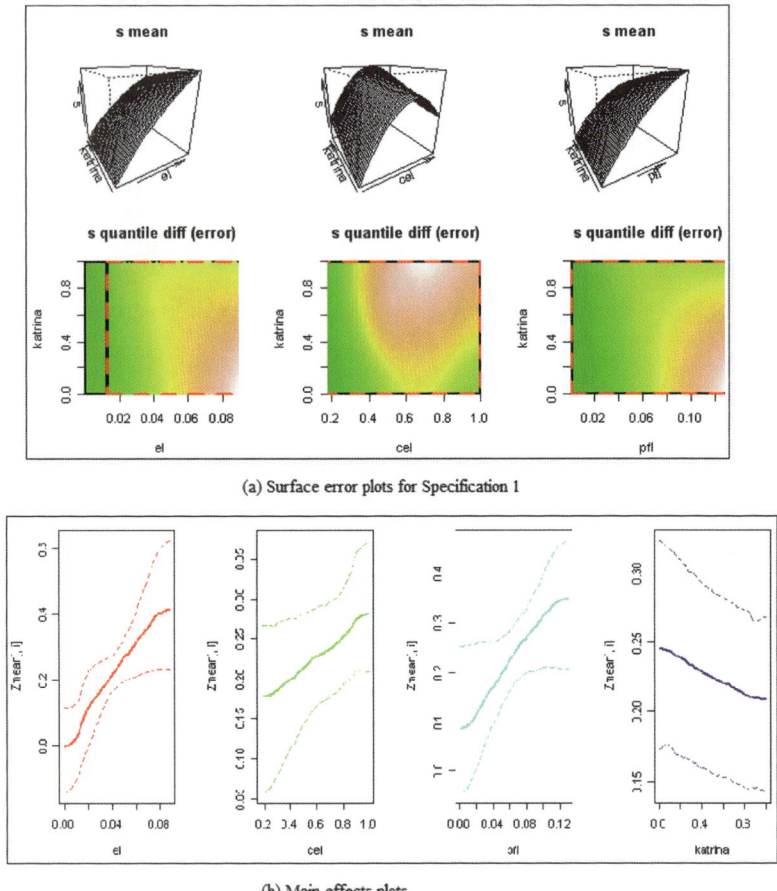

(a) Surface error plots for Specification 1

(b) Main effects plots

Fig. 4.4 Test of the Supply Hypothesis under Specification 1. Panel (**a**) shows surface and error plots, while Panel (**b**) illustrates the main effects of the variables on the predictive posterior mean

the whole range of *pfl* before Katrina, while the impact of *cel* increases dramatically with *pfl* after Katrina. The main effect graphs in both specifications show an increased slope in *cel*. However, the sensitivity indices imply that the increased impact results from an increase in interaction with other measures, rather than from an increase in variance. This indicates the change in perception of importance given to the *cel* measure. Because of space limitations we only show a selection of figures.

In the second group, Specifications 4 and 5, the trigger type is the hurricane peril which is set to either 0 or 1 in the base model, and is conditioned on Katrina. Figure 4.6 shows a change in the direction of influence on the plots of *el* versus *cel*. Panel (a) of Fig. 4.6 illustrates the relationship between *el*, *cel*, and *pfl* with hurricane exposure only; Specification 4 refers to the base models with Hurricane Peril = 1 and augmented by the Katrina dummy, whereas Specification 5

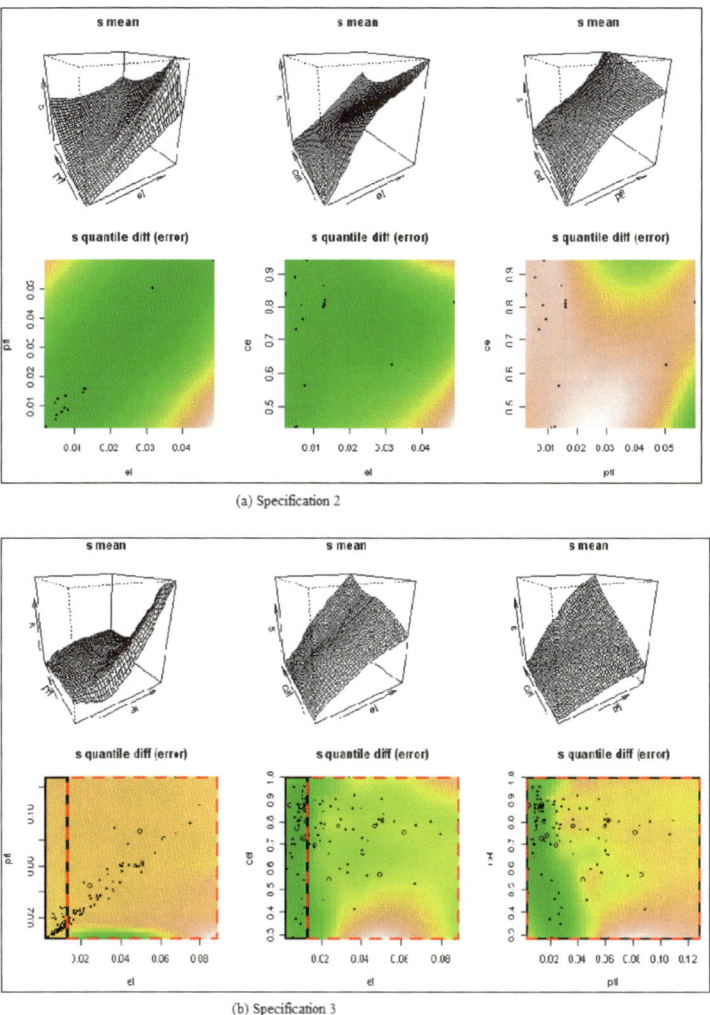

Fig. 4.5 Test of the Perception Hypothesis under Specifications 2 and 3. Panel (**a**) shows the surface and error plots for Specification 2, whereas Panel (**b**) illustrates the plots for Specification 3

in Panel (b) of Fig. 4.6 refers to the base models with Hurricane Peril = 0 and the Katrina dummy. The dataset that includes only the hurricane sample shows an increase in spread, along with increases in both el and cel; whereas for the dataset without hurricanes, the spread decays when the other two variables increase. The sensitivity indices for specification 4 show that cel barely contributes to the spread variance, which results in a low interaction with other variables. The expected loss and the attachment probability thus do not interact with other variables, which results in the variability of the spread. Contrary to Specification 4, Specification 5

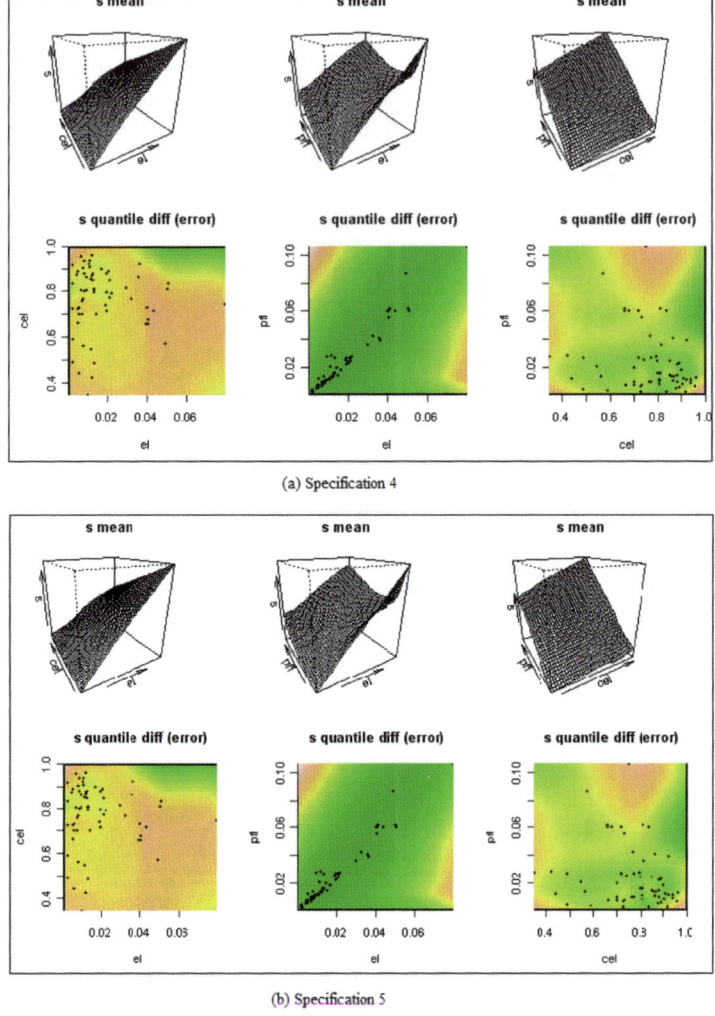

Fig. 4.6 Test of the Perception Hypothesis under Specifications 4 and 5. Panel (**a**) shows the surface and error plots for Specification 4, whereas Panel (**b**) illustrates the plots for Specification 5

indicates higher interactions among all variables, as well as the domination of the variance of expected loss over the variance of spread.

4.4.5 Impact of Ratings

According to Slovic [21], individuals often perceive events with catastrophic potential to be very risky, regardless of the low probability of their occurrence.

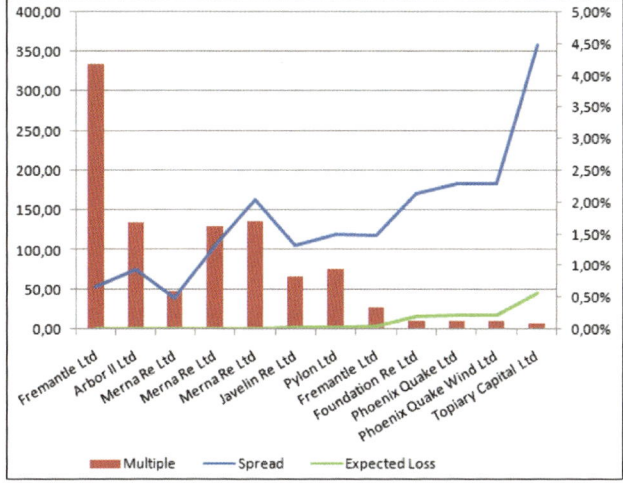

(a) Multiple bonds above BBB rating

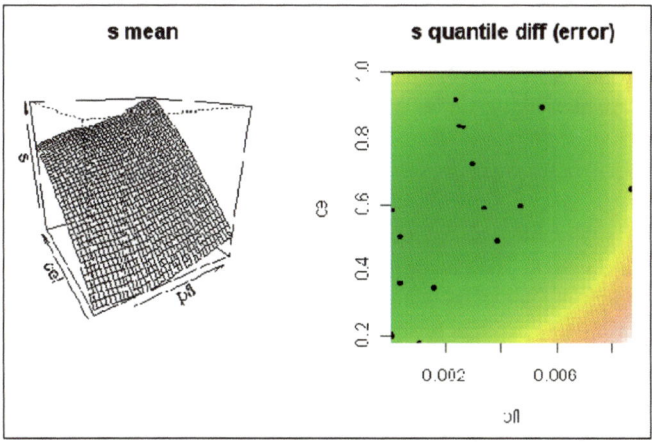

(b) Surface and error plots

Fig. 4.7 Impact of rating BBB or above on CAT bonds. Panel (**a**) shows the graphs of spread, expected loss and multiple, while Panel (**b**) illustrates surface and error plots

Thus, investors may worry about even the small possibility of losing money due to a catastrophic event. They may prefer a LIBOR investment, which is perceived as more secure, unless the bonds pay significantly more than LIBOR, which leads to a high multiple even with low spreads. Thus, CAT bonds have unusually high multiples when premiums approach LIBOR. Panel (a) of Fig. 4.7 shows all observations rated at least BBB, sorted by expected loss. The highest multiples are clearly seen in a region where the expected loss is hardly distinguishable from zero.

Table 4.2 Descriptive statistics with and without investment grade

Variable	No. Obs.	Mean	Std. Dev.	Min	Max
Complete data without investment grade					
s	128	0.06252	0.0395	0.00483	0.17824
el	128	0.02057	0.0183	0.00002	0.0881
Multiple	128	9.946	32.5535	1.1712	334.5588
Reduced data: rating BBB and above					
s	20	0.0185	0.0088	0.0048	0.0448
el	20	0.0017	0.0016	0.00002	0.0056
Multiple	20	52.8247	80.7349	6.6816	334.5588

Twenty bonds in the dataset are rated BBB and above. Table 4.2 lists the two hypotheses "Supply" and "Perception" to determine the impact of Hurricane Katrina and the corresponding specifications of the base model. From the graphs, we can conclude that such investment-grade bonds yield smaller expected losses. However, the variable multiple exhibits the opposite, i.e. the higher the rating the higher the relative risk load. This implies that investors who demanded highly rated bonds may be more concerned than junk bond investors about possible losses.

The CAT bonds rating is one of the factors that influences the market price, especially before or at issue [2]. The key rating factors for CAT bonds are: (1) the risk characteristics of the underlying perils in terms of attachment probabilities, (2) the covered peril(s) as well as the granularity of the exposure data used for modeling the underlying risk measures, (3) the catastrophe models used by the modeling agent and the results of stress tests on key modeling factors, (4) the basis risk, if applicable, according to the trigger mechanism, or the existence of a multiple event trigger, (5) the credit risk of the sponsor, paying the premiums to the special purpose vehicles (SPV), and the credit risk of the swap counterparties, (6) miscellaneous factors, such as the experience of the sponsor, the covered risk period, annual risk resetting, or extension options. Given the factors above, the LFC model obviously accounts for the same factors as the rating. Therefore, if the assigned rating significantly impacts the shape of the posteriors, there should be other unlisted factors not covered by the LFC model. When we consider the subinvestment data, the relationships do not change much compared to either the base model or the post-Katrina data. The majority of the observations in our dataset stem from post-Katrina, and exhibit sub-investment ratings.

To analyze the impact of investment-grade ratings on conditional expected loss and attachment probability, we must draw surface and error plots for the LFC model for the reduced data. Panel (b) of Fig. 4.7 shows that the conditional expected loss already has a positive impact on the spread at the low values of *pfl*. The main effect of *cel* increases, but the proportion of explained variance due to *cel* does not. We further find that an investment-grade rating increases the impact of *cel*.

Table 4.3 Out-of-sample forecast performance

Statistic	LFC based	LLFC based	Katrina	Rating	Both
RMSE	0.0172	0.5407	0.0246	0.0197	0.0244
MAPE	0.2945	0.1181	0.2977	0.2939	0.3749

4.4.6   Accuracy

To facilitate out-of-sample forecasting, we exclude a sample of thirty observations from the posterior estimation. Although the selection is carried out randomly, checks on all relevant domains of predictor variables are performed as there may be different parameter values for the prediction of the posterior mean over the whole predictor space.

We use the values from all partitions to cross-validate the fit of the augmented models relative to the base model and calculate the RMSE and the MAPE. To check if the augmented models provide a better forecast than the base model (LFC-based), we consider three other specifications (Katrina, the Rating, and Both) and the linearized version of the LFC model (LLFC-based). The Katrina impact corresponds to Specification 1, while Rating corresponds to investment-grade.

Table 4.3 shows the out-of-sample forecast by using all partitions to cross-validate the fit of the augmented models relative to the base model. It can be seen that the LFC-based model performs very well compared to its augmented relatives. LFC-based model yields smaller RMSE than the LLFC-based model. Note also that the LLFC-based model yields the highest RMSE, but the lowest MAPE compared to the other specifications.

4.5   Concluding Remarks

This study aims to develop an econometric model to analyze the impact of the 2005 hurricane season, particularly Hurricane Katrina, on CAT bond pricing and its components. Bayesian treed Gaussian estimation technique is implemented to estimate the parameters of the model proposed by Lane [13]. Data on CAT bonds from 2003 to 2008 are used to derive parameter estimates of the model under assumptions on the factors which have direct and indirect effect on pricing.

The empirical results show that the severity component of the spread has an increasing impact, which indicates either a shift in the investor perceptions or a change in weight during the pricing process. After the 2005 hurricane season, an increase in the supply of CAT bonds led to a reduction in spreads. The level of this supply increase that is attributable to Katrina is not distinguishable. However, an impact on the increase in conditional expected loss is recognizable. It is found that Katrina did not directly contribute through variance, but through the interaction between expected loss and attachment probability. It is also argued that the 2005 hurricane season shifted investor's perceptions and catastrophe models. Thus, during the investment decision-making process, investors ultimately weighted the expected severity of CAT bonds higher. Also, during the catastrophe model adjustment process, the implied correlation structure was changed. Additionally, an investment-grade rating is found to increase the impact of the conditional expected loss. We posit that this is because investors who demand highly rated bonds may be more concerned on possible losses than junk bond investors. The forecast evaluation of the models illustrates that accounting for Katrina and the rating does not improve forecasting performance. However, the linearized version of the LFC model does perform better than the non-linear Cobb-Douglas form.

Acknowledgements The authors are grateful to the Department of Empirical Research and Applied Econometrics at Albert Ludwigs University Freiburg, Germany for the support received during completion of this study and Willis, U.K. for enabling us the dataset and for the permission to use it for academical purposes.

References

1. Bantwal, V.J., Kunreuther, H.C.: A cat bond premium puzzle? J. Psychol. Financ. Markets **1**(1), 76–91 (2000)
2. Best, A.M.: Rating Natural Catastrophe Bonds. A.M. Best's Quick Reference. http://www.ambest.com/ratings/methodology/QuickRef-CatBonds.pdf January (2008)
3. Chipman, H.A., George, E.I., McCulloch, R.E.: Bayesian CART model search. J. Am. Stat. Assoc. **93**(443), 935–960 (1998)
4. Chipman, H.A., George, E.I., McCulloch, R.E.: Bayesian treed models. Mach. Learn. **48**, 299–320 (2002)
5. Cox, S.H., Pederson, H.W.: Catastrophe risk bonds. N. Am. Actuarial J. **4**(4), 56–82 (2000)
6. Cummins, D.J.: CAT bonds and other risk-linked securities: state of the market and recent developments. Risk Manag. Insur. Rev. **11**(1), 23–47 (2008)
7. Egami, M., Young, V.R.: Indifference prices of structured catastrophe (CAT) bonds. Insur. Math. Econ. **42**(2), 771–778 (2008)
8. Froot, K.A., Murphy, B., Stern, A., Usher, S.: The emerging asset class: insurance risk. Special Report from Guy Carpenter and Company, Inc. Reprinted in Viewpoint **24**(3), 19–28. http://www.nber.org/reporter/summer99/froot.html (July 1995)
9. Gramacy, R.B.: tgp: an R package for Bayesian nonstationary, semiparametric nonlinear regression and design by treed Gaussian process models. J. Stat. Software 19(9), 1–46 (2007)
10. Gramacy, R.B., Lee, H.K.H.: Bayesian treed Gaussian process models with an application to computer modeling. J. Am. Stat. Assoc. **103**(483), 1119–1130 (2008)
11. Kleindorfer, P.R., Kunreuther, H.C.: Challenges facing the insurance industry in managing catastrophic risks. In: Froot. K.A. (ed.) The Financing of Catastrophe Risk, A National Bureau of Economic Research Project Report, University of Chicago Press, pp. 149–194 (1999)
12. Kull, A.: A unifying approach to pricing insurance and financial risk. Casualty Actuarial Society (ed.) Winter forum 2003, data management, quality, and technology call papers and ratemaking discussion papers, Arlington, pp. 317–350 (2003)
13. Lane, M.N.: Rationale and results with the LFC cat bond pricing model. Discussion paper, Lane Financial LLC, Wilmette (2003)
14. Lane, M.N.: Over the top, but not off the boil. Discussion paper, Lane Financial LLC, Wilmette (2006)

15. Lane, M.N.: What were we thinking? Discussion paper, Lane Financial LLC, Wilmette (2008)
16. Lane, M.N., Beckwith, R.G.: How high is up?: the 2006 review of the insurance securitization market. Discussion paper, Lane Financial LLC, Wilmette (2006)
17. Lane, M.N., Beckwith, R.G.: The 2008 review of ILS transaction – what price ILS?: a work in progress. Discussion paper, Lane Financial LLC, Wilmette (2008)
18. Lane, M.N., Mahul, O.: Catastrophe Risk Pricing: An Empirical Analysis. Policy Research Working Paper, vol. 4765. The World Bank, Washington, DC (2008)
19. Munich Re: Zwischen Hoch und Tief: Wetterrisiken in Mitteleuropa. Discussion paper, Munich (2007)
20. Saltelli, A.: Making best use of model evaluations to compute sensitivity indices. Comput. Phys. Commun. **145**, 280–297 (2002)
21. Slovic, P.: Perception of risk. Science **236**, 280–285 (1987)
22. Tilley, J.A.: The Securitization of Catastrophic Property Risks in ASTIN/AFIR. Colloquia. International Actuarial Association ed., Cairns (1997)
23. Wang, S.S.: CAT bond pricing using probability transforms. The Geneva Papers on Risk and Insurance: Études et Dossiers, vol. 278, pp. 19–29 (2004)
24. Zweifel, P., Eisen, R.: Versicherungsökonomie: Mit 65 Tabellen. Springer, Berlin (2003)

Chapter 5
Properties and Comparative-Static Effects in Models of Decision Under Uncertainty: Applications to the Theory of the Firm

Alberto A. Álvarez-López, Inmaculada Rodríguez-Puerta,
Francisco Sebastiá-Costa, and Mónica Buendía

5.1 Introduction

Many models of decision under uncertainty, especially from the theory of the firm under uncertainty, can be seen as particular cases of a general formulation: an agent maximizes the expected utility of a random "wealth", this wealth being the balance of an "income" and a "cost", where the income can be defined as the product of a "production" and a "price" (interpreting these three elements: production, price and cost, in a wide sense).

In this paper, we formulate a decision model following this pattern, with only one decision variable, and only one source of uncertainty that is independent of the decision variable. Due to this simplification, the model is simple and easy to handle. Moreover, it remains quite general. In fact, we postulate the wealth in a quite general form in what concerns to the decision variable, so that the formulation is indeed easily applicable to models from different situations.

For this general model, we present a method to study its properties, and especially comparative-static effects, in a systematic manner. This method is analytic, rather than geometric as usual in many other models. One of the key points of the method

A.A. Álvarez-López (✉) • F. Sebastiá-Costa
Department of Quantitative Applied Economics II, UNED, Paseo Senda del Rey, 11, Madrid 28040, Spain
e-mail: aalvarez@cee.uned.es; fransecos@yahoo.es

I. Rodríguez-Puerta
Department of Economics, Quantitative Methods and Economic History, Pablo de Olavide University, Carretera de Utrera Km. 1, Sevilla 41013, Spain
e-mail: irodpue@upo.es

M. Buendía
Department of Economics, ESERP, C/ Costa Rica, 9, Madrid 28016, Spain
e-mail: monicabuen@gmail.com

A.A. Pinto and D. Zilberman (eds.), *Modeling, Dynamics, Optimization and Bioeconomics I*, Springer Proceedings in Mathematics & Statistics 73, DOI 10.1007/978-3-319-04849-9__5,
© Springer International Publishing Switzerland 2014

is a lemma (Lemma 5.1 in Appendix) that gives us useful bounds for products of random variables. This lemma generalizes one from [5] that is barely used in the literature.

In Sect. 5.2, we formulate the model and present the method.[1] We find a characterization for corner solution, and compare the optimal solution with the corresponding solution in absence of uncertainty. We also examine comparative-static effects. As an application of this, we consider a proportional wealth tax and study the effect of a variation in its rate.

In Sect. 5.3, we apply the previous results to two models from the theory of the firm under uncertainty. For both of them, we easily obtain their key results. In particular, for the second model, the corresponding result about the effect of a proportional wealth tax is new in the literature.

Finally, Sect. 5.4 concludes the paper.

5.2 The General Model and Its Properties

5.2.1 The Model

We consider an agent who faces some kind of uncertainty. There is only one variable of decision: x. For each possible value of x, a random wealth $W(x)$ is defined. The agent seeks to maximize the expected utility of this wealth.

We assume that the variable x belongs to an interval $I \subseteq \mathbb{R}_+ = \{x \in \mathbb{R} \mid x \geq 0\}$, bounded or not, such that $0 \in I$. For each $x \in I$, we consider that the wealth $W(x)$ is given by[2]

$$W(x) = f(x)\, Z - g(x),$$

where $Z \geq 0$ is a non-degenerate random variable with expectation $\mu > 0$, and f and g are real (non-random) functions of class $\mathbb{C}^2$ on I such that $f' > 0$, $g' > 0$, and f is concave and g is convex (on I). In addition, we assume that $f(0) = 0$, and thus $f(x) > 0$ if $x \neq 0$.

The agent's attitude towards risk is modeled by a BERNOULLI utility function u, regular enough (at least of class $\mathbb{C}^2$ on $\mathbb{R}$) and such that $u' > 0$ and $u'' < 0$. In particular, the agent is risk averse.

For each $x \in I$, the expected utility of choosing x is given by $U(x) \equiv E[u(W(x))]$. The agent maximizes this function U on the interval I:

---

[1]Some aspects in this section were presented, in a preliminary version, in [2]. On the other hand, the model formulated here is similar to that in [4], but our method to analyze the problem and our results are different.

[2]In order to simplify the notation, if possible we will write simply W instead of $W(x)$.

$$\max_{x \in I} U(x). \tag{5.1}$$

The first and second derivatives of U are:

$$U'(x) = \mathsf{E}\big[u'(W)\big(f'(x)\, Z - g'(x)\big)\big],$$

$$U''(x) = \mathsf{E}\big[u''(W)\big(f'(x)\, Z - g'(x)\big)^2 + u'(W)\big(f''(x)\, Z - g''(x)\big)\big].$$

According to the hypotheses, we obtain that $U'' < 0$ on I, and thus the function U is strictly concave on I: if the maximization problem (5.1) admits a solution, this solution is unique. Notice that the equality $U'(x) = 0$ is a sufficient condition for a point $x \in I$ to be the (unique) solution of the problem.

From now on, we consider the function $h \equiv g'/f'$. In terms of this function h and the expectation μ of the random variable Z, the following proposition gives us both a sufficient condition for the existence of a solution and a characterization for a corner solution.

Proposition 5.1. *If there exists some $x_0 \in I$ such that $\mu \leqslant h(x_0)$, or if the interval I is bounded, then the maximization problem (5.1) has a unique solution. In addition, if we do know that there exists a unique solution for the problem (5.1), then this solution will be positive if and only if $\mu > h(0)$.*

Proof. Given that $\mu \leqslant h(x_0)$ for some $x_0 \in I$, applying Lemma 5.1 to the random variable $X = f'(x_0)Z - g'(x_0)$, and the real functions

$$\psi \equiv 1 \quad \text{and} \quad \phi(s) = u'\big([s + g'(x_0)][f(x_0)/f'(x_0)] - g(x_0)\big),$$

as the function ϕ is positive and decreasing, we can write:

$$\mathsf{E}\big[u'\big(W(x_0)\big)\big(f'(x_0)\, Z - g'(x_0)\big)\big] \leqslant \phi(0)\, \mathsf{E}\big[f'(x_0)\, Z - g'(x_0)\big]$$

$$= \phi(0)\, f'(x_0)\, \big(\mu - h(x_0)\big);$$

that is: $U'(x_0) \leqslant 0$; thus: $U(x_0) > U(x)$ if $x > x_0$, and the unique solution of the problem (5.1) is precisely that of the problem $\max_{x \in [0, x_0]} U(x)$, which exists because of the continuity of U.

On the other hand, note that $W(0) = -g(0)$ is non-random, and we have:

$$U'(0) = u'\big(-g(0)\big) \cdot \big(f'(0)\, \mu - g'(0)\big) = u'\big(-g(0)\big) \cdot f'(0) \cdot \big(\mu - h(0)\big),$$

and thus $\mu > h(0)$ is equivalent to $U'(0) > 0$. If we know that there exists a unique solution x^*, the inequality $U'(0) > 0$ is equivalent to $x^* > 0$. $\qquad\square$

Henceforth, we assume that there exists a unique, interior solution x^* for the maximization problem (5.1). This solution is thus characterized by the equality $U'(x^*) = 0$.

5.2.2 An Auxiliary Function

The following function will be useful in our analysis:

$$F(x) = \frac{\mathsf{E}\big[u'\big(W(x)\big)\,Z\big]}{\mathsf{E}\big[u'\big(W(x)\big)\big]}, \qquad x \in I,$$

with derivative:

$$F'(x) = \frac{\mathsf{E}\big[u''\big(W(x)\big)\big(Z - h(x)\big)\big(Z - F(x)\big)\big]\,f'(x)}{\mathsf{E}\big[u'\big(W(x)\big)\big]}.$$

The function F and the marginal utility U' are closely related:

$$U'(x) = \mathsf{E}\big[u'(W)\big]\,f'(x)\,\big(F(x) - h(x)\big),$$

so that the optimal solution is characterized by this equality:

$$F(x^*) = h(x^*). \tag{5.2}$$

We also have:

$$F(0) = \mu, \qquad \text{and} \qquad F(x) < \mu \quad \text{if } x \in I - \{0\}. \tag{5.3}$$

The latter can be proved by applying Lemma 5.1 with the random variable $X = Z - \mu$, and the real functions $\psi \equiv 1$ and $\phi(s) = u'(f(x)(s + \mu) - g(x))$: as $f(x) > 0$, the function ϕ is strictly decreasing, and we obtain:

$$\mathsf{E}\big[u'(W)(Z - \mu)\big] < \phi(0) \cdot \mathsf{E}[Z - \mu] = 0,$$

and hence $F(x) < \mu$.

An Interpretation. We can give an interpretation of the number $F(x^*)$. For the moment, we focus our attention on a non-random maximization problem: that resulting from writing a positive number z instead of the random variable Z in (5.1); that is:

$$\max_{x \in I} u\big(f(x)\,z - g(x)\big). \tag{5.4}$$

We can see that, if this problem admits an interior solution x_z, then this solution is unique (the objective function is strictly concave) and is characterized by the condition: $f'(x_z)\,z - g'(x_z) = 0$, or equivalently: $h(x_z) = z$.

Therefore, recalling the characterization (5.2) of the optimal solution x^* of problem (5.1), we claim that x^* is also the solution of a non-random problem: the problem (5.4) with $z = F(x^*)$, or equivalently: the problem resulting from writing $F(x^*)$ instead of Z in (5.1).

In other words: *if it were possible for the agent to operate in a certainty environment, $F(x^*)$ would be the value of Z for which the agent would choose exactly the level x^*.*

5.2.3 The Corresponding Certainty Problem

We wish to compare the solution of problem (5.1) to the one of problem (5.4) with $z = \mu$:

$$\max_{x \in I} u\big(f(x)\,\mu - g(x)\big), \qquad (5.5)$$

provided this solution indeed exists. We refer to this problem as the *corresponding certainty problem*. The following proposition establishes the result of comparison of both solutions. Recall that we are assuming that the solution x^* of problem (5.1) is interior, so that the equality (5.2) holds.

Proposition 5.2. *If the problem (5.5) admits a solution x_c^*, then $x^* < x_c^*$.*

Proof. We already know that the solution x_c^* is unique; we can assume that it is interior. Indeed. Set: $U_c(x) = u\big(f(x)\,\mu - g(x)\big)$. As x^* is itself interior, we have: $\mu > h(0)$ (cf. Proposition 5.1), and thus: $U_c'(0) > 0$; that is: $x_c^* > 0$. Also, if the interval I were bounded and closed, x_c^* being its right extreme, then the thesis of the proposition would hold, because x^* would be smaller than that right extreme. Therefore, we can assume that the solution x_c^* is characterized by the condition $U_c'(x_c^*) = 0$, or equivalently: $h(x_c^*) = \mu$.

Furthermore, the function h is strictly increasing: the sign of h' is the same as that of $g'' f' - g' f''$, which is positive according to the hypotheses assumed for f and g.[3] Recalling (5.3), we can write:

$$h(x_c^*) = \mu > F(x^*) = h(x^*),$$

and hence $x^* < x_c^*$.   □

That is: *the optimal solution under uncertainty is less than the optimal solution under certainty.*

In this general model, we have just proved that uncertainty *reduces* the optimal amount of the decision variable chosen by the agent.

---

[3] Actually, we would only have that $g'' f' - g' f'' \geqslant 0$, because $f'' \leqslant 0$ and $g'' \geqslant 0$. But if f'' and g'' were simultaneously null (so that the function h turns out to be constant), the problem (5.5) would admit the solution x_c^* if and only if $I = [0, x_c^*]$, and in this case the result would also hold.

5.2.4 Comparative-Static Effects

The function F can also be used to study comparative-static effects. If ν denotes a parameter of the model, we can write $F(x; \nu)$ and $h(x; \nu)$ instead of $F(x)$ and $h(x)$, respectively, to stand for this further dependence on ν of these functions, and we can write $\mathrm{d}x^*/\mathrm{d}\nu$ to stand for the corresponding comparative-static effect.

Proposition 5.3. *The sign of* $\dfrac{\mathrm{d}x^*}{\mathrm{d}\nu}$ *is the same as that of* $F'_\nu(x^*; \nu) - h'_\nu(x^*; \nu)$.

Proof. From the characterization $F(x^*; \nu) = h(x^*; \nu)$ [cf. (5.2)], we can write:

$$\frac{\mathrm{d}x^*}{\mathrm{d}\nu} = -\frac{F'_\nu(x^*; \nu) - h'_\nu(x^*; \nu)}{F'_x(x^*; \nu) - h'_x(x^*; \nu)},$$

and the denominator is negative. Indeed: we already know that $h' > 0$, and also:

$$F'(x^*) = \frac{\mathsf{E}\big[u''(W^*)\,(Z - h(x^*))\,(Z - F(x^*))\big]\,f'(x^*)}{\mathsf{E}\big[u'(W^*)\big]}$$

$$= \frac{\mathsf{E}\big[u''(W^*)\,(Z - F(x^*))^2\big]\,f'(x^*)}{\mathsf{E}\big[u'(W^*)\big]} < 0,$$

where $W^* \equiv W(x^*)$. □

An Application: A Proportional Wealth Tax. In order to illustrate this result, let us assume that the wealth function is given by:

$$W(x) = (1 - \tau)\big(f_0(x)\,Z - g_0(x)\big),$$

where $0 < \tau < 1$ and f_0 and g_0 are in the same conditions as f and g, respectively. We can interpret this wealth as the result of a proportional wealth tax at rate τ.[4] What we wish to study is the effect, in the optimal value x^*, of a variation in the tax rate τ.

Note that $h = g'_0/h'_0$ does not depend on τ, and hence the sign of $\mathrm{d}x^*/\mathrm{d}\tau$ is simply that of $F'_\tau(x^*; \tau)$ (cf. Proposition 5.3). We have: $F(x^*; \tau) = \mathsf{E}\big[u'(W^*)\,Z\big] / \mathsf{E}\big[u'(W^*)\big]$, where $W^* = W(x^*) = (1 - \tau)\big(f_0(x^*)\,Z - g_0(x^*)\big)$; and we obtain (recall (5.2)):

$$F'_\tau(x^*; \tau) = -\frac{\mathsf{E}\big[u''(W^*)\,W^*\,(Z - h(x^*))\big]}{(1 - \tau)\,\mathsf{E}\big[u'(W^*)\big]}.$$

---

[4]We are implicitly assuming that there is a full loss offset. See [7, p. 70] for a discussion about the plausibility of this hypothesis at least in the case of a firm.

If R_u denotes the ARROW–PRATT measure of relative risk aversion for the agent, we can write:

$$-E[u''(W^*) W^* (Z - h(x^*))] = E[R_u(W^*) u'(W^*) (Z - h(x^*))].$$

Considering the real functions:

$$\psi(s) = u'\Big((1 - \tau)[f(x^*)(s + h(x^*)) - g(x^*)]\Big),$$

$$\phi(s) = R_u\Big((1 - \tau)[f(x^*)(s + h(x^*)) - g(x^*)]\Big),$$

we find that $\psi > 0$, and that ϕ is increasing, constant or decreasing depending on whether the agent exhibits IRRA, CRRA or DRRA,[5] respectively. In the IRRA case, for instance, taking the random variable $X = Z - h(x^*)$, the application of Lemma 5.1 yields:

$$E[R_u(W^*) u'(W^*) (Z - h(x^*))] \geq \phi(0) \, E[u'(W^*) (Z - h(x^*))] = 0,$$

where the last equality follows from the first order condition of the maximization problem. For the CRRA and DRRA cases, we can proceed, *mutatis mutandis*, in the same manner.

Finally, we can assert that, *as the tax rate increases, the optimal level x^* increases, remains constant or reduces depending on whether the agent exhibits IRRA, CRRA or DRRA, respectively.*

5.3 Application to the Theory of the Firm Under Uncertainty

In this section, we apply the results of Sect. 5.2 to two models from the theory of the firm under uncertainty. The first one, due to SANDMO, is one of the seminal models of the theory; we easily obtain some of its most important results. The second one, due to DALAL and ALGHALITH, is an important model of the theory with the feature, unlike SANDMO's model, of considering two sources of uncertainty; for this model, we also obtain its main result in a easy way, and derive a new property.

5.3.1 About SANDMO's Model

In SANDMO's model, as presented in [7], it is considered a competitive firm that produces a single output and faces uncertainty in the price at which this output will

---

[5]As usual in the literature, these acronyms mean *i*ncreasing, *c*onstant or *d*ecreasing (respectively) *r*elative *r*isk *a*version.

be sold. The firm decides the amount of output to be produced *before* the sale date, and hence before knowing the spot price.

From the firm's point of view, the price of the output is a non-degenerate random variable $P \geqslant 0$ with expectation $\mu > 0$, and the total cost of choosing an amount of output $q \geqslant 0$ is $C(q) + B$, $C(q)$ being the variable cost (with $C(0) = 0$) and B being the fixed cost. The firm's attitude facing uncertainty is modeled by a BERNOULLI utility function u. For the functions C and u, it is assumed that $C' > 0$, $C'' \geqslant 0$, $u' > 0$ and $u'' < 0$, so that the firm is risk averse. The firm's profit of producing q units is given by $\pi(q) = Pq - C(q) - B$, and the firm seeks a level of output that maximizes the expected utility of this profit.

This model can be considered as a particular case of the general model presented in Sect. 5.2, just taking $Z = P$, $x = q$, $I = \mathbb{R}_+$, $f(q) = q$ and $g(q) = C(q) + B$, and $W = \pi$. Notice that $h = C'$; that is, the function h is precisely the marginal cost.

SANDMO assumes that there exists an optimal solution q^* for the corresponding maximization problem, and discusses the possibility of a corner solution. Our Proposition 5.1 establishes that this optimal solution will be positive if and only if the expected price is bigger than the marginal cost of the null level of production: $\mu > C'(0)$. Remark also that, when the optimal solution q^* is positive, this solution is characterized by the condition $F(q^*) = C'(q^*)$ [cf. (5.2)]; thus, if the firm could sell its product at a certain price, the number $F(q^*)$ would be the value of that price for which the firm would choose to produce exactly q^* units of output.

On the other hand, the main result obtained by SANDMO is this: "under price uncertainty, output is smaller than the certainty output" (cf. [7, pp. 66–67]). More precisely: the firm reduces its output with respect to the firm under certainty for which the output price equals the expected price. In the new context from Sect. 5.2, this key result follows as a simple consequence of Proposition 5.2.

Finally, among others comparative-static effects, SANDMO studies the influence, on the optimal solution q^*, of a proportional profit tax and of a further variation of its rate. He finds that "increasing the tax rate will increase, leave constant or reduce output according as relative risk aversion is increasing, constant, or decreasing" (cf. [7, p. 70]). This result can easily be obtained from the statement about this effect in Sect. 5.2.4.

5.3.2 A Model with Two Sources of Uncertainty: A New Result

Now we apply the previous results to a model from the theory of the firm under uncertainty due to DALAL and ALGHALITH (see [3]). This model has become a relevant reference in the theory, since it is a very complete study of a firm with two simultaneous sources of uncertainty: output price and production. In spite of introducing two sources of uncertainty, the model can also be considered as a particular case of the framework presented in Sect. 5.2.

The wealth of the firm is given by[6]:

$$W(y) = pvy - c(y) + W_0,$$

where y, the decision variable, is the target level of production; v is a random factor so that the product vy can be interpreted as the *actual* level of production; p is the random output price; c is the cost function; and W_0 is an initial level of wealth. It is assumed that $E[v] = 1$, and that $c' > 0$ and $c'' > 0$. Also, the initial wealth W_0 is such that $W(y) \geq 0$ for all y. And the expected output price is denoted by $\bar{p}$: $E[p] = \bar{p}$. The firm seeks to maximize the expected utility of its wealth, considering that the attitude towards risk is modeled by a BERNOULLI utility function u such that $u' > 0$ and $u'' < 0$.

We consider this model in the context of Sect. 5.2, with $Z = pv$, $f(y) = y$, and $g(y) = c(y) - W_0$, which results in $h(y) = c'(y)$. We are especially interested in applying Proposition 5.2, which compares the optimal level of production y^* in the original problem (under uncertainty in both variables) with the optimal level of production y_c^* in the corresponding certainty problem (in the sense given in Sect. 5.2.3). This corresponding certainty problem results from writing $E[pv]$ instead of pv. With Proposition 5.2, we deduce that $y^* < y_c^*$.

But DALAL and ALGHALITH focus their attention on a different certainty problem: that resulting from writing $E[v] = 1$ instead of v, and writing $\bar{p}$ instead of p; they denote its solution as y_c. Note that y_c is characterized by the condition $h(y_c) = \bar{p}$, and according to (5.2) and (5.3) we have that $h(y^*) < E[pv]$. Under the hypothesis that $cov(p, v) \leq 0$, we have: $E[pv] = cov(p, v) + \bar{p} \leq \bar{p}$, and we can write:

$$h(y^*) < E[pv] \leq \bar{p} = h(y_c),$$

and hence: $y^* < y_c$. This is exactly the assertion of Proposition 1 in [3].

Proposition 2 in [3], about uncertainty in production but certainty in price, is also an immediate consequence of our analysis.

Finally, on translating, to this model, the comparative-static result about a tax rate (see Sect. 5.2.4), we obtain a new property of the model. Consider that the wealth is given by:

$$W(y) = (1 - \tau)(pvy - c(y) + W_0),$$

where $0 < \tau < 1$, so that the firm's final wealth is reduced by a proportional tax at rate τ. We can assert that *the tax rate increases, leaves constant, or reduces the optimal decision level of production y^* depending on whether the firm exhibits IRRA, CRRA or DRRA, respectively.*

---

[6]We specify here a slightly simplified version of the original model presented in [3]. For our purposes, that will be enough.

5.4 Concluding Remarks

We consider a simple one-variable model of decision under uncertainty. The specification of the wealth (the expected utility of which is to be maximized) is quite general. We propose a method to obtain several properties for this model. This method requires to consider the function $F(x) = \mathsf{E}\big[u'(W)\,Z\big]\big/\mathsf{E}\big[u'(W)\big]$, which lets us establish a simple characterization of the optimal solution. It is a remarkable fact that the number $F(x^*)$, where x^* is the optimal solution, lets the agent—at least theoretically—consider a decision problem without uncertainty that is equivalent to the original one under uncertainty, in the sense that both problems share the optimum at the same point.

Furthermore, we consider the so-called corresponding certainty problem, which results from writing, instead of the random variable Z, its expectation μ. We find that the optimal solution of the original problem is less than that of this non-random problem. That is: uncertainty reduces the optimal amount chosen by the agent.

From the characterization of the optimal solution, it is also possible to obtain comparative-static results. In order to illustrate this tool, we consider a proportional wealth tax and study the effect, on the optimal solution, of a variation in the tax rate.

The use of the method presented here is illustrated in two models of the theory of the firm under uncertainty. For the first one (due to SANDMO, see [7], one of the seminal models of the theory), the method easily lets us obtain the key results. The second model (due to DALAL and ALGHALITH, see [3], a relevant model of the theory) presents the remarkable feature of considering two simultaneous sources of uncertainty (output price and production). Formally, there are now two random variables, but they appear as a product, in such a way that it is also possible to analyze the problem in the context presented in Sect. 5.2. We readily obtain an important property (the first key result of that paper), namely: under uncertainty, the optimal amount of the decision variable (the "target" production) is reduced. Finally, we translate, to this model, the general result about the effect of a variation in a tax rate, what lets us establish a property that had not been studied by the authors in their paper.

Acknowledgements ÁLVAREZ-LÓPEZ would like to thank the financial support provided by the Spanish Interministerial Commission of Science and Technology (CICYT: Comisión Interministerial de Ciencia y Tecnología), under the Projects with the reference numbers ECO2008-06395-C05-03 and ECO2012-39553-C04-01.

Appendix

The following lemma slightly generalizes a result taken from [5]:

Lemma 5.1. *Let ψ and ϕ be two real functions defined on $\mathbb{R}$ such that $\psi > 0$ and ϕ is increasing. If $\xi = \psi \cdot \phi$, and X is a real random variable such that the* *expectation $E[X \, \psi(X)]$ is finite, and such that the probability of the set $\{X \neq 0\}$ is positive, then:*

$$E[X \, \xi(X)] \geq \phi(0) \, E[X \, \psi(X)] \,,$$

and the contrary inequality holds when ϕ is decreasing. In addition, if ϕ is strictly increasing or strictly decreasing, the corresponding inequality also holds strictly.

Proof. See [6]: in Lemma 1, write $F \equiv 1$ and $Z = X$. See also [1]. $\square$

References

1. Álvarez-López, A.A., Rodríguez-Puerta, I.: Teoría de la empresa bajo incertidumbre con mercado de futuros: el papel de los costes fijos y de un impuesto sobre los beneficios. Rect@ **10**(1), 253–265 (2009)
2. Álvarez-López, A.A., Rodríguez-Puerta, I.: A methodological contribution in the theory of the firm under uncertainty. In: Peixoto, M.M., Pinto, A.A., Rand, D.A. (eds.) Dynamics, Games and Science II, chap. 7. Springer, Berlin (2011)
3. Dalal, A.J., Alghalith, M.: Production decisions under joint price and production uncertainty. Eur. J. Oper. Res. **197**, 84–92 (2009)
4. Feder, G.: The impact of uncertainty in a class of objective functions. J. Econ. Theory **16**(2), 504–512 (1977)
5. Lippman, S.A., McCall, J.J.: The economics of uncertainty: selected topics and probabilistic methods. In: Arrow, K.J., Intriligator, M.J. (eds.) Handbook of Mathematical Economics, vol. 1, chap. 6. North-Holland, Amsterdam (1982)
6. Rodríguez-Puerta, I., Sebastiá-Costa, F., Álvarez-López, A.A., Buendía, M.: Una herramienta de análisis teórico en la teoría de la empresa bajo incertidumbre. Revista de Métodos Cuantitativos para la Economía y la Empresa **11**, 33–40 (2011)
7. Sandmo, A.: On the theory of the competitive firm under price uncertainty. Am. Econ. Rev. **61**(1), 65–73 (1971)

Chapter 6
On Sensitive Dependence on Initial Conditions and Existence of Physical Measure for 3-Flows

Vítor Araújo

6.1 Introduction

The development of the theory of dynamical systems has shown that models involving expressions as simple as quadratic polynomials (as the *logistic family* or *Hénon attractor*, see e.g. [8] for a gentle introduction), or autonomous ordinary differential equations with a hyperbolic equilibrium of saddle-type accumulated by regular orbits, as the *Lorenz flow* (see e.g. [3, 12, 30]), exhibit *sensitive dependence on initial conditions*, a common feature of *chaotic dynamics*: small initial differences are rapidly augmented as time passes, causing two trajectories originally coming from practically indistinguishable points to behave in a completely different manner after a short while. Long term predictions based on such models are unfeasible since it is not possible to both specify initial conditions with arbitrary accuracy and numerically calculate with arbitrary precision. For an introduction to these notions see [8, 28].

Formally the definition of sensitivity is as follows for a flow X^t on some compact manifold M: a X^t-invariant subset Λ is *sensitive to initial conditions* or has *sensitive dependence on initial conditions*, or simply *chaotic* if, for every small enough $r > 0$ and $x \in \Lambda$, and for any neighborhood U of x, there exists $y \in U$ and $t \neq 0$ such that $X^t(y)$ and $X^t(x)$ are r-apart from each other: $\mathrm{dist}\big(X^t(y), X^t(x)\big) \geq r$. See Fig. 6.1. An analogous definition holds for diffeomorphism f of some manifold, taking $t \in \mathbb{Z}$ and setting $f = X^1$ in the previous definition.

Using some known results on robustness of attractors from Mañé [18] and Morales, Pacifico and Pujals [21] together with observations on their proofs, we

V. Araújo
Instituto de Matemática, Universidade Federal da Bahia

Av. Adhemar de Barros, S/N , Ondina, 40170-110, Salvador, Brazil
e-mail: vitor.araujo.im.ufba@gmail.com

A.A. Pinto and D. Zilberman (eds.), *Modeling, Dynamics, Optimization and Bioeconomics I*, Springer Proceedings in Mathematics & Statistics 73, DOI 10.1007/978-3-319-04849-9_6,

Fig. 6.1 Sensitive
dependence on initial
conditions

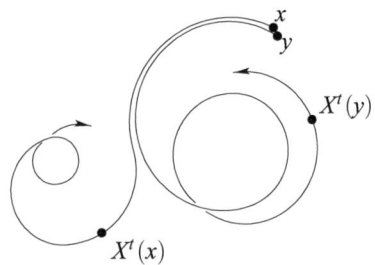

show that for attractors of three-dimensional flows, robust chaotic behavior (in the
above sense of sensitiveness to initial conditions) is equivalent to the existence
of certain hyperbolic structures. These structures, in turn, are associated to the
existence of physical measures. In short *in low dimensions, robust chaotic behavior
ensures the existence of a physical measure.*

6.2 Preliminary Notions

Here and throughout the text we assume that M is a three-dimensional compact
connected manifold without boundary endowed with some Riemannian metric
which induces a distance denoted by dist and a volume form Leb which we
name *Lebesgue measure* or *volume*. For any subset A of M we denote by $\overline{A}$ the
(topological) closure of A.

We denote by $\mathfrak{X}^r(M), r \geqslant 1$ the set of C^r smooth vector fields X on M endowed
with the C^r topology. Given $X \in \mathfrak{X}^r(M)$ we denote by X^t, with $t \in \mathbb{R}$, the flow
generated by the vector field X. Since we assume that M is a compact manifold
the flow is defined for all time. Recall that the flow $(X^t)_{t \in \mathbb{R}}$ is a family of C^r
diffeomorphisms satisfying the following properties:

1. $X^0 = Id \cdot M \to M$ is the identity map of M;
2. $X^{t+s} = X^t \circ X^s$ for all $t, s \in \mathbb{R}$,

and it is *generated by the vector field X* if

3. $\frac{d}{dt} X^t(q)\big|_{t=t_0} = X\big(X_{t_0}(q)\big)$ for all $q \in M$ and $t_0 \in \mathbb{R}$.

We say that a compact X^t-invariant set Λ is *isolated* if there exists a neighbor-
hood U of Λ such that $\Lambda = \cap_{t \in \mathbb{R}} X^t(U)$. A compact invariant set Λ is *attracting*
if $\Lambda_X(U) := \cap_{t \geq 0} X^t(U)$ equals Λ for some neighborhood U of Λ satisfying
$\overline{X^t(U)} \subset U$, for all $t > 0$. In this case the neighborhood U is called an *isolating
neighborhood* of Λ. Note that $\Lambda_X(U)$ is in general different from $\cap_{t \in \mathbb{R}} X^t(U)$, but
for an attracting set the extra condition $\overline{X^t(U)} \subset U$ for $t > 0$ ensures that every
attracting set is also isolated. We say that Λ is *transitive* if Λ is the closure of both
$\{X^t(q) : t > 0\}$ and $\{X^t(q) : t < 0\}$ for some $q \in \Lambda$. An *attractor* of X is a

transitive attracting set of X and a *repeller* is an attractor for $-X$. We say that Λ is a *proper* attractor or repeller if $\emptyset \neq \Lambda \neq M$.

An *equilibrium* (or *singularity*) for X is a point $\sigma \in M$ such that $X^t(\sigma) = \sigma$ for all $t \in \mathbb{R}$, i.e. a fixed point of all the flow maps, which corresponds to a zero of the associated vector field X: $X(\sigma) = 0$. An *orbit* of X is a set $\mathcal{O}(q) = \mathcal{O}_X(q) = \{X^t(q) : t \in \mathbb{R}\}$ for some $q \in M$. A *periodic orbit* of X is an orbit $\mathcal{O} = \mathcal{O}_X(p)$ such that $X^T(p) = p$ for some minimal $T > 0$. A *critical element* of a given vector field X is either an equilibrium or a periodic orbit.

We recall that a X^t-invariant probability measure μ is a probability measure satisfying $\mu(X^t(A)) = \mu(A)$ for all $t \in \mathbb{R}$ and measurable $A \subset M$. Given an invariant probability measure μ for a flow X^t, let $B(\mu)$ be the *(ergodic) basin* of μ, i.e., the set of points $z \in M$ satisfying for all continuous functions $\varphi : M \to \mathbb{R}$

$$\lim_{T \to +\infty} \frac{1}{T} \int_0^T \varphi(X^t(z)) \, dt = \int \varphi \, d\mu.$$

We say that μ is a *physical* (or *SRB*) measure for X if $B(\mu)$ has positive Lebesgue measure: $\text{Leb}(B(\mu)) > 0$.

The existence of a physical measures for an attractor shows that most points in a neighborhood of the attractor have well defined long term statistical behavior. So, in spite of chaotic behavior preventing the exact prediction of the time evolution of the system in practical terms, we gain some statistical knowledge of the long term behavior of the system near the chaotic attractor.

6.3 Chaotic Systems

We distinguish between forward and backward sensitive dependence on initial conditions. We say that an invariant subset Λ for a flow X^t is *future chaotic with constant $r > 0$* if, for every $x \in \Lambda$ and each neighborhood U of x in the ambient manifold, there exists $y \in U$ and $t > 0$ such that $\text{dist}(X^t(y), X^t(x)) \geq r$. Analogously we say that Λ is *past chaotic with constant r* if Λ is future chaotic with constant r for the flow generated by $-X$. If we have such *sensitive dependence both for the past and for the future*, we say that Λ is *chaotic*. Note that in this language sensitive dependence on initial conditions is weaker than chaotic, future chaotic or past chaotic conditions.

An easy consequence of chaotic behavior is that it prevents the existence of sources or sinks, either attracting or repelling equilibria or periodic orbits, inside the invariant set Λ. Indeed, if Λ is future chaotic (for some constant $r > 0$) then, were it to contain some attracting periodic orbit or equilibrium, any point of such orbit (or equilibrium) would admit no point in a neighborhood whose orbit would move away in the future. Likewise, reversing the time direction, a past chaotic invariant set cannot contain repelling periodic orbits or repelling equilibria. As an almost reciprocal we have the following.

Lemma 6.1. *If* $\Lambda = \cap_{t \in \mathbb{R}} X^t(U)$ *is a compact isolated proper subset for* $X \in \mathfrak{X}^1(M)$ *with isolating neighborhood* U *and* Λ *is* not future chaotic *(respective not past chaotic), then* $\Lambda_X^-(U) := \overline{\cap_{t>0} X^{-t}(U)}$ *(respective* $\Lambda_X^+(U) := \overline{\cap_{t>0} X^t(U)}$*) has non-empty interior.*

Proof. If Λ is not future chaotic, then for every $r > 0$ there exists some point $x \in \Lambda$ and a neighborhood V of x such that $\text{dist}\big(X^t(y), X^t(x)\big) < r$ for all $t > 0$ and each $y \in V$. If we choose $0 < r < \text{dist}(M \setminus U, \Lambda)$ (we note that if $\Lambda = U$ then Λ would be open and closed, and so, by connectedness of M, Λ would not be a proper subset), then we deduce that $X^t(y) \in U$, that is, $y \in X^{-t}(U)$ for all $t > 0$, hence $V \subset \Lambda_X^-(U)$. Analogously if Λ is not past chaotic, just by reversing the time direction. $\square$

In particular *if an invariant and isolated set* Λ *with isolating neighborhood* U *is given such that the volume of both* $\Lambda_X^+(U)$ *and* $\Lambda_X^-(U)$ *is zero, then* Λ *is chaotic.*

Sensitive dependence on initial conditions is part of the definition of *chaotic system* in the literature, see e.g. [8]. It is an interesting fact that sensitive dependence is a consequence of another two common features of most systems considered to be chaotic: existence of a dense orbit and existence of a dense subset of periodic orbits.

Proposition 6.1. *A compact invariant subset* Λ *for a flow* X^t *with a dense subset of periodic orbits and a dense (regular and non-periodic) orbit is chaotic.*

A short proof of this proposition can be found in [5]. An extensive discussion of this and related topics can be found in [10].

6.4   Lack of Sensitiveness for Flows on Surfaces

We recall the following celebrated result of Mauricio Peixoto in [25, 26] (and for a more detailed exposition of this results and sketch of the proof see [12]) built on previous work of Poincaré [27] and Andronov and Pontryagin [1], that characterizes structurally stable vector fields on compact surfaces.

Theorem 6.1 (Peixoto). *A* C^r *vector field,* $r \geq 1$, *on a compact surface* S *is structurally stable if, and only if:*

1. *the number of critical elements is finite and each is hyperbolic;*
2. *there are no orbits connecting saddle points;*
3. *the non-wandering set consists of critical elements alone.*

Moreover if S *is orientable, then the set of structurally stable vector fields is open and dense in* $\mathfrak{X}^r(S)$.

In particular, this implies that for a structurally stable vector field X on S there is an open and dense subset B of S such that the positive orbit $X^t(p), t \geq 0$ of $p \in B$ converges to one of finitely many attracting equilibria. Therefore *no sensitive dependence on initial conditions arises for an open and dense subset of all vector fields in orientable surfaces.*

The extension of Peixoto's characterization of structural stability for C^r flows, $r \geq 1$, on non-orientable surfaces is known as *Peixoto's Conjecture*, and up until now it has been proved for the projective plane $\mathbb{P}^2$ [23], the Klein bottle $\mathbb{K}^2$ [19] and $\mathbb{L}^2$, the torus with one cross-cap [13]. Hence for these surfaces we also have no sensitiveness to initial conditions for most vector fields.

This explains in part the great interest attached to the Lorenz attractor which was one of the first examples of sensitive dependence on initial conditions.

6.5 Robustness and Volume Hyperbolicity

Related to chaotic behavior is the notion of *robust dynamics*. We say that an attracting set $\Lambda = \Lambda_X(U)$ for a three-flow X and some open subset U is *robust* if there exists a C^1 neighborhood $\mathcal{U}$ of X in $\mathfrak{X}^1(M)$ such that $\Lambda_Y(U)$ is transitive for every $Y \in \mathcal{U}$.

The following result obtained by Morales, Pacifico and Pujals in [21] characterizes robust attractors for three-dimensional flows.

Theorem 6.2. *Robust attractors for flows containing equilibria are singular-hyperbolic sets for X.*

We remark that robust attractors cannot be C^1 approximated by vector fields presenting either attracting or repelling periodic points. This implies that, on three-manifolds, any periodic orbit inside a robust attractor is hyperbolic of saddle-type.

We now define the concept of singular-hyperbolicity. A compact invariant set Λ of X is *partially hyperbolic* if there are a continuous invariant tangent bundle decomposition $T_\Lambda M = E_\Lambda^s \oplus E_\Lambda^c$ and constants $\lambda, K > 0$ such that

- E_Λ^c (K, λ)-*dominates* E_Λ^s, i.e. for all $x \in \Lambda$ and for all $t \geq 0$

$$\|DX^t(x) \mid E_x^s\| \leq \frac{e^{-\lambda t}}{K} \cdot m(DX^t(x) \mid E_x^c); \qquad (6.1)$$

- E_Λ^s is (K, λ)-*contracting*: $\|DX^t \mid E_x^s\| \leq Ke^{-\lambda t}$ for all $x \in \Lambda$ and for all $t \geq 0$.

For $x \in \Lambda$ and $t \in \mathbb{R}$ we let $J_t^c(x)$ be the absolute value of the determinant of the linear map $DX^t(x) \mid E_x^c : E_x^c \to E_{X^t(x)}^c$. We say that the sub-bundle E_Λ^c of the partial hyperbolic set Λ is (K, λ)-*volume expanding* if

$$J_t^c(x) = \left| \det(DX^t \mid E_x^c) \right| \geq Ke^{\lambda t},$$

for every $x \in \Lambda$ and $t \geq 0$.

We say that a partially hyperbolic set is *singular-hyperbolic* if its singularities are hyperbolic and it has volume expanding central direction.

A *singular-hyperbolic attractor* is a singular-hyperbolic set which is an attractor as well: an example is the (geometric) Lorenz attractor presented in [11, 17]. Any equilibrium σ of a singular-hyperbolic attractor for a vector field X is such that $DX(\sigma)$ has only real eigenvalues $\lambda_2 \leq \lambda_3 \leq \lambda_1$ satisfying the same relations as in the Lorenz flow example:

$$\lambda_2 < \lambda_3 < 0 < -\lambda_3 < \lambda_1, \tag{6.2}$$

which we refer to as *Lorenz-like equilibria*. We recall that an compact X^t-invariant set Λ is *hyperbolic* if the tangent bundle over Λ splits $T_\Lambda M = E^s_\Lambda \oplus E^X_\Lambda \oplus E^u_\Lambda$ into three DX^t-invariant subbundles, where E^s_Λ is uniformly contracted, E^u_Λ is uniformly expanded, and E^X_Λ is the direction of the flow at the points of Λ. It is known, see [21], that a partially hyperbolic set for a three-dimensional flow, with volume expanding central direction and without equilibria, is hyperbolic. Hence the notion of singular-hyperbolicity is an extension of the notion of hyperbolicity.

Recently in a joint work with Pacifico, Pujals and Viana [4] the following consequence of transitivity and singular-hyperbolicity was proved.

Theorem 6.3. *Let* $\Lambda = \Lambda_X(U)$ *be a singular-hyperbolic attractor of a flow* $X \in \mathfrak{X}^2(M)$ *on a three-dimensional manifold. Then* Λ *supports a unique physical probability measure* μ *which is ergodic and its ergodic basin covers a full Lebesgue measure subset of the topological basin of attraction, i.e.* $B(\mu) = W^s(\Lambda)$ *Lebesgue mod 0. Moreover the support of* μ *is the whole attractor* Λ.

It follows from the proof in [4] that the singular-hyperbolic *attracting set* $\Lambda_Y(U)$ for all $Y \in \mathfrak{X}^2(M)$ which are C^1-close enough to X *admits finitely many physical measures whose ergodic basins cover* U *except for a zero volume subset.*

6.5.1 Absence of Sinks and Sources Nearby

The proof of Theorem 6.2 given in [21] uses several tools from the theory of normal hyperbolicity developed first by Mañé in [18] together with the low dimension of the flow. Indeed, going through the proof in [21] we can see that the arguments can be carried through assuming that

1. Λ is an attractor for X with isolating neighborhood U such that every equilibria in U is hyperbolic with no resonances;
2. there exists a C^1 neighborhood $\mathcal{U}$ of X such that for all $Y \in \mathcal{U}$ every periodic orbit and equilibria in U is hyperbolic of saddle-type.

The condition on the equilibria amounts to restricting the possible three-dimensional vector fields in the above statement to an open a dense subset of all C^1 vector fields. Indeed, the hyperbolic and no-resonance condition on a equilibrium σ means that:

- either $\lambda \neq \mathfrak{R}(\omega)$ if the eigenvalues of $DX(\sigma)$ are $\lambda \in \mathbb{R}$ and $\omega, \overline{\omega} \in \mathbb{C}$;
- or σ has only real eigenvalues with different norms.

Indeed, conditions (1) and (2) ensure that no bifurcations of periodic orbits or equilibria leading to sinks or sources are allowed for any nearby flow in U. This implies, by now standard arguments, that the flow on Λ must have a dominated splitting which is *volume hyperbolic*: both subbundles of the splitting must contract/expand volume. For a three-dimensional flow one of the subbundles is one-dimensional, and so we deduce singular-hyperbolicity either for X or for $-X$. If Λ has no equilibria, then Λ is uniformly hyperbolic. Otherwise, it follows from the arguments in [21] that all singularities of Λ are Lorenz-like and this shows that Λ must be singular-hyperbolic for X.

We note that the second condition above is a consequence of any one of the following assumptions on U:

Robust Chaoticity   for every $Y \in \mathcal{U}$ the maximal invariant subset $\Lambda_Y(U)$ is chaotic;

Zero Volume and Future Chaoticity   for every $Y \in \mathcal{U}$ the maximal invariant subset $\Lambda_Y(U)$ has zero volume and is future chaotic;

Zero Volume and Robust Positive Lyapunov Exponent   for every $Y \in \mathcal{U}$ the maximal invariant subset $\Lambda_Y(U)$ has zero volume and there exists a full Lebesgue measure subset P_Y of U such that

$$\limsup \frac{1}{n} \sum_{i=0}^{n-1} \log \|DY_x^i\| > 0, \quad x \in P_Y. \tag{6.3}$$

The result of Mañé analogous to Theorem 6.2 in [18]

Theorem 6.4. *Robust attractors for surface diffeomorphisms are hyperbolic.*

also follows from the absence of sinks and sources for all C^1 close diffeomorphisms in a neighborhood of the attractor.

Extensions of these results to higher dimensions for diffeomorphisms, by Bonatti et al. in [6], show that robust transitive sets always admit a volume hyperbolic splitting of the tangent bundle. Vivier in [31] extends previous results of Doering [9] for flows, showing that a C^1 robustly transitive vector field on a compact boundaryless n-manifold, with $n \geq 3$, admits a global dominated splitting. Metzger and Morales extend the arguments in [21] to homogeneous vector fields (inducing flows allowing no bifurcation of critical elements, i.e. no modification of the index of periodic orbits or equilibria) in higher dimensions leading to the concept of two-sectional expanding attractor in [20].

6.5.2   Robust Chaoticity, Volume Hyperbolicity and Physical Measure

The preceding observations allows us to deduce that robust chaoticity is a sufficient conditions for singular-hyperbolicity of a generic attractor.

Corollary 6.1. *Let Λ be an attractor for $X \in \mathfrak{X}^1(M^3)$ such that every equilibrium in its trapping region is hyperbolic with no resonances. Then Λ is singular-hyperbolic if, and only if, Λ is robustly chaotic.*

This means that *if we can show that arbitrarily close orbits, in an isolating neighborhood of an attractor, are driven apart, for the future as well as for the past, by the evolution of the system, and this behavior persists for all C^1 nearby vector fields, then the attractor is singular-hyperbolic.*

To prove the necessary condition on Corollary 6.1 we use the concept of expansiveness for flows, and through it show that singular-hyperbolic attractors for 3-flows are robustly expansive and, as a consequence, robustly chaotic also. This is done in the last Sect. 6.6.

We recall the following conjecture of Viana, presented in [29].

Conjecture 6.1. If an attracting set $\Lambda(U)$ of smooth map/flow has a non-zero Lyapunov exponent at Lebesgue almost every point of its isolated neighborhood U [i.e. it satisfies (6.3) with $P_Y \subset U$], then it admits some physical measure.

From the preceding results and observations we can give a partial answer to this conjecture for 3-flows in the following form.

Corollary 6.2. *Let $\Lambda_X(U)$ be an attractor for a flow $X \in \mathfrak{X}^1(M)$ such that*

- *the divergence of X is negative in U;*
- *the equilibria in U are hyperbolic with no resonances;*
- *there exists a neighborhood $\mathcal{U}$ of X in $\mathfrak{X}^1(M)$ such that for $Y \in \mathcal{U}$ one has (6.3) almost everywhere in U.*

Then there exists a neighborhood $\mathcal{V} \subset \mathcal{U}$ of X in $\mathfrak{X}^1(M)$ and a dense subset $\mathcal{D} \subset \mathcal{V}$ such that

1. *$\Lambda_Y(U)$ is singular-hyperbolic for all $Y \in \mathcal{V}$;*
2. *there exists a physical measure μ_Y supported in $\Lambda_Y(U)$ for all $Y \in \mathcal{D}$.*

Indeed, item (2) above is a consequence of item (1), the denseness of $\mathfrak{X}^2(M)$ in $\mathfrak{X}^1(M)$ in the C^1 topology, together with Theorem 6.3 and the observation following its statement.

Item (1) above is a consequence of Corollary 6.1 and the observations of Sect. 6.5.1, noting that negative divergence on the isolating neighborhood U ensures that the volume of $\Lambda_Y(U)$ is zero for Y in a C^1 neighborhood $\mathcal{V}$ of X.

6.6 Expansive Systems

Here we explain why robust chaotic behavior necessarily follows from singular-hyperbolicity in an attractor, completing the proof of Corollary 6.1. For this we need the concept and some properties of expansiveness for flows.

A concept related to sensitiveness is that of expansiveness, which roughly means that points whose orbits are always close for all time must coincide. The concept of expansiveness for homeomorphisms plays an important role in the study of transformations. Bowen and Walters [7] gave a definition of expansiveness for flows which is now called *C-expansiveness* [15]. The basic idea of their definition is that two points which are not close in the orbit topology induced by $\mathbb{R}$ can be separated at the same time even if one allows a continuous time lag—see below for the technical definitions. The equilibria of C-expansive flows must be isolated [7, Proposition 1] which implies that the Lorenz attractors and geometric Lorenz models are not C-expansive.

Keynes and Sears introduced [15] the idea of restriction of the time lag and gave several definitions of expansiveness weaker than C-expansiveness. The notion of *K-expansiveness* is defined allowing only the time lag given by an increasing surjective homeomorphism of $\mathbb{R}$. Komuro [16] showed that the Lorenz attractor and the geometric Lorenz models are not K-expansive. The reason for this is not that the restriction of the time lag is insufficient, but that the topology induced by $\mathbb{R}$ is unsuited to measure the closeness of two points in the same orbit.

Taking this fact into consideration, Komuro [16] gave a definition of *expansiveness* suitable for flows presenting equilibria accumulated by regular orbits. This concept is enough to show that two points which do not lie on a same orbit can be separated.

Let $C\left(\mathbb{R}, \mathbb{R}\right)$ be the set of all continuous functions $h : \mathbb{R} \to \mathbb{R}$ and let us write $C\left((\mathbb{R}, 0), (\mathbb{R}, 0)\right)$ for the subset of all $h \in C\left(\mathbb{R}, \mathbb{R}\right)$ such that $h(0) = 0$. We define

$$\mathcal{K}_0 = \{h \in C\left((\mathbb{R}, 0), (\mathbb{R}, 0)\right) : h(\mathbb{R}) = \mathbb{R}, \ h(s) > h(t) , \forall s > t\},$$

and

$$\mathcal{K} = \{h \in C\left(\mathbb{R}, \mathbb{R}\right) : h(\mathbb{R}) = \mathbb{R}, \ h(s) > h(t) , \forall s > t\},$$

A flow X is *C-expansive* (*K-expansive* respectively) on an invariant subset $\Lambda \subset M$ if for every $\epsilon > 0$ there exists $\delta > 0$ such that if $x, y \in \Lambda$ and for some $h \in \mathcal{K}_0$ (respectively $h \in \mathcal{K}$) we have

$$\text{dist}\left(X^t(x), X^{h(t)}(y)\right) \leqslant \delta \quad \text{for all} \quad t \in \mathbb{R}, \tag{6.4}$$

then $y \in X^{[-\epsilon,\epsilon]}(x) = \{X^t(x) : -\epsilon \leqslant t \leqslant \epsilon\}$.

We say that the flow X is *expansive* on Λ if for every $\epsilon > 0$ there is $\delta > 0$ such that for $x, y \in \Lambda$ and some $h \in \mathcal{K}$ (note that now we do not demand that 0 be fixed by h) satisfying (6.4), then we can find $t_0 \in \mathbb{R}$ such that $X^{h(t_0)}(y) \in X^{[t_0-\epsilon, t_0+\epsilon]}(x)$.

Observe that expansiveness on M implies sensitive dependence on initial conditions for any flow on a manifold with dimension at least 2. Indeed if ϵ, δ satisfy the expansiveness condition above with h equal to the identity and we are given a point $x \in M$ and a neighborhood U of x, then taking $y \in U \setminus X^{[-\epsilon,\epsilon]}(x)$ (which always exists since we assume that M is not one-dimensional) there exists $t \in \mathbb{R}$ such that $\text{dist}\left(X^t(y), X^t(x)\right) \geqslant \delta$. The same argument applies whenever we have expansiveness on an X-invariant subset Λ of M containing a dense regular orbit of the flow.

Clearly C-expansive $\Longrightarrow$ K-expansive $\Longrightarrow$ expansive by definition. When a flow has no fixed point then C-expansiveness is equivalent to K-expansiveness [22, Theorem A]. In [7] it is shown that on a connected manifold a C-expansive flow has no fixed points. The following was kindly communicated to us by Alfonso Artigue from IMERL, the proof can be found in [2].

Proposition 6.2. *A flow is C-expansive on a manifold M if, and only if, it is K-expansive.*

We will see that singular-hyperbolic attractors are expansive. In particular, the Lorenz attractor and the geometric Lorenz examples are all expansive and sensitive to initial conditions. Since these families of flows exhibit equilibria accumulated by regular orbits, we see that expansiveness is compatible with the existence of fixed points by the flow.

6.6.1 Singular-Hyperbolicity and Expansiveness

The full proof of the following given in [4]. We provide a sketch of the proof in Sect. 6.7.2.

Theorem 6.5. *Let Λ be a singular-hyperbolic attractor of $X \in \mathfrak{X}^1(M)$. Then Λ is expansive.*

The reasoning is based on analyzing Poincaré return maps of the flow to a convenient (δ-adapted) cross-section. We use the family of adapted cross-sections and corresponding Poincará maps R, whose Poincaré time $t(\cdot)$ is large enough, obtained assuming that the attractor Λ is singular-hyperbolic. These cross-sections have a co-dimension one foliation, which are dynamically defined, whose leaves are uniformly contracted and invariant under the Poincaré maps. In addition R is uniformly expanding in the transverse direction and this also holds near the singularities.

From here we argue by contradiction: if the flow is not expansive on Λ, then we can find a pair of orbits hitting the cross-sections infinitely often on pairs of points uniformly close. We derive a contradiction by showing that the uniform expansion in the transverse direction to the stable foliation must take the pairs of points apart, unless one orbit is on the stable manifold of the other.

This argument only depends on the existence of finitely many Lorenz-like singularities on a compact partially hyperbolic invariant attracting subset $\Lambda = \Lambda_X(U)$, with volume expanding central direction, and of a family of adapted cross-sections with Poincaré maps between them, whose derivative is hyperbolic. It is straightforward that if these conditions are satisfied for a flow X^t of $X \in \mathfrak{X}^1(M^3)$, then the same conditions hold for all C^1 nearby flows Y^t and for the maximal invariant subset $\Lambda_Y(U)$ *with the same family of cross-sections* which are also adapted to $\Lambda_Y(U)$ (as long as Y is C^1-close enough to X).

Corollary 6.3. *A singular-hyperbolic attractor $\Lambda = \Lambda_X(U)$ is robustly expansive, that is, there exists a neighborhood $\mathcal{U}$ of X in $\mathfrak{X}^1(M)$ such that $\Lambda_Y(U)$ is expansive for each $Y \in \mathcal{U}$, where U is an isolating neighborhood of Λ.*

Indeed, since transversality, partial hyperbolicity and volume expanding central direction are robust properties, and also the hyperbolicity of the Poincaré maps depends on the central volume expansion, all we need to do is to check that a given adapted cross-section Σ to X is also adapted to $Y \in \mathfrak{X}^1$ for every Y sufficiently C^1 close to X. But $\Lambda_X(U)$ and $\Lambda_Y(U)$ are close in the Hausdorff distance if X and Y are close in the C^0 distance, by the following elementary result.

Lemma 6.2. *Let Λ be an isolated set of $X \in \mathfrak{X}^r(M)$, $r \geq 0$. Then for every isolating block U of Λ and every $\epsilon > 0$ there is a neighborhood $\mathcal{U}$ of X in $\mathfrak{X}^r(M)$ such that $\Lambda_Y(U) \subset B(\Lambda, \epsilon)$ and $\Lambda \subset B(\Lambda_Y(U), \epsilon)$ for all $Y \in \mathcal{U}$.*

Thus, if Σ is an adapted cross-section we can find a C^1-neighborhood $\mathcal{U}$ of X in $\mathfrak{X}^1$ such that Σ is still adapted to every flow Y^t generated by a vector field in $\mathcal{U}$.

6.6.2 Singular-Hyperbolicity and Chaotic Behavior

We already know that expansiveness implies sensitive dependence on initial conditions. An argument with the same flavor as the proof of expansiveness provides the following, whose proof also sketch in the following Sect. 6.7.2.

Theorem 6.6. *A singular-hyperbolic isolated set $\Lambda = \cap_{t \in \mathbb{R}} \overline{X^t(U)}$ is robustly chaotic, i.e. there exists a neighborhood $\mathcal{U}$ of X in $\mathfrak{X}^1(M)$ such that $\cap_{t \in \mathbb{R}} \overline{Y^t(U)}$ is chaotic for each $Y \in \mathcal{U}$, where U is an isolating neighborhood of Λ.*

This completes the argument proving that robust chaoticity is a necessary property of singular-hyperbolicity, in Corollary 6.1.

6.7 Sketch of the Proof of Expansiveness and of Chaotic Behavior

We need the following notions to understand the proof of Theorems 6.5 and 6.6 as a consequence.

6.7.1 Adapted Cross-Sections and Poincaré Maps

To help explain the ideas of the proofs we give here a few properties of *Poincaré maps*, that is, continuous maps $R : \Sigma \to \Sigma'$ of the form $R(x) = X^{t(x)}(x)$ between cross-sections Σ and Σ' of the flow near a singular-hyperbolic set. We always assume that the Poincaré time $t(\cdot)$ is large enough as explained in what follows.

We assume that Λ is a compact invariant subset for a flow $X \in \mathfrak{X}^1(M)$ such that Λ is a singular-hyperbolic attractor. Then every equilibrium in Λ is Lorenz-like.

6.7.1.1 Stable Foliations on Cross-Sections

We start recalling standard facts about uniformly hyperbolic flows from e.g. [14].

An embedded disk $\gamma \subset M$ is a (local) *strong-unstable manifold*, or a *strong-unstable disk*, if $\mathrm{dist}(X^{-t}(x), X^{-t}(y))$ tends to zero exponentially fast as $t \to +\infty$, for every $x, y \in \gamma$. Similarly, γ is called a (local) *strong-stable manifold*, or a *strong-stable disk*, if $\mathrm{dist}(X^t(x), X^t(y)) \to 0$ exponentially fast as $n \to +\infty$, for every $x, y \in \gamma$. It is well-known that every point in a uniformly hyperbolic set possesses a local strong-stable manifold $W_{loc}^{ss}(x)$ and a local strong-unstable manifold $W_{loc}^{uu}(x)$ which are disks tangent to E_x and G_x at x respectively with topological dimensions $d_E = \dim(E)$ and $d_G = \dim(G)$ respectively. Considering the action of the flow we get the (global) *strong-stable manifold*

$$W^{ss}(x) = \bigcup_{t>0} X^{-t}\left(W_{loc}^{ss}(X^t(x))\right)$$

and the (global) *strong-unstable manifold*

$$W^{uu}(x) = \bigcup_{t>0} X^t\left(W_{loc}^{uu}(X_{-t}(x))\right)$$

for every point x of a uniformly hyperbolic set. These are immersed submanifolds with the same differentiability of the flow. We also consider the *stable manifold* $W^s(x) = \cup_{t \in \mathbb{R}} X^t\left(W^{ss}(x)\right)$ and *unstable manifold* $W^u(x) = \cup_{t \in \mathbb{R}} X^t\left(W^{uu}(x)\right)$ for x in a uniformly hyperbolic set, which are flow invariant.

Now we recall classical facts about partially hyperbolic systems, especially existence of strong-stable and center-unstable foliations. The standard reference is [14].

We have that Λ is a singular-hyperbolic isolated set of $X \in \mathfrak{X}^1(M)$ with invariant splitting $T_\Lambda M = E^s \oplus E^{cu}$ with $\dim E^{cu} = 2$. Let $\tilde{E}^s \oplus \tilde{E}^{cu}$ be a continuous extension of this splitting to a small neighborhood U_0 of Λ. For convenience we take U_0 to be forward invariant. Then $\tilde{E}^s$ may be chosen invariant under the derivative: just consider at each point the direction formed by those vectors which are strongly contracted by DX^t for positive t. In general $\tilde{E}^{cu}$ is not invariant. However we can consider a cone field around it on U_0

$$C_a^{cu}(x) = \{v = v^s + v^{cu} : v^s \in \tilde{E}_x^s \text{ and } v^{cu} \in \tilde{E}_x^{cu} \text{ with } \|v^s\| \leq a \cdot \|v^{cu}\|\}$$

which is forward invariant for $a > 0$:

$$DX^t(C_a^{cu}(x)) \subset C_a^{cu}(X^t(x)) \quad \text{for all large } t > 0. \tag{6.5}$$

Moreover we may take $a > 0$ arbitrarily small, reducing U_0 if necessary. For notational simplicity we write E^s and E^{cu} for $\tilde{E}^s$ and $\tilde{E}^{cu}$ in all that follows.

From the standard normal hyperbolic theory, there are locally strong-stable and center-unstable manifolds, defined at every regular point $x \in U_0$ and which are embedded disks tangent to $E^s(x)$ and $E^{cu}(x)$, respectively. Given any $x \in U_0$ define the strong-stable manifold $W^{ss}(x)$ and the stable-manifold $W^s(x)$ as for an hyperbolic flow (see the beginning of this section).

Denoting $E_x^{cs} = E_x^s \oplus E_x^X$, where E_x^X is the direction of the flow at x, it follows that

$$T_x W^{ss}(x) = E_x^s \quad \text{and} \quad T_x W^s(x) = E_x^{cs}.$$

We fix ϵ once and for all. Then we call $W_\epsilon^{ss}(x)$ the local *strong-stable manifold* and $W_\epsilon^{cu}(x)$ the local *center-unstable manifold* of x.

Now let Σ be a *cross-section* to the flow, that is, a C^1 embedded compact disk transverse to X: at every point $z \in \Sigma$ we have $T_z\Sigma \oplus E_z^X = T_zM$ (recall that E_z^X is the one-dimensional subspace $\{s \cdot X(z) : s \in \mathbb{R}\}$). For every $x \in \Sigma$ we define $W^s(x, \Sigma)$ to be the connected component of $W^s(x) \cap \Sigma$ that contains x. This defines a foliation $\mathcal{F}_\Sigma^s$ of Σ into co-dimension 1 sub-manifolds of class C^1.

Given any cross-section Σ and a point x in its interior, we may always find a smaller cross-section also with x in its interior and which is the image of the square $[0, 1] \times [0, 1]$ by a C^1 diffeomorphism h that sends horizontal lines inside leaves of $\mathcal{F}_\Sigma^s$. In what follows we assume that cross-sections are of this kind, see Fig. 6.2.

We also assume that each cross-section Σ is contained in U_0, so that every $x \in \Sigma$ is such that $\omega(x) \subset \Lambda$.

On the one hand $x \mapsto W_\epsilon^{ss}(x)$ is usually not differentiable if we assume that X is only of class C^1, see e.g. [24]. On the other hand, assuming that the cross-section is small with respect to ϵ, and choosing any curve $\gamma \subset \Sigma$ crossing transversely every leaf of $\mathcal{F}_\Sigma^s$, we may consider a Poincaré map

$$R_\Sigma : \Sigma \to \Sigma(\gamma) = \bigcup_{z \in \gamma} W_\epsilon^{ss}(z)$$

with Poincaré time close to zero, see Fig. 6.2. This is a homeomorphism onto its image, close to the identity, such that $R_\Sigma(W^s(x, \Sigma)) \subset W_\epsilon^{ss}(R_\Sigma(x))$. So, identifying the points of Σ with their images under this homeomorphism, we pretend that indeed $W^s(x, \Sigma) \subset W_\epsilon^{ss}(x)$.

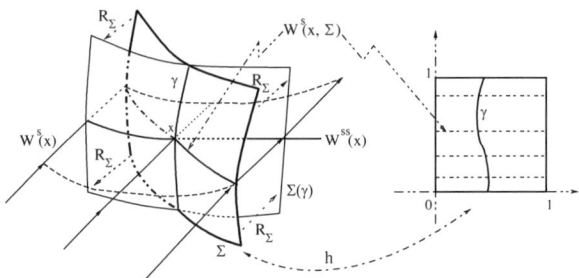

Fig. 6.2 The sections Σ, $\Sigma(\gamma)$, the manifolds $W^s(x)$, $W^{ss}(x)$, $W^s(x, \Sigma)$ and the projection R_{Σ}, on the *right*. On the *left*, the square $[0, 1] \times [0, 1]$ is identified with Σ through the map h, where $\mathscr{F}_{\Sigma}^s$ becomes the horizontal foliation and the curve γ is transverse to the horizontal direction. *Solid lines with arrows* indicate the flow direction

6.7.1.2 Hyperbolicity of Poincaré Maps

Let Σ be a small cross-section to X and let $R : \Sigma \to \Sigma'$ be a Poincaré map $R(y) = X^{t(y)}(y)$ to another cross-section Σ' (possibly $\Sigma = \Sigma'$). Here R needs not correspond to the first time the orbits of Σ encounter Σ'.

The splitting $E^s \oplus E^{cu}$ over U_0 induces a continuous splitting $E_{\Sigma}^s \oplus E_{\Sigma}^{cu}$ of the tangent bundle $T\Sigma$ to Σ (and analogously for Σ'), defined by

$$E_{\Sigma}^s(y) = E_y^{cs} \cap T_y\Sigma \quad \text{and} \quad E_{\Sigma}^{cu}(y) = E_y^{cu} \cap T_y\Sigma. \tag{6.6}$$

We now show that if the Poincaré time $t(x)$ is sufficiently large then (6.6) defines a hyperbolic splitting for the transformation R on the cross-sections restricted to Λ.

Proposition 6.3. *Let $R : \Sigma \to \Sigma'$ be a Poincaré map as before with Poincaré time $t(\cdot)$. Then $DR_x(E_{\Sigma}^s(x)) = E_{\Sigma'}^s(R(x))$ at every $x \in \Sigma$ and $DR_x(E_{\Sigma}^{cu}(x)) = E_{\Sigma'}^{cu}(R(x))$ at every $x \in \Lambda \cap \Sigma$.*

Moreover for every given $0 < \lambda < 1$ there exists $T_1 = T_1(\Sigma, \Sigma', \lambda) > 0$ such that if $t(\cdot) > T_1$ at every point, then

$$\|DR \mid E_{\Sigma}^s(x)\| < \lambda \quad \text{and} \quad \|DR \mid E_{\Sigma}^{cu}(x)\| > 1/\lambda \quad \text{at every } x \in \Sigma.$$

Given a cross-section Σ, a positive number ρ, and a point $x \in \Sigma$, we define the unstable cone of width ρ at x by

$$C_{\rho}^u(x) = \{v = v^s + v^u : v^s \in E_{\Sigma}^s(x), \ v^u \in E_{\Sigma}^{cu}(x) \text{ and } \|v^s\| \leqslant \rho\|v^u\|\}. \tag{6.7}$$

Let $\rho > 0$ be any small constant. In the following consequence of Proposition 6.3 we assume the neighborhood U_0 has been chose sufficiently small.

Corollary 6.4. *For any* $R : \Sigma \rightarrow \Sigma'$ *as in Proposition 6.3, with* $t(\cdot) > T_1$, *and any* $x \in \Sigma$, *we have* $DR(x)(C_\rho^u(x)) \subset C_{\rho/2}^u(R(x))$ *and*

$$\|DR_x(v)\| \geq \frac{5}{6}\lambda^{-1} \cdot \|v\| \quad \textit{for all} \quad v \in C_\rho^u(x).$$

As usual a *curve* is the image of a compact interval $[a, b]$ by a C^1 map. We use $\ell(\gamma)$ to denote its length. By a *cu-curve* in Σ we mean a curve contained in the cross-section Σ and whose tangent direction is contained in the unstable cone $T_z\gamma \subset C_\rho^u(z)$ for all $z \in \gamma$. The next lemma says that *the length of cu-curves linking the stable leaves of nearby points* x, y *must be bounded by the distance between* x *and* y.

Lemma 6.3. *Let us we assume that* ρ *has been fixed, sufficiently small. Then there exists a constant* κ *such that, for any pair of points* $x, y \in \Sigma$, *and any cu-curve* γ *joining* x *to some point of* $W^s(y, \Sigma)$, *we have* $\ell(\gamma) \leq \kappa \cdot d(x, y)$.

Here d is the intrinsic distance in the C^2 surface Σ, that is, the length of the shortest smooth curve inside Σ connecting two given points in Σ.

In what follows we take T_1 in Proposition 6.3 for $\lambda = 1/3$.

6.7.1.3 Adapted Cross-Sections

Now we exhibit stable manifolds for Poincaré transformations $R : \Sigma \rightarrow \Sigma'$. The natural candidates are the intersections $W^s(x, \Sigma) = W_\epsilon^s(x) \cap \Sigma$ we introduced previously. These intersections are tangent to the corresponding sub-bundle E_Σ^s and so, by Proposition 6.3, they are contracted by the transformation. For our purposes it is also important that the stable foliation be invariant:

$$R(W^s(x, \Sigma)) \subset W^s(R(x), \Sigma') \qquad \text{for every } x \in \Lambda \cap \Sigma. \tag{6.8}$$

In order to have this we restrict our class of cross-sections whose center-unstable boundary is disjoint from Λ. Recall that we are considering cross-sections Σ that are diffeomorphic to the square $[0, 1] \times [0, 1]$, with the horizontal lines $[0, 1] \times \{\eta\}$ being mapped to stable sets $W^s(y, \Sigma)$. The *stable boundary* $\partial^s \Sigma$ is the image of $[0, 1] \times \{0, 1\}$. The *center-unstable boundary* $\partial^{cu} \Sigma$ is the image of $\{0, 1\} \times [0, 1]$. The cross-section is δ-*adapted* if

$$d(\Lambda \cap \Sigma, \partial^{cu} \Sigma) > \delta,$$

where d is the intrinsic distance in Σ, see Fig. 6.3.

Lemma 6.4. *Let* $x \in \Lambda$ *be a regular point, that is, such that* $X(x) \neq 0$. *There exists* $\delta > 0$ *such that there exists a* δ-*adapted cross-section* Σ *at* x.

Fig. 6.3 An adapted
cross-section for Λ

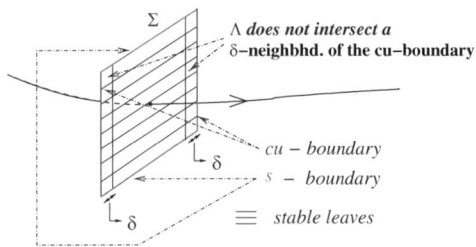

We are going to show that if the cross-sections are adapted, then we have the invariance property (6.8). Given $\Sigma, \Sigma' \in \Xi$ we set $\Sigma(\Sigma') = \{x \in \Sigma : R(x) \in \Sigma'\}$ the domain of the return map from Σ to Σ'.

Lemma 6.5. *Given $\delta > 0$ and δ-adapted cross-sections Σ and Σ', there exists $T_2 = T_2(\Sigma, \Sigma') > 0$ such that if $R : \Sigma(\Sigma') \to \Sigma'$ defined by $R(z) = R_{t(z)}(z)$ is a Poincaré map with time $t(\cdot) > T_2$, then*

1. $R\big(W^s(x, \Sigma)\big) \subset W^s(R(x), \Sigma')$ for every $x \in \Sigma(\Sigma')$, and also
2. $d(R(y), R(z)) \leq \frac{1}{2} d(y, z)$ for every $y, z \in W^s(x, \Sigma)$ and $x \in \Sigma(\Sigma')$.

Clearly we may choose $T_2 > T_1$ so that all the properties of the Poincaré maps obtained up to here are valid for return times greater than T_2.

6.7.2 The Proof of Expansiveness

Here we sketch the proof of Theorem 6.5. The proof is by contradiction: let us suppose that there exist $\epsilon > 0$, a sequence $\delta_n \to 0$, a sequence of functions $h_n \in \mathcal{H}$ and sequences of points $x_n, y_n \in \Lambda$ such that

$$d\left(X^t(x_n), X^{h_n(t)}(y_n)\right) \leq \delta_n \quad \text{for all } t \in \mathbb{R}, \tag{6.9}$$

but

$$X^{h_n(t)}(y_n) \notin X^{[t-\epsilon, t+\epsilon]}(x_n) \quad \text{for all } t \in \mathbb{R}. \tag{6.10}$$

The main step of the proof is a reduction to a forward expansiveness statement about Poincaré maps which we state in Theorem 6.7 below.

We are going to use the following observation: there exists some regular (i.e. non-equilibrium) point $z \in \Lambda$ which is accumulated by the sequence of ω-limit sets $\omega(x_n)$. To see that this is so, start by observing that accumulation points do exist, since Λ is compact. Moreover, if the ω-limit sets accumulate on a singularity then they also accumulate on at least one of the corresponding unstable branches which, of course, consists of regular points. We fix such a z once and for all. Replacing our sequences by subsequences, if necessary, we may suppose that for every n there exists $z_n \in \omega(x_n)$ such that $z_n \to z$.

Let Σ be a δ-adapted cross-section at z, for some small δ. Reducing δ (but keeping the same cross-section) we may ensure that z is in the interior of the subset

$$\Sigma_\delta = \{y \in \Sigma : d(y, \partial\Sigma) > \delta\}.$$

By definition, x_n returns infinitely often to the neighborhood of z_n which, on its turn, is close to z. Thus dropping a finite number of terms in our sequences if necessary, we have that the orbit of x_n intersects Σ_δ infinitely many times. Let t_n be the time corresponding to the nth intersection.

Replacing x_n, y_n, t, and h_n by $x^{(n)} = X^{t_n}(x_n)$, $y^{(n)} = X^{h_n(t_n)}(y_n)$, $t' = t - t_n$, and $h_n'(t') = h_n(t' + t_n) - h_n(t_n)$, we may suppose that $x^{(n)} \in \Sigma_\delta$, while preserving both relations (6.9) and (6.10). Moreover there exists a sequence $\tau_{n,j}$, $j \geq 0$ with $\tau_{n,0} = 0$ such that

$$x^{(n)}(j) = X^{\tau_{n,j}}(x^{(n)}) \in \Sigma_\delta \quad \text{and} \quad \tau_{n,j} - \tau_{n,j-1} > \max\{t_1, t_2\} \qquad (6.11)$$

for all $j \geq 1$, where t_1 is given by Proposition 6.3 and t_2 is given by Lemma 6.5.

Theorem 6.7. *Given $\epsilon_0 > 0$ there exists $\delta_0 > 0$ such that if $x \in \Sigma_\delta$ and $y \in \Lambda$ satisfy*

(a) there exist τ_j such that

$$x_j = X^{\tau_j}(x) \in \Sigma_\delta \quad \text{and} \quad \tau_j - \tau_{j-1} > \max\{t_1, t_2\} \quad \text{for all } j \geq 1;$$

(b) $\mathrm{dist}\left(X^t(x), X^{h(t)}(y)\right) < \delta_0$, *for all $t > 0$ and some $h \in \mathcal{H}$;*

then there exists $s = \tau_j \in \mathbb{R}$ *for some* $j \geq 1$ *such that* $X^{h(s)}(y) \in W^{ss}_{\epsilon_0}(X^{[s-\epsilon_0, s+\epsilon_0]}(x))$.

The proof of Theorem 6.7 will not be given here, and can be found in [4]. We explain why this implies Theorem 6.5. We are going to use the following observation.

Lemma 6.6. *There exist $\rho > 0$ small and $c > 0$, depending only on the flow, such that if z_1, z_2, z_3 are points in Λ satisfying $z_3 \in X^{[-\rho, \rho]}(z_2)$ and $z_2 \in W^{ss}_\rho(z_1)$, with z_1 away from any equilibria of X, then*

$$\mathrm{dist}(z_1, z_3) \geq c \cdot \max\{\mathrm{dist}(z_1, z_2), \mathrm{dist}(z_2, z_3)\}.$$

This is a direct consequence of the fact that the angle between E^{ss} and the flow direction is bounded from zero which, on its turn, follows from the fact that the latter is contained in the center-unstable sub-bundle E^{cu}.

We fix $\epsilon_0 = \epsilon$ as in (6.10) and then consider δ_0 as given by Theorem 6.7. Next, we fix n such that $\delta_n < \delta_0$ and $\delta_n < c\rho$, and apply Theorem 6.7 to $x = x^{(n)}$

Fig. 6.4 Relative positions of the strong-stable manifolds and orbits

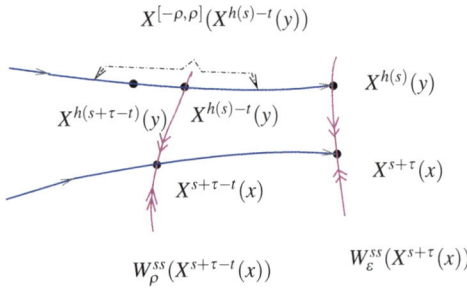

and $y = y^{(n)}$ and $h = h_n$. Hypothesis (a) in the theorem corresponds to (6.11) and, with these choices, hypothesis (b) follows from (6.9). Therefore we obtain that $X^{h(s)}(y) \in W_\epsilon^{ss}(X^{[s-\epsilon,s+\epsilon]}(x))$. Equivalently there is $|\tau| \leq \epsilon$ such that $X^{h(s)}(y) \in W_\epsilon^{ss}(X^{s+\tau}(x))$. Condition (6.10) then implies that $X^{h(s)}(y) \neq X^{s+\tau}(x)$. Hence since strong-stable manifolds are expanded under backward iteration, there exists $\theta > 0$ maximum such that

$$X^{h(s)-t}(y) \in W_\rho^{ss}(X^{s+\tau-t}(x)) \quad \text{and} \quad X^{h(s+\tau-t)}(y) \in X^{[-\rho,\rho]}(X^{h(s)-t}(y))$$

for all $0 \leq t \leq \theta$, see Fig. 6.4. Moreover $s = \tau_j$ for some $j \geq 1$ so that x is close to cross-section of the flow which we can assume is uniformly bounded away from the equilibria, and then we can assume that $\|X(X^t(x))\| \geq c$ for $0 \leq t \leq \theta$. Since θ is maximum

$$\text{either dist}\left(X^{h(s)-t}(y), X^{s+\tau-t}(x)\right) \geq \rho$$

$$\text{or dist}\left(X^{h(s+\tau-t)}(y), X^{h(s)-t}(y)\right) \geq c_0\rho$$

for $t = \theta$, because $\|X(X^t(x))\| \geq c_0 > 0$ for $0 \leq t \leq \theta$. Using Lemma 6.6, we conclude that $\text{dist}(X^{s+\tau-t}(x), X^{h(s+\tau-t)}(y)) \geq c\rho > \delta_n$ which contradicts (6.9). This contradiction reduces the proof of Theorem 6.5 to that of Theorem 6.7.

6.7.3 Singular-Hyperbolicity and Chaotic Behavior

Here we explain why singular-hyperbolic attractors, like the Lorenz attractor, are necessarily robustly chaotic.

Proof (of Theorem 6.6). The assumption of singular-hyperbolicity on an isolated proper subset Λ with isolating neighborhood U ensures that the maximal invariant subsets $\cap_{t \in \mathbb{R}} \overline{Y^t(U)}$ for all C^1 nearby flows Y are also singular-hyperbolic. Therefore to deduce robust chaotic behavior in this setting it is enough to show that a proper isolated invariant compact singular-hyperbolic subset is chaotic.

Let Λ be a singular-hyperbolic isolated proper subset for a C^1 flow. Then there exists a strong-stable manifold $W^{ss}(x)$ through each of its points x. We claim that this implies that Λ is past chaotic. Indeed, assume by contradiction that we can find $y \in W^{ss}(x)$ such that $y \neq x$ and dist$\left(X^{-t}(y), X^{-t}(x)\right) < \epsilon$ for every $t > 0$, for some small $\epsilon > 0$. Then, because $W^{ss}(x)$ is uniformly contracted by the flow in positive time, there exists $\lambda > 0$ such that

$$\text{dist}(y, x) \leqslant Const \cdot e^{-\lambda t} \text{ dist} \left(X^{-t}(y), X^{-t}(x)\right) \leqslant Const \cdot \epsilon e^{-\lambda t}$$

for all $t > 0$, a contradiction since $y \neq x$. Hence for any given small $\epsilon > 0$ we can always find a point y arbitrarily close to x (it is enough to choose y is the strong-stable manifold of x) such that its past orbit separates from the orbit of x.

To obtain future chaotic behavior, we argue by contradiction: we assume that Λ is not future chaotic. Then for every $\epsilon > 0$ we can find a point $x \in \Lambda$ and an open neighborhood V of x such that the future orbit of each $y \in V$ is ϵ-close to the future orbit of x, that is, dist $\left(X^t(y), X^t(x)\right) \leqslant \epsilon$ for all $t > 0$.

First, x is not a singularity, because all the possible singularities inside a singular-hyperbolic set are hyperbolic saddles and so each singularity has a unstable manifold. Likewise, x cannot be in the stable manifold of a singularity. Therefore $\omega(x)$ contains some regular point z. Let Σ be a transversal section to the flow X^t at z.

Hence there are infinitely many times $t_n \to +\infty$ such that $x_n := X^{t_n}(x) \in \Sigma$ and $x_n \to z$ when $n \to +\infty$. Taking Σ sufficiently small looking only to very large times, the assumption on V ensures that each $y \in V$ admits also an infinite sequence $t_n(y) \xrightarrow[n \to +\infty]{} +\infty$ satisfying

$$y_n := X^{t_n(y)}(y) \in \Sigma \quad \text{and} \quad \text{dist}(y_n, x_n) \leqslant 10\epsilon.$$

We can assume that $y \in V$ does not belong to $W^s(x)$, since $W^s(x)$ is a C^1 immersed sub-manifold of M. Hence we consider the connected components $\gamma_n := W^s(x_n, \Sigma)$ and $\xi_n := W^s(y_n, \Sigma)$ of $W^s(x) \cap \Sigma$ and $W^s(y) \cap \Sigma$, respectively. We recall that we can assume that every y in a small neighborhood of Λ admits an invariant stable manifold because we can extend the invariant stable cone fields from Λ to a small neighborhood of Λ. We can also extend the invariant center-unstable cone fields from Λ to this same neighborhood, so that we can also define the notion of cu-curve in Σ in this setting.

The assumption on V ensures that there exists a cu-curve ζ_n in Σ connecting γ_n to ξ_n, because $X^{t_n}(V) \cap \Sigma$ is an open neighborhood of x_n containing y_n. But we can assume without loss of generality that $t_{n+1} - t_n > \max\{t_1, t_2\}$, forgetting some returns to Σ in between if necessary and relabeling the times t_n. Thus Proposition 6.3 applies and the Poincaré return maps associated to the returns to Σ considered above are hyperbolic.

Fig. 6.5 Expansion between
visits to a cross-section

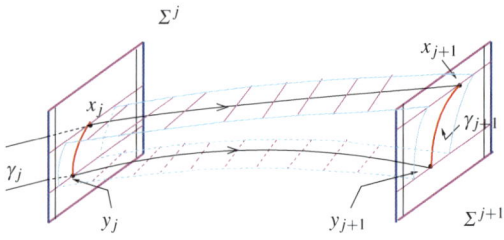

The same argument as in the proof of expansiveness guarantees that there exists a
flow box connecting $\{x_n, y_n\}$ to $\{x_{n+1}, y_{n+1}\}$ and sending ζ_n into a *cu*-curve $R(\zeta_n)$
connecting γ_{n+1} and ξ_{n+1}, for every $n \geqslant 1$.

The hyperbolicity of the Poincaré return maps ensures that the length of $R(\zeta_n)$
grows by a factor greater than one, see Fig. 6.5. Therefore, since y_n, x_n are uniformly
close, this implies that the length of ζ_1 and the distance between γ_1 and ξ_1 must be
zero. This contradicts the choice of $y \neq W^s(x)$.

This contradiction shows that Λ is future chaotic, and concludes the proof.   □

Acknowledgements Author was partially supported by CNPq, FAPERJ and PRONEX-
Dynamical Systems (Brazil).

References

1. Andronov, A., Pontryagin, L.: Systèmes grossiers. Dokl. Akad. Nauk. USSR **14**, 247–251 (1937)
2. Araujo, V., Pacifico, M.J.: Three Dimensional Flows. XXV Brazillian Mathematical Collo-quium. IMPA, Rio de Janeiro (2007)
3. Araújo, V., Pacifico, M.J.: Three-Dimensional Flows. Ergebnisse der Mathematik und ihrer Grenzgebiete. 3. Folge. A Series of Modern Surveys in Mathematics [Results in Mathematics and Related Areas. 3rd Series. A Series of Modern Surveys in Mathematics], vol. 53. Springer, Heidelberg (2010). With a foreword by Marcelo Viana
4. Araújo, V., Pujals, E.R., Pacifico, M.J., Viana, M.: Singular-hyperbolic attractors are chaotic. Trans. A.M.S. **361**, 2431–2485 (2009)
5. Banks, J., Brooks, J., Cairns, G., Davis, G., Stacey, P.: On Devaney's definition of chaos. Am. Math. Mon. **99**(4), 332–334 (1992)
6. Bonatti, C., Díaz, L.J., Pujals, E.: A C^1-generic dichotomy for diffeomorphisms: weak forms of hyperbolicity or infinitely many sinks or sources. Ann. Math. **157**(2), 355–418 (2003)
7. Bowen, R., Walters, P.: Expansive one-parameter flows. J. Differ. Equat. **12**, 180–193 (1972)
8. Devaney, R.: An Introduction to Chaotic Dynamical Systems, 2nd edn. Addison-Wesley, New York (1989)
9. Doering, C.I.: Persistently transitive vector fields on three-dimensional manifolds. In: Procs. on Dynamical Systems and Bifurcation Theory, vol. 160, pp. 59–89. Pitman, London (1987)
10. Glasner, E., Weiss, B.: Sensitive dependence on initial conditions. Nonlinearity **6**(6), 1067–1075 (1993)
11. Guckenheimer, J.: A strange, strange attractor. In: The Hopf Bifurcation Theorem and Its Applications, pp. 368–381. Springer, New York (1976)

12. Guckenheimer, J., Holmes, P.: Nonlinear Oscillations, Dynamical Systems and Bifurcation of Vector Fields. Springer, New York (1983)
13. Gutiérrez, C.: Structural stability for flows on the torus with a cross-cap. Trans. Am. Math. Soc. **241**, 311–320 (1978)
14. Hirsch, M., Pugh, C., Shub, M.: Invariant Manifolds. Lecture Notes in Mathematics, vol. 583. Springer, New York (1977)
15. Keynes, H.B., Sears, M.: F-expansive transformation groups. Gen. Topology Appl. **10**(1), 67–85 (1979)
16. Komuro, M.: Expansive properties of Lorenz attractors. In: Kawakami, H. (ed.) The Theory of Dynamical Systems and Its Applications to Nonlinear Problems, pp. 4–26. World Scientific Publishing Co., Singapure (1984). Papers from the meeting held at the Research Institute for Mathematical Sciences, Kyoto University, Kyoto, July 4–7, 1984
17. Lorenz, E.N.: Deterministic nonperiodic flow. J. Atmosph. Sci. **20**, 130–141 (1963)
18. Mañé, R.: An ergodic closing lemma. Ann. Math. **116**, 503–540 (1982)
19. Markley, N.G.: The Poincaré-Bendixson theorem for the Klein bottle. Trans. Am. Math. Soc. **135**, 159–165 (1969)
20. Metzger, R., Morales, C.: Sectional-hyperbolic systems. Ergod. Theor Dyn. Syst. **28**, 1587–1597 (2008)
21. Morales, C.A., Pacifico, M.J., Pujals, E.R.: Robust transitive singular sets for 3-flows are partially hyperbolic attractors or repellers. Ann. Math. **160**(2), 375–432 (2004)
22. Oka, M.: Expansiveness of real flows. Tsukuba J. Math. **14**(1), 1–8 (1990)
23. Palis, J., de Melo, W.: Geometric Theory of Dynamical Systems. Springer, New York (1982)
24. Palis, J., Takens, F.: Hyperbolicity and Sensitive-Chaotic Dynamics at Homoclinic Bifurcations. Cambridge University Press, Cambridge (1993)
25. Peixoto, M.M.: On structural stability. Ann. Math. **69**(2), 199–222 (1959)
26. Peixoto, M.M.: Structural stability on two-dimensional manifolds. Topology **1**, 101–120 (1962)
27. Poincaré, H.: Les méthodes nouvelles de la mécanique céleste. Tome I. Les Grands Classiques Gauthier-Villars. [Gauthier-Villars Great Classics]. Librairie Scientifique et Technique Albert Blanchard, Paris (1987) Solutions périodiques. Non-existence des intégrales uniformes. Solutions asymptotiques [Periodic solutions. Nonexistence of uniform integrals. Asymptotic solutions]. Reprint of the 1892 original, With a foreword by J. Kovalevsky, Bibliothèque Scientifique Albert Blanchard [Albert Blanchard Scientific Library]
28. Robinson, C.: An Introduction to Dynamical Systems: Continuous and Discrete. Pearson Prentice Hall, Upper Saddle River (2004)
29. Viana, M.: Dynamics: a probabilistic and geometric perspective. In: Proceedings of the International Congress of Mathematicians, vol. I. (Berlin), number I in extra volume, pp. 557–578 (1998, electronic)
30. Viana, M.: What's new on Lorenz strange attractor. Math. Intel. **22**(3), 6–19 (2000)
31. Vivier, T.: Flots robustement transitifs sur les variétés compactes. C. R. Math. Acad. Sci. Paris **337**(12), 791–796 (2003)

Chapter 7
Thermodynamic Formalism for the General One-Dimensional *XY* Model: Positive and Zero Temperature

A.T. Baraviera, L. Ciolleti, A.O. Lopes, J. Mengue, J. Mohr, and R.R. Souza

7.1 Introduction

This is a survey paper on the general one-dimensional *XY* model. The proofs of the results presented here appear in two papers which are [2] and [18]. In the last mentioned work it is consider a more general setting where the state space is a compact metric space and the a-priory probability is any fixed probability on the metric space.

Let S^1 the unitary circle, d_1 the metric induced by the usual Riemmanian structure and the metric in $(S^1)^{\mathbb{N}}$ given by: $d(x, y) = \sum_{n=1}^{\infty} \frac{1}{2^n} d_1(x_n, y_n)$, where $x = (x_1, x_2, \ldots)$ and $y = (y_1, y_2, \ldots)$. Note that $\mathscr{B} := (S^1)^{\mathbb{N}}$ is compact by Tychonoff's theorem.

We denote by H_α the set of α-Hölder functions $A : \mathscr{B} \to \mathbb{R}$ with the norm $\|A\|_\alpha = \|A\| + |A|_\alpha$, where $\|A\| = \sup_{x \in \mathscr{B}} |A(x)|$ and $|A|_\alpha = \sup_{x \neq y} \frac{|A(x) - A(y)|}{d(x,y)^\alpha}$. $\sigma : \mathscr{B} \to \mathscr{B}$ denotes the shift map which is defined by $\sigma(x_1, x_2, x_3, \ldots) = (x_2, x_3, x_4, \ldots)$.

We point out that a Hölder potential A defined on $(S^1)^{\mathbb{Z}}$ is coboundary with a potential in $(S^1)^{\mathbb{N}}$ (same proof as in [20]). In this way the Statistical Mechanics of interactions on $(S^1)^{\mathbb{Z}}$ can be understood via the analysis of the similar problem in $(S^1)^{\mathbb{N}}$.

Let $\mathscr{C}$ be the space of continuous functions from $\mathscr{B}$ to $\mathbb{R}$, and we will fix the a-priori probability measure dx acting the Borel sigma algebra over S^1.

A.T. Baraviera • A.O. Lopes (✉) • J. Mengue • J. Mohr • R.R. Souza
Instituto de Matemática, UFRGS, Porto Alegre, Brazil
e-mail: atbaraviera@gmail.com; arturoscar.lopes@gmail.com; jairokras@gmail.com; joanamohr@gmail.com; rafars.mat@gmail.com

L. Ciolleti
Departamento de Matemática, UNB, Brasilia, Brazil
e-mail: leandro.mat@gmail.com

A.A. Pinto and D. Zilberman (eds.), *Modeling, Dynamics, Optimization and Bioeconomics I*, Springer Proceedings in Mathematics & Statistics 73, DOI 10.1007/978-3-319-04849-9__7,
© Springer International Publishing Switzerland 2014

For a fixed potential $A \in H_\alpha$ we define a Transfer Operator (also called Ruelle operator) $\mathscr{L}_A : \mathscr{C} \to \mathscr{C}$ by the rule

$$\mathscr{L}_A(\varphi)(x) = \int_{S^1} e^{A(ax)} \varphi(ax) da ,$$

where $x \in \mathscr{B}$ and $ax = (a, x_1, x_2, \ldots)$ denote a pre-image of x with $a \in S^1$.

The so called one-dimensional XY model (see [8, 16, 24]) is considered in several applications to real problems in Physics. The spin in each site of the lattice is described by an angle from $[0, 2\pi)$. In the Physics literature, as far as we know, the potential A depends on two coordinates. A well known example in applications is the potential $A(x) = A(x_0, x_1) = \cos(x_1 - x_0 - \alpha) + \gamma \cos(2x_0)$. We consider here potentials which can depend on the all string $x = (x_1, x_2, \ldots)$ but which are in the Hölder class. We call this setting the general one-dimensional XY model.

There are several possible points of view for understanding Gibbs states in Statistical Mechanics (see [21,22] for interesting discussions). We prefer the transfer operator method because we believe that the eigenfunctions and eigenprobabilities (which can be derived from the theory) allow a more deep understanding of the problem. For example, the information one can get from the main eigenfunction (defined in the whole lattice) is worthwhile, mainly in the limit when temperature goes to zero.

In Sect. 7.3 we consider the entropy, pressure and Variational Principle and its relations with eigenfunctions and eigenprobabilities of the Ruelle operator. This setting, as far as we know, was not considered before. In this case the entropy, by its very nature, is always a nonpositive number. Invariant probabilities with support in a fixed point will have entropy equal to minus infinity. The infinite product of dx on $(S^1)^{\mathbb{N}}$ will have zero entropy. We point out that, although at first glance, the fact that the entropy we define here is negative may look strange, our definition is the natural extension of the concept of Kolomogorov entropy. In the classical case, the entropy is positive because the a-priori measure is not a probability: is the counting measure.

Among other things we consider in Sect. 7.4 the case where temperature goes to zero and show some selection results related with the Ergodic Optimization (see [7, 10, 12, 23]). Using the variational principle we obtain a simple proof of the fact that Gibbs states converge to maximizing measures when the temperature goes to zero (a question not discussed in [2]).

An important issue that does not appear in the classical Thermodynamic Formalism (in the sense of [20] and [11]) is the differentiable structure. We will show in Sect. 7.5 that, in the case A is smooth, then, the associated main eigenfunction is also smooth.

7.2   The Ruelle Operator

Let a^n be an element of $(S^1)^n$ having coordinates $a^n = (a_n, a_{n-1}, \ldots, a_2, a_1)$, we denote by $a^n x \in \mathcal{B}$ the concatenation of $a^n \in (S^1)^n$ with $x \in \mathcal{B}$, i.e., $a^n x = (a_n, \ldots, a_1, x_1, x_2, \ldots)$. In the case of $n = 1$ we will write $a := a^1 \in S^1$, and $ax = (a, x_1, x_2, \ldots)$.

The n-th iterate of $\mathcal{L}_A$ has the following expression

$$\mathcal{L}_A^n(\varphi)(x) = \int_{(S^1)^n} e^{S_n A(a^n x)} \varphi(a^n x) (da)^n,$$

where $S_n A(a^n x) = \sum_{k=0}^{n-1} A(\sigma^k(a^n x))$.

Theorem 7.1. *Let us fix $A \in H_\alpha$, then there exists a strictly positive Hölder eigenfunction ψ_A for $\mathcal{L}_A : \mathcal{C} \to \mathcal{C}$ associated to a strictly positive eigenvalue λ_A. This eigenvalue is simple, which means the eigenfunction is unique (modulo multiplication by constant).*

We say that a potential B is normalized if $\mathcal{L}_B(1) = 1$, which means it satisfies $\int_{S^1} e^{B(ax)} da = 1$, $\forall x \in \mathcal{B}$.

Let $A \in H_\alpha$, ψ_A and λ_A given by Theorem 7.1, it is easy to see that

$$\int_{S^1} \frac{e^{A(ax)} \psi_A(ax)}{\lambda_A \psi_A(x)} \, da = 1, \ \forall x \in \mathcal{B}. \tag{7.1}$$

Therefore we define the normalized potential $\bar{A}$ associated to A, as

$$\bar{A} := A + \log \psi_A - \log \psi_A \circ \sigma - \log \lambda_A, \tag{7.2}$$

where $\sigma : \mathcal{B} \to \mathcal{B}$ is the shift map. As $\psi_A \in H_\alpha$ we have that $\bar{A} \in H_\alpha$.

We say a probability measure μ is invariant, if for any Borel set B, we have that $\mu(B) = \mu(\sigma^{-1}(B))$. We denote by $\mathcal{M}_\sigma$ the set of invariant probability measures.

We note that $\mathcal{B}$ is a compact metric space and by the Riesz Representation Theorem, a probability measure on the Borel sigma-algebra is identified with a positive linear functional $L : \mathcal{C} \to \mathbb{R}$ that sends the constant function 1 to the real number 1. We also note that $\mu \in \mathcal{M}_\sigma$ if and only if, for any $\psi \in \mathcal{C}$ we have $\int_{\mathcal{B}} \psi \, d\mu = \int_{\mathcal{B}} \psi \circ \sigma \, d\mu$.

We define the dual operator $\mathcal{L}_A^*$ on the space of Borel measures on $\mathcal{B}$ as the operator that sends a measure μ to the measure $\mathcal{L}_A^*(\mu)$, defined by $\int_{\mathcal{B}} \psi \, d\mathcal{L}_A^*(\mu) = \int_{\mathcal{B}} \mathcal{L}_A(\psi) \, d\mu$, for any $\psi \in \mathcal{C}$.

Theorem 7.2. *Let A be a Hölder continuous potential, not necessarily normalized, ψ_A and λ_A the eigenfunction and eigenvalue given by the Theorem 7.1. We associate to A the normalized potential $\bar{A} = A + \log \psi_A - \log \psi_A \circ \sigma - \log \lambda_A$. Then*

(a) *there exists an unique fixed point μ_A for $\mathscr{L}_{\bar{A}}^*$, which is a σ-invariant probability measure;*

(b) *the measure $\rho_A = \frac{1}{\psi_A} \mu_A$ satisfies $\mathscr{L}_A^*(\rho_A) = \lambda_A \rho_A$. Therefore, ρ_A is an eigenmeasure for $\mathscr{L}_A^*$;*

(c) *for any Hölder continuous function $w : \mathscr{B} \to \mathbb{R}$, we have that, in the uniform convergence topology, $\frac{\mathscr{L}_A^n(w)}{(\lambda_A)^n} \to \psi_A \int_{\mathscr{B}} w \, d\rho_A$ and $\mathscr{L}_{\bar{A}}^n \omega \to \int_{\mathscr{B}} \omega d\mu_A$, where $\mathscr{L}_A^n$ denotes the n-th iterate of the operator $\mathscr{L}_A : H_\alpha \to H_\alpha$.*

We call μ_A the **Gibbs probability** (or, Gibbs state) for A. We will leave the term **equilibrium probability** (or, equilibrium state) for the one which maximizes pressure. As we will see, this invariant probability measure over $\mathscr{B}$ describes the statistics in equilibrium for the interaction described by the potential A. The assumption that the potential is Hölder implies that the decay of interaction is fast.

Proposition 7.1. *The only Hölder continuous eigenfunction ψ of $\mathscr{L}_A$ which is totally positive is ψ_A.*

Proposition 7.2. *Suppose $\bar{A}$ is normalized, then the eigenvalue $\lambda_{\bar{A}} = 1$ is maximal. Moreover, the remainder of the spectrum of $\mathscr{L}_{\bar{A}} : H_\alpha \to H_\alpha$ is contained in a disk centered at zero with radius $\lambda_{\bar{A}}^{\frac{1}{A}}$ strictly smaller than one.*

Proposition 7.3. *If $v, w \in \mathscr{L}^2(\mu_A)$ are such that w is Hölder and $\int w \, d\mu_A = 0$, then, there exists $C > 0$ such that for all n $\int (v \circ \sigma^n) w \, d\mu_A \leq C \, (\lambda_{\bar{A}}^{\frac{1}{A}})^n$. In particular μ_A is mixing and therefore ergodic.*

7.3 Entropy and Variational Principle

In this section (which was taken from [18]) we will introduce a notion of entropy. Initially, this will be done only for Gibbs probabilities, and then we will extend this definition to invariant probabilities. After that we prove that the Gibbs probability obtained in the general setting above satisfies a variational principle.

We point out that any reasonable concept of entropy must satisfy two principles: the entropy of probabilities with support in periodic orbits should be minimal and the entropy of the independent probability should be maximal. This will happen for our definition.

Definition 7.1. We denote by $\mathscr{G}$ the set of Gibbs measures, which means the set of $\mu \in \mathscr{M}_\sigma$, such that, $\mathscr{L}_B^*(\mu) = \mu$, for some normalized potential $B \in H_\alpha$. We define the entropy of $\mu \in \mathscr{G}$ as $h(\mu) = -\int_{\mathscr{B}} B(x) d\mu(x)$.

One can show that $-\int B \, d\mu$ is the infimum of $\left\{ -\int A \, d\mu + \log(\lambda_A) \ : \ A \in H_\alpha \right\}$.

The above definition which appears in [18] is different from the one briefly mentioned in Sect. 3 in [2].

Proposition 7.4. *If $\mu \in \mathscr{G}$, then we have $h(\mu) \leq 0$.*

This follow from Jensen's inequality.

It is easy to see that the Gibbs state $(dx)^N$ has zero entropy.

Now we state a lemma that is used to prove the main result of this section, namely, the variational principle of Theorem 7.3. This lemma was shown to be true in the classical Bernoulli case in [13].

Lemma 7.1. *Let us fix a Hölder continuous potential A and a measure* $\mu \in \mathcal{G}$ *with associated normalized potential B. We call* $\mathcal{C}^+$ *the space of continuous positive functions on* $\mathcal{B}$. *Then, we have*

$$h(\mu) + \int_{\mathcal{B}} A(x)d\mu(x) = \inf_{u \in \mathcal{C}^+} \left\{ \int_{\mathcal{B}} \log\left(\frac{\mathcal{L}_A u(x)}{u(x)}\right) d\mu(x) \right\}.$$

Definition 7.2. Let μ be an invariant measure. We define the entropy of μ as

$$h(\mu) = \inf_{A \in H_\alpha} \left\{ -\int_{\mathcal{B}} A \, d\mu + \log \lambda_A \right\},$$

where λ_A is the maximal eigenvalue of $\mathcal{L}_A$, given by Theorem 7.1.

This value is non positive and can be $-\infty$ as we will se later.

Definition 7.3. Given a Hölder potential A we call the pressure of A the value

$$P(A) = \sup_{\mu \in \mathcal{M}_\sigma} \left\{ h(\mu) + \int_{\mathcal{B}} A(x)d\mu(x) \right\}.$$

A probability which attains such maximum value is called equilibrium state for A.

Theorem 7.3 (Variational Principle). *Let* $A \in H_\alpha$ *be a Hölder continuous potential and* λ_A *be the maximal eigenvalue of* $\mathcal{L}_A$, *then*

$$\log \lambda_A = P(A) = \sup_{\mu \in \mathcal{M}_\sigma} \left\{ h(\mu) + \int_{\mathcal{B}} A(x)d\mu(x) \right\}.$$

Moreover the supremum is attained on the Gibbs measure, i.e. the measure μ_A *that satisfies* $\mathcal{L}_{\bar{A}}^*(\mu_A) = \mu_A$.

Therefore, the Gibbs state and the equilibrium state for A are given by the same measure μ_A, which is the unique fixed point for the dual Ruelle operator associated to the normalized potential $\bar{A}$.

Theorem 7.4 (Pressure as Minimax). *Given a Hölder potential A*

$$P(A) = \sup_{\mu \in \mathcal{M}_\sigma} \left[\inf_{u \in \mathcal{C}^+} \left\{ \int_{\mathcal{B}} \log\left(\frac{\mathcal{L}_A u(x)}{u(x)}\right) d\mu(x) \right\} \right].$$

Remark. The entropy of a probability measure supported on periodic orbit can be $-\infty$. Indeed, suppose $S^1 \sim (0, 1]$, and $A_c : (S^1)^{\mathbb{N}} \to \mathbb{R}$ given by $A_c(x) = \log\left(\frac{c}{1-e^{-c}}e^{-cx_1}\right)$. We have that for each $c > 0$, the function A_c is a C^1 normalized potential (therefore belongs to H_α), which depends only on the first coordinate of x. Note that $\mathscr{L}_{A_c}(1) = 1$. Let μ be the Dirac Measure on 0^∞. We have $h(\mu) \leq -\int A_c d\mu = -A_c(0^\infty) = -\log\left(\frac{c}{1-e^{-c}}\right) \to -\infty$ when $c \to \infty$. This shows that $h(\mu) = -\infty$. An easy adaptation of the arguments can be done to prove that, in this setting, invariant measures supported on periodic orbits have entropy $-\infty$.

7.4 Zero Temperature

Consider a fixed Hölder potential A and a real variable $\beta > 0$. We denote, respectively, by $\psi_{\beta A}$ and $\mu_{\beta A}$, the eigenfunction for the Ruelle operator associated to βA and the equilibrium measure (Gibbs) for βA. We would like to investigate general properties of the limits of $\mu_{\beta_n A}$ and of $\frac{1}{\beta_n} \log \psi_{\beta_n A}$ when $\beta_n \to \infty$. Some results of this section are generalizations of the ones in [2]. It is well known that the parameter β represents the inverse of the temperature.

It is fair to call "Gibbs state at zero temperature for the potential A" any of the weak limits of convergent subsequences $\mu_{\beta_n A}$. Even when the potential A is Hölder, Gibbs state at zero temperature do not have to be unique. In the case there exist the weak limit $\mu_{\beta A} \rightharpoonup \mu$, $\beta \to \infty$, we say that there exists selection of Gibbs state for A at temperature zero.

Remark. Given β and A, the Hölder constant of $u_{\beta A} = \log(\psi_{\beta A})$, depends on the Hölder constant for βA, and is given by $\beta \frac{2^\alpha}{2^\alpha-1} Hol_A$ (see [2]). As we normalize $\psi_{\beta A}$ assuming that $\max \psi_{\beta A} = 1$, the family of functions $\frac{1}{\beta} \log(\psi_{\beta A})$, $\beta > 0$, is uniformly bounded. Note that when we normalize $\psi_{\beta A}$ the Hölder constant of $\log(\psi_{\beta A})$ remains unchanged, which assures the family $\frac{1}{\beta} \log(\psi_{\beta A})$, $\beta > 0$, is equicontinuous.

Therefore, there exists a subsequence $\beta_n \to \infty$, and V Hölder, such that, on the uniform convergence topology $V := \lim_{n\to\infty} \frac{1}{\beta_n} \log(\psi_{\beta_n A})$.

Remember that we denote by $\mathscr{M}_\sigma$ the set of σ invariant Borel probability measures over $\mathscr{B}$. As $\mathscr{M}_\sigma$ is compact, given A, there always exists a subsequence β_n, such that $\mu_{\beta_n A}$ converges to an invariant probability measure.

The limits of $\mu_{\beta A}$ are related (see below) with the following problem: given $A : \mathscr{B} \to \mathbb{R}$ Hölder, we want to find probabilities that maximize, over $\mathscr{M}_\sigma$, the value

$$\int_{\mathscr{B}} A(x) \, d\mu(\mathbf{x}).$$

We define

$$m(A) = \max_{\mu \in \mathcal{M}_\sigma} \left\{ \int_{\mathcal{B}} A d\mu \right\} .$$

Any of the probability measures which attains the maximal value will be called a maximizing probability measure, which will be sometimes denoted generically by μ_∞. As $\mathcal{M}_\sigma$ is compact, there exist always at least one maximizing probability measure. It is also true that there exists ergodic maximizing probability measures. Indeed, the set of maximizing probability measures is convex, compact and the extreme probability measures of this convex set are ergodic (can not be expressed as convex combination of others [11]). Results obtained in this setting belong to what is called Ergodic Optimization Theory [3, 10].

The possible limits of $\frac{1}{\beta_n} \log \psi_{\beta_n A}$ are related (see below) with the following concept:

Definition 7.4. A continuous function $u : \mathcal{B} \to \mathbb{R}$ is called a *calibrated subaction* for $A : \mathcal{B} \to \mathbb{R}$, if, for any $y \in \mathcal{B}$, we have

$$u(y) = \max_{\sigma(x)=y} [A(x) + u(x) - m(A)]. \tag{7.3}$$

This can also be expressed as

$$m(A) = \max_{a \in M} \{A(ay) + u(ay) - u(y)\}.$$

Note that for any $x \in \mathcal{B}$ we have

$$u(\sigma(x)) - u(x) - A(x) + m(A) \geq 0.$$

If u is a calibrated subaction, then $u + c$, where c is a constant, is also a calibrated subaction. An interesting question is when such calibrated subaction u is unique up to an additive constant (see [16] and [9]).

Remember that if μ is σ-invariant, then for any continuous function $u : \mathcal{B} \to \mathbb{R}$ we have

$$\int_{\mathcal{B}} [u(\sigma(x)) - u(x)] d\mu = 0.$$

Therefore if μ_∞ is a maximizing probability measure for A and u is a calibrated subaction for A, then (see for instance [7, 10, 23] for a similar result) for any x in the support of μ_∞, we have

$$u(\sigma(x)) - u(x) - A(x) + m(A) = 0. \tag{7.4}$$

In this way if we know the value $m(A)$, then a calibrated subaction u for A can help us to identify the support of maximizing probabilities. The above equation can be eventually true outside the union of the supports of the maximizing probabilities (see an interesting example due to Leplaideur [6]).

One can show that in the case there exists a subsequence $\beta_n \to \infty$, such that on the uniform convergence $V := \lim_{n \to \infty} \frac{1}{\beta_n} \log(\psi_{\beta_n A})$, then such V is a calibrated subaction for A. When there exists a V which is the limit $V := \lim_{\beta \to \infty} \frac{1}{\beta} \log(\psi_{\beta A})$, (not just via a subsequence) we say we have selection of subaction at temperature zero. Positive results in this direction are presented in [4, 15] and [16].

Proposition 7.5. *Given a potential A Hölder continuous, we have:*
i) $\lim_{\beta \to \infty} \frac{1}{\beta} \log \lambda_\beta = m(A)$. *ii) Any limit, in the uniform topology,*

$$V := \lim_{n \to \infty} \frac{1}{\beta_n} \log(\psi_{\beta_n A}),$$

is a calibrated subaction for A.

Now we return to study the Gibbs measures at zero temperature.

Theorem 7.5. *Consider a Hölder potential A. Suppose that for some subsequence we have $\mu_{\beta_n A} \rightharpoonup \mu_\infty$. Then μ_∞ is a maximizing probability, i.e.,*
$\int_{\mathscr{B}} A(x) d\mu_\infty(x) = m(A).$
In the case the maximizing probability for A is unique, we have selection of Gibbs probability at temperature zero.

Questions related to the Large Deviation property on the XY model, when $\beta \to \infty$, are considered in [15]. The existence of a calibrated subaction plays an important role in this kind of result.

7.5   The Differentiable Structure and the Involution Kernel

We consider in this section (which was taken from [18]) the differentiability structure of the general one-dimensional XY model.

Let $\mathscr{B}^* = \{(\ldots, y_2, y_1) \in (S^1)^{\mathbb{N}}\}$, and we denote by the pair $(y|x) = (\ldots, y_2, y_1|x_1, x_2 \ldots)$, the general element of $\hat{\mathscr{B}} := \mathscr{B}^* \times \mathscr{B} = (S^1)^{\mathbb{Z}}$.

We denote by $\hat{\sigma}$ the shift on $\hat{\mathscr{B}}$, i.e. $\hat{\sigma}(\ldots, y_2, y_1|x_1, x_2, \ldots) = (\ldots, y_2, y_1, x_1|x_2, x_3, \ldots)$.

Definition 7.5. Let $A : \mathscr{B} \to \mathbb{R}$ be a continuous potential (considered as a function on $\hat{\mathscr{B}}$). A continuous function $W : \hat{\mathscr{B}} \to \mathbb{R}$ is called an involution kernel, if

$$A^* := A \circ \hat{\sigma}^{-1} + W \circ \hat{\sigma}^{-1} - W$$

depends only on the variable y.

Let us fix $x' \in \mathcal{B}$ and A a Hölder continuous potential, then we define

$$W(y|x) = \sum_{n \geq 1} A(y_n, \ldots, y_1, x_1, x_2, \ldots) - A(y_n, \ldots, y_1, x'_1, x'_2, \ldots). \qquad (7.5)$$

An easy calculation shows that $W(y|x)$ is a involution kernel (see [1]).

Let $A : \mathcal{B} \to \mathbb{R}$ be a Hölder continuous potential and $W : \hat{\mathcal{B}} \to \mathbb{R}$ an involution kernel, then for any $a \in S^1$, $x \in \mathcal{B}$ and $y \in \mathcal{B}^*$, we have

$$(A^* + W)(ya|x) = (A + W)(y|ax). \qquad (7.6)$$

Questions related to Ergodic Transport Theory and the involution kernel are analyzed in [6, 17, 19], and [14].

Remember that $m(A) = \sup_{\mu \text{ is } \sigma-\text{invariant}} \int A d\mu$, and, in an analogous way we define $m(A^*) = \sup_{\mu \text{ is } \sigma^*-\text{invariant}} \int A^* d\mu$.

The next result is an adaptation to the present setting of a result in [1].

Lemma 7.2. *Let $\mathcal{L}_A$ and $\mathcal{L}_{A^*}$ be the Ruelle operators defined on $\mathcal{B}$ and $\mathcal{B}^*$, and $W(y|x)$ an involution kernel.*

Then, for any $x \in \mathcal{B}$, $y \in \mathcal{B}^*$, and any function $f : \hat{\mathcal{B}} \to \mathbb{R}$

$$\mathcal{L}_{A^*} \left(f(\cdot|x) \, e^{W(\cdot|x)} \right)(y) = \mathcal{L}_A \left(f \circ \hat{\sigma}(y|\cdot) \, e^{W(y|\cdot)} \right)(x). \qquad (7.7)$$

Let ρ_A and ρ_A^* the eigenmeasures for $\mathcal{L}_A^*$ and $\mathcal{L}_{A^*}^*$, given in Theorem 7.2. Suppose c is such that $\iint e^{W(y|x)-c} \, d\rho_A^*(y) d\rho_A(x) = 1$.

Proposition 7.6. *Suppose $K(y|x) = e^{W(y|x)-c}$. Then, $d \hat{\mu}_A = K(y|x) d\rho_A^*(y) d\rho_A(x)$ is invariant for $\hat{\sigma}$ and is the natural extension of the Gibbs measure μ_A.*

The function $\psi_A(x) = \int_{\mathcal{B}^*} K(y|x) d\rho_A^*(y)$ is the main eigenfunction for $\mathcal{L}_A$, and the function $\psi_{A^*}(y) = \int_{\mathcal{B}} K(y|x) d\rho_A(x)$ is the main eigenfunction for $\mathcal{L}_{A^*}$. Furthermore $\lambda_A = \lambda_{A^*}$.

Remark. Note that, as $\lambda_{\beta A} = \lambda_{\beta A^*}$, we have that $m(A) = \lim_{\beta \to \infty} \dfrac{1}{\beta} \log \lambda_{\beta A} = m(A^*)$.

Suppose that A is differentiable in each coordinate of $x \in \mathcal{B}$, and given $\epsilon > 0$, there exists $H_\epsilon > 0$, such that, if $|h| < H_\epsilon$, then

$$\left| \frac{A(x + h e_j) - A(x)}{h} - D_j A(x) \right| \leq \frac{\epsilon}{2^j} \quad \forall j \in \mathbb{N}, \qquad (7.8)$$

where $D_j A(x)$ denote the derivative of A with respect to the j-th coordinate.

Proposition 7.7. *Suppose A is differentiable in the sense defined above. Then, for any j*

$$\frac{\partial}{\partial x_j} W(y|x) = \sum_{n \geq 1} D_{n+j} A(y_n, \ldots, y_1, x_1, x_2, \ldots)$$

Proposition 7.8. *Let $\psi_A(x) = \int_{\mathscr{B}} e^{W(y|x)-c} d\rho_A^*(y)$, and suppose A is differentiable. Then, the eigenfunction ψ_A is differentiable, moreover*

$$\frac{\partial}{\partial x_j} \psi_A(x) = \int_{\mathscr{B}} e^{W(y|x)-c} \sum_{n \geq 1} D_{n+j} A(y_n, \ldots, y_1, x_1, x_2, \ldots) d\rho_A^*(y).$$

We remark that the conditions necessary to interchange the integral and the derivative are satisfied.

Remark. In the case where A depends only on the two first coordinates, we have that

$$\psi_A(x_1) = \frac{1}{\lambda_A} \int_{S^1} e^{A(y_1, x_1)} \psi_A(y_1) d\nu(y_1).$$

Hence, ψ_A satisfies the equation

$$\frac{\partial}{\partial x_1} \psi_A(x_1) = \frac{1}{\lambda_A} \int_{S^1} e^{A(y_1, x_1)} D_2 A(y_1, x_1) \psi_A(y_1) d\nu(y_1).$$

References

1. Baraviera, A., Lopes, A.O., Thieullen, P.: A large deviation principle for equilibrium states of Hölder potentials: the zero temperature case. Stochast. Dyn. **6**, 77–96 (2006)
2. Baraviera, A.T., Cioletti, L., Lopes, A.O., Mohr, J., Souza, R.R.: On the general one-dimensional XY model: positive and zero temperature, selection and non-selection. Rev. Math. Phys. **23**(10), 1063–1113 (2011). 82Bxx
3. Baraviera, A., Leplaideur, R., Lopes, A.: Ergodic optimization, zero temperature limits and the max-plus algebra, mini-course in XXIX Coloquio Brasileiro de Matemática - IMPA (2013)
4. Baraviera, A.T., Lopes, A.O., Mengue, J.K.: On the selection of subaction and measure for a subclass of potentials defined by P. Walters. Ergod. Theory Dyn. Syst. **33**(05), 1338–1362 (2013)
5. Chazottes, J.R. Gambaudo, J.M., Ulgade E.: Zero-temperature limit of one dimensional Gibbs states via renormalization: the case of locally constant potentials. Ergod. Theory Dyn. Syst. **31**(4), 1109–1161 (2011)
6. Contreras, G., Lopes, A., Oliveira, E.: Ergodic transport theory, periodic maximizing probabilities and the twist condition. In: Zilberman, D., Pinto, A. (eds.) Modeling, Dynamics, Optimization and Bioeconomics. Springer (2014)
7. Contreras, G., Lopes, A.O., Thieullen, P.H.: Lyapunov minimizing measures for expanding maps of the circle. Ergod. Theory Dyn. Syst. **21**, 1379–1409 (2001)

8. Fukui, Y., Horiguchi, M.: One-dimensional Chiral XY model at finite temperature. Interdiscip. Inf. Sci. **1**(2), 133–149 (1995)
9. Garibaldi, E., Lopes, A.O.: On Aubry-Mather theory for symbolic dynamics. Ergod. Theory Dyn. Syst. **28**(3), 791–815 (2008)
10. Jenkinson, O.: Ergodic optimization. Discrete Continuous Dyn. Syst. Ser. A **15**, 197–224 (2006)
11. Keller, G.: Gibbs States in Ergodic Theory. Cambrige Press, Cambridge (1998)
12. Leplaideur, R.: A dynamical proof for convergence of Gibbs measures at temperature zero. Nonlinearity **18**(6), 2847–2880 (2005)
13. Lopes, A.O.: An analogy of charge distribution on Julia sets with the Brownian motion. J. Math. Phys. **30**(9), 2120–2124 (1989)
14. Lopes, A., Mengue, J.: Duality theorems in ergodic transport. J. Stat. Phys. **149**(5), 921–942 (2012)
15. Lopes, A., Mengue, J.: Selection of measure and a large deviation principle for the general one-dimensional XY model. Dyn. Syst. **29**, 24–39 (2014)
16. Lopes, A.O., Mohr, J., Souza, R., Thieullen, P.H. Negative entropy, zero temperature and stationary Markov chains on the interval. Bull. Braz. Math. Soc. **40**, 1–52 (2009)
17. Lopes, A.O., Oliveira, E.R., Thieullen, P.H.: The dual potential, the involution kernel and transport in ergodic optimization. Arxiv (2011)
18. Lopes, A.O., Mengue, J.K., Mohr, J., Souza, R.R.: Entropy and variational principle for one-dimensional lattice systems with a general a-priori measure: positive and zero temperature, Arxiv (2012)
19. Lopes, A., Oliveira, E.R., Smania, D.: Ergodic transport theory and piecewise analytic subactions for analytic dynamics. Bull. Braz. Math. Soc. **43**(3), 467–512 (2012)
20. Parry, W., Pollicott, M.: Zeta functions and the periodic orbit structure of hyperbolic dynamics. Asterisque 187–188 (1990)
21. Ruelle, D.: Thermodynamic Formalism, 2nd edn. Cambridge University Press, Cambridge (2004)
22. Sarig, O.: Lecture Notes on Thermodynamic Formalism for Topological Markov Shifts. Pennsylvania State University, Pennsylvania (2009)
23. Souza, R.R.: Sub-actions for weakly hyperbolic one-dimensional systems. Dyn. Syst. **18**(2), 165–179 (2003)
24. Thompson, C.: Infinite-spin Ising model in one dimension. J. Math. Phys. **9**(2), 241–245 (1968)

Chapter 8
Impact of Political Economy and Logistical Constraints on Assessments of Biomass Energy Potential: New Jersey as a Case Study

Margaret Brennan-Tonetta, Gal Hochman, and Brian Schilling

8.1 Introduction

Methodologies used to estimate biomass energy potential have evolved over the last several decades. Early studies made simple assumptions regarding land use and the amounts of residue recoverable from agricultural production and forestry management for use in bioenergy production (McKeever 2002) [7, 14, 15, 17, 20, 21]. More recent studies employ more sophisticated models to explicitly simulate various scenarios that include environmental, technical, and economic assumptions, such as energy prices, food costs, land use and other environmental factors [1, 3–6, 8–10, 19]. In the recent United Kingdom Energy Research Centre report, *Energy from Biomass: the Size of the Global Resource* [16], ninety global studies on biomass potential were reviewed. Many of the studies reviewed in this report impose assumptions governing land or land use. To a lesser extent these studies also incorporate assumptions or constraints regarding yield and production capacity, sustainability, environmental impacts, and economic and market factors. However, realistic political economy and logistical constraints such as institutional factors, existing policy frameworks and collection infrastructure that impact the utilization of biomass feedstocks and the feasibility of installing alternative biomass-to-energy conversion technologies, are absent.

In this chapter, we review a state-level biomass-to-energy assessment conducted in New Jersey to demonstrate the effect of political economic and logistical considerations on estimates of energy generation potential [2]. The assessment is expansive, incorporating thirteen biomass-to-energy conversion technologies and forty potential biomass feedstocks. Biomass-to-energy generation simulations were

M. Brennan-Tonetta • G. Hochman (✉) • B. Schilling
Rutgers University, New Brunswick, NJ, USA
e-mail: brennan@AESOP.Rutgers.edu; gal.hochman@rutgers.edu;
schilling@AESOP.Rutgers.edu

A.A. Pinto and D. Zilberman (eds.), *Modeling, Dynamics, Optimization
and Bioeconomics I*, Springer Proceedings in Mathematics & Statistics 73,
DOI 10.1007/978-3-319-04849-9_8,
© Springer International Publishing Switzerland 2014

supported by the development of a novel Bioenergy Calculator.[1] This tool allows decision makers to evaluate the sensitivity of energy generation estimates to various political economy, logistical and technological parameters affecting the availability and conversion of biomass feedstocks into energy. Political economy constraints in the model encompass assumptions regarding existing policies and logistical (i.e., feedstock collection, aggregation and distribution) or institutional barriers that limit or support the diversion of biomass resources into energy production systems. Qualitative assumptions are also made regarding the commercialization status or market readiness of alternative energy production technologies.

The physical quantities of biomass feedstocks were estimated (on a dry tonnage basis) at the state level and for each of New Jersey's 21 counties based on the collection of extensive secondary data. These estimates represent theoretical maxima in the Bioenergy Calculator or, in other words, the upper bounds for feedstock quantities potentially suitable for bioenergy generation. Subsequent calculations discounted these maxima by considering political economy and logistical barriers that limit or preclude the use of each biomass feedstock for bioenergy production. The resulting feedstock balances, termed "practically recoverable feedstocks", total only 65% of the theoretical dry tonnage maxima. The New Jersey case study therefore demonstrates the sensitivity of biomass evaluations to assumptions regarding institutional capacity, feedstock aggregation and distribution infrastructure, and regulatory/policy considerations.

To set the context for the New Jersey biomass-to-energy assessment, the following section provides a brief policy background on the role of bioenergy in the state's recent energy master plan. This is followed by a more detailed description of the data collection methodologies, structure of the Bioenergy Calculator and estimates of New Jersey's biomass energy potential after consideration of political economy and logistical constraints. The chapter concludes with a discussion of the results and policy recommendations.

8.2 The Role of Renewable Energy in Achieving New Jersey's Energy Goals

New Jersey has taken significant steps to increase the development and utilization of renewable energy resources through legislation, mandatory greenhouse gas (GHG) emission reduction programs, incentives, and regional collaborations. In February 2007, Governor Jon Corzine issued Executive Order 54, requiring that 20% of the state's electricity be derived from Class 1[2] renewable resources by

---

[1] The Bioenergy Calculator is accessible at: http://bioenergy.rutgers.edu/biomass-energy-potential/default.asp.

[2] Class One Renewable Energy definition as provided by NJDEP: Class I renewable energy is defined as electricity derived from solar energy, wind energy, wave or tidal action, geothermal

2020. The executive order also outlined targets for reducing GHG emissions [18]. In 2011, the New Jersey State Energy Master Plan (EMP), for the first time, incorporated renewable energy strategies, including biomass utilization [12]. These strategies are designed to meet several goals, including the adoption of a Renewable Portfolio Standard (RPS) requiring that 22.5 % of the state's energy needs be met from renewable sources by 2021. The EMP further specifies a target of 900 MW of biopower as part of the State's 2020 RPS.[3] The EMP further recognizes the importance of enabling public policies to achieve these goals, stating that "both State and federal mandates regarding the use of renewables are predicated on the need to establish worthwhile public policy goals to support renewable energy technology."[4]

Within the context of New Jersey's total energy generation portfolio, renewable energy generation lags far behind other energy resources (i.e., nuclear, natural gas and coal). In 1997, renewable energy sources accounted for 2.8 % of New Jersey's total energy generation. In 2011, renewables accounted for only 1.2 %, demonstrating that growth in the renewable sector has not kept pace with the increased energy demands of the state [23]; New Jersey ranks 46th in the country in terms of renewable electricity net generation [24]. To meet future demand, New Jersey's growth rate of renewable energy will therefore have to increase substantially over current levels. As noted in the EMP, meeting renewable energy goals will require policies that ensure significant energy conservation, energy efficiencies, and high growth in renewable energy sources.[5]

New Jersey has established economic incentives along with renewable energy standards, to encourage the development and utilization of alternative energy. The New Jersey Board of Public Utilities (NJBPU) Office of Clean Energy establishes and implements the majority of renewable energy programs for the state. From 2009–2012, BPU approved $1.2 billion for energy-efficiency and renewable energy initiatives (an increase of $475 million over the 2005–2008 allocation). In 2011, the Office of Clean Energy expended over $380 million to provide New Jersey residents, municipalities and businesses with incentives to install energy efficient and renewable energy technologies [13]. New Jersey also offers numerous financial incentives for alternative fuel and alternative fuel vehicles in the form of rebates and tax exemptions. In addition to the energy programs operated by BPU, the New Jersey Economic Development Authority offers a suite of "Clean Energy Solutions" programs[6] that offer grants and low-cost financing.

---

energy, landfill gas, anaerobic digestion, fuel cells using renewable fuels, and, with written permission of the New Jersey Department of Environmental Protection (DEP), certain other forms of sustainable biomass.

[3] New Jersey Energy Master Plan [12, p. 1].

[4] New Jersey Energy Master Plan [12, p. 86].

[5] New Jersey Energy Master Plan [12, p. 103].

[6] Available at http://www.njeda.com/web/Aspx_pg/Templates/Pic_Text.aspx?Doc_Id=1080&topid=722&midid=1357.

New Jersey currently has no financial incentives to support the development of alternative fuel infrastructure to serve the general public, which will be needed if the state intends to effectuate the transition to alternative fuel vehicles and associated infrastructure. The Alternative Fuel Infrastructure Rebate provides only limited funding to eligible local governments, state colleges and universities, school districts, and governmental authorities for the purchase and installation of refueling infrastructure for alternative fuels. Lack of public access to alternative fuels will impede the state's goals of alternative fuel use and greenhouse gas reductions. However, New Jersey has instituted a number of regulations designed to support emissions reductions and use of alternative fuel and vehicles which include biodiesel fuel use rebates, emissions reductions requirements, low emission vehicle standards and plug-in hybrid electric vehicle promotion.

To advance biomass energy use in an environmentally responsible and sustainable manner, and as mandated by existing legislation and NJBPU regulations, the Department of Environmental Protection is developing an objective and systematic process of sustainability determinations that will facilitate environmental permitting of qualified biomass projects meeting the sustainability criteria of: (1) superior environmental performance (including meeting state-of-the-art (SOTA) air quality standards); (2) socioeconomic sustainability; and (3) reduction of greenhouse gas emissions.[7]

8.3 Assessment of Biomass Energy Potential for New Jersey

The assessment of biomass energy potential for New Jersey was completed in 2007 and provided critical information to the NJBPU as it incorporated bioenergy potential into the state's energy master plan. The 2011 EMP was the first version to contain information and recommendations on the development and utilization of bioenergy to meet state energy demands, necessitating baseline information on biomass feedstocks and practical estimates of potential bioenergy generation. The objectives of the biomass assessment were therefore to:

(1) assess the characteristics and quantity of New Jersey's biomass resources;
(2) assess technologies (commercially or near commercially available) that are capable of producing biopower or biofuels from New Jersey's biomass resources;
(3) develop the first statewide mapping of waste/biomass resources and bioenergy potential; and,
(4) develop policy recommendations for creating a bioenergy industry in New Jersey [2].

---

[7]New Jersey Energy Master Plan [12, pp. 102–103].

Since no previous assessment of biomass feedstocks and bioenergy potential had ever been conducted for New Jersey, a method was required to guide the collection and analysis of a substantial amount of data. In addition, a decision support tool was needed to calculate energy potential based: (1) estimated biomass feedstock quantities, (2) availability of suitable bioenergy generation technologies, and (3) previously described political economic and logistical constraints. These were foundational elements in the development of the Bioenergy Calculator and are detailed in the following sections.

8.3.1 Feedstock Assessment

The feedstock assessment efforts concentrated on collecting existing data on quantity and location for each type of biomass identified as a potentially significant feedstock for energy production. The data was evaluated for validity and completeness, and additional data collected as needed to fill existing gaps. This data was then used for assessing conversion technologies to identify the most promising and efficient technologies in terms of feedstock needs and cost per unit of fuel/energy production.

The methodology for identification of the type and quantity of biomass resources involved the following steps:

1. Identify all types of biomass resources in New Jersey that are suitable for bioenergy production.
2. Identify, by county, the quantity and location of the resources
3. Place these resources into categories that are identified by similar feedstock characteristics and common type of energy conversion technologies.
4. Determine factors that could reduce the availability of these resources for use as a bioenergy feedstock.
5. Create a screening function in the database to eliminate those resources or percentages of resources that are not realistically able to be used.

Public data (as of September 2006) on biomass resources for each New Jersey county (21) was collected to determine an estimated total biomass quantity for the state of New Jersey. The characteristics and quantities of forty of New Jersey's biomass resources were divided into five categories based on their physical characteristics: sugars/starches, lignocellulosic biomass, bio-oils, solid wastes, and other waste (i.e. animal waste). Sugars/starches are traditional agricultural crops and food processing residues suitable for fermentation using first generation technologies. Lignocellulosic biomass is clean, woody and herbaceous material from a variety of sources including: agricultural residues, cellulosic energy crops, food processing residues, forest residues, mill residues, urban wood waste (wood from urban forests, used pallets) and yard waste. Bio-oils are traditional edible oil crops and waste oils suitable for conversion to biodiesel, including soybean oil, used cooking oil ("yellow grease") and grease trap waste ("brown grease"). Solid wastes are primar-

County	Sugar/ Starch	Ligno	Bio-Oils	Solid Waste Recycled	Solid Waste Landfilled Biomass	C&D non-recycled	Other Wastes	Totals (Tons)
Atlantic	3,170	108,957	1,179	31,919	115,217	25,602	30,315	316,358
Bergen	4	87,455	3,779	169,401	294,436	69,209	115,775	740,060
Burlington	29,787	255,697	23,040	60,576	149,554	32,570	130,609	681,833
Camden	2,477	118,822	2,550	29,799	39,659	41,743	34,565	269,615
Cape May	831	145,752	851	24,249	42,421	24,471	8,925	247,500
Cumberland	26,681	216,226	10,823	54,495	56,829	13,574	42,461	421,088
Essex	-	37,392	3,313	76,587	87,559	71,750	40,251	316,851
Gloucester	15,206	173,089	11,462	27,420	15,704	20,022	58,327	321,229
Hudson	-	7,949	2,527	109,051	191,915	41,639	19,328	372,410
Hunterdon	25,370	138,574	5,985	11,304	42,090	56,986	31,986	312,295
Mercer	9,306	80,835	8,101	75,089	113,978	25,883	12,200	325,393
Middlesex	11,212	95,451	8,216	169,437	260,179	81,044	52,927	678,466
Monmouth	11,537	151,043	8,639	92,865	199,296	49,677	54,940	567,996
Morris	4,429	114,985	2,431	71,636	165,620	38,695	33,375	431,170
Ocean	2,239	156,619	2,833	85,768	221,097	43,008	17,981	529,543
Passaic	6	52,724	2,090	94,517	177,172	38,164	3,308	367,980
Salem	59,560	135,424	18,675	5,396	17,035	14,625	37,777	288,492
Somerset	9,267	67,465	2,282	40,404	104,843	1,482	14,546	240,289
Sussex	6,796	160,795	653	17,667	40,322	11,216	35,978	273,427
Union	5	42,242	2,225	46,261	60,536	48,164	10,022	209,455
Warren	48,006	135,236	5,014	10,588	11,150	7,822	53,302	271,117
TOTALS	265,887	2,482,731	126,666	1,304,429	2,406,613	757,346	838,899	8,182,570

©2007 New Jersey Agricultural Experiment Station

Fig. 8.1 Theoretical biomass potential

ily lignocellulosic biomass that may be contaminated or commingled with other biomass types. This category includes the biomass component of municipal solid waste (MSW), construction and demolition (C&D) wood, food wastes, non-recycled paper and recycled materials. Other Wastes are biomass wastes that are generally separate from the solid waste stream, including animal waste, wastewater treatment biogas and landfill gas. This data was entered into the biomass resource database which forms the foundation of the bioenergy calculator.

In addition, the geographic distribution of biomass feedstocks were mapped using GIS technology to facilitate analysis of the logistics and practicality of collection and aggregation. The assessment concluded that an estimated 8.2 million dry tons of biomass is produced in New Jersey annually, representing the theoretical or upper bound for biomass feedstocks available for bioenergy generation (see Fig. 8.1). The most abundant feedstocks in the state are from lignocellulosic sources and landfilled biomass. These feedstocks are also the most difficult to collect (lignocellulosic) and the most bound by political economic restraints (landfilled biomass).

8.3.2 Technology Assessment

There are many bioenergy technologies currently available or in the pipeline that are capable of converting New Jersey biomass resources into bioenergy. For the purposes of this study only biofuel and biopower were considered. Bio-heat applications were not included in the assessment since these are typically "captive" opportunities in biomass based industries like forest products, and are therefore limited in New Jersey.

Biofuel technologies are often classified as either first or second generation technologies. Ethanol and biodiesel, as examples, are considered first generation biofuels. A significant advantage of these fuels is the ease of their incorporation into existing petroleum infrastructure. Currently, they are the only renewable option for liquid transport fuels. The downside with first generation biofuels is limited scalability, economically unviable, impacts on grain supply for food, and uncertainty of indirect land use change effects. Second generation biofuel R&D efforts are focused on increasing the range of feedstocks from which to produce biofuels and reducing the biomass to liquid fuel conversion costs. Three second generation technology platforms are under development:

- Biochemical pathway: conversion of cellulose to fermentable sugars to multiple alcohol fuels
- Thermochemical pathway: conversion of biomass to syngas and synthesis to multiple fuels
- Purification of biogas (landfill gas and anaerobic digester gas) into biomethane for transportation fuels (as a compressed or liquefied gas)

These technologies require significant private and public money investment in R&D, but they offer great potential for fossil fuel displacement.

Four conversion technologies are generally considered the most appropriate for biopower applications. These are as follows:

- **Direct combustion** is the primary form of biomass utilization for power generation. It is a mature technology that is applied broadly in industrial combined heat and power (CHP) and stand-alone grid power applications.
- **Gasification** has received significant public and private sector investment and numerous technologies are commercially available. Although this technology is much less widely deployed relative to direct combustion, it is considered a major technology platform for future biopower development.
- **Pyrolysis** is less developed than either direct combustion or gasification, but is the subject of moderate technology development and commercialization activities. One company (DynaMotive) is constructing a 200 tpd power plant in Canada.
- **Anaerobic Digestion** is commonly practiced in wastewater treatment plants and increasingly on animal farms. Landfill gas is also a product of natural anaerobic digestion in landfills. Power and CHP are the most common applications.

In an effort to identify the technologies that would be most applicable to New Jersey's biomass resources, a technology evaluation analysis was conducted using the following criteria (Fig. 8.2):

- Certain technologies are not well developed yet and/or are likely to be applicable primarily to niche applications. These were excluded from detailed analysis.

R&D	Demonstration			Market Entry	Market Penetration	Market Maturity
	Initial System Prototypes		Commercial Prototypes			
• Research on component technologies • General assessment of market needs • Assess general magnitude of economics	• Integrating component technologies • Initial system prototype for debugging • Monitoring "Policy & Market developments	• Ongoing development to reduce costs or for other needed improvements • "Technology" (systems) demonstrations • Some small-scale "commercial" demonstrations	• Commercial demonstration • Full size system in commercial operating environment • Communicate program results to early adopters/ selected niches	• Commercial orders • Early movers or niche segments • Product reputation is initially established • Business concept implemented • Market support usually needed to address high cost production	• Follow-up orders based on need and product reputation • Broad(er) market penetration • Infrastructure developed • Full-scale manufacturing	• Roll-out of new models, upgrades • Increased scale drives down costs and results in learning
10+ years	4 - 8 years			1 - 3 years	5-10 years	Ongoing

Fig. 8.2 Technology commercialization timeline (the time required to pass through any given stage can vary considerably. The values shown here are representative of a technology that passes successfully from one stage to the next without setbacks)

- Though there are many biomass feedstocks that could be used with a particular conversion technology, in practice, certain feedstocks are better suited to certain conversion processes. Technologies that could more efficiently utilize New Jersey feedstocks were given priority.
- Given the wide range of technologies within a particular "platform" (e.g., types of biomass gasification reactors), the analysis focused on broad technology platforms with similar characteristics. Representative feedstock-conversion-end use pathways were selected for the economic analysis.
- Technologies were assessed on a Market Readiness scale to determine their location along the development/commercialization continuum. The scale was defined as five points (1) Research and Development stage; (2) Demonstration stage; (3) Market Entry stage; (4) Market Penetration stage; and (5) Market Maturity stage.

The evaluation process yielded thirteen technologies deemed feasible for the conversion of New Jersey's biomass feedstocks into biofuel or biopower. The selected technologies were divided into five major categories: Direct Combustion, Thermo-Chemical Conversion, Fermentation, Anaerobic Digestion and Physio-Chemical Conversion (see Fig. 8.3).

Energy generation data for the thirteen selected bioenergy technologies, which takes into consideration advances in energy output and efficiency over time, was calculated. Estimated energy potential included energy produced using current or near-term technologies appropriate for each feedstock resource. All the resource and technology data was integrated with other information (e.g. technology process efficiencies and yields). A unique bioenergy calculator was then developed to aggregate all biomass and technology information and to automatically calculate

	Core technology platforms and applications				
Application	Direct Combustion	Thermo-chemical Conversion	Fermentation	Anaerobic Digestion	Physio-chemical Conversion
Power/CHP	1.Stand-alone rankine (steam) cycle plant 2.Small-scale rankine cycle CHP plant 3.Biomass co-firing with coal	4. Stand-alone BIGCC plant 5. Small-scale gasification-IC engine CHP plant 6. Stand-alone pyrolysis plant		11.Food waste anaerobic digester with IC engine CHP plant/ Landfill gas with microturbine	
Heat Only	• Discussed qualitatively and shown in context of CHP applications above.				
Transportation Fuels		7. .Biomass-to-liquids plant (Fischer-Tropsch) 8..Dilute acid hydrolysis for biofuels production[1]	9. Corn-ethanol dry mill 10.Cellulosic ethanol plant	12. CNG or LNG from landfill gas/AD gas	13.Transester-fication Biodiesel

©2007 New Jersey Agricultural Experiment Station

Fig. 8.3 Core technology platforms and applications

energy generation potential for each county in New Jersey.[8] The database is designed to analyze the biomass resource data and technology assessment data in an interactive fashion and can be updated and modified. A screening tool embedded in the database (described in next section) allows for sensitivity analyses to be conducted on the estimates of recoverable biomass and energy potential. This also tempered estimates of energy generation potential through incorporation of existing political economic and logistical constraints.

8.3.3 Political Economic and Logistical Screening Process

Considering a variety of political economy factors to arrive at an estimate of the realistic quantity of biomass in the state available for energy conversion was an important step in the New Jersey assessment. First, all biomass currently going to the state's seven incinerators were removed from the assessment, reflecting a state policy priority to maintain the flow of feedstocks into these facilities. Keeping the incinerators functioning to the end of their useful life was a priority in terms of maintaining a viable financial model. In addition, incinerated waste is considered a Class 2 resource and the focus of state biomass policy is only on utilization of Class

---

[8]The calculator programming was developed by Navigant Consulting.

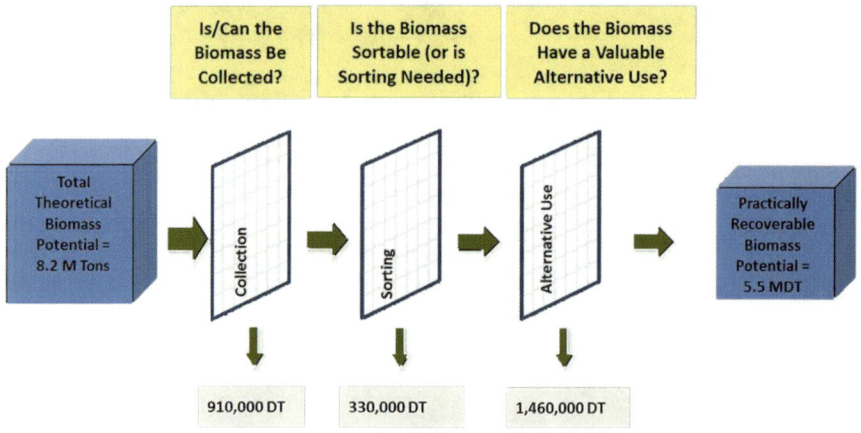

Fig. 8.4 Biomass screening process

1 resources. The total amount of waste going to incinerators per year is 688,012 dry tons and was not included in the analysis.

A screening process that considered political economy/logistics criteria for Class 1 biomass resources was incorporated to determine the proportions of each existing biomass resource that are "practically recoverable" for energy production based on several considerations (Fig. 8.4). First, *is/can the biomass be collected?* A well established collection and delivery infrastructure is in place to manage New Jersey's municipal solid waste. However, for some biomass resources (i.e., agricultural and forestry residues) new or significantly modified collection systems would be required. The development of additional waste management systems would require economic incentives to encourage adaptation of existing waste collection operations or the development of new systems. Economic incentives would also be needed to incentivize the owners of collection operations to add to or divert a portion of their fleet for these purposes. This screen removed 910,000 dry tons of biomass.

The second screening criterion focused on *is the biomass sortable (or is sorting needed)?* Further source separation policies are needed if New Jersey is to take advantage of wastes that are now not fully separated, such as food waste and C&D wood. This will require a change in policy and behavior for businesses and residents which may be difficult to achieve. The elimination of 330,000 dry tons of biomass resources due to inadequate or non-existent sorting logistics does not, of course, preclude their inclusion in the future should policy or behavioral adaptations promote needed waste separation.

The final screening criterion was *does the biomass have a valuable alternative use?* As a prime example, much of New Jersey's urban waste biomass is recycled and used in alternative markets. These markets are well established and may, at least in the present day, provide higher economic returns than would be realized if the biomass was diverted into bioenergy production. This screen resulted in the largest quantity of biomass being removed—1,460,000 dry tons.

8.4 Results

Of the estimated 8.2 million dry tons (MDT) of Class 1 biomass produced annually in New Jersey (see Table 9.1), nearly[9] three-quarters of New Jersey's biomass is produced directly by the state's population in the form of solid waste (e.g. municipal waste). The majority of New Jersey's biomass is concentrated in the counties of central and northeastern New Jersey due to the large populations in those counties. Landfill gas is considered a Class 1 resource and the major source of renewable energy in the state.

Based on the screening process described previously, of the 8.2 MDT of biomass, approximately 5.5 MDT (65 %) could ultimately be available to produce energy, in the form of power or transportation fuels. The 35 % reduction in biomass volume reflects the exclusion of materials due to political economic or logistical constraints and demonstrates the clear importance of evaluating existing infrastructure limitations, policy priorities, and logistical realities during the evaluation and formation of bioenergy production alternatives.

Energy generation estimates were developed that account for the "practically recoverable" amounts of biomass available in each feedstock category after the screening process, and capable of being converted to energy by current or near-term technologies appropriate for each resource. All resource and technology data were integrated with other information (e.g. technology process efficiencies and yields) within the Bioenergy Calculator. The calculator is designed to analyze the biomass resource and technology assessment data in an interactive fashion and can be updated and modified. The screening tool embedded in the database allows for sensitivity analyses to be conducted on the estimates of recoverable biomass and energy potential based on political economy changes or scenario alternatives.

The Bioenergy Calculator yields projected biopower and biofuel estimates for 2007, 2010, 2015, and 2020. It is currently being updated to 2025. "Typical" moisture and energy content and/or yield assumptions for each resource were developed and used to calculate dry weights needed to estimate total bioenergy potential. The results showed that New Jersey's 5.5 MDT of available Class 1 biomass resources could deliver up to 1,124 MW of power (equivalent to approximately 9 % of New Jersey's current electricity consumption) or 311 million gallons of gasoline equivalent (roughly 5 % of current transportation fuel consumed), if appropriate

---

[9] This does NOT include biomass that is currently used for incineration or sewage sludge because these are not classified as Class I renewable feedstocks in New Jersey.

policies, technologies and infrastructure were in place to produce the bioenergy. This is currently not the case, and at this time only 1.2 % of the state's energy is generated from biomass resources [22].

8.5 Policy Discussion and Conclusions

Creating an effective institutional, regulatory and feedstock supply infrastructure, as well as comprehensive strategic and tactical industry development plans is vital to the successful achievement of any state's renewable energy goals. Based on an examination of the impacts of political economic and logistical constraints in the utilization of the state's biomass resources, the State of New Jersey needs to establish the capacity and infrastructure required for rapid biofuels and biorefinery development and to create sustainable markets for bioenergy products. A key step for achieving the bioenergy goals and targets is for policy to focus on institutional infrastructure, regulations, market-based incentives and market transformation through technological innovation. Market transformation will take place once the technological and infrastructure capabilities exist and can function in an economically viable and environmentally sustainable fashion. Institutional infrastructure capable of supporting the development of a renewable energy industry in New Jersey is essential to achieve the goals of the state's energy master plan. These include the establishment/appointment of a state agency with primary responsibility for the development and support of the emerging renewable energy industry. Furthermore, policy harmonization must be facilitated across all state agencies so that the state's renewable energy goals can be successfully achieved. This effort will need to be fully integrated, include public and private partnerships, and incorporate comprehensive research, policy and marketing plans. And finally, regional partnerships with surrounding states must be built to take advantage of related programs, maximize utilization of biomass feedstock, coordinate research activities and share expertise.

Furthermore, since the assessment results showed that 75 % of New Jersey's biomass is produced directly by the state's population, in the form of solid waste (e.g. municipal waste), this leads to a recommendation that New Jersey pursue the development of energy from waste industry. Energy from waste in New Jersey is particularly attractive because waste disposal costs are high and waste collection and consolidation infrastructure is already in place. Agriculture and forestry management comprise the majority of the remaining biomass produced in New Jersey and, therefore, are also important potential energy sources. The biomass from agricultural sources includes both crops and crop residues. The use of agricultural crops for energy production would require the decision to convert the current food supply chain into energy production, which could have other major policy implications. Crop residues, however, are generally underutilized and undervalued, which should allow for an easier decision to use these resources. In the case of energy crop production, New Jersey would need to decide whether to maintain current crop varieties (i.e. corn, soybean, hay, etc.), or introduce new crops that would be better suited for energy production (e.g. Poplar or switchgrass).

The results of the New Jersey bioenergy analysis demonstrates the importance of considering the influence of political economy and logistical constraints when developing and optimizing assessment models. It illuminates potentially large biases that may result from the exclusion of these constraints, thus generating dramatically misleading results.

References

1. Birur, D.K., Hertel, T.W., Tyner, W.E.: The biofuels boom: implications for world food markets. Center for Global Trade Analysis, Department of Agricultural Economics, Purdue University, 2007
2. Brennan, M.F., Schilling, B.J., et al.: Assessment of Biomass Energy Potential in New Jersey, 2007-1. New Jersey Agricultural Experiment Station, New Brunswick (2007)
3. Capehart, T., Richardson, J.: Food Price Inflation: Causes and Impacts. Congressional Research Service, The Library of Congress, Washington, DC (2008)
4. Elobeid, A., Hart, C.: Ethanol expansion in the food versus fuel debate: how will developing countries fare? J. Agric. Food Ind. Organ. 5(2), 86 (2007)
5. Gulen, G., Shenoy, B.: Supporting biofuels: a case study on the law of unintended consequences. Working draft, center for energy economics, bureau of economic geology, Jackson School of Geosciences, University of Texas at Austin (2007)
6. Helbling, T., Mercer-Blackman, V., Cheng, K.: Commodities boom, riding a wave. Finance Dev. 45(1), 10–15 (2008)
7. Hitzhusen, F., Jeanty, P.W.: Inventory and Economic Assessment of Ohio Biomass for Energy. Renewable Energy Conference, Piketon, Ohio (2007): McKeever, D.B.: Inventories of Woody Residues and Solid Wood Waste in the United States. USDA Forest Service, Forest Products Laboratory, Madison (2002)
8. Laurance, W.F.: Switch to Corn Promotes Amazon Deforestation. Science 318:1721b (2007)
9. Lazear, E.P.: Testimony of Edward P. Lazear Chairman, Council of Economic Advisors, before the Senate Foreign Relations Committee. http://foreign.senate.gov/testimony/2008/LazearTestimony080514a.pdf (2008). Accessed 14 May 2008
10. Masters, M.W.: Testimony of Michael W. Masters, Managing Member/Portfolio Manager, Masters Capital Management, LLC, before the Committee on Homeland Security and Governmental Affairs, United States Senate. http://hsgac.senate.gov/public/files/052008Masters.pdf(accessedJuly2008) (2008). Accessed 20 May 2008
11. McKeever, D.B.: Inventories of Woody Residues and Solid Wood Waste in the United States. USDA Forest Service, Forest Products Laboratory (2002)
12. New Jersey Energy Master Plan: http://nj.gov/emp/docs/pdf/2011FinalEnergyMasterPlan.pdf (2011)
13. New Jersey Clean Energy Program: http://www.njcleanenergy.com/files/file/Library/2012%20Final%20Budget%20Order%2012-14-11.pdf (2012)
14. Pennsylvania Biomass Working Group: Pennsylvania Biomass Assessment. http://www.pabiomass.org/inventory.html (2008). Accessed 7 Nov 2008
15. Sherman, A.R., Montpelier, V.: Vermont Wood Fuel Supply Study: An Examination of the Availability and Reliability of Wood Fuel for Biomass Energy in Vermont. Biomass Energy Resource Center, Montpelier (2007)
16. Slade, R., Saunders, R., Gross, R., Bauen, A.: Energy from biomass: the size of the global resource. Imperial College Centre for Energy Policy and Technology and UK Energy Research Centre, London. www.ukerc.ac.uk/support/tiki-downloadfile.php (2011).

17. Smith, B.W., Patrick D.M., John S.V., Scott A.P.: Forest Resources of the United States. General Technical Report NC-241, Forest Service, North Central Research Station, St. Paul, MN: USDA, 2003
18. State of New Jersey: http://www.state.nj.us/infobank/circular/eojsc54.htm (2007)
19. The World Bank: Addressing the Food Crisis: The Need for Rapid and Coordinated Action, 2008
20. U.S. Department of Energy: Roadmap for Agriculture Biomass Feedstock Supply in the United States. U.S. Dept. of Energy, Washington, DC (2003)
21. United States Department of Agriculture and the United States Department of Energy: Biomass as feedstock for a bioenergy and bioproducts industry: the technical feasibility of a billion-ton annual supply. http://www.osti.gov/bridge (2005)
22. U.S. Department of Energy, Energy information administration. Available at: http://www.eia.gov/renewable/state/newjersey/ (viewed October, 2012)
23. U.S. Department of Energy, Energy Information Administration: http://www.eia.gov/renewable/state/newjersey/pdf/newjersey.pdf (2012)
24. U.S. Department of Energy, Energy Information Administration: http://www.eia.gov/renewable/state/ (2012)

Chapter 9
Modelling Decentralized Interaction in a Monopolistic Competitive Market

Juan Gabriel Brida and Nicolás Garrido

9.1 Introduction

In this paper, we assume that an economy or a market jumps through a countable infinitum space of states. Every state is identified by a tuple of observable variables. The market cannot move from a given state to any other state. Instead of this, there is a connected graph that links all the states specifying the possible changes in the observable variables that the agents of the market are able to cause within a given institutional framework.

Agents have incentives to move the economy from one state to another according to the utility or profit that they obtain in the current state. Indeed, when an agent is not in his most desirable state, he tries to make decisions that improve his *payoff*. The farthest he is from the desirable state the higher the intensity to move the market toward the desirable one. The interaction of all the incentives, within the institutional framework, makes the economy travel through the network of states, drawing different trajectories on the graph.

J.G. Brida (✉)
Competence Centre in Tourism Management and Tourism Economics (TOMTE)
School of Economics and Management, Free University of Bolzano, Via Universita 1,
I-39100 Bolzano, Italy
e-mail: JuanGabriel.Brida@unibz.it

N. Garrido
Universidad Diego Portales, Departamento de Economía, Nucleous of the Millennium Science
Initiative "Regional Science and Public Policy", Santiago, Chile
e-mail: nicogarrido@gmail.com

A.A. Pinto and D. Zilberman (eds.), *Modeling, Dynamics, Optimization and Bioeconomics I*, Springer Proceedings in Mathematics & Statistics 73, DOI 10.1007/978-3-319-04849-9_9,

The aim of this paper is double. On the one hand it describes the states in which an economy spends more time, when it is inhabited by selfish agents, having imperfect information and interacting within a monopolistic competitive market. On the other hand, it explores whether the resulting steady states change as consequence of two variations; first, a change in the firm learning algorithms and the release of the monopolistic competitive price rule, and second a different specification of the agent "intensities". The first part of the paper is studied using Markov random fields, whereas the second part, is explored with computational models.

Following [15], individual choice and the institutional constraints are integrated in order to study how the agents are channeled through the space of states. In our model, a monopolistic competitive market is the institutional framework in which agents with imperfect information and bounded procedural capacities attempt to make optimal choices.

In recent years there has been an increasing use of Markov random fields to analyze stochastic interaction in the social sciences. Some references are Aoki [2, 3], Aoki and Shirai [4], Blume [5], Blume and Durlauf [6,7], and Wolfgang [19]. The main concept is that interaction among agents produces aggregate effects, i.e. mean fields, that affect their individual behavior. Thus, there is a feedback between the individual behavior and the aggregated result obtained as consequence of the individual interaction.

The interaction using mean fields has also been combined with network theory. Dalle [9], assuming that there is a network of heterogeneous agents, shows through the use of Gibbs random fields, how, according to the network externalities among the neighbors, different landscapes of technology are found in an industry.

Instead of having a network of states, Kauffman et al. [11] introduce a technology network (i.e. technological landscape) where each node has an economic value according to the efficiency of the technology that represents. In this landscape it is modeled a technological search, driven by the decisions made for optimizing labor efficiency firms, constrained by the search cost.

The framework presented here is similar to [8] because firms employ relatively simple behavioral routines, without making conjectures about other firms and using only internally available information. Moreover, firms make decision on more than one variable; price and production capacities.

The introduction of a graph to connect the state of the economy is not a necessary formalism to solve the model. However, it is useful to makes clear that, due to the bounded rationality of the agents and physical and institutional constraints in the economy it is not possible to jump from one state to any other state.

The work is organized as follows: in the next section we develop the model explaining how firms and consumers act in the market. Afterward, using the master equation, the mean trajectory is presented as a dynamical system and the steady state solution is found. In the next section, we simulate the model with firms learning through an evolutionary algorithm. Finally, the conclusions are presented.

9.2 Model

There are N consumers and F firms buying and selling respectively an homogeneous perishable good. Every time that a consumer goes shopping he buys one unit of the good from one of the F active brands in the market.

Consumers are indistinguishable, therefore the state of the demand is characterized by the number of consumers buying from the different firms. On the other hand the state of the supply is described by two vectors; current prices and produced quantities.

Defining the vector $\mathbf{n} = \{n_0, n_1 \ldots n_F\}$, with $N = \sum_{i=0}^{F} n_i$, as the distribution of consumers buying in the different firms, the vector $\mathbf{p} = \{p_1, \ldots, p_F\}$ as the current prices in the market and $\mathbf{q} = \{q_1, \ldots, q_F\}$ as the production of the firms, it is possible to describe the state of the market through the following state vector of observable variables,

$$V = \{\mathbf{n}, \mathbf{p}, \mathbf{q}\} \tag{9.1}$$

The state vector V can take an infinite number of states, however, in reality is observed just a small sample of them. This sample of most visited states can be computed as consequence of the specification of the stochastic transitional dynamics that *move* the system through the whole space of states. This dynamic in turn, is determined as consequence of the satisfaction that the agents, either firms or consumers, have in a given state.

An agent, entrepreneur or consumer, translate his optimal bounded rational decision into an adaptive rule. The rule is adaptive because drives an agent from the current less desirable state toward his most desirable one in a finite number of steps. The rule is applied by the agent with a probability proportional to the distance between his optimal state and the current one. In other words, dissatisfied agents are trying to apply their adaptive rule to improve their state with high frequency.

Every agent attempts to reach his most desirable state. However, there is no social agreement on which is the social desirable state. Accordingly, the interaction among all the agents results in a state (or a subset of states) where the intensities of all the agents to leave it are compensated. For instance, assume that the states of a market is described by the values of one price. In this market, the "equilibrium" price, i.e. the state where the market stay locked-in, is the state where the incentives to reduce the price by the consumers and to increase the price by the firms is compensated.

In the following subsections the transitional probabilities and the adaptive rules that determines the dynamics of the demand, the price and the quantity vectors are specified. Afterwards, we identify the most visited states of the market.

9.2.1 Demand

The demand is composed by N consumers that periodically buy the unique homogeneous good available in the economy.

We can identify the consumers in two groups: consumers with and without a good. Having the ownership means that the consumer has a unit of the good produced by one of the firms in his shelf. In the other case, if the shelf is empty the good has to be replaced, so the consumer is ready to buy a unit the next time that he goes shopping.

Thus, in any instant of time the vector $\mathbf{n} = \{n_0, n_1, \ldots, n_F\}$ describes the number of consumers with ownership of a good and the number of consumers that are shopping. The former is represented by n_j, i.e. the number of consumers that have in their shelves the good from the brand j, and n_0 represents the number of consumers ready to go shopping. The vector $\mathbf{n}$ characterize the state of all the consumers in the market; in other words $N = \sum_{j=0}^{F} n_j$.

There are two different transitional dynamics to be specified on the demand side; first, how a consumer decide from which firm to buy given that they have his shelf empty, and second how a consumer "consume" the good in his shelf. In other words, it has to be explained how consumers move from n_0 to n_j and from n_j to n_0, where $j = 1..F$.

9.2.1.1 How Consumers Buy

Each consumer i has a stochastic utility function U_i defined over the good offered by the firms.

Following [1], the utility of the consumer i defined over the good of firm j is defined as,

$$U_{i,j} = u_{i,j} + \varepsilon_{i,j} \tag{9.2}$$

where $u_{i,j}$ corresponds to the deterministic utility of the individual and $\varepsilon_{i,j}$ is the portion of the individual behavior that can not be appreciated by the modeler. It could be interpreted as idiosyncratic differences within the population of consumers N.[1] Without loss of generality we assume homogeneity in the deterministic component of the utility, so it is possible to avoid the sub index i in the previous specification.

The probability that an individual chosen at random from N buys from the firm j is given by

$$P_F(j) = \Pr\left(S_j = \max_{f=1..F} S_f\right)$$

where $S_j = u_j - p_j$ is the surplus that the consumer obtains when buying from j, p_j is the price of firm j and $\Pr\left(S_j = \max_{f=1..F} S_f\right)$ represents the probability that the surplus of the firm j is the maximum surplus among all the firms in the

---

[1]Blume and Durlauf [6] attribute this randomness to some form of bounded rationality.

market. Considering that the good sold by the firms is homogeneous, it is possible to normalize the deterministic utility for all the brands, such that $u_j = 0 \; \forall j$.

Theorem 2.2 in [1], proves that the random utility with appropriate assumptions on ε_j, generates a multinomial logit model where the probability that a consumers, randomly chosen, buys from firm j at price p_j is given by

$$P_F(j) = \frac{e^{-p_j}}{\sum_{f=1}^{F} e^{-p_f}} \tag{9.3}$$

$P_F(0, j)$ defines the transition probability that a consumer buys one unit from the firm j, given the vector of prices $\mathbf{p}$.

We do not assume neither, inertial demand nor switching cost. Accordingly, for a given consumer, the probability of buying from a firm j is independent from the brand that the consumer previously had. Moreover, notice that all the firms have a positive demand. Indeed, while firms fixing low prices have high demand, there are firms with high prices having a positive demand.

Following [12] the probabilistic transition rate is the probability generated per unit of time to the occupation of a new state. The relationship between the transition rate and transition probabilities is $P(j, k) = \frac{\omega_{k,j}}{\omega_j}$, where $\omega_{k,j}$ is the transition rate from j to k and $\omega_j = \sum_{k=1}^{F} \omega_{k,j}$.

Using Eq. (9.3), the transition rate for a consumer of moving from the state of not having the ownership of a good to buy it from firm j is, $\omega_{j,0} = \upsilon e^{-p_j}$. Considering that there are n_0 agents without the good the transition rate becomes,

$$\omega_{j0}^c = \upsilon e^{-p_j} n_o \tag{9.4}$$

where υ represents the intensity in an instant of time at which the consumers n_0, decide to go shopping. To make simple the interpretation of υ it can be considered as the inverse of a measure of the period of time that a consumer wait until he goes shopping. Thus, longer the period of time, lower is the intensity of υ, therefore there is a lower probability that the consumers in n_0 will go shopping in a given instant of time. The value of ω_{j0}^c can be interpreted as the realized demand for the firm j.

9.2.1.2 How Consumers Consume

As time goes by, consumers exhaust the unit that they have in their shelf so they have to go shopping again. In other words they move from n_j to n_0 with $j = 1..F$.

It is assumed that consumers have the good in their shelves during a period of time ρ', so that $\rho = 1/\rho'$ represents the number of consumers that loose the ownership of the good in a unit of time. According to this the transition rate is given by,

$$\omega_{0j}^c = \rho n_j \tag{9.5}$$

Therefore, $\sum_{j=1}^{F} \omega_{0j}^{c}$ can be interpreted as the number of consumers that, in an instant of time, consume the good from their shelves.

9.2.2 Firms

Firms operate in the market according to price and production plans that they revise periodically. The revision are due to differences between their plans and the results obtained from the market. When the difference between plans and results is high, firms revise their plans and adopt corrective actions frequently. If the difference is negligible, firms are lazy to correct it.

Firms revise their plans asynchronically, in the sense that if firm j decide to revise its production plan, it does not mean neither, it revises its price plan nor that other firms revise their production plans.

It is also assumed that firms adjust their plans slowly, in the sense that they can change either price or production decision just through the possible path defined by local network of states; there are not shortcuts between two states that are not linked. Even though this assumption seems to be restrictive, this is not necessarily the case, because the further a firm is from its desirable state the higher the frequency with which it decides to revise its plans.

9.2.2.1 Price Dynamics

The firm's profit is given by,

$$\Pi_j = p_j q_j - c(q_j) \tag{9.6}$$

where $c(q_j)$ and $p_j q_j$ represent the cost function and the revenue for the firm j respectively. We assume that $c(0) = 0$ and $c'(\cdot) = \bar{c} > 0$.

Even though firms can survive in the market having different prices, changes on the price of a firm affect the demand of other firms. This characterize the dynamic of the monopolistic competitive market interaction. Moreover, as in the long run firms prevent new entries, every firms follows the plan of equalizing price to average cost, $p_j = \bar{c}(q_j)$.[2]

Thus the adaptive rule to update the price is: when the price is below the average cost increases it, in the opposite case reduces it. In any case the price is adjusted in one unit.[3]

---

[2]Notice that the average cost and the marginal cost in this case coincide.

[3]This assumption about the size of the price adjustment $\Delta p = 1$ does not affect the general result. The size of the jumps can be reduced below one but greater than zero without modifying the main results.

On the other hand, a firm revises its price with a frequency proportional to both: the difference between its price and average cost, and its price. A firm is reactive to change its price if it realizes that it is far from the average cost because if the price is too high, the high profits could produce new entries. If the price is below the average cost, the firm is having negative profits, therefore it has to change its price soon.

The additional awareness to revise firms' plans when prices are high capture the idea that firms having high prices have a low demand, therefore they have to be ready to change when there is a deviation from the plan very soon.

Thus, the frequency with which firm j wants to revise its plan is given by,

$$X_j = (\bar{c}_j(q_j) - p_j)p_j \tag{9.7}$$

The sign of X_j specifies the direction in which the price has to be adjusted.

Any price revision by firm j can be from p_j to $p_j + 1$ in the case that $X_j > 0$ or to $p_j - 1$ when $X_j < 0$. Formally the transition rate for the price revision is given by,

$$w^p_{j+} = \beta X_j \vartheta(X_j) \tag{9.8}$$

$$w^p_{j-} = \beta(-X_j)\vartheta(-X_j) \tag{9.9}$$

where the function $\vartheta(X)$ is 1 in case that $X_j \geq 0$, and 0 otherwise, β measure the intensity of the transition rates, according to the same interpretation of the parameters υ and ρ in the consumer analysis. Thus, if firms wait an infinitum period of time to adjust their plans, this mean $\beta = 0$. In this case, prices remain the same.

9.2.2.2 Production Dynamics

The revision in the production plans is done according to differences between the production and the demand; higher the excess or the short demand of the firm, higher will be the probability of making revisions in the production plan. Formally,

$$Z_j = (d_j - q_j)q_j \tag{9.10}$$

where d_j is the demand that the firm receives at any time, $d_j = w^c_{j0}n_0$.

As in the case of price revision, it is assumed that when the level of production is higher firms are more sensitive to make an adjustment in their production plan.

According to this, the transition rates are given by

$$w^q_{j+} = \alpha Z_j \vartheta(Z_j)$$

$$w^q_{j-} = \alpha(-Z_j)\vartheta(-Z_j)$$

where α is the intensity parameter that represents how sensitive firms are to a revision of the production plan and the function $\vartheta(X)$ is the same as before.

Assuming that β is higher than α, means that to an equal difference in the price and the production plan firms are more willing to revise their prices.

9.2.3 Most Visited Nodes of the Network of States

The transitions rates constitute the basic ingredient for the equation of probabilistic motion of the system through the network of states.

The state vector V in (9.1) characterizes the system in a specific instant of time, whereas the transition rates ω^c, ω^p and ω^q specify the change in probability that the state vector has of moving from a given state V to a neighbor V'.

A neighborhood of V in the network is defined according to the three dimensions that the state vector describes: demand, price and production.

Given an initial state $V = \{\mathbf{n}, \mathbf{p}, \mathbf{q}\}$ a demand neighborhood V^n is the state vector in which $\mathbf{n} = \{n_0 \pm 1, \ldots, n_j \mp 1\}$ with $j \in [1, F]$ and the vector describing price and production are exactly the same as in V. In other words a demand neighbor of V is the state in which one consumer either moves from having the ownership to go shopping or vice versa.

In the same way a price or production neighborhood of V is the state in which one firm increases or decreases its price or production by one unit.

Thus, the dynamics of the model, is specified in terms of the transition probabilities of the outflows and inflows transitional probabilities from the reference state V. Transitions from V toward any of the possible neighbors are outflows and the transition rates coming from the neighbors toward the reference state are inflows.

The probabilistic transitions rate between neighbor states, either outflows or inflows, capture the micro dynamic of the model and how the aggregate state affects it, i.e. mean fields. For instance the multinomial model of demand, is the mean field that aggregate the interaction of the decision of the consumers and firms. Indirectly this field will affect the production plans of the firms and the decisions that the consumers make to buy a good.

The main assumption to obtain the solution of the model is that there is a distribution that describes the probability of observing at time t the system in a given state V. Such a probability satisfies,

$$\Pr(V, t) \geq 0$$

$$\sum_{V \in S} \Pr(V, t) = 1$$

where the summation is over all the possible states in the space of state S.

The equation that represents the dynamic evolution of the probability in the system, is the master equation,[4]

$$\frac{d \Pr(V, t)}{dt} = \sum_{V'} inflows \left(V' \to V\right) - \sum_{V'} outflows \left(V \to V'\right) \qquad (9.11)$$

where V' represents all the neighbors' state of V.

Thus, the system is in the stationary state when the flow of probability of leaving is equal to the flow of probability of reaching the state V.

The stationary state solution of this equation gives information about the most probable state or subset of states. If we denote the stationary state probability as $\Pr^*$, the subset of most visited spaces V^* is the solution to $\arg \max_{V \in S} \Pr^* (V)$. This subset of states represents the demands, prices and quantities that are found with higher probability in our monopolistic competitive market.

The specification of the master equation for our model is given by:

$$\frac{\partial \Pr(V, t)}{\partial t} = \sum_{j=1}^{F} \omega_{0j}^{c} \, \Pr\left(\{(n_0 - 1, n_1, \ldots, n_j + 1, \ldots, n_F), \mathbf{p}, \mathbf{q}\}, t\right) \quad (9.12)$$

$$- \sum_{j=1}^{F} \omega_{j0}^{c} \Pr\left(\{\mathbf{n}, \mathbf{p}, \mathbf{q}\}, t\right)$$

$$+ \sum_{j=1}^{F} \omega_{j0}^{c} \Pr\left(\{(n_0 + 1, n_1, \ldots, n_j - 1, \ldots, n_F), \mathbf{p}, \mathbf{q}\}, t\right)$$

$$- \sum_{j=1}^{F} \omega_{0j}^{c} \Pr\left(\{\mathbf{n}, \mathbf{p}, \mathbf{q}\}, t\right)$$

$$+ \sum_{j=1}^{F} \omega_{j+}^{p} \Pr(\{\mathbf{n}, (p_1, \ldots, p_j - 1, \ldots, p_F), \mathbf{q}\}, t)$$

$$- \sum_{j=1}^{F} \omega_{j-}^{p} \Pr\left(\{\mathbf{n}, \mathbf{p}, \mathbf{q}\}, t\right)$$

$$+ \sum_{j=1}^{F} \omega_{j-}^{p} \Pr(\{\mathbf{n}, (p_1, \ldots, p_j + 1, \ldots, p_F), \mathbf{q}\}, t)$$

---

[4]For more details on the master equation see Chap. 10 of [19]

$$-\sum_{j=1}^{F} \omega_{j+}^{p} \Pr\left(\{\mathbf{n}, \mathbf{p}, \mathbf{q}\}, t\right)$$

$$+\sum_{j=1}^{F} \omega_{j+}^{q} \Pr(\{\mathbf{n}, \mathbf{p}, \{q_1, \ldots, q_j - 1, \ldots, q_F\}\}, t)$$

$$-\sum_{j=1}^{F} \omega_{j-}^{q} \Pr\left(\{\mathbf{n}, \mathbf{p}, \mathbf{q}\}, t\right)$$

$$+\sum_{j=1}^{F} \omega_{j-}^{q} \Pr(\{\mathbf{n}, \mathbf{p}, \{q_1, \ldots, q_j + 1, \ldots, q_F\}\}, t)$$

$$-\sum_{j=1}^{F} \omega_{j+}^{q} \Pr\left(\{\mathbf{n}, \mathbf{p}, \mathbf{q}\}, t\right)$$

The first and second rows of the differential equation represent the flow of probability between the reference state $V = \{\mathbf{n}, \mathbf{p}, \mathbf{q}\}$ and all its possible demand neighborhoods. For instance, the first term on the right side of the first row, is the inflow probability that the system moves from a demand neighbor into the state V. On the other hand, the second term represents the outflow probability that the system moves from V to a neighbor state.

The third and fourth lines of the equation represent the flow of probability between a price neighbor state V^p and the state of reference V.

Finally the fifth and sixth rows represent the flow of probability among quantity neighbor states and the reference state V.

9.2.4 Master Equation Solution

Suppose that it is possible to produce a bundle of experiments using the model previously defined with different initial conditions. Accordingly, suppose that we collect information about the evolution of V for each experiment.

The solution to (9.13) represents the mean trajectory thorough the graph of states that the market would follow over all the experiments.

We use the quasi-mean value equations as in [3, 12, 19] to describe the mean trajectory through the following dynamical system,

This approach is better suited for social systems where there is no chance of having a big number of trajectories to describe the complete behavior of the momentum of its underlying distribution. Instead of this, the researcher only dispose of one or at best of a few comparable systems to describe.

$$\frac{dn_j(t)}{dt} = \omega_{j0}^c - \omega_{0j}^c = ve^{-p_j}n_0 - \rho n_j \tag{9.13}$$

$$\frac{dn_0(t)}{dt} = \sum_{j=1}^{F} \omega_{0j}^c - \sum_{j=1}^{F} \omega_{j0}^c = -\sum_{j=1}^{F} \frac{dn_j(t)}{dt} \tag{9.14}$$

$$\frac{dp_j(t)}{dt} = \omega_{j+}^p - \omega_{j-}^p = \beta X_j \vartheta(X_j) - \beta(-X_j)\vartheta(-X_j) \tag{9.15}$$

$$\frac{dq_j(t)}{dt} = \omega_{j+}^q - \omega_{j-}^q = \alpha Z_j \vartheta(Z_j) - \alpha(-Z_j)\vartheta(-Z_j) \tag{9.16}$$

for $j = 1..F$.

Replacing the transition rate with the formula derived previously and given the complementary between the two demand equations, the system which determines the solution is given by,

$$\frac{dn_j(t)}{dt} = ve^{-p_j}n_0 - \rho n_j \tag{9.17}$$

$$\frac{dp_j(t)}{dt} = \beta(\bar{c}_j(q_j) - p_j)p_j \tag{9.18}$$

$$\frac{dq_j(t)}{dt} = \alpha(d_j - q_j)q_j \tag{9.19}$$

$$\frac{dn_0(t)}{dt} = -\sum_{j=1}^{F} \frac{dn_j(t)}{dt} \tag{9.20}$$

which is a system with $3F + 1$ equations, that for a relatively big number of firms it becomes hard to deal with.

9.2.5 Homogeneous Oligopoly Case

In order to simplify the solution of the dynamical system (9.17) we use the *slaving principle*[5] to express the system as a function of order parameters.

We assume that the total number of consumers N does not growth, the incumbents have production capacities to satisfy all consumers and that firms are capable of adjusting their prices without cost.

---

[5]The slaving principle comes from the synergetic. The idea is that in a dynamic system there are fast and slow variables. The dynamics of the fast variables is driven by the slow variables. Thus, it is possible to analyze the system assuming that the fast variables are in their steady state, and that the stable state is driven by the order parameters of the slow variables.

If firms can adjust prices and productions to the demand, the order parameter of our model is the demand. In other words, prices and productions are fast variables that firms adjust, following the slow changes in the demand. Thus,

$$0 = \beta(\bar{c}_j(q_j) - p_j)p_j$$
$$0 = \alpha(d_j - q_j)q_j$$

which means that firms set price equal to the average cost and that firms produce exactly what is demanded in the market at t. In other words,

$$p_j = \bar{c}_j(q_j)$$
$$q_j = ve^{-p_j}n_0$$

for $j = 1..F$.

Given the fixed number of consumers, the dynamic of the demand equation can be expressed as,

$$\frac{dn_j(t)}{dt} = ve^{-p_j}(N - \sum_{i=1}^{F} n_i) - \rho n_j \tag{9.21}$$

which after the normalization of $\mathbf{n}$ into $\mathbf{x}$, with $\sum_{i=0}^{F} x_i = 1$, we obtain,

$$\frac{dx_j(t)}{dt} = ve^{-p_j}(1 - \sum_{i=1}^{F} x_i) - \rho x_j \text{ for } j = 1..F \tag{9.22}$$

Thus, the system of F Eq. (9.22) represents the dynamic of the quasi-mean demand for every firm of the monopolistic competitive market of the model under the assumption that price and production decisions are variables that follow the demand.

In steady state (i.e. $\frac{dx_i(t)}{dt} = \dot{x}_i - 0$) the solution to the system is given by,

$$x_j = \frac{e^{-p_j}}{\frac{\rho}{v} + \sum_{i=1}^{F} e^{-p_i}} \text{ for } j = 1..F \tag{9.23}$$

where price is equal to average cost and ρ and v represent the reciprocal of the periods that consumers have the good in their shelves and the periods that they wait before of going shopping once they exhausted the good in their shelves.[6]

---

[6]**Stability Analysis**: We assume that the average cost of every firm is constant. Thus, the Jacobian of the system (9.22) is given by

From Eq. (9.23) it is possible to compute the proportion of consumers that are shopping in any period of time,

$$x_0 = \frac{\rho/v}{\rho/v + \sum_{i=1}^{F} e^{-p_i}} \tag{9.27}$$

The proportion of consumers shopping increases with any of these three factors: (1) the time that the consumers have the good in their stack is short, so the parameter ρ is high, (2) the time that the consumers spend shopping is long, so that v is low, and (3) when the prices are high.

If there is no market, i.e. $F = 0$, no consumer has a good $x_0 = 1$. In the competitive market, where the number of firms setting the market price is infinitum all the consumers has a good in their stack, $x_0 = 0$.

Equation (9.27) can be used to explain why there is a big number of retailers and few shops selling durable goods. Food is consumed relatively fast, so ρ is high, therefore the population of consumers shopping is high. On the other hand durable goods are consumed during a long period of time, therefore all else equal, there are few consumers buying durable goods. However, the durable case has a caveat: when the cost of producing a particular durable good is reduced, its price fall, therefore the proportion of consumers shopping increases. Thus, the number of shops selling the durable good with low price increase as well.

The solution obtained in (9.23) differs from the standard oligopoly solutions á la Bertrand and Cournot, because in steady state prices and production can be different. The asymmetry comes from an unobserved consumer component in his utility function. This stochastic component, specified in (9.2), represents an idiosyncratic bias in the consumer utility function.

Our model differ from spatial models of product differentiation as [16] because the interaction between firms is not local. In our model, the interaction is among all the firms in the market. Our model is close to the monopolistic competitive literature as in [10]. The interaction is global and there is a low level of strategic interaction,

$$\begin{bmatrix} \psi_1 - \rho & \psi_2 & \cdots & \psi_F \\ \psi_1 & \psi_2 - \rho & \cdots & \psi_F \\ \cdots & \cdots & \cdots & \cdots \\ \psi_1 & \psi_2 & \cdots & \psi_F - \rho \end{bmatrix} \tag{9.24}$$

where $\psi_i = -ve^{-p_i}$. Notice that when firms have the same average cost, this means that all the ψ_i are equal. The generic solution of the characteristic polynomial of this matrix is given by

$$\lambda_i = -\rho \ i = 1..F - 1 \tag{9.25}$$

$$\lambda_F = -\sum_{i=1}^{F} \psi_i - \rho \tag{9.26}$$

Notice that in any case the first $F - 1$ roots are real and negatives, therefore the system is asymptotically stable.

"in that the strategies of any particular firm do not affect the payoff of any other firm" [14, p. 400].

The model of this paper is close to models of bounded rationality in [17], because all the agents use private information. Moreover, the agents do not use any "proxy" of the market behavior, as the mean demand or the mean price. In all the cases, agents use adaptive rules, and the asymmetric equilibrium is explained as consequence of the interaction among all the agents.

9.2.6 Heterogeneous Duopoly

In this section ρ is assumed to be a proxy of the quality of the goods produced by the firms. We assume that high quality goods stay longer in consumer's shelves than low quality goods.

Assume that in the economy there are two firms producing a good with different quality. The good of firm 1 has higher quality than the good of firm 2, i.e. $\rho_1 < \rho_2$.

With these assumptions and using the slave principle, the solution to (9.17) is given by.

$$x_1 = \frac{\rho_2 e^{-p_1}}{\frac{\rho_1 \rho_2}{v} + \rho_1 e^{-p_2} + \rho_2 e^{-p_1}}$$

$$x_2 = \frac{\rho_1 e^{-p_2}}{\frac{\rho_1 \rho_2}{v} + \rho_1 e^{-p_2} + \rho_2 e^{-p_1}}$$

$$x_0 = \frac{\frac{\rho_1 \rho_2}{v}}{\frac{\rho_1 \rho_2}{v} + \rho_1 e^{-p_2} + \rho_2 e^{-p_1}}$$

Assuming that firms do not modify their prices when the quality changes, and setting $v = 1$, we can compute how the shares changes as the quality of the good of firm 1 improve, i.e. ρ_1 falls.

$$\frac{dx_1}{d\rho_1} = -\frac{\rho_2 e^{-p_1} (e^{-p_2} + \rho_2)}{(\rho_1 \rho_2 + \rho_2 e^{-p_1} + \rho_1 e^{-p_2})^2} < 0$$

$$\frac{dx_2}{d\rho_1} = \frac{\rho_2 e^{-(p_1+p_2)}}{(\rho_1 \rho_2 + \rho_2 e^{-p_1} + \rho_1 e^{-p_2})^2} > 0$$

$$\frac{dx_0}{d\rho_1} = \frac{\rho_2^2 e^{-p_2}}{(\rho_1 \rho_2 + \rho_2 e^{-p_1} + \rho_1 e^{-p_2})^2} > 0$$

An improve on the quality of the good produced by firm 1 increases the market share of firm 1, reduces the share of firm 2 and the proportion of consumers shopping x_0. Moreover, notice that $\frac{dx_0}{d\rho_1} + \frac{dx_1}{d\rho_1} + \frac{dx_2}{d\rho_1} = 0$.

9.3 Computational Agent Model

The model of the previous sections, shows the aggregated macro behavior that results as consequence of the probabilistic micro behavior of the agents within a given institutional settings. Two characteristics of our institutional setting identify it with a monopolistic competitive market: first, the interaction is global with a low level of strategic interaction and second, in order to prevent new entries firms set price equal to average cost.

In this section we construct a model with computational agents of our previous specification.

As in the analytical model, each of the N customers buys and consumes a homogeneous good sold by one of the F firms. Every firm i chooses price and quantity to maximize its profit, $\pi_i^e = p_i q_i^e - C(q_i^e)$, where q_i^e is the expected demand for the firms and the cost function is $C(q) = cq$.

The dynamic of the system is represented by a set of events that modify the state of the system. The system change when either of the following four types of events happen; (1) a consumer goes shopping, (2) a firm revises price, (3) a firm revises the produced quantity, or (4) a customer consumes the good that he bought previously.

In every period t, a number E_t of events may occur, with $0 \leq E_t \leq N+2F$. If the system does not change in a period t, it means that $E_t = 0$. On the other extreme, the maximum number of events occurs in the unlikely case where all these events happen in a period t: each of the $n_0(t)$ consumers buys a good from one of the firms, each of the $N - n_0(t)$ consumers consumes the good that they bought in previous periods, the F firms revise their prices, and the F firms revise their production plans. Notice that the sum of all these events is $E_t = n_0(t) + (N - n_0(t)) + F + F = N + 2F$.

In every event an agent, consumer of firm, executes an action or change his state. Events are asynchronics because the agents are not coordinated to make decisions. For instance, when a firm revises its price it does not mean that any other firm has to revise its price.

Thus, in the specification of the model to be simulated the following two characteristics have to be described; first, when the events of making an action is triggered and second, how the agents execute the specific action.

9.3.1 Consumers

Consumers may be in two different states during a simulation; *consuming* the good that they have in their shelves or *shopping* a unit of good because they do not have the good anymore.

9.3.1.1 Consuming

The good sold by the firms is perishable. In fact, once a costumer i acquire a good he spends a period of time τ_i^c consuming it. Every consumer takes a different period of time to consume its good. This depends on a set of unobservable, for the modeler, characteristics.

We assume that the periods to consume a good is independently among the consumers and that in average a consumer spend $1/\hat{\rho}$ periods of time with a good. In order to simulate this behavior we use two variations: first, every time that a consumer i buys a good we draw the value of τ_i^c from an exponential distribution with mean $1/\hat{\rho}$. Thus, the consumer i moves to the group of consumers n_0 after τ_i^c periods. The second strategy is that in every period, the consumers in n_j move to n_0 with probability $\hat{\rho}$.

When a consumer having a good from firm j consumes his good, he becomes a consumer ready to go shopping again. This means, that the number n_0 of customers ready to go shopping is increased by one unit and the number of customers from a firm j is reduced by one unit.

9.3.1.2 Shopping

The consumer search in the market before deciding from which firm to buy. Indeed, we assume that the agent i spends τ_i^s periods of time searching before of buying the good. We assume that the agents spend in average $1/\hat{v}$ periods of time searching. We use two strategies to modelling this, according to the strategy follows to model the consuming time. With the first strategy, the amount of periods τ_i^s for the agent i is drawn from an exponential distribution with mean $1/\hat{v}$. With the second strategy the value of $\hat{v}$ is used to compute the probability that every agent from n_0 stop searching and decide to buy.

After the searching process elapses, a consumer buys from firm j with probability $\Pr(j) = \frac{e^{-p_j}}{\sum_{i=1}^{F} e^{-p_i}}$, where p_j is the price of firm j. We rationalize this decision as in Sect. 9.2.1.1

The parameters $\rho, v, \hat{\rho}$ and $\hat{v}$ capture the flow intensity of the transitions from consuming to shopping and vice versa. In the analytical model the influence is through the intensity of the transition rates. In the computational model we explore two different alternatives to model the intensity. In the first one, we use the parameters with the same effect of the analytical model: intensity of the transition rate.

However, transition rates is an infinitesimal concept hard to be measured in reality. Thus, we use the computational model to explore whether the predicted results of the analytical model change with a different specification of the flow intensity of the transitions from consuming to shopping and viceversa. In the second strategy we specifically model the time. We assume that the parameters are mean times of a distribution. In particular we use a negative exponential distribution[7] to model the time that every consumer spend consuming and shopping.

9.3.2 Firms

We assume that every firm follows a price and produces a quantity that is fixed in its plan. A firm revise its plan according to its performance. Indeed, plans with low performance are revised frequently while good plans are revised with less frequency.

We explore two specifications about how firms revise their plans; in the benchmark case, every firm has a plan for the price and a plan for the quantity whereas in the evolutionary model, firms have a single plan for both, price and quantity.

In the benchmark case firms set a plan and they adapt their current price or quantity to their plan using adaptive rules. In the evolutionary case firms use an evolutionary algorithm to select the best plan.

In both cases firms attempt to maximize their profit.

9.3.2.1 Production Decision in the Benchmark Case

In this case, a firm has more incentive to adjust its production plan if it is far from the demand that it received during the current period. Thus, the probability that a firm i revises its production plan at the end of the period t increases according to $|\beta(d_{i,t} - q_{i,t})q_{i,t}|$, where $d_{i,t}$ represents the current demand, $q_{i,t}$ is the quantity produced according to the last production plan, β is a normalization parameter and $|\cdot|$ represent the absolute value of $\cdot$.

Firms cannot adjust their production instantaneously. When a firm revises his production plan, it adjust the production on one unit according to whether it receives an excess or not of demand.

In other words, firms will adjust the production according to the following adaptive rule,

$$q_{i,t+1} = q_{i,t} + sign(d_{i,t} - q_{t,i}) \tag{9.28}$$

where $sign(.)$ is the sign function.[8]

---

[7]The negative exponential distribution applies frequently in the simulation of the interarrival and interdeparture times of service facilities. For more details see [13].

[8]The sign function is defined as,

$$sign(x) = \begin{cases} 1 & x > 0 \\ 0 & x = 0 \\ -1 & x < 1 \end{cases} \tag{9.29}$$

Notice that if the current production is far from the current demand the adjustment on q_i are frequent. The distance is measured by the normalization parameter β. If $|\beta(d_{i,t} - q_{i,t})q_{i,t}| \geq 1$ the production is adjusted in every period.

9.3.2.2 Price Adjustment in the Benchmark Case

Following the setting of the analytical model, in the benchmark case firms set price equal to average cost.

The further the current price is from the current average cost, the higher will be the incentives to adjust the price. Thus a firm j revises its price, at the end of period t, with a probability given by $|\alpha(\bar{c}(d_{j,t}) - p_{j,t})p_{j,t}|$ where $d_{j,t}$ is the demand in the current period, $p_{j,t}$ is the price of the current period, $\bar{c}(d_j)$ is the average cost and α is a normalization parameter.

Firms adjust prices according to the following adaptive rule,

$$p_{j,t+1} = p_{j,t} + sign(\bar{c}(d_j) - p_j) \tag{9.30}$$

9.3.2.3 Evolutionary Firms: Production and Price Decision Made Simultaneously

In this case, we assume that firms revise their prices and productions plans every Λ periods of time in order to obtain the maximum profit,

$$\pi_j = min(d_j, q_j)p_j - C(q_j) \tag{9.31}$$

where the $min(d_j, q_j)$ means the minimum between the demand and the production.

Every firm has a set of plans L. Every plan has a price and a production decision. A firm explore different plans until it manages to obtain the most profitable one. Every plan produces a profit, as consequence of interacting with the plans of all the other firms.

To obtain the best plan that gives the maximum profit every firm *evolve* its set of plans, discarding low performance plans and using good performance plans to generate new ones. The performance of a plan is measured by its profit.

Every firm apply every Λ periods, three genetic operators over its set of plans; selection, crossover and mutation.[9] Selection selects the best plans, and crossover and mutation generate a set of new one.

Using the evolutionary algorithm to simulate how firms make their price and quantity decisions, introduces two modifications in the original assumptions of the analytical model: on the one hand it changes the structure of neighborhood of one

---

[9]See [18] for more details on Genetic Algorithms.

states; a firm can change its price or its production in more than one unit. The neighborhood of one state is now is defined by the mixing matrices in Vose's model of simple genetic algorithms (see [18]).[10] The second deviation from the original assumption is that firms do not follow the rule of price equal to average cost.

With the goal of observing the consequences of this modification of assumptions in the next section we simulate our computational specifications.

9.3.3 Experiments

We simulate the computational model using the following base line parameters: there are $N = 1,000$ consumers and $F = 10$ firms interacting during $T = 5,000$ periods. The 5,000 periods are in all the cases enough to reach in every case the steady state.

The normalization parameters for the price and production revisions are $\alpha = \beta = 0.2$. This means that if the difference between the price and the average cost, or between demand and production for a firm is equal or greater than 5, the firm revises his price or production plan with probability 1.

For the evolutionary firms we assume that every firm has a set of $L = 30$ plans. Every $\Lambda = 30$ periods each firm evolve its set of plans.[11]

In order to explore the effect of heterogeneity in the cost function, two cases are studied: the homogeneous case, as a benchmark where all the firms have the same cost function and the heterogeneous case where there are two groups of firms differentiated by their cost function.

In the homogeneous case all the firms have the same cost function $C(q) = 2.5q$. In the heterogeneous case half the firms, type 1, have the cost function $C(q) = 2.5q$ and half, type 2, $C(q) = 3q$.

The two different cost structures together with the two different model of firms, produce four simulations to be tested,

1. Benchmark firms with homogeneous in their cost function.
2. Benchmark firms with heterogeneous cost function.
3. Evolutionary firms with homogeneous cost function.
4. Evolutionary firms with heterogeneous cost function.

According to the analytical results obtained in Sect. 9.2.3, and using the parameters described in this section, the steady state of the system (i.e. the most probable result observed) for each case is given in Table 9.1. Thus, for instance when all the firms have the same cost function, each firm will have 7.6 % of the market, and it will price the good to 2.5 monetary units.

---

[10]The mixing matrices define all the possible states that can be visited as consequence of applying the mutation and crossover operator to the active population of plans that the firms have.

[11]We select 25 % of the plans using roulette. Moreover mutation is applied with 0.01 probability.

Table 9.1 Proportion of
agents consuming a given
brand and prices according to
the analytical model

	Homogeneous	Heterogeneous	
		Type 1	Type 2
Quantity	7.6 %	9.02 %	5.4 %
Price	2.5	2.5	3

These analytical results are useful as benchmarks to compare the simulated experiments.

For each of the four experiments we run 250 simulations. Each of the 1,000 simulations is started with a different initial conditions. Thus, at time $t = 0$ consumers are distributed among firms uniformly and initial prices are set equal to $c_i (1 + e_i)$ with c_i the average cost of firm i and e an independent draw from a normal distribution $N (0, 1)$.

Consumers spend $1/\hat{\rho} = 40$ periods consuming a unit of the good and they spend searching for a new unit $1/\hat{\upsilon} = 10$ periods. In Sect. 9.3.1 we specify two strategies to model the transitions from shopping to consuming and vice versa: the transition rate probability and the modelling of time strategy respectively. We explore the results using both strategies and we do not distinguish differences in the outcomes. Therefore, in order to reduce the size of the paper, we present only results based on the modelling of time strategy. In other words, $\hat{\rho}$ and $\hat{\upsilon}$ are parameters of a negative exponential distribution. We simulate the periods of time that a consumer spend shopping or consuming through a random draw from a negative exponential distribution with parameters $\hat{\upsilon}$ and $\hat{\rho}$ respectively.

In the first row of Table 9.2 we plot the mean quantity in the period 5,000 for the benchmark.[12] The first bar is the proportion $\bar{x}_0$. of consumers shopping. The next ten bars represent the proportion of consumers that have in their shelves a good of one of the firms. Notice that for the benchmark with homogeneous and heterogeneous costs the results are close to the analytical one. The small fluctuations are consequence of the stochasticity of the model.

In Table 9.3 we plot the average price for the benchmark case in the first row again. In both cases, homogeneous and heterogeneous cost structure, the average cost is reached with high precision.

The results for the evolutionary case are reported in the second row of Tables 9.2 and 9.3. In this case the outcomes require an additional analysis. Although firms use a completely diverse learning mechanism and set simultaneously price and quantity, the quantities of steady state are close to the benchmark, but prices are different.

There are two differences between the prices in the benchmark and the evolutionary case: first the prices are higher than the average costs and second, the prices have more variability.

---

[12]Each bar in the plot represents the mean of 250 simulations. For instance, in the benchmark with homogenous cost the first bar is $\bar{x}_0 = \frac{1}{250} \sum_{j=1}^{250} x_{0,5000,j}$. Where the subindex j counts the simulations.

Table 9.2 Proportion of agents shopping and consuming

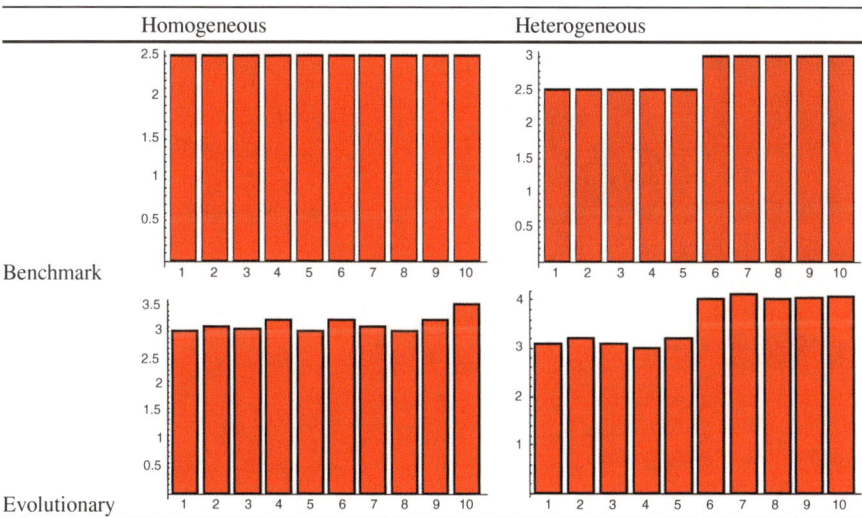

Table 9.3 Firm's prices

The additional variations is explained because in the optimization algorithm there is a population of firms exploring different plans. This process of searching converge to select the best plans. Although after 5,000 periods there are few plans left to be explored and the best plans have been selected, the mutations from the genetic algorithm keeps introducing new combinations of price and quantity. Thus, this is translated into the observed variation.

The intuition is that even when the market is in equilibrium, there are entrepreneurs searching for new plans. This exploration introduces the additional variation.

The increase in the prices is explained by the (low) strategic interaction between the firms. In the evolutionary case, firms do not follow the rule of price equal to average cost. In this case there is a fix number of firms maximizing their profits. The demand is not infinitely inelastic. Instead, every firm has a "local" elasticity consequence of the global interaction and of the assumption that consumers do not switch to the firm offering the lowest price.

Indeed, in the evolutionary case the maximization of profit for every firm is when they set their marginal revenue equal to their marginal cost. Using our cost function we have,

$$p_i \left(1 + \frac{1}{\varepsilon_i} \right) = c_i \text{ for } i=1..F \tag{9.32}$$

where p_i is the price, ε_i is the price elasticity and c_i is the marginal cost for firm i.

We use the analytical model to compute the elasticity $\varepsilon_i = -p_i \left(1 - \frac{ve^{-p_i}}{\rho + v \sum_{j=1}^{F} e^{-p_j}} \right)$.

We are not able to provide a close form for the F prices that solve (9.32). However, using the same parameters of the computational model we obtain numerical solutions for the prices.

When the cost structure is the same for all the firms, the homogeneous case, firms set price $p_i = 3.51$. The mean of prices obtained from the simulations is $\bar{p} = 3.39$.

When the cost functions are heterogeneous, the price for the firms with low cost is $p_i = 3.51$, whereas the price for the firms of high cost is $p_i = 4.01$. The mean prices that we obtain for every group with the evolutionary firms, are $\beta p = 3.24$ and $\bar{p} = 3.93$ for the low and high cost respectively.

Notice that the price values are close in every case. The small difference between the theoretical values and the simulated ones are due to the variation introduced by the optimization algorithm.

9.4   Conclusion

The aim of this paper is double: first, we present a model where consumers and firms interact according to their bounded rational adaptive rules and to the constrain established by an institution. We aim to capture the massive interaction that is present in the market through an stochastic model. The second goal of the paper is to construct a computational model with the same assumptions of the analytical one. This comparison of methods to modelling the same problem allows us to understand better some features of the analytical model.

We learnt from the analytical model that the time that consumers spend shopping and consuming is helpful to approximate the population of consumers shopping.

When the goods are homogeneous, either one of this two effects: short period of time that consumers spend consuming a good or the long periods that they spend searching (shopping) for a good produce the same result: the proportion of consumers shopping increases. Moreover, if the prices of the firms are raised, all else equal, there will be a higher proportion of consumer shopping. The price effect is explained by the consumers gathering more information about the goods when prices are high.

Using the analytical model we also describe the effect of increasing the quality of the good of a firm. Thus, in a duopoly, if a firm increases the quality of its good there is an increase in the proportion of consumers buying it. The increase is explained with both, a reduction in the proportion of consumers buying the other good, and a reduction in the proportion of consumers doing shopping.

From the computational models we learnt two things. First, the result of the analytical model is conditioned by the price equal to average cost rule. This rule is exogenously introduced in the analytical model with the assumption that firms do it to prevents the entry of new firms. In the simulated model, when we released that institutional constraint, the results change. This time, as the number of firms is fixed and each firm optimizes they set prices higher than the average cost. We shown that as consequence of introducing evolutionary firms the outcomes of the simulation are close to an equilibrium where all the firms set their marginal revenue equal to marginal cost.

The second thing that we learnt from the implementation of the computational model is about the interpretations of the parameters ρ and ν. Using the computational models, we validate the interpretation of those parameters as periods of time that consumers spend either consuming or shopping. Moreover, we model the time evolution as random draws from a negative exponential distribution, using ρ and ν as parameters. This last result, within the assumptions of our analytical model, allows us to estimate the proportion of customers doing shopping, consuming and buying using market prices and the mean period of time that a consumer spend consuming and shopping.

References

1. Anderson, S., De Palma, A., Thisse, J.-F.: Discrete Choice Theory of Product Differentiation. MIT Press, Cambridge (1992)
2. Aoki, M.: Simple model of asymmetrical business cycles: interactive dynamics of a large number of agents with discrete choice. Macroeconomic Dyn. **2**, 427–442 (1998)
3. Aoki, M.: Cluster size distributions of economic agents of many types in a market. J. Math. Anal. Appl. **249**, 35–52 (2000)
4. Aoki, M., Shirai, Y.: A new look at the Diamond search model: stochastic cycles and equilibrium selection in search equilibrium. Macroeconomic Dyn. **4**, 487–505 (2000)
5. Blume, L.: Population games. In: Arthur, W.B., Durlauf, S., Lane, D. (eds.) The Economy as a Complex Evolving System II. Addison-Wesley, Redding (1995)
6. Blume, L., Durlauf, S.: Equilibrium concepts for social interaction models. Int. Game Theory Rev. **5**(3), 193–209 (2002)

7. Blume, L., Durlauf, S.: The interactions-based approach to socioeconomic behavior. In: Durlauf, S., Young, H.P. Social Dynamics. MIT/Brooking Institution Press, Washington, DC/Cambridge (2000)
8. Currie, M., Metcalfe, S.: Firms routines, customer switching and market selection under duopoly. J. Evol. Econ. **11**, 433–456 (2001)
9. Dalle, J.-M.: Heterogeneity vs. externalities in technological competition: a tale of possible technological landscape. J. Evol. Econ. **7**, 395–413 (1997)
10. Hart, O.D.: Monopolistic competition in the spirit of Chamberlin: a general model. Rev. Econ. Stud. **52**, 529–46 (1985)
11. Kauffman, S., Lobo, J., Macready, W.: Optimal search on a technology landscape. J. Econ. Behav. Organ. **43**, 141–166 (2000)
12. Kelly, F.P.: Reversibility and Stochastic Networks. Wiley, New York (1979)
13. Law, A., Kelton, W.: Simulation Modeling and Analysis. McGraw-Hill Book Company, New York (1982)
14. Mas Colell, A., Whinston, M., Green, J.: Microeconomic Theory. Oxford University Press, New York (1995)
15. North, D.: Institutions, Institutional Change and Economic Performance. Cambridge University Press, Cambridge (1990)
16. Salop, S.: Monopolistic competition with outside goods. Bell J. Econ. **10**, 141–156 (1979)
17. Sargent, T.: Bounded Rationality in Macroeconomics. Oxford University Press, Oxford (1993)
18. Vose, M.: The Simple Genetic Algorithm: Foundation and Theory. MIT Press, Cambridge (1999)
19. Wolfgang, W.: Sociodynamics: A Systematic Approach to Mathematical Modeling in the Social Sciences. Harwood Academic Publishers, Amsterdam (2000)

Chapter 10
Measuring the Effectiveness of an E-Commerce Site Through Web and Sales Activity

Ana Ribeiro Carneiro, Alípio Mário Jorge, Pedro Quelhas Brito, and Marcos Aurélio Domingues

10.1 Introduction

There are currently several ways of measuring the success of e-commerce sites. Many companies specialize in analyzing e-commerce and each of them uses different methods and techniques to solve the same problem: Is this Website successful? Why or why not? How do you make it successful?

In this chapter a case study is presented where the effectiveness of an e-commerce site is studied using Web server log file (Web access logs) data and commercial data (sales figures). The Web data and selling data are combined to determine the Website's ability to achieve the goal for which it was created: to sell. For this we will use clustering techniques and values obtained for site success metrics by adapting the work developed by Spiliopoulou [10]. The final goal is to provide the company under analysis with objective information that can be used to improve the site, raise marketer's awareness of these factors and increase sales. This process combines technical Web data analysis and marketing analysis.

A.R. Carneiro
FEP, Universidade do Porto, Porto, Portugal
e-mail: rianabeiro@hotmail.com

A.M. Jorge (✉)
LIAAD-INESC TEC, FCUP, Universidade do Porto, Porto, Portugal
e-mail: amjorge@fc.up.pt

P.Q. Brito
LIAAD-INESC TEC, FEP, Universidade do Porto, Porto, Portugal
e-mail: pbrito@fep.up.pt

M.A. Domingues
University of São Paulo, São Paulo, Brazil
e-mail: mmad@icmc.usp.br

A.A. Pinto and D. Zilberman (eds.), *Modeling, Dynamics, Optimization and Bioeconomics I*, Springer Proceedings in Mathematics & Statistics 73, DOI 10.1007/978-3-319-04849-9__10,
© Springer International Publishing Switzerland 2014

Firstly theoretical concepts, such as the concept of e-loyalty and how to measure the success of a site are presented. This is followed by an introduction to the problem and the company behind this case study. How the company's Website's buying process is organized is analyzed along with how the Web server log data are pre-processed and stored in a data warehouse. The Web server log results are measured, the selling results are evaluated and a clustering analysis is performed. The Website success is then measured for each of the clusters discovered. Finally, the main conclusions for this case study are presented along with proposals for future work.

10.2   Electronic Commerce on the Web

Electronic commerce may be defined as the process of buying, selling, or exchanging products, services, or information via computer networks [13]. E-commerce has definitely been changing the world in terms of the way people interact and schedule their time, in the way companies reorganize their selling processes and human resources, in the way governments relate to the people, to companies and to other countries. The economy, markets, society, the labor market and industry have all been and still are being shaken by e-commerce.

10.2.1   Electronic Commerce from a Marketing Perspective

> "The ability to track a user's browsing behavior down to the individual mouse clicks has brought the vendor and end customer closer than ever before. It is now possible for a vendor to personalize their product message for individual customers on a massive scale; a phenomenon that is being referred to as mass customization" [11].

Electronic commerce can be a huge source of income. Analyzing the data generated by Web usage in electronic commerce sites provides important added value. Web usage mining [4, 10, 11] can contribute to optimizing a company's Websites, meeting the needs of users/customers and accomplishing the aims of the owners of the sites. For marketers, it is clear that they are most concerned with the return on investment. For Travis [12], approaching the World Wide Web in the right way brings considerable advantages; "There are four key benefits from a customer-centred approach: higher revenues, loyal customers, improved brand volume and process improvement."

10.2.2   The Customer and E-Loyalty

Understanding consumer behavior on the World Wide Web is essential for the success of a business. With this knowledge, marketers will be able to respond to their consumers' needs on time. After all, competitors are just one click away. Moreover, customers' tolerance of inconsistency and mediocrity is rapidly disappearing [8].

Turban et al. [13] define e-loyalty as the customer's loyalty to an entity that sells online, be it apparel, music, books or any kind of service. Customer acquisition and retention is a critical success factor in e-commerce. Liao et al. [6] warn that the maturity and low cost of technology in this business lowers the entrance threshold for new competitors. Furthermore, the transparency of information makes the business model easy for competitors to mimic.

After acquiring a new customer how can we gain his/her loyalty? "To gain the loyalty of customers, you must first gain their trust" [8]. According to Liao et al. [6] in order to maintain customers' trust, you have to constantly improve the usability of your Website. The Website is the only way the customer has of getting to know its supplier. Therefore, the sites' usability is of extreme importance in terms of deciding whether to trust the supplier or not.

The basis for loyalty is not technological; loyalty is based on old-fashioned customer service basics like: quality customer support, on-time delivery, compelling product presentations, convenient and reasonably priced shipping and handling and clear and trustworthy privacy policies. What is actually changing is the rhythm at which economies are played out and the need for speed in improving products and services [8]. Companies must constantly deliver a total customer experience to their customers. Seybold et al. [9] define a total customer experience as: "A consistent representation and flawless execution, across distribution channels and interaction touchpoints, of the emotional connection and relationship you want your customers to have with your brand."

When customers identify themselves with a brand it is more likely that they become and remain loyal to it. Every time they come into contact with the brand, if they have a positive experience they are reinforcing that loyalty [9].

A company must define what kind of customers they are willing to attract because the site design strongly defines this. When defining what kind of customers to attract and which ones to avoid, the company must be aware of the different categories of on-line customers. There are types of loyalty-oriented customers and types of customers who flit—like butterflies—from site to site seeking bargains [8].

For companies that use both channels for business (the traditional and the Web) it is important to balance both in terms of human resources. A company should not view the Web channel as a mere way to reduce costs by bypassing its commissioned sales force. Reichheld and Schefter [8] give the example of a successful company that seamlessly integrates its Web channel with its traditional channel. This company pays sales commissions independently of the channel which was used to sell, because this means that the sales representatives direct customers to the most convenient channel. Windham and Orton [14] conclude that if Web retailers wish to establish Web brand loyalty and remain competitive, they must provide the components that create a good consumer experience.

10.3 How to Measure the Success of a Site?

"For a business deploying a personalization system, accuracy of the system will be little solace if it does not translate into an increase in quantitative business metrics such as profits or qualitative metrics such as customer loyalty" [1].

Website usage has been monitored one way or the other by examining Web server logs. A large number of success measures have been tried and developed since the first Websites were created and many programs are available for this end. Berthon et al. [2] modeled the flow of surfer activity on a Website as a six-stage process; this process has six indexes that measure: the awareness efficiency, the locability/attractability efficiency, the contact efficiency, the conversion efficiency and the retention efficiency. In 2001, Spiliopoulou and Pohle [10] used two of Berthon's measures of a site's success: the contact efficiency and the conversion efficiency. The first measure assesses how effectively the organization transforms Website hits into visits, which means the number of users that spent at least a user-defined minimum amount of time exploring the site. The second measures the ability to turn visitors into purchasers, which is measured by the ratio of users that after exploring the site also made a purchase. In this study, Spiliopoulou and Pohle [10] defined their goal as being not only to measure but also to improve a Website's success. Mafe and Navarre [7] have made further developments by defining an online consumer typology. This is made by segmenting consumers according to their Web behavior and Web purchases. Their study intends to make the Website's marketing actions more profitable and obtain a competitive advantage.

10.4 The B2B Portal Case

Starting with the measures mentioned above, the online activity of the e-commerce site owned by INTROduxi, a Portuguese company (www.introduxi.pt) was analyzed with the aim of identifying its strengths and weakness and taking into account its customers' profiles. The core business of this Website is selling hardware and software products to retailers. It is a business to business Website. In order to be a registered user of this Website, the client must be a company with an activity related to selling computing material or providing assistance in the area. The company has a policy of selling through the online channel as well as through the call centre placed at the headquarters (this channel will be called the "classic channel"). As a consequence, the Website not only aims to offer electronic commerce but it also makes it possible for registered users to check which products exist and if they are available, using the call centre afterwards. As a matter of fact, there are some products for which the price is not available even when added to the shopping cart. In these situations the user is invited to contact their account manager at the call centre.

INTROduxi was formed in 1995 and is now one of the main players in the Portuguese computer market. The company aims to make their Website more successful with respect to either direct sales (selling through the Web channel)

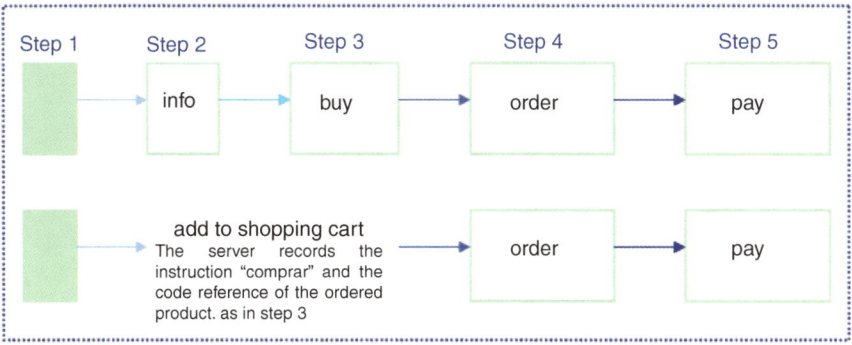

Fig. 10.1 The two different ways of completing a buying process on the INTROduxi Website

Table 10.1 Success measures

Contact efficiency of an action page A	The proportion of sessions containing page A with respect to all sessions in the log in the considered period of time
Relative contact efficiency of an action page A	The proportion of sessions containing page A with respect to the number of logged active sessions, called *aSessions. Active sessions* are sessions that contain at least one action page
Conversion efficiency of a page P to a target page T	This measure estimates the success of an arbitrary page in helping/ guiding the users towards a target page; the proportion of active sessions containing pages P and T (the access to page T is after the access to page P) with respect to active sessions containing P

or indirect sales (using the Web channel to promote traditional sales). Although the activity of the site can be analyzed using Web analytics tools such as Google Analytics [5], such a solution has limitations that the study presented in this chapter proposes to overcome. For example, we present a cross analysis between Web accesses and offline sales.

The buying process is the process that leads the user to reach the site's goal by confirming and concluding the order; it is also the focus of this analysis. The buying process on INTROduxi's Website takes five steps (Fig. 10.1) from entering the site (step 1) to actually ordering the product (step 5). In the buying process there are two turning points that deserve attention. One is when the user enters an *action page* (a page which indicates that the user is potentially interested in purchasing) (step 2). The other turning point corresponds to accessing a *target page*, the page that corresponds to an actual purchase (step 5). It is important to note that it is possible to skip step 2 when a specific product is directly added to the shopping cart. In this case, after selecting the kind of product the user is interested in (step 1), the user immediately adds it to the shopping cart (step 3) instead of having a closer look at the specific product of choice (step 2).

The meaning of the site's success measures are now defined more precisely: the *contact efficiency* of an action page A, the *relative contact efficiency* of an action page A and the *conversion efficiency* of a page P with respect to a target page T. The definitions presented in Table 10.1 are adapted from Spiliopoulou et al. [10]:

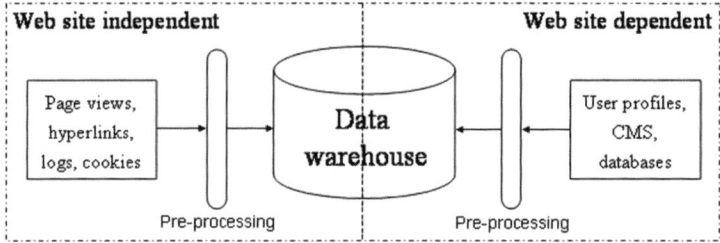

Fig. 10.2 Architecture of the data warehouse

The pre-processed Web server log data are stored in a data warehouse and the measures are implemented using SQL queries as described in the next section.

10.4.1 Pre-processing and Storage of Web Server Log Data

The ETL (extraction–transformation–loading) process and the data warehouse proposed in [3] were used to pre-process and store the Web server log data. The general architecture of such a data warehouse is presented in Fig. 10.2. In [3] the goal is to use the architecture with the widest possible applicability. For this case study the architecture was adapted according to our needs.

Only the site independent part of our data warehouse is used here. We exploit usage data extracted from Web access logs corresponding to 1 month of activity. The general model of the data warehouse follows a star scheme that is represented by centralized fact tables which are connected to multiple dimension tables (Fig. 10.3). The characters "#" and "*" indicate that the field is a primary or a foreign key in the table, respectively. The tables *Parameter Page* and *Parameter Referer* store the name and value of the parameters of an URI. These are neither fact nor dimension tables, they are just normalized relational tables to make the usage of parameters of an URI easier.

The *Fact Table Usage* is filled with data about accesses/requests to pages of the Website. The ETL process is described in Fig. 10.4 and it is implemented as a composition of different existing tools.

In the extraction step, the process creates a local version of the remote Website and access logs. This local version is stored in the Data Staging Area (DSA), a simple directory in the file system. *Wget*[1] and *Scp*[2] were used for this task.

---

[1] http://www.gnu.org/software/wget/.

[2] http://www.openssh.org/.

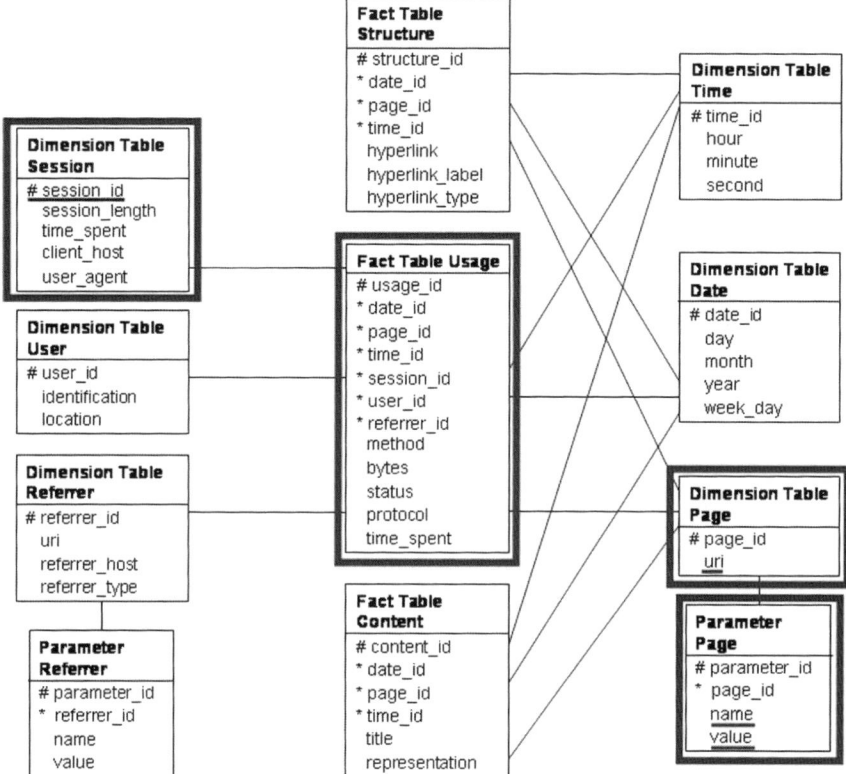

Fig. 10.3 Star schema of the data warehouse

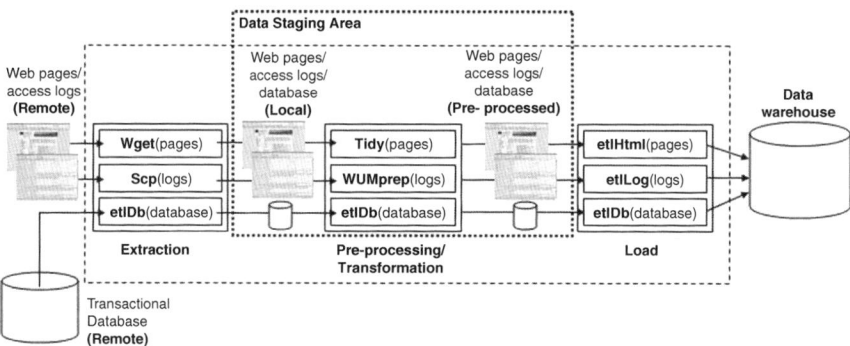

Fig. 10.4 The process of extracting, transforming and loading (ETL) the Web data into the data warehouse

In the transformation step, the local version of the site and logs are pre-processed and transformed into useful information that is ready to be loaded into the data warehouse. The pre-processing of the access logs consists of merging the log files, removing irrelevant requests and/or data fields, removing robot requests and identifying users and sessions for the local version of the access logs. *WUMPrep*,[3] a collection of Perl programs supporting data preparation for data mining of web logs was used. For the loading step, two components were implemented, *etlHtml* and *etlLog*, they use simple SQL commands to load data into the data warehouse. Additionally, to handle data collected from a transactional database, a component was developed, called *etlDb*, to select data, pre-process and load them into the data warehouse.

Once the data has been stored in the data warehouse, the success measures can be calculated. To do that, only the highlighted fields in Fig. 10.3 were used. Using simple SQL queries, the numerator and denominator values can be calculated for the three measures. For example, the Contact Efficiency measure can be calculated as follows:

Numerator: Query to count all different sessions containing action pages about laptops.

```
SELECT COUNT(DISTINCT session.session_id) FROM session,

usage, page, parameter_page WHERE session.session_id =

usage.session_id AND usage.page_id = page.page_id AND

page.uri LIKE '%product_info.asp%' AND page.page_id =

parameter_page.page_id AND parameter_page.name =

'product' AND parameter_page.value = 'laptop'
```

Denominator: Query to count all different sessions.

```
SELECT COUNT(DISTINCT session.session_id) FROM session
```

10.5   Case Study Results

10.5.1   Contact Efficiency

The contact efficiency of A is the percentage of sessions in which an attempt to reach the site's goal has been made using action page A. By computing this value for each action page, it is possible to (1) identify the impact of each page on the

---

[3]http://hypknowsys.sourceforge.net/wiki/Web-LogPreparationwithWUMprep/.

Table 10.2 Efficiency results per product type

Product type	Contact efficiency (%)	Relative contact efficiency (%)	Conversion efficiency (%)
ACESSORIOS	9.21	14.41	5.10
COMPONENTES	20.97	32.82	4.56
COMPUTADORES	8.26	12.92	0.99
CONSUMIVEIS	2.26	3.53	4.52
IMAGEM	2.55	3.99	3.11
PERIFERICOS	16.83	26.33	4.36
REDES/COMUNICACAO	6.57	10.28	3.58
SOFTWARE	1.35	2.11	3.89

overall success of a site in engaging visitors and (2) detect pages with low contact efficiency. In the case of INTROduxi, the action page and the target page were chosen for each of the 81 product families. Each page is identified in the access log through a specific substring in the URI, as described in the example above. Table 10.2 depicts an aggregate of the results for the three measures for each type of product (accessories, components, computers, consumables, images, peripherals, networks and communications, software).

The overall results for the contact efficiency measure are necessarily low. This is an expected implication of the fact that there are many action pages that the user can go to. In other words, a user is not likely to visit such a number of different products at every session. However, it can be observed that different product types can have very different contact efficiency. This can be a measure of the popularity of the product or product type. Components and peripherals are in fact the products sold in the highest quantities. The relative contact efficiency is a measure appropriate for sites with many action pages, which is the case, or with a large number of inactive sessions, which is also the case.

10.5.2 Conversion Efficiency

The conversion efficiency estimates the success of a given page in helping/guiding users towards a target page. With this measure, it is possible to study the impact of each page on the success of the site and to identify pages that have low conversion efficiency and require improvement. The action page for Computer (Computador), for example has a very low conversion efficiency (the lowest of the 8) for a page with such high contact efficiency (the fourth highest). This kind of discrepancy can be monitored and the sales force can be informed. However, as we will see, such differences can also be explained by different prices and product prevalence.

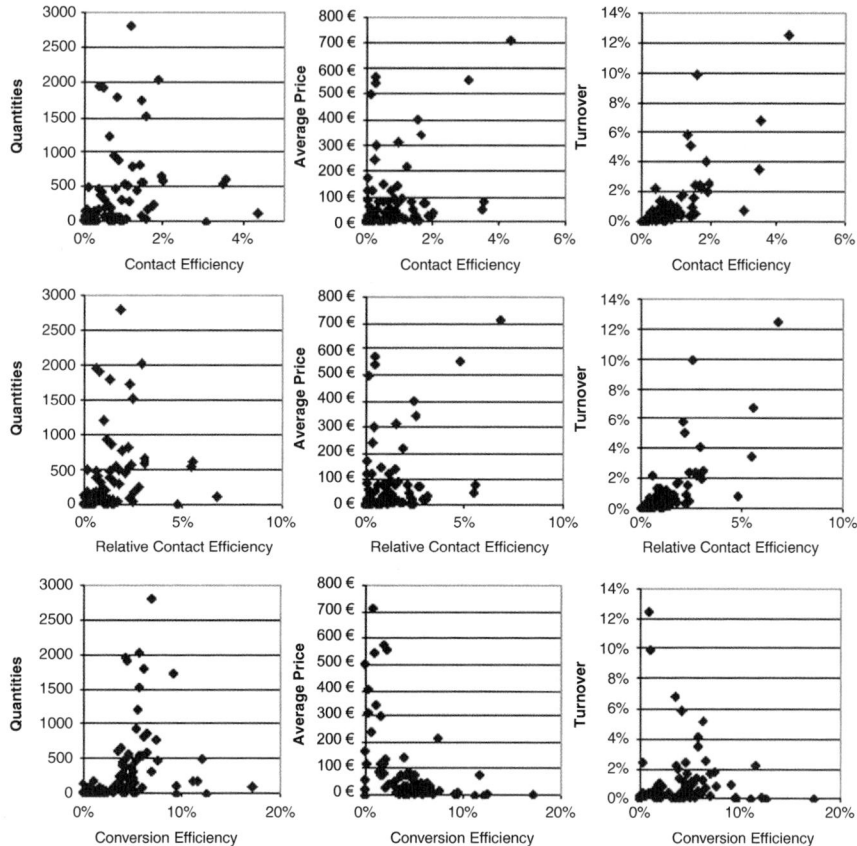

Fig. 10.5 Combining success measures and sales figures per family of products

10.5.3 Log Based Metrics vs. Sales Metrics

As seen above, the efficiency of a page may be very different for different products or product types. Although this variation may be caused by usability problems, most of it is probably explained by the popularity and the price of the products. Products sold in large quantities also get more visits to their pages. Expensive products may have lower conversion efficiency. This hypothetical relationship will be studied in this section by relating the success measures obtained from the access log with the sales figures. The sales numbers consider three determinant variables for the success of the business: the quantities sold, the average price of the products and the turnover for each product. The success metrics come from the calculation of the contact efficiency, relative contact efficiency and conversion efficiency measures considering the entire log.

In Fig. 10.5 each point represents a family of products. The first observation is that there are different shapes of point clouds. Some tend to follow a diagonal line where the y axis increases with the x axis. Others follow a kind of power law. After a detailed look, it is possible to see that contact efficiency tends to increase with average price and turnover. Conversion efficiency has the opposite behaviour. In other words, it can be confirmed that expensive products have low conversion efficiency. Inexpensive products with low conversion efficiency should therefore be analysed in terms of the usability of their representing Web pages.

The relation between contact efficiency and quantity sold is more surprising, since there are highly visited products sold in relatively low quantities and vice versa. This may be explained by direct sales external to the Web channel.

10.5.4 Segmented Analysis

The next question is how usage behavior changes with customer segments. For this, INTROduxi's customer profiles have been grouped according to their click stream behaviour and selling information. Therefore, contrary to the above analysis, which was performed on all users of the site, independently of being buying users or not, this clustered analysis was performed on customers only (users who actually made a purchase). The variables chosen were the following: the *number of page views*, the *average page views per session*, the *average time per session*, the *customer share*, the *total number of units ordered* and the *average price per order*. Using the SPAD software[4] seven clusters were found which can be characterized according to the relative contact efficiency results, as follows:

Cluster 1: "Low-price I" (557 customers)—the action pages or the type of products that are relatively more important on the site for these customers are the ones with the lowest average prices.

Cluster 2: "Good share & Poor navigation performance" (64 customers)—the Componentes and Perifericos (components and peripherals) type of products are the most important on the site, although the majority of these types of products are cheap, this cluster leans towards the products of greater value within this range.

Cluster 3: "Big value orders" (eight customers)—the most important type of product for this cluster is Computador (computer), this makes a lot of sense since they are known for the incredibly high average price of their orders.

Cluster 4: "Online all day long" (ten customers)—the type of products relatively more important to the customers within this cluster are Perifericos (peripherals). They visualize more pages per session than the others and have an extreme relative contact efficiency result of 100 %.

---

[4]http://eng.spad.eu.

Table 10.3 Relative contact efficiency results per cluster and by type of product

	Cluster						
Relative contact efficiency	1 (%)	2 (%)	3 (%)	4 (%)	5 (%)	6 (%)	7 (%)
ACESSORIOS	25.10	16.54	28.57	58.33	36.36	24.96	18.64
COMPONENTES	38.23	44.88	14.29	25.00	76.66	56.45	52.54
COMPUTADORES	4.58	19.69	100.00	0.00	9.37	9.38	7.63
CONSUMIVEIS	3.96	6.30	0.00	33.33	6.61	4.36	3.39
IMAGEM	3.33	5.51	14.29	16.67	5.23	4.36	3.81
PERIFERICOS	34.27	40.94	42.88	100.00	52.82	39.53	33.05
REDES/COMUNICACAO	13.13	10.24	0.00	25.00	17.91	15.41	12.71
SOFTWARE	1.25	9.45	0.00	0.00	2.20	4.02	2.54

Cluster 5: "Low-price II" (197 customers)—as in cluster 1, the type of products that are relatively more important on the site for these customers are the ones with the lowest average prices.

Cluster 6: "Good share & Good performance" (140 customers)—the Componentes and Perifericos (components and peripherals) type of products are the most important within the site for this cluster; it offsets the low value of the products ordered with the volume of quantities ordered.

Cluster 7: "Top share customers" (30 customers)—the action pages relatively more important on the site are the ones concerning the type of product Componentes (components), again a type of product with a low average price; this cluster totally wins the ordering championship, achieving the highest customer share via quantity.

Table 10.3 shows how the action pages for each type of product are visited by each cluster. The pages about computers, accessories and peripherals are successful within cluster 3. Otherwise the pages for computers do not have a great impact. With this information, the owners of the site can focus on particular groups when they analyze the success of each part of the site. In other words, the pages for a certain product type are only expected to be successful for certain groups of customers. It must be noted that this analysis can be performed with a finer level of detail, such as on product family or even product. Another interesting example is cluster 4, where peripherals are very important. If, during the continuous monitoring of these measures, this value drops for these customers, then something wrong is happening on the site.

Table 10.4 depicts the conversion efficiency for the Web pages of each type of product under each segment. It is interesting to see, for example, that despite the huge difference in terms of contact efficiency of the pages on computers for clusters 3 and 2 (100 vs. 19.69 %), the difference is very small when it comes to conversion (71.43 vs. 60 %). This means that something on the action pages for selling computers is not working as expected for cluster 3.

Clusters 1 "Low-price I" and 5 "Low-price II" showed unexpectedly good conversion efficiency results for most types of products. It could be interesting

Table 10.4 Conversion efficiency results per cluster and by type of product

Conversion efficiency	Cluster						
	1 (%)	2 (%)	3 (%)	4 (%)	5 (%)	6 (%)	7 (%)
ACESSORIOS	56.43	66.67	50.00	85.71	50.00	50.34	65.91
COMPONENTES	65.40	42.11	100.00	0.00	60.22	49.85	62.10
COMPUTADORES	27.27	60.00	71.43	0.00	17.65	19.64	44.44
CONSUMIVEIS	57.89	75.00	0.00	50.00	79.17	73.08	50.00
IMAGEM	62.50	71.43	100.00	0.00	57.89	42.31	55.56
PERIFERICOS	62.31	51.92	0.00	58.33	61.78	48.31	61.54
REDES/COMUNICACAO	59.52	38.46	0.00	33.33	47.69	40.22	63.33
SOFTWARE	91.67	83.33	0.00	0.00	37.50	41.67	33.33

to develop a targeted marketing campaign for the customers of these clusters. The purpose would be to increase these customers' interest in higher added value products similarly to cluster 2 or to raise the number of orders among these customers similarly to cluster 6. Since the behaviour of these two clusters is very similar it would be better to test this marketing campaign on only one of the clusters. It could make more sense to start with cluster 5 "Low-price II" since its navigation performance results are better than those from cluster 1 "Low-price I".

10.6   Conclusions and Future Work

In this chapter we have objectively measured the success of an e-commerce Website from a Portuguese company. We have adapted the Website success measures proposed in the literature to the specific buying process under study. We have observed that products with relatively high contact efficiency can have low conversion efficiency and vice versa. This highlights the pages which need more attention and the ones which are already successful.

Furthermore, we have combined the analysis of the access log data with the selling data available, obtaining further validation information on the marketing success of the Web pages. We conclude that, despite the fact that we have observed major tendencies; some product pages do not follow the general laws. Further investigation into the relation between the Web success measures and the commercial activity measures would be required.

We have performed clustering to segment clients according to sales and navigational behavior and studied the success of the site by segment. This makes it possible to see the success of parts of the site for different user groups. We can therefore design different action pages for different customer segments. We can also ignore expected low Web performance for some pages within some groups, as well as high demand performance within other groups.

Following the work presented in this chapter we would like to study the dynamics of Website usability, collecting contact and conversion efficiency measures over a period of time and determining maximum and minimum threshold values that can be used to monitor the success of the site and its parts. At the same time we are interested in performing a similar study for each of the segments identified.

It would be interesting to relate sudden variations in the success of the site and of its pages to decisions made by the marketing team, or to outside events. For example, if a new campaign increases contact but reduces conversion, this could be studied.

We would also like to implement a tool that collects data and continuously compiles the measures and makes it available to the site's management. This way the management could make decisions in real time, based on how the customers are behaving on their Web site.

Acknowledgements SIBILA Project "NORTE-07-0124-FEDER-000059" is financed by the North Portugal Regional Operational Programme (ON.2—O Novo Norte), under the National Strategic Reference Framework (NSRF), through the European Regional Development Fund (ERDF), and by national funds, through the Portuguese funding agency, Fundação para a Ciência e a Tecnologia (FCT). Grant 2012/13830-9, São Paulo Research Foundation (FAPESP).

References

1. Anand, S.S., Mobasher, B.: Intelligent techniques for web personalization. In: Lecture Notes in Artificial Intelligence, vol. 3169, pp. 1–36. Springer, Berlin (2005)
2. Berthon, P., Pitt, L.F., Watson, R.T.: The world wide web as an advertising medium. J. Advert. Res. **36**, 43–54 (1996)
3. Domingues, M.A., Jorge, A.M., Soares, C., Leal, J.P., Machado, P.: A data warehouse for web intelligence. In: Proceedings of the Thirteenth Portuguese Conference on Artificial Intelligence, pp. 487–499 (2007)
4. Fayyad, U., Piatetsky-Shapiro, G., Smyth, P.: The kdd process for extracting useful knowledge from volumes of data. Commun. ACM **39**(11), 27–34 (1996)
5. Ledford, J.L., Tyler, M.E.: Google Analytics 2.0. Wiley, Indianapolis (2007)
6. Liao, C., Palvia, P., Lin, H.: The roles of habit and web site quality in e-commerce. Int. J. Inf. Manag. **26**, 469–483 (2006)
7. Ruiz Mafe, C., Lassala Navarre, C.: Segmenting consumers by e-shopping behaviour and online purchase intention. J. Internet Bus. **7**, (2006)
8. Reichheld, F.F., Schefter, P.: E-loyalty : Your secret weapon on the web. Harv. Bus. Rev. **78**(4), 105–113 (2000)
9. Seybold, P.B., Marshak, R.T., Jeffrey, J.M.: The Customer Revolution - How to Thrive When Your Customers Are in Control. Business Books, London (2002)
10. Spiliopoulou, M., Pohle, C.: Data mining for measuring and improving the success of web sites. Data Min. Knowl. Discov. **5**, 85–114 (2001)
11. Srivastava, J., Cooley, R., Deshpande, M., Tan, P.: Web usage mining: discovery and applications of usage patterns from web data. ACM SIGKDD **1**(2), 12–23 (2000)
12. Travis, D.: E-Commerce Usability - Tools and Techniques to Perfect the On-line Experience. Taylor & Francis, London (2003)
13. Turban, E., King, D., McKay, J., Marshall, P., Lee, J., Viehland, D.: Electronic Commerce 2008 - A Managerial Perspective. Pearson/Prentice Hall, Upper Saddle River (2008)
14. Windham, L., Orton, K.: The Soul of the New Consumer - The Attitudes, Behaviors, and Preferences of E-Customers. Allworth Press, New York (2000)

Chapter 11
Worldwide Survey of Biodegradable Feedstocks, Waste-to-Energy Technologies, and Adoption of Technologies

Mike Centore, Gal Hochman, and David Zilberman

11.1 Introduction

The generation of waste is an unavoidable consequence of most human activity. During the economic life cycle of the production and consumption of goods and services, undesirables are discarded from start to finish. These undesirables, if not managed properly, pose a serious environmental and public health risk. One category of waste—biodegradable waste—originates from plant or animal sources. Despite emanating from organic sources, improper handling of such waste can threaten air, water, and land resources. Globally, as both population and affluence increase, production of biodegradable waste is also likely to increase. Properly managing such waste as to reduce pollution and limit exposure while promoting nutrient recycling will be an important undertaking for both developed and developing countries in the years to come.

This book chapter has three main purposes: (1) to survey six broad categories of biodegradable waste; (2) to identify and analyze four major types of waste management technologies; and (3) to examine global adoption of such technologies. The six categories of waste (from here on, "waste" will refer to only biodegradable waste) to be appraised are manure and animal waste, food waste, crop residues, sewage sludge, and a general category of other industrial, municipal, and residential waste. Next, the four technologies to be studied are landfilling, composting, incineration, and anaerobic digestion. Afterwards, we conclude with discussion of international adoption of such technologies.

M. Centore • G. Hochman (✉)
Rutgers University, New Brunswick, NJ, USA
e-mail: mcentore@Eden.Rutgers.edu; gal.hochman@rutgers.edu

D. Zilberman
University of California, Berkeley, CA, USA
e-mail: zilber11@berkeley.edu

A.A. Pinto and D. Zilberman (eds.), *Modeling, Dynamics, Optimization and Bioeconomics I*, Springer Proceedings in Mathematics & Statistics 73, DOI 10.1007/978-3-319-04849-9_11,
© Springer International Publishing Switzerland 2014

11.2 Survey of Biodegradable Feedstocks

The waste categories surveyed below—manure and animal waste, food waste, crop residues, sewage sludge, and other waste—differ in terms of generation rates, worldwide distribution, and biological characteristics.

11.2.1 Manure and Animal Waste

Most animals, which are produced for human consumption, are produced in concentrated animal feeding operations or CAFOs [5]. According to a 2004 EPA study, "One animal facility with a large population of animals can easily equal a small city in terms of waste production," and a dairy farm with 2,500 cattle produces a similar amount of waste to a city of 411,000 people [27, 28]. The abundance of CAFOs today presents challenges for animal waste management. The traditional method of manure management is land deposition, which simply involves dispersing the manure onto the fields as fertilizer, either in a solid or liquid form. This method is adequate for traditional farms that have small concentrations of livestock, but because of the size and concentration of today's farms, large quantities of waste are produced in a relatively small area. The great volumes of waste produced often exceed the assimilative capacity of the land for any land within an economically feasible transport distance [27, 28]. One hundred and thirty-three million dry tons of manure is produced in the United States each year, and there is a question of how to effectively treat and dispose of this waste. Animal manure is high in organics, nitrogen, and phosphorus, which make it useful as a fertilizer or conditioner of farmland—however, if manure is over-applied to the land, the soil will be overloaded with nutrients and heavy metals [5].

There is a number of other environmental, public health, and water quality issues related to the treatment of manure from CAFOs. In the US, human waste is treated before disposal into the environment, but animal waste is not treated or minimally treated before disposal, such as in land deposition [27, 28]. There are possible environmental contaminants in animal waste from CAFOs, including: nutrients, pathogens, veterinary pharmaceuticals, heavy metals (especially zinc and copper), and naturally excreted hormones. There are also over 100 microbial pathogens in swine waste that can cause human illness and disease. Parasites, viruses, and bacteria are present in animal wastes at the level of 1 billion per gram. Animal waste is also high in biochemical oxygen-demanding materials (BODs)—namely, in the quantity of oxygen used by microorganisms (e.g., aerobic bacteria) in the oxidation of organic matter. Swine waste slurry contains 20,000–30,000 mg of BODs per liter. Treated human sewage contains 20–60 mg of BODs per liter, and raw human sewage contains 300–400 mg of BODs per liter [5].

Laws regarding the treatment and disposal of animal waste differ between countries. In Norway, disinfection of sludge has been mandatory since 1995. In the EU, laws regarding disinfection are being proposed. Land deposition of animal waste is the most common method of disposal in the US [27, 28]. In Europe, 35–45 % of animal waste is deposited on land, but land deposition is being phased out. In Norway, the use of disinfected biosoils is prohibited in areas where vegetables, berries, potatoes, or fruits are produced. In the EU, use in forests is prohibited. For treating pathogens in animal waste, there are a number of effective methods, but in the case of unexpected waste spills the high concentrations of nutrients, heavy metals, and pathogens present in waste can contaminate surface and ground waters.

The properties of manure differ between animals, and the differences in manure types can affect the type of waste treatment or waste storage method used. Manure from swine, dairy cows, and layer chickens are typically handled as a liquid, which requires liquid storage methods. Manure from broiler chickens, turkeys, and beef cows is typically handled as a solid. In general, liquid wastes are more suitable for anaerobic digestion and solid wastes are more suitable for dry composting. Poultry manure has seen the largest increases for any category of animal waste: From 1982 to 2001, manure from poultry increased by more than 80 %. Poultry manure is higher in phosphorus than other animal wastes, which has implications for treatment and deposition of the waste, because certain nitrogen-to-phosphorus ratios are desirable for plant growth [27, 28]. Manure also changes characteristics after leaving the animal. The most prominent of these changes is the loss of nitrogen as ammonia into the air. By the time the manure is applied to soil, nitrogen losses may be 90 %. These changes adversely affect the fertilizer value of manure. Loss of nitrogen to the air may also be a pollution concern if it gets redeposited in other watersheds, and airborne ammonia will cause bad odors, which becomes an annoyance to people in the community. The method of land deposition affects how much nitrogen is lost to the air—the loss is maximized with sprayers and minimized with direct injections into the soil. Phosphorus does not get lost to the air as nitrogen does, so the method of land deposition affects the nitrogen-to-phosphorus ratio as well [27, 28].

Besides manure, effluents from dairy and meat processing factories can also be included in the category of animal waste. Large quantities of water are used in dairy and meat processing operations. This water is used to clean the outsides and insides of carcasses, and for cleaning equipment and facilities before and after killing. There are three types of wastewater from dairy processing: sanitary wastewater, process wastewater, and cleaning wastewater. Sanitary wastewater is disposed of directly into municipal sewer systems. Processing wastewater is mainly used for cooling and heating. This water does not normally have pollutants and can also be disposed of in sewer drains. Cleaning water—used for cleaning facilities and equipment—regularly comes in contact with, and is mixed with, milk products. This type of wastewater is high in organics, is mixed with cleaning compounds, and is high in sodium [19]. In meat processing plants, wastewater comes from stockyard washdowns and stock watering, rendering operations, fecal removal from intestines, and hair scalding. Effluents from these plants contains high organic compounds, high fat concentrations, high levels of N, P and Na, and it fluctuates in pH and temperature This wastewater effluent, like manure, also contains high levels of nitrogen and phosphorus. For farmers, the effluent can be a free source of nutrients, but use of the effluent on farmland carries with it environmental dangers such as dissolved salts causing soil salinity and groundwater contamination. These effluents can be treated in a two-stage process—first with screens to remove large debris, then with anaerobic or aerobic digestion.

11.2.2 Food Waste

Food waste in the US has increased approximately 50 % since 1974, reaching 1,400 kcal per person per day or 150 trillion kcal per year. Food waste is the biggest category of waste in the US. An average US family of 2.63 persons generates 100 kg of wet food waste in slightly over a year [11]. Including restaurants and institutions as well as households, US food waste amounts to 43.6 million tons per year. In the US, over one quarter of total freshwater supply is used on wasted food, and wasted food accounted for 4 % of total US oil consumption in 2003 [15]. Food waste is 30 % solids and 70 % water. The solid portion of the waste is 95 % decomposable and 5 % ash. The energy contained in wet food waste is 4,650 kJ/kg [11]. This waste can be converted into energy in a variety of ways, including incineration with energy recovery, pyrolysis and gasification, and anaerobic digestion.

Food waste can be divided into three main categories: post-harvest losses of perishable and non-perishable crops, waste from food processing and retail, and post-consumer waste. Different countries, and different types of economies, have different sources of food waste. In developed economies, post-consumer food waste accounts for the most losses, and "as much as half of all food grown is lost or wasted before and after it reaches the consumer". In industrialized countries, post-harvest losses of non-perishable food crops are not significant. In other countries, 15 % of grain may be lost in the post-harvest system. The post-harvest losses of perishable food crops are greater than non-perishable losses in both developing and developed countries. In the US, these losses are estimated to be (depending on the commodity) between 2 and 23 %. The overall average post-harvest food loss for perishable crops in the US is 12 %. In the UK, these losses are estimated to be 9 %. In developing countries, current methods of fruit harvesting (leading to bruised fruit) may lead to greater losses if supply chains get longer; with current short supply chains, the bruised fruit always finds a consumer [24]. For example, 30 % of fresh fruit and vegetable production in India is lost through lack of a temperature-controlled supply chain.

In the UK, food and drink waste is estimated to be approximately 14 megatons. The largest source of waste in the UK is from households. For the food and drink manufacturing and processing sector, food waste is approximately 2.6 megatons. At the retail and distribution stage, the number is 366 kilotons per year [24].

In the UK, Post-consumer food waste amounts to 8.3 megatons of food and drink each year. This waste has a retail value of 12.2 billion pounds (in 2008 prices). This waste is equivalent to over 20 megatons of carbon dioxide emissions. The total percentage of food wasted in the UK is estimated at 25 % of food purchased. In the US, Kantor et al. estimated that 25 % of food is wasted. The EPA estimated that food waste was 12.7 % of municipal waste, or 31.79 megatons. In Australia, food waste is estimated at 15 % of the 20 megatons of municipal waste that goes to landfills each year. Dutch consumers throw away 8–11 % of food purchased. A study by REFORSK found that food waste from households in six municipalities in Sweden in 1998 was 40.4 % of total household waste (by weight). Other studies found that food waste was 14.8 % of total household waste. This amounted to 177,682 dry tons/year. In a South Korean study of municipal waste, food waste was found to be 26–27 % of household waste [24].

Several trends affect worldwide food waste. Urbanization in developing countries requires that new food supply chains be developed. The structure of these supply chains will affect future food waste. In BRIC countries (Brazil, Russia, India and China), diets are also changing to include less starchy food and more meat, dairy, and fish. These foods are more perishable. Also, transitioning to longer food supply chains may bring more waste; in the UK, "contractual penalties, product take-back clauses and poor demand forecasting had a combined influence that drove 10 % over-production and high levels of wastage in the UK FSC [food supply chain]" [24].

It cannot be assumed that all food waste goes into municipal solid waste. Household sink food disposers or "garbage disposals" divert the solid waste into wastewater systems. Of the 100 kg of wet food waste that an average US family generates in slightly over a year, 75 % can be disposed of in a food waste disposer. One thousand and thirty-one kilograms of water is needed to flush 100 kg of food waste through the food waste disposer, and this is added to the food waste total [11].

Waste cooking oil is a source of food waste used to produce biodiesel Waste cooking oil is high in abundance (~9 pounds/person/year) and low in cost (between $0.09 and $0.20/lb)— [16]. Its properties, however, differ based on cooking process, source collection, and storage conditions. When compared to petroleum-based diesel, Kulkarni and Dalai [16] argue that emissions of HC, CO, NO, SO_2 and CO_2 decrease but NOx increase observed. Further, although engine performance is the same, biodiesel consumption is slightly higher (while the volumetric energy content of biodiesel reported in the literature is 33.3–35.7 MJ/L, petroleum based diesel is 40.3 MJ/L), but that blend of biodiesel and petroleum-based diesel (75:25 and 50:50) maintains balance for fuel consumption and emissions.

Another source of oil used to produce biodiesel is restaurant Waste Lipids [7]. Canakci analysis suggests that 1 lb of fats and oils can be converted to 1 lb of biodiesel (1:1 ratio), and that the samples of feedstocks analyzed had moisture levels, which varied from 0.01 to 18.06 %, and Free Fatty Acids content that varied from 0.7 to 41.8 %. Yet another source used to produce biodiesel is soapstock, a byproduct of edible oil refining, which some argue is less expensive than edible-grade refined oils—it costs $0.11/kg, 1/5 price of crude soybean oil.

11.2.3 Crop Residues

Crop residues are a major byproduct of agriculture. These residues are cellulosic plant material such as corn stalks, corncobs, wheat straw, and rice chaff. Corncobs make up 20 % of corn crop residue by weight. It is estimated that US production could produce 40–50 million tons of corn cobs per year. For the entire world, the UNEP states that "140 billion metric tons of biomass is generated every year from agriculture". Biomass is defined as all agricultural wastes, manures, and forestry scraps. This biomass is equivalent to approximately 50 billion tons of oil and could provide energy to 1.6 billion people in developing countries. In these countries, agricultural biomass is mostly left to rot in fields or burned.

Currently, biomass contributes 9–13 % of the world's energy supply. There has recently been great interest in converting energy-crops into biofuels and other types of energy, but their efficacy is limited by space constraints, growing conditions, and competition with food crops. Crop residues may be a better option—they do not interfere with food production, they can be harvested on a large scale in different climates, and "they can be harvested sustainably without affecting soil quality" [12]. Agricultural residues such as corn stover, crop straw, and sugar cane bagasse show great potential for the production of energy. Corn stover particularly is predicted to play an important role in bioethanol production. According to the US National Renewable Energy Laboratory, it is possible to produce 288–447 l of ethanol per dry ton of corn stover.

A simple method of using crop residues for energy is to burn them in a boiler to create steam and electric power. However, the efficiency of this process is limited by the high water content of the biomass. The most versatile method, according to [12], is gasification and pyrolysis. The authors designate three different products generated from this process: high-pressure steam, electricity, and steam for the conversion of char into activated carbon. Pyrolysis involves subjecting the biomass to high temperatures in an oxygen-deficient environment. Activated carbon is in high demand for industrial applications, water purification, and air pollutant removal.

From the 5 % of corn left on fields, 9.3 gigaliters of ethanol could be produced. Furthermore, the dry milling process of producing bioethanol produces dry grains that can be used as animal feed. These grains would replace a certain amount of corn as animal feed. If this corn were used for bioethanol instead of animal feed, another 5.3 gigaliters would be produced. This could replace. Ninety-three percent of world gasoline consumption. Allowing for 60 % ground cover, the remaining available corn stover could produce 58.6 gigaliters of bioethanol. This could replace 3.8 % of world gasoline consumption.

In the same way, bioethanol from barley waste and residue could replace 1.3 % of global gasoline consumption. Oat waste and oat straw could produce 3.16 gigaliters of bioethanol, or 0.2 % of global gasoline consumption. Energy produced from rice waste would amount to 14.3 % global gasoline consumption, from wheat and wheat straw 7.5 %, from sorghum 0.3 %, and from sugar cane waste and sugar cane bagasse 3.4 %. Furthermore, lignin residues from the process of creating ethanol could be used for generating electricity and steam. This electricity would amount to 0.7 % of total global electricity generation. The total potential replacement of global gasoline consumption from crop residues is 32 %, and the potential replacement of global electricity production is 3.6 %.

The use of crop residues for energy production is not without potential environmental issues. Soil conservation is a major consideration. Conservation tillage practices state that at least 30 % of soil surface be covered with crop residues. In the US today, 90 % of corn stover is left in the fields. Some authors argue that the best use of crop residues is to leave them on the fields, due to the nutrients and erosion-management benefits they provide to the environment. It is argued that the crop residues will nourish the soil and increase crop yields in the following season. Others estimate that 30 % of crop residue on the field can be harvested without affecting soil quality [12].

11.2.4   Sewage Waste

Sewage sludge is a mixture of biosolids and liquids from wastewater treatment plants. Sludge is an unwanted byproduct of wastewater treatment, and it presents the challenges of disposal and disinfection. However, sewage sludge also has the potential to be used for beneficial means, such as fertilizer or energy production. Its high organic, nitrogen, and phosphorus content makes it suitable for this purpose. In the US, over half of biosolids that are a product of municipal wastewater treatment go toward fertilizing or conditioning land, and the remaining portion is either incinerated or landfilled. Dumping sewage sludge into the ocean is no longer allowed [27, 28]. In the EU, policies exist to enhance the use of sewage sludge in agriculture. However, there are regulations concerning acceptable levels of certain compounds in sewage sludge, depending on the country. The amount of sewage sludge produced is relatively constant—it is estimated to be about 50g of dry matter per person per day. The EU member states produce 8 million metric tons of sewage sludge each year. The presence of sewage sludge, and the treatment of it, will continue to pose environmental challenges into the future.

The use of sewage sludge as fertilizer is potentially very beneficial, as the sludge has the ability to put nutrients into the soil and replace some amount of chemical fertilizers. However, sludge carries a wide variety of pathogens that are infectious to humans as well as various species of animals and plants. The treatment and disinfection of sewage sludge is therefore an important process, but it can be costly. Sewage sludge treatment accounts for over half of total wastewater treatment costs [25].

There has been a large amount of interest over the past two decades in converting sewage wastewater effluent to energy in various forms. The main two options for producing energy from sewage sludge are incineration with energy recovery and anaerobic digestion. Heat from the incineration process can be recovered and used as energy. In anaerobic digestion, methane is produced which can be used to power generators to produce electricity.

11.2.5 Other Organic Industrial, Municipal, and Residential Waste

Municipal solid waste is composed of organic and non-organic portions. Paper and organic waste are the major components of the municipal solid waste stream by weight. Plastics make up 10 % of the waste weight but 40 % of the volume. The amount and composition of solid waste differs between developed and developing countries. According to one estimate, developed countries produce an average of 1.5 kg of municipal solid waste per person per day. In developed Asian countries, solid waste generation is estimated at 0.4–0.6 kg per person per day. In developing Asian countries, the number is 0.7–0.8 kg per person per day. Also, recyclable waste is more common in developed countries, and organic waste is more common in developing countries [23]. Global MSW is estimated at 1,100 million tons per year.

The methane emissions from municipal solid waste in landfills are a major concern if they are not trapped and used as fuel. Global average emissions of methane into the atmosphere from MSW alone are estimated at 34 million tons per year.

Landfill emissions are estimated to be in the range of 6–20 % of total world methane emissions (anthropogenic or otherwise). In the EU, landfill methane is required to be either used as fuel or flared. Flaring refers to the burning of methane, which converts it into carbon dioxide—a less potent greenhouse gas. Besides the greenhouse gas potential of methane from landfills, improper disposal of waste can also have far-reaching environmental and health effects such as ground and surface water contamination, soil and air pollution, spreading of diseases, and bad odors [23].

11.3 Waste Management Technologies

A waste hierarchy is often used in waste policy. This hierarchy is: (1) Waste reduction, (2) Reuse, (3) Recycle, (4) Incinerate with heat recovery, (5) Landfill. This hierarchy can be seen in certain pieces of legislation, such as the German "Law on the Prevention and Disposal of Waste" (1986), followed by the "Closed Loop Economy Law" (1994); the US "Pollution Prevention Act" (1990); and in Denmark, the "Government Action Plan on Waste and Recycling" (1993), which set targets for recycling, incineration, and landfill. After waste reduction, the priorities of waste treatment are often contested [13].

11.3.1 Anaerobic Digestion

There is a question of where to place anaerobic digestion and composting in this hierarchy. Anaerobic digestion is an effective and widely used treatment strategy, which utilizes microorganisms to digest waste without the presence of oxygen. Four types of microorganisms are used in anaerobic digestion: hydrolytic, fermentative, acetogenic, and methanogenic [3]. Anaerobic digestion produces methane (CH_4) and carbon dioxide (CO_2), as well as trace amounts of other gases. Anaerobic digestion produces 60–70 % methane and 30–40 % carbon dioxide. Anaerobic digestion is also effective in disinfecting waste, which is a mandatory process in countries such as Norway, and could become mandatory in the EU. The biogas products of anaerobic digestion can be used to generate heat or electricity, and the resulting digestate can be used as fertilizer, which can be applied in agriculture. Mata-Alvarez et al. [20] analysis suggests that while emissions of VOC from composting are 747.4 g/ton biowaste, anaerobic digestion is only 100.6 g/ton biowaste. Furthermore, they document a reduction of emissions with anaerobic digestion of CO_2 emissions of 25–67 % than composting.

Methods of anaerobic digestion include covered lagoons, complete mix digesters, and plug-flow digesters [27, 28]. Covered lagoons are a wet digestion system that involves digesting the waste underneath a covering which traps biogas emissions. Complete mix digesters allow inflows of new material to be mixed with the partially digested material and plug-flow digesters do not allow new material to be mixed with the biomass undergoing digestion. After digestion, the digestate is usually refined for use on farmland or compost, and some portion of the liquid effluent will go back into the digestion process as inoculum [3].

11.3.1.1 Animal Waste

With anaerobic digestion, farmers of livestock can obtain energy from biogas, fertilizer and, in countries such as Brazil, even carbon credits. Anaerobic digestion is also used to treat effluent wastewater from dairy and meat processing factories. Anaerobic lagoons are common where land is available for their construction and where the climate is warm. They are not common in urban areas. Suspended biomass reactors include anaerobic contact reactors and suspended biomass reactors. Anaerobic contact reactors are considered dated technology but are still used in older and smaller factories throughout the world [19]. Anaerobic sequencing batch reactors have shown high efficiency in treating wastewater. Immobilized cell reactors—i.e., cells that are entrapped within or associated with an insoluble matrix—have been successfully applied in full-scale operations for almost two decades [19].

Around the world, the most common disposal strategy for animal manure is land deposition. In Brazil, the most common treatment strategy is anaerobic digestion.

11.3.1.2 Food Waste

For the most part, food waste in the US goes to landfills, but energy-from-waste is becoming more economically viable. Due to the high moisture content of food waste, anaerobic digestion is preferred to combustion or gasification [29]. Zhang et al. [29] concluded that "...food waste contained the required nutrients for anaerobic microorganisms." and that it was "a highly desirable feedstock for anaerobic digestion."

In one study by Finnveden et al. [13], electricity consumed in the anaerobic digestion process was 31 MJ per ton of organic household waste (dry weight). Dry weight is 30 % of wet weight. The plant's heat consumption in the study was 495 MJ per ton organic household waste (dry weight). This heat was generated from the methane gas produced. The digestion residue contained 7.6 kg/ton nitrogen and 1.1 kg/ton phosphorus (dry weight). Describing the energy needed for spreading the produced residue from anaerobic digestion and composting, the authors say "The energy required for spreading is approximately 20 MJ diesel/ton digestion residue and 15 MJ/ton compost" [13]. From 1 ton food waste, anaerobic digestion produces 858 kg residue (14.2 % weight loss, wet weight). Composting produces 500 kg (50 % weight loss, wet weight). The avoided energy costs of producing and spreading artificial fertilizers must also be taken into account in the life-cycle analysis. The energy produced from methane from food waste is 3,743 MJ/ton food waste, and 495 MJ of that is used for the plant itself [13].

It is also possible to co-digest food waste and yard waste, which possibly gives greater biogas yields. Co-digestion of other wastes (food waste with dairy manure) has been shown to increase total methane production. 30.9 million metric tons/year of yard waste (grass, leaves, wood chips) is available for co-digestion. The literature has shown that, at F/E ratio of 2, increased methane production was seen when food waste was 10 % of the substrate. At F/E ratio of 1, increased methane production was seen when food waste was 20 % of the substrate.

Oil waste, a type of food waste, can be used to produce biodiesel, and several alternative technologies are discussed in the literature. One of those technologies is Supercritical Methanol—a catalyst free technology of biodiesel production under high temperature and pressure. Different from existing technologies, it is insensitive to water and Free Fatty Acid content. When using this technology, the feedstocks can account for 80 % of total costs of production, biodiesel produced using Supercritical Methanol is similar to commercial diesel fuel, and the reaction time is relatively short [9]. Lee et al. [17] evaluated three biodiesel processes:

1. Alkali-catalyzed, virgin oil (Alakai-FVO);
2. Two-step process, WCO (Alkali-WVO); and
3. Supercritical Methanol process, WCO (SC-WVO).

Their analysis suggests that energy consumption of alkali-FVO of 2,349 kW, alkali-WVO of 5,258 kW, and SC-WVO of 3,927 kW. Those authors conclude that energy consumption of SC comparable to Alkali-FVO, and that the most significant variable affecting production cost is feedstock (64–84 %). Using co-solvent in Supercritical Methanol, Cao et al. [8] conclude that because Supercritical Methanol requires temperatures of 350–400 °C and pressures of 45–65 MPa, this process is one with high production costs and energy consumption. The analysis presented by those authors suggests that using propane as a co-solvent reduces severity of necessary conditions: At 280 °C and 12.8 MPa, they obtained methanol/oil ratio of 24 and 98 % yield within 10 min. Thus, the authors conclude that the Supercritical Methanol with co-solvent is superior to Supercritical Methanol by itself, and that when using co-solvent lower temperatures and pressures is required, and that less energy required for process.

When using Waste Cooking Oil, Kulkarni and Dalai [16] argue that Chemical Transesterification Processes: (1) Alkali-Catalyzed Transesterification has fast reaction rate but is inefficient when Free Fatty Acid content is high; and (2) Acid-Catalyzed Transesterification has a slow reaction rate but works better if Free Fatty Acid content is high. On the other hand, a Two-Step Acid and Alkali Transesterification, whereby

1. Stage 1: pretreatment process (acid-catalyzed)
2. Stage 2: transesterification (alkali-catalyzed),

has problems with pretreatment of feedstock, recovery of glycerol, removal of catalyst, and energy intensity. Other Transesterification Processes surveyed, included Enzyme-Catalyzed Transesterification, where the authors argued that this process has no generation of by-products, easy product recovery, mild reaction condition, and is insensitive to FFA and moisture content. Another process discussed and analyzed by the authors was a Catalyst-Free Technology, i.e., the Supercritical methanol process—an emerging technology that results in higher yield of biodiesel even with high Free Fatty Acid and moisture content. However, scaling up and commercialization has yet to be proven. Testing of Biodiesel Produced from Waste Cooking Oil, the presented analysis suggested estimated costs for biodiesel from:

1. Soybean oil—$0.418/L
2. Yellow grease—$0.317/L; and
3. Brown grease—$0.241/L.

The author concluded that chemical processes are not recommended due to high costs. But that enzyme transesterification may be a good option that should be developed. The authors also concluded that Supercritical methanol method has potential but scaling up is required. Canakci [7], while focusing on Restaurant Waste Lipids concludes that because moisture and Free Fatty Acid content vary widely, the Transesterification processes needs to account for this, and that Two-step process (alkali and acid transesterification) is recommended.

Diaz-Felix et al. [10] looked at Yellow Grease, and seek to quantify output based upon Free Fatty Acid content of feedstock and methanol use during reaction. They also advocate for a two-step process, with pre-treatment followed by transesterification. Their analysis suggests that while using molar ratio of 18:1 and 95 % conversion, the Fatty Free Acid content were reduced to 0.4 %. The authors further argues that this process can be further optimized as only 85 % conversion is necessary for reduction in Fatty Free Acid content to 2 %, and that the alkaline-catalyzed transesterification successful at 2 % content. Haas [14] also argued for the two-step process, while focusing on Soapstock–a byproduct of edible oil refining. Haas argued that the Saponification process

1. Step 1: alkali-catalyzed hydrolysis increases FFA; and
2. Step 2: acid-catalyzed transesterification produces FAME,

is efficient and that it met ASTM specifications for biodiesel and the product was comparable to biodiesel from soy oil. It costs $0.41/L to produce biodiesel from soapstock, which is 25 % less than production from soy oil.

11.3.1.3 Crop Residues

In Europe, biogas plants are mainly fed with animal waste effluents and dedicated energy crops. As the number of biogas plants has increased, the amount of dedicated energy crops has also increased. There are concerns about world food supply in this system. As an alternative, crop residues can be used, and there will be no food supply concerns (setting aside the argument that taking crop residues off the fields can reduce subsequent crop yields). If the residues are harvested sustainably, leaving enough on the fields for effective nutrient deposition and erosion control, then this will not be a concern.

Combustion-based plants are rare in Italy due to the fact that the production of thermal energy is not incentivized, but the government incentivizes electricity production. Thus, the thermal energy is converted into electricity by heating water to create steam to power turbines. The efficiency of this process is less than using methane-powered engines which would be used at biogas plants. Combustion plants also require flue gas purification, but anaerobic digestion biogas plants do not. And the digestate from biogas plants can be used as fertilizer. Using crop by-products instead of energy crops also reduces greenhouse gas emissions [21].

As the result of a study by Menardo and Balsari [21] which compared methane yields of different types of crop residues, maize stalks produced the lowest yields. Rice chaff, wheat straw, kiwi, and onion were not significantly different in yields. Dairy products produced the most biogas and methane yields due to high protein and fat content—however, dairy products are not available in large amounts, and would not be applicable on a large scale. Dry bread biogas yields were not significantly different from rice chaff, maize stalks and wheat straw but had a high methane percentage compared to other gases. The lignocellulosic compounds in corn stalks and wheat straw negatively affected methane yields, but the huge amounts of these resources that is available for use makes them attractive energy sources nonetheless. Rice chaff also has high availability in Italy and it is also an attractive potential source of biogas. Mechanical shredding can be used as pre-treatment for these types of crops to prepare them better for digestion [21].

11.3.1.4 Sewage Waste and Municipal Solid Waste

In a typical situation, sewage waste is mechanically and biologically treated with sedimentation followed by anaerobic digestion. This process involves three steps: a hydrolysis step, an acidification step, and a third step in which biogas is produced. Anaerobic digestion is used to stabilize the sewage and convert it into biogas. Menger-Krug et al. [22] state that, "Currently, anaerobic digestion of sewage sludge is mainly applied at large and medium-sized wastewater treatment plants. However, also, a growing interest is observed in the application of anaerobic treatment in small-sized plants." Pretreating the sewage in a variety of ways can also increase the amount of biogas produced.

The portions of municipal solid waste that are suitable for anaerobic digestion are paper, green waste or yard waste and food waste. The organic fraction of municipal solid waste is approximately 50 % [3]. The typical method of disposal for municipal solid waste is incineration or landfill. Anaerobic digestion has been used for over 100 years to treat sewage sludge, but experience with digesting the organic portion of municipal solid waste is more recent. Typically, anaerobic digestion is most appropriate when the waste is mostly food waste, and aerobic composting is most appropriate when it is mostly yard waste. Anaerobic digestion provides a good solution for treating the wet fraction of municipal solid waste which is too high in moisture for incineration. This method also requires less land than aerobic composting. Furthermore, because anaerobic digestion can be applied to many different types of waste, it is possible to co-digest the organic portion of municipal solid waste with other wastes such as agricultural manure [3]. The efficiency of the digestion process is even increased in some cases when wastes are co-digested [4].

In order for municipal solid waste to undergo anaerobic digestion, it must be pre-treated, which includes sorting and separation of differently sized particles. In wet systems, sorting and separation can occur simultaneously. For dry digestion systems, rotating drums are typically used. For wet digestion systems, pulpers are used [3].

11.3.2 Waste Incineration

11.3.2.1 Sewage Waste

Incineration with energy recovery is a process that can be applied to mechanically-dewatered or dried sludge. The sludge before dewatering is 98 % water. The dewatering techniques include drying beds, belt filter presses, plate and frame presses, and centrifuges. Pretreatment strategies can also be used before dewatering, which involves mixing the sewage sludge with lime, ferric chloride, or other chemicals [27, 28]. The dewatering process adds to the cost of incineration.

There are potential environmental problems with incineration, involving emissions of pollutants and quality of ash produced. There are systems (scrubbers) to eliminate the harmful pollutants from the emissions, but they are costly. This is one of the main reasons why incineration is an expensive option compared to other treatment methods. The heavy metals in the ash are not an environmental problem because, due to the high heat treatment, they are immobilized and not prone to leaching. Currently, incineration of sludge is mainly used for generating heat (steam) or electricity. Incineration is used worldwide and on a large scale. The performance largely depends on the water content and the performance of the drying process. To reduce costs, the dried sludge can be co-incinerated in an existing coal-fired power plant. The combustion and the gas treatment systems will be already in place, so initial costs will be lowered. This option is already applied in practice [25].

The ash from the incineration process can be used in the production of various products such as bricks or portland cement. This process is used today in Japan. The energy efficiency of the production process is not very high; it has high costs, and the current practical applications of this technology are limited [25].

11.3.2.2 Food Waste

Incineration is expanding in Sweden: "The amount of waste incinerated has more than doubled since 1980 and the energy production has increased almost fourfold during the same period" [13]. In 1995, 1.8 megatons of waste were incinerated, and the energy produced from this waste was 5 terawatt-hours [13]. It is possible that incineration of food waste offers no net energy benefit due to the high moisture content in food waste, which requires energy for drying or dewatering. However, some authors argue that the moisture can be evaporated off efficiently and that net energy benefits can be attained [2].

11.3.3 Municipal Solid Waste

Incineration is an option for managing municipal solid waste. Incineration is effective in reducing the volume of the waste to be disposed and sterilizing it. Volume reductions of the waste are up to 90 %. However, strict emissions regulations have made this option more costly. Nonetheless, some countries are attempting to reduce the amount of organic content of post-recycled waste destined for landfill and promoting incineration as an alternative. If the infrastructure for incineration and emissions purification are already in place, such as with co-incineration inside coal-fired power plants, then this option becomes more economical.

Many countries have deemed "bottom ash" (the ash collected during incineration—contrasted with the finer and more leaching-prone "fly ash") suitable for disposal in landfills. But it is also used in construction materials. Germany, Denmark, and Sweden use 60–90 % of bottom ash in road construction or in asphalt/concrete. The fly ash, by contrast, is considered hazardous and needs to be disposed separately. However, in the US, the fly ash and bottom ash are combined and disposed of together [26].

11.3.4 Composting

Composting can be used to treat a variety of organic wastes, including animal manure, sewage waste, the organic portion of municipal solid waste, food waste and yard waste. Composting is commonly used to treat these wastes in the US. Composting reduces the volume of the waste as well as disinfecting and stabilizing it. Reduced volume and weight leads to easier and more economical transportation of the waste product. After the composting process, the compost can be used to fertilize land or be placed in a landfill. However, composting does not result in usable amounts of methane. The process is aerobic and mostly results in the production of carbon dioxide. From 1990 to 2010, the amount of waste composted in the US rose 392 %. This increase is mainly attributable to population growth and legislation in the 1990s to reduce the amount of yard trimming disposed in landfills [27,28]. Sakai et al. [26] stated that biological treatment methods were reemerging as commercially viable, which had led toward composting garden, kitchen, and commercial food wastes, and away from mixed solid waste processing.

Large amounts of composting can be done through aerated static pile, aerated windrow, or in-vessel composting. Aerated static pile composting simply involves one large pile of waste mixed together with layers of bulking agents such as wood chips, which allow air to reach the bottom of the compost pile. Aerated static pile composting works well for yard waste, food waste and paper, but not for animal waste. Aerated windrow composting is different from static pile composting in that the waste is placed in rows which get turned over periodically. In-vessel composting involves depositing the waste in large drums, silos, or trenches and controlling the moisture and aeration of the waste mixture. This method of composting works well for all types of organic waste, although, like all aerobic composting, it does not have the potential to produce useable amounts of methane from the waste.

Composting (especially in-vessel composting) solves many problems for farmers who need to manage large amounts of solid or liquid manure. Composting transforms liquid manure into solid, which reduces the volume and weight of the manure and lessens its transportation cost. Transportation cost is one of the main considerations in animal waste management because the waste must be transported off the farm site due to the inability of the on-site farmland to assimilate the nutrients present in the waste. Composting is also effective for disinfecting waste. The common methods of waste disinfection are as follows:

- Lime treatment
- Dry composting
- Wet composting
- Wet composting + anaerobic digestion
- Pre-pasteurization + anaerobic digestion
- Anaerobic digestion + thermal drying

Composting for disinfection purposes can be performed in either a wet or dry state, or combined with anaerobic digestion.

In the composting of food waste, there are commonly three stages: mechanical pre-treatment, composting, and manufacturing of soil products. The soil products produced can be used for fertilizer. The composting process on a large scale has implicit costs associated with it. One study found that electricity consumed in the process is 54.4 MJ per dry ton of food waste. Also, consumption of diesel in the process is 555.5 MJ per dry ton of food waste. The compost residue contained 8.3 kg of nitrogen per dry ton and 2.0 kg of phosphorus per dry ton [13].

11.3.5 Landfilling

Landfills are the third-largest source of human-related methane emissions in the US. The EPA estimates landfill methane emissions to be 16.4 % of total US methane emissions in 2010 [27, 28]. This figure is contrasted with the percentage of methane emissions from wastewater treatment (2.5 %) and composting (<1 %). Clearly, landfills present a huge environmental issue: how best to deal with these emissions of methane, which is a greenhouse gas with 25 times the potency of carbon dioxide.

There are about 1,900 landfills operating in the US today. Over the past 20 years, the number of landfills has decreased and the average landfill size has increased. The process of decomposition in landfills is initially an aerobic one, but then turns anaerobic when oxygen is depleted. Eventually, this process leads to the production of methane and carbon dioxide in approximately equal parts. From 1990 to 2010, net methane emissions from landfills decreased 27 %, even though municipal solid waste deposition in landfills increased 23 %. This change is due to several factors. Even though the amount of waste increased, the amount of decomposable waste decreased by 21 %. Methane-recovery in landfills also increased. From 2009 to 2010, there was an increase in methane recovery of 5 %. In 2010, 54 new landfill gas-to-energy projects began. It is estimated, in part because of the rising US population, that municipal solid waste will increase in the coming years—but the percentage landfilled may decrease if more waste is recycled and composted [27, 28].

Food waste in the US, for the most part, goes to landfills, but energy-from-waste is becoming more economically viable [29]. Landfilling is also one of the most widely used options in Sweden with 300 municipal landfill plants. In 1998, 4.8 million tons of waste were landfilled. Energy gained from landfills is used for 60 % heat and 30 % energy, and 10 % is lost. There are also more efficient types of landfills known as biocells [13].

Municipal solid waste landfills receive approximately 69 % of the solid waste generated in the US. The remaining waste is sent to industrial landfills. However, the municipal solid waste landfills accounted for 94 % of the total methane emissions. The US has banned many polluting compounds, and banned disposal of certain items like car batteries in landfills.

11.4   Discussion and Concluding Remarks

Many facets of converting biodegradable waste to energy and the associated technologies are mature and used extensively in some regions but not in others. These technologies can potentially reduce greenhouse gases and help countries achieve their Kyoto Protocol goals. However, the realm of adoption of biodegradable technologies is one of divergent experiences. Within Europe, Germany and Denmark were the early adopters of the anaerobic technology in the early 1990s, with much of Europe following. The US, however, only began to experience growth starting in 2003. Interesting but Europe does not have a comparative advantage in either technology or feedstock availability, although work has detected large differences in policy and especially in use of the feed-in-tariffs overtime [1].

More generally, the literature has shown that adoption is impacted by lack of access to credit, and that these constraints may hamper the introduction of otherwise profitable technologies [30]. Furthermore, financial incentives have a positive effect on adoption of energy conservation technologies but that the elasticity of adoption in response to financial incentives is low [18]. These conclusions can be explained using the putty-clay nature of capital-intensive assets and the long-term commitments associated with adoption of capital goods, where response to financial incentives that results in the adoption of new technologies is associated with the response to new entrants. Further, research has concluded that capital investments are barriers to adoption and use of agricultural anaerobic digestion among farmers (e.g., [6]).

Experiences in Europe as well as to the west of the Atlantic indicate policy stability in the form of financial support is the most important factor in promoting the adoption of technologies converting biodegradable waste to energy. As more countries acknowledge the multitude of benefits from these technologies, adoption will likely increase. In future work we plan on further understanding the importance of policy and further explain differences in patterns of adoption of these technologies among countries and regions.

References

1. Bangalore, M., Hochman, G., Zilberman, D.: Differences in the Adoption of Agricultural Anaerobic Digestion in Europe and the United States (2012)
2. Bernstad, A., la Cour Jansen, J.: Review of comparative LCAs of food waste management systems: Current status and potential improvements. Waste Manag. **32**, 2439–2455 (2012)
3. Braber, K.: Anaerobic digestion of municipal solid waste: a modern waste disposal option on the verge of breakthrough. Biomass Bioenergy **9**(1), 365–376 (1995)
4. Brown, D., Li, Y.: Solid state anaerobic co-digestion of yard and food waste for biogas production. Bioresour. Technol. **127**, 275–280 (2012)
5. Burkholder, J., Libra, B., Weyer, P., Heathcote, S., Kolpin, D., Thorne, P.S., Wichman, M.: Impacts of waste from concentrated animal feeding operations on water quality. Environ. Health Perspect. **115**(2), 308–312 (2007). http://www.ncbi.nlm.nih.gov/pmc/articles/PMC1817674/
6. Bywater, A.: A review of anaerobic digestion plants on UK farms – barriers. Roy. Agr. Soc. Engl. (2011)
7. Canakci, M.: The potential of restaurant waste lipids as biodiesel feedstocks. Bioresour. Technol. **98**(1), 183–190 (2007)
8. Cao, W., Han, H., et al.: Preparation of biodiesel from soybean oil using supercritical methanol and co-solvent. Fuel **84**(4), 347–351 (2005)
9. Demirbas, A.: Biodiesel from waste cooking oil via base-catalytic and supercritical methanol transesterification. Energ Convers. Manag. **50**(4), 923–927 (2009)
10. Diaz-Felix, W., Riley, M.R., et al.: Pretreatment of yellow grease for efficient production of fatty acid methyl esters. Biomass Bioenergy **33**(4), 558–563 (2009)
11. Diggelman, C., Ham, R.K.: Household food waste to wastewater or to solid waste? That is the question. Waste Manag. Res. **21**(6), 501–514 (2003)
12. Donaldson, A.A., Kadakia, P., et al.: Production of energy and activated carbon from agri-residue: sunflower seed example. Appl. Biochem. Biotechnol. **168**, 1–9 (2011)
13. Finnveden, G.R., Johansson, J., et al.: Life cycle assessment of energy from solid waste part 1: general methodology and results. J. Cleaner Production **13**(3), 213–229 (2005)
14. Haas, M.J.: Improving the economics of biodiesel production through the use of low value lipids as feedstocks: vegetable oil soapstock. Fuel Process. Technol. **86**(10), 1087–1096 (2005)
15. Hall, K.D., Guo, J., et al.: The progressive increase of food waste in America and its environmental impact. PLoS One **4**(11), e7940 (2009)
16. Kulkarni, M.G., Dalai, A.K.: Waste cooking oil an economical source for biodiesel: a review. Ind. Eng. Chem. Res. **45**(9), 2901–2913 (2006)
17. Lee, S., Posarac, D., Ellis, N.: Process simulation and economic analysis of biodiesel production processes using fresh and waste vegetable oil and supercritical methanol. Chem. Eng. Res. Design **89**(12), 2626–2642 (2011)
18. Linn, J.: Energy prices and the adoption of energy-saving technology. Econ. J. **118**(553), 1986–2012 (2008)
19. Liu, Y.-Y., Haynes, R.: Origin, nature, and treatment of effluents from dairy and meat processing factories and the effects of their irrigation on the quality of agricultural soils. Crit. Rev. Environ. Sci. Technol. **41**(17), 1531–1599 (2011)
20. Mata-Alvarez, J., Mace, S., et al.: Anaerobic digestion of organic solid wastes. An overview of research achievements and perspectives. Bioresour. Technol. **74**(1), 3–16 (2000)
21. Menardo, S., Balsari, P.: An analysis of the energy potential of anaerobic digestion of agricultural by-products and organic waste. BioEnergy Res. **5**, 759–767 (2012)
22. Menger-Krug, E., Niederste-Hollenberg, J., et al.: Integration of microalgae systems at municipal wastewater treatment plants: implications for energy and emission balances. Environ. Sci. Technol. **46**(21), 11505–11514 (2012)
23. Othman, S.N., Noor, Z.Z., et al.: Review on life cycle assessment of integrated solid waste management in some Asian countries. J. Cleaner Production **41**, 251–262 (2012)

24. Parfitt, J., Barthel, M., et al.: Food waste within food supply chains: quantification and potential for change to 2050. Philos. Trans. R. Soc. B Biol. Sci. **365**(1554), 3065–3081 (2010)
25. Rulkens, W.: Sewage sludge as a biomass resource for the production of energy: overview and assessment of the various options. Energy Fuels **22**(1), 9–15 (2007)
26. Sakai, S., Sawell, S., et al.: Waste Manag. **16**(5), 341–350 (1996)
27. U.S. Environmental Protection Agency: primer for municipal wastewater treatment systems. http://www.epa.gov/npdes/pubs/primer.pdf (2004)
28. U.S. Environmental Protection Agency secretariat: risk assessment evaluation for concentrated animal feeding operations. The U.S. Environmental Protection Agency (2004)
29. Zhang, R., El-Mashad, H.M., et al.: Characterization of food waste as feedstock for anaerobic digestion. Bioresour. Technol. **98**(4), 929–935 (2007)
30. Zilberman, D., Zhao, J., Heiman, A.: Adoption versus adaptation, with emphasis on climate change. Annu. Rev. Resource Econ. **4**, 27–53 (2012)

Chapter 12
Ergodic Transport Theory, Periodic Maximizing Probabilities and the Twist Condition

G. Contreras, A.O. Lopes, and E.R. Oliveira

12.1 Introduction

We will state in this section the mathematical definitions and concepts we will consider in this work. We denote by T the action of the shift in the Bernoulli space $\{1, 2, 3, .., d\}^{\mathbb{N}} = \Sigma$.

Definition 12.1.1. Denote

$$m(A) = \max_{\nu \text{ an invariant probability for } T} \int A(x) \, d\nu(x),$$

and, $\mu_{\infty,A}$, any probability which attains the maximum value. Any one of these probabilities $\mu_{\infty,A}$ is called a maximizing probability for A.

We will assume here that the maximizing probability is unique and has support in a periodic orbit. An important conjecture claimed that this property is generic (see [11] for partial results). Recently this conjecture was proved by Contreras (see [9])

The analysis of this kind of problem is called Ergodic Optimization Theory [2,3, 5,8,11,12,15,17–19,29]. A generalization of such problems from the point of view of Ergodic Transport can be found in [21].

We denote the set of α-Hölder potentials on Σ by $C^\alpha(\Sigma, \mathbb{R})$.

If A is Hölder, and, the maximizing probability for A is unique, then the probability $\mu_{\infty,A}$ is the limit of the Gibbs states $\mu_{\beta A}$, for the potentials βA, when

G. Contreras
CIMAT, Guanajuato, CP: 36240, Mexico
e-mail: gonzalo@cimat.mx

A.O. Lopes (✉) • E.R. Oliveira
Instituto de Matemática, UFRGS, 91509-900, Porto Alegre, Brazil
e-mail: arturoscar.lopes@gmail.com; oliveira.elismar@gmail.com

A.A. Pinto and D. Zilberman (eds.), *Modeling, Dynamics, Optimization*
and Bioeconomics I, Springer Proceedings in Mathematics & Statistics 73,
DOI 10.1007/978-3-319-04849-9__12,
© Springer International Publishing Switzerland 2014

$\beta \to \infty$ [11, 13]. Therefore, our analysis concerns Gibbs probabilities at zero temperature ($\beta = 1/T$).

The metric d in Σ is defined by

$$d(\omega, v) := \lambda^N, \qquad N := \min\{ k \in \mathbb{N} \mid \omega_k \neq v_k \}.$$

The norm we consider in the set $C^\alpha(\Sigma, \mathbb{R})$ of α-Hölder potentials A is

$$||A||_\alpha = \sup_{d(x,y)<\varepsilon} \frac{|A(x) - A(y)|}{d(x, y)^\alpha} + \sup_{x \in \Sigma} |A(x)|.$$

Contreras showed that for a generic set of α-Hölder potentials in this topology the maximizing probability for A has support in a periodic orbit (see [9]).

One can also consider Lipchitz potentials and the theory works in the same way. We denote $\hat{\Sigma} = \Sigma \times \Sigma = \{1, 2, 3, .., d\}^{\mathbb{Z}}$, and, we use the notation

$$\bar{x} = (\ldots x_{-2}, x_{-1} \mid x_0, x_1, x_2, ..) \in \hat{\Sigma},$$

$w = (x_0, x_1, x_2, ..) \in \Sigma$, $x = (x_{-1}, x_{-2}, ..) \in \Sigma$. Sometimes we denote $\bar{x} = (w, x)$. We say $w = (x_0, x_1, x_2, ..)$ are the future coordinates of $\bar{x}$ and $x = (x_{-1}, x_{-2}, ..) \in \Sigma$ are the past coordinates of $\bar{x}$. We also use the notation: σ is the shift acting in the past coordinates w and T is the shift acting in the future coordinates x. Moreover, $\mathbb{T}$ is the (right side)-shift on $\hat{\Sigma}$, that is,

$$\mathbb{T}^{-1}(\ldots x_{-2}, x_{-1} \mid x_0, x_1, x_2, ..) = (\ldots x_{-2}, x_{-1} \, x_0 \mid x_1, x_2, ..).$$

As we said before we denote the action of the shift in the coordinates x by T, that is,

$$T(x_{-1}, x_{-2}, x_{-2}..) = (x_{-2}, x_{-3}, x_{-4}..).$$

For $w = (x_0, x_1, x_2, ..) \in \Sigma$, $x = (x_{-1}, x_{-2}, ..) \in \Sigma$, denote $\tau_w(x) = (x_0, x_{-1}, x_{-2}, ..)$.

We use sometimes the simplified notation $\tau_w = \tau_{x_0}$, because the dependence on w is just on the first symbol x_0.

Using the simplified notation $\bar{x} = (w, x)$, we have

$$\mathbb{T}^{-1}(w, x) = (\sigma(w), \tau_w(x)).$$

We also define $\tau_{k,a} x = (a_k, a_{k-1}, \ldots, a_1, x_0, x_1, x_2, \ldots)$, where $x = (x_0, x_1, x_2, \ldots)$, $a = (a_1, a_2, a_3, \ldots)$.

Bellow we consider that A acts on the past coordinate x and A^* acts on the future coordinate w.

Definition 12.1.2. Consider $A : \Sigma \to \mathbb{R}$ Hölder. We say that a Hölder continuous function $W : \hat{\Sigma} \to \mathbb{R}$ is a involution kernel for A, if there is a Hölder function $A^* : \Sigma \to \mathbb{R}$, such that,

$$A^*(w) = A \circ \mathbb{T}^{-1}(w, x) + W \circ \mathbb{T}^{-1}(w, x) - W(w, x).$$

We say that A^* is a dual potential of A, or, that A and A^* are in involution.

The above expression can be also written as

$$A^*(w) = A(\tau_w(x)) + W(\sigma(w), \tau_w(x)) - W(w, x).$$

Theorem 1 ([1]). *Given A Hölder there exist a Hölder W which is an involution kernel for A.*

We call μ_{∞, A^*} the maximizing probability for A^* (it is unique if A has a unique maximizing $\mu_{\infty A}$, as one can see in [1]).

To consider a dual problem is quite natural in our setting (see also [27]). Note that $m(A) = m(A^*)$ (see [1]).

In [3] it is presented several examples where one can get explicitly the involution kernel and the dual potential.

We denote by $\mathbb{K} = \mathbb{K}(\mu_{\infty, A}, \mu_{\infty, A^*})$ the set of probabilities $\hat{\eta}(w, x)$ on $\hat{\Sigma}$, such that $\pi_x^*(\hat{\eta}) = \mu_{\infty, A}$, and also that $\pi_y^*(\hat{\eta}) = \mu_{\infty, A^*}$.

We are interested in the solution $\hat{\mu}$ of the Kantorovich Transport Problem for $-W$, that is, the solution of

$$\inf_{\hat{\eta} \in \mathbb{K}} \int \int -W(w, x) \, d\, \hat{\eta}.$$

Note that in the definition of W we use the dynamics of $\mathbb{T}$. Note also that if we consider a new cost of the form $c(x, w) = -W(x, w) + \varphi(w)$, instead of $-W$, where φ bounded and measurable, then we **do not** change the original minimization problem.

Definition 12.1.3. A calibrated sub-action $V : \Sigma \to \mathbb{R}$ for the potential A, is a continuous function V such that

$$\sup_{y \text{ such that } T(y)=x} \{ V(y) + A(y) - m(A) \} = V(x).$$

We denote by V^* a calibrated subaction for A^*.

We denote by $\hat{\mu}$ the minimizing probability over $\hat{\Sigma} = \{1, 2, 3, .., d\}^{\mathbb{Z}}$ for the natural Kantorovich Transport Problem associated to the $-W$, where W is the involution kernel for A (see [1]).

We call $\hat{\mu}_{max}$ the natural extension of $\mu_{\infty, A}$ as described in [1, 28].

In [22] was shown (not assuming the maximizing probability is a periodic orbit) that:

Theorem 2. *Suppose the maximizing probability for A Hölder is unique (not necessarily a periodic orbit). Then, the minimizing Kantorovich probability $\hat{\mu}$ on $\hat{\Sigma}$ associated to $-W$, where W is the involution kernel for A, is $\hat{\mu}_{max}$.*

Moreover, it was shown in [22]:

Theorem 3. *Suppose the maximizing probability is unique (not necessarily a periodic orbit). If V is the calibrated subaction for A, and V^* is the calibrated subaction for A^*, then, the pair $(-V, -V^*)$ is the dual $(-W)$-Kantorovich pair of $(\mu_{\infty,A}, \mu_{\infty,A^*})$.*

The solution cames from the so called complementary slackness condition [7,30, 31] which were obtained in Proposition 10 (1) [1]. We can assume that γ in [1] is equal to zero.

One can consider in the Bernoulli space $\Sigma = \{0, 1\}^{\mathbb{N}}$ the lexicographic order. In this way, $x < z$, if and only if, the first element i such that, $x_j = z_j$ for all $j < i$, and $x_i \neq z_i$, satisfies the property $x_i < z_i$. Moreover, $(0, x_1, x_2, \ldots) < (1, x_1, x_2, \ldots)$.

One can also consider the more general case $\Sigma = \{0, 1, \ldots, d-1\}^{\mathbb{N}}$, but in order to simplify the notation and to avoid technicalities, we consider only the case $\Sigma = \{0, 1\}^{\mathbb{N}}$. We also suppose, from now on, that $\hat{\Sigma} = \Sigma \times \Sigma$.

Definition 12.1.4. We say a continuous $G : \hat{\Sigma} = \Sigma \times \Sigma \to \mathbb{R}$ satisfies the twist condition on $\hat{\Sigma}$, if for any $(a, b) \in \hat{\Sigma} = \Sigma \times \Sigma$ and $(a', b') \in \Sigma \times \Sigma$, with $a' > a$, $b' > b$, we have

$$G(a, b) + G(a', b') < G(a, b') + G(a', b). \tag{12.1}$$

Definition 12.1.5. We say a continuous $A : \Sigma \to \mathbb{R}$ satisfies the twist condition, if its involution kernel W satisfies the twist condition.

Examples of twist potentials are presented in [22].
One of the main results in [22] is:

Theorem 4. *Suppose the maximizing probability is unique (not necessarily a periodic orbit) and W satisfies the twist condition on $\hat{\Sigma}$, then, the support of $\hat{\mu}_{max} = \hat{\mu}$ on $\hat{\Sigma}$ is a graph (up to one possible orbit).*

Given $A : \Sigma = \{1, 2, .., d\}^{\mathbb{N}} \to \mathbb{R}$ Hölder, the Ruelle operator $\mathscr{L}_A : C^0(\Sigma) \to C^0(\Sigma)$ is given by $\mathscr{L}_A(\phi)(x) = \psi(x) = \sum_{T(z)=x} e^{A(z)} \phi(z)$.

We also consider $\mathscr{L}_{A^*}$ acting on continuous functions in $\Sigma = \{1, 2, .., d\}^{\mathbb{N}}$.

We also denote by ϕ_A and ϕ_{A^*} the corresponding eigen-functions associated to the main common eigenvalue $\lambda(A)$ of the operators $\mathscr{L}_A$ and $\mathscr{L}_{A^*}$ (see [26]).

ν_A, and, ν_{A^*} are respectively the eigen-probabilities for the dual of the Ruelle operator $\mathscr{L}_A$ and $\mathscr{L}_{A^*}$.

Finally, $\mu_A = \nu_A \phi_A =$ and $\mu_{A^*} = \nu_{A^*} \phi_{A^*}$ are the invariant probabilities such that they are solution of the respective pressure problems for A and A^*. The probability μ_A is called the Gibbs measure for the potential A.

If the maximizing probability is unique, then, it is easy to see that considering for any real β the potential βA and the corresponding $\mu_{\beta A}$, then, $\mu_{\beta A} \to \mu_{\infty,A}$, when $\beta \to \infty$.

In the same way, if we take any real β, the potential βA^*, and, the corresponding $\mu_{\beta A^*}$, then $\mu_{\beta A^*} \to \mu_{\infty,A^*}$, when $\beta \to \infty$.

In Statistical Mechanics $\beta = \frac{1}{T}$ where T is temperature. Then, $\mu_{\infty,A}$ is the version of the Gibbs probability at temperature zero.

One can choose c (a normalization constant) such that

$$\int \int e^{W(w,x)-c} \, d\nu_{A^*}(w) \, d\nu_A(x) = 1. \tag{12.2}$$

A calibrated sub-action V can obtained as the limit [11, 13]

$$V(x) = \lim_{\beta \to \infty} \frac{1}{\beta} \log \phi_{\beta A}(x). \tag{12.3}$$

In the same way we can get a calibrated sub-action V^* for A^* by taking

$$V^*(w) = \lim_{\beta \to \infty} \frac{1}{\beta} \log \phi_{\beta A^*}(w) . \tag{12.4}$$

Moreover, by [1]

$$\phi_{A^*}(w) = \int e^{W_A(w,x)-c} \, d\nu_A(x), \tag{12.5}$$

and,

$$\phi_A(x) = \int e^{W_A(w,x)-c} \, d\nu_{A^*}(w) = \int e^{W_A(w,x)-c} \frac{1}{\phi_{A^*}} d\mu_{A^*}(w). \tag{12.6}$$

Note that if W and A^* define an involution for A, then, given any real β, we have that βW and βA^* define an involution for the potential βA. The normalizing constant c, of course, changes with β (see [1]).

We say a family of probabilities μ_β, $\beta \to \infty$, satisfies a Large Deviation Principle, if there is function $I : \Sigma \to \mathbb{R} \cup \infty$, which is non-negative, lower semi-continuous, and such that, for any cylinder $K \subset \Sigma$, we have

$$\lim_{\beta \to \infty} \frac{1}{\beta} \log(\mu_\beta(K)) = - \inf_{z \in K} I(z).$$

In this case we say the I is the deviation function. The function I can take the value ∞ in some points.

In [1] a Large Deviation Principle is described for the family $\mu_\beta = \mu_{\beta A}$ of equilibrium states for βA, under the assumption that the maximizing probability for A is unique (see also [20] for a different case where it is not assumed uniqueness). The function I is zero on the support of the maximizing probability for A. We point out that there are examples where I can be zero outside the support (even when the maximizing probability is unique) as we will show bellow in an example due to Leplaideur.

Applying the same to A^* we get a deviation function $I^* : \Sigma \to \mathbb{R} \cup \{\infty\}$.

The function I^* is defined by

$$I^*(w) = \sum_{n \geq 0} \left(V^* \circ \sigma - V^* - A^* \right) \circ \sigma^n(w),$$

where V^* is a fixed calibrated subaction.

We point out that in fact the claim of Theorem 3 should say more precisely that the pair $(-V, -V^*)$ is the dual $(-W + I^*)$-Kantorovich pair of $(\mu_{\infty,A}, \mu_{\infty,A^*})$.

Using the property (12.6) above, and, adapting Varadhan's Theorem [14] to the present setting, it is shown in [22], that given any $x \in \Sigma$, then, it is true the relation

$$V(x) = \sup_{w \in \Sigma} \left[W(w, x) - V^*(w) - I^*(w) \right].$$

For each x we get one (or, more than one) w_x such attains the supremum above. Therefore,

$$V(x) = W(w_x, x) - V^*(w_x) - I^*(w_x).$$

A pair of the form (x, w_x) is called an optimal pair. We can also say that w_x is optimal for x.

Given A the involution kernel W and the dual potential A^* are not unique. But, the above maximization problem is intrinsic on A. That is, if we take another A^* and the corresponding W, there is some canceling, and we get the same problem as above (given x the optimal w_x does not change).

It is also true that (see [22]) there exists γ such that for any w

$$\gamma + V^*(w) = \sup_{x \in \Sigma} \left[W(w, x) - V(x) - I(x) \right].$$

Given w, a solution of the above maximization is denoted by x_w. **We denote the pair (x_w, w) a ∗-optimal pair.**

We assume without lost of generality that $\gamma = 0$, by adding $-\gamma$ to the function W.

Definition 12.1.6. The set of all (x, w_x), is called the optimal set for A, and, denoted by $\mathbb{O}(A)$.

Remark 1. Note that in [22] it was shown that, under the twist assumption, the support of the optimal transport periodic probability $\hat{\mu}_A$, for the cost $-W$, is a graph, that is, for each x, there is only one w such that (x, w) is in the support of $\hat{\mu}_A$. But, nothing is said about the graph property of the set $\mathbb{O}(A)$. Note that the support of $\hat{\mu}_A$ is contained in $\mathbb{O}(A)$.

Remark 2. Note also that the minimal transport problem for the cost $-W$, or the cost $-W + I^*$ is the same (see [22]).

Our main result is:

Theorem 5. *Generically, in the set of potentials Hölder potentials A that satisfy*

(i) *the twist condition,*
(ii) *uniqueness of maximizing probability which is supported in a periodic orbit, the set of possible optimal w_x, when x covers the all range of possible elements x in $\in \Sigma$, is finite.*

We point out that this is a result for points outside the support of the maximizing probability.

In the first two sections we will show that under certain conditions the set of possible optimal w_x is finite, for any x. In Sects. 12.3 and 12.4 we will show that these conditions are generic.

In [24] it is also considered the twist condition and results for the analytic setting are obtained.

12.2   The Twist Property

In order to simplify the notation we assume that $m(A) = m(A^*) = 0$.

Given A, we denote

$$\Delta(x, x', y) = \sum_{n \geq 1} A \circ \tau_{y,n}(x) - A \circ \tau_{y,n}(x').$$

The involution kernel W can be computed for any (ω, x) by $W(\omega, x) = \Delta_A(\omega, \overline{x}, x)$, where we choose a point $\overline{x}$ for good.

It is known the following relation: for any $x, x', w \in \Sigma$, we have that $W(w, x) - W(w, x') = \Delta(w, x, x')$ (see [1]).

We assume from now on that A satisfies the twist condition. It is known in this case (see [4, 23]), that $x \to w_x$ (can be multivaluated) is monotonous decreasing.

Proposition 12.2.1. *If A is twist, then* $x \to w_x$ *is monotonous decreasing.*

Proof. See [24]. □

We define R by the expression $R(x) = V(\sigma(x)) - V(x) - A(x) \geq 0$, and, we define R^* by $R^*(w) = V^*(\sigma(w)) - V^*(w) - A^*(w) \geq 0$.

Note that given y, there is a preimage x of y, such that, $R(x) = 0$. The analogous property is true for R^*.

Given A (and a certain choice of A^* and W) the next result (which does not assume the twist condition) claims that the dual of R is R^*, and the corresponding involution kernel is $(V^* + V - W)$.

Proposition 12.2.2 (Fundamental Relation (FR)).

$$R(\tau_w x) = (V^* + V - W)(x, w) - (V^* + V - W)(\tau_w x, \sigma(w)) + R^*(w).$$

Proof. see [24]. □

We know that the calibrated subaction satisfies

$$V(x) = \max_{w \in \Sigma}(-V^* - I^* + W)(x, w).$$

Then, we define

$$b(x, w) = (V^* + V + I^* - W)(x, w) \geq 0,$$

and,

$$\Gamma_V = \{(x, w) \in \Sigma \times \Sigma \,|\, V(x) = (-V^* - I^* + W)(x, w)\},$$

which can be written in an equivalent form

$$\Gamma_V = \{(x, w) \in \Sigma \times \Sigma \,|\, b(x, w) = 0\}.$$

Given x, this maximum at w_x can not be realized where $I^*(w)$ is infinity.

Remark 3. Note, that $b(x, w) = 0$, if and only if, (x, w) is an optimal pair. We are not saying anything for $*$-optimal pairs.

If we use $R^*(w) = I^*(w) - I^*(\sigma w)$, the FR becomes

$$R(\tau_w x) = (V^* + V - W)(x, w) - (V^* + V - W)(\tau_w x, \sigma(w)) + I^*(w) - I^*(\sigma(w)),$$

or

$$R(\tau_w x) = b(x, w) - b(\tau_w x, \sigma(w)) \qquad \text{FR1.}$$

From this main equation we get:

Lemma 1. *If* $\mathbb{T}^{-1}(x, w) = (\tau_w x, \sigma(w))$, *then*

a) $b - b \circ \mathbb{T}^{-1}(x, w) = R(\tau_w x)$;
b) *The function b it is not decreasing in the trajectories of $\mathbb{T}$;*
c) Γ_V *is backward invariant;*
d) *when (x, w) is optimal then $R(\tau_w(x)) = 0$.*

Proof. The first claim is a trivial consequence of the definition of $\mathbb{T}^{-1}$. The second one it is a consequence of $R \geq 0$:

$$b - b \circ \mathbb{T}^{-1}(x, w) = R(\tau_w x) \geq 0$$

$$b(x, w) \geq b \circ \mathbb{T}^{-1}(x, w).$$

In order to see the third part we observe that

$$(x, w) \in \Gamma_V \Leftrightarrow b(x, w) = 0$$

Since

$$b(x, w) \geq b \circ \mathbb{T}^{-1}(x, w) \geq 0$$

we get $b(\tau_w x, \sigma(w)) = 0$ thus $(\tau_w x, \sigma(w)) \in \Gamma_V$. $\square$

From the above we get that in the case (x, w) is optimal, then, $\mathbb{T}^{-1}(x, w)$ is also optimal. Indeed, we have that

$$b(x, w) = 0 \Rightarrow b(\tau_w x, \sigma(w)) = 0.$$

This is equivalent to

$$V(x) = -V^*(w) - I^*(w) - W(x, w) \Rightarrow$$
$$V(\tau_w x) = -V^*(\sigma(w)) - I^*(\sigma(w)) - W(\tau_w x, \sigma(w)).$$

In this way $\mathbb{T}^{-n}$ spread optimal pairs.
We denote by M the support of the maximizing probability periodic orbit.
Consider the compact set of points $P = \{w \in \Sigma,$ such that $\sigma(w) \in M$, and w is not on $M\}$.

Definition 12.2.1. We say that A is good if, for each $w \in P$, we have that $R^*(w) > 0$.

We alternatively, say sometimes that R^* is good for A^*.
We point out that there are examples of potentials A^* (with a unique maximizing probability) where the corresponding R^* is not good (see Example 1 in the end of the present section).

Remember that ,

$$I^*(w) = \sum_{n \geq 0} \left(V^* \circ \sigma - V^* - A^* \right) \circ \sigma^n(w) = \sum_{n \geq 0} R^*(\sigma^n(w)).$$

In [23] Sect. 5 it is shown that if $I^*(w)$ is finite, then

$$\lim_{n \to \infty} \frac{1}{n} \sum_{j=0}^{n-1} \delta_{\sigma^j(w)} \to \mu_\infty^*.$$

One important assumption here is that R^* is good. We will show later that this property is true for generic potentials A.

Proposition 12.2.3. *Assume A is good. If $I^*(w) < \infty$, then, w is in the preimage of the maximizing probability for A^*.*

For a proof see Sect. 5 in [23].

Proposition 12.2.4. *Suppose (x, w_x) is an optimal pair, and, A is good, then, there exists k, such that, $\tilde{w}_k = \sigma^k(w_x)$ is in the support of the maximizing probability for A^*. Moreover, for such k, we have that $\mathbb{T}^{-k}(x, w_x)$ is an optimal pair.*

Proof. Suppose (x, w_x) is optimal. Therefore, $I^*(w_x) < \infty$. Then, by a previous proposition, w_x is in the pre-image of the maximizing probability for A^*, that is, there exists k such that $\tilde{w} = \sigma^k(w_x)$ is in the support of the maximizing probability for A^*.

Moreover, $b(\mathbb{T}^{-k}(x, w_x)) = (x_k, \sigma^k(w_x)) = (x_k, \tilde{w}_k)$ is also optimal. $\square$

Proposition 12.2.5. *Assume A is good, then, the set of w such that $I^*(w) < \infty$ is countable.*

Definition 12.2.2. We say a continuous $A : \Sigma \to \mathbb{R}$ satisfies the the countable condition, if there are a countable number of possible optimal w_x, when x ranges over the interval Σ.

Remark. If A is good, then, it satisfies the countable condition.

We showed before that the twist property implies that for $x < x'$, if $b(x, w) = 0$ and $b(x', w') = 0$, then $w' < w$, which means that the optimal sequences are monotonous not increasing. Thus, we define the **"turning point c"** as being the maximum of the point x that has his optimal sequence starting in 1:

$$c = \sup\{x \mid b(x, w) = 0 \Rightarrow w = (1\, w_1\, w_2 \ldots)\}.$$

The main criteria is the following:
"If $x \in \sigma$ has the optimal sequence $w = (w_0\, w_1\, w_2\, \ldots)$ then

$$w_0 = \begin{cases} 1, & if \ x \in [0^\infty, c] \\ 0, & if \ x \in (c, 1^\infty]'' \end{cases}$$

Starting from (x_0, w_0) we can iterate FR1 by $\mathbb{T}^{-n}(x, w) = (x_n, w_n)$ in order to obtain new points $w_1, w_2 \ldots \in \Sigma$. Unless the only possible optimal point w_x, for all x, is a fixed point for σ, then, $0 < c < 1$.

Note that for c there are two optimal pairs (c, w) and (c, w'), where the first symbol of w is zero, and, the first symbol of w' is one.

We denote

$$B(w) = \{x \mid b(x, w) = 0\}.$$

Lemma 2 (Characterization of Optimal Change). *Let $c \in (0^\infty, 1^\infty)$ be the turning point then, for any $x < x'$ and $b(x, w) = 0$ and $b(x', w') = 0$, we have $w \neq w'$ if, and only if, there exists $n \geq 0$ such that $T^n(c) \in [x, x']$. Moreover, if x, x' are such that w_x and w'_x are identical until the n coordinate, then, $T^n(c) \in (x, x')$.*

Each set $B(w) = [a, b]$ is such that $a = T^n(c)$, or, a it is accumulated by a subsequence of $T^j(c)$ from the left side. Similar property is true for b (accumulated by the right side).

Proof. Step 0   If $x < x' \leq c$ then $w_0 = w'_0 = 1$ else if $c < x < x'$ then $w_0 = w'_0 = 0$. Suppose $w_0 = w'_0 = \in \{0, 1\}$ then applying FR1 we get $\tau_i x < \tau_i x'$ and $b(\tau_i x, (w_1 w_2 \ldots)) = 0$ and $b(\tau_i x', (w'_1 w'_2 \ldots)) = 0$.

Step 1   If $\tau_i x < \tau_i x' \leq c$ then $w_1 = w'_1 = 1$ else if $c < \tau_i x < \tau_i x'$ then $w_1 = w'_1 = 0$. Otherwise if $\tau_1 x < c < \tau_1 x'$ we can use the monotonicity of T in each branch in order to get $x < T(c) < x'$. Thus

$$w_1 \neq w'_1 \Leftrightarrow x < T(c) < x'.$$

The conclusion comes by iterating this algorithm.

The last claim is obvious from the above.

A point x is called pre-periodic (or, eventually periodic) if there is $n \neq m$, such that, $T^n(x) = T^m(x)$.

We denote $[a, b]$, $a, b \in \Sigma$, $a \leq b$, the set of all points w in Σ such $a \leq w \leq b$. We call $[a, b]$ the interval determined by a and b. Each interval $[a, b]$, with $a < b$, is not countable.

Lemma 3. *The set*

$$B(w) = \{x \mid b(x, w) = 0\}$$

is an interval (can eventually be a single point). More specifically, if $B(w) = [a, b]$, then, a and b are adherence points of the orbit of c.

In particular, if c is pre-periodic, then, for any non-empty $B(w)$, there exists n, m such that $B(w) = [T^n(c), T^m(c)]$ (unless $B(w)$ is of the form $[0, b]$, or $[a, 1]$.

For a proof see [24].

Lemma 4. *Let $c \in \Sigma$ be the turning point. Let us suppose the c is isolated from his orbit, which means that, there is d, e, w^- and w^+, such that, $b(x, w^-) = 0$, for any $x \in (d, c]$, and, $b(x, w^+) = 0$, for any $x \in [c, e)$, then, there is no accumulation points of the orbit of c. In this case c is eventually periodic.*

For a proof see [24].

Remark 4. The main problem we have to face is the possibility that the orbit of c is dense in $[0, 1]$.

If c is eventually periodic there exist just a finite number intervals $B(w)$ with positive length. The other $B(w)$ are reduced to points and they are also finite.

Lemma 5. *Suppose A satisfies the twist and the countable condition. Then there is at least one $B(w)$ with positive length of the form $(T^n(c), T^m(c))$. Moreover, for each subinterval (a, b) there exists at least one $B(w)$ with positive length of the form $(T^n(c), T^m(c))$ contained on it. Therefore, there exists an infinite number of such intervals.*

Denote the possible w, such that, $I^*(w) < \infty$, by w_j, $j \in \mathbb{N}$.

For each w^j, $j \in \mathbb{N}$, denote $I_j = B(w^j)$, the maximal interval where for all $x \in I_j$, we have that, (x, w^j) is an optimal pair. Some of these intervals could be eventually a point, but, an infinite number of them have positive length, because the set Σ is not countable. We consider from now on just the ones with positive length.

Note that by the same reason, in each subinterval (e, u), there exists an infinite countable number of $B(w)$ with positive length.

We suppose, by contradiction, that each interval $B(w) = [a, b]$, with positive length is such that, each side is approximated by a sub-sequence of points $T^j(c)$.

Take one interval (a_1, b_1) with positive length inside Σ. There is another one (a_2, b_2) inside $(0^\infty, a_1)$, and one more (a_3, b_3) inside $(b_1, 1^\infty)$.

If we remove from the interval Σ these three intervals we get four intervals. Using our hypothesis, we can find new intervals with positive length inside each one of them. Then we do the same removal procedure as before. This procedure is similar to the construction of the Cantor set. If we proceed inductively on this way, the set of points x which remains after infinite steps is not countable. An uncountable number of such x has a different w_x. This is not possible because the optimal w_x are countable.

Then, the first claim of the lemma is true.

Given an interval $(a, b) \subset \Sigma$, we can do the same and use the fact that (a, b) is not countable. $\square$

Lemma 6. *Under the twist and the countable condition the turning point c is eventually periodic.*

Proof. Denote by w^j, $j \in \mathbb{N}$, the countable set of pre-images of the periodic orbit maximizing A^*.

For each w^j, $j \in \mathbb{N}$, denote $I_j = B(w^j)$, the maximal interval where for all $x \in I_j$, we have that, (x, w^j) is an optimal pair. Some of these intervals could be, eventually, a point, but, an infinite number of them have positive length, because the set Σ is not countable.

We will suppose c is not eventually periodic, and, we will reach a contradiction. Therefore, if $T^n(c) = T^m(c)$, then $m = n$.

From, now on we consider just j, such that, for the corresponding w_j, the interval I_j has positive length, and, it is of the form $[T^n(c), T^m(c)]$, $m, n \geq 1$. From last lemma there exist an infinite number of them.

Denote by $I_j = [a_j, b_j]$. We denote I_0 the interval of the form $[0^\infty, b_0]$, and, I_1 the interval of the form $[a_0, 1^\infty]$. From last lemma, for $j \neq 0, 1$, there is n_j and m_j, such that $a_j = T^{n_j}(c)$ and $b_j = T^{m_j}(c)$.

Consider the inverse branch $\tau_{i_1^j}$, where $i_1^j = i_1$ is such that $\tau_{i_1}((T^{n_j})(c)) = T^{n_j-1}(c)$. This i_1 **do not have to be the first symbol of the optimal** w for $T^{n_j}(c)$. Then, $\tau_{i_1}(I_j)$ is another interval, which is strictly inside a domain of injectivity of T, does not contain any forward image of c, and in its left side we have the point $T^{n_j-1}(c)$.

Then, repeating the same procedure inductively, we get i_2, such that $\tau_{i_2}((T^{n_j-1})(c)) = T^{n_j-2}(c)$, determining another interval which does not contain any forward image of c, and in his left side we have the point $T^{n_j-2}(c)$. Repeating the reasoning over and over again, always taking the same inverse branch which contain $T^n(c)$, $0 \leq n \leq n_j$, after n_j times we arrive in an interval of the form (c, r_j). Note that each inverse branch preserves order. It is not possible to have an iterate $T^k(c)$, $k \in \mathbb{N}$, inside this interval (c, r_j) (by the definition of I_j). Then, the optimal w for x in this interval (c, r_j) is a certain $\tilde{w}_j$ which can be different of $\sigma^{n_j}(w_j)$.

Suppose now that $n_j > m_j$. Using the analogous procedure we get that there exists r^j, such that the optimal w_x for x in the interval (r^j, c) is $\sigma^{n_j}(w_j)$.

If both cases happen, then c is eventually periodic.

The trouble happens when just one type of inequality is true. Suppose without lost of generality that we have always $n_j < m_j$, for all possible j.

Let's fix for good a certain j.

Therefore, all we can get with the above procedure is that c is isolated by the right side.

In the procedure of taking pre-image of $T^{n_j}(c)$, always following the forward orbit $T^n(x)$, $0 \leq n \leq n_j$, we will get a sequence of $i_1, i_2, \ldots, i_{n_j}$. In the first step we have two possibilities: $\tau_{i_1}(T^{m-j}(c)) = T^{m_j-1}(c)$, or not.

If it happens the second case, we are done. Indeed, the interval $\tau_{i_1}[T^{n_j}(c), T^{m_j}(c)]$ does not contain forward images of c (otherwise $[T^{n_j}(c), T^{m_j}(c)]$ would also have). Now we follow the same procedure as before, but, this time following the branches which contains the orbit of $T^m(c)$, $0 \leq m \leq m_j$. In this way, we get that c is isolated by the left side.

Suppose $\tau_{i_1}(T^{m_j}(c)) = T^{m_j-1}(c)$. Consider the interval, $[T^{n_j-1}(c), T^{m_j-1}(c)]$, which do not contain forward images of c.

Now, you can ask the same question: $\tau_{i_2}(T^{m_j-1}(c)) = T^{m_j-2}(c)$? If this do not happen (called the second option), then, in the same way as before, we are done (c is also isolated by the right side). If the expression is true, then, we proceed with the same reasoning as before.

We proceed in an inductive way until time n_j. If in some time we have the second option, we are done, otherwise, we show that any $x \in (c, T^{m_j-n_j}(c))$ has a unique optimal w_x (there is no forward image of c inside it).

Denote $k = m_j - n_j$ for the j we fixed.

From the above we have that for any $B(w)$, which is an interval of the form $[T^{n_i}(c), T^{m_i}(c)]$, for any possible i, it is true that $m_i - n_i = k$. There are an infinite number of intervals of this form.

We claim that the set of points x which are extreme points of any $B(\tilde{w})$, and, such that x can be approximated by the forward orbit of c is finite. Suppose without lost of generality that x is the right point of a $B(w) = (z, x)$.

If the above happens, then, by the last lemma, we have an infinite sequence of intervals of the form $[T^{n_i}(c), T^{n_i+k}(c)]$, such that $T^{n_i}(c) \to x$, as $n_i \to \infty$. Therefore, x is a periodic point of period k. There are a finite number of points of period k. This shows our main claim. Finally, c is eventually periodic. $\square$

Definition 12.2.3. We denote by $\mathbb{G}$ the set of Hölder potentials such that

1. the maximizing probability is unique, and it is a periodic orbit;
2. The potential A satisfies the twist condition;
3. R^* is good for A^*.

From the above one can get:

Theorem 6. *Suppose A is in $\mathbb{G}$, then, for any x, there exists a finite number of possible w_x such that (x, w_x) are optimal pairs. If we denote by $I_j = (a_j, b_j)$, $j = 1, 2, ..., n$, the maximal open intervals where for $x \in I_j$ the w_x is constant, then, just on the points $x = a_j$, or $x = b_j$, we can get two different w_x, which define points (x, w_x) in the optimal set $\mathbb{O}(A)$.*

Note that if x is in the maximizing orbit for $A \in \mathbb{G}$, then, at least one of the optimal w_x is in the support of the maximizing probability for A^*. This point x can be eventually in the extreme of one of this maximal intervals I_j. This do not contradicts the graph property.

Definition 12.2.4. We denote by $\mathbb{A}$ the open set of Hölder potentials such that

1) the maximizing probability is unique, and it is a periodic orbit;
2) The potential A satisfies the twist condition.

The next theorem shows the class we consider above is large.

Theorem 7. *The set $\mathbb{G}$ is generic in the open set $\mathbb{A}$.*

The proof of this result will be done in the next two sections (see Theorem 12.4.1 bellow).

Corollary 1. *For any $A \in \mathbb{G}$, the value $c = c_A$ is given by the expression*

$$c = \inf\{x \mid V(x) - V(\tau_1 x) - A(\tau_1(x)) > 0\},$$

where V is any calibrated subaction for A. Moreover, c_A is locally constant as a function of A.

Proof. The first claim follows from the fact that we have to use 0 as first symbol of the optimal w_x, when x is on the right of c.

If R^* is good the point c is in the pre-image of the support of the maximizing probability (which is locally constant by the continuous varying support property [11]). Therefore, the possible c are in a countable set.

Note that under the uniqueness hypothesis of the maximizing probability for A, the sub-action $V = V_A$ can be chosen in a continuous fashion with A. From this follows the last claim of the corollary. $\square$

Now we will provide a counterexample.

Example 1. The following example is due to Leplaideur.

We will show an example on the shift where the maximizing probability for a certain Lipschitz potential $A^* : \{0, 1\}^\mathbb{N} \to \mathbb{R}$ is a unique periodic orbit γ of period two, denoted by $p_0 = (01010101\ldots)$, $p_1 = (10101010\ldots)$, but for a certain point, namely, $w_0 = (110101010\ldots)$, which satisfies $\sigma(w_0) = p_1$, we have that $R^*(w_0) = 0$.

The potential A^* is given by $A^*(w) = -d(w, \gamma \cup \Gamma)$, where d is the usual distance in the Bernoulli space. The set Γ is described later.

For each integer n, we define a $2n + 3$-periodic orbit $z_n, \sigma(z_n), \ldots, \sigma^{2n+2}(z_n)$ as follows:

we first set

$$b_n = (\underbrace{01010101\ldots01}_{2n} 1\,01),$$

and the point z_n is the concatenation of the word b_n: $z_n = (b_n, b_n, \ldots)$.

The main idea here is to get a sequence of periodic points which spin around the periodic orbit $\{p_0, p_1\}$ during the time $2n$, and then pass close by w_0 (note that $d(\sigma^{2n}(z_n), w_0) = 2^{-2(n+1)}$).

Denote γ_n the periodic orbit $\gamma_n = \{z_n, \sigma(z_n), \sigma^2(z_n), \ldots, \sigma^{2n+2}(z_n)\}$.

Consider the sequence of Holder (could be also Lipschitz) potentials $A_n^*(w) = -d(w, \gamma_n \cup \gamma)$. The support of the maximizing probability for A_n^* is $\gamma_n \cup \gamma$. Moreover

$$0 = m(A_n^*) = \max_{\substack{\nu \text{ an invariant probability for } \sigma}} \int A_n^*(w)\, d\nu(w).$$

Denote by V_n^* a Holder (or Lipschitz) calibrated subaction for A_n^* such that $V_n^*(w_0) = 0$. In this way, for all w

$$R_n^*(w) = (V_n^* \circ \sigma - V_n^* - A_n^*)(w) \geq 0,$$

and for $w \in \gamma_n \cup \gamma$ we have that $R_n^*(w) = 0$.

We know that R_n^* is zero on the orbit γ_n, because γ_n is included in the Masur set. Note that we not necessarily have $R_n^*(w_0) = 0$.

By construction, the Lipschitz constant for A_n^* is 1. This is also true for V_n^*. Hence the family of subactions (V_n^*) is a family of equicontinuous functions. Let us denote by V^* any accumulation point for (V_n^*) for the $\mathscr{C}^0$-topology. Note that V^* is also 1-Lipschitz continuous. For simplicity we set

$$V^* = \lim_{k \to \infty} V_{n_k}^*.$$

We denote by Γ the set which is the limit of the sets γ_n (using the Hausdorff distance). $\gamma \cup \Gamma$ is a compact set. Note that Γ is not a compact set, but the set of accumulation points for Γ is the set γ. We now consider $A^*(w) = -d(w, \gamma \cup \Gamma)$.

As any accumulation point of Γ is in γ, any maximizing probability for the potential A^* has support in γ. On the contrary, the unique σ-invariant measure with support in γ is maximizing for A^*.

Remember that for any n we have $V_n^*(w_0) = 0$. We also claim that we have $A_{n_k}^*(w_0) \to 0$ and $V_{n_k}^*(\sigma(w_0)) \to 0$, as $k \to \infty$.

For each fixed w we set

$$R_{n_k}^*(w) = (V_{n_k}^* \circ \sigma - V_{n_k}^* - A_{n_k}^*)(w) \geq 0.$$

The right hand side terms converge (for the $\mathscr{C}^0$-topology) as k goes to $+\infty$. Then $R_{n_k}^*$ converge, and we denote by R^* its limit. Then for every w we have:

$$R^*(w) = (V^* \circ \sigma - V^* - A^*)(w) \geq 0.$$

This shows that V^* is a subaction for A^*. Note that $R^*(w_0) = 0$. From the uniqueness of the maximizing probability for A^* we know that there exists a unique calibrated subaction for A^* (up to an additive constant).

Consider a fixed w and its two preimages w_a and w_b. For any given n, one of the two possibilities occur $R_n^*(w_a) = 0$ or $R_n^*(w_b) = 0$, because V_n^* is calibrated for A_n^*.

Therefore, for an infinite number of values k either $R_{n_k}^*(w_a) = 0$ or $R_{n_k}^*(w_b) = 0$.

In this way the limit of $V_{n_k}^*$ is unique (independent of the convergent subsequence) and equal to V^*, the calibrated subaction for A^* (such that $V^*(w_0) = 0$).

Therefore,

$$R^*(w_0) = (V^* \circ \sigma - V^* - A^*)(w_0) = 0,$$

and V^* is a calibrated subaction for $A^*(w) = d(w, \gamma \cup \Gamma)$.

12.3 Generic Continuity of the Aubry Set

In this section and in the next we will present the proof of the generic properties we mention before.

In order to do that we will need several general properties in Ergodic Optimization.

Remember that we will denote the action of the shift in the points x by T, and, we leave σ for the action of the shift in the coordinates w.

We will present our main results in great generality. First, in this section, we analyze the main properties of sub-actions and its dependence on the potential A.

First we will present the main definitions we will consider here.

$\mathscr{F} \subset C^0(\Sigma, \mathbb{R})$ denotes a complete metric space with a (topology finer than) metric larger than $d_{C^0}(f, g) = \|f - g\|_0 := \sup_{x \in \Sigma} |f(x) - g(x)|$; (for instance, Hölder functions, Lispchitz functions, etc.) AND such that

$$\forall K \subset \Sigma \text{ compact}, \ \exists \psi \in \mathscr{F} \text{ s.t. } \psi \leq 0, \ [\psi = 0] = \{x \mid \psi(x) = 0\} = K. \tag{12.7}$$

Given $A \in \mathscr{F}$ and F a calibrated sub-action for A, remember that its *error* is denoted by $R = R_A : \Sigma \to [0, +\infty[$:

$$R(x) := F(T(x)) - F(x) - A(x) + m_A \geq 0.$$

$\mathscr{S}(A)$ denotes the set of Hölder calibrated sub-actions (it is not empty [11, 13]). Given A, the *Mañé action potential* is:

$$S_A(x, y) := \lim_{\varepsilon \to 0} \left[\sup \left\{ \sum_{i=0}^{n-1} [A(T^i(z)) - m_A] \mid n \in \mathbb{N}, \ T^n(z) = y, \ d(z, x) < \varepsilon \right\} \right].$$

Given x and y the above value describe the A-cost of going from x to y following the dynamics.

The *Aubry set* is $\mathbb{A}(A) := \{x \in \Sigma \mid S_A(x, x) = 0\}$.

The terminology is borrowed from the Aubry-Mather Theory [10].

For any $x \in \mathbb{A}(A)$, we have that $S_A(x, .)$ is a sub-action (in particular, in this case, $S_A(x, y) > -\infty$, for any y), see Proposition 23 in [11].

The set of maximizing measures is

$$\mathscr{M}(A) := \{ \mu \in \mathscr{M}(T) \mid \int A \, d\mu = m_A \}.$$

If $F \in C^\alpha(\Sigma, \mathbb{R})$ is a Hölder function define

$$|F|_\alpha := \sup_{x \neq y} \frac{|F(x) - F(y)|}{d(x, y)^\alpha}.$$

Define the *Mañé set* as

$$\mathbb{N}(A) := \bigcup_{F \in \mathscr{S}(A)} I_F^{-1}\{0\},$$

where the union is among all the α-Hölder calibrated sub-actions F for A and

$$I_F(x) = \sum_{i=0}^{\infty} R_F(T^i(x)).$$

$I_F(x)$ is the deviation function we considered before.
For $A \in \mathscr{F}$ define the *Mather set* as

$$\mathbb{M}(A) := \bigcup_{\mu \in \mathscr{M}(A)} \mathrm{supp}(\mu).$$

The *Peierls barrier* is

$$h_A(x, y) := \lim_{\varepsilon \to 0} \limsup_{k \to +\infty} S_A(x, y, k, \varepsilon),$$

where $S_A(x, y, k, \varepsilon) := \sup \left\{ \sum_{i=0}^{n-1} \left[A(T^i(z)) - m_A \right] \, \Big| \, n \geq k, \ T^n(z) = y, \right.$

$$\left. d(z, x) < \varepsilon \right\}.$$

Several properties of the Mañé potential and the Peierls barrier are similar (but not all, see Sect. 4 in [16]). We will present proofs for one of them and the other case is similar.

Lemma 7.

1. If μ is a minimizing measure then

$$\mathrm{supp}(\mu) \subset \mathbb{A}(A) = \{ x \in \Sigma \mid S_A(x, x) = 0 \}.$$

2. $S_A(x, x) \leq 0$, for every $x \in \Sigma$.
3. For any $z \in \Sigma$, the function $F(y) = h_A(z, y)$ is Hölder continuous.
4. If, $a \in \mathbb{A}(A)$, then, $h_A(a, x) = S_A(a, x)$, for all $x \in \Sigma$.
     In particular, $F(y) = S_A(a, x)$ is continuous, if $a \in \mathbb{A}(A)$.
5. If $S_A(w, y) = h_A(w, y)$ then the function $F(y) = S_A(w, y)$ is continuous at y.
6. If $S(x_0, T^{n_k}(x_0)) = \sum_{j=0}^{n_k-1} A(T^j(x_0))$, $\lim_k T^{n_k}(x_0) = b$ and $\lim_k n_k = +\infty$,
   then

$$\lim_k S(x_0, T^{n_k}(x_0)) = S(x_0, b).$$

Item (1) follows from Atkinson-Mañé's lemma which says that if μ is ergodic for μ-almost every x and every $\varepsilon > 0$, the set

$$N(x, \varepsilon) := \left\{ n \in \mathbb{N} \mid \left| \sum_{j=0}^{n-1} A(T^j(x)) - n \int A \, d\mu \right| \leq \varepsilon \right\}$$

is infinite (see Lemma 2.2 [25] (which consider non-invertible transformation, [10], [11] or [15] for the proof). We will show bellow just the items which are not proved in the mentioned references.

The problem with the discontinuity of $F(y) = S_A(w, y)$ is when the maximum is obtained at a *finite* orbit segment (i.e. when $S_A(w, y) \neq h_A(w, y)$), the hypothesis in item (5).

Proof. By adding a constant we can assume that $m_A = 0$.

(2). Let F be a continuous sub-action for A. Then

$$-R_F = A + F - F \circ T \leq 0.$$

Given $x_0 \in \Sigma$, let $x_k \in \Sigma$ and $n_k \in \mathbb{N}$ be such that $T^{n_k}(x_k) = x_0$, $\lim_k x_k = x_0$ and

$$S(x_0, x_0) = \lim_k \sum_{j=0}^{n_k-1} A(T^j(x_k)).$$

We have

$$\sum_{j=0}^{n_k-1} A(T^j(x_k)) = \left[\sum_{j=0}^{n_k-1} (A + F - F \circ T)(T^j(x_k)) \right] + F(x_0) - F(x_k)$$

$$\leq F(x_0) - F(x_k).$$

Then

$$S(x_0, x_0) = \lim_k \sum_{j=0}^{n_k-1} A(T^j(x_k)) \leq \lim_k \left[F(x_0) - F(x_k) \right] = 0.$$

The proofs of (3)–(5) can be found in [11, 15].

(6). Let τ_k be the branch of the inverse of T^{n_k} such that $\tau_k(T^{n_k}(x_0)) = x_0$. Let $b_k = \tau_k(b)$ for k sufficiently large. Then, by the expanding property of the shift, there is $\lambda < 1$, such that,

$$d(x_0, b_k) \leq \lambda^{n_k} d(T^{n_k}(x_0), b) \xrightarrow{k} 0,$$

$$\left| \sum_{i=0}^{n_k-1} A(T^i(x_0)) - \sum_{i=0}^{n_k-1} A(T^i(b_k)) \right| \leq \frac{\|A\|_\alpha}{1 - \lambda^\alpha} d(T^{n_k}(x_0), b)^\alpha.$$

Write $Q := \frac{\|A\|_\alpha}{1-\lambda^\alpha}$, then

$$S(x_0, b) \geq \limsup_k \sum_{i=0}^{n_k-1} A(T^i(b_k))$$

$$\geq \limsup_k S(x_0, T^{n_k}(x_0)) - Q \, d(T^{n_k}(x_0), b)^\alpha$$

$$\geq \limsup_k S(x_0, T^{n_k}(x_0)).$$

Now for $\ell \in \mathbb{N}$ let $b_\ell \in \Sigma$ and $m_\ell \in \mathbb{N}$ be such that $\lim b_\ell = x_0$, $T^{m_\ell}(b_\ell) = b$ and

$$\lim_\ell \sum_{j=0}^{m_\ell-1} A(T^j(b_\ell)) = S(x_0, b).$$

Let $\hat{\tau}_l$ be the branch of the inverse of T^{m_ℓ} such that $\hat{\tau}_l(b) = b_\ell$. Let $x_\ell := \hat{\tau}_l(T^{n_k}(x_0))$. Then

$$d(x_\ell, x_0) \leq d(x_0, b_\ell) + d(b_\ell, x_\ell)$$

$$\leq d(x_0, b_\ell) + \lambda^\ell d(T^{n_k}(x_0), b) \xrightarrow{\ell} 0.$$

$$\left| \sum_{j=0}^{m_\ell-1} A(T^j(x_\ell)) - \sum_{j=0}^{m_\ell-1} A(T^j(b_\ell)) \right| \leq Q \, d(T^{n_k}(x_0), b)^\alpha.$$

Since $x_\ell \to x_0$ and $T^{m_\ell}(x_\ell) = T^{n_k}(x_0)$, we have that

$$S(x_0, T^{n_k}(x_0)) \geq \limsup_\ell \sum_{j=0}^{m_\ell - 1} A(T^j(x_\ell))$$

$$\geq \limsup_\ell \sum_{j=0}^{m_\ell - 1} A(T^j(b_\ell)) - Q \, d(T^{n_k}(x_0), b)^\alpha$$

$$\geq S(x_0, b) - Q \, d(T^{n_k}(x_0), b)^\alpha.$$

And hence

$$\liminf_k S(x_0, T^{n_k}(x_0)) \geq S(x_0, b).$$

$\square$

Proposition 12.3.1. *The Aubry set is*

$$\mathbb{A}(A) = \bigcap_{F \in \mathscr{S}(A)} I_F^{-1}\{0\},$$

where the intersection is among all the α-Hölder calibrated sub-actions for A.

Proof.   By adding a constant we can assume that $m_A = 0$.

We first prove that $\mathbb{A}(A) \subset \bigcap_{F \in \mathscr{S}(A)} I_F^{-1}\{0\}$.

Let $F \in \mathscr{S}(A)$ be a Hölder sub-action and $x_0 \in \mathbb{A}(A)$. Since $S_A(x_0, x_0) = 0$ then there is $x_k \to x_0$ and $n_k \uparrow \infty$ such that $\lim_k T^{n_k}(x_k) = x_0$ and $\lim_k \sum_{j=0}^{n_k - 1} A(T^j(x_k)) = 0$. If $m \in \mathbb{N}$ we have that

$$F(T^{m+1}(x_0)) \geq F(T^m(x_0)) + A(T^m(x_0))$$

$$\geq F(T^{m+1}(x_k)) + \sum_{j=m+1}^{n_k + m - 1} A(T^j(x_k)) + A(T^m(x_0))$$

$$\geq F(T^{m+1}(x_k)) + \sum_{j=0}^{n_k - 1} A(T^j(x_k)) - \sum_{j=0}^{m} |A(T^j(x_k)) - A(T^j(x_0))|$$

$$(12.8)$$

When $k \to \infty$ the right hand side of (12.8) converges to $F(T^{m+1}(x_0))$, and hence all those inequalities are equalities. Therefore $R_F(T^m(x_0)) = 0$ for all m and hence $I_F(x_0) = 0$.

Now let $x_0 \in \bigcap_{F \in \mathscr{S}(A)} I_F^{-1}\{0\}$. Since, Σ is compact there is $n_k \overset{k}{\to} +\infty$ such that the limits $b = \lim_k T^{n_k}(x_0) \in \Sigma$ and $\mu = \lim_k \mu_k \in \mathscr{M}(T)$,

$\mu_k := \frac{1}{n_k} \sum_{i=0}^{n_k-1} \delta_{T^i(x_0)}$ exist and $b \in \text{supp}(\mu)$. Let G be a Hölder calibrated sub-action. For $m \geq n$ we have

$$G(T^n(x_0)) + S_A(T^n(x_0), T^m(x_0)) \geq G(T^n(x_0)) + \sum_{j=n}^{m-1} A(T^j(x_0))$$

$$= G(T^m(x_0)) \qquad\qquad [\text{because } I_G(x_0) = 0]$$

$$\geq G(T^n(x_0)) + S_A(T^n(x_0), T^m(x_0)).$$

Then they are all equalities and hence for any $m \geq n$

$$S_A(T^n(x_0), T^m(x_0)) = \sum_{j=n}^{m-1} A(T^j(x_0)).$$

Since

$$0 = \lim_k \frac{1}{n_k} S_A(T^n(x_0), T^m(x_0)) = \lim_k \frac{1}{n_k} \sum_{j=n}^{m-1} A(T^j(x_0)) = \int A d\mu,$$

μ is a minimizing measure. By Lemma 7.(1), $b \in \mathbb{A}(A)$.

Let $F : \Sigma \to \mathbb{R}$ be $F(x) := S_A(b, x)$. Then F is a Hölder calibrated sub-action. By hypothesis $I_F(x_0) = 0$ and then

$$F(T^n(x_0)) = F(x_0) + S_A(x_0, T^{n_k}(x_0)).$$

$$S_A(b, T^{n_k}(x_0)) = S_A(b, x_0) + S_A(x_0, T^{n_k}(x_0)).$$

By Lemma 7.(4) and 7.(6), taking the limit on k we have that

$$0 = S_A(b, b) = S_A(b, x_0) + S_A(x_0, b) = 0.$$

$$0 \geq S_A(x_0, x_0) \geq S_A(x_0, b) + S_A(b, x_0) = 0$$

Therefore $x_0 \in \mathbb{A}(A)$. $\qquad\qquad\qquad\qquad\qquad\qquad\qquad\qquad\qquad\qquad\qquad\qquad$ □

We want to show the following result which will require several preliminary results.

Theorem 12.3.1. *The set*

$$\mathscr{R} := \{ A \in C^\alpha(\Sigma, \mathbb{R}) \mid \mathscr{M}(A) = \{\mu\}, \ \mathbb{A}(A) = supp(\mu) \} \qquad (12.9)$$

contains a residual set in $C^\alpha(\Sigma, \mathbb{R})$.

The proof of the bellow lemma (Atkinson-Mañé) can be found in [25] and [10].

Lemma 8. *Let $(X, \mathcal{B}, \nu)$ be a probability space, f an ergodic measure preserving map and $F : X \to \mathbb{R}$ an integrable function. Given $A \in \mathcal{B}$ with $\nu(A) > 0$ denote by $\hat{A}$ the set of points $p \in A$ such that for all $\varepsilon > 0$ there exists an integer $N > 0$ such that $f^N(p) \in A$ and*

$$\left| \sum_{j=0}^{N-1} F\big(f^j(p)\big) - N \int F \, d\nu \right| < \varepsilon.$$

Then $\nu(\hat{A}) = \nu(A)$.

Corollary 2. *If besides the hypothesis of Lemma 8, X is a complete separable metric space, and $\mathcal{B}$ is its Borel σ-algebra, then for a.e. $x \in X$ the following property holds: for all $\varepsilon > 0$ there exists $N > 0$ such that $d(f^N(x), x) < \varepsilon$ and*

$$\left| \sum_{j=0}^{N-1} F\big(f^j(x)\big) - N \int F \, d\nu \right| < \varepsilon$$

Proof. Given $\varepsilon > 0$ let $\{V_n(\varepsilon)\}$ be a countable basis of neighborhoods with diameter $< \varepsilon$ and let $\hat{V}_n$ be associated to V_n as in Lemma 8. Then the full measure subset $\bigcap_m \bigcup_n \hat{V}_n(\frac{1}{m})$ satisfies the required property. □

Lemma 9. *Let $\mathcal{R}$ be as in Theorem 12.3.1. Then if $A \in \mathcal{R}$, $F \in \mathcal{S}(A)$ we have*

1. If $a, b \in \mathbb{A}(A)$ then $S_A(a,b) + S_A(b,a) = 0$.
2. If $a \in \mathbb{A}(A) = \mathrm{supp}(\mu)$ then $F(x) = F(a) + S_A(a,x)$ for all $x \in \Sigma$.

Proof.

(1). Let $a, b \in \mathbb{A}(A) = \mathrm{supp}(\mu)$. Since μ is ergodic, by Corollary 2 there are sequences $\alpha_k \in \Sigma$, $m_k \in \mathbb{N}$ such that $\lim_k m_k = \infty$, $\lim_k \alpha_k = a$, $\lim_k d(T^{m_k}(\alpha_k), \alpha_k) = 0$,

$$\sum_{j=0}^{m_k-1} A(T^j(\alpha_k)) \geq \frac{1}{k}, \quad \text{and writing} \quad \mu_k := \frac{1}{m_k} \sum_{j=1}^{m_k-1} \delta_{T^{m_k}(\alpha_k)}, \quad \lim_k \mu_k = \mu.$$

Since $b \in \mathrm{supp}(\mu)$ there are $n_k \leq m_k$ such that $\lim_k T^{n_k}(\alpha_k) = b$.

Let τ_k be the branch of the inverse of T^{n_k} such that $\tau_k(T^{n_k}(\alpha_k)) = \alpha_k$. Let $b_k := \tau_k(b)$. Then $T^{n_k}(b_k) = b$ and

$$d(b_k, a) \leq d(b_k, \alpha_k) + d(\alpha_k, a)$$
$$\leq \lambda^{n_k} d(b, T^{n_k}(\alpha_k)) + d(\alpha_k, a)$$
$$\leq d(b, T^{n_k}(\alpha_k)) + d(\alpha_k, a) \xrightarrow{k} 0.$$

We have that

$$\left| \sum_{j=0}^{n_k-1} A(T^j(b_k)) - \sum_{j=0}^{n_k-1} A(T^j(\alpha_k)) \right| \le \frac{\|A\|_\alpha}{1-\lambda^\alpha} \, d(T^{n_k}(\alpha_k), b)^\alpha.$$

$$S(a,b) \ge \limsup_k \sum_{j=0}^{n_k-1} A(T^j(b_k))$$

$$\ge \limsup_k \sum_{j=0}^{n_k-1} A(T^j(\alpha_k)) - Q \, d(T^{n_k}(\alpha_k), b)^\alpha.$$

Let τ_k be the branch of the inverse of $T^{m_k-n_k}$ such that $\tau_k(T^{m_k}(\alpha_k)) = T^{n_k}(\alpha_k)$. Let $a_k := \tau_k(a)$ Then $T^{m_k-n_k}(a_k) = a$ and

$$d(b, a_k) \le d(b, T^{n_k}(\alpha_k)) + d(T^{n_k}(\alpha_k), a_k)$$

$$\le d(b, T^{n_k}(\alpha_k)) + \lambda^{m_k-n_k} d(T^{m_k}(a_k), a)$$

$$\le d(b, T^{n_k}(\alpha_k)) + d(T^{m_k}(a_k), a) \xrightarrow{k} 0.$$

Also

$$\left| \sum_{j=0}^{m_k-n_k-1} A(T^j(a_k)) - \sum_{j=n_k}^{m_k-1} A(T^j(\alpha_k)) \right| \le \frac{\|A\|_\alpha}{1-\lambda^\alpha} \, d(a, T^{m_k}(\alpha_k)).$$

$$S(a,b) \ge \limsup_k \sum_{j=0}^{m_k-n_k-1} A(T^j(a_k))$$

$$\ge \limsup_k \sum_{j=n_k}^{m_k-1} A(T^j(\alpha_k)) - Q \, d(a, T^{n_k}(\alpha_k))$$

Therefore

$$0 \ge S(a,a) \ge S(a,b) + S(b,a)$$

$$\ge \limsup_k \sum_{j=0}^{n_k-1} A(T^j(\alpha_k)) + \limsup_k \sum_{j=n_k}^{m_k-1} A(T^j(\alpha_k))$$

$$\geq \limsup_k \left[\sum_{j=0}^{n_k-1} A(T^j(\alpha_k)) + \sum_{j=n_k}^{m_k-1} A(T^j(\alpha_k)) \right]$$

$$\geq \limsup_k \frac{1}{k}$$

$$\geq 0.$$

(2). We first prove that if for some $x_0 \in \Sigma$ and $a \in \mathbb{A}(A)$ we have

$$F(x_0) = F(a) + S_A(a, x_0), \tag{12.10}$$

then Eq. (12.10) holds for every $a \in \mathbb{A}(A)$. If $b \in \mathbb{A}(A)$, using item (1) we have that

$$F(x_0) \geq F(b) + S(b, x_0)$$
$$\geq F(a) + S_A(a, b) + S_A(b, x_0)$$
$$\geq F(a) + S_A(a, b) + S_A(b, a) + S_A(a, x_0)$$
$$= F(a) + S_A(a, x_0)$$
$$= F(x_0).$$

Therefore $F(x_0) = F(b) + S(b, x_0)$.

It is enough to prove that given any $x_0 \in \Sigma$ there is $a \in \mathbb{A}(A)$ such that the equality (12.10) holds. since F is calibrated there are $x_k \in \Sigma$ and $n_k \in \mathbb{N}$ such that $T^{n_k}(x_k) = x_0$, $\exists \lim_k x_k = a$ and for every $k \in \mathbb{N}$,

$$F(x_0) = F(x_k) + \sum_{j=0}^{n_k-1} A(T^j(x_k)).$$

We have that

$$S_A(a, x_0) \geq \limsup_k \sum_{j=0}^{n_k-1} A(T^j(x_k))$$

$$= \limsup_k F(x_0) - F(x_k)$$

$$= F(x_0) - F(a)$$

$$\geq S(a, x_0).$$

Therefore equality (12.10) holds.

It remains to prove that $a \in \mathbb{A}(A)$, i.e. that $S_A(a, a) = 0$. We can assume that the sequence n_k is increasing. Let $m_k = n_{k+1} - n_k$. Then $T^{m_k}(x_{k+1}) = x_k$. Let τ_k

be the branch of the inverse of T^{m_k} such that $\tau_k(x_k) = x_{k+1}$ and $a_{k+1} := \tau_k(a)$. We have that

$$\left| \sum_{j=0}^{m_k-1} A(T^j(a_{k+1})) - \sum_{j=0}^{m_k-1} A(T^j(x_{k+1})) \right| \leq \frac{\|A\|_\alpha}{1 - \lambda^\alpha} \, d(a, x_k)^\alpha.$$

Since $x_k \to a$ we have that

$$\begin{aligned}
d(a_{k+1}, a) &\leq d(a_{k+1}, x_{k+1}) + d(x_{k+1}, a) \\
&\leq \lambda^{m_k} \, d(x_k, a) + d(x_{k+1}, a) \\
&\leq d(x_k, a) + d(x_{k+1}, a) \xrightarrow{k} 0.
\end{aligned}$$

Therefore

$$\begin{aligned}
0 \geq S_A(a, a) &\geq \limsup_k \sum_{j=0}^{m_k-1} A(T^j(a_{k+1})) \\
&\geq \limsup_k \sum_{j=0}^{m_k-1} A(T^j(x_{k+1})) - Q \, d(a, x_k)^\alpha \\
&= \limsup_k F(x_k) - F(x_{k+1}) - Q \, d(a, x_k)^\alpha \\
&= 0.
\end{aligned}$$

$\square$

The above result (2) is true for F only continuous.

Corollary 3. *Let $\mathscr{R}$ be as in Theorem 12.3.1. Then if $A \in \mathscr{R}$, $F \in \mathscr{S}(A)$ we have*

1. If $x \notin \mathbb{A}(A)$ then $I_F(x) > 0$.
2. If $x \notin \mathbb{A}(A)$ and $T(x) \in \mathbb{A}(A)$ then $R_F(x) > 0$.

Proof. **(1).** By Lemma 9.(2) modulo adding a constant there is only one Hölder calibrated sub-action F in $\mathscr{S}(A)$. Then by Proposition 12.3.1, $\mathbb{A}(A) = [I_F = 0]$. Since $I_F \geq 0$, this proves item (1).
(2). Since $T(x) \in \mathbb{A}(A)$

$$I_F(x) = \sum_{n \geq 1} R_F(T^n(x)) = 0.$$

Since $x \notin \mathbb{A}(A)$, by item (2) and Proposition 12.3.1, $\mathbb{A}(A) = [I_F = 0]$. Then

$$I_F(x) = \sum_{n \geq 0} R_F(T^n(x)) > 0.$$

Hence $R_F(x) > 0$.

□

Lemma 10. *1. $A \mapsto m_A$ has Lipschitz constant 1.*
2. Fix $x_0 \in \Sigma$. The set $\mathscr{S}(A)$ of α-Hölder calibrated sub-actions F for A with $F(x_0) = 0$ is an equicontinuous family. In fact

$$\sup_{F \in \mathscr{S}(A)} |F|_\alpha < \infty.$$

3. The set $\mathscr{S}(A)$ of α-Hölder continuous calibrated sub-actions is closed under the C^0 topology.
4. If $\#\mathscr{M}(A) = 1$, $A_n \overset{n}{\to} A$ uniformly, $\sup_n |A_n|_\alpha < \infty$, and, $F_n \in \mathscr{S}(A_n)$, then, $\lim_n F_n = F$ uniformly.
5. $A \leq B$ & $m_A = m_B \implies S_A \leq S_B$.
6. $\limsup\limits_{B \to A} \mathbb{N}(B) \subseteq \mathbb{N}(A)$, where

$$\limsup_{B \to A} \mathbb{N}(B) = \Big\{ \lim_n x_n \mid x_n \in \mathbb{N}(B_n),\ B_n \overset{n}{\to} A,\ \exists \lim_n x_n \Big\}$$

7. If $A \in \mathscr{R}$ then

$$\lim_{B \to A} d_H(\mathbb{A}(B), \mathbb{A}(A)) = 0,$$

where d_H is the Hausdorff distance.
8. If $A \in \mathscr{R}$ with $\mathscr{M}(A) = \{\mu\}$ and $\nu_B \in \mathscr{M}(B)$ then

$$\lim_{B \to A} d_H(supp(\nu_B), supp(\mu)) = 0.$$

9. If $A \in \mathscr{R}$ then

$$\lim_{B \to A} d_H(\mathbb{M}(B), \mathbb{A}(A)) = 0.$$

If X, Y are two metric spaces and $\mathbb{F} : X \to 2^Y = \mathbb{P}(Y)$ is a set valued function, define

$$\limsup_{x \to x_0} \mathbb{F}(x) = \bigcap_{\varepsilon > 0} \bigcap_{\delta > 0} \bigcup_{d(x,x_0) < \delta} V_\varepsilon(\mathbb{F}(x)),$$

$$\liminf_{x \to x_0} \mathbb{F}(x) = \bigcap_{\varepsilon > 0} \bigcup_{\delta < 0} \bigcap_{d(x,x_0) < \delta} V_\varepsilon(\mathbb{F}(x)),$$

where

$$V_\varepsilon(C) = \bigcup_{y \in C} \{ z \in Y \mid d(z, y) < \varepsilon \}.$$

Proof.

(1). We have that $A \le B + \|A - B\|_0$, then

$$\int A \, d\mu \le \int B \, d\mu + \|A - B\|_0, \qquad \forall \mu \in \mathscr{M}(T),$$

$$\int A \, d\mu \le \sup_{\mu \in \mathscr{M}(T)} \int B \, d\mu + \|A - B\|_0 = m_B + \|A - B\|_0,$$

$$m_A \le m_B + \|A - B\|_0.$$

Similarly $m_B \le m_A + \|A - B\|_0$ and then $|m_A - m_B| \le \|A - B\|_0$.
See also [18] and [11] for a proof.

(2). Let $\varepsilon > 0$ and $0 < \lambda < 1$ be such that for any $x \in \Sigma$ there is an inverse branch τ of T which is defined on the ball $B(T(x), \varepsilon) := \{ z \in \Sigma \mid d(z, T(x)) < \varepsilon \}$, has Lipschitz constant λ and $\tau(T(x)) = x$.

Let $F \in \mathscr{S}(A)$. Let

$$K := |F|_\alpha := \sup_{d(x,y)<\varepsilon} \frac{|F(x) - F(y)|}{d(x, y)^\alpha}, \quad a := |A|_\alpha := \sup_{d(x,y)<\varepsilon} \frac{|A(x) - A(y)|}{d(x, y)^\alpha}$$

be Hölder constants for F and A. Given $x, y \in \Sigma$ with $d(x, y) < \varepsilon$ let τ_i, $i = 1, \ldots, m(x) \le M$ be the inverse branches for T about x and let $x_i = \tau_i(x)$, $y_i = \tau_i(y)$. We have that

$$|F(x_i) - F(y_i)| \, K; \lambda^\alpha \, d(x, y)^\alpha, \qquad |A(x_i) - A(y_i)| \alpha \, \lambda^\alpha \, d(x, y)^\alpha.$$

$$F(x_i) + A(x_i) \le F(y_i) + A(y_i) + (K + a) \, \lambda^\alpha \, d(x, y)^\alpha,$$

$$\max_i \left[F(x_i) + A(x_i) - m_A \right] \le \max_i \left[F(y_i) + A(y_i) - m_A \right] + (K + a) \, \lambda^\alpha \, d(x, y)^\alpha,$$

$$F(x) \le F(y) + (K + a) \, \lambda^\alpha \, d(x, y)^\alpha,$$

Then $|F|_\alpha \le \lambda^\alpha (|F|_\alpha + |A|_\alpha)$ and hence

$$|F|_\alpha \le \frac{\lambda^\alpha}{(1 - \lambda^\alpha)} |A|_\alpha. \tag{12.11}$$

This implies the equicontinuity of $\mathscr{S}(A)$.

The proof of the above result could be also get if we just assume that F is continuous.

(3). It is easy to see that uniform limit of calibrated sub-actions is a sub-action, and it is calibrated because the number of inverse branches of T is finite, i.e. $\sup_{y \in \Sigma} \#T^{-1}\{y\} < \infty$. By (2) all C^α calibrated sub-actions have a common Hölder constant, the uniform limits of them have the same Hölder constant.

(4). The family $\{F_n\}$ satisfies $F_n(x_0) = 0$ and by inequality (12.11)

$$|F_n|_\alpha < \frac{\lambda^\alpha}{(1 - \lambda^\alpha)} \sup_n |A_n|_\alpha < \infty.$$

Hence $\{F_n\}$ is equicontinuous. By Arzelá-Ascoli theorem it is enough to prove that there is a unique $F(x) = S_A(x_0, x)$ which is the limit of any convergent subsequence of $\{F_n\}$. Since $\sup |A_n|_\alpha < \infty$, by inequality (12.11), any such limit is α-Hölder. Since by Lemma 9.(2), $\mathscr{S}(A) \cap [F(x_0) = 0] = \{F(x) = S_A(x_0, x)\}$, it is enough to prove that any limit of a subsequence of $\{F_n\}$ is a calibrated sub-action. But this follows form the continuity of $A \mapsto m_A$, the equality

$$F_n(x) = \max_{T(y)=x} F_n(y) + A_n(y) - m_{A_n}$$

and the fact $\sup_{x \in \Sigma} \#(T^{-1}\{x\}) < \infty$.

(5). The proof follows from the expression

$$S_A(x, y) := \lim_{\varepsilon \to 0} \left[\sup \left\{ \sum_{i=0}^{n-1} \left[A(T^i(z)) - m_A \right] \,\middle|\, n \in \mathbb{N}, \ T^n(z) = y, \right.\right.$$

$$\left.\left. d(z, x) < \varepsilon \right\} \right].$$

(6). Let $x_n \in B_n \to A$ be such that $x_n \to x_0$. Let $F_n \in \mathscr{S}(A)$ be such that $I_{F_n}(x_n) = 0$. Adding a constant we can assume that $F_n(x_0) = 0$ for all n. By (2), taking a subsequence we can assume that $\exists F = \lim_n F_n$ in the C^0 topology. Then F is a C^α calibrated sub-action for A. Also $R_{F_n} \to R_F$ uniformly and there is a common Hölder constant C for all the R_{F_n}. We have that

$$|R_{F_n}(T^k(x_n)) - R_F(T^k(x_0))| \leq$$

$$\left| R_{F_n}(T^k(x_n)) - R_{F_n}(T^k(x_0)) \right| + \left| R_{F_n}(T^k(x_0)) - R_F(T^k(x_0)) \right|$$

$$\leq C \, d(T^k(x_n), T^k(x_0))^\alpha + \| R_{F_n} - R_F \| \xrightarrow{n} 0$$

Since for all n, k, $R_{F_n}(T^k(x_n)) = 0$, we have that $R_F(T^k(x_0)) = 0$ for any k. Hence $I_F(x_0) = 0$ and then $x_0 \in \mathbb{N}(A)$.

(7). By Lemma 9.(2), there is only one calibrated sub-action modulo adding a constant. Then by Proposition 12.3.1, $\mathbb{A}(A) = \mathbb{N}(A)$. Then by (6) $\limsup_{B \to A} \mathbb{A}(B) \subset \mathbb{A}(A)$. It is enough to prove that for any $x_0 \in \mathbb{A}(A)$ and $B_n \to A$, there is $x_n \in \mathbb{A}(B_n)$ such that $\lim_n x_n = x_0$. Let $\mu_n \in \mathcal{M}(B_n)$. Then $\lim_n \mu_n = \mu$ in the weak* topology. Given $x_0 \in \mathbb{A}(A) = \text{supp}(\mu)$ we have that

$$\forall \varepsilon > 0 \quad \exists N = N(\varepsilon) > 0 \quad \forall n \geq N : \quad \mu_n(B(x_0, \varepsilon)) > 0.$$

We can assume that for all $m \in \mathbb{N}$, $N(\frac{1}{m}) < N(\frac{1}{m+1})$. For $N(\frac{1}{m}) \leq n < N(\frac{1}{m+1})$ choose $x_n \in \text{supp}(\mu_n) \cap B(x_0, \frac{1}{m})$. Then $x_n \in \mathbb{A}(B_n)$ and $\lim_n x_n = x_0$.

(8). For any $B \in \mathcal{F}$ We have that

$$\text{supp}(\nu_B) \subseteq \mathbb{A}(B) \subseteq \mathbb{N}(B).$$

By item (7),

$$\limsup_{B \to A} \text{supp}(\nu_B) \subseteq \mathbb{A}(A) = \text{supp}(\mu).$$

It remains to prove that

$$\liminf_{B \to A} \text{supp}(\nu_B) \supseteq \text{supp}(\mu).$$

But this follows from the convergence $\lim_{B \to A} \nu_B = \mu$ in the weak* topology.

(9). Write $\mathcal{M}(A) = \{\mu\}$. By items (7) and (8) we have that

$$\limsup_{B \to A} \mathbb{M}(B) \subseteq \limsup_{B \to A} \mathbb{A}(B) \subseteq \mathbb{A}(B),$$

$$\mathbb{A}(A) = \text{supp}(\mu) \subseteq \liminf_{B \to A} \mathbb{M}(B).$$

$\square$

Proof of Theorem 12.3.1. The set

$$\mathcal{D} := \{ A \in \mathcal{F} \mid \#\mathcal{M}(A) = 1 \}$$

is dense (c.f. [11]). We first prove that $\mathcal{D} \subset \overline{\mathcal{R}}$, and hence that $\mathcal{R}$ is dense.

Given $A \in \mathcal{D}$ with $\mathcal{M}(A) = \{\mu\}$ and $\varepsilon > 0$, let $\psi \in \mathcal{F}$ be such that $\|\psi\|_0 + \|\psi\|_\alpha < \varepsilon$ $\psi \leq 0$, $[\psi = 0] = \text{supp}(\mu)$. It is easy to see that $\mathcal{M}(A + \psi) = \{\mu\} = \mathcal{M}(A)$. Let $x_0 \notin \text{supp}(\mu)$. Given $\delta > 0$, write

$$S_A(x_0, x_0; \delta) := \sup \left\{ \sum_{k=0}^{n-1} A(T^k(x_n)) \mid T^n(x_n) = x_0, \ d(x_n, x_0) < \delta \right\}.$$

If $T^n(x_n) = x_0$ is such that $d(x_n, x_0) < \delta$ then

$$\sum_{k=0}^{n-1}(A + \psi)(T^k(x_n)) \leq S_A(x_0, x_0; \delta) + \sum_{k=0}^{n-1} \psi(T^k(x_n))$$

$$\leq S_A(x_0, x_0; \delta) + \psi(x_n).$$

Taking $\lim \sup_{\delta \to 0}$,

$$S_{A+\psi}(x_0, x_0) \leq S_A(x_0, x_0) + \psi(x_0) \leq \psi(x_0) < 0.$$

Hence $x_0 \notin \mathbb{A}(A + \psi)$. Since by Lemma 7.(1), $\mathrm{supp}(\mu) \subset \mathbb{A}(A + \psi)$, then $\mathbb{A}(A + \psi) = \mathrm{supp}(\mu)$ and hence $A + \psi \in \mathscr{R}$.

Let

$$\mathscr{U}(\varepsilon) := \{ A \in \mathscr{F} \mid d_H(\mathbb{A}(A), \mathbb{M}(A)) < \varepsilon \}.$$

From the triangle inequality

$$d_H(\mathbb{A}(B), \mathbb{M}(B)) \leq d_H(\mathbb{A}(B), \mathbb{A}(A)) + d_H(\mathbb{A}(A), \mathbb{M}(B))$$

and items (7) and (9) of Lemma 10, we obtain that $\mathscr{U}(\varepsilon)$ contains a neighborhood of $\mathscr{D}$. Then the set

$$\mathscr{R} = \bigcap_{n \in \mathbb{N}} \mathscr{U}(\tfrac{1}{n})$$

contains a residual set.   □

12.4   Duality

In this section we have to consider properties for A^* which depends of the initial potential A.

We will consider now the specific example described before. We point out that the results presented bellow should hold in general for natural extensions.

We will assume that T and σ are topologically mixing.

Remember that $\mathbb{T} : \hat{\Sigma} \to \hat{\Sigma}$,

$$\mathbb{T}(x, \omega) = (T(x), \tau_x(\omega)), \qquad \mathbb{T}^{-1}(x, \omega) = (\tau_\omega(x), \sigma(\omega))$$

Given $A \in \mathscr{F}$ define $\Delta_A : \Sigma \times \Sigma \times \Sigma \to \mathbb{R}$ as

$$\Delta_A(x, y, \omega) := \sum_{n \geq 0} A(\tau_{n,\omega}(x)) - A(\tau_{n,\omega}(y))$$

where

$$\tau_{n,\omega}(x) = \tau_{\sigma^n\omega} \circ \tau_{\sigma^{n-1}\omega} \circ \cdots \circ \tau_\omega(x).$$

Fix $\overline{x} \in \Sigma$ and $\overline{w} \in \Sigma$.

The involution W-kernel can be defined as $W_A : \Sigma \times \Sigma \to \mathbb{R}$, $W_A(x,\omega) = \Delta_A(x,\overline{x},\omega)$. Writing $\mathsf{A} := A \circ \pi_1 : \mathbb{T} \to \mathbb{R}$, we have that

$$W(x,\omega) = \sum_{n\geq 0} \mathsf{A}(\mathbb{T}^{-n}(x,\omega)) - \mathsf{A}(\mathbb{T}^{-n}(\overline{x},\omega))$$

We can get the *dual function* $A^* : \Sigma \to \mathbb{R}$ as

$$A^*(\omega) := (W_A \circ \mathbb{T}^{-1} - W_A + A \circ \pi_1)(x,\omega).$$

Remember that we consider here the metric on Σ defined by

$$d(\omega,v) := \lambda^N, \qquad N := \min\{k \in \mathbb{N} \mid \omega_k \neq v_k\}$$

Then λ is a Lipschitz constant for both τ_x, and τ_ω and also for $\mathbb{T}|_{\{x\}\times\Sigma}$ and $\mathbb{T}^{-1}|_{\Sigma\times\{\omega\}}$.

Write $\mathscr{F} := C^\alpha(\Sigma,\mathbb{R})$ and $\mathscr{F}^* := C^\alpha(\Sigma,\mathbb{R})$. Let $\mathscr{B}$ and $\mathscr{B}^*$ be the set of coboundaries

$$\mathscr{B} := \{u \circ T - u \mid u \in C^\alpha(\Sigma,\mathbb{R})\},$$
$$\mathscr{B}^* := \{u \circ \sigma - u \mid u \in C^\alpha(\Sigma,\mathbb{R})\}.$$

Remember that

$$\|z\|_\alpha = \|z\|_0 + |z|_\alpha.$$

We also use the notation $[z]_\alpha = \|z\|_\alpha$.

Lemma 11.

1. $z \in \mathscr{B} \iff z \in C^\alpha(\Sigma,\mathbb{R})$ & $\forall \mu \in \mathcal{M}(T)$, $\int z \, d\mu = 0$.
2. *The linear subspace* $\mathscr{B} \subset C^\alpha(\Sigma,\mathbb{R})$ *is closed.*
3. *The function*

$$[z + \mathscr{B}]_\alpha = \inf_{b\in\mathscr{B}} [z+b]_\alpha$$

is a norm in $\mathscr{F}/\mathscr{B}$.

Proof.

(1). This follows[1] from [6], Theorem 1.28 (ii) $\implies$ (iii).

(2). We prove that the complement $\mathscr{B}^c$ is open. If $z \in C^\alpha(\Sigma, \mathbb{R}) \setminus \mathscr{B}$, by item (1), there is $\mu \in \mathscr{M}(T)$ such that $\int z \, d\mu \neq 0$. If $u \in C^\alpha(\Sigma, \mathbb{R})$ is such that

$$\|u - z\|_0 < \frac{1}{2} \left| \int z \, d\mu \right|$$

then $\int u \, d\mu \neq 0$ and hence $u \notin \mathscr{B}$.

(3). This follows from item (2).

$\square$

Lemma 12.

1. *If A is C^α then A^* is C^α.*
2. *The linear map $L : C^\alpha(\Sigma, \mathbb{R}) \to C^\alpha(\Sigma, \mathbb{R})$ given by $L(A) = A^*$ is continuous.*
3. *$\mathscr{B} \subset \ker L$.*
4. *The induced linear map $L : \mathscr{F}/\mathscr{B} \to \mathscr{F}^*/\mathscr{B}^*$ is continuous.*
5. *Fix one $\overline{\omega} \in \Sigma$. Similarly the corresponding linear map $L^* : \mathscr{F}^* \to \mathscr{F}$, given by*

$$L^*(\psi) = W_\psi^* \circ \mathbb{T} - W_\psi + \psi \circ \pi_2$$

$$= \sum_{n \geq 0} \Psi(\mathbb{T}^n(x, \overline{\omega})) - \Psi(\mathbb{T}^n(Tx, \overline{\omega}))$$

$$= \Psi(x, \overline{\omega}) + \sum_{n \geq 0} \Psi(\mathbb{T}^n(Tx, \tau_x\overline{\omega})) - \Psi(\mathbb{T}^n(Tx, \overline{\omega})),$$

with $\Psi = \psi \circ \pi_2$, is continuous and induces a continuous linear map $L^* : \mathscr{F}^*/\mathscr{B}^* \to \mathscr{F}/\mathscr{B}$, which is the inverse of $L : \mathscr{F}/\mathscr{B} \to \mathscr{F}^*/\mathscr{B}^*$.

Proof.

(1) and (2). We have that

$$A^*(\omega) = \sum_{n \geq 0} A(\mathbb{T}^{-n}(\overline{x}, \omega)) - A(\mathbb{T}^{-n}(\overline{x}, \sigma\,\omega))$$

$$= A(\overline{x}) + \sum_{n \geq 0} A(\mathbb{T}^{-n}(\tau_\omega\,\overline{x}, \sigma\,\omega)) - A(\mathbb{T}^{-n}(\overline{x}, \sigma\,\omega))$$

Since $d(\mathbb{T}^{-n}(\tau_\omega\,\overline{x}, \sigma\,\omega), \mathbb{T}^{-n}(\overline{x}, \sigma\,\omega)) \leq \lambda^n \, d(\tau_\omega\,\overline{x}, \overline{x}) \leq \lambda^n$ and $\|A\|_\alpha = \|A \circ \pi_1\|_\alpha = \|A\|_\alpha$, we have that

$$\|A^*\|_0 \leq \|A\|_0 + \frac{\|A\|_\alpha}{1 - \lambda^\alpha}.$$

---

[1]Theorem 1.28 of Bowen [6] asks for T to be topologically mixing.

Also if $m := \min\{k \geq 0 \mid w_k \neq v_k\}$

$$A^*(\omega) - A^*(v) = \sum_{n \geq m-1} \mathsf{A}(\mathbb{T}^{-n}(\tau_\omega \bar{x}, \sigma \omega)) - \mathsf{A}(\mathbb{T}^{-n}(\bar{x}, \sigma \omega))$$

$$- \sum_{n \geq m-1} \mathsf{A}(\mathbb{T}^{-n}(\tau_v \bar{x}, \sigma v)) - \mathsf{A}(\mathbb{T}^{-n}(\bar{x}, \sigma v))$$

$$|A^*(\omega) - A^*(v)| \leq 2 \|A\|_\alpha \frac{\lambda^{(m-1)\alpha}}{1 - \lambda^\alpha} = \frac{2 \|A\|_\alpha \lambda^{-\alpha}}{1 - \lambda^\alpha} d(\omega, v)^\alpha.$$

$$\|A^*\|_\alpha \leq \frac{2 \|A\|_\alpha}{\lambda^\alpha (1 - \lambda^\alpha)}.$$

(3). If $u \in \mathscr{F}$ and $\mathbb{U} := u \circ \pi_1$ from the formula for L (in the proof of item [1]) we have that

$$L(u \circ T - u) = \mathbb{U}(\mathbb{T}(\bar{x}, \omega)) - \mathbb{U}(\mathbb{T}(\bar{x}, \sigma \omega))$$

$$= u(T \bar{x}) - u(T \bar{x}) = 0.$$

(4). Item (4) follows from items (2) and (3).

(5). We only prove that for any $A \in \mathscr{F}$, $L^*(L(A)) \in A + \mathscr{B}$. Write

$$L^*(L(A)) = \left(W_{A^*}^* \circ \mathbb{T} - W_{A^*}^* + \mathsf{A}^*\right)(\cdot, \overline{\omega})$$

$$= (W_{A^*}^* \circ \mathbb{T} - W_{A^*}^* + W_A \circ \mathbb{T}^{-1} - W_A) + A$$

Write

$$B := W_{A^*}^* \circ \mathbb{T} - W_{A^*}^* + W_A \circ \mathbb{T}^{-1} - W_A. \tag{12.12}$$

Since $A, L^*(L(A)) \in \mathscr{F} = C^\alpha(\Sigma, \mathbb{R})$, then $B \in C^\alpha(\Sigma, \mathbb{R})$.

Following Bowen, given any $\mu \in \mathscr{M}(T)$ we construct an associated measure $\nu \in \mathscr{M}(\mathbb{T})$. Given $z \in C^0(\Sigma \times \Sigma, \mathbb{R})$ define $z^\sharp \in C^0(\Sigma, \mathbb{R})$ as $z^\sharp(x) := z(x, \overline{\omega})$. We have that

$$\left\| (z \circ \mathbb{T}^n)^\sharp \circ T^m - (z \circ \mathbb{T}^{n+m})^\sharp \right\|_0 \leq \mathrm{var}_n z \xrightarrow{n} 0,$$

where

$$\mathrm{var}_n z = \sup\{ |z(a) - z(b)| \mid \exists x \in \Sigma, \; a, b \in \mathbb{T}^n(\{x\} \times \Sigma) \}$$

$$\leq \sup\{ |z(a) - z(b)| \mid d_{\Sigma \times \Sigma}(a, b) \leq \lambda^n \} \xrightarrow{n} 0,$$

$$d_{\Sigma \times \Sigma} = d_\Sigma \circ (\pi_1, \pi_1) + d_\Sigma \circ (\pi_2, \pi_2).$$

Then

$$\left| \mu((z \circ \mathbb{T}^n)^\sharp) - \mu((z \circ \mathbb{T}^{n+m})^\sharp) \right| = \left| \mu((z \circ \mathbb{T}^n)^\sharp \circ T^m) - \mu((z \circ \mathbb{T}^{n+m})^\sharp \right| \le \mathrm{var}_n z.$$

Therefore $\mu((z \circ \mathbb{T}^n)^\sharp)$ is a Cauchy sequence in $\mathbb{R}$ and hence the limit

$$v(z) := \lim_n \mu((z \circ \mathbb{T}^n)^\sharp)$$

exists. By the Riesz representation theorem v defines a Borel probability measure in $\Sigma \times \Sigma$, and it is invariant because

$$v(z \circ \mathbb{T}) = \lim_n \mu((z \circ \mathbb{T}^{n+1})^\sharp) = v(z).$$

Now let $B := L^*(L(A)) - A$ and $\mathbb{B} := B \circ \pi_1$. By formula (12.12) we have that $\mathbb{B}$ is a coboundary in $\Sigma \times \Sigma$. Since $\pi_1 \circ T^n = T^n$ we have that

$$0 = v(\mathbb{B}) = \lim_n \mu((\mathbb{B} \circ \mathbb{T}^n)^\sharp)$$

$$= \lim_n \mu(B \circ T^n)$$

$$= \mu(B).$$

Since this holds for every $\mu \in \mathcal{M}(T)$, by Lemma 11.(1), $B \in \mathcal{B}$ and then $(L^+ \circ L)(A + \mathcal{B}) \subset A + \mathcal{B}$.   □

Theorem 12.4.1.   *There is a residual subset $\mathcal{Q} \subset C^\alpha(\Sigma, \mathbb{R})$ such that if $A \in \mathcal{Q}$ and $A^* = L(A)$ then*

$$\mathcal{M}(A) = \{\mu\}, \qquad \mathbb{A}(A) = supp(\mu),$$
$$\mathcal{M}(A^*) = \{\mu^*\}, \qquad \mathbb{A}(A^*) = supp(\mu^*). \tag{12.13}$$

In particular

$$I_A(x) > 0 \qquad if \qquad x \notin supp(\mu),$$
$$I_A(\omega) > 0 \qquad if \qquad \omega \notin supp(\mu^*),$$

and

$$R_A(x) > 0 \qquad if \qquad x \notin supp(\mu) \quad and \quad T(x) \in supp(\mu),$$
$$R_A(\omega) > 0 \qquad if \qquad \omega \notin supp(\mu^*) \quad and \quad \sigma(\omega) \in supp(\mu).$$

Proof. Observe that the subset $\mathcal{R}$ defined in (12.9) in Theorem 12.3.1 is invariant under translations by coboundaries, i.e. $\mathcal{R} = \mathcal{R} + \mathcal{B}$. Indeed if $B = u \circ T - u \in \mathcal{B}$, we have that

$$\int (A + B)\, d\mu = \int A\, d\mu, \qquad \forall \mu \in \mathscr{B},$$

$$S_{A+B}(x, y) = S_A(x, y) + B(x) - B(y), \qquad \forall x, y \in \Sigma.$$

Then the Aubry set and the set of minimizing measures are unchanged: $\mathscr{M}(A + B) = \mathscr{M}(A)$, $\quad \mathbb{A}(A + B) = \mathbb{A}(A)$.

For the dynamical system (Σ, σ) let

$$\mathscr{R}^* = \{ \psi \in C^\alpha(\Sigma, \mathbb{R}) \mid \mathscr{M}(\psi) = \{\mu\}, \ \mathbb{A}(\psi) = \mathrm{supp}(\nu) \}$$

By Theorem 12.3.1, the subset $\mathscr{R}^*$ contains a residual set in $\mathscr{F}^* = C^\alpha(\Sigma, \mathbb{R})$ and it is invariant under translations by coboundaries: $\mathscr{R}^* = \mathscr{R}^* + \mathscr{B}^*$.

By Lemma 12 the linear map $L : \mathscr{F}/\mathscr{B} \to \mathscr{F}^*/\mathscr{B}^*$ is a homeomorphism with inverse L^*. Then the set $\mathscr{Q} := \mathscr{R} \cap L^{-1}(\mathscr{R}^*) = \mathscr{R} \cap (L^*(\mathscr{R}^*) + \mathscr{B})$ contains a residual subset and satisfies (12.13).

By Corollary 3 the other properties are automatically satisfied.   □

From this last theorem it follows our main result about the generic potential A to be in $\mathbb{G}$.

References

1. Baraviera, A., Lopes, A.O., Thieullen, P.: A large deviation principle for equilibrium states of Hölder potencials: the zero temperature case. Stoch. Dyn. **6**, 77–96 (2006)
2. Baraviera, A.T., Cioletti, L.M., Lopes, A.O., Mohr, J., Souza, R. R.: On the general one-dimensional XY Model: positive and zero temperature, selection and non-selection. Rev. Mod. Phys. **23**(10), 1063–1113 (2011)
3. Baraviera, A., Leplaideur, R., Lopes, A.: Ergodic optimization, zero temperature limits and the max-plus algebra. In: Mini-course in XXIX Coloquio Brasileiro de Matemática-IMPA, Rio de Janeiro (2013)
4. Bhattacharya, P., Majumdar, M.: Random Dynamical Systems. Cambridge University Press, Cambridge (2007)
5. Bissacot, R., Garibaldi, E.: Weak KAM methods and ergodic optimal problems for countable Markov shifts. Bull. Braz. Math. Soc. **41**(3), 321–338 (2010)
6. Bowen, R.: Equilibrium states and the ergodic theory of Anosov diffeomorphisms. In: Lecture Notes in Mathematics, vol. 470. Springer, Berlin (1975)
7. Boyd, S.: Convex Optimization. Cambridge University Press, Cambridge (2004)
8. Collier, D., Morris, I.D.: Approximating the maximum ergodic average via periodic orbits. Ergod. Theor Dyn. Syst. **28**(4), 1081–1090 (2008)
9. Contreras, G.: Ground States are generically a Periodic Orbit (2013). Arxiv
10. Contreras, G., Iturriaga, R.: Global minimizers of autonomous Lagrangians, 22° Colóquio Brasileiro de Matemática. IMPA (1999)
11. Contreras, G., Lopes, A.O., Thieullen, P.: Lyapunov minimizing measures for expanding maps of the circle. Ergod. Theor Dyn. Syst. **21**, 1379–1409 (2001)
12. Conze, J.-P., Guivarc, Y.: Croissance des sommes ergodiques, manuscript, circa (1993)
13. Conze, J.P., Guivarc'h, Y.: Croissance des sommes ergodiques et principe variationnel, manuscript circa (1993)

14. Dembo, A., Zeitouni, O.: Large Deviation Techniques and Applications. Springer, Berlin (1998)
15. Garibaldi, E., Lopes, A.O.: On the Aubry-Mather theory for symbolic dynamics. Ergod. Theor Dyn. Syst. **28**(3), 791–815 (2008)
16. Garibaldi, E., Lopes, A.O., Thieullen, Ph.: On calibrated and separating sub-actions. Bull. Braz. Math. Soc. **40**(4), 577–602 (2009)
17. Hunt, B.R., Yuan, G.C.: Optimal orbits of hyperbolic systems. Nonlinearity **12**, 1207–1224 (1999)
18. Jenkinson, O.: Ergodic optimization. Discrete Continuous Dyn. Syst. Ser. A **15**, 197–224 (2006)
19. Leplaideur, R.: A dynamical proof for the convergence of Gibbs measures at temperature zero. Nonlinearity **18**(6), 2847–2880 (2005)
20. Lopes, A.O., Mengue, J.: Zeta measures and thermodynamic formalism for temperature zero. Bull. Braz. Math. Soc. **41**(3), 449–480 (2010)
21. Lopes, A.O., Mengue, J.: Duality theorems in ergodic transport. J. Stat. Phys. **149**(5), 921–942 (2012)
22. Lopes, A.O., Oliveira, E.R., Thieullen, P.: The dual potential, the involution kernel and transport in ergodic optimization. Preprint (2008)
23. Lopes, A.O., Mohr, J., Souza, R., Thieullen, Ph.: Negative entropy, zero temperature and stationary Markov chains on the interval. Bull. Braz. Math. Soc. **40**, 1–52 (2009)
24. Lopes, A.O., Oliveira, E.R., Smania, D.: Ergodic transport theory and piecewise analytic subactions for analytic dynamics. Bull. Braz. Math. Soc. **43**(3) (2012)
25. Mañé, R.: Generic properties and problems of minimizing measures of Lagrangian systems. Nonlinearity **9**, 273–310 (1996)
26. Parry, W., Pollicott, M.: Zeta Functions and the Periodic Orbit Structure of Hyperbolic Dynamics, pp. 187–188. Societe Mathematique de France, Astérisque (1990)
27. Pinto, A.A., Rand, D.: Existence, uniqueness and ratio decomposition for Gibbs states via duality. Ergod. Theor Dyn. Syst. **21**(2), 533–543 (2001)
28. Pollicott, M., Yuri, M.: Dynamical Systems and Ergodic Theory. Cambridge University Press, Cambridge (1998)
29. Tal, F.A., Zanata, S.A.: Maximizing measures for endomorphisms of the circle. Nonlinearity **21**, 2347–2359 (2008)
30. Villani, C.: Topics in Optimal Transportation. AMS, Providence (2003)
31. Villani, C.: Optimal Transport: Old and New. Springer, Berlin (2009)

Chapter 13
Measuring the Welfare Impact of Biofuel Policies: A Review of Methods and Findings from Numerical Models

Christine L. Crago

13.1 Introduction

The growth of the biofuel sector in the United States (US) and other parts of the world has been dramatic in the past several years. Buoyed by support policies in the form of subsidies, mandates, and trade protection, US production increased from 175 million gallons in 1980 to 1.8 billion gallons in 2000, and over 13 billion gallons in 2011. Rationales for these support policies include: increased energy security, reduced greenhouse gas (GHG) emissions, and strengthened rural economies. Although the industry has benefited from subsidies since the early 1990s, it is only recently that much attention has been given to evaluating the social welfare and environmental impacts of biofuel support policies. This is due in part to the increase in government expenditures needed to finance these policies, as well as the widening reach of these policies in terms of what markets are affected. In particular, the role of biofuels in increasing food prices [1, 24, 40], and its contribution to global deforestation [27, 29] by inducing an expansion in global cropland acreage have heightened scrutiny toward biofuel policies.

This review article discusses different modeling frameworks for assessing the welfare impacts of biofuel policies and presents an overview of the findings of studies that examine the welfare consequences of biofuel policies in the US. Determinants of the magnitude of welfare impacts are also discussed and gaps in current research are identified. Although this research area is relatively new, much ground has been covered in the last several years. Analyses and modeling efforts have also evolved to adapt to changes in regulation and in response to new findings about the impact of biofuels and biofuel policies on related markets.

C.L. Crago (✉)
Department of Resource Economics, University of Massachusetts Amherst, MA, USA
e-mail: ccrago@resecon.umass.edu

A.A. Pinto and D. Zilberman (eds.), *Modeling, Dynamics, Optimization and Bioeconomics I*, Springer Proceedings in Mathematics & Statistics 73, DOI 10.1007/978-3-319-04849-9_13,
© Springer International Publishing Switzerland 2014

Studies of biofuel policies use models that differ in scope and modeling framework, which has sometimes led to differences in their conclusions about the impact of biofuel policies and whether these policies increase or decrease social welfare in the US. Numerical models used in these studies show that biofuel policies lead to allocative inefficiencies, but these losses can be partially or fully offset by gains from improved terms of trade in corn and oil markets. While current literature has made strides in quantifying the social surplus and environmental impacts of biofuel policies in the US, more research is needed to address impacts of biofuel policies on water use, agricultural pollution, and biodiversity. Further research is also needed to address the long term impact of biofuel policies on energy security. Extending current methods to incorporate long-run energy security benefits requires considering the role of biofuel policies in inducing technological change, the potential share of biofuels in the overall energy mix, and future prices and availability of other energy resources.

The chapter is organized as follows: Sect. 13.2 provides a background on the evolution of biofuel policies in the US. Section 13.3 discusses the general framework used to examine the welfare impact of biofuel policies and the factors that affect the magnitude of these impacts, and presents the different modeling approaches used to quantify the welfare impact of biofuel policies. Section 13.4 discusses quantitative findings of selected studies analyzing the welfare impact of biofuel policy. Section 13.5 concludes.

13.2 Background

Ethanol from corn has received government support since the early 1900s. However, this support did not receive much scrutiny until the last several years. The technology to produce ethanol and use it as transportation fuel has been around since the early 1900s. Indeed, Ford's Model T car and Fordson (a tractor) was designed to run on both gasoline and ethanol. One of the first federal incentives for ethanol was signed by President Theodore Roosevelt in 1906, exempting ethanol from the $2.08 per gallon tax on alcohol imposed to generate revenue during the Civil War. However, even with this subsidy, the growth of the ethanol sector was limited by its high production cost relative to gasoline. During the Great Depression, ethanol was seen as a means to support agricultural income, but support quickly faded once agricultural income rose [39]. Steady support for the ethanol industry started after the oil shocks of the 1970s. The Energy Tax Act of 1978 exempted ethanol from the $0.04 exercise tax on fuel. In 1980, the Energy Security Act added a $0.54 per gallon tariff to the already existing 2.5 % ad-valorem tariff on imported ethanol. Between 1980 and 2004, ethanol tax exemptions ranged from $0.5 to $0.6 per gallon, in addition to other government supports in the form of loan guarantees for ethanol production plants, and research and development funding. In 2004, the $0.51 tax exemption for ethanol began to be administered as a tax credit [30, 38]. Federal support for ethanol received little attention prior to mid-2000s because the total support package was not substantial. The subsidy per gallon in 2004 was $0.51. With production at 3.4 billion gallons, the total subsidy bill was about $1.8 billion or less than a third of subsidy expenditures in 2011.

Starting in the mid-2000s, several factors led to the rapid growth of the ethanol sector, with production growing threefold from 2004 to 2011. First, the widespread ban on methyl tertiary butyl ether (MTBE) as a fuel oxygenate created a steady demand of about six billion gallons per year. MTBE was the preferred oxygenate by the petroleum industry, but findings by the Environmental Protection Agency (EPA) that MTBE contaminates groundwater led to a sharp decline in its use [10]. Second, crude oil prices started to increase, fueling fears of energy insecurity and securing support for home-grown fuels such as ethanol. Finally, concerns about climate change came to the fore, and ethanol received support as a cleaner alternative to gasoline owing to its lower direct GHG emissions compared to gasoline. Increasing energy security and curbing GHG emissions, along with the desire to support rural communities motivated legislation to impose mandates on annual ethanol consumption. The Energy Policy Act of 2005 established the Renewable Fuel Standard (RFS) that mandates biofuel consumption of 7.5 billion gallons by 2015. The mandate was later increased in the Energy Independence and Security Act of 2007 (EISA) to 36 billion gallons by 2022. EISA has additional provisions about the mix of feedstocks that should be used in biofuel production, as well as minimum standards for GHG emissions from biofuels [9].

With the expansion of biofuel production and consumption, the total expenditure to support the industry (over $6 billion in 2011), as well as the support policies' distortionary effect on fuel and agricultural markets increased, leading to concerns about the social welfare impacts of these policies. One of the first analyses was by [10] who examine the impact of the first RFS mandate. This was followed by a number of studies, as analyses attempted to keep pace with changes in biofuel policies as well as emerging findings about which markets are impacted by biofuel policies and what the environmental impact of these policies are.

Early studies focused on the interaction of ethanol and corn markets. In particular, whether increasing ethanol production and increasing demand for corn, which resulted in increased price of corn, lowered the expenditures needed to finance price-contingent farm subsidies (e.g. loan rate program). Gardner [11] compares the deadweight loss of a deficiency payment given directly to corn farmers to an ethanol subsidy. De Gorter and Just [7] examine the interaction between the ethanol subsidy and a loan rate program for corn. Historically, Brazil was the largest producer and exporter of ethanol. Between 2004 and 2007, sugarcane ethanol imports from Brazil were 2–8 % of US corn ethanol production [33]. Sugarcane ethanol produced in Brazil was also historically cheaper than corn ethanol produced in the US. Elobeid and Tokgoz [8] and Lasco and Khanna [22] focus on the welfare impact of removing the ethanol tariff. Studies by [21] and [2] focus on the impact of biofuel policies on the market for fuel and vehicle miles traveled. The second RFS stipulated that more than 50 % of biofuels used to meet the mandate must be derived from feedstocks other than corn, such as sugarcane, crop and forest residues, woody biomass, and dedicated biomass crops. Huang et al. [18] incorporate the use of these feedstocks in their analysis of the welfare effect of biofuel policies. Hertel et al. [14] examine the welfare effect of combined US and EU policies on US markets, as they interact with the rest of the world commodity market, while [25] examine the impact of policies in the US and Brazil. Some studies emphasize the effect of biofuel policies on environmental externalities. Assuming monetary values for the marginal cost of these externalities, they include changes in the level of externalities in their calculation of the welfare effect of biofuel policies. Cui et al. [5] and Lasco and Khanna [22] include externality from CO_2 emissions. Khanna et al. [21] and Ando et al. [2] include the impact of biofuel policies on miles externalities such as congestion.

The next section discusses the general framework used to examine the welfare impact of biofuel policies as well as differences in modeling approaches. Factors that affect the magnitude of biofuel policies' impacts on affected markets are also discussed.

13.3   Modeling Approaches to Measure the Welfare Effect of Biofuel Policy

Welfare impacts of a policy are typically measured by changes in social surplus (the sum of producer surplus and consumer surplus) due to the policy change [19]. Thus, it is important to accurately identify markets affected by biofuel policies and measure impacts (i.e. changes in prices and quantities) in those markets. Because biofuels are produced from agricultural feedstock, but work as a substitute for fuel in the transportation market, biofuels have created a link among agriculture, transportation, and energy markets. To examine the impact of biofuel policies, a model has to capture the dynamic interaction among agricultural and energy markets, both domestically and internationally. Ideally, these interactions will have a spatial and temporal dimension. The cost and location of biofuel production will partly be determined (and limited) by land availability constraints. Since policies may span many years, analyses should consider how the affected sectors will evolve over the lifespan of the policies considered.

Models used to quantify the impact of biofuel policies differ in their modeling framework, assumptions about baselines used to compare biofuel policies against, and the range of policies they analyze.

Modeling Framework. Numerical models used to quantify the impact of biofuel policies vary in scope and sophistication, and can be categorized into partial equilibrium and general equilibrium models. Under these categories, models can be further characterized as static and dynamic. Partial equilibrium models consider markets that are closely impacted by biofuel policies, and do not include other inter-sectoral linkages or impacts to government budget. Some static partial equilibrium models include only a few markets, usually fuel (ethanol and gasoline), corn, and oil markets [2, 7, 11, 21, 22]. Others like the Food and Agricultural Policy Research

Institute (FAPRI) model are more extensive and include the major agricultural commodities as well as the livestock sector [8]. It also links to a model of world commodity markets and Brazil's ethanol market. The Biofuels and Environmental Policy Analysis (BEPAM) model used by [18] is a dynamic partial equilibrium model, with a rolling time horizon. It also includes the major agricultural crops, as well as a detailed representation of non-corn ethanol biofuels like cellulosic ethanol from miscanthus, switchgrass and residues, as well as biodiesel. In addition, it models the existing vehicle fleet by considering different types of vehicles: conventional vehicles, flex-fuel vehicles, hybrid vehicles, and diesel vehicles. The advantage of focusing on a limited number of markets is the ability to model production activities in great detail, and achieve fine-scale spatial resolution. For example, models like BEPAM have detailed representation of different biofuel supply chains, and model variation in land characteristics by crop reporting districts. General equilibrium models include stylized models consisting of several markets directly related to the biofuel sector [5], as well as computable general equilibrium models like the Global Trade Analysis Project (GTAP) model with numerous commodities linked to defined regions of the world market via trade [14]. General equilibrium models usually consider broader inter-sectoral linkages and include interactions with factor markets and government budget. The drawback of broader coverage is the lack of detailed representation of production activities, and fine-scale spatial resolution, as the world economy is often divided into large countries and regional economies.

Baselines and Policy Mix. Studies differ in the mix of policies they evaluate, and the baseline those policies are compared against. Different biofuel policies, such as subsidies and mandates have different market impacts when implemented on their own. Furthermore, the combined effect of support policies can be quite different from their individual effects, and in most cases lead to greater welfare losses. For example, the negative effect of a subsidy is further exacerbated when it is used in conjunction with a mandate because the subsidy results in subsidizing fuel consumption when the mandate is binding [6]. In addition to biofuel policies, other policies in the fuel market also affect biofuels. A carbon tax on fuels and a low carbon fuel standard will shift consumption toward biofuel if it is less carbon-intensive relative to gasoline potentially increasing the welfare cost of biofuel policies [17,18]. The examination of multiple policies in tandem is expected because biofuels have been supported by several different policies.

Not only is the policy mix different across studies, the baseline is also different. Some studies evaluate the impact of biofuel policies against a no-market intervention baseline, while others include non-biofuels policies, such as a fuel tax in the baseline. Studies that consider externalities also define an optimal scenario, in which optimal tax rates that account for externalities (carbon emissions, miles externalities) are in place [2, 5, 18, 21, 22]. Cui et al. [5] also include optimal tariffs in the optimal scenario. It is important to take these differences into account when evaluating findings about the impact of biofuel policies. For example, in [5] an ethanol subsidy along with a fuel tax increases welfare when compared to a no-ethanol policy scenario, but decreases welfare compared to a scenario with optimal taxes and tariffs.

The remainder of this section focuses on the different markets affected by biofuel policy, and the determinants of the magnitudes of the biofuel policies' impacts on those markets.

13.3.1 Agricultural Markets

The most significant impact of biofuel policies on the agricultural sector is in the corn market. As of 2012, over 90 % of biofuel produced in the US is ethanol from corn, and 40 % of the corn crop goes to ethanol production. Corn feedstock costs account for 65 % of the cost of ethanol production [4]. Increased corn demand for biofuel production increases the price of corn, increasing the producer surplus of corn producers, while decreasing the consumer surplus of those buying corn in the world market and livestock producers that use corn as feed. Higher corn prices also reduce the welfare of consumers that purchase products using corn as input (cereal, meat). The increase in the price of corn also has important feedback effects to the ethanol sector. An increase in corn prices also increases the production cost of ethanol. Constraints on agricultural land, and cropland for corn in particular, limit the growth of ethanol production.

The impact of biofuel on the agricultural sector is affected by the types of feedstock used to produce biofuel. Feedstocks vary in their land requirement and carbon footprint. For example, using Miscanthus grass as feedstock can yield up to 9,782 L of cellulosic ethanol per hectare, while one hectare of corn will yield at most 4,452 L of ethanol [20]. Producing biofuel from feedstocks with lower land requirement will minimize the need to displace crops for food and feed with crops going to fuel production. High yielding feedstocks are also typically grasses that can be grown on marginal lands [13, 34], which further minimizes the impact of biofuels on agriculture. Different feedstocks also have different environmental impacts. Corn ethanol has been shown to reduce direct GHG emissions by 20–40 % [23, 36], while cellulosic ethanol from grasses and residues reduce GHG emissions relative to gasoline by up to 90 % [37]. Furthermore, since cellulosic ethanol has a smaller impact on the agricultural sector, it also minimizes the potential for GHG emissions due to deforestation in other countries, as increased world crop prices induces farmers to deforest in order to extend the agricultural frontier. High prices of grains induce more production, and induce technological change. As demand for feedstocks grows, the desire to maximize profits will incentivize farmers to increase yields by applying more non-land inputs such as fertilizer and developing higher-yielding varieties. The degree to which yield increases on land currently used for biofuel feedstock, as well as on land that is brought in for additional agricultural production will affect the impact of biofuel on agricultural markets. A greater increase in yields to meet greater demand will lessen the impact of biofuels on grain prices.

To the extent that the increased demand for corn changes the crop mix, biofuel policy affects the overall agricultural market, including the livestock sector that derives its inputs from grain markets. The production of ethanol also creates by-products such as dried distillers grains (DDGS) and corn oil. Failing to consider by-products will lead to an overestimate of the negative impact of diverting corn for feed to corn for ethanol, because part of the output of corn ethanol production is DDGS which substitutes for corn and soy meal as animal feed. Using the GTAP model, Taheripour et al. [31] show that omission of by-products leads to an overestimation of cropland conversion due to US and EU mandates by 27 %.

Since the US is a large agricultural exporter, the impact of biofuel policies could extend to world commodity markets. Changes in the world commodity markets will then have feedback effects on the biofuel sector. For example, the extent to which grain prices increase due to decreased exports by the US would determine the increase in production cost of ethanol due to rising corn prices. For models with linkages to world markets, assumptions about the ease of price transmission from domestic markets like the US, from which a biofuel policy shock originates, to world markets affect the magnitude of policies' impact on world commodity markets. The ease of price transmission is related to how easy it is for goods to be traded among countries. Two approaches are currently used. The Armington approach has a tendency to preserve the status quo because it differentiates otherwise homogenous goods by country of origin. Countries first decide on trading partners before determining traded goods and their quantities. In contrast, the integrated world market (IWM) framework assumes that goods will be produced where it is least costly to do so. If there are omitted factors in the model, such as political stability, the IWM assumption can produce unrealistic predictions about trade flows [12]. For example, the model may predict that a country that has very low production cost but is traditionally a closed economy due to political reasons will be exporting significant amounts of biofuel.

13.3.2 Fuel Market

The impact of biofuel policies on the fuel market depends on the price elasticity of miles demand, substitutability between different fuels, and the extent to which biofuels change oil prices. Biofuel policies affect the price of fuel and also the price of miles. A greater price elasticity of miles demand increases the impact of biofuel policies on the quantity of fuel consumed. The impact of biofuels on the fuel mix is also determined by the substitutability between gasoline and biofuel. Although gasoline and biofuel are perfect substitutes as energy sources, demand and supply side constraints such as fleet structure and distribution facilities prevent them from being blended as such in practice. Assumptions of perfect [7] and imperfect [22] substitution are used in current studies. Assuming greater substitutability will lead to a larger share of biofuels relative to gasoline, given a policy change supporting biofuels.

Biofuel policies could also impact world oil markets. Earlier studies assume that the price of oil or petroleum is exogenous and unaffected by the quantities of ethanol in the market [6]. However, studies like [28] show that biofuels can decrease the domestic gasoline price. If a significant portion of gasoline is produced from imported crude oil, reduced demand for crude oil can also affect world oil prices because the US is a major importer of oil [16]. Recent work has endogenized the price of oil or gasoline [5, 6, 18]. With the assumption that gasoline price is exogenous, a policy such as the RFS that drives up the price of ethanol increases overall fuel prices. However with the price-endogenous assumption, gasoline price could decrease as its demand decreases. Thus, even with an increased use of higher-priced ethanol, overall fuel prices could still fall. The size of the rebound effect (increase in gasoline consumption due to price decrease) will have a significant effect on how much fuel (and miles) consumption increases, which in turn has important implications for the impact of biofuels on externalities related to fuel use, such as GHG emissions and importantly, miles-related externalities such as congestion, accidents and air pollution that account for a significant share (over 95 %) of the total external cost of driving [26]. The world oil market also affects the impact of biofuel policies on the domestic fuel market. If the price of oil is low, biofuels will find it difficult to compete with gasoline, and the level of subsidies needed to induce adoption of biofuels will be higher.

13.4 Empirical Results

This section discusses the quantitative results obtained by selected studies examining the welfare impact of biofuel policies.

13.4.1 Ethanol and Corn Market Interaction

Gardner [11] examine the impact of a $0.51 per gallon ethanol subsidy on ethanol and corn markets, and compare its impact to a corn production subsidy in the form of a deficiency payment equal to the difference between the market price and a set target price for corn. For an equal amount of government expenditure ($2.6 billion), Gardner [11] finds that the deadweight loss of the ethanol subsidy is 5–18 times greater than that of the deficiency payment. In the short run (1 year),the deadweight loss from the subsidy is $91 million compared to $18 million for the deficiency payment. In the long run (10 years), the deadweight loss from the subsidy is $665 million compared to $37 million for the deficiency payment. If the demand elasticity for non-ethanol use of corn is around −1 or smaller, corn producers gain more in the long run from the ethanol subsidy compared to the deficiency payment. On the other hand, if demand for non-ethanol corn was more elastic (−2), corn producers would prefer the deficiency payment, since a more elastic corn demand would lead to a greater price increase with the deficiency payment. Also, because corn supply is expected to be less elastic in the long run than ethanol supply (owing to inelastic land supply), corn farmers capture most of the benefits of the subsidy or deficiency payment compared to ethanol producers.

De Gorter and Just [7] focus on the interaction of the biofuel subsidy (tax credit) with price-contingent farm subsidies in the corn market. They assume that gasoline prices are exogenous and constant, although the price of fuel changes due to changes in ethanol price. Gasoline and ethanol are perfect substitutes. They convert the per gallon ethanol subsidy of $0.51 to a per bushel corn subsidy, which equals $2.04 after accounting for by-products of ethanol production. They examine the impact of the tax credit on the tax costs and deadweight costs of the loan rate program. In their numerical simulation for the years 2001–2007, they find that the combined tax credit and loan rate program leads to an average annual social welfare loss of $1.3 billion. The tax credit on its own or the loan rate program on its own leads to a smaller welfare loss of $913 million and $613 million, respectively. The tax credit reduces the cost of an existing loan rate program by increasing the price of corn for non-ethanol consumers, and financing a portion of the loan rate payment to corn that goes to ethanol production. However, the cost of the tax credit program itself is not fully offset by the reduction in the cost of the loan rate program. 'Rectangular deadweight' loss that represents the costs of the tax credit program that are not offset by savings in the costs of the loan rate program averaged $1.5 billion per year.

The results from the studies by [11] and [7] suggest that the ethanol tax credit program is less efficient than the loan rate (or deficiency payment) program for corn, and that the combined effect of the two policies lead to even greater welfare losses.

13.4.2 Ethanol Trade With Brazil

Elobeid and Tokgoz [8] examine the impact of removing the US ethanol subsidy ($0.51 per gallon) and tariff on imported ethanol ($0.54 per gallon, plus 2.5 % ad-valorem) using the FAPRI model, a multi-market open-economy model consisting of the markets for US ethanol and its by-products, oil refining into gasoline, and agricultural crops. These markets are linked to an international sugar market and Brazil's domestic ethanol market. Exports of ethanol from CBI countries are also included but play a small role in the analysis. They find that the removal of the tariff increases US consumption of ethanol by 3.75 %, and increases imports by 199 %. The authors assume that ethanol and gasoline have a dominant complementary relationship. Thus, gasoline consumption also increases by 0.11 % when the tariff is removed. When both the subsidy and tariff are removed, imports increase by 137 %. The subsidy's removal dominates the result, so that production and consumption decrease by 10 and 2 % respectively, while gasoline consumption decreases by 0.06 %. Elobeid and Tokgoz [8] also report that the combined welfare in the US ethanol and corn markets decreases under the removal of the tariff (or tariff and subsidy), due to the decline in ethanol and corn prices. This result suggests that the tariff and subsidy policies cause wealth transfers from producers of livestock that use corn as input and producers of other crops that compete with corn for land, to corn and ethanol producers.

Lasco and Khanna [22] also examine the joint impact of the ethanol subsidy and tariff on welfare and GHG emissions. Unlike [8], they assume that gasoline and ethanol are imperfect substitutes. They also link the demand for fuels to the demand for miles. Using a stylized partial equilibrium model of the US fuel and miles markets linked to Brazil's ethanol market, they compare the impact of a $0.51 per gallon ethanol subsidy and a $0.54 per gallon, plus 2.5 % ad valorem tariff on imported ethanol mostly from Brazil, to a baseline scenario without these policies. They find that the subsidy decreases welfare by $2.9 billion if the US is a price taker in the ethanol market, and $3.47 billion if the US has market power (i.e. can influence the price) in the world ethanol market. Combined with the tariff, welfare losses increase to $3.51 billion under the price-taker assumption, and decrease to $3.2 billion under the market-power scenario. The tariff increases US welfare if the US has market power because it improves terms of trade of the US relative to Brazil, as the tariff decreases the price received by ethanol exporters. Both policies provide negligible (in some cases negative) GHG reduction benefits.

13.4.3   Impact on Environmental Externalities

Gallagher et al. [10] analyze the welfare implications of a national ban on the use of MTBE as a fuel additive and the imposition of a Renewable Fuel Standard of five billion gallons of ethanol by 2015 relative to the baseline of existing EPA regulations, including a 2 % oxygenate standard. They use a static partial equilibrium model that includes markets for petroleum, gasoline, petroleum refining byproducts, corn, corn refining by-products, and fuel additives. Gallagher et al. [10] show that ethanol price increases while gasoline price decreases, leading to an overall fuel price increase and a decrease in fuel consumption. The combined MTBE ban and RFS reduce welfare by $14.7 billion (excluding excise tax payments, in 2000 dollars), owing to reduced consumer surplus in the fuel market and reduced profits in the petroleum refining and gasoline retail sectors. Corn producers, however benefit from both policies. They also find that emissions of EPA clean air criteria pollutants (volatile organic compounds (VOC), toxics, and nitrous oxide (NOX)) decrease with the MTBE ban and RFS, although these environmental benefits are not monetized.

Khanna et al. [21] examine the impact of an ethanol subsidy on environmental externalities due to the consumption of fuel and vehicle miles traveled, namely GHG emissions and congestion. They assume that consumers derive utility from miles driven and miles are produced using gasoline and ethanol. Consumers also experience disutility from externalities such as congestion and GHG emissions. Gasoline and ethanol are imperfect substitutes in the production of miles, and both have upward sloping supply curves. The optimal policy to address externalities is to set a Pigouvian tax on fuels equal to their marginal external damage. They compare the welfare effect of the ethanol subsidy, relative to the optimal scenario in which Pigouvian taxes are levied on miles, gasoline and ethanol. The optimal taxes are: $0.085 per mile, $0.08 per gallon gasoline, and $0.04 per gallon ethanol. They find that compared to the optimal policy, the ethanol subsidy of $0.51 per gallon increases consumption of fuels and leads to a welfare loss of $19 billion (2007) dollars per year. Compared to the status quo with the current fuel tax ($0.39 per gallon), the welfare loss is $1.5 billion, due to the deadweight loss of the subsidy. Miles consumption is greater in the scenario with the subsidy compared to the optimal and status quo scenarios because the subsidy decreases the overall price of fuel. They also find that the subsidy increases GHG emissions by 20.5 % compared to the optimal scenario, and decreases emissions by 0.09 % compared to the status quo.

Ando et al. [2] extend the framework of [21] to examine the impact of mandates. Specifically they consider the effect of stacking the EISA RFS that requires the blending of 36 billion gallons of corn ethanol and other advanced biofuels by 2022. By 2015, which is the time period considered in their analysis, the projected biofuel mix is 15 billion gallons of corn ethanol and 5.5 billion gallons of cellulosic ethanol. Cellulosic ethanol receives a tax credit of $1.01 per gallon, while corn ethanol receives $0.51 per gallon. They first compare the status quo policy of a $0.39 per gallon fuel tax and $0.51 corn ethanol subsidy with the optimal scenario where emissions and miles externalities are taxed according to their marginal external damage at $0.025 per mile and $0.085 per kg CO_2. Compared to the optimal scenario, the status quo policy leads to a welfare loss of $52 billion. Relative to the scenario with a fuel tax and corn ethanol subsidy (status quo), the RFS reduces the price of gasoline and increases the price of ethanol; the overall effect is lower cost of fuel and miles. Consumers of gasoline and producers of ethanol see an increase in welfare whereas producers of gasoline see a decrease in producer surplus. The net welfare effect of adding the RFS and subsidizing cellulosic ethanol at $1.01 relative to the status quo is a 16 % greater loss in welfare. Differences in GHG emissions are negligible between the RFS and status quo scenarios, but the social cost of miles externalities are greater with the RFS.

Cui et al. [5] use an open-economy general equilibrium model to examine the welfare impact of alternative policies in the US fuel sector, in contrast to an optimal policy scenario. The optimal policy scenario includes an optimal carbon tax on fuel, as well as an optimal import tariff on oil and export tariff on corn. Alternative policies include combinations of fuel taxes and ethanol policies. They find that compared to the equilibrium with no policies and a scenario with the current fuel tax but no ethanol policy, policy alternatives that include ethanol support policies increase social welfare, although they do not always reduce GHG emissions. Most of the increase in welfare associated with ethanol policies is owing to its favorable effect on terms of trade in oil and corn markets.

Results from the studies that examine the impact of biofuel policies on environmental externalities, notably emissions, suggest that at best the effect is modest, and there are cases when emissions may even increase. Considering the welfare costs of biofuel policies, the emissions savings from biofuel policies come at great costs. Importantly, if fuel prices decrease due to biofuel policies, miles consumption could increase, leading to large welfare losses due to increases in miles externalities such as congestion, air pollution, and traffic accidents. Current studies do not include the impact of biofuel policies on water use, agricultural pollution and biodiversity. More research is needed to quantify the impact of biofuel policies in these areas.

13.4.4 Interaction with Fuel Market Policies

Huang et al. [18] use BEPAM, a multi-market, dynamic, price-endogenous, non-linear mathematical programming model of the US fuel and agricultural sectors, linked by trade to the rest of the world. Unlike other studies that focus on corn ethanol, the model includes several feedstocks for biofuel including forest and agricultural residues, dedicated biomass such as miscanthus and switchgrass, as well as corn, soybeans, and sugarcane. They examine the impacts of the RFS, and its combination with a low carbon fuel standard and carbon tax from 2005 to 2035, relative to a business-as-usual (BAU) scenario without those policies. They find that the increase in biofuel demand owing to these policies increase agricultural prices and lower fuel prices (as gasoline demand decreases, so does its price). Agricultural consumers and fuel producers incur losses while fuel consumers and agricultural producers gain. Government revenue increases due to more fuel tax revenue from higher fuel consumption. The net social surplus effect of these policies are positive because the reduction in fuel price offsets the negative effect of rising agricultural prices. The main impact of the LCFS is to increase the share of cellulosic biofuel relative to corn ethanol. The carbon tax increases both fuel and agricultural prices, while reducing the price received by producers. However government revenue increases, offsetting the losses in the agricultural and fuel markets. GHG emissions from the food and agricultural sectors decrease under all these scenarios with the reduction largest for the policy with a carbon tax. The RFS alone decreases emissions by 4.8 % relative to BAU.

13.4.5 World Biofuel Policies and Commodity Markets

Hertel et al. [14] examine the impact of biofuel mandates in the US and in the European Union (EU). Using the GTAP model, they simulate the effect of the US mandate that increases corn ethanol consumption to 15 billion gallons, and the EU mandate that increases the share of renewable fuels by 6.25 % by 2015. They report changes in bilateral trade in coarse grains, oilseeds, and other food products as well as changes in land cover and aggregate welfare for the different world regions represented in the model. They find that the combined US and EU biofuel policies decrease allocative efficiency both in the US and EU due to the mandated increase in biofuel quantities in the absence of high oil prices that would make biofuels competitive with gasoline. US welfare changes by −$15 billion from 2006 to 2015 due to reduction in allocative efficiency. On the other hand, the US gains $7 billion from improved terms of trade, as reduced demand for petroleum depresses the world price of oil. Thus, the total welfare change from the US and EU mandates is −$8 million in the US. Including net welfare effect in the EU (−$25 million), welfare gains of exporting countries like Brazil, and welfare losses of oil exporters, total welfare loss in the world market is $43 billion.

Nuñez et al. [25] examine the impact of biofuel policies in the US and Brazil on economic surplus, GHG emissions and land use. They use a static partial equilibrium model that includes the fuel and agricultural sectors of the US, Brazil, China, Argentina, and the rest-of-the-world. Sources of biofuels include corn ethanol and cellulosic ethanol produced in the US, and sugarcane ethanol produced in Brazil. They find that relative to the baseline case without biofuel support policies, the RFS, corn ethanol subsidy, and import tariff increase Brazil's social surplus, while having a minimal effect on the social surplus of the US. Removing the subsidy and tariff while keeping the RFS leads to a greater social surplus in Brazil, compared to the scenario with the subsidy and tariff in place.

The studies discussed above suggest that biofuel policies lead to welfare losses due to allocative inefficiencies. On the other hand, they can improve terms of trade in markets where changes in US excess demand and supply can affect world market prices. Hertel et al. [14] find that terms of trade gains do not fully compensate for welfare losses from inefficiencies introduced by biofuel policies, while [18] and [5] find that terms of trade gains dominate.

13.5 Conclusion

In evaluating biofuel policies, it is important to consider the stated goals of biofuel policy: increase energy security, reduce GHG emissions and support farm incomes. While current studies on the welfare effect of biofuel policies have made significant progress toward incorporating aspects of these objectives in their analysis, further research is needed to take into account energy security benefits. Future benefits from diversified energy sources and reduced dependence on foreign oil are hard to quantify. Studies have used expenditures to secure oil supplies from the Middle East to estimate benefits from displacing gasoline with ethanol [35]. Will it be welfare increasing in the long run to continue supporting the biofuel industry? The production cost of corn ethanol has decreased by 43–60 % from the early 1980s to 2005 due to learning-by-doing and technological change, and is expected to further decrease in the future [3, 15]. Without policy supports for corn ethanol, these cost reductions may not have been realized. Cellulosic ethanol currently requires large subsidies to be competitive, additional support for research and development, and loan guarantees to induce production at a large scale. It is hard to predict whether these investments will pay off in the form of cheap, domestically produced renewable energy with a small carbon footprint. It will depend on what technological breakthroughs will be achieved, and what future costs of other energy sources will be.

Even with the uncertainty in the future energy picture, some clear guidelines emerge that would enable the US to move toward a future of diversified energy sources and reduced dependence on foreign oil. Current analyses of biofuel policies have identified mandates to be the least distorting form of support for the biofuel sector. It appears that policy makers have heeded this advice as the corn ethanol subsidy of $0.45 per gallon and the ethanol tariff of $0.54 per gallon were both allowed to expire in January 2012. It is also clear that the ideal feedstock for biofuel is not one that would compete with food [32]. These include perennial grasses that can be grown on marginal land, residues, and waste materials. Studies have also shown that biofuel from cellulosic feedstock have a much greater potential to reduce GHG emissions than ethanol from corn [36]. Thus, there is some rationale for continuing to support the development of the cellulosic biofuel industry. There is currently a debate about whether it is worthwhile to keep the mandate for cellulosic biofuels even with the lack of commercial scale production. Some have argued that mandates should be waived, while others argue that the biofuel sector will be stalled indefinitely without certainty in the regulatory environment. Further research needs to be undertaken to determine what support mechanism will strike a balance between minimizing inefficiency, and enabling the biofuel sector to grow.

References

1. Abbott, P.C., Hurt, C., Tyner, W.E.: What's driving food prices? March 2009 update. Technical report, Farm Foundation (2009)
2. Ando, A.W., Khanna, M., Taheripour, F.: Market and Social Welfare Effects of the Renewable Fuels Standard. Handbook of Bioenergy Economics and Policy, pp. 233–250. Springer, New York (2010)
3. Chen, X., Khanna, M.: Explaining the reductions in US corn ethanol processing costs: Testing competing hypotheses. Energy Pol. **44**, 153–159 (2012)
4. Crago, C.L., Khanna, M., Barton, J., Giuliani, E., Amaral, W.: Competitiveness of brazilian sugarcane ethanol compared to us corn ethanol. Energy Pol. **38**(11), 7404–7415 (2010)
5. Cui, J., Lapan, H., Moschini, G.C., Cooper, J.: Welfare impacts of alternative biofuel and energy policies. Am. J. Agr. Econ. **93**(5), 1235–1256 (2011)
6. De Gorter, H., Just, D.R.: The economics of a blend mandate for biofuels. Am. J. Agr. Econ. **91**(3), 738–750 (2009)
7. De Gorter, H., Just, D.R.: The welfare economics of a biofuel tax credit and the interaction effects with price contingent farm subsidies. Am. J. Agr. Econ. **91**(2), 477–488 (2009)
8. Elobeid, A., Tokgoz, S.: Removing distortions in the us ethanol market: what does it imply for the united states and brazil? Am. J. Agr. Econ. **90**(4), 918–932 (2008)
9. EPA: Regulation of fuels and fuel additives: changes to renewable fuel standard program. Fed. Regist. **75**, 14669–15320 (2010)
10. Gallagher, P.W., Shapouri, H., Price, J., Schamel, G., Brubaker, H.: Some long-run effects of growing markets and renewable fuel standards on additives markets and the us ethanol industry. J. Pol. Model. **25**(6), 585–608 (2003)

11. Gardner, B.: Fuel ethanol subsidies and farm price support. J. Agr. Food Ind. Organ. **5**(2), Article 4 (2007)
12. Golub, A., Hertel, T.W., Taheripour, F., Tyner, W.E.: Modeling biofuels policies in general equilibrium: insights, pitfalls, and opportunities. Fron. Econ. Globalization **7**, 153–187 (2010)
13. Heaton, E.A., Dohleman, F.G., Long, S.P.: Meeting us biofuel goals with less land: the potential of miscanthus. Global Change Biol. **14**(9), 2000–2014 (2008)
14. Hertel, T.W., Tyner, W.E., Birur, D.K.: The global impacts of biofuel mandates. Energy J. **31**(1), 75 (2010)
15. Hettinga, W.G., Junginger, H.M., Dekker, S.C., Hoogwijk, M., McAloon, A.J., Hicks, K.B.: Understanding the reductions in us corn ethanol production costs: an experience curve approach. Energy Pol. **37**(1), 190–203 (2009)
16. Hochman, G., Rajagopal, D., Zilberman, D.: The effect of biofuels on crude oil markets. AgBioForum **13**(2), 112–118 (2010)
17. Holland, S.P., Hughes, J.E., Knittel, C.R.: Greenhouse gas reductions under low carbon fuel standards? Am. Econ. J. Econ. Pol. **1**(1), 106–146 (2009)
18. Huang, H., Khanna, M., Önal, H., Chen, X.: Stacking low carbon policies on the renewable fuels standard: economic and greenhouse gas implications. Energy Pol. **56**, 5–15 (2013)
19. Just, R.E., Hueth, D.L., Schmitz, A.: The Welfare Economics of Public Policy. E. Elgar, Cheltenham (2004)
20. Khanna, M., Crago, C.L.: Measuring indirect land use change with biofuels: implications for policy. Annu. Rev. Resource Econ. **4**, 161–184 (2012)
21. Khanna, M., Ando, A.W., Taheripour, F.: Welfare effects and unintended consequences of ethanol subsidies. Appl. Econ. Perspect. Pol. **30**(3), 411–421 (2008)
22. Lasco, C., Khanna, M.: Us–Brazil Trade in Biofuels: Determinants, Constraints, and Implications for Trade Policy. Handbook of Bioenergy Economics and Policy, pp. 251–266. Springer, New York (2010)
23. Liska, A.J., Yang, H.S., Bremer, V.R., Klopfenstein, T.J., Walters, D.T., Erickson, G.E., Cassman, K.G.: Improvements in life cycle energy efficiency and greenhouse gas emissions of corn-ethanol. J. Ind. Ecol. **13**(1), 58–74 (2009)
24. Mitchell, D.: A note on rising food prices. Technical report, World Bank Policy Research Working Paper Series (2008)
25. Nuñez, H.M., Onal, H., Khanna, M.: A prospective analysis of brazilian biofuel economy: land use and infrastructure development, 2012. Selected Paper prepared for presentation at the International Association of Agricultural Economists (IAAE) Triennial Conference, Foz do Iguaçu, 18–24 August 2012.
26. Parry, I.W.H., Small, K.A.: Does Britain or the United States have the right gasoline tax? Am. Econ. Rev. **95**(4), 1276–1289 (2005)
27. Plevin, R.J., Jones, M.O.H.A.D., Torn, M.S., Gibbs, H.K.: Greenhouse gas emissions from biofuels indirect land use change are uncertain but may be much greater than previously estimated. Environ. Sci. Technol. Columbus **44**(21), 8015 (2010)
28. Rajagopal, D., Sexton, S.E., Roland-Holst, D., Zilberman, D.: Challenge of biofuel: filling the tank without emptying the stomach? Environ. Res. Lett. **2**(4), 044004 (2007)
29. Searchinger, T.D.: Biofuels and the need for additional carbon. Environ. Res. Lett. **5**(2), 024007 (2010)
30. Solomon, B.D., Barnes, J.R., Halvorsen, K.E.: Grain and cellulosic ethanol: History, economics, and energy policy. Biomass Bioenergy **31**(6), 416–425 (2007)
31. Taheripour, F., Hertel, T.W., Tyner, W.E., Beckman, J.F., Birur, D.K.: Biofuels and their by-products: Global economic and environmental implications. Biomass Bioenergy **34**(3), 278–289 (2010)
32. Tilman, D., Socolow, R., Foley, J.A., Hill, J., Larson, E., Lynd, L., Pacala, S., Reilly, J., Searchinger, T., Somerville, C., et al.: Beneficial biofuels – The food, energy, and environment trilemma. Science **325**(5938), 270–271 (2009)
33. United States International Trade Commission (USITC): Interactive Tariff and Trade Dataweb. http://dataweb.usitc.gov/ (2011)

34. Varvel, G.E., Vogel, K.P., Mitchell, R.B., Follett, R.F., Kimble, J.M.: Comparison of corn and switchgrass on marginal soils for bioenergy. Biomass Bioenergy **32**(1), 18–21 (2008)
35. Vedenov, D., Wetzstein, M.: Toward an optimal US ethanol fuel subsidy. Energy Econ. **30**(5), 2073–2090 (2008)
36. Wang, M., Wu, M., Huo, H.: Life-cycle energy and greenhouse gas emission impacts of different corn ethanol plant types. Environ. Res. Lett. **2**(2), 024001 (2007)
37. Wu, M., Wang, M., Hou, H.: Fuel-cycle assessment of selected bioethanol production. Technical report, ANL/ESD/06-7, Argonne National Laboratory (ANL) (2007)
38. Yacobucci, B.D.: Biofuels incentives: a summary of federal programs. Technical report, Congressional Research Service (2011)
39. Yergin, D.: The Quest: Energy, Security, and the Remaking of the Modern World. Penguin Books, New York (2011)
40. Zilberman, D., Hochman, G., Rajagopal, D., Sexton, S., Timilsina, G.: The impact of biofuels on commodity food prices: assessment of findings. Am. J. Agr. Econ. **95**(2), 275–281 (2013)

Chapter 14
Advanced Mathematical and Statistical Tools in the Dynamic Modeling and Simulation of Gene-Environment Regulatory Networks

Özlem Defterli, Vilda Purutçuoğlu, and Gerhard-Wilhelm Weber

14.1 Introduction

In order to deeply understand and examine in details the whole structure of a biological network on a system-level approach rather than single-cell components, some technologies are recently developed. In this direction, there is an essential need for the methods to model and analyze the regulatory networks [1]. With the help of biochemical and genetics studies and user friendly computer tools, various mathematical models have been already constructed to describe gene interactions and to make predictions precisely and representatively [2]. Additionally, the development of new mathematical methods for the analysis of these highly interconnected systems allows a deeper understanding in the dynamic behavior and the topological aspects of complex regulatory systems appearing not only in biology, but also in finance, engineering and environmental sciences [2–5].

There are three main *deterministic* approaches as listed below to model the biological networks. These are *modeling by graphs*, *Boolean networks* and *dynamic models*. Moreover, *Bayesian networks* is another modeling technique which considers stochastic effects.

Modeling by graphs is one of the well known way of modeling the gene regulatory networks by representing as directed graphs. A directed graph or network

Ö. Defterli (✉)
Department of Mathematics and Computer Science, Çankaya University, 06810, Ankara, Turkey
e-mail: defterli@cankaya.edu.tr

V. Purutçuoğlu
Department of Statistics, Middle East Technical University, 06531, Ankara, Turkey
e-mail: vpurutcu@metu.edu.tr

G.-W. Weber
Institute of Applied Mathematics, Middle East Technical University, 06531, Ankara, Turkey
e-mail: gweber@metu.edu.tr

A.A. Pinto and D. Zilberman (eds.), *Modeling, Dynamics, Optimization and Bioeconomics I*, Springer Proceedings in Mathematics & Statistics 73, DOI 10.1007/978-3-319-04849-9_14,
© Springer International Publishing Switzerland 2014

is expressed by $G = (V, A)$ components which consist of a finite set of nodes, i.e. vertices and denoted by V, and a subset A of arcs of the cross product $(V \times V)$ [6, 7]. Here, an arc or edge $a \in A$ is defined as an ordered pair of distinct nodes $(v_1, v_2) \in (V \times V)$ in the directed graph. In a genetic network, the nodes refer to genes and the arcs refer to the relationships between the genes, i.e., gene interactions. Hereby, the pair $a = (v_1, v_2)$ of nodes indicates an arc which connects two nodes (genes), v_1 and v_2, and weighted with a value where it represents the influence of the first node v_1 (gene 1) on the second node v_2 (gene 2). This associated weight can be positive or negative which stands for an activation or inhibition effect, respectively, between gene 1 and gene 2 [8].

On the other hand the *Boolean networks*, firstly studied by Kauffman [9], are another sort of dynamic approaches in order to model the gene regulation [10–12]. The advantage of the Boolean networks over its alternatives is its ability to present a rich variety of dynamic behaviors such as convergence to a stable steady-state, multi-stationarity, oscillations, switch-like behavior or hysteresis activation in a system [13]. But as they are in the class of deterministic models, they cannot represent the stochastic nature of gene expressions and do not account for noise in the measurements. Whereas many extensions of these networks have been proposed in order to overcome some of these limitations [1]. Among alternatives, the *probabilistic Boolean networks* can be seen the most important method since they enable us to include the stochastic fluctuations of the regulation process [1].

As an alternative way of modeling, *Bayesian networks* initially used by Murphy and Mian [14] to deterministically construct the gene interactions by using expression data [15]. Getting information for a Bayesian network from dataset is related with the estimation of the joint probability distribution which defines the structure of the corresponding directed acyclic graph [1]. Therefore the Bayesian networks are probabilistic systems whereas they are static and so disable to model complex phenomena like the oscillations, multi-stationarity, etc. [3, 13]. But in order to deal with these restrictions, the *dynamic Bayesian networks* are developed [1, 14, 15].

Finally, *modeling with ordinary differential equations* is a recently suggested approach to represent the dynamic behavior of gene regulatory networks quantitatively. This way of modeling enables us to understand better the underlying mechanisms causing certain kinds of dynamic behaviors rather than Boolean and Bayesian networks. In this approach, different parametrizations of the right-hand side function are proposed to solve the problem of initialization of the system. These parameters appearing in the model refer directly to reaction rates, binding affinities or degradation rates, which are useful for both a reasonable restriction of the parameter space and the interpretation of inference results [1]. A linear version of this model has been firstly proposed by Chen et al. [16] and its extensions which include piecewise linear and nonlinear approaches, hybrid system approach [17–22], stochastic kinetic approaches, partial differential equations, and delay differential equations [23] have been further suggested. In [19], the mixed continuous-discrete model has been introduced to contain the most relevant regulating interactions in a cell as a complementary approach to the one developed by [16] and later on by [17, 18, 24] where the dynamics is obtained by a hybrid system.

The common challenge in all these listed modeling approaches for the network inference is to select a model that can fit better with the given data which are generally sparse. This sparsity means that the number of network components p is large with respect to the number of different conditions or time points N, i.e., $p >> N$. Hence, there is a fitting problem of a high-dimensional function due to the measurements for a few data points. Therefore, the optimization problems are ill-posed. In order to overcome this challenge, various regularization methods such as the application of distinct model selection criteria are developed. The Akaike information criterion (AIC) and the Bayesian information criterion (BIC), which take into account the negative logarithm of the likelihood function and penalized term having a large number of parameters, are the most popular ones among alternatives [8, 19, 25–27]. These type of criterias restrict the parameter space by considering the biological knowledge in the optimization process. This can be done by introducing constraints into the optimization problem like upper bounds for single parameters or by adding penalty terms in the associated likelihood function [1].

Hereby, in this study, we particularly deal with the inference of the system via the ordinary differential equations' and Gaussian graphical models' approaches among other listed alternatives since both of them enable us to get high accuracy in the construction of the biological network. So, in the following part we explain the mathematical details of each model comprehensively. Then, in Sect. 14.4 we present simulation of the system to obtain future states via the estimated structure of the network by the discretization of system of ordinary differential equations. In this simulation, recently developed numerical methods, namely a class of explicit Runge-Kutta approximations, are reviewed briefly. The evaluation of the new results are given in Sect. 14.5 on an artificial dataset where the network structure is estimated by two different model, i.e. ordinary differential equations (ODEs) based model and Gaussian graphical model (GGM). Correspondingly, the graphical representation of these two different inference are obtained and presented. Additionally, Sect. 14.5 contains the previously obtained simulation results for the forecasting of future behaviour of the system are presented, only for the inference results of the ODE based model. Finally, we conclude and discuss the findings in Sect. 14.6 with the statement of our future work.

14.2 Modeling of Gene-Environment Regulatory Networks via Differential Equations

The differential equations (DE) are commonly used for modeling the biological systems. The main advantage of this approach is it gains in the accuracy of the physical systems and it has ability for the possible extensions like the dynamic system theory in analyzing and capturing the dynamic behaviours of the system. Moreover, considering that the activation of the biological systems is continuous in time, it is preferred to use this approach since it enables us to do instantaneous changes in their right-hand side expression [28].

In this model, a differential relation among n variables of gene networks is simply described by the following expression

$$\frac{dx_i}{dt} = f_i(\mathbf{x}), \tag{14.1}$$

where $i = 1, \ldots, n$, and each function $f_i : \mathbb{R}^n \to \mathbb{R}$ is nonlinear. Furthermore, the vector $\mathbf{x} = (x_1, \ldots, x_n)^T$ stands for the positive concentrations of proteins, mRNAs, or small components. In the literature, there are a number of DE methods which can represent the behaviour of the system under the assumption of continuous in time.

14.2.1 Gene-Environment Networks and Time-Continuous Model Class

Among many ordinary differential equations' approaches, the first time-continuous model is suggested as a system of time-autonomous ODEs that can represent the dynamic structure of the gene-environment network via the expression below

$$\dot{\mathbf{E}} = \mathbf{F}(\mathbf{E}), \tag{14.2}$$

where $\mathbf{E} = (E_1, \ldots, E_d)^T$ shows the d-dimensional vector of positive concentration levels of proteins (or mRNAs or small components) under certain levels of environmental factors. Here, $\mathbf{E} = \mathbf{E}(t)$ in which the time t is in the interval I where $I = (a, b) \subseteq \mathbb{R}$. Here, the first n components of the vector $\mathbf{E}$ refer to the genes, whereas, the remaining $(d - n)$ components denotes the environmental factors. Moreover, $\dot{\mathbf{E}}(= \frac{dE}{dt})$ presents a continuous change in the gene-expression data and $\mathbf{F}_i : \mathbb{R}^d \to \mathbb{R}$ are nonlinear coordinate functions of $\mathbf{F}$, i.e., $\mathbf{F} = (F_1(E), \ldots, F_d(E))^T$ [8, 16, 24, 25, 27, 29, 30]. The associated parameters appearing in the expression of $\mathbf{F}$ can be estimated by considering the experimental data vectors $\bar{\mathbf{E}}$ which are obtained from microarray experiments and environmental measurements at the sample times. Indeed, the vectors $\bar{\mathbf{E}}$ are approximated values of the actual states $\mathbf{E}$, thereby may contain some errors, noise and uncertainties coming from these measurements [20,21,31–33]. Here, $\mathbf{E}(t_0) = \mathbf{E}_0$ refers the initial values, where $\mathbf{E}_0 = \bar{\mathbf{E}}_0$. Furthermore, $E_i(t)$ denotes the gene-expression level, i.e., concentration rate, of the ith-gene at time t, and $E_i(t)$ describes the first n coordinates in the d-dimensional vector $\mathbf{E}$ of genetic and environmental states.

In [8, 16–19, 24–26, 30, 34–36], various dynamical model types are presented as improved versions of the first model given by Eq. (14.2). These improved model types have the following forms [37, 38]:

1. Chen et al. [16] propose the first time-continuous model consisting of system of the first-order ODE's to model time-series gene expression patterns. This model is formulated as

$$\dot{\mathbf{E}} = \mathbf{M}\mathbf{E}, \tag{14.3}$$

where $\mathbf{M}$ is an $(n \times n)$-dimensional constant matrix as a transition matrix representing regulatory interactions for both genes and proteins, and $\mathbf{E}$ indicates the $(n \times 1)$-dimensional vector showing the expression level of individual genes. Here n denotes the total number of genes.

On the other hand, Hoon et al. [29, 30] use similar version of this model but they consider only the mRNA concentrations and the AIC value to find the places and the number of nonzero parameters in $\mathbf{M}$, respectively. Later, Sakamoto and Iba [24] develop a more flexible model as given below:

$$\dot{E}_j = F_j(E_1, \ldots, E_n)^T. \tag{14.4}$$

Here F_j $(j = 1, \ldots, n)$ denotes the function of $\mathbf{E} = (E_1, \ldots, E_n)^T$ obtained by the genetic programming and least-squares methods.

2. The models suggested by part (1) are improved by the following aspects [8, 25, 26, 35, 39]: Gebert et al. [8] considers a constant gene interaction matrix, $\mathbf{M}$, in the model given by Eq. (14.3) and apply the discrete least-squares approximation [40] to find the parameters in the regulatory relations. On the other hand, in [25], the following nonlinear, also called quasi-linear, model is proposed which has a matrix multiplicative form

$$\dot{\mathbf{E}} = \mathbf{M}(\mathbf{E})\mathbf{E}. \tag{14.5}$$

In this expression, the interaction matrix $\mathbf{M}$ depends on the current metabolic state $\mathbf{E}$ which allows nonlinear interactions between network variables to be taken into account. But the solution space is restricted by imposing bounds on the number of regulating factors for each gene so that the regulatory network can be uniquely identified.

This dynamical system is represented by the n genes and their associated interactions. Hereby, the matrix $\mathbf{M}$ owns an $(n \times n)$-dimensional matrix with entries as the functions of polynomials, exponential, trigonometric, splines or wavelets containing some parameters to be optimized. In the matrix $\mathbf{M}$, each row and column correspond to one gene in the genetic network, thereby, the value of each entry denotes the interactions between genes. If these entries are zero, they imply no interaction between the genes. Moreover, the smaller the absolute value of the entries of the matrix, the less is the influence or interaction between genes [8].

3. In the study of Yılmaz [18], an extended version of the model given by Eq. (14.5) is derived in order to emphasize the nonlinear interactions with the environment by the following model

$$\dot{\mathbf{E}} = \mathbf{F}(\mathbf{E}), \tag{14.6}$$

in which $\mathbf{F}(\mathbf{E})$, given as $\mathbf{F} = (F_1, \ldots, F_n)^T$, consists of a sum of quadratic functions as shown below while $j = 1, \ldots, n$.

$$\mathbf{F}_j(\mathbf{E}) = f_{j,1}(E_1) + f_{j,2}(E_2) + \ldots + f_{j,n}(E_n). \tag{14.7}$$

Later, the affine linear shifts terms are included into to the model.

4. In order to preserve the idea of recursive iteration that is given in [26] by the underlying affine linear shifts, Eq. (14.5) is reconstructed as in Eq. (14.8) [17,34]:

$$\dot{\mathbf{E}} = \mathbf{M}(\mathbf{E})\mathbf{E} + \mathbf{C}(\mathbf{E}). \tag{14.8}$$

The additional column vector $\mathbf{C}(\mathbf{E})$ stands for the environmental perturbations or contributions and provides more accurate data fitting (see [18, 35] for the case of a constant $\mathbf{C}$). The shift term $\mathbf{C}(\mathbf{E})$ does not need to reveal $\mathbf{E}$ as a factor. In the extended model [17, 18, 34–36], represented by Eq. (14.8), the dimension of the vector $\mathbf{E}$ becomes $(n + m)$ by considering the m-dimensional vector $\check{\mathbf{E}}(t) = (\check{E}_1(t), \ldots, \check{E}_m(t))^T$, which indicates m environmental factors affecting the gene-expression levels and their variations. In order to represent the weights of the effect of the jth-environmental factor $\check{E}_j$ on the gene-expression data E_i, the $(n \times m)$-dimensional weight matrix (also called gene-environment matrix) $\check{\mathbf{M}}(\mathbf{E})$ is introduced so that the vector $\mathbf{C}(\mathbf{E})$ can be written as $\mathbf{C}(\mathbf{E}) = \check{\mathbf{M}}(\mathbf{E})\check{\mathbf{E}}$, where

$$\check{\mathbf{M}}(\mathbf{E}) = \begin{bmatrix} c_{11}(\mathbf{E}) & \cdots & c_{1m}(\mathbf{E}) \\ \vdots & \ddots & \vdots \\ c_{n1}(\mathbf{E}) & \cdots & c_{nm}(\mathbf{E}) \end{bmatrix}. \tag{14.9}$$

Hence, the gene-environment network described by the dynamic in Eq. (14.8) becomes

$$\dot{\mathbf{E}} = \mathbf{M}(\mathbf{E})\mathbf{E} + \check{\mathbf{M}}(\mathbf{E})\check{\mathbf{E}}. \tag{14.10}$$

The extended initial value problem can finally be written in a multiplicative form as follows:

$$\dot{\mathbb{E}} = \mathbb{M}(\mathbb{E})\mathbb{E}, \qquad \mathbb{E}_0 = \begin{bmatrix} \mathbf{E}_0 \\ \check{\mathbf{E}}_0 \end{bmatrix}, \tag{14.11}$$

where

$$\mathbb{E} := \begin{bmatrix} \mathbf{E} \\ \check{\mathbf{E}} \end{bmatrix}, \quad \mathbb{M}(\mathbb{E}) := \begin{pmatrix} \mathbf{M}(\mathbf{E}) & \check{\mathbf{M}}(\mathbf{E}) \\ \mathbf{0} & \mathbf{0} \end{pmatrix}, \tag{14.12}$$

are an $(n+m)$-dimensional vector and $[(n + m) \times (n + m)]$-dimensional matrix, respectively. The different kinds of extension of the gene-environment network described in Eq. (14.8) are studied in [20, 31, 36, 41].

As a system of ordinary differential equations, Eqs. (14.3)–(14.8), hence, Eq. (14.11) are all autonomous, which means that the right-hand side depends on the state $\mathbb{E}$ only, but not on time t. This implies that trajectories do not cross themselves [25].

The models given by the continuous dynamical equations in items (1)–(4) can be written in general via

$$\dot{\mathbb{E}} = \mathbb{M}(\mathbb{E})\mathbb{E}, \tag{14.13}$$

where $\mathbb{M}(\mathbb{E})$ is a $(d \times d)$-matrix which is estimated to compute the gene network based on the gene expression dataset [42, 43] and $\mathbb{E}$ denotes the d-dimensional concentration vector of the genes throughout the time [20, 31, 33, 36]. Hereby the initial value is equal to $\mathbb{E}_0 = \mathbb{E}(t_0)$. Accordingly the entries of $\mathbb{M}(\mathbb{E})$, which can be polynomial, trigonometric, exponential, logarithmic, hyperbolic, spline, etc., by standing for the growth, cyclicity or other kinds of changes in the genetic or environmental concentration rates. The form of system in Eq. (14.13) allows a time-discretization such that the dynamics is given by a step-wise matrix multiplication. This recursive property is an important advantage based on the algorithmic stability analysis [17, 34].

On the other hand, in inference of the network matrix $\mathbb{M}(\mathbb{E})$ appearing in the ODE based model, the following discrete least-squares optimization problem has to be solved:

$$\min_{\mathbf{y}} \sum_{k=0}^{N-1} \left\| \mathbb{M}_{\mathbf{y}}(\bar{\mathbb{E}}^{(k)})\bar{\mathbb{E}}^{(k)} - \dot{\bar{\mathbb{E}}}^{(k)} \right\|_2^2 . \tag{14.14}$$

In this expression we aim to find the matrix $\mathbb{M}(\mathbb{E})$ which enables us to obtain the distance between the forecasted and actual observed values as small as possible with respect to the $\| \cdot \|_2$ norm. Here, $\mathbf{y}$ is the vector of a subset of all the parameters and N refers to the number of biological measurements (observations). $\dot{\bar{\mathbb{E}}}^{(k)}$ describe the difference quotients based on the kth-experimental data $\bar{\mathbb{E}}^{(k)}$ with step lengths $\bar{h}_k := \bar{t}_{k+1} - \bar{t}_k$ between neighbouring sampling times $\bar{t}_{k+1}$ and $\bar{t}_k$ [8, 19, 25, 32, 33, 37, 38]. The forward and central difference approximations are the common choices for approximating $\dot{\mathbb{E}}^{(k)}$ as presented via

$$\dot{\mathbb{E}}^{(k)} := \left\{ \begin{array}{ll} (\mathbb{E}^{(k+1)} - \mathbb{E}^{(k)})/\bar{h}_k, & \text{if} \quad k \in \{0, 1, \ldots, N - 1\}, \\ (\mathbb{E}^{(N)} - \mathbb{E}^{(N-1)})/\bar{h}_k, & \text{if} \quad k = N \end{array} \right\} . \tag{14.15}$$

For the equidistant step-size, $\bar{h}_k \equiv c \ (c > 0)$, $\dot{\mathbb{E}}^{(k)} := (\mathbb{E}^{(k+1)} - \mathbb{E}^{(k-1)})/2c$ is a common choice [19, 20, 25, 33].

Hereby in Eq. (14.14), the estimation of the entries of $\mathbb{M}(\mathbb{E})$ can be calculated by the least-squares methods of linear and nonlinear regression in such a way that the set of a given experimental data can fit to the model and the statistical properties of the estimates can be characterized. In this optimization problem, a set of constraints can be imposed in order to restrict the solution space because of the high complexity and dimension of the problem. Here, the choice of constraints and their bounds are related with the decision making, multi-criteria optimization, rarefication and regularization [44–46]. The bounds to be added to the problem in (14.14) will form it to the mixed integer problem [47], given in Sect. 14.5, and they regularize and rarefy the gene-network by selecting the most important and meaningful elements [8, 25, 26, 38].

14.3 Modeling via Gaussian Graphical Models

The Gaussian graphical model (GGM) is a dynamic modeling of the biological system under the assumption of the multivariate normality of the data with mean vector μ and covariance matrix Σ for p nodes or genes, i.e., $N_p(\mu, \Sigma)$. In this approach we consider that the direct association or interaction between any two genes can be described by the partial correlation coefficient between the genes conditional on all other genes in the system [6]. In inference this conditional correlation can be found from the inverse of Σ, also called precision matrix Σ^{-1}. But, particularly, in genomic data since the number of observations for each node N is much smaller than the number of dimensions p, that is $N << p$, the empirical covariance matrix can be singular [48]. In the study of Schäfer and Strimmer [49], the underlying challenge is solved via the *shrinkage approach* which combines the maximum likelihood estimator of Σ, i.e., denoted by Σ_{ML}, with the constrained estimator of Σ_C such that

$$\hat{\Sigma}^* = \gamma \hat{\Sigma}_{ML} + (1 - \gamma)\hat{\Sigma}_C, \tag{14.16}$$

where γ describes the shrinkage indicator and satisfies $0 \leq \gamma \leq 1$. In Eq. (14.16), γ which minimizes the mean squared error between the unconstrained high-dimensional model and the constrained low-dimensional correspondence as seen in Eq. (14.17), can be found from the Ledoit and Wolf lemma [50]. Indeed as given in Eq. (14.17), this is an optimization problem for γ in the sense that we need to select a best fitted model among the unconstrained estimates of the actual model with large variance and more model parameters, and the constrained estimates of the actual model with lower variance and bias factor:

$$\min_{0<\gamma<1} \text{MSE}(\gamma) := E\left[\sum_{i=1}^{p}(\hat{\Sigma}^* - \hat{\Sigma}_{C_i}^*)^2\right], \tag{14.17}$$

in which min MSE(γ) shows the mean squared error function of γ that is minimized. $\hat{\boldsymbol{\Sigma}}^*$ represents the estimated $\boldsymbol{\Sigma}$ according to Eq. (14.16). In this expression $\boldsymbol{\Sigma}$ is the covariance structure of the actual p-dimensional multivariate normal model where the associated means from each dimension are equal. Hereby, in Eq. (14.17), the optimal γ, i.e., γ^*, which minimizes MSE(γ) is obtained via

$$\gamma^* = \frac{\sum_{i=1}^{p} V(\hat{\boldsymbol{\Sigma}}_i) - Cov(\hat{\boldsymbol{\Sigma}}_{C_i}, \hat{\boldsymbol{\Sigma}}_i) - B(\hat{\boldsymbol{\Sigma}}_i) E(\hat{\boldsymbol{\Sigma}}_{C_i} - \hat{\boldsymbol{\Sigma}}_i)}{\sum_{i=1}^{p} E[(\hat{\boldsymbol{\Sigma}}_{C_i} - \hat{\boldsymbol{\Sigma}}_i)^2]}, \qquad (14.18)$$

where $B(\hat{\boldsymbol{\Sigma}}_i)$ describes the bias due to $\hat{\boldsymbol{\Sigma}}_i$. In this expression the estimated γ exits always and is unique under $p \ll N$ [49]. Whereas the underlying approach can produce invertible $\boldsymbol{\Sigma}$ under $p < N$. Moreover, the calculation of the optimal γ can be computationally demanding. In order to simplify the problem for practical purposes such as the inference in high-dimensional genomic data, it is assumed that the nodes are pairwise uncorrelated, but can have unequal variances. But once the interactions between the nodes are estimated, the results are tested for their statistical significance. For this purpose different strategies can be followed. Schäfer and Strimmer [51] suggest to test the distribution of the observed partial correlations across edges, which is taken as beta density, via the false discovery rate. As an alternative of this testing procedure, the model check can be also done by regressing each gene i to the remaining set of $p - 1$ variables and the repetition of this process in turn for every gene [52]. In this calculation, which is called the *lasso regression*, the inference of γ is found from the following expression.

$$\hat{\gamma}_i = 2\sqrt{\frac{s_{ii}}{N}} \Phi^{-1} \left(1 - \frac{\alpha}{2p^2}\right), \qquad (14.19)$$

where $\hat{\gamma}_i$ refers to the estimated γ for each regressed variable or node g_i ($i = 1, \ldots, p$). Φ is the cumulative density of the standard normal and α indicates the constant that controls the probability of false positives, i.e., the probability of the falsely connected two nodes while there is no connection between them. Finally s_{ii} presents the maximum likelihood estimate of the variance of the ith node. In the lasso regression model, the matrix of regression coefficients indicates the precision matrix, thereby the interpretation of the conditional independence structure can be also obtained from the regression coefficients. From the analysis it is shown that the underlying strategy is successful if N goes to infinity, that is when the system is large [53]. But it does not guarantee the symmetry of the estimated $\boldsymbol{\Sigma}$. In order to unravel this limitation, Friedman et al. [54] propose the maximization of the penalized likelihood approach as

$$\max_{\boldsymbol{\Theta} \in \mathbb{R}^{p \times p}} \log [\det(\boldsymbol{\Theta})] - \text{tr}(\boldsymbol{S}\boldsymbol{\Theta}) - \gamma \|\boldsymbol{\Theta}\|_1, \qquad (14.20)$$

in which $\det(\cdot)$ and $\text{tr}(\cdot)$ stand for the determinant and trace of the underlying matrix, respectively, and $\boldsymbol{S}$ displays the estimate of the covariance matrix of the data. Moreover $||\Theta||_1$ describes the L_1-norm of the inverse of Σ, i.e., $\Theta = \Sigma^{-1}$. $\Theta \in \mathbb{R}^{p \times p}$ matrix where Θ belongs to the class of symmetric, invertible, and non-negative matrices [6]. L_1-norm represents the sum of the absolute values of the entries for the given matrix.

This optimization problem, as seen in Eq. (14.20), is also known as the graphical lasso in literature and this method has two major advantages over its alternatives: (1) the solution space is always positive definite for all γ's under both singular and nonsingular S, (2) the estimate of Θ is sparse even if γ is large [55]. In order to solve the underlying graphical lasso challenge by maintaining a block diagonal matrix, various approaches can be applicable. Hence in the calculation to guarantee the presence of results with block diagonal, the following two conditions are required [55]. These conditions enable us to resolve a smaller graphical lasso problem on each block in the estimate of Σ.

1. If the inverse of Σ is block diagonal, Eq. (14.20) can be handled by separately maximizing this expression for each block in such a way that Σ^{-1} and S are replaced by Σ_i^{-1} ($i = 1, \ldots, p$ for the p nodes) and S_i where Σ_i^{-1} and S_i imply the submatrices of Σ^{-1} and S, respectively.
2. $|S_{jj'}| < \gamma$ for all j belonging to Σ_j ($j \in \Sigma_j$, $j' \in \Sigma_j$ and $j \neq j'$). Here, Σ_j refers to the partition of the p variables for $j = 1, \ldots, k$, as used previously. k is the total number of block diagonals in this partitioning.

On the other hand in the computation of the solution, different methods can be used. The coordinate descent method [53] is one of these alternatives. In this approach, briefly, a loss function, which is the negative of the log-likelihood function, is minimized with respect to the precision, Σ^{-1}. The step of the algorithm can be summarized as follows:

(a) The diagonal elements of Σ^{-1}, denoted by $W_d(0)$, are initialized as 1, the off-diagonal terms, i.e., $W_o(0)$, and the counter of the iteration, shown by k, are set to zero.
(b) The negative gradient, $g(k)$, is computed via $g(k) = -\partial l / \partial W_o$ for the current W_o and W_d where $l(W_o, W_d) = -W(W_o, W_d)$ represents the log-likelihood function and $W(W_o, W_d)$ is found via

$$W(W_o, W_d) = \frac{N}{2} \log |\Sigma^{-1}| - \frac{1}{2} \sum_{i=1}^{N} X_i' \Sigma^{-1} X_i, \qquad (14.21)$$

in which X_i denotes the ith observation ($i = 1, \ldots, N$).
(c) $f_j(k) = I\{|g_j(k)| \geq \gamma \times \max_{1 \leq j \leq q} |g_j(k)|\}$ is calculated. Here $I(\cdot)$ displays the indicator function and γ refers to the threshold parameter, $\gamma \in [0, 1]$, by adjusting the diversity of the values of $f_j(k)$. When $\gamma = 0$, the estimate of Σ^{-1} becomes sparser, whereas, while $\gamma = 1$, the elements of Σ^{-1} have the most diversity. In other words as γ changes, the path of the precision alters from the maximum likelihood estimate of the precision to the precision matrix surface defined by S^{-1}, that is the estimate of sample covariance matrix. Therefore γ

can be seen the parameter which can control the sparsity in W_o and penalty in the graphical lasso problem. Finally, $q = p(p-1)/2$ for the p-dimensional Σ shows the number of off-diagonal elements of Σ^{-1}. Hence, $h(k)$ is computed as the following way:

$$h(k) = \sum_{j=1}^{q} f_j(k)g_j(k), \qquad (14.22)$$

where it presents the tangent vector in each step of the algorithm by a descent direction.

(d) The off-diagonal elements of W, i.e., W_o, is calculated by the gradient descent step as

$$W_o(k + \Delta k) = W_o(k) + \Delta k h(k), \qquad (14.23)$$

where $k := k + \Delta k$ and Δk displays the infinitesimal increment in each iteration of k.

(e) The W_d parameters are updated by maximizing the log-likelihood via the Newton-Raphson algorithm while W_o is taken as fixed during $W_o(k + \Delta k)$.

(f) The process is repeated from step b until the convergence is obtained from the Newton-Raphson iterations [56].

In this calculation since the performance of the method is closely related to the selection of γ, i.e., degree of the penalty, different strategies can be used. The control of the false positive rate, that is the proportion of the falsely detected interaction, is one of these alternatives. In order to find the optimal γ under this condition, the ROC (receiving operating characteristic) curve can be implemented by checking γ that maximizes the sensitivity of the model, i.e., the true positive rate. Moreover the k-fold cross-validation method can be also performed for the detection of the optimal γ [54]. In this approach the observation matrix is separated into two parts, namely, training and validation sets, in such a way that the data are randomly partitioned into k subsamples. Then in each iteration one of these subsamples is assigned as the validation set to test the performance of the fitted model and the remaining $k - 1$ subsamples are retained as the training set to get the fitted model. Accordingly, in the calculation initially the precision matrix is estimated by the graphical model via the training set. This precision is then used for the computation of the likelihood of the validation set. The same procedure is repeated for all the folds iteratively in order to compute the overall likelihood. Finally, the penalty γ which maximizes this likelihood is chosen for the network construction. On the other hand, apart from this idea, distinct model selection criteria such as the BIC can be also applied to select the best fitted models among various choices of γ [53]. Whereas these alternatives are computationally demanding for large systems. To solve this problem by guaranteeing the two sufficient conditions as presented above, Witten et al. [55] suggest two approaches which are based on searching the parameters' space that gives the block diagonal in the estimate of Σ^{-1}. In the first approach, initially, the fully unconnected nodes in the graphical lasso solution from Eq. (14.20) are identified such that $|S_{ii'}| < \gamma$ for $i' = 1, \ldots, i-1, i+1, \ldots, p$, under totally p number of nodes. Then the underlying fully unconnected nodes, which are totally q amount, and afterward $p - q$ remaining nodes are ordered. Finally the estimate of Σ^{-1} is found as

$$
\Sigma^{-1} = \begin{pmatrix} \frac{1}{S_{11}+\gamma} & \cdots & 0 & 0 \\ 0 & \ddots & 0 & 0 \\ \vdots & & & \vdots \\ 0 & \cdots & \frac{1}{S_{qq}+\gamma} & 0 \\ 0 & \cdots & 0 & \Sigma^{-1}_{q+1} \end{pmatrix}, \tag{14.24}
$$

where Σ^{-1}_{q+1} denotes the graphical lasso problem implemented to the square symmetric $(p-q) \times (p-q)$ submatrix of S whose nodes are not fully unconnected to the all other nodes in the system. On the other side in the second approach this algorithm is further adjusted in such a way that the second condition can be included in the iterations too [55].

14.4 Simulation of Gene-Environment Regulatory Networks

In the generation of the system to capture its behavior, the starting time point t is set to t_0 with a given initial value $\mathbb{E}^{(0)}$. Then the states of the system, which are taken as the solution at discrete time points, is approximated during the given time interval. This approximate value of the solution at a state is simulated from the value at the previous states iteratively. Here, the main concerns are the stability and precision of these generated approximate results [17, 57].

By using discretization methods, we can transform a continuous process to obtain the approximate solutions $\mathbb{E}^{(0)}, \mathbb{E}^{(1)}, \ldots, \mathbb{E}^{(N-1)}$ at discrete time points $t_0, t_1, \ldots, t_{N-1}$. Then, we can compare them with the given experimental values $\bar{\mathbb{E}}^{(0)}, \bar{\mathbb{E}}^{(1)}, \ldots, \bar{\mathbb{E}}^{(N-1)}$. The comparison of Euler's and Runge-Kutta methods, as discretization methods (schemes) which are based on the ordinary differential equations, is studied in [57] for a concrete example of the differential equations' system. Among alternatives, the Euler's method is the simplest discretization scheme and depends on the first-order Taylor series expansion, resulting in the approximation of the function by a straight line [17, 57]. From the study of Dubois and Kalisz [57], it is reported that this scheme may produces unstable results compared with the exact solution. Moreover, it is slow and inaccurate.

On the other hand in [17, 18, 20, 31, 33, 34, 36], the Euler's method and the second-order Heun's method, also known as the second-order Runge-Kutta method, are applied for the time-discrete dynamics of gene-environment, also called target-environment, regulatory systems given in Sect. 14.2.1 together with the derived matrix algebra correspondingly.

Considering the most general form of gene-environment network modeled by Eq. (14.13), the Euler's method is applied to discretize the time-continuous process as follows (see [17, 18, 34, 35, 41] and the references therein):

$$\dot{\mathbb{E}}^{(k)} := \frac{\mathbb{E}^{(k+1)} - \mathbb{E}^{(k)}}{h_k} = \mathbb{M}(\mathbb{E}^{(k)})\mathbb{E}^{(k)} \quad \text{and} \tag{14.25}$$

$$\mathbb{E}^{(k+1)} = (I + h_k \mathbb{M}(\mathbb{E}^{(k)}))\mathbb{E}^{(k)}, \tag{14.26}$$

where $\dot{\mathbb{E}}(t_k) \approx \dot{\mathbb{E}}^{(k)}$ for all $k \in \mathbb{N}_0$, $h_k = t_{k+1} - t_k$ is the step-size and $t_k < t_{k+1}$.

Correspondingly, the time-discrete dynamics are obtained by the following equation

$$\mathbb{E}^{(k+1)} = \mathbb{M}^{(k)}\mathbb{E}^{(k)} \quad (k \in \mathbb{N}_0). \tag{14.27}$$

The next state can be iteratively generated from the previous states approximately as below

$$\hat{\mathbb{E}}^{(k)}(:= \mathbb{E}^{(k)}) = \mathbb{M}^{(k-1)}(\mathbb{M}^{(k-2)} \cdots (\mathbb{M}^{(1)}(\mathbb{M}^{(0)}\bar{\mathbb{E}}^{(0)}))) \quad (k \in N_0), \tag{14.28}$$

for a given initial value $\bar{\mathbb{E}}^{(0)} = \mathbb{E}^{(0)}$, where $\bar{\mathbb{E}}^{(0)}, \bar{\mathbb{E}}^{(1)}, \ldots, \bar{\mathbb{E}}^{(N-1)}$ denote the provided experimental values and $\hat{\mathbb{E}}^{(0)}, \hat{\mathbb{E}}^{(1)}, \ldots, \hat{\mathbb{E}}^{(N-1)}$ refer to the approximated, i.e., estimated values. This multiplicative form provides an important advantage both computationally and analytically.

In the derivation of iterative formula of second-order Heun's method for the dynamics of gene-environment networks and the corresponding matrix algebra, we refer to [17, 18, 20, 31, 33, 34, 36, 41, 58].

On the other hand, *explicit higher order Runge-Kutta methods*, namely the third-order Heun's method and the fourth-order classical Runge-Kutta method, are newly derived and studied in details for the construction of the discrete dynamics of gene-environment/target-environment regulatory networks by Defterli in [32, 37, 38, 59] in junction with their corresponding matrix algebra. The applications of third-order Heun's method and the fourth-order classical Runge-Kutta method on a set of artificial data and real-world data are newly studied in [32, 37, 38, 59] within a comparison process of these methods with Euler's and second-order Heun's methods. The performance of these methods are investigated with different step-sizes and also their sensitivity with respect to various perturbations is tested in [38].

14.5 Application and Comparison of Inference Results

For the comparison of both ODE based modeling and graphical modeling approaches, we use the bench-mark artificial dataset in Table 14.1 [25] which has been applied previously in the literature [25] in order to assess different time-discretization models based on the fixed time interval between the states.

Table 14.1 Expression
scores of the genes A, B, C
and D at four time points [25]

Time/genes	A	B	C	D	
1	255	250	0	255	$= (\bar{\mathbf{E}}^{(1)})^T$
2	255	200	50	0	$= (\bar{\mathbf{E}}^{(2)})^T$
3	255	180	70	255	$= (\bar{\mathbf{E}}^{(3)})^T$
4	255	170	80	0	$= (\bar{\mathbf{E}}^{(4)})^T$

In inference, initially the ODE technique is implemented. Hereby, by formulating
the system via the differential equation

$$\dot{\mathbf{E}} = \mathbf{M}\mathbf{E},\tag{14.29}$$

where the network matrix $\mathbf{M}$ contains some unknown parameters to be estimated. By
solving the following nonlinear mixed-integer problem, that is defined and solved
in [19,25,26,32,37,38,59], the network is inferred by obtaining $\mathbf{M}$ in various ways
corresponding to the given gene-expression data and approximated derivative data.
The mentioned nonlinear mixed-integer problem is formulated as follows [19, 25,
26,32,37,38,59]:

$$\min_{\mathbf{M}=(m_{ij})} \sum_{k=1}^{N} \left\| \mathbf{M}\bar{\mathbf{E}}^{(k)} - \dot{\bar{\mathbf{E}}}^{(k)} \right\|_2^2,$$

subject to

$$m_{ij} \geq \begin{cases} -\lambda(i), & i = j, \\ 0, & i \neq j, \end{cases}$$

$$(1 - y_{ij}) \cdot m_{ij} = 0, \quad \forall i, j \in G,$$

$$\sum_{j \in G} y_{ij} \leq deg_{max,i} \quad \forall i \in G,\tag{14.30}$$

where $G = \{1, 2, 3, 4\}$ is the set of genes in the considered network, $\lambda(i)$ is the
degradation rate and $deg_{max,i} \in \mathbb{Z}_+$ is a bound on the indegree of each gene for
$i \in G$ [19,25,26]. These bounds for the imposed constraints are biologically given
as $\lambda(i) = 2$, $deg_{max,i} = 2$, for $i = 1, \ldots, 4$ [19,25,26].

In [32,37,38,59], the mixed-integer nonlinear programming (MINLP) problem
stated by Eq. (14.30) is solved for the gene-expression data in Table 14.1 and for
different kinds of the derivative data. The problem is solved for the gene derivative
data approximated by third-order Heun's approximation in [32,37,38,59] and also
by Euler's approximation in [37,38]. Here, we present in below the inference results
for $\mathbf{M}$ where Euler's approximation is used for the derivative data $\dot{\bar{\mathbf{E}}}^{(1)}, \dot{\bar{\mathbf{E}}}^{(2)}, \dot{\bar{\mathbf{E}}}^{(3)}$ (see
[37,38]):

$$M = \begin{pmatrix} 0 & 0 & 0 & 0 \\ 0 & -0.20 & 0.38 & 0 \\ 0.19 & 0 & -0.58 & 0 \\ 1 & 0 & 0 & -2 \end{pmatrix}. \tag{14.31}$$

On the other hand, as the second modeling approach for the data in Table 14.1, we implement the L_1-penalized likelihood approach. In the selection of the penalty value, γ, we perform the twofold cross validation method. From the calculation, we find that $\gamma = 2$ is the best fitted penalty term which maximize the likelihood function of the time-course observation in Table 14.1. Indeed such a small dataset, we can expect large values of γ as computed here. Whereas if the dataset becomes larger, the values found by the cross-validation can tend to small values of γ which indicates highly connected networks.

The estimated precision matrix, $\hat{\Theta} = \hat{\Sigma}^{-1}$, which corresponds to the matrix of the regression coefficient in the lasso-based regression model is obtained and presented in Eq. (14.32). In the entries of $\hat{\Theta}$ we consider that the values less that 10^{-3} can be taken as the low strengths between the genes, thereby, setting to zero.

$$\hat{\Theta} = \begin{pmatrix} 0.5 & 0 & 0 & 0 \\ 0 & 0.13 & 0.12 & 0 \\ 0 & 0.12 & 0.13 & 0 \\ 0 & 0 & 0 & 0 \end{pmatrix}. \tag{14.32}$$

Moreover, in this calculation the estimated $\hat{\Theta}$, can be seen the equivalent matrix M in Eq. (14.13) based on the ODE approach. But, in order to compare the final structures of the network from both graphical and ODE models, we add the matrix $2I$, where $I = I_{4 \times 4}$ is the identity matrix, to $\hat{\Theta}$. The reason is that, the network matrix in the ODE model given by Eq. (14.29) is estimated by solving a constrained optimization problem described in Eq. (14.30). Among these constraints,

$$m_{ij} \geq \begin{cases} -2, & i = j, \\ 0, & i \neq j, \end{cases} \tag{14.33}$$

imposes a bound on self-degradation of the genes by the value -2 and off-diagonal genes are prevented from negative regulation. Therefore, in order to compare the two inferred network matrices M in (14.31) and $\hat{\Theta}$ in (14.32) which are obtained from ODE based model and GGM, respectively, we have to define the following network matrix

$$\hat{\Theta} := M + 2I = \begin{pmatrix} 2 & 0 & 0 & 0 \\ 0 & 1.80 & 0.38 & 0 \\ 0.19 & 0 & 1.42 & 0 \\ 1 & 0 & 0 & 0 \end{pmatrix}. \tag{14.34}$$

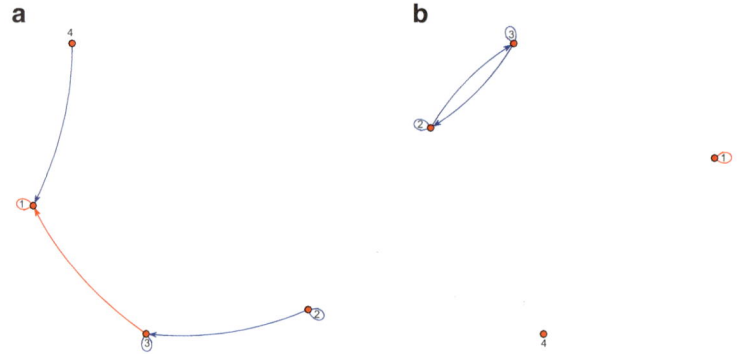

Fig. 14.1 Estimated structure of the gene-environment network via (**a**) ordinary differential equations via MINLP solutions and (**b**) graphical lasso with L_1-penalized likelihood approach. In the graph the nodes 1, 2, 3, and 4 represent the gene A, B, C, and D, respectively

In the Fig. 14.1a, b, we draw the corresponding directed graphs of the networks $\hat{\Theta}$ given in (14.34) and $\hat{\Theta}$ in (14.32), respectively, to figure out the estimated interactions between genes A, B, C, and D.

According to the estimated network matrices given above, the gene-expression vectors $\mathbf{E}^{(k)}$ at the future time levels t_k $(k = N, N + 1, \ldots)$ can be generated iteratively by obtaining the time-discrete dynamics with a numerical approximation scheme as mentioned in Sect. 14.4. As a class of explicit Runge-Kutta methods, Euler's method, second-order Heun's method, third-order Heun's methods and fourth-order classical Runge-Kutta method are studied in details and applied in [25, 32, 37, 38, 59] for forecasting the future behaviour of the network. In this respect, Euler's method is studied in [25], second-order Heun's method, third-order Heun's method and fourth-order classical Runge-Kutta method are examined with a comparative study in [32, 37, 38, 59].

Here, we chose the results of the ODE approach given in Eq. (14.31) and present the simulation of the estimated system by the Euler's method, second-order Heun's method, third-order Heun's method for a fixed step-size h, for the fixed gene data in Table 14.1 and correspondingly approximated derivative data in [37, 38]. These simulation results are obtained in [37, 38] in order to show the performance of these numerical schemes. The following simulations, given by Figs. 14.2, 14.3, 14.4, and 14.5, presents the generated time series results for the gene-expression values that are calculated using these three different discretization schemes [37, 38].

14.6 Conclusion

The modeling approaches representing the dynamic nature of gene-environment regulatory systems are presented. Various discretization methods are reviewed in order to explain the way of obtaining corresponding time discrete dynamics of the

Fig. 14.2 Forecasting of Gene A using different methods for fixed data [37, 38]

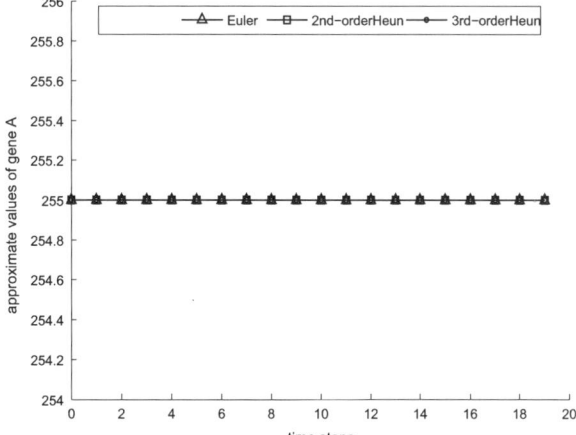

Fig. 14.3 Forecasting of Gene B using different methods for fixed data [37, 38]

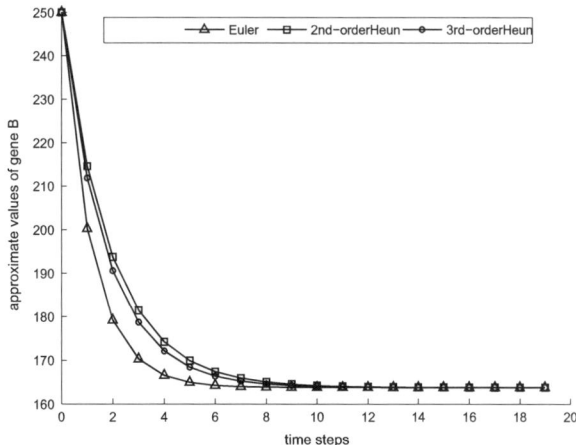

system and also predicting the future behaviour of the system. The inference of a gene-environment network by considering two different modeling approaches, namely, ODE based modeling with mixed-integer optimization method and graphical modeling by GGM is studied. The results of the inference problem is presented by the corresponding graphs. Then, for the forecasting of the future behaviour of the network by three different numerical methods, only previously obtained time-series results of the network estimated by ODE based modeling is given. From the findings presented by Figs. 14.2, 14.3, 14.4, and 14.5 [37, 38], it is observed that the higher-order approximations in these methods outperform in convergence and smoothing with respect to the results of the Euler scheme, as expected, for the generation of future states.

Fig. 14.4 Forecasting of
Gene C using different
methods for fixed data
[37, 38]

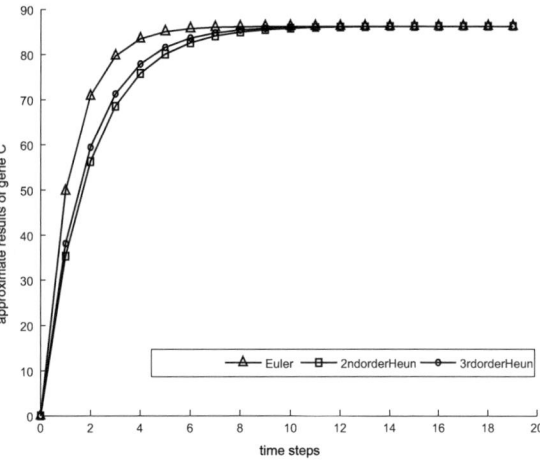

Fig. 14.5 Forecasting of
Gene D using different
methods for fixed data
[37, 38]

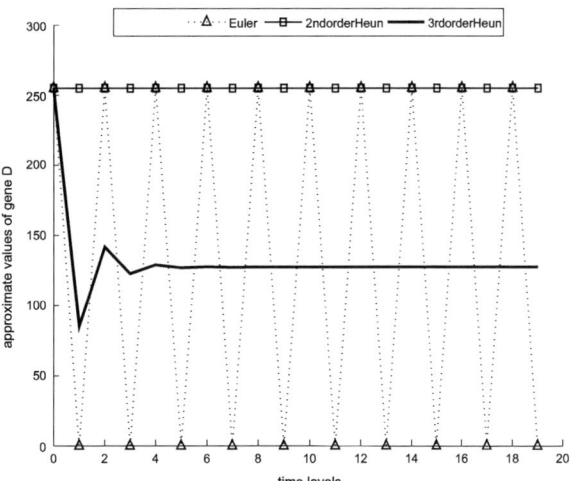

In our further studies, we will examine the future behaviour of a huge network
estimated by GGM by implementing the mentioned numerical method class and
compare these results with the forecasting results of the ODE based model.

Acknowledgements This study is a part of Ph.D. thesis of Özlem Defterli at the Department of
Mathematics in Middle East Technical University (METU). Her work is partially supported by the
Scientific and Technical Research Council of Turkey. Moreover, Vilda Purutçuoğlu and Gerhard-
Wilhelm Weber thank to the EU 7th Framework Programme Project PATHOSYS (No: 260429) for
their financial support in the computational equipment.

References

1. Radde, N.: Modeling non-linear dynamic phenomena in biochemical networks. Ph.D. thesis, Faculty of Mathematics and Natural Sciences, University of Köln (2007)
2. Jong, H.D.: Modeling and simulation of genetic regulatory systems: a literature review. J. Comput. Biol. **9**(1), 67–103 (2002)
3. Hasty, J., McMillen, D., Isaacs, F., Collins, J.J.: Computational studies of gene regulatory networks: in numero molecular biology. Nat. Rev. Genet. **2**, 268–279 (2001)
4. Smolen, P., Baxter, D.A., Byrne, J.H.: Modeling transcriptional control in gene networks - methods, recent results, and future directions. Bull. Math. Biol. **62**, 247–292 (2000)
5. Werhli, A., Grzegorczyk, M., Husmeier, D.: Comparative evaluation of reverse engineering gene regulatory networks with relevance networks, graphical Gaussian models and Bayesian networks. Bioinformatics **22**(20), 2523–2531 (2006)
6. Whittaker, J.: Graphical Models in Applied Multivariate Statistics. Wiley, Chichester (1990)
7. Ahuja, R.K., Magnanti, T.L., Orlin, J.B.: Network Flow: Theory, Algorithms and Applications. Prentice Hall, New Jersey (1993)
8. Gebert, J., Lätsch, M., Ming Poh Quek, E., Weber, G.-W.: Analyzing and optimizing genetic network structure via path-finding. J. Comput. Technol. **9**(3), 3–12 (2004)
9. Kauffman, S.: Metabolic stability and epigenesis in randomly constructed genetic nets. J. Theor. Biol. **22**, 437–467 (1969)
10. Bornholdt, S.: Less is more in modeling large genetic networks. Science **310**(5747), 449–451 (2005)
11. Li, F., Long, T., Lu, Y., Ouyangm, Q., Tang, C.: The yeast cell-cycle network is robustly designed. Proc. Natl. Acad. Sci. **101**, 4781–4786 (2004)
12. Thieffry, D., Thomas, R.: Qualitative analysis of gene networks. Pac. Symp. Biocomput. **3**, 77–88 (1998)
13. Thomas, R., D'Ari, R.: Biological Feedback. CRC Press, Boca Raton (1990)
14. Murphy, K., Mian, S.: Modelling gene expression data using dynamic Bayesian networks. Technical report, Computer Science Division, University of California, Berkeley (1999)
15. Husmeier, D.: Sensitivity and specificity of inferring genetic regulatory interactions from microarray experiments with dynamic Bayesian networks. Bioinformatics **19**(17), 2271–2282 (2003)
16. Chen, T., He, H.L., Church, G.M.: Modeling gene expression with differential equations. Proc. Pac. Symp. Biocomput. **4**, 29–40 (1999)
17. Taştan, M.: Analysis and prediction of gene expression patterns by dynamical systems, and by a combinatorial algorithm. M.Sc. thesis, Institute of Applied Mathematics, Middle East Technical University, Ankara (2005)
18. Yılmaz, F.B.: A mathematical modeling and approximation of gene expression patterns by linear and quadratic regulatory relations and analysis of gene networks. M.Sc. thesis, Institute of Applied Mathematics, Middle East Technical University, Ankara (2004)
19. Gebert, J., Radde, N., Weber, G.-W.: Modelling gene regulatory networks with piecewise linear differential equations. Eur. J. Oper. Res. **181**(3), 1148–1165 (2007)
20. Weber, G.-W., Uğur, Ö., Taylan, P., Tezel, A.: On optimization, dynamics and uncertainty: a tutorial for gene-environment networks. Discrete Appl. Math. **157**(10), 2494–2513 (2009)
21. Weber, G.-W., Kropat, E., Tezel, A., Belen, S.: Optimization applied on regulatory and eco-finance networks-survey and new developments. Pac. J. Optim. **6**(2), 319–340 (2010)
22. Weber, G.-W., Kropat, E., Akteke-Öztürk, B., Görgülü, Z.K.: A survey on OR and mathematical methods applied on gene-environment networks. Cent. Eur. J. Oper. Res. **17**(3), 315–341 (2009)
23. Kaderali, L., Radde, N.: Inferring gene regulatory networks from expression data. Studies in Computational Intelligence, vol. 1, chapter 2. Springer, Berlin (2007)
24. Sakamoto, E., Iba, H.: Inferring a system of differential equations for a gene regulatory network by using genetic programming. In: Proceedings of Congress on Evolutionary Computation, pp. 720–726 (2001)

25. Gebert, J., Lätsch, M., Pickl, S.W., Weber, G.-W., Wünschiers, R.: Genetic networks and anticipation of gene expression patterns. In: Computing Anticipatory Systems: CASYS(92)03 - Sixth International Conference. AIP Conference Proceedings, vol. 718, pp. 474–485 (2004)
26. Gebert, J., Lätsch, M., Pickl, S.W., Weber, G.-W., Wünschiers, R.: An algorithm to analyze stability of gene-expression pattern. Discrete Appl. Math. **154**(7), 1140–1156 (2006)
27. Gebert, J., Pickl, S., Shokina, N., Weber, G.-W., Wünschiers, R.: Algorithmic analysis of gene expression data with polyhedral structures. In: Kröplin, B., Rudolph, S., Häcker, J. (eds.) Proceedings of Similarity Methods (5th International Workshop), pp. 79–87 (2001). ISBN: 3-930683-47-4
28. Weber, G.-W., Taylan, P., Akteke-Öztürk, B., Uğur, Ö.: Mathematical and data mining contributions to dynamics and optimization of gene-environment networks. Electron. J. Theor. Phys. **4**(16(II)), 115–146 (2007)
29. Hoon, M., Imoto, S., Miyano, S.: Inferring gene regulatory networks from time-ordered gene expression data using differential equations. Discov. Sci. 267–274 (2002)
30. Hoon, M.D., Imoto, S., Kobayashi, K., Ogasawara, N., Miyano, S.: Inferring gene regulatory networks from time-ordered gene expression data of Bacillus Subtilis using differential equations. Proc. Pac. Symp. Biocomput. **8**, 17–28 (2003)
31. Weber, G.-W., Taylan, P., Alparslan Gök, S.Z., Özöğür, S., Akteke Öztürk, B.: Optimization of gene-environment networks in the presence of errors and uncertainty with Chebychev approximation. TOP **16**(2), 284–318 (2008)
32. Weber, G.-W., Defterli, O., Kropat, E., Alparslan-Gök, S.Z.: Modeling, inference and optimization of regulatory networks based on time series data. Eur. J. Oper. Res. **211**(1), 1–14 (2011)
33. Uğur, Ö., Weber, G.-W.: Optimization and dynamics of gene-environment networks with intervals. J. Ind. Manag. Optim. **3**(2), 357–379 (2007)
34. Taştan, M., Ergenç, T., Pickl, S.W., Weber, G.-W.: Stability analysis of gene expression patterns by dynamical systems and a combinatorial algorithm. In: Proceedings of International Symposium on Health Informatics and Bioinformatics, pp. 67–75 (2005)
35. Yılmaz, F.B., Öktem, H., Weber, G.-W.: Mathematical modeling and approximation of gene expression patterns and gene networks. In: Fleuren, F., den Hertog, D., Kort, P. (eds.) Operations Research Proceedings, pp. 280–287 (2005)
36. Weber, G.-W., Tezel, A., Taylan, P., Soyler, A., Çetin, M.: Mathematical contributions to dynamics and optimization of gene-environment networks. Optimization **57**(2), 353–377 (2008)
37. Defterli, O., Fügenschuh, A., Weber, G.-W.: Modern tools for the time-discrete dynamics and optimization of gene-environment networks. Commun. Nonlin. Sci. Numer. Simulat. **16**(12), 4768–4779 (2011)
38. Defterli, Ö.: Modern mathematical methods in modeling and dynamics of regulatory systems of gene-environment networks. Ph.D. thesis in Graduate School of Natural and Applied Sciences, Department of Mathematics, Middle East Technical University (METU), Ankara (August 2011)
39. Akhmet, M.U., Gebert, J., Öktem, H., Pickl, S.W., Weber, G.-W.: An improved algorithm for analytical modelling and anticipation of gene expression patterns. J. Comput. Technol. **10**(4), 3–20 (2005)
40. Isaacson, E., Keller, H.B.: Analysis of Numerical Methods. Wiley, New York (1966)
41. Taştan, M., Pickl, S.W., Weber, G.-W.: Mathematical modeling and stability analysis of gene-expression patterns in an extended space and with Runge-Kutta discretization. In: Proceedings of Operations Research, Bremen, September 2005, pp. 443–450 (2006)
42. Aster, R.C., Borchers, B., Thurber, C.H.: Parameter Estimation and Inverse Problems. Academic, New York (2004)
43. Hastie, T.J., Tibshirani, R.J., Friedman, J.: The Elements of Statistical Learning, Data Mining, Inference and Prediction. Springer, New York (2001)
44. Özmen, A.: Robust conic quadratic programming applied to quality improvement - A robustification of CMARS. M.Sc. thesis, Institute of Applied Mathematics, Middle East Technical University, Ankara (2010)

45. Yerlikaya, F.: A new contribution to nonlinear robust regression and classification with MARS and its application to data mining for quality control in manufacturing. M.Sc. thesis at the Institute of Applied Mathematics, Middle East Technical University, Ankara (2008)
46. Özmen, A., Weber, G-W., Batmaz, I.: The new robust CMARS (RCMARS) method. In: ISI Proceedings of 24th MEC-EurOPT 2010-Continuous Optimization and Information-Based Technologies in the Financial Sector, Izmir, pp. 362–368 (2010). ISBN: 978-9955-28-598-4
47. Fügenschuh, A., Martin, A.: Computational integer programming and cutting planes. In: Aardal, K., Nemhauser, G., Weismantel, R. (eds.) Handbooks in Operations Research and Management Science, Handbook on Discrete Optimization, vol. 12, pp. 69–122. Elsevier, Amsterdam (2005)
48. Grzegorczyk, M., Husmeier, D., Werhli, A.V.: Reverse engineering gene regulatory networks with variaous machine learning methods. In: Emmert-Streib, E., Dehmer, M. (eds.) Analysis of Micoarray Data, a Network-Based Approach. Wiley-VCH Verlag, Weinheim (2008)
49. Schäfer, J., Strimmer, K.: A shrinkage approach to large-scale covariance matrix estimation and implications for functional genomics. Stat. Appl. Genet. Mol. Biol. 4(1), 1–29 (2005)
50. Ledoit, O., Wolf, W.: Improved estimation of the covariance matrix of stock returns with an application to portfolio selection. J. Empir. Finance 10, 603–621 (2003)
51. Schäfer, J., Strimmer, K.: An empirical Bayes approach to inferring large-scale gene association networks. Bioinformatics 21, 754–764 (2005)
52. Meinshausen, N., Bühlmann, P.: High dimensional graphs and variable selection with the Lasso. Ann. Stat. 34(3), 1436–1462 (2006)
53. Li, H.: Statistical methods for inference of genetic networks and regulatory modules. In: Emmert-Streib, E., Dehmer, M. (eds.) Analysis of Micoarray Data, a Network-Based Approach. Wiley-VCH Verlag, Weinheim (2008)
54. Friedman, J., Hastie, T., Tibshirani, R.: Sparse inverse covariance estimation with the graphical lasso. Biostatistics 9(3), 432–441 (2008)
55. Witten, D.M., Friedman, J.H., Simon, N.: New insights and faster computationas for the graphical lasso. J. Comput. Graph. Stat. 20(4), 892–900 (2011)
56. Hastie, T., Tibshirani, R., Friedman, J.: The Elements of Statistical Learning: Data Mining, Inference, and Prediction. Springer, New York (2009)
57. Dubois, D.M., Kalisz, E.: Precision and stability of Euler, Runge-Kutta and incursive algorithm for the harmonic oscillator. Int. J. Comput. Anticipatory Syst. 14, 21–36 (2004)
58. Ergenç, T., Weber, G.-W.: Modeling and prediction of gene-expression patterns reconsidered with Runge-Kutta discretization. J. Comput. Technol. 9(6), 40–48 (2004)
59. Defterli, O., Fügenschuh, A., Weber, G.-W.: New discretization and optimization techniques with results in the dynamics of gene-environment networks. In: Barsoum, N., Vasant, P., Habash, R. (eds.) Proceedings of the 3rd Global Conference on Power Control&Optimization, Gold Coast, 2–4 February 2010. CD-ISBN: 978-983-44483-1-8

Chapter 15
BHP Universality in Energy Sources

Helena Ferreira, Rui Gonçalves, and Alberto Adrego Pinto

15.1   Introduction

Modeling the time series of the prices of different energy sources is important in economics, finance and energy market, and it is essential in the management of large stock portfolios [5–7,9,18–20,22–25]. In this paper we analyze the daily price of north american biofuel ethanol and the plant from which ethanol is produced, specifically, corn. Herein we present the results from two data sets, ethanol and corn daily prices, from May 2005 to August 2013. Let $Y(t)$ be the energy source (ES) price or adjusted close value at day t. We define the *ES daily return* on day t by

$$r(t) = \frac{Y(t) - Y(t-1)}{Y(t-1)}$$

H. Ferreira (✉)
LIAAD—INESC TEC, Porto, Portugal
e-mail: helenaisafer@gmail.com.

R. Gonçalves
LIAAD—INESC TEC, Porto, Portugal

Faculty of Engineering, Section of Mathematics,University of Porto, R. Dr. Roberto Frias s/n, 4200-465 Porto, Portugal
e-mail: rjasg@fe.up.pt

A.A. Pinto
LIAAD—INESC TEC, Porto, Portugal

Faculty of Sciences, Department of Mathematics, University of Porto, Rua do Campo Alegre, 687, 4169-007, Porto, Portugal
e-mail: aapinto@fc.up.pt

A.A. Pinto and D. Zilberman (eds.), *Modeling, Dynamics, Optimization and Bioeconomics I*, Springer Proceedings in Mathematics & Statistics 73, DOI 10.1007/978-3-319-04849-9__15,
© Springer International Publishing Switzerland 2014

We define the α *re-scaled ES daily positive returns* $r(t)^\alpha$, for $r(t) > 0$, that we call, after normalization, the α *positive fluctuations*. We define the α *re-scaled ES daily negative returns* $(-r(t))^\alpha$, for $r(t) < 0$, that we call, after normalization, the α *negative fluctuations*. We analyze separately the α positive and α negative daily fluctuations that can have different statistical and economic natures (see, for example, [1, 2, 16, 17, 21]). Our aim is to find the values of α that optimize the data collapse of the histogram of the α positive and negative fluctuations to the Bramwell-Holdsworth-Pinton (truncated BHP) probability density function (pdf) f_{BHP} truncated to the support range of the data (see Bramwell et al. [3, 4]). Our approach is to apply the Kolmogorov-Smirnov (K-S) statistic test as a method to find the values of α that optimize the data collapse. The α values represent a new measure that allows the comparison between the intensity of gains and losses of market activity in different energy sources prices. Using this data collapse we do a change of variable that allows us to compute the analytical approximations of the pdf of the normalized positive and negative ES daily returns in terms of the BHP pdf f_{BHP}. Similar results have been observed for some other energy prices, exchange rates, commodity prices as well as different indices with different time scales (see [13–15]). Since the BHP probability density function appears in several other dissimilar phenomena (see, for example, [8, 10–12]), our result reveals a universal feature of the prices of energy sources. Furthermore, these results lead to the construction of a new qualitative and quantitative econophysics model for the stock market based on the two-dimensional spin model (2dXY) at criticality and a new stochastic differential equation model for the stock exchange market indices that provides a better understanding of several stock exchange crisis.

15.2 Energy Sources and BHP

15.2.1 Energy Source Positive Returns

Let T^+ be the set of all days t with positive returns, i.e.

$$T^+ = \{t : r(t) > 0\}.$$

Let n^+ be the cardinal of the set T^+. The α *re-scaled energy source daily positive returns* are the returns $r(t)^\alpha$ with $t \in T^+$. The *mean* μ_α^+ of the α re-scaled energy source daily positive returns is given by

$$\mu_\alpha^+ = \frac{1}{n^+} \sum_{t \in T^+} r(t)^\alpha \qquad (15.1)$$

The *standard deviation* σ_α^+ of the α re-scaled energy source daily positive returns is given by

$$\sigma_\alpha^+ = \sqrt{\frac{1}{n^+} \sum_{t \in T^+} r(t)^{2\alpha} - (\mu_\alpha^+)^2} \qquad (15.2)$$

We define the α *positive fluctuations* by

$$r_\alpha^+(t) = \frac{r(t)^\alpha - \mu_\alpha^+}{\sigma_\alpha^+} \qquad (15.3)$$

for every $t \in T^+$. Hence, the α *positive fluctuations* are the normalized α re-scaled energy source daily positive returns. Let L_α^+ be the *smallest* α positive fluctuation, i.e.

$$L_\alpha^+ = \min_{t \in T^+} \{r_\alpha^+(t)\}.$$

Let R_α^+ be the *largest* α positive fluctuation, i.e.

$$R_\alpha^+ = \max_{t \in T^+} \{r_\alpha^+(t)\}.$$

We denote by $F_{\alpha,+}$ the *probability distribution of the α positive fluctuations*. Let the *truncated BHP probability distribution* $F_{BHP,\alpha,+}$ be given by

$$F_{BHP,\alpha,+}(x) = \frac{F_{BHP}(x)}{F_{BHP}(R_\alpha^+) - F_{BHP}(L_\alpha^+)}$$

where F_{BHP} is the BHP probability distribution (see Bramwell et al. [4] and [14]). We apply the K-S statistic test to the null hypothesis claiming that the probability distributions $F_{\alpha,+}$ and $F_{BHP,\alpha,+}$ are equal. The Kolmogorov-Smirnov P *value* P_α^+ is plotted in Fig. 15.1. We observe that α_{BHP}^+ is the point where the P value $P_{\alpha_{BHP}^+}^+$ attains its maximum.

It is well-known that the Kolmogorov-Smirnov P value P_α^+ decreases with the distance

$$D_{\alpha,+} = \|F_{\alpha,+} - F_{BHP,\alpha,+}\|$$

between $F_{\alpha,+}$ and $F_{BHP,\alpha,+}$. In Fig. 15.2, we plot

$$D_{\alpha_{BHP}^+,+}(x) = \left| F_{\alpha_{BHP}^+,+}(x) - F_{BHP,\alpha_{BHP}^+,+}(x) \right|$$

and we observe that $D_{\alpha_{BHP}^+,+}(x)$ for ethanol and corn attains its maximum value for the α^+ positive fluctuations below the mean of the probability distribution.

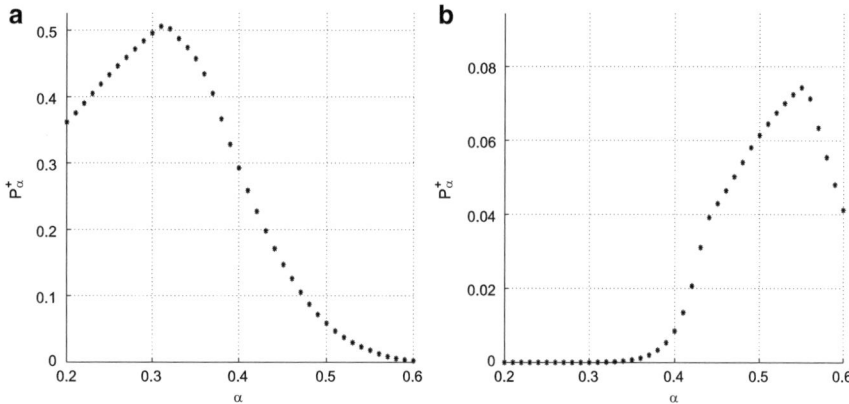

Fig. 15.1 The Kolmogorov-Smirnov P value P_α^+ for values of α in the range $[0.2, 0.6]$, for daily returns. (**a**) Ethanol. (**b**) Corn

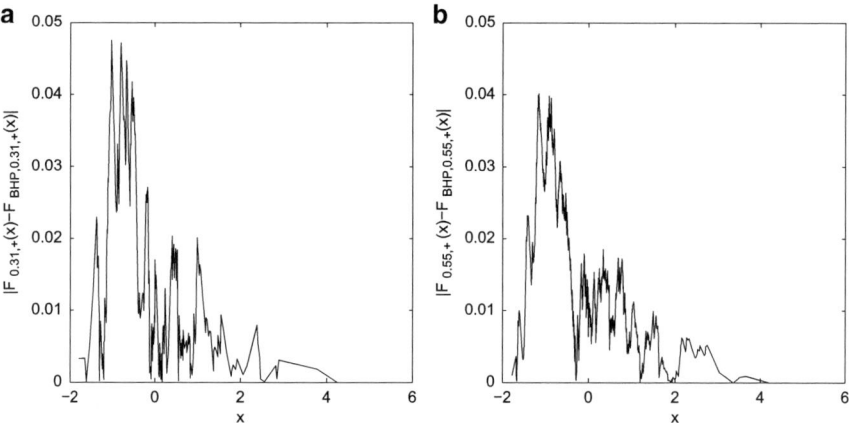

Fig. 15.2 The map $D_{\alpha_{BHP}^+,+}(x) = |F_{\alpha_{BHP}^+,+}(x) - F_{BHP,\alpha_{BHP}^+,+}(x)|$, for daily returns. (**a**) Ethanol. (**b**) Corn

In Fig. 15.3, we show the data collapse of the histogram $f_{\alpha_{BHP}^+,+}$ of the α_{BHP}^+ positive fluctuations to the truncated BHP pdf $f_{BHP,\alpha_{BHP}^+,+}$.

We assume that the probability distribution of the α_{BHP}^+ positive fluctuations $r_{\alpha_{BHP}^+}^+(t)$ is approximated by $F_{BHP,\alpha_{BHP}^+,+}$. The pdf of the energy source positive returns $r(t)$ is approximated by (see [14])

$$f_{BHP,ES,+}(x) = \frac{\alpha_{BHP}^+ x^{\alpha_{BHP}^+ - 1} f_{BHP}\left(\left(x^{\alpha_{BHP}^+} - \mu_{\alpha_{BHP}^+}^+\right)/\sigma_{\alpha_{BHP}^+}^+\right)}{\sigma_{\alpha_{BHP}^+}^+\left(F_{BHP}\left(R_{\alpha_{BHP}^+}^+\right) - F_{BHP}\left(L_{\alpha_{BHP}^+}^+\right)\right)}.$$

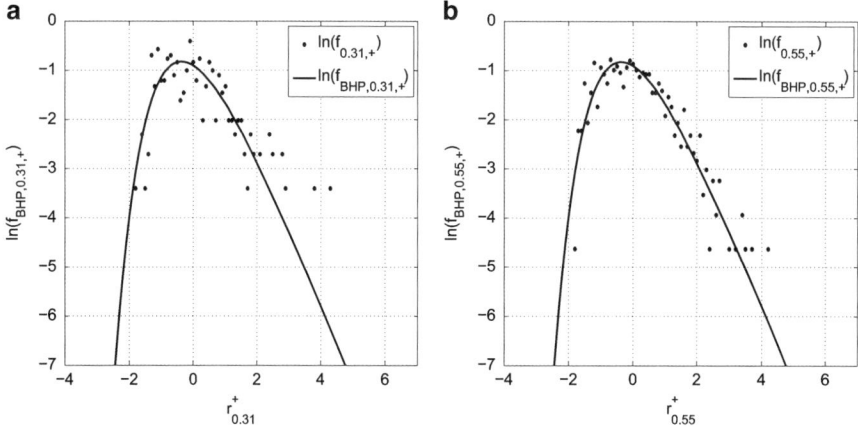

Fig. 15.3 The histogram of the α_{BHP}^{+} fluctuations with the truncated BHP pdf $f_{BHP,\alpha,+}$ on top, in the semi-log scale, for daily returns. (**a**) Ethanol. (**b**) Corn

Hence, we get

$$f_{BHP,Ethanol,+}(x) = 1.02x^{-0.69} f_{BHP}(10.40x^{0.31} - 3.17).$$

and

$$f_{BHP,Corn,+}(x) = 5.53x^{-0.45} f_{BHP}(20.53x^{0.55} - 1.77).$$

If $\alpha < \alpha'$ then the probability density of the returns near zero scale with order $x^{\alpha-1}$ and $x^{\alpha'-1}$ near zero. Hence the returns near probability zero have higher probability for the density with α than with α'. Therefore the exponent α is a new measure of intensity of the market.

We denote that $x^{\alpha_{BHP}^{+}-1}$ is the intensity term of $f_{BHP,ES,+}(x)$ at zero.

15.2.2   Energy Source Daily Negative Returns

Let T^{-} be the set of all days t with negative returns, i.e.

$$T^{-} = \{t : r(t) < 0\}.$$

Let n^{-} be the cardinal of the set T^{-}.

The α *re-scaled energy source daily negative returns* are the returns $(-r(t))^{\alpha}$ with $t \in T^{-}$. We note that $-r(t)$ is positive. The *mean* μ_{α}^{-} of the α re-scaled energy source daily negative returns is given by

$$\mu_\alpha^- = \frac{1}{n^-} \sum_{t \in T^-} (-r(t))^\alpha \qquad (15.4)$$

The *standard deviation* σ_α^- of the α re-scaled energy source daily negative returns is given by

$$\sigma_\alpha^- = \sqrt{\frac{1}{n^-} \sum_{t \in T^-} (-r(t))^{2\alpha} - (\mu_\alpha^-)^2} \qquad (15.5)$$

We define the α *negative fluctuations* by

$$r_\alpha^-(t) = \frac{(-r(t))^\alpha - \mu_\alpha^-}{\sigma_\alpha^-} \qquad (15.6)$$

for every $t \in T^-$. Hence, the α *negative fluctuations* are the normalized α re-scaled energy source daily negative returns. Let L_α^- be the *smallest* α negative fluctuation, i.e.

$$L_\alpha^- = \min_{t \in T^-} \{r_\alpha^-(t)\}.$$

Let R_α^- be the *largest* α negative fluctuation, i.e.

$$R_\alpha^- = \max_{t \in T^-} \{r_\alpha^-(t)\}.$$

We denote by $F_{\alpha,-}$ the *probability distribution of the α negative fluctuations*. Let the *truncated BHP probability distribution* $F_{BHP,\alpha,-}$ be given by

$$F_{BHP,\alpha,-}(x) = \frac{F_{BHP}(x)}{F_{BHP}(R_\alpha^-) - F_{BHP}(L_\alpha^-)}$$

where F_{BHP} is the BHP probability distribution. We apply the K-S statistic test to the null hypothesis claiming that the probability distributions $F_{\alpha,-}$ and $F_{BHP,\alpha,-}$ are equal. The Kolmogorov-Smirnov P *value* P_α^- is plotted in Fig. 15.4. Hence, we observe that α_{BHP}^- is the point where the P value $P_{\alpha_{BHP}^-}^-$ attains its maximum. It is well-known that the Kolmogorov-Smirnov P value P_α^- decreases with the distance

$$D_{\alpha,-} = \|F_{\alpha,-} - F_{BHP,\alpha,-}\|$$

between $F_{\alpha,-}$ and $F_{BHP,\alpha,-}$. In Fig. 15.5, we plot

$$D_{\alpha_{BHP}^-,-}(x) = \left| F_{\alpha_{BHP}^-,-}(x) - F_{BHP,\alpha_{BHP}^-,-}(x) \right|$$

and we observe that $D_{\alpha_{BHP}^-,-}(x)$ attains its maximum value, in ethanol and corn prices, for the α_{BHP}^- negative fluctuations below the mean of the probability

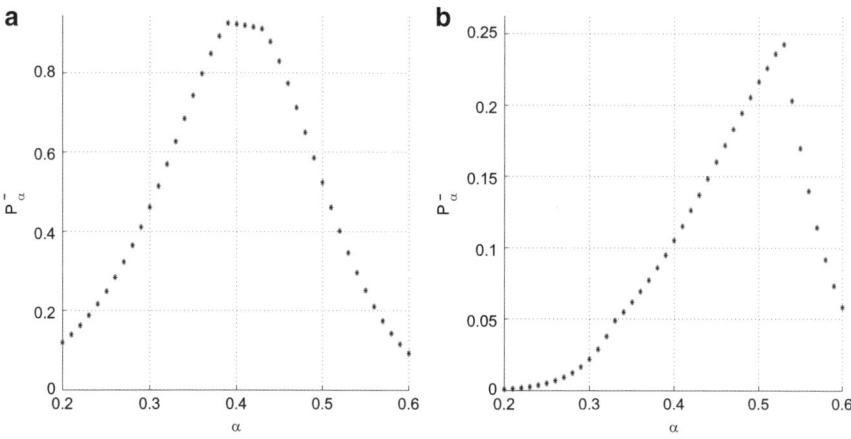

Fig. 15.4 The Kolmogorov-Smirnov P value P_α^- for values of α in the range $[0.2, 0.6]$, for daily returns. (**a**) Ethanol. (**b**) Corn

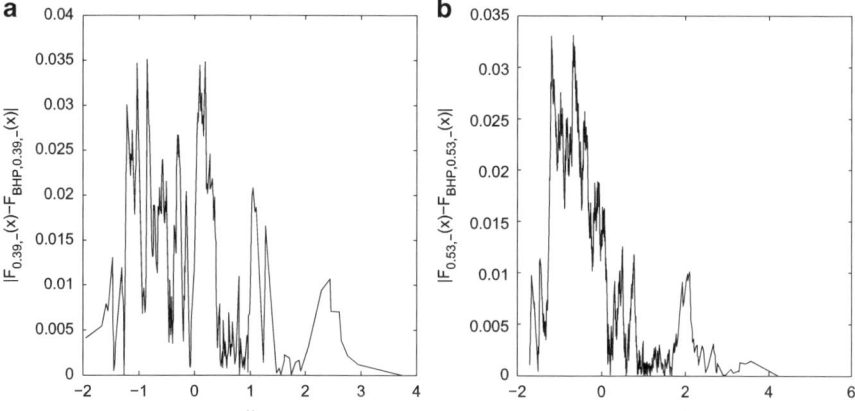

Fig. 15.5 The map $D_{\alpha_{BHP}^-}, -(x) = |F_{\alpha_{BHP}^-,-}(x) - F_{BHP,\alpha_{BHP}^-,-}(x)|$, for daily returns. (**a**) Ethanol. (**b**) Corn

distribution. In Fig. 15.6, we show the data collapse of the histogram $f_{\alpha_{BHP}^-,-}$ of the α_{BHP}^- negative fluctuations to the truncated BHP pdf $f_{BHP,\alpha_{BHP}^-,-}$.
We assume that the probability distribution of the α_{BHP}^- negative fluctuations $r_{\alpha_{BHP}^-}^-(t)$ is approximated by $F_{BHP,\alpha_{BHP}^-,-}$. The pdf of the energy source daily (symmetric) negative returns $-r(t)$, with $T \in T^-$, is approximated by (see [14])

$$f_{BHP,Energysource,-}(x) = \frac{\alpha_{BHP}^- x^{\alpha_{BHP}^- 1} f_{BHP}\left(\left(x^{\alpha_{BHP}^-} - \mu_{\alpha_{BHP}^-}^-\right) / \sigma_{\alpha_{BHP}^-}^-\right)}{\sigma_{\alpha_{BHP}^-}^-\left(F_{BHP}\left(R_{\alpha_{BHP}^-}^-\right) - F_{BHP}\left(L_{\alpha_{BHP}^-}^-\right)\right)}.$$

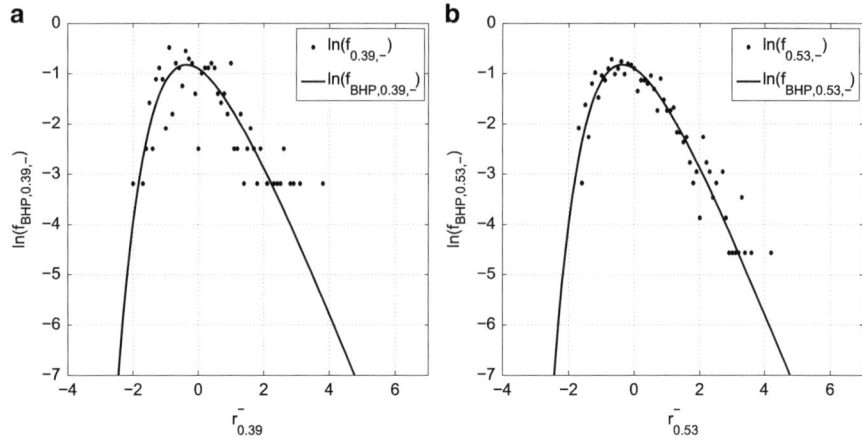

Fig. 15.6 The histogram of the α_{BHP}^- fluctuations with the truncated BHP pdf $f_{BHP,\alpha,-}$ on top, in the semi-log scale, for daily returns. (**a**) Ethanol. (**b**) Corn

Hence, we get

$$f_{BHP,Ethanol,-}(x) = 1.67x^{-0.61} f_{BHP}(10.72x^{0.39} - 2.51)$$

$$f_{BHP,Corn,-}(x) = 5.02x^{-0.47} f_{BHP}(20.07x^{0.53} - 2.12)$$

We denote that $x^{\alpha_{BHP}^- - 1}$ is the intensity term of $f_{BHP,ES,-}(x)$ at zero.

15.2.3 Ethanol and Corn Prices Gains and Losses

In the presented data analysis all the P values are always higher that 0.01, which can indicate universality of the values in these energy source prices. For daily returns, in ethanol and corn prices, we observe that $P_{BHP}^- > P_{BHP}^+$.

Considering the exponent α as new measure of intensity of the market, we observe that in daily returns, in the period of 8 years considered, in ethanol prices, $\alpha_{BHP}^+ < \alpha_{BHP}^-$ which can indicate that the market was more intense or more active in the losses than in the gains.

A smaller value of α causes a higher blowup of the returns near zero and the fact that, in corn daily prices, $\alpha_{BHP}^+ > \alpha_{BHP}^-$, shows that the market was less intense or less active in losses (Fig. 15.7).

Fig. 15.7 Values of α_{BHP}^{+} and α_{BHP}^{-}, for daily returns

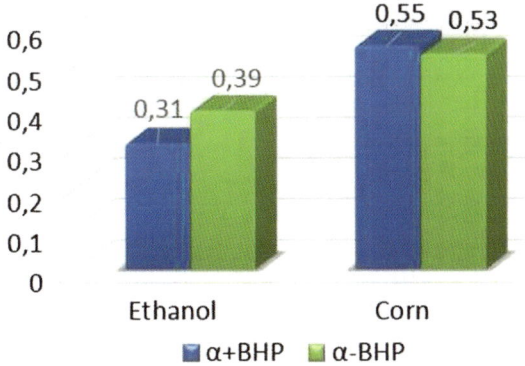

15.3 Conclusions

We computed the analytical approximations of the pdf of the distinct normalized daily positive and negative returns for specific energy sources, in terms of the truncated BHP pdf. We showed that the data collapse of the histogram of the positive and negative returns supports our proposed theoretical pdfs. We presented a measure of the intensity of gains and losses of different energy sources prices.

Acknowledgements We thank Nico Stollenwerk, Jason Gallas, Peter Holdsworth, Imre Janosi and Henrik Jensen for showing us the relevance of the Bramwell-Holdsworth-Pinton distribution. We thank Susan Jenkins for the helpful discussions and comments. A previous version of this work was presented in ICIAM 2011. We acknowledge the financial support of LIAAD-INESC TEC through 'Strategic Project-LA 14-2013–2014' with reference PEst-C/EEI/LA0014/2013, USP-UP project, IJUP, Faculty of Sciences, University of Porto, Calouste Gulbenkian Foundation, FEDER, POCI 2010 and COMPETE Programmes and Fundação para a Ciência e a Tecnologia (FCT) through Project "Dynamics and Applications", with reference PTDC/MAT/121107/2010.

References

1. Andersen, T.G., Bollerslev, T., Frederiksen, P., Nielse, M.: Continuos-time models, realized volatilities and testable distributional implications for daily stock returns. J. Appl. Econom. **25**(2), 233–261 (2010)
2. Barnhart, S.W., Giannetti, A.: Negative earnings, positive earnings and stock return predictability: an empirical examination of market timing. J. Empir. Finance **16**, 70–86 (2009)
3. Bramwell, S.T., Holdsworth, P.C.W., Pinton, J.F.: Universality of rare fluctuations in turbulence and critical phenomena. Nature **396**, 552–554 (1998)
4. Bramwell, S.T., Fortin, J.Y., Holdsworth, P.C.W., Peysson, S., Pinton, J.F., Portelli, B., Sellitto, M.: Magnetic fluctuations in the classical XY model: the origin of an exponential tail in a complex system. Phys. Rev E **63**, 041106 (2001)
5. Chowdhury, D., Stauffer, D.: A generalized spin model of financial markets. Eur. Phys. J. **B8**, 477–482 (1999)

6. Cont, R., Potters, M., Bouchaud, J.P.: Scaling in stock market data: stable laws and beyond. In: Dubrulle, B., Groner, F., Sornette, D. (eds.) Scale Invariance and Beyond. Centre de Physique des Houches, vol. 7, pp. 75–85. Springer, Berlin (1997)
7. Cutler, D.M., Poterba, J.M., Summers, L.H.: What moves stock prices? J. Portfolio Manag. **15**(3), 4–12 (1989)
8. Dahlstedt, K., Jensen, H.J.: Universal fluctuations and extreme-value statistics. J. Phys. A Math. Gen. **34**, 11193–11200 (2001)
9. Gabaix, X., Parameswaran, G., Plerou, V., Stanley, E.: A theory of power-law distributions in financial markets. Nature **423**, 267–270 (2003)
10. Gonçalves, R., Pinto, A.A.: Negro and Danube are mirror rivers. Special issue dynamics & applications in honor of Mauricio Peixoto and David Rand. J. Differ. Equat. Appl. **16**(12), 1491–1499 (2010)
11. Gonçalves, R., Ferreira, H., Pinto, A.A., Stollenwerk, N.: Universality in nonlinear prediction of complex systems. Special issue in honor of Saber Elaydi. J. Differ. Equat. Appl. **15**(11–12), 1067–1076 (2009)
12. Gonçalves, R., Pinto, A.A., Stollenwerk, N.: Cycles and universality in sunspot numbers fluctuations. Astrophys. J. **691**, 1583–1586 (2009)
13. Gonçalves, R., Ferreira, H., Stollenwerk, N., Pinto, A.A.: Universal fluctuations of the AEX index. Phys. A Stat. Mech. Appl. **389**, 4776–4784 (2010)
14. Gonçalves, R., Ferreira, H., Pinto, A.A.: Universality in the stock exchange market. Special issue: dynamics, games and applications I in honour of Mauricio Peixoto and David Rand. J. Differ. Equat. Appl. **17**(7), 1049–1063 (2011)
15. Gonçalves, R., Ferreira, H., Pinto A.A.: Universality in PSI20 fluctuations. In: Peixoto, M., Pinto, A.A., Rand, D. (eds.) Dynamics, Game and Science I. Springer Proceedings in Mathematics, vol. 1, pp. 405–420. Springer, Berlin (2011)
16. Kantz, H., Schreiber, T.: Nonlinear Time Series Analysis. Cambridge University Press, Cambridge, UK (1997)
17. Landau, L.D., Lifshitz, E.M.: Statistical Physics, vol. 1. Pergamon Press, Oxford (1980)
18. Lillo, F., Mantegna, R.: Statistical properties of statistical ensembles of stock returns. Int. J. Theor. Appl. Finance **3**, 405–408 (2000)
19. Lillo, F., Mantegna, R.: Ensemble properties of securities traded in the Nasdaq market. Phys. A **299**, 161–167 (2001)
20. Mandelbrot, B.: The variation of certain speculative prices. J. Bus. **36**, 392–417 (1963)
21. Mandelbrot, B.: Fractals and Scaling in Finance. Springer, Berlin (1997)
22. Mantegna, R., Stanley, E.: Scaling behaviour in the dynamics of a economic index. Nature **376**, 46–49 (2001)
23. Pagan, A.: The econometrics of financial markets. J. Empir. Finance **3**, 15–102 (1996)
24. Plerou, V., Amaral, L., Gopikrishnan, P., Meyer, M., Stanley, E.: Universal and nonuniversal properties of cross correlations in financial time series. Phys. Rev. Lett. **83**(7), 1471–1474 (1999)
25. Stanley, H., Plerou, V., Gabix, X.: A statistical physics view of financial fluctuations: evidence for scaling and universality. Phys. A **387**, 3967–3981 (2008)

Chapter 16
Dynamical Phase Transition in Slowed Exclusion Processes

Tertuliano Franco, Patrícia Gonçalves, and Adriana Neumann

16.1 Introduction

A central question in Statistical Mechanics consists in obtaining the scaling limits of interacting particle systems: given a microscopical interaction in a time-evolving particle system, properly rescaled, what is the limiting behavior of the system? Here, we are concerned about a one-dimensional particle system with a random evolution where the interaction dynamics is given by the exclusion rule, namely, at most one particle can occupy a fixed state (the so-called *fermions* in Physics) and all the bonds have a constant rate of passage of particles, except a finite number of bonds, whose rate is slowed down in order to difficult the passage across them.

The exclusion rule (Exclusion Process) is a standard model in Probability and Statistical Physics, with wide literature about it. The special bonds slowing down the passage of particles, denominated by *slow bonds*, have been considered in [5] and have origin in the works [2, 4] and [3].

T. Franco (✉)
Instituto de Matemática, Universidade Federal da Bahia, Av. Adhemar de Barros S/N, Ondina, CEP 40170-110, Salvador, Brazil
e-mail: tertu@impa.br

P. Gonçalves
PUC-RIO, Departamento de Matemática, Rua Marquês de São Vicente, no. 225, 22453-900, Gávea, Rio de Janeiro, Brazil

Centro de Matemática da Universidade do Minho, Campus de Gualtar,
4710-057 Braga, Portugal
e-mail: patricia@mat.puc-rio.br

A. Neumann
Universidade Federal do Rio Grande do Sul, Porto Alegre, Brazil
e-mail: aneumann@impa.br

A.A. Pinto and D. Zilberman (eds.), *Modeling, Dynamics, Optimization and Bioeconomics I*, Springer Proceedings in Mathematics & Statistics 73, DOI 10.1007/978-3-319-04849-9__16,
© Springer International Publishing Switzerland 2014

The scaling limit considered here is the *hydrodynamical limit* (see [8] for a reference on the subject) for a one-dimensional particle system, where the spatial mesh of the discrete lattice is taken as N^{-1} and particles can evolve at a scaling time given by N^2. We further assume that at the initial time, the density of particles is approximated by a continuous profile $\gamma(\cdot)$. The parameter characterizing the intensity of the rate of passage at the slow bonds is taken as $N^{-\beta}$, where $\beta \in [0, \infty]$. It is understood here that $N^{-\infty} = 0$. In order to keep notation simple we suppose the presence of a single slow bond. The extension to a finite number of slow bonds is straightforward, see [5] for details.

The main result of this paper consists on establishing a *dynamical phase transition* that depends on the parameter $\beta \in [0, \infty]$, for the hydrodynamic limit of exclusion processes with a finite number of slow bonds. In other words, we prove that, when $N \to \infty$, the time trajectory of the spatial density of particles converges to a space-time function $\rho(t, x)$ that is the weak solution of a certain partial differential equation, depending on the chosen regime of β. More precisely, if $\beta \in [0, 1)$, then $\rho(t, x)$ is the unique weak solution of the well known *heat equation on the torus* $\mathbb{T}$:

$$\begin{cases} \partial_t \rho = \partial_x^2 \rho, \\ \rho(0, \cdot) = \gamma(\cdot). \end{cases}$$

meaning that, although the rate of passage of particles across the slow bond goes to zero, its microscopical effect is not strong enough in order to have any consequence in the continuum.

On the other hand, at the critical value $\beta = 1$, $\rho(t, x)$ is the unique weak solution of the *heat equation on the torus* $\mathbb{T}$, *with Robin's boundary conditions at the origin*:

$$\begin{cases} \partial_t \rho = \partial_x^2 \rho \\ \rho(0, \cdot) = \gamma(\cdot) \\ \partial_x \rho_t(1) = \partial_x \rho_t(0) = \rho_t(0) - \rho_t(1). \end{cases}$$

We suppose that the slow bond is close to the origin, otherwise if it is close to some point uN in the one-dimensional discrete torus $\mathbb{T}_N$, then the boundary condition is given at $u \in \mathbb{T}$. It is possible to recognize above the Fick's Law (or the Fourier's Law for the heat conduction), which states that the passage of mass across an interface is proportional to the difference of the concentration (or the temperature in the Fourier's Law).

Now, if the rate of passage is slowed down in such a way that $\beta \in (1, \infty]$, then $\rho(t, x)$ is the solution of the *heat equation on the torus* $\mathbb{T}$, *with Neumann's boundary conditions*:

$$\begin{cases} \partial_t \rho = \partial_x^2 \rho \\ \rho(0, \cdot) = \gamma(\cdot) \\ \partial_x \rho_t(1) = \partial_x \rho_t(0) = 0. \end{cases}$$

In the case $\beta \in (1, \infty)$, the intensity of the slow bond is big enough in order to prevent the passage of mass in the continuum. In the microscopic scenario, for each fixed N, it is possible to observe particles crossing the slow bond, nevertheless in the macroscopic limit, the corresponding boundary is isolated, as predicted by the Neumann's boundary conditions. Microscopically, the case $\beta = \infty$ denotes a forbidden passage of particles across the slow bond and the same behavior is reflected at the continuum, but in this case the system will evolve in a finite box instead of the discrete torus $\mathbb{T}_N$.

The paper is organized as follows. In Sect. 16.2, we define our microscopic dynamics as an exclusion type model with a finite number of slow bonds whose rate of passage is given by $N^{-\beta}$ with $\beta \in [0, \infty]$. In Sect. 16.3, we present the partial differential equations that we obtain for the different regimes of β and we define what we mean by weak solutions of each one of those equations. In Sect. 16.4, we formally define the concept of hydrodynamic limit and in Sect. 16.5 we state the main result of this paper, namely, the *dynamical phase transition* at the level of the hydrodynamic limit. We finish the paper, presenting in Sect. 16.6, some extensions to higher dimensions and some future open problems.

16.2 Microscopic Dynamics

Let $\mathbb{T}_N = \mathbb{Z}/N\mathbb{Z} = \{1, \ldots, N\}$ be the one-dimensional discrete torus with N points. We consider a microscopic dynamics of exclusion type, at each site $x \in \mathbb{T}_N$ there can be at most one particle. As a consequence of this exclusion rule, our Markov process has state space $\{0, 1\}^{\mathbb{T}_N}$, its configurations being denoted by the Greek letter $\eta \in \{0, 1\}^{\mathbb{T}_N}$. The occupation variable at the site x is defined in such a way that $\eta(x) = 1$, if the site x is occupied, otherwise $\eta(x) = 0$ (Fig. 16.1).

Now, we define the dynamics of our interacting particle system. For that purpose, suppose that initially we have particles distributed in $\mathbb{T}_N$ as:

Let $p : \mathbb{T}_N \times \mathbb{T}_N \to [0, 1]$ be a probability measure such that for all $x, y \in \mathbb{T}_N$ it holds that $p(x, y) := p(y - x)$.

At each site $x \in \mathbb{T}_N$, there exists a random clock with exponential law of parameter $\lambda(x)$ (usually this parameter is equal to one), which is independent of the random clocks at the other sites. When one clock rings, if there is a particle at that site, then it jumps to a site y with probability $p(y - x)$. By the exclusion rule, particles can only jump to empty sites (Figs. 16.2 and 16.3). For instance, this jump is allowed by the dynamics:

while this other jump is forbidden:

We suppose that jumps are performed to the nearest-neighbors, namely, two sites $x, y \in \mathbb{T}_N$ are nearest-neighbors, that we denote by $x \sim y$, if $|x - y| = 1$.

We also assume that there are a finite number of slow bonds whose rate of passage of particles is decreased in such a way that those bonds act somehow as a *barrier* to the movement of particles, see Fig. 16.4:

To each pair of sites $x, y \in \mathbb{T}_N$ such that $x \sim y$, we associate a number $\xi_{x,y}^N = \xi_{y,x}^N > 0$, usually called *conductance*. At the slow bonds the conductance is a smaller

Fig. 16.1 One possible initial configuration

Fig. 16.2 Since the destination site is empty the particle can jump

Fig. 16.3 Since the destination site is occupied the particle does not move

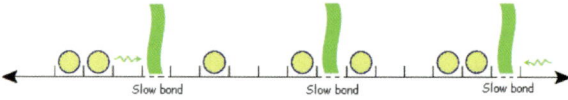

Fig. 16.4 Exclusion process with three slow bonds acting as a barrier. The rate at the slow bonds is defined in such a way that the passage of particles across them is more difficult than in the other bonds

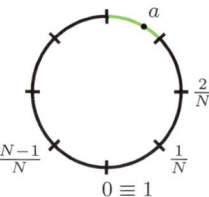

Fig. 16.5 Discrete torus $\frac{1}{N}\mathbb{T}_N$ embedded in continuous torus $\mathbb{T}$. The slow bond is the *green* one, containing the macroscopical point $a \in \mathbb{T}$

in comparison with the value of the conductance at other bonds. The conductances are related to the parameter of the exponential clock as follows: for each site $x \in \mathbb{T}_N$, $\lambda(x) = \sum_{y;\, y \sim x} \xi_{x,y}^N$.

We consider a finite number of slow bonds, each one associated to a point $b_1, \ldots, b_k \in \mathbb{T}$. Having a bond associated to the macroscopic point $a \in \mathbb{T}$, means that this bond contains the point a in the natural embedding of the discrete torus $\mathbb{T}_N$ in the continuous torus $\mathbb{T}$, $\frac{1}{N}\mathbb{T}_N \subset \mathbb{T}$, see Fig. 16.5:

Given $b_1, \ldots, b_k \in \mathbb{T}$, we consider the following conductances:

$$\xi_{x,x+1}^N = \begin{cases} N^{-\beta}, & \text{if } \{b_1, \ldots, b_k\} \cap \left(\frac{x}{N}, \frac{x+1}{N}\right] \neq \varnothing, \\ 1, & \text{otherwise}. \end{cases}$$

The rate of the conductances are chosen in such a way that particles cross bonds at rate one, except k particular bonds in which the dynamics is slowed down by a factor $N^{-\beta}$, with $\beta \in (0, \infty]$. Each one of these particular bonds contains the macroscopic point $b_i \in \mathbb{T}$; or b_i coincides with some vertex $\frac{x}{N}$ and the slow bond is chosen as the bond to the left of $\frac{x}{N}$.

Denote by $\{\eta_t := \eta_{tN^2} : t \geq 0\}$ the Markov process on $\{0, 1\}^{\mathbb{T}_N}$ whose interacting dynamics is described above, *speeded up* by N^2. Although η_t depends on N and β, we are not indexing it on that, in order not to overload notation. Formally, this Markov process has generator given on local functions $f : \{0, 1\}^{\mathbb{T}_N} \to \mathbb{R}$ by

$$\mathcal{L}_N f(\eta) = \sum_{\substack{x, y \in \mathbb{T}_N \\ x \sim y}} \xi^N_{x,y} (f(\eta^{x,y}) - f(\eta)),$$

where $\eta^{x,y}$ is the configuration obtained from η by exchanging the variables $\eta(x)$ and $\eta(y)$.

Let $D(\mathbb{R}_+, \{0, 1\}^{\mathbb{T}_N})$ be the path space of càdlàg trajectories with values in $\{0, 1\}^{\mathbb{T}_N}$. For a measure μ^N on $\{0, 1\}^{\mathbb{T}_N}$, denote by $\mathbb{P}^\beta_{\mu^N}$ the probability measure on $D(\mathbb{R}_+, \{0, 1\}^{\mathbb{T}_N})$ induced by the initial state μ^N and the Markov process $\{\eta_t : t \geq 0\}$.

16.3 Macroscopic Hydrodynamic Equations

In this section we present the partial differential equations governing the evolution of the density profile for the different regimes of β and we define the notion of weak solutions of each one of these equations.

Denote by ρ_t a function $\rho(t, \cdot)$ and for an integer n denote by $C^n(\mathbb{T})$ the set of continuous functions from $\mathbb{T}$ to $\mathbb{R}$ and with continuous derivatives of order up to n. For $\mathcal{I}$ an interval of $\mathbb{T}$, here and in the sequel we use the notation $C^{n,m}([0, T] \times \mathcal{I})$ for the set of functions defined on the domain $[0, T] \times \mathcal{I}$, that are of class C^n in time and C^m in space, for n, m integers.

Fix a bounded density profile $\gamma : \mathbb{T} \to \mathbb{R}$.

Definition 16.1. A bounded function $\rho : [0, T] \times \mathbb{T} \to \mathbb{R}$ is said to be a weak solution of the parabolic differential equation:

$$\begin{cases} \partial_t \rho = \partial^2_u \rho \\ \rho(0, \cdot) = \gamma(\cdot) \end{cases} \tag{16.1}$$

if, for $t \in [0, T]$ and $H \in C^{1,2}([0, T] \times \mathbb{T})$, $\rho(t, \cdot)$ satisfies the integral equation

$$\int_{\mathbb{T}} \rho(t, u) H(t, u) \, du - \int_{\mathbb{T}} \gamma(u) H(0, u) \, du$$

$$- \int_0^t \int_{\mathbb{T}} \rho(s, u) \{\partial^2_u H(s, u) + \partial_s H(s, u)\} \, du \, ds = 0.$$

We repeat here the definition of the Sobolev Space from [1].

Definition 16.2 (Sobolev space). For $a, b \in \mathbb{T}$, the Sobolev space $\mathcal{H}^1(a, b)$ consists of all locally summable functions $\zeta : (a, b) \to \mathbb{R}$ such that there exists $\partial \zeta \in L^2(a, b)$ satisfying

$$\int_{\mathbb{T}} \partial_u G(u) \zeta(u) \, du = - \int_{\mathbb{T}} G(u) \partial \zeta(u) \, du,$$

for all $G \in C^\infty(a, b)$ with compact support. For $\zeta \in \mathcal{H}^1(a, b)$, we define the norm

$$\|\zeta\|_{\mathcal{H}^1(a,b)} = \|\partial \zeta\|_{L^2(a,b)}.$$

Definition 16.3. The space $L^2(0, T; \mathcal{H}^1(a, b))$ consists of all measurable functions $\xi : [0, T] \to \mathcal{H}^1(a, b)$ with

$$\|\xi\|_{L^2(0,T;\mathcal{H}^1(a,b))} := \left(\int_0^T \|\xi_t\|_{\mathcal{H}^1(a,b)}^2 \, dt \right)^{1/2} < \infty.$$

Definition 16.4. Let $\{b_1, \ldots, b_k\} \subset \mathbb{T}$. A bounded function $\rho : [0, T] \times \mathbb{T} \to \mathbb{R}$ is said to be a weak solution of the following parabolic differential equation with Robin's boundary conditions at the points $\{b_1, \ldots, b_k\} \subset \mathbb{T}$:

$$\begin{cases} \partial_t \rho = \partial_u^2 \rho \\ \rho(0, \cdot) = \gamma(\cdot) \\ \partial_u \rho_t(b_i^+) = \partial_u \rho_t(b_i^-) = \rho_t(b_i^+) - \rho_t(b_i^-), \ \forall t \in [0, T], \forall i = 1, \ldots, k \end{cases} \tag{16.2}$$

if the following two conditions are fulfilled:

(1) $\rho \in L^2(0, T; \mathcal{H}^1(\mathbb{T}\backslash\{b_1, \ldots, b_k\}))$;
(2) For all functions $H \in C^{1,2}([0, T] \times \mathbb{T}\backslash\{b_1, \ldots, b_k\})$ and for all $t \in [0, T]$, $\rho(t, \cdot)$ satisfies the integral equation

$$\int_{\mathbb{T}} \rho(t, u) H(t, u) \, du - \int_{\mathbb{T}} \gamma(u) H(0, u) \, du$$

$$- \int_0^t \int_{\mathbb{T}} \rho(s, u) \{\partial_u^2 H(s, u) + \partial_s H(s, u)\} \, du \, ds$$

$$- \sum_{i=1}^k \int_0^t \{\rho(s, b_i^+) \partial_u H(s, b_i^+) - \rho(s, b_i^-) \partial_u H(s, b_i^-)\} \, ds \tag{16.3}$$

$$+ \sum_{i=1}^k \int_0^t \{\rho(s, b_i^+) - \rho(s, b_i^-)\}\{H(s, b_i^+) - H(s, b_i^-)\} \, ds = 0.$$

Definition 16.5. Let $\{b_1, \ldots, b_k\} \subset \mathbb{T}$. A bounded function $\rho : [0, T] \times \mathbb{T} \to \mathbb{R}$ is said to be a weak solution of the following parabolic differential equation with *Neumann's boundary conditions* at the points $\{b_1, \ldots, b_k\} \subset \mathbb{T}$:

$$\begin{cases} \partial_t \rho = \partial_u^2 \rho \\ \rho(0, \cdot) = \gamma(\cdot) \\ \partial_u \rho(t, b_i^-) = \partial_u \rho(t, b_i^+) = 0, \ \forall t \in [0, T], \forall i = 1, \ldots, k \end{cases} \tag{16.4}$$

if the following two conditions are fulfilled:

(1) $\rho \in L^2(0, T; \mathcal{H}^1(\mathbb{T} \backslash \{b_1, \ldots, b_k\}))$;
(2) For all functions $H \in C^{1,2}([0, T] \times \mathbb{T} \backslash \{b_1, \ldots, b_k\})$ and for all $t \in [0, T]$, $\rho(t, \cdot)$ satisfies the integral equation

$$\int_{\mathbb{T}} \rho(t, u) \, H(t, u) \, du - \int_{\mathbb{T}} \gamma(u) \, H(0, u) \, du$$

$$- \int_0^t \int_{\mathbb{T}} \rho(s, u) \, \{\partial_u^2 H(s, u) + \partial_s H(s, u)\} \, du \, ds \tag{16.5}$$

$$- \sum_{i=1}^k \int_0^t \{\rho(s, b_i^+) \partial_u H(s, b_i^+) - \rho(s, b_i^-) \partial_u H(s, b_i^-)\} \, ds = 0.$$

For classical results about Sobolev spaces, we refer the reader to [1] and [9]. Since in Definitions 16.4 and 16.5 we imposed $\rho \in L^2(0, T; \mathcal{H}^1(\mathbb{T} \backslash \{b_1, \ldots, b_k\}))$, the integrals above are well-defined at the boundary points.

Heuristically, in order to establish an integral equation for the weak solution of the heat equation with Robin's or Neumann's boundary conditions as above, one should multiply both sides of (16.2) or (16.4) (respectively) by a test function H, integrate in space and time and then perform twice a formal integration by parts, obtaining the equation

$$\int_{\mathbb{T}} \rho(t, u) \, H(t, u) \, du - \int_{\mathbb{T}} \gamma(u) \, H(0, u) \, du$$

$$- \int_0^t \int_{\mathbb{T}} \rho(s, u) \, \{\partial_u^2 H(s, u) + \partial_s H(s, u)\} \, du \, ds$$

$$- \sum_{i=1}^k \int_0^t \{\rho(s, b_i^+) \partial_u H(s, b_i^+) - \rho(s, b_i^-) \partial_u H(s, b_i^-)\} \, ds$$

$$+ \sum_{i=1}^k \int_0^t \{\partial_u \rho(s, b_i^+) H(s, b_i^+) - \partial_u \rho(s, b_i^-) H(s, b_i^-)\} \, ds = 0.$$

Applying the formal boundary conditions on ρ, one gets to (16.3) or (16.5), respectively. Besides that, any strong solution of (16.2) or (16.4) is a weak solution of (16.2) or (16.4), respectively.

16.4 Hydrodynamic Limit

In this section we define formally the hydrodynamic limit for the processes we described above. For that purpose, we define the empirical measure by:

$$\pi_t^N(du) = \pi^N(\eta_t, du) = \frac{1}{N} \sum_{x \in \mathbb{T}_N} \eta_t(x)\delta_{\frac{x}{N}}(du),$$

where δ_u denotes the Dirac measure at $u \in \mathbb{T}$. As mentioned in the introduction we assume that at the initial time the density of particles is approximated by a given profile. Now we define exactly what is the assumption we need on the initial distribution of the system.

Fix a continuous density profile $\gamma : \mathbb{T} \to [0, 1]$ and denote by $(\mu^N)_N$ a sequence of probability measures on $\{0, 1\}^{\mathbb{T}_N}$.

Definition 16.6. A sequence $(\mu^N)_N$ is *associated* to an initial profile $\gamma(\cdot)$, if for every continuous function $H : \mathbb{T} \to \mathbb{R}$ and for every $\delta > 0$

$$\lim_{N \to +\infty} \mu^N \left[\left| \frac{1}{N} \sum_{x \in \mathbb{T}_N} H\left(\frac{x}{N}\right)\eta(x) - \int_{\mathbb{T}} H(u)\gamma(u)du \right| > \delta \right] = 0. \qquad (16.6)$$

We can translate the definition above by saying that a sequence of measures $(\mu^N)_N$ is associated to a profile $\gamma(\cdot)$ if a Law of Large Number (in the weak sense) holds for the empirical measure at time $t = 0$ under the probability μ^N. We can rewrite (16.6) as

$$\lim_{N \to +\infty} \mu^N \left[\left| \int_{\mathbb{T}} H(u)\pi_0^N(du) - \int_{\mathbb{T}} H(u)\gamma(u)du \right| > \delta \right] = 0.$$

The goal in hydrodynamic limit consists in showing that if at time $t = 0$ the empirical measures are associated to some initial profile $\gamma(\cdot)$, then at time t they are associated to a profile ρ_t, where ρ_t is the solution of the corresponding hydrodynamic equation. So, the goal is to show that *if a Law of Large Numbers holds for the empirical measure at time $t = 0$ then it holds at any time t.*

In order to prove this result, we follow the Entropy Method which was introduced by Guo, Papanicolau and Varadhan in [6]. This method requires the uniqueness of weak solutions of the hydrodynamic equation. We notice that all the partial differential equations (16.1), (16.2) and (16.4), have a unique weak solution. For details we refer the reader to [5].

16.5   The Main Result

Here, we state the main result of the article in which we establish the hydrodynamic limit for the exclusion process with slow bonds depending on the regime of β:

Theorem 1 (Franco, Gonçalves and Neumann [5]). *Fix* $\beta \in [0, \infty]$. *Consider the exclusion process with k slow bonds corresponding to macroscopic points $b_1, \ldots, b_k \in \mathbb{T}$ and with conductance $N^{-\beta}$ at each one of these bonds. Fix a continuous profile $\gamma : \mathbb{T} \to [0, 1]$. Let $(\mu^N)_N$ be a sequence of probability measures on $\{0, 1\}^{\mathbb{T}_N}$ associated to $\gamma(\cdot)$.*

Then, for any $t \in [0, T]$, for every $\delta > 0$ and every $H \in C(\mathbb{T})$, it holds that

$$
\lim_{N \to \infty} \mathbb{P}^{\beta}_{\mu^N} \left\{ \eta_{\cdot} : \left| \frac{1}{N} \sum_{x \in \mathbb{T}_N} H\left(\frac{x}{N}\right) \eta_t(x) - \int_{\mathbb{T}} H(u)\, \rho(t, u)du \right| > \delta \right\} = 0,
$$

where :

- *if $\beta \in [0, 1)$, $\rho(t, \cdot)$ is the unique weak solution of (16.1);*
- *if $\beta = 1$, $\rho(t, \cdot)$ is the unique weak solution of (16.2);*
- *if $\beta \in (1, \infty]$, $\rho(t, \cdot)$ is the unique weak solution of (16.4).*

16.6   Future Problems

We finish the paper by presenting some extensions to higher dimensions and some future problems.

The extension of the results presented here to higher dimensional cases can be obtained as in [4]. In that paper the slow bonds have a spatial position associated to a smooth closed surface, modeling a membrane slowing down the passage of particles. There, the authors study the case $\beta = 1$ and it is an interesting problem to obtain the *dynamical phase transition* as in [5], for any d-dimensional exclusion process with slow bonds. Also, it would be a challenging problem to extend the result to particle systems without exclusion constrains, as for example, the generalized exclusion or the zero-range process.

One interesting problem that we are studying is related to the *dynamical phase transition* at the level of the fluctuations or at the limit distributions of a tagged particle. In the former case, we fix a profile $\gamma(\cdot)$ and we suppose the system to start from the Bernoulli product measure $\nu_{\gamma(\cdot)}$, with parameter $\gamma(x/N)$ at the site $x \in \mathbb{T}_N$. We consider the density fluctuation field defined as:

$$
\mathcal{Y}^N_t := \frac{1}{\sqrt{N}} \sum_{x \in \mathbb{T}_N} \delta_{x/N} \left\{ \eta_t(x) - \mathbb{E}_{\nu_{\gamma(\cdot)}}[\eta_t(x)] \right\}.
$$

The goal is to obtain the limit $\mathcal{Y}$. of the density field as defined above and to characterize it as a solution to some stochastic partial differential equation. As in the hydrodynamic limit scenario, this equation will depend on the regime of β. For $\beta = 0$, it was proved in [10] that $\mathcal{Y}$. is an Ornstein-Uhlenbeck process with certain characteristics depending on the non-equilibrium thermodynamical quantities of the underlying system. The goal is to characterize the limit process for $\beta \neq 0$. In the case of the tagged particle, at the initial time we fix one of the particles that are distributed according to $\nu_{\gamma(\cdot)}$ and we follow its trajectory. We want to establish a *dynamical phase transition* at the level of the limiting distributions of this particle. For $\beta = 0$, it is proved in [7] that the limit distributions are given by a fractional Brownian motion. We want to characterize the limit when $\beta \neq 0$. This is a step towards characterizing the limit behavior of exclusion processes with slow bonds.

Acknowledgements The authors thank FCT (Portugal) and Capes (Brazil) for the financial support through the research project "Non-Equilibrium Statistical Mechanics of Stochastic Lattice Systems". PG thanks FCT (Portugal) for support through the research project "Non-Equilibrium Statistical Physics" PTDC/MAT/109844/2009. PG thanks the Research Centre of Mathematics of the University of Minho, for the financial support provided by "FEDER" through the "Programa Operacional Factores de Competitividade COMPETE" and by FCT through the research project PEst-C/MAT/UI0013/2011.

References

1. Evans, L.: Partial differential equations. In: Graduate Studies in Mathematics. American Mathematical Society, Providence (1998)
2. Faggionato, A., Jara, M., Landim, C.: Hydrodynamic behavior of one dimensional subdiffusive exclusion processes with random conductances. Probab. Theor Relat. Field **144**(3–4), 633–667 (2008)
3. Franco, T., Landim, C.: Hydrodynamic limit of gradient exclusion processes with conductances. Arch. Ration. Mech. Anal. **195**, 409–439 (2010)
4. Franco, T., Neumann, A., Valle, G.: Hydrodynamic limit for a class of exclusion type process in dimension ≥ 2. J. Appl. Probab. **49.2**, 333–351 (2011)
5. Franco, T., Gonçalves, P., Neumann, A.: Hydrodynamical behavior of symmetric exclusion with slow bonds. Annales de l'Institut Henri Poincaré Probab. Stat. B **49**(2), 402–427 (2013)
6. Guo, M.Z., Papanicolau, G.C., Varadhan, S.R.S.: Nonlinear diffusion limit for a system with nearest neighbor interactions. Commun. Math. Phys. **118**, 31–59 (1988)
7. Jara, M., Landim, C.: Non equilibrium central limit theorem for a tagged particle in symmetric simple exclusion. Annales de l'Institute Henri Poincaré Probab. Stat. **42**(5), 567–577 (2006)
8. Kipnis, C., Landim, C.: Scaling Limits of Interacting Particle Systems. Springer, New York (1999)
9. Leoni, G.: A First Course in Sobolev Spaces. In: Graduate Studies in Mathematics. American Mathematical Society, Providence (2009)
10. Ravishankar, K.: Fluctuations from the hydrodynamical limit for the symmetric simple exclusion in $\mathbb{Z}^d$. Stoch. Process. Appl. **42**, 31–37 (1992)

Chapter 17
Cooperative Ellipsoidal Games: A Survey

S.Z. Alparslan Gök and G.-W. Weber

17.1 Introduction

Uncertainty affects our decision making activities on a daily basis. There are many sources of uncertainty in the real world such as: noise in observation and experimental design, incomplete information, vagueness in preference structures and decision making. The important issue of whether and why individuals and organizations choose to cooperate (or not) when faced with uncertainty on outcomes or costs has generated a productive line of research in recent years. Crises in the world and the need that players act in the directed manner of, e.g., the joint implementation of the Kyoto Protocol, has brought the necessity of collaboration into public awareness [5–7].

Cooperative game theory in coalitional form is a popular research area with many new developments in the last few years. Several models are introduced in this area such as the classical model of cooperative games with transferable utility (TU-games), cooperative game theory in which the players have the possibility to cooperate partially, namely, fuzzy games, multi-choice games and cooperative interval games. In a cooperative TU-game, the agents are either fully involved or not involved at all in cooperation with some other agents, while in a fuzzy game players are allowed to cooperate with infinitely many different participation levels, varying from non-cooperation to full cooperation. A multi-choice game describes an intermediate case in the sense that each player may have a fixed finite number of

S.Z. Alparslan Gök (✉)
Faculty of Arts and Sciences, Department of Mathematics, Süleyman Demirel University,
32 260 Isparta, Turkey
e-mail: zeynepalparslan@yahoo.com

G.-W. Weber
Institute of Applied Mathematics, Middle East Technical University, METU, Ankara, Turkey
e-mail: gweber@metu.edu.tr

A.A. Pinto and D. Zilberman (eds.), *Modeling, Dynamics, Optimization and Bioeconomics I*, Springer Proceedings in Mathematics & Statistics 73, DOI 10.1007/978-3-319-04849-9_17,
© Springer International Publishing Switzerland 2014

activity levels. A cooperative interval game fits all the situations where participants consider cooperation and know with certainty only the lower and upper bounds of all potential revenues or costs generated via cooperation. For details of these models we refer reader to [3].

The structure of interval data and the model of cooperative interval games [1] have had broad applicability in Operational Research, climate negotiations and policy, environmental management and pollution control, etc. However, the interval-valued games model treats all its items separately, i.e., even if we have confidence intervals given (as data) or to be estimated (as unknown parameters and levels of states), there is not yet any stochastic kind of dependence, correlation, implied into this model. For this reason and since the interval calculus is not able to represent the players' mutual dependencies, similarities and possible affinities to collaborate, Alparslan Gök and Weber [2] recently introduced cooperative ellipsoidal games, where ellipsoids were associated with clusters of players which were considered to share such kinds of correlations. We notice that when changing to new/more general structures (such as ellipsoids), we regularized—meaning: compromised between the gain in accuracy with the forthcoming difficulties/sensitivities of the model and computations. Ellipsoids go beyond intervals (and their Cartesian products: cubes or parallelepipeds) in that they allow to include correlation between the items (genes, actors, players) interpreted as dependences, common interests, etc.; intervals mean parallelism—no correlations included. Cooperative ellipsoidal games are cooperative games where the worth of each coalition is an ellipsoid. In [8], the ellipsoidal core was introduced to solve the difficult problem of distribution of the ellipsoidal value of the grand coalition over the players and in [9] a necessary condition for the non-emptiness of the ellipsoidal core is given and the basic properties of the ellipsoidal core are studied. Motivation for the model of cooperative ellipsoidal games and for the ellipsoidal core relies on the need to collaborate, e.g., for overcoming environmental challenges, in the presence of uncertainty that exist in the real world, especially in gene-environment and eco-finance networks.

In the aforementioned studies, the research started with fully deterministic models, in the form of networks, their dynamics and optimization. Since, however, the real world is characterized by uncertainty in the form of noise in the observations and a lack of knowledge about how the items of the models interact, the authors left the deterministic *real*-valuedness of the models and turn to interval-valued models. In fact, the given (data) and predicted values of the biological and entire environmental information can hardly be identified by single scalar numbers, but they can be easily hosted in some confidence intervals. The same can be said for the levels of interactions of all these items. When aggregating all the intervals of data vectors, state vectors or vectors of parameters using Cartesian products, we obtain (confidence) parallelepipeds. Those parallelepipeds and intervals usually come from a perspective where functional dependencies among any two of the errors made in the measurements of the gene-environmental levels are not taken into account explicitly.

The paper is organized as follows. Section 17.2 gives ellipsoid calculus needed for handling ellipsoidal coalition values and payoff vectors whose components are ellipsoids. The model of cooperative ellipsoidal games is given in Sect. 17.3. Section 17.4 deals with ellipsoidal solution concepts and related results for cooperative ellipsoidal games. An economic application related with this model is given in Sect. 17.5. Finally, we indicate some topics for further research.

17.2 Ellipsoid Calculus

In this section, some preliminaries from ellipsoidal calculus are given based on [4, 8].

An ellipsoid in $\mathbb{R}^p$ will be parameterized in terms of its center $c \in \mathbb{R}^p$ and a symmetric non-negative definite configuration matrix $\Sigma \in \mathbb{R}^{p \times p}$ as

$$\mathcal{E}(c, \Sigma) = \{\Sigma^{1/2} u + c \mid \|u\| \leq 1\},$$

where $\Sigma^{1/2}$ is any matrix square root satisfying $\Sigma^{1/2}(\Sigma^{1/2})^T = \Sigma$ and $\|\cdot\|$ representing the Euclidean norm. When Σ is of full rank, the non-degenerate ellipsoid $\mathcal{E}(c, \Sigma)$ may be expressed as

$$\mathcal{E}(c, \Sigma) = \{x \in \mathbb{R}^p \mid (x - c)^T \Sigma^{-1}(x - c) \leq 1\}.$$

We denote by Ξ the set of all ellipsoids. The null ellipsoid will be denoted by 0. Now, we introduce the basic operations needed to deal with ellipsoidal uncertainty such as sums, intersections (fusions) and affine-linear transformations of ellipsoids. The family of ellipsoids in $\mathbb{R}^p$ is closed with respect to affine-linear transformations but neither the sum nor the intersection is generally ellipsoidal, so both must be approximated by ellipsoidal sets. Given two non-degenerate ellipsoids $\mathcal{E}_1 = \mathcal{E}(c_1, \Sigma_1)$ and $\mathcal{E}_2 = \mathcal{E}(c_2, \Sigma_2)$, their geometric (Minkowksi) sum $\mathcal{E}_1 + \mathcal{E}_2 = \{z_1 + z_2 \mid z_1 \in \mathcal{E}_1, z_2 \in \mathcal{E}_2\}$ is not generally an ellipsoid. However, it can be tightly approximated by parameterized families of external ellipsoids. The range of values of $\mathcal{E}_1 + \mathcal{E}_2$ is contained in the ellipsoid $\mathcal{E}_1 \oplus \mathcal{E}_2 := \mathcal{E}(c_1 + c_2, \Sigma(s))$ for all $s > 0$, where $\Sigma(s) = (1 + s^{-1})\Sigma_1 + (1 + s)\Sigma_2$.

For a minimal and unique external ellipsoidal approximation an additional condition has to be fulfilled. The value of s is commonly chosen to minimize either the trace or the determinant of $\Sigma(s)$.

The family of ellipsoids is closed with respect to affine transformations. Given an ellipsoid $\mathcal{E}(c, \Sigma) \subset \mathbb{R}^p$, a matrix $A \in \mathbb{R}^{m \times p}$ and a vector $b \in \mathbb{R}^m$ we get $A\mathcal{E}(c, \Sigma) + b = \mathcal{E}(Ac + b, A\Sigma A^T)$. Thus, ellipsoids are preserved under affine transformation. If the rows of A are linearly independent (which implies $m \leq p$), and $b = 0$, the affine transformation is called a projection.

The (maximal in size) inner and (minimal in size) outer approximations of ellipsoids by parallelepipeds and the projections of such parallelepipeds on the coordinate axes provide a criterion to compare ellipsoids. We introduce binary relation $\triangleright$ on Ξ as follows.

Let $\mathscr{E}_1, \mathscr{E}_2 \in \varXi$ be non-degenerated ellipsoids whose interior parallelepipeds are disjoint. We say that $\mathscr{E}_1 \rhd \mathscr{E}_2$ if and only if $d_k^1 \geq d_k^2$ for each $k = 1, \ldots, p$, where d_k^1 and d_k^2 are the coordinate wise sizes of the projected inner parallelepipeds approximating $\mathscr{E}_1$ and $\mathscr{E}_2$, respectively. Note that the relation $\rhd$ is a partial order, that is a reflexive and transitive binary relation.

17.3   The Model

A cooperative ellipsoidal game [2] is an ordered pair $< N, \tilde{w} >$, where $N = \{1, 2, \ldots, n\}$ is the set of players, and $\tilde{w} : 2^N \to \varXi$ is the characteristic function which assigns to each coalition $S \in 2^N$ an ellipsoid such that $\tilde{w}(\emptyset) = 0$, where $\varXi$ is the family of all ellipsoids.

Reward/cost sharing problems in situations modeled by cooperative ellipsoidal games can be addressed with ellipsoidal solution concepts, i.e., solutions which associate to each cooperative ellipsoidal game a set consisting of vectors of ellipsoids. Such solutions inform the players about the ranges of individual payoffs generated by cooperation within the grand coalition, and are helpful post cooperation to distribute among the players the outcome of the grand coalition when the ellipsoidal uncertainty is resolved.

The following example illustrates the concept of an ellipsoidal game.

Example 17.1 (Ellipsoid Glove Game [2]). Consider the ellipsoidal game with three players with $\tilde{w}(1, 3) = \tilde{w}(2, 3) = \tilde{w}(1, 2, 3) = \mathscr{E}(c, \varSigma)$ and $\tilde{w}(S) = 0$, otherwise.

Recall that a map $\lambda : 2^N \setminus \{\emptyset\} \to \mathbb{R}_+$ is called a balanced map [3] if $\sum_{S \in 2^N \setminus \{\emptyset\}} \lambda(S) e^S = e^N$. Here, e^S is the characteristic vector for coalition S with

$$e_i^S = \begin{cases} 1 & \text{if } i \in S \\ 0 & \text{if } i \in N \setminus S. \end{cases}$$

Let $< N, \tilde{w} >$ be a cooperative ellipsoidal game. We say that $< N, \tilde{w} >$ is E-balanced if for each balanced map $\lambda : 2^N \setminus \{\emptyset\} \to \mathbb{R}_+$ we have $\tilde{w}(N) \rhd \sum_{S \in 2^N \setminus \{\emptyset\}} \lambda(S) \tilde{w}(S)$.

17.4   Ellipsoidal Solution Concepts

Let $< N, \tilde{w} >$ be a cooperative ellipsoidal game. Its ellipsoidal core is defined by

$$\mathbb{C}(\tilde{w}) = \left\{ (\mathscr{E}_1, \ldots, \mathscr{E}_n) \in \varXi \,\Big|\, \sum_{i \in N} \mathscr{E}_i = \tilde{w}(N), \sum_{i \in S} \mathscr{E}_i \rhd \tilde{w}(S), \forall S \in 2^N \setminus \{\emptyset\} \right\}.$$

Example 17.2. The ellipsoidal core of Example 17.1 is $\mathbb{C}(\tilde{w}) = \{(0, 0, \mathscr{E}(c, \Sigma))\}$.

In [9] a necessary condition for the non-emptiness of the ellipsoidal core is given as follows:

Proposition 17.1. *If a cooperative ellipsoidal game has a non-empty ellipsoidal core then it is E-balanced.*

The following basic properties of the ellipsoidal core are proved in [9].

Proposition 17.2. *Let $\tilde{w}$ be a cooperative ellipsoidal game. Then, the ellipsoidal core $\mathbb{C}(\tilde{w})$ of $\tilde{w}$ is a convex set.*

Proposition 17.3. *Let $\tilde{w}$ be a cooperative ellipsoidal game. Then, the ellipsoidal core $\mathbb{C}(\tilde{w})$ is relative invariant with respect to strategic equivalence.*

Proposition 17.4. *Let $\tilde{w}$ be a cooperative ellipsoidal game. Then, the ellipsoidal core $\mathbb{C}(\tilde{w})$ of $\tilde{w}$ is a superadditive map.*

17.5   An Economic Application

In this section we illustrate an economic application of cooperative ellipsoidal games to motivate our work.

Example 17.3 ([8]). Three countries, say Australia, New Zealand and Indonesia, consider cooperation for reducing their greenhouse gas emission because they do not have enough money to do this separately. Australia and New Zealand have much in common from cultural viewpoint and industrialization degree, while Indonesia is rather complementary in technological sense. A joint CO_2 reduction by the cluster {Australia, New Zealand} has a positive influence on the cluster {Indonesia} and vice versa. The technological complementary of Indonesia and the group consisting of Australia and New Zealand together with the use of the TEM model allow us to associate with this situation an ellipsoidal game $< N, \tilde{w} >$, where $N = \{1, 2, 3\}$, with Australia as player 1, New Zealand as player 2 and Indonesia as player 3 whose characteristic function $\tilde{w}$ is given by $\tilde{w}(1, 3) = \tilde{w}(2, 3) = \tilde{w}(1, 2, 3) = \mathscr{E}_1(c, \Sigma_1)$, $\tilde{w}(S) = \mathscr{E}_2(c, \Sigma_2)$, otherwise, with $\mathscr{E}_1(c, \Sigma_1) \rhd \mathscr{E}_2(c, \Sigma_2)$. Any ellipsoidal core element gives equal ellipsoidal payoffs to players 1 and 2 and a better ellipsoidal payoff to player 3 that owns the scarce good. We emphasize that the establishment of the ellipsoids needs real-world data, the use of modern clustering theory and further data mining and statistics.

17.6 Final Remarks

In this paper we give a survey on the class of cooperative ellipsoidal games. Now, we indicate some topics for further research regarding the ellipsoidal core such as: to find conditions guaranteeing the non-emptiness of the interval core of a cooperative ellipsoidal game, to design efficient algorithms for determining ellipsoidal core elements, to characterize the ellipsoidal core on suitable classes of cooperative ellipsoidal games. We invite the interested reader to participate in this exciting new research programme and its modern applications.

Acknowledgements The authors thank Rodica Branzei and Stef Tijs for their valuable contributions to the research.

References

1. Alparslan Gök, S.Z.: Cooperative Interval Games: Theory and Applications. LAP-Lambert Academic Publishing house, Germany (2010). PROJECT-ID:5124, ISBN-NR: 978-3-8383-3430-1
2. Alparslan Gök, S.Z., Weber, G.-W.: Cooperative games under ellipsoidal uncertainty. In: Proceedings of PCO 2010, 3rd Global Conference on Power Control and Optimization, Gold Coast, Queensland, Australia, 2–4 February 2010. ISBN: 978-983-44483-1-8
3. Branzei, R., Dimitrov, D., Tijs, S.: Models in Cooperative Game Theory. Springer, Berlin (2008)
4. Kropat, E., Weber, G.-W., Sekhar, P.C.: Regulatory networks under ellipsoidal uncertainty: optimization theory and dynamical systems. Preprint no. 138, Institute of Applied Mathematics, METU (2009)
5. Pickl, S., Weber, G.-W.: Optimization of a time-discrete nonlinear dynamical system from a problem of ecology: an analytical and numerical approach. J. Comput. Technol. **6**(1), 43–52 (2001)
6. Weber, G.-W., Alparslan-Gök, S.Z., Dikmen, N.: Environmental and life sciences: gene-environment networks: optimation, games and control: a survey on recent achievements, invited paper. Special issue, DeTombe, D. (guest ed.). J. Organisational Transformation Soc. Change **5**(3), 197–233 (2008)
7. Weber, G.-W., Alparslan Gök, S.Z., Söyler, B.: A new mathematical approach in environmental and life sciences: gene-environment networks and their dynamics. Environ. Model. Assess. **14**(2), 267—288 (2009)
8. Weber, G.-W., Branzei, R., Alparslan-Gök, S.Z.: On cooperative ellipsoidal games. In: Proceedings of 24th MEC-EurOPT 2010 Mini EURO Conference: Continuous Optimization and Information-Based Technologies in the Financial Sector, Izmir-Turkey, 23–26 June 2010
9. Weber, G.-W., Branzei, R., Alparslan-Gök, S.Z.: On the ellipsoidal core for cooperative games under ellipsoidal uncertainty. In: Proceedings of 2nd International Conference on Engineering Optimization, Lisbon, Portugal, September 6–9 2010

Chapter 18
Zeta Functions and Continuous Time Dynamics

Paolo Giulietti

18.1 Introduction

In the following we will address the following questions: "What is it known about dynamical zeta functions for continuous dynamical systems? Why do we care at all about zeta functions?"

As a brief comment to the skeptical reader, zeta functions are interesting for their ability to encode statistical properties of physical systems, such as the abundance and distribution of orbits. For example, in [25, 36], it is shown that the study of the scattering of a point particle on a plane arising from the presence of three hard disks can be carried out using considering orbits and zeta functions; moreover the scattering poles are efficiently computed.

While in the last 40 years there were some advances in understanding zeta functions as invariants of dynamical systems, the search for a unifying framework is still not concluded; however, we can now recognize few recurrent patterns.

This survey takes firmly the continuous dynamic viewpoint, close in spirit to [71]. In fact, if the reader is more interested in the discrete case, he can read through the many excellent introductions available [11, 12, 80] and follow the references contained therein.

Proofs are omitted and the exposition is limited to outline roughly the ideas involved; we hope that a casual reader will appreciate the elegance of the ideas, while those who are already acquainted with the techniques can use this survey as a reference to original papers.

Most of the definitions are grouped in Sect. 18.2, while few others are introduced along the related results. In Sects. 18.3, 18.4 and 18.5, we present, mostly in a

P. Giulietti (✉)
Paolo Giulietti Universidade Federal do Rio Grande do Sul, Porto Alegre, RS, Brasil
e-mail: paologiulietti.math@gmail.com

A.A. Pinto and D. Zilberman (eds.), *Modeling, Dynamics, Optimization and Bioeconomics I*, Springer Proceedings in Mathematics & Statistics 73, DOI 10.1007/978-3-319-04849-9__18,
© Springer International Publishing Switzerland 2014

time-line fashion, available results. We do not claim to have collected exhaustively all the available results, the choices made reflect the author[1] taste.

18.2   General Framework

A continuous dynamical systems is, to the extent of this presentation, an action of $\mathbb{R}$ on a smooth manifold[2] M; we will denote our flows by ϕ_t.

There are three main ideas which are central to our study: that of an orbit of a dynamical system, that of correlation between observables and that of entropy. Let us begin by fixing some notation.

Definition 18.1. τ will be an orbit, τ_p be the prime orbit associated to it. $\lambda(\tau)$ will indicate the length of a orbit; $\mu(\tau)$ is its multiplicity with respect to its prime orbit so that $\lambda(\tau) = \mu(\tau)\lambda(\tau_p)$. $\mathcal{T}$ will indicate the set of orbits and $\mathcal{T}_p$ will indicate the set of prime orbits. γ will be reserved to closed geodesics.

Definition 18.2. If ϕ_t preserves a measure μ we can define the correlation function for two observables f, g in some reasonable class as

$$\rho_{f,g}(t) \doteq \int_M f(\phi_t(x))g(x)d\mu(x) - \left(\int_M f(x)d\mu(x) \right) \cdot \left(\int_M g(x)d\mu(x) \right).$$

(18.1)

Note that choosing the "most adequate" class of observables will always be a critical part of our work.

If $\rho(t) \to 0$ for $t \to \infty$ we say that the flow is mixing. The flow is weak mixing if the equation $f \circ \phi_t = e^{iat} f$ has no solutions for $f \in C^1(M, \mathbb{C})$ and $a > 0$.[3] Given a parametric Banach norm $\| \cdot \|_\alpha, \alpha \in (0, 1)$ for our class of observables, if there exists $C_\alpha > 0$ and $\sigma_\alpha > 0$ such that $|\rho_t(x)| \leq \|f\|_\alpha \|g\|_\alpha e^{-\sigma_\alpha t}$ we say that the flow is exponentially mixing. Last note that, given at least a weak-mixing flow, its Fourier transform is well defined and we set

$$\hat{\rho}(\omega) \doteq \int_{-\infty}^{+\infty} e^{i\omega t} \rho(t)dt,$$

(18.2)

whose poles are called "resonances".

Definition 18.3. Let X be a nonempty compact metric space with its distance $d(\cdot, \cdot)$ and T a continuous map on X. A set $A \subseteq X$ is said to be (n, ϵ)-separated, if for

---

[1] Arguably bad

[2] The attention here is restricted to smooth manifolds, but in some occasions one could relax such hypothesis or deal with a compact metric space X.

[3] Note that if $a = 0$, by transitivity, any solution f must be constant.

all $x, y \in A$ with $x \neq y$ the $d(T^i x, T^i y) \geq \epsilon$ for some $i < n$. Let $\omega(n, \epsilon)$ be the maximal cardinality of a (n, ϵ)-separated set in X. The topological entropy is defined as

$$h_{\text{top}}(T) \doteq \lim_{\varepsilon \to 0} \limsup_{n \to \infty} \frac{\log(\omega(n, \epsilon))}{n}.$$

By $h_{\text{top}}(\phi_1)$ we mean the topological entropy of the flow at time one (see [65] for more details). For a discussion on equivalent definitions of topological entropy see Bowen [18]. Conjugated to the notion of topological entropy there is, obviously, the notion of measure of maximal entropy, for the definition and the relation with topological entropy see [57–59].

Remark 18.1. From now on C is a running constant, in particular it could change value even within the same sentence. C_a is a running constant, dependent from the parameter a. Numbered constants C_i, $i \in \mathbb{N}$, have fixed values.

18.3   The Prototype

Given a surface M of nonpositive sectional curvature one is led to ask which are the properties of the geodesic flow that lives on it. The study of the geodesic flow steams from the physical perspective that such flow describe the motion of a particle constrained on the surface when no other force acts on it (for an elegant and extensive introduction to geodesic flows as physically interesting objects see [4]).

In 1956 Selberg [84] highlighted the relation between the objects we are interested in. Let M be a Riemann surface of curvature $k = -1$ and let

$$\zeta_{\text{Selberg}}(s) \doteq \prod_{\gamma} \prod_{k=0}^{\infty} \left(1 - e^{-(s+k)\lambda(\gamma)}\right). \tag{18.3}$$

Selberg was able to study its properties by using the trace formula which bears his name and the following elegant strategy. First he showed that each factor $\prod_{k=0}^{\infty} \left(1 - e^{-(s+k)\lambda(\gamma)}\right)$ converges for $s \in \mathbb{C}$ sufficiently large. Since formally we have

$$\frac{d}{ds} \log \left(\zeta_{\text{Selberg}}(s)\right) = \sum_{\gamma} \sum_{k=0}^{\infty} \frac{d}{ds} \log \left(1 - e^{-(s+k)\lambda(\gamma)}\right) = \sum_{\gamma} \sum_{k=0}^{\infty} \frac{\lambda(\gamma)}{1 - e^{-(s+k)\lambda(\gamma)}}, \tag{18.4}$$

we can proceed to interpret the rightmost element as following. Let Δ be the Laplacian of the surface, let $\{\lambda_j\}_{j \in \mathbb{N}}$ be its positive discrete spectrum, and let

$\phi_j \in C^\infty(\Gamma/\mathbb{H}^2)$ the related eigenfunctions. The set $\{\phi_j\}$ is a basis of $L^2(\Gamma/\mathbb{H}^2)$ and satisfies the equation

$$(\Delta + \lambda_j)\phi_j = 0.$$

We let $\rho_j \doteq \sqrt{\lambda_j - \frac{1}{4}}$ for $-\pi/2 \le \arg \rho_j < \pi/2$, then

Theorem 18.1 (Selberg [84]). *Let $h : \mathbb{C} \to \mathbb{C}$ be a suitable test function such that $h(s) = h(-s)$, $h(s)$ is holomorphic in a strip $\Im(s) \le \frac{1}{2} + \epsilon$ for $\epsilon > 0$ and $|h(s)| \le a(1 + |s|^2)^{-1-\epsilon}$ for some $a > 0$. Then*

$$\sum_{j=0}^{\infty} h(\rho_j) - \frac{Area(M)}{4\pi} \int_{-\infty}^{\infty} h(\rho)\tanh(\pi\rho)\rho d\rho = \sum_{\gamma \in \mathcal{T}_p} \sum_{n=1}^{\infty} \frac{\lambda(\gamma)\widehat{n\lambda(\gamma)}}{2\sinh(n\lambda(\gamma)/2)}$$

$$(18.5)$$

where $\hat{f}$ indicates the Fourier transform of f. Thus while keeping some freedom in the choice of the function h, by using the right hand sides of Eqs. (18.5) and (18.4) we establish a straightforward connection between the set of orbit and the set of eigenvalues.

Different h can be used to achieve different purposes: here given a parameter $k \in \mathbb{R}$ we let $h_k(\rho_j) = (k^2 - \rho_j^2)^{-1}$ and by doing so we formally obtain

$$\sum_{j=0}^{\infty} h_{\lambda^2 + \frac{1}{4}}(\rho_j) = \sum_{j=0}^{\infty} (\lambda - \lambda_j)^{-1} = \text{Tr}\left((\Delta + \lambda)^{-1}\right). \qquad (18.6)$$

Thus, we can study the property of $\zeta_{Selberg}$ by studying the trace of the resolvent of the Laplacian, which can be shown, in a regularized sense, to extend to the whole $\mathbb{C}$. Selberg also proved that

Theorem 18.2 (Selberg [84]). *$\zeta_{Selberg}$ converges absolutely for $\Re(s) > 1$ and can be analytically extended to an entire function on the whole $\mathbb{C}$. Moreover $\zeta_{Selberg}$ has non-trivial zero precisely at $s = h_{top}(\phi_1) = 1$ and at $s = \frac{1}{2} \pm i\rho_j$ where ρ_j are as above. Last $\zeta_{Selberg}$ satisfies the following functional equation*

$$\zeta_{Selberg}(s) = \zeta_{Selberg}(1 - s)\exp\left(Area(M)\int_0^{s-\frac{1}{2}} u\tan(\pi u)du\right).$$

Thus the functional equation allows us to compute the values of $\zeta_{Selberg}$ at any point, by first computing the values in the region of granted convergence. Next the region free of zeros grants the following result on the distribution of orbits based on a classical tauberian theorem approach. In fact, with such result at hand, one obtains (see [43]) that the number $\Pi(L)$ of closed geodesics of length less than L, $L \to \infty$, satisfies

$$\Pi(L) = \sum_{j=0}^{\bar{j}} \int_1^L dt \frac{e^{(1/2+i\rho_j)t}}{t} + \mathcal{O}\left(\frac{e^{\frac{3}{4}L}}{\sqrt{L}}\right)$$

where $\bar{j}$ is such that $\rho_0, \ldots \rho_{\bar{j}}$ satisfy $\Im(\rho_j) < 0$.

Later results shows that ζ_{Selberg} can be decomposed into a product formula over its zeros, as it is usually done in number theory with the Hadamard product. In the development of such formulas the main ingredients are the study of a functional determinants, which encode the spectrum of the Laplacian, and the employment of Barnes double gamma function [83, 94, 95].

See [60] for an overall presentation of the Selberg trace formula and its implications.

18.4 Anosov Flows

It is important to recall that geodesic flows on a surface of constant negative curvature are ergodic, mixing, and have been the model out of which Anosov flows [75, 86] were defined.

Definition 18.4. An Anosov flow is a flow such that there exists a $D\phi_t$-invariant continuous splitting $TM = E^0 \oplus E^s \oplus E^u$, constants $C_1 > 1$ and $\lambda > 0$, such that for $t \geq 0$, E^0 is the one-dimensional subspace tangent to the flow and for all $x \in M$

$$\|D\phi_t(v)\| \leq C_1\|v\|e^{-\lambda t} \text{ if } t \geq 0, v \in E^s$$
$$\|D\phi_{-t}(v)\| \leq C_1\|v\|e^{-\lambda t} \text{ if } t \geq 0, v \in E^u.$$

For such flows it holds

Theorem 18.3 (Anosov [3]). *An Anosov flow has only one of the following two properties, either it has a strong stable and strong unstable manifold everywhere dense or it is a suspension of an Anosov diffeomorphism by a constant roof function.*

Note that the geodesic flow on a surface falls in the first case. Moreover, note that, in the second case even a smooth perturbation[4] forces the perturbed flow to fall into the first class.

There was a natural attempt to generalize Selberg's results by considering Anosov flows on a manifold of variable negative curvature. However such results were not attainable at first, due to the fact that the trace formula relies critically on the constant curvature assumption and on the means of representation theory.

---

[4] For example the perturbation introduced by a smooth reparametrization obtained by choosing two points which have irrational ratio.

However, one is naturally led to ask if by replacing some of the elements involved in such scheme one could obtain the same cascade of results.

In 1976, Ruelle [76] proposed a dynamical zeta function that reads as

$$\zeta_{\text{Ruelle}}(z) \doteq \prod_{\tau \in T_p} \left(1 - e^{-z\lambda(\tau)}\right)^{-1}, z \in \mathbb{C}. \tag{18.7}$$

Since, at the time, Margulis already showed that

Theorem 18.4 (Margulis [57–59]). *Given a geodesic flow on a surface of negative curvature one has for $C_2 > 0$ and $L > 0$*

$$\#\{\tau \in \mathcal{T} : \lambda(\tau) < L\} \sim \frac{e^{C_2 L}}{C_2 L},$$

by using the properties of the horocycle foliations (improving a previous result by Sinai [87]), Ruelle zeta is convergent for z sufficiently large in the same geometrical setting. An alternative proof which also showed that the constant of Theorem 18.4 is exactly $h_{\text{top}}(\phi_1)$ was found later on by Parry and Pollicott [64,65] and it is based on transfer operators and symbolic coding.

The definition (18.7) arose in Ruelle work in a natural way given the works of Lang [51], Artin and Mazur [5], Smale [88] which all dealt, in different frameworks, with counting periodic points for a discrete transformation. Moreover note that whenever ζ_{Ruelle} and ζ_{Selberg} are both defined we coherently have

$$\zeta_{\text{Ruelle}}(z) = \frac{\zeta_{\text{Selberg}}(z+1)}{\zeta_{\text{Selberg}}(z)} \quad , \quad \zeta_{\text{Selberg}}(z) = \prod_{i=0}^{\infty} \zeta_{\text{Ruelle}}(z+i)^{-1}.$$

First of all we want to highlight that, while for the geodesic flow the meromorphic extension of (18.7) is granted to the whole $\mathbb{C}$, in the context of flows of limited regularity, it is not the case. Such result is coherent with examples by Gallavotti [35], Ruelle [77] and Pollicott [67], where one finds an essential singularity for the suspension flows constructed with an arbitrarily slow rates of decay correlations.

Thus, following the strategy of Selberg, we look for a suitable related operator. The connection between our objects is highlighted in the following formal computations. Begin by noticing that we can turn the product formula into a sum formula via

$$\zeta_{\text{Ruelle}}(z) = \prod_{\tau \in \mathcal{T}_p} \left(1 - e^{-z\lambda(\tau)}\right)^{-1} = \exp\left(\sum_{\tau \in \mathcal{T}_p} \sum_{m=1}^{\infty} \frac{1}{m} e^{-zm\lambda(\tau)}\right)$$

$$= \exp\left(\sum_{\tau \in \mathcal{T}} \frac{1}{\mu(\tau)} e^{-z\lambda(\tau)}\right). \tag{18.8}$$

Then, following Ruelle, we use the linear algebra identity for the linear finite dimensional operator A given by

$$\det(\mathbb{1} - A) = \sum_{\ell=0}^{\dim A} (-1)^\ell \, \mathrm{tr}(\wedge^\ell A).$$

Thus, as long as $\det(\mathbb{1} - A) \neq 0$, we obtain

$$\zeta_{\text{Ruelle}}(z) = \exp\left(\sum_{\ell=0}^{\dim A} \sum_{\tau \in \mathcal{T}} \frac{1}{\mu(\tau)} \frac{(-1)^\ell \, \mathrm{tr}(\wedge^\ell A)}{\det(\mathbb{1} - A)} e^{-z\lambda(\tau)} \right). \qquad (18.9)$$

On the other hand we define the operator

Definition 18.5. Let $g : M \to R$ a continuous weight function. Let f be a function in C^r (even C^∞). Then we define a family of transfer operators as

$$\mathcal{L}_{t,g} f(x) \doteq g(x) f(\phi_{-t}(x)).$$

Thus we define its resolvent as

$$R(z)f = \int_0^\infty dt \, e^{-zt} \mathcal{L}_{t,g}(f).$$

Such resolvent is in general not in trace class, however it can be constructed a regularized trace for which holds

$$\mathrm{Trace}\,(R) \sim \sum_{i=0}^\infty (1 - \lambda_{g,i})^{-1} \qquad (18.10)$$

where $\lambda_{g,i}$ are the eigenvalues of the transfer operator on a suitable functional space. Thus we formally write a trace for the resolvent as

$$\int_M dx\, \delta(x)(R(z)\delta(x)) = \int_M dx\, \delta(x) \int_0^\infty dt\, e^{-zt} g(x)\delta(\phi_{-t}(x)) \qquad (18.11)$$

where we forced $R(z)$ to act on δ as if it was a function, implying that we used some regularization scheme.

Next we can decompose the space M in its orbits by considering $\delta(x - \phi_{-t}(x))$, where δ is the usual delta function. Thus if $U_\tau \subset M$ is a sufficiently small neighborhood of a single orbit we obtain

$$\int_{U_\tau} dx\, g(x)\delta(x - \phi_{-t}(x)) = \frac{g_\tau}{|\det(\mathbb{1} - D_\tau \phi_{-\lambda(\tau)})|}$$

if g depends only on the orbit and not on a specific point. Thus we can write

$$\int_M dx \int_0^\infty dt \, e^{-zt} g(x)\delta(x - \phi_{-t}(x)) = \sum_{\tau \in \mathscr{T}_p} \frac{e^{-z\lambda(\tau)}g_\tau}{|\det(\mathbb{1} - D_\tau \phi_{-\lambda(\tau)})|}. \tag{18.12}$$

which can be used[5] with (18.10) to provide at the same time a representation and an estimate of the trace of the resolvent.

Thus by using Eq. (18.12) in (18.9) we can obtain (18.8), provided the g-s can be chosen for each ℓ to compensate for the weights induced by exterior algebra. Such regularized traces have been used before, and are traditionally called "flat traces", they are analogous of those obtained by Atiyah-Bott [6–8] and thus inherited the same nomenclature.

Hence the core of the work lies in giving mean to such flat trace and to define a space of function on which the resolvent is (at least) quasi-compact, i.e. its spectrum is given by an isolated eigenvalue at one, a spectral gap, some point spectrum and an essential spectrum with radius strictly smaller then one. Such strategy has been fruitful in a variety of situations.

In the case of expanding flows with real analytic stable and unstable foliations[6] Ruelle already showed that ζ_{Ruelle} has a meromorphic extension to $\mathbb{C}$ and that it is the quotient between two entire functions d_1, d_2 such that $|d_i(z) - 1| \leq e^{-C\Re(z)}$. He was able to do so by coding the flow by a Markov partition à la Bowen [19] and then studying the action of a suitable transfer operator on Hölder observables. In the case of expanding semiflows (constructed as suspension of expanding maps, without the analytic assumption) Ruelle [79] later showed that if one starts with[7] $f, r \in C^{(k,\alpha)}$, where $f : M \to M$ is an expanding map and $r : M \to \mathbb{R}^+$ is a roof function, ζ_{Ruelle} is meromorphic in the half-plane $\Re(s) > \eta$ where $\eta < h_{\text{top}}(\phi_1)$ is the unique number such that $P(f, -\eta \cdot r) = \log \theta^{-(k+\alpha)}$ where P is the topological pressure and θ is the expansion coefficient. In such context the extension is optimal (see [10]).

The meromorphic extension to $\mathbb{C}$ was obtained by Rugh [81], for generic smooth Anosov flows on three dimensional manifolds, and later by Fried [33,34] in arbitrary dimension but still assuming analyticity of the flow.

In a series of paper by Parry and Pollicott [64,68,69] it was proved that for weak mixing Anosov flows, $\zeta_{\text{Ruelle}}(z)$ is analytic and non zero for $\Re(z) \geq h_{\text{top}}(\phi_1)$ apart for a single pole at $z = h_{\text{top}}(\phi_1)$. Moreover if ϕ_t is topologically weak mixing then

---

[5]Modulo few nontrivial caveats tied to the orientation of E_s along orbits, the choice of a suitable g, the choice of how approximate the δ.

[6] Recall that by real analytic function in a neighborhood of a point we mean that in such neighborhood the function is infinitely many time differentiable and the Taylor expansion of $f(x)$ converges to $f(x)$. Thus, a real-analytic manifold is a manifold such that the charts are real-analytic and a real-analytic foliation is a foliation such that the map from each leaf to $\mathbb{R}^n$ is real-analytic.

[7] $C^{(k,\alpha)}$ being the class of function k-times differentiable such that the k-th derivative has Hölder exponent α.

there are no other poles on the line $\Re(s) = h_{\text{top}}(\phi_1)$. On the other hand if ϕ_t is not topologically weak mixing then it has poles at $s = h_{\text{top}}(\phi_1) + iak, k \in \mathbb{Z}, a \in \mathbb{R}$. The monograph [65] contains all the details and the references on such results. At the moment the best available result is the following

Theorem 18.5 (Giulietti-Liverani-Pollicott [37]). *Given a compact, connected and orientable $\mathscr{C}^\infty$ Riemannian manifold M, for any $\mathscr{C}^r$ Anosov flow ϕ_t with $r > 2$, $\zeta_{Ruelle}(z)$ is meromorphic in the region*

$$\Re(z) > h_{\text{top}}(\phi_1) - \frac{\lambda}{2} \left\lfloor \frac{r-1}{2} \right\rfloor$$

where λ is determined by the Anosov splitting, and $\lfloor x \rfloor$ denotes the integer part of x. Moreover, $\zeta_{Ruelle}(z)$ is analytic for $\Re(z) > h_{top}(\phi_1)$ and non zero for $\Re(z) > \max\{h_{\text{top}}(\phi_1) - \frac{\lambda}{2}\lfloor\frac{r-1}{2}\rfloor, h_{\text{top}}(\phi_1) - \lambda\}$. If the flow is topologically mixing then $\zeta_{Ruelle}(z)$ has no poles on the line $\{h_{top}(\phi_1) + ib\}_{b\in\mathbb{R}}$ apart from a single simple pole at $z = h_{top}(\phi_1)$.

Corollary 18.1. *For any $\mathscr{C}^\infty$ Anosov flow the zeta function $\zeta_{Ruelle}(z)$ is meromorphic in the entire complex plane.*

The whole strategy relied, as in the original Ruelle [76] paper, on studying a generalization of transfer operators when applied to forms. The ability of disposing of Markov coding, thus of the analyticity hypothesis, is obtained by choosing suitable Banach spaces of forms, on which one could study the resolvent of the generator of the flow, refining what was done before in a variety of (discrete or continuous) situations [15–17, 20, 38, 39, 50, 53, 54]. The second part, i.e. relating "flat traces" to transfer operators, also implements ideas presents in the literature [9, 13, 16, 55]. It is likely that "another" combination of the above techniques could produce optimal results, at the price of lengthier computations.

Much less, despite the effort, is known about finer properties of ζ_{Ruelle}, such as precise location of zeros and poles; nevertheless some estimates are available in specific cases. Recall that such estimates, by means of well established tauberian theorems, readily give results for the number of orbits of a given length.

In 1998, Chernov [21] showed that the rate of mixing for sufficiently regular Anosov flows on three dimensional manifolds is, at worst, stretched exponentially. Dolgopyat [15], proved exponential mixing for $C^{2+\varepsilon}$ weak-mixing Anosov ow with C^1 stable and unstable foliations. Dolgopyat, as Chernov before him, did not translate his papers into the language of zeta functions. By having exponential decay at hand, one obtains that a dynamical zeta function is analytic, except as usual at $h_{\text{top}}(\phi_1)$, in a region determined by $|\Re(z) - h_{\text{top}}| \leq |\Im(z)|^{-c}$ for some $c > 0$ i.e. there is a small strip free of zeros at the left of the half plane of convergence. Thus, by tauberian theorems, on a negatively curved Riemannian surface [27, 72] we have

$$\pi(L) = \text{li}(e^{h_{\text{top}}(\phi_1)L}) + O(e^{c_3 L}) \tag{18.13}$$

where $c_3 < h_{top}(\phi_1)$, $li(x) = \int_2^x (\ln(s))^{-1} ds$. For weak-mixing transitive [73], the error term estimate was later improved, for flows which satisfy a really weak Diophantine condition, since it was proven that there exists a $\delta > 0$ such that

$$\pi(L) = \frac{e^{h_{top}(\phi_1)L}}{h_{top}(\phi_1)L} \left(1 + \mathcal{O}\left(\frac{1}{T^\delta}\right)\right).$$

Also there are some available results on resonances (see Eq. (18.2)). For real analytic flows, Ruelle [78] showed that $\hat{\rho}$ is meromorphic in a strip $|\Im\omega| < \delta$ and the related resonances and their residues can be understood as some special Gibbs distribution. Note that such poles are often computable [24, Chap. 17]. In the context of real analytic suspension semiflows over uniformly expanding real-analytic map of the interval of Naud [63] proves that $\hat{\rho}$ extends meromorphically to the whole complex plane and that one can find infinitely many resonances in a strip $\{-2h-\epsilon \leq \Re(z) \leq 0\}$, where h is the measure entropy with respect to an equilibrium measure and a chosen potential.

In the analytic framework, it is known that both the topological entropy and the metric entropy[8] vary under an analytical perturbation in a real analytical sense [48, 70]. Such phenomena is captured by ζ_{Ruelle}, and is reflected by the location of the entropy pole, which moves accordingly to the perturbation.

Remark 18.2 (Contact Anosov flows). The estimate (18.13) is available not only when the horocycle foliation is $\mathcal{C}^1$ [72, 90] but also for weaker conditions, and it is conjectured to hold for all contact Anosov flows. Note that the first paper on Anosov flows where one finds exponential decay of correlation, without requiring the extra regularity of foliations, restricts the attention to contact Anosov flow [53]. In this sense the best estimate available at this time is obtained in [37], following and adapting strategies developed elsewhere [14, 52–54]; a Dolgopyat type estimate is established for d_s-forms, where d_s is the dimension of the strong stable manifold. The extension of the geometric part of the original Dolgopyat argument to the framework of [37] required a pinching condition on top of the contact requirement, to counteract for the fact that, morally integration is performed with respect to the measure of maximal entropy. It is likely that recent work on contact flows, which allowed to study directly the transfer operator associated to the time one map of the flow [30, 31, 91–93] could allow a generalization of such estimate.

Remark 18.3 (L-functions). The importance of studying generalization of zeta functions, such as L-functions, it is clear in the paper of Katsuda and Sunada [49]. In fact, one construct a complex function associated to a unitary character $\chi : H_1(X, \mathbb{Z}) \to U(1)$ as

---

[8] The metric entropy is defined as $h(\phi, \mu) = \int E^u(x) d\mu(x)$ where μ is the SRB measure related of ϕ_t. See [98] for an introduction to SRB measures.

$$L(s, \chi) = \prod_{\tau_p} \left(1 - \chi([\tau_p]) e^{-s\lambda(\tau_p)}\right)^{-1}.$$

Such definition is natural if one wants to keep track of the homological distribution of orbits, in fact, Katsuda and Sunada prove an equidistribution theorem by perturbation, by carefully studying the location of zeros and poles, near the real axis and near the line defined by $\Re(z) = h_{\text{top}}(\phi_1)$.

Remark 18.4 (Torsion). It is possible to connect the geodesic flows and torsion, either à la Reidemeister or à la Ray-Singer, since they are known to be equivalent. Fried [32] shows that the value of $\zeta_{\text{Ruelle}}(0)$ is equivalent to the value of torsion for a closed oriented hyperbolic manifold.

Following the idea of exploiting analytic flows, and the results of Rugh and Fried, Morgado [82] proves a stronger version of the theorem which relates the torsion with the zeros of the zeta function already obtained by Fried. With this new approach he is able to get rid of the requirement of the extra regularity on the foliations.

Theorem 18.6 (Morgado [82], Fried [32]). *Let ϕ_t be an analytic transitive Anosov flow on an orientable closed 3-manifold M. Let $\rho : \pi_1(M) \to U(n)$ be an acyclic representation. Suppose there is a periodic orbit τ such that 1 and the holonomy of τ are not eigenvalues of $\rho(\tau)$. Then the L-function is regular at the origin and $\text{Torsion}_\rho(M) = |L_{\phi,\rho}(0)|$.*

Anantharaman [1] uses a dynamical zeta to construct an asymptotic expansion of the functions which counts closed geodesics under cohomological constraint on surfaces of negative curvature, she is able to do so by using the result of Dolgopyat and Chernov-Dolgopyat on standard pairs.

Let $\xi \in H_1(M, \mathbb{R})$, $\alpha \in H_1(M, \mathbb{Z})$ and $\delta > 0$. Let the "integral part" map be $[\cdot] : H_1(M, \mathbb{R}) \to H_1(M, \mathbb{Z})$ and let

$$\pi(\xi, \alpha, \delta, T) \doteq \{\gamma, T \leq \lambda(\gamma) \leq T + \delta, [\gamma] = \alpha + [T\xi]\}.$$

Let $H : \xi \to \sup_{m \in [\xi]} h(m)$ i.e. H maps the element ξ to the supremum of the metric entropy for the winding cycles in the class of ξ. Then we have

Theorem 18.7 (Anantharaman [1]). *Let M be a three dimensional manifold and consider an Anosov flow on it. Suppose moreover that the characteristic foliations are of class C^1 and uniformly jointly non integrable. Then there are analytic functions $c_n, n \in \mathbb{N}, n \neq 0$ in $D \times \mathbb{R}^{d+1}$ such that for a $\delta > 0$*

$$\pi(\xi, \alpha, \delta, T) = \frac{e^{TH(\xi^T) - \langle \nabla H | \alpha \rangle}}{T^{d/2+1}} \left(c_0(\xi^T, \delta) + \sum_{k=i}^{N} \frac{c_k(\xi^T, \alpha, \delta)}{T^k} + \mathcal{O}(T^{1-n}) \right).$$

(18.14)

Note that the requirements over the foliations are coherent with the Anosov alternative and necessary to the proof. The strategy of the proof requires mainly two ingredient. First by using the regularity of the foliations to ensure the regularity of the partitions as Dolgopyat does, the proofs exploit the cancellation effect. Second a zeta function is introduced naturally as the inverse of the Laplace transform both for $\lambda(\gamma) - T$ and for the homology class of γ. Stretching further these ideas, in [2] the authors show that there is relation between Wigner distributions and Patterson-Sullivan distributions i.e. residues of weighted dynamical zeta functions for a compact hyperbolic surface. In fact they show that such distributions are asymptotically the same, in the region where they are both defined, on any line parallel to the imaginary axis.

Remark 18.5 (Algebraic flavor). On a different direction, Mayer develops further the algebraic side of the problem, in particular we find that given the Gauss map $G(x) = x^{-1} \mod 1$ on the unit interval and the geodesic flow of a modular surface[9] one has the following theorem

Theorem 18.8 (Mayer [61]). *$\zeta_{Selberg}$ for the geodesic flow related to the modular group $PSL(2, \mathbb{Z})$ can be written as $\zeta_{Selberg}(z) = \det(1 - L_z) \det(1 + L_z)$ with L_z the transfer operator of the Gauss map. Moreover $\zeta_{Selberg}$ is meromorphic in the entire complex plane with (possibly removable) singularities at the points $z_k = (1 - k)/2$, for $k \in \mathbb{N}$.*

18.5   Semi-Classical Approximation

Zeta functions have proven to be quite successful in the study of semiclassical regimes, i.e when one can relate periodic orbits in a classical system to the energy levels of the corresponding quantum system. If the underlying geometry is $\mathbb{R}^n$ or $\mathbb{S}^n$, the solutions of the wave equation

$$\frac{\partial^2 u}{\partial t^2} = \Delta u$$

for suitable boundary conditions and a class of functions u which satisfy rapid decay are well understood. Such solutions are recovered by first reducing the problem, through a change of variable, to a pair of transport equations. The properties of Δu are then studied through the resolvent of the associated semigroup. Note that if one is given a sufficiently regular potential V, the solutions of $(\Delta + V)u = \lambda u$ are obtained through the analysis of its scattering matrix.[10] Such problem becomes extremely complex if we move to richer geometries, for example by studying spaces of negative curvatures or by adding obstacles throughout the space. In such cases the

---

[9]The surface of constant negative curvature constructed from a modular group.

[10] It is an operator. Such nomenclature comes from the one-dimensional framework, where it is actually the refraction/transmission matrix associated to the potential.

scattering matrix should extends to a meromorphic operator on a wider region, and it is a widespread feeling that the poles of the scattering matrix should approximately determine the singularities of a "semiclassical" ζ_{SM}. The explicit expression of ζ_{SM} is usually fairly complex (compared to (18.7) or (18.3)), we will show later on an example of it.

It is common for zeta functions to have both sum and products representations (for example as we have seen in Eq. (18.8)) and as a matter of fact is often easier to study the properties of the elements of the sum. Nevertheless it is usual to study the "determinant"

$$\mathscr{D}_{SM}(s) \doteq \sum_{\tau \in \mathscr{T}} \frac{S(\tau) e^{-s\lambda(\tau)}}{|\det(I - P_\tau)|^{\frac{1}{2}}} \tag{18.15}$$

where $S(\tau)$ embeds the properties of the system studied (for example the geometry of the boundary or the geometry of the scatterers) and P_τ is a representation of the Poincaré return map along the orbit τ. Note that the usual $\det(I - P_\tau)^{-1}$ in expressions like (18.9), is replaced here by $\det(I - P_\tau)^{-1/2}$ to compensate for the fact that we are dealing with a wave amplitude, as can be seen from formal computations similar to those of Eqs. (18.11) and (18.12) applied to waves.

A stepping stone to understand the properties of (18.15) is the Gutzwiller-Voros trace formula [40, 96], which has the ability of tyeing eigenenergies of the wave operator to $\mathscr{D}_{SM}$; such trace is formally similar to that of Selberg (18.5).

Thus, one is allowed to think of zeta functions as a good tool to numerically compute eigenenergies, though only in an aggregate manner, since such computations provide a good degree of accuracy even when truncated and evaluated through a small number of orbits.

Recall that, generically, the study of a billiard is the study of the trajectory of particles with given energy in a limited region, such that a particle obeys to the laws of optical reflections whenever it encounters an obstacle (either a scatterer or the border of its space).

In a series of a papers Harayama, Shudo and Tasaki [41,42,85] studied dispersing strongly chaotic billiards.[11] In such setting it is possible to define a Fredholm determinant[12] starting from the boundary element method. Here the authors are able to show that if one chooses the symbolic dynamic associated to the flow in a opportune manner, then such determinant agrees completely with the zeta function defined by Gutzwiller-Voros. If we enrich the geometry with several strictly convex obstacles, Ikawa [44–47] shows that $\mathscr{D}_{SM}$ with an opportune weight gives rises to the expected zeta function which has zeros and poles directly related to that of the scattering matrix for the related perturbed symbolic flows. Stoyanov [89] is able to

---

[11] By dispersing we mean that the boundary is strictly concave inward at every smooth point of the boundary. By strongly chaotic we mean hyperbolic, ergodic, mixing, and Bernoulli.

[12] In the sense of defining $\det(I + A)$ by setting $\det(I + A) \doteq \sum_0^\infty \operatorname{Tr}\Lambda^k(A)$.

show that on $\mathbb{R}^3$, under suitable hypothesis on the scatterers, ζ_{SM} shows an infinite number of poles on a strip near the real axis.

For dispersing billiards on $\mathbb{R}^3$, Dahlqvist [26] approximates the zeta functions by determinants with different weights and shows that the related traces can be dominated by isolated zeros or by the continuous spectra. He also notes that there is a phase transition between exponential decay and polynomial decay (reflected by the properties of the zeta functions) for different values of the largest eigenvalue of the weighted operator. His scheme seems numerically stable, and it can be used to compute topological entropy and other dynamical features, such as Lyapunov exponents or rate of decay of correlations.

In Naud [62], we find that for an open billiard flow in $\mathbb{R}^3$ there is a generic Diophantine condition[13] which grants an analytic extension of ζ_{SM} on a strip to the left of topological entropy of width polynomially decreasing. In particular assume ϕ_t has two primitive orbits $\tau_{p,1}$ and $\tau_{p,2}$ such that $\frac{\lambda(\tau_{p,1})}{\lambda(\tau_{p,2})}$ is a Diophantine number and recall that the condition on the orbits is met with ease since can be deduced if we have three obstacles such that the relative distances are Diophantine. In this context, we define the two determinants

$$Z_d(s) = \sum_{\tau \in \mathcal{T}} \frac{(-1)^{m_\tau} \lambda(\tau_p) e^{-s\lambda(\tau)}}{|\det(I - P_\tau)|^{\frac{1}{2}}} \qquad Z_0(s) = \sum_{m=1}^{\infty} \sum_{\tau_p \in \mathcal{T}} (-1)^{m r_\tau} \lambda(\tau_p) e^{-s\lambda(\tau_p) + \delta_\tau}$$

where m_τ are the number of reflections, $r_\tau = 0$ if $\lambda(\tau_p)$ has an even number of reflections, 1 otherwise and $\delta_\tau = -\frac{1}{2} \log(e_1 e_2)$ where e_1 and e_2 are eigenvalues of P_τ. Thus he can set $Z_0(s) = -\zeta'_{SM}(s)$ and obtain

$$\zeta_{SM}(s) = \sum_{m=1}^{\infty} \frac{1}{m} \sum_{\sigma^m x = x} e^{-sf^m(x) + g^m(x) + i\pi m}$$

and he can show meromorphic continuation of such zeta by introducing symbolic dynamics into the billiard flow along with the irrationality condition. Moreover he proves that there exist $C, \rho > 0$ such that Z_d has an analytic continuation up to the domain $\{\sigma + it \in \mathbb{C} : |t| \geq 1, x_c - \frac{C}{|t|^\rho} \leq \sigma \leq x_c\}$. The strategy is to estimate the resolvent of the transfer operator using the regularity of the Gibbs measure, following of Dolgopyat [27], and then using the irrationality condition to prove that $(\xi I - \mathcal{L}_t)$ is invertible.

Last we want to mention that Petkov and Stoyanov [66] show that it is possible to compute an estimate on equidistribution of lengths of periodic orbits in no-eclipse billiards on the plane. No-eclipse billiards are billiards where the convex hull of any two scatterers has empty intersection with any other scatterer. Their results are close

---

[13] An irrational number $x \in \mathbb{R}$ is Diophantine if there exist $\nu > 0$ and $M > 0$ such that for all $(p, q) \in \mathbb{Z} \times \mathbb{N}^*$ we have $\left| x - \frac{p}{q} \right| > \frac{M}{q^{2+\nu}}$.

in spirits to the theoretical ones obtained by Pollicott and Sharp [74] for negatively curved surfaces. In the last mentioned paper the cancellation of oscillatory integrals plays a remarkable role in proving estimates concerning the distribution of pairs of closed geodesic on a compact surface of negative curvature whose difference in length is given.[14]

Remark 18.6 (Casimir Effect).   The Casimir effect, that is the attractive or repulsive effect observed when two neutral metallic plates are pulled very close to each other, can be computed from a quantum-mechanical billiard-type framework [97]. Its strength can be deduced from the appropriately weighted zeta function, by means of a trace formula for transfer operators. It is shown that one could apply such calculations to any length of given billiards, though good results are obtained only for medium to large separations of scatterers.

18.6   Conclusions

We hope that we do not need to give any other motivation to persuade the reader of the general interest of such questions. It is a pity that so far we have not been able to find a better framework which encloses all the situations above, and we hope that the above survey stimulated the reader curiosity. Nevertheless we would like to conclude this survey by the following extraordinary path. Connes [22, 23] suggests that one could use a suitable transfer operator to study the action of an opportune "Riemann flow". In fact, in his context he is able to define a suitable trace, similar to what we presented so far which coherently relies on what we called dynamical determinant. Next he guesses that if we could find a replacement for the standard Selberg trace formula for such Riemann flow, we could probably be on the right track to reformulate the Riemann hypothesis in dynamical terms.

Acknowledgements I would like to thank V. Baladi, C. Liverani and M. Tsujii for helpful discussions and comments along the years. I also thank the anonymous referee for pointing out a shameful quotation error. Partially supported by ERC Advanced Grant MALADY (246953) and CNPq Brazil.

References

1. Anantharaman, N.: Precise counting results for closed orbits of Anosov flows. Ann. Sci. École Norm. Sup. (4) **33**(1), 33–56 (2000)
2. Anantharaman, N., Zelditch, S.: Patterson-Sullivan distributions and quantum ergodicity. Ann. Henri Poincaré **8**(2), 361–426 (2007)

---

[14] One should think of this as the problem of describing the set of prime numbers such that $p, p+2$ are both prime.

3. Anosov, D.V.: Geodesic flows on Riemann manifolds with negative curvature. In: Proceedings of the Steklov Institute of Mathematics, No. 90. American Mathematical Society, Providence (1967). Translated from the Russian by S. Feder (1969)

4. Arnold, V.: Sur la géométrie différentielle des groupes de Lie de dimension infinie et ses applications à l'hydrodynamique des fluides parfaits. Ann. Inst. Fourier (Grenoble) 16(fasc. 1), 319–361 (1966)

5. Artin, M., Mazur, B.: On periodic points. Ann. Math. 81(2), 82–99 (1965)

6. Atiyah, M.F., Bott, R.: Notes on the Lefschetz fixed point formula for elliptic complexes, vol. 2. Bott's Collected Papers, Harvard University (1964)

7. Atiyah, M.F., Bott, R.: A Lefschetz fixed point formula for elliptic differential operators. Bull. Am. Math. Soc. 72, 245–250 (1966)

8. Atiyah, M.F., Bott, R.: A Lefschetz fixed point formula for elliptic complexes, I. Ann. Math 86, 374–407 (1967)

9. Baillif, M.: Kneading operators, sharp determinants, and weighted lefschetz zeta functions in higher dimensions. Duke Math. J. 124, 145–175 (2004)

10. Baladi, V.: Optimality of ruelle's bound for the domain of meromorphy of generalized zeta functions. Port. Math. 49, 69–83 (1992)

11. Baladi, V.: Positive transfer operators and decay of correlations. In: Advanced Series in Nonlinear Dynamics, vol. 16. World Scientific, River Edge (2000)

12. Baladi, V.: Dynamical zeta functions and kneading operators. http://www.math.ens.fr/~baladi/kyoto.ps (2002)

13. Baladi, V., Baillif, M.: Kneading determinants and spectra of transfer operators in higher dimensions, the isotropic case. Ergod. Theor Dyn. Syst. 25, 1437–1470 (2005)

14. Baladi, V., Liverani, C.: Exponential decay of correlations for piecewise cone hyperbolic contact flows. Commun. Math. Phys. 314(3), 689–773 (2012)

15. Baladi, V., Tsujii, M.: Anisotropic hölder and sobolev spaces for hyperbolic diffeomorphisms.c Annales de l'Institut Fourier, Grenoble 57, 127–154 (2007)

16. Baladi, V., Tsujii, M.: Dynamical determinants and spectrum for hyperbolic diffeomorphisms. In: Geometric and probabilistic structures in dynamics. Contemporary Mathematics series, vol. 469, pp. 29–68. American Mathematical Society, Providence (2008)

17. Blank, M., Keller, G., Liverani, C.: Ruelle-Perron-Frobenius spectrum for Anosov maps. Nonlinearity 15(6), 1905–1973 (2002)

18. Bowen, R.: Periodic points and measures for axiom A diffeomorphisms. Trans. Am. Math. Soc. 154, 377–397 (1971)

19. Bowen, R.: Symbolic dynamics for hyperbolic flows. Am. J. Math. 95, 429–460 (1973)

20. Butterley, O., Liverani, C.: Smooth Anosov flows: correlation spectra and stability. J. Mod. Dyn. 1(2), 301–322 (2007)

21. Chernov, N.I.: Markov approximations and decay of correlations for Anosov flows. Ann. Math. 147(2), 269–324 (1998)

22. Connes, A.: Noncommutative geometry and the riemann zeta function. http://www.alainconnes.org (1998)

23. Connes, A.: Trace formula in noncommutative geometry and the zeros of the riemann zeta function. http://www.alainconnes.org (1999)

24. Cvitanović, P., Artuso, R., Mainieri, R., Tanner, G., Vattay, G.: Chaos: Classical and Quantum. Niels Bohr Institute, Copenhagen (2009)

25. Cvitanovic, P., Eckhardt, B.: Periodic-orbit quantization of chaotic systems. Phys. Rev. Lett. 63(8), 823–826 (1989)

26. Dahlqvist, P.: Approximate zeta functions for the Sinai billiard and related systems. Nonlinearity 8(1), 11–28 (1995)

27. Dolgopyat, D.: On decay of correlations in Anosov flows. Ann. Math. 147, 357–390 (1998)

28. Dolgopyat, D.: Prevalence of rapid mixing for hyperbolic flows. Ergod. Theor Dyn. Syst. 18, 1097–1114 (1998)

29. Dolgopyat, D.: Prevalence of rapid mixing 2: topological prevalence. Ergod. Theor Dyn. Syst. 20, 1045–1059 (2000)

30. Faure, F., Sjöstrand, J.: Upper bound on the density of Ruelle resonances for Anosov flows. Commun. Math. Phys. **308**, 325–364 (2011)
31. Faure, F., Tsujii, M.: Prequantum transfer operator for symplectic Anosov diffeomorphism. ArXiv e-prints, arXiv:1206.0282v2 (2013)
32. Fried, D.: Analytic torsion and closed geodesics on hyperbolic manifolds. Invent. Math. **84**(3), 523–540 (1986)
33. Fried, D.: The zeta functions of Ruelle and Selberg: I. Ann. Scientifiques de L' É.N.S. **19**(4), 491–517 (1986)
34. Fried, D.: Meromorphic zeta functions for analytic flows. Commun. Math. Phys. **174**, 161–190 (1995)
35. Gallavotti, G.: Zeta functions and basic sets. Atti Accademia Nazionale dei Lincei **61**, 309–317 (1976)
36. Gaspard, P., Rice, S.: Semiclassical quantization of the scattering from a classically chaotic repellor. J. Chem. Phys. **90**, 2242–2254 (1989)
37. Giulietti, P., Liverani, C., Pollicott, M.: Anosov flows and dynamical zeta functions. Ann. Math. (2) **178**(2), 687–773 (2013)
38. Gouëzel, S., Liverani, C.: Banach spaces adapted to anosov systems. Ergod. Theor Dyn. Syst. **26**, 189–217 (2006)
39. Gouëzel, S., Liverani, C.: Compact locally maximal hyperbolic sets for smooth maps: fine statistical properties. J. Differ. Geom. **79**, 433–477 (2008)
40. Gutzwiller, M.C.: Periodic orbits and classical quantization conditions. J. Math. Phys. **12**, 343–358 (1971)
41. Harayama, T., Shudo, A.: Zeta function derived from the boundary element method. Phys. Lett. A **165**(5–6), 417–426 (1992)
42. Harayama, T., Shudo, A., Tasaki, S.: A functional equation for semiclassical fredholm determinant for strongly chaotic billiards. Progress Theor. Phys. Suppl. **139**, 460–469 (2000)
43. Hejhal, D.A.: The selberg trace formula and the riemann zeta function. Duke Math. J. **43**, 441–482 (1976)
44. Ikawa, M.: On the existence of the poles of the scattering matrix for several convex bodies. Proc. Jpn. Acad. Ser. A Math. Sci. **6**(4), 91–93 (1988)
45. Ikawa, M.: Singular perturbation of symbolic flows and poles of the zeta function. Osaka J. Math. **27**, 281–300 (1990)
46. Ikawa, M.: Singular perturbation of symbolic flows and poles of the zeta function. Osaka J. Math. **27**, 161–174 (1992)
47. Ikawa, M.: On zeta functions and the scattering poles for several complex bodies. J. Equ. aux Derivees Partielles **2** (1994)
48. Katok, A., Knieper, G., Pollicott, M., Weiss, H.: Differentiability of entropy for Anosov and geodesic flows. Bull. Am. Math. Soc. **22**(2), 285–293 (1990)
49. Katsuda, A., Sunada, T.: Closed orbits in homology classes. Publications Mathématiques de L'IHÉS **71**(1), 5–32 (1990)
50. Kitaev, A.Y.: Fredholm determinants for hyperbolic diffeomorphisms of finite smoothness. Nonlinearity **12**, 141–179 (1999)
51. Lang, S.: Sur les séries L d'une variété algébrique. Bull. Soc. Math. France **84**, 385–407 (1956)
52. Liverani, C.: Decay of correlation. Ann. Math. **142**, 239–301 (1995)
53. Liverani, C.: On contact Anosov flows. Ann. Math. **159**, 1275–1312 (2004)
54. Liverani, C.: Fredholm determinants, Anosov maps and Ruelle resonances. Discrete Contin. Dyn. Syst. **13**(5), 1203–1215 (2005)
55. Liverani, C., Tsujii, M.: Zeta functions and dynamical systems. Nonlinearity **19**(10), 2467–2473 (2006)
56. MacPherson, R.D. (ed.): Raoul Bott Collected Papers, vol. 2. Chapter Notes on the Lefschetz Fixed Point Theorem for Elliptic Complexes. Contemporary Mathematicians. Birkhäuser, Boston (1994)
57. Margulis, G.A.: Certain applications of ergodic theory to the investigation of manifolds of negative curvature. Funkcional. Anal. i Priložen. **3**(4), 89–90 (1969)

58. Margulis, G.A.: Certain measures that are connected with U-flows on compact manifolds. Funkcional. Anal. i Priložen. **4**(1), 62–76 (1970)
59. Margulis, G.A.: On some aspects of the theory of Anosov systems. In: Springer Monographs in Mathematics. With a survey by R. Sharp: Periodic Orbits of Hyperbolic Flows, (trans: Russian by Szulikowska, V.V). Springer, Berlin (2004)
60. Marklof, J.: Selberg's trace formula: an introduction. In: Bolte, J., Steiner, F. (eds.) Hyperbolic Geometry and Applications in Quantum Chaos and Cosmology. London Mathematical Society Lecture Notes Series, vol. 397, pp. 83–119. Cambridge University Press, Cambridge (2011)
61. Mayer, D.H.: The thermodynamic formalism approach to Selberg's zeta function for PSL(2, **Z**). Bull. Am. Math. Soc. (N.S.) **25**(1), 55–60 (1991)
62. Naud, F.: Analytic continuation of a dynamical zeta function under a Diophantine condition. Nonlinearity **14**(5), 995–1009 (2001)
63. Naud, F.: Entropy and decay of correlations for real analytic semi-flows. Ann. Henri Poincaré **10**(3), 429–451 (2009)
64. Parry, W., Pollicott, M.: An analogue of the prime number theorem for closed orbits of shifts of finite type and their suspensions. Ann. Math. **118**, 573–591 (1983)
65. Parry, W., Pollicott, M.: Zeta functions and the periodic orbit structure of hyperbolic dynamics. In: Astérisque, vol. 187–188. Société mathématique de France, Paris (1990)
66. Petkov, V.: Dynamical zeta function for several strictly convex obstacles. Can. Math. Bull **51**(1), 100–113 (2008)
67. Pollicott, M.: A complex Ruelle-Perron-Frobenius theorem and two counterexamples. Ergod. Theor Dyn. Syst. **4**, 135–146 (1984)
68. Pollicott, M.: On the rate of mixing of Axiom A flows. Invent. Math. **81**, 413–426 (1985)
69. Pollicott, M.: Meromorphic extensions of generalised zeta functions. Invent. Math. **85**, 147–164 (1986)
70. Pollicott, M.: Zeta functions and analyticity of metric entropy for Anosov systems. Isr. J. Math. **76**(3), 257–263 (1991)
71. Pollicott, M.: Dynamical zeta functions and closed orbits for geodesic and hyperbolic flows. In: Frontiers in Number Theory, Physics, and Geometry. I, pp. 379–398. Springer, Berlin (2006)
72. Pollicott, M., Sharp, R.: Exponential error terms for growth functions on negatively curved surfaces. Am. J. Math **120**, 1019–1042 (1998)
73. Pollicott, M., Sharp, R.: Error terms for closed orbits of hyperbolic flows. Ergod. Theor Dyn. Syst. **21**, 545–562 (2001)
74. Pollicott, M., Sharp, R.: Correlations for pairs of closed geodesics. Invent. Math. **163**, 1–24 (2006)
75. Ratner, M.: The rate of mixing for geodesic and horocycle flows. Ergod. Theor Dyn. Syst. **7**, 267–288 (1987)
76. Ruelle, D.: Zeta-functions for expanding maps and Anosov flows. Invent. Math. **34**, 231–242 (1976)
77. Ruelle, D.: Flots qui ne melangént pas exponentiellement. C. R. Acad. Sci. Paris Ser. I Math. **296** (1983)
78. Ruelle, D.: Resonances for Axiom A flows. J. Differ. Geom. **25**(1), 99–116 (1987)
79. Ruelle, D.: An extension of the theory of Fredholm determinants. Publications mathématique de l'I.H.É.S. **72**, 175–193 (1991)
80. Ruelle, D.: Dynamical zeta functions and transfer operators. Not. Am. Math. Soc. **49**(8), 887–895 (2002)
81. Rugh, H.H.: Generalized Fredholm determinants and Selberg zeta functions for Axiom A dynamical systems. Ergod. Theor Dyn. Syst. **16**(4), 805–819 (1996)
82. Sanchez-Morgado, H.: R-torsion and zeta functions for analytic Anosov flows on 3-manifolds. Trans. Am. Math. Soc. **348**(3), 963–973 (1996)
83. Sarnak, P.: Determinants of laplacian. Commun. Math. Phys. **110**, 113–120 (1987)
84. Selberg, A.: Harmonic analysis and discontinuous groups in weakly symmetric Riemannian spaces with applications to Dirichlet series. J. Indian Math. Soc. (N.S.) **20**, 47–87 (1956)

85. Shudo, A., Harayama, T., Tasaki, S.: Semiclassical Fredholm determinant for strongly chaotic billiards. Nonlinearity **12**(4), 1113 (1999)
86. Sinaĭ, J.G.: Geodesic flows on compact surfaces of negative curvature. Soviet Math. Dokl. **2**, 106–109 (1961)
87. Sinai, Y.G.: The asymptotic behaviour of the number of closed geodesics on a compact manifold of negative curvature. Trans. Am. Math. Soc. **73**, 227–250 (1968)
88. Smale, S.: Differentiable dynamical systems. Bull. Am. Math. Soc. **73**(6), 747–817 (1967)
89. Stoyanov, L.: Scattering resonances for several small convex bodies and the Lax-Phillips conjecture. Mem. Am. Math. Soc. **199**(933) (2009)
90. Stoyanov, L.: Regular decay of ball diameters and spectra of ruelle operators for contact anosov flows. ArXiv e-prints (2011)
91. Tsujii, M.: Decay of correlations in suspension semi-flows of angle-multiplying maps. Ergod. Theor Dyn. Syst. **28**(1), 291–317 (2008)
92. Tsujii, M.: Quasi-compactness of transfer operators for contact Anosov flows. Nonlinearity **23**(7), 1495–1545 (2010)
93. Tsujii, M.: Contact Anosov flows and the Fourier-Bros-Iagolnitzer transform. Ergod. Theory Dyn. Syst. **32**(6), 2083–2118 (2012)
94. Voros, A.: The hadamard factorization of the Selberg zeta function on a compact Riemann surface. Phys. Lett. B **180**(3), 245–246 (1986)
95. Voros, A.: Spectral functions, special functions and the Selberg zeta function. Commun. Dyn. Syst. **110**, 439–465 (1987)
96. Voros, A.: Unstable periodic orbits and semiclassical quantisation. J. Phys. A Math. Gen. **21**(3), 685–692 (1988)
97. Wirzba, A.: The Casimir effect as a scattering problem. J. Phys. A Math. Theor. **41**(16) (2008)
98. Young, L-S.: What are SRB measures, and which dynamical systems have them? J. Stat. Phys. **108**, 733–754 (2002)

Chapter 19
Diffusion Dynamics in Economics:
An Application to the Effects of Fiscal Policy

Orlando Gomes

19.1 Introduction

Mainstream macroeconomic analysis typically adopts a rationality concept according to which each agent is endowed with an automatic capacity to maximize intertemporal utility subject to some resource constraints. Recent work searches for reasonable departures from this paradigm of full rationality in order to better replicate observable phenomena. 'Bounded rationality', 'near-rational behavior/expectations' or 'limited rationality' are some of the designations that have been progressively introduced in the macroeconomic lexicon in order to characterize more or less significant departures from the idyllic setting in which every agent possesses the capabilities needed to compute optimal solutions in every circumstance and in an instantaneous way. The voluminous literature that touches the mentioned subjects includes the following relevant references: [1, 4–8].

With this study, we intend to contribute to the literature on bounded rationality in macroeconomics by imposing the following central assumption: only a small share of educated and well informed agents in the economy will be able or willing to solve the optimization problem they face (these can be called the 'innovators'). All the other agents will just imitate the behavior of the first group, but with a time lag that varies across individuals, following a diffusion process (these are the 'adopters' or 'imitators').

The general setting is a trivial intertemporal utility maximization problem where three types of taxes influence the choices of private agents: labor income taxes, taxes on asset revenues and taxes over the acquisition of consumption goods. The

O. Gomes
Instituto Superior de Contabilidade e Administração de Lisboa (ISCAL) and Unidade de
Investigação em Desenvolvimento Empresarial—Economics Research Center
[UNIDE/ISCTE—ERC]. Av. Miguel Bombarda 20, 1069-035 Lisbon, Portugal
e-mail: omgomes@iscal.ipl.pt

A.A. Pinto and D. Zilberman (eds.), *Modeling, Dynamics, Optimization
and Bioeconomics I*, Springer Proceedings in Mathematics & Statistics 73,
DOI 10.1007/978-3-319-04849-9_19,
© Springer International Publishing Switzerland 2014

optimization problem is solved, allowing for presenting a trivial long-term outcome with constant values for the assumed endogenous variables (in the case, wealth, consumption and time allocation between labor and leisure). The benchmark setting for the analysis is this steady-state, which can be disturbed through changes in the way the fiscal policy is conducted. Any perturbation on one of the assumed tax rates will trigger a re-optimization process. According to the bounded rationality assumption, only a small share of agents (which in aggregate terms might correspond to an infinitesimal share) will instantly update their behavior and compute the new optimal solution; every other agent will follow in time according to some diffusion pattern.

The diffusion dynamics are adapted from [9], who considers three categories of sources of diffusion: contagion, social influence and social learning. We will study the three, for different kinds of policy changes, in order to understand how consumption and labor-leisure choices are eventually affected on the aggregate. We will encounter inertia on the time paths of macro variables following the policy shocks: there is not an instantaneous jump from one equilibrium point to another one generated by the disturbance, but a gradual adjustment. That is, after a tax change, the large majority of the agents (or even the entire population) will end up by adopting the same behavior as the individuals with the instantaneous re-optimization ability, however this will not occur for everyone at the same time; ranther, a sluggish adjustment will be observed. Assuming some standard distribution of agents' attentiveness (e.g., a normal distribution) as we will do in the cases of social influence and social learning, or a simple logistic pattern, as in the contagion scenario, we will encounter a diffusion path for the evolution of the assumed endogenous variables or, which is the same, a hump-shaped evolution for the rate of change of the mentioned variables.

The central piece of the diffusion analysis will be the time allocation choice. The problem involves a state variable, wealth, that reacts immediately to tax changes, and two control variables, consumption and time allocation between labor and leisure. When a fiscal policy disturbance takes place, agents will take time (following the diffusion process) to apprehend such change in terms of their labor-leisure decision; as a result, some agents will continue, for some time length, to offer an amount of labor supply that is compatible with the pre-perturbation optimum but that is sub-optimal for the new tax rante. Consumption levels will be the direct outcome of financial and labor income, and therefore a non optimal labor-leisure choice will conduct to an amount of consumption that does not allow for maximizing the steady-state aggregate level of utility, as long as the diffusion process is not extinguished.

An important remark relates to the causes of sluggishness in the labor market: we are not departing from a perfectly competitive environment, i.e., the labor market is fully flexible. The causes for stickiness in the behavior of workers relate to their incapacity to immediately adjust to a change in the relevant parameters: only when observing the behavior of others, will the majority of the agents progressively adapt their decisions to the new economic conditions. In other words, it will be less costly for most of the agents to 'adopt' an existing behavior with some delay than to engage in a systematic re-optimization process (although we do not directly consider re-optimization costs, they will implicitly exist; they are the reason for the diffusion process to take place).

The remainder of the chapter is organized as follows: Sect. 19.2 describes the structure of the model and Sect. 19.3 characterizes the trivial steady-state of the problem. In Sect. 19.4, we discuss types of diffusion processes, which are applied in Sects. 19.5 and 19.6 to address transitional dynamics under diffusion, after a policy shock. The first of these two sections concentrantes on the impact of policy measures over the evolution of endogenous variables while the latter focuses its attention on long-run utility or welfare implications. Section 19.7 concludes.

19.2 The Optimization Problem

Consider an economy populated by a large number of individual households. They all face the same optimization problem, which consists in maximizing intertemporal utility, given an infinite horizon. The arguments of the instantaneous utility function are consumption, c_t, and the amount of time allocated to leisure. Normalizing the amount of available time to the unity, $\ell_t \in (0, 1)$ will represent the share of time allocated to working hours, while $1 - \ell_t$ will be the percentage of the household's time destined to leisure. Both variables, c_t and ℓ_t, are control variables for the agent, meaning that the problem faced by households consists in evaluating consumption—savings and labor—leisure trande-offs that best serve the goal of utility maximization. Let $\beta \in (0, 1)$ represent the intertemporal discount factor. The individual agent maximizes the value of the following objective function:

$$V_0 = \sum_{t=0}^{\infty} \beta^t u(c_t, 1 - \ell_t) \tag{19.1}$$

Utility function $u(\cdot)$ is continuous, twice differentiable, increasing on its two arguments and subject to decreasing marginal returns, also relatively to both arguments, i.e., the following derivative signs must hold: $u_c > 0, u_{cc} < 0$ and $u_{1-\ell} > 0, u_{1-\ell,1-\ell} < 0$. One also considers that the utility function is additively separable, that is, $u_{c,1-\ell} = u_{1-\ell,c} = 0$.

The optimization problem is subject to a resource constraint. This constraint reflects the accumulation of wealth. Households increase their wealth levels through labor participation, which pays a fixed wage rate, and through the accumulation of financial assets. In order to obtain a steady-state result such that wealth, consumption and labor and leisure shares are kept constant, we assume that assets' accumulation is subject to decreasing returns: additional returns on financial investments fall with the amount of wealth already applied in the markets (this view can be supported on the idea that more profitable investment opportunities are the first to be seized).

Three types of taxes will be considered in this economy: taxes over consumption, taxes over labor income and taxes over the returns on financial wealth. The corresponding rates are: $\tau_c, \tau_\ell, \tau_s \in (0, 1)$. These rates will appear on the resource constraint, which takes the form

$$a_{t+1} - a_t = (1 - \tau_\ell)w\ell_t + (1 - \tau_s)ra_t^\alpha - (1 + \tau_c)c_t, \ a_0 \text{ given} \qquad (19.2)$$

In difference equation (19.2), a_t represents the level of wealth held by the households at time t, w translates a constant wage rate, r is a parameter that allows to measure the returns on financial wealth and elasticity $\alpha \in (0, 1)$ reflects the degree of decreasing returns on the accumulation of wealth. Increasing consumption levels or assuming higher taxes (independently of their type) constitutes a penalty over the amount of available wealth (it will suffer a negative change).

It is straightforward to solve the optimization problem *Max V_0* subject to resource constraint (19.2). The computation of first order conditions requires writing the Hamiltonian function,

$$H(a_t, c_t, \ell_t) = u(c_t, 1 - \ell_t) +$$
$$\beta p_{t+1}\left[(1 - \tau_\ell)w\ell_t + (1 - \tau_s)ra_t^\alpha - (1 + \tau_c)c_t\right] \qquad (19.3)$$

with p_t the co-state variable associated to the state variable a_t.

To derive optimality conditions we resort to the Pontryagin's principle; the necessary conditions for optimality are:

$$\frac{\partial H}{\partial c} = 0 \Rightarrow u_c = (1 + \tau_c)\beta p_{t+1} \qquad (19.4)$$

$$\frac{\partial H}{\partial \ell} = 0 \Rightarrow u_\ell = -(1 - \tau_\ell)w\beta p_{t+1} \qquad (19.5)$$

$$\beta p_{t+1} - p_t = -\frac{\partial H}{\partial a_t} \Rightarrow \left[1 + \alpha(1 - \tau_s)ra_t^{-(1-\alpha)}\right]\beta p_{t+1} = p_t \qquad (19.6)$$

The transversality condition $\lim_{t \to \infty} p_t \beta^t a_t = 0$ needs to be imposed in order to avoid the possibility of perpetual debt.

To develop further the optimality relations, we need to specify a functional form for the utility function. A shape that obeys to the properties one has imposed to this function is the following: $u(c_t, 1 - \ell_t) = \ln\left[c_t (1 - \ell_t)^m\right]$, where $m > 0$ is the parameter reflecting the relevance that the household attributes to the utility of leisure relatively to consumption utility. For the chosen function, $u_c = 1/c_t$ and $u_{1-\ell} = m/(1 - \ell_t)$.

Optimality conditions (19.4) and (19.6) allow for presenting an equation of motion for consumption,

$$c_{t+1} = \beta \left[1 + \alpha(1 - \tau_s) r a_{t+1}^{-(1-\alpha)} \right] c_t \tag{19.7}$$

By combining (19.4) and (19.5), one also encounters a static relation between share ℓ_t and the value of the consumption variable:

$$\ell_t = 1 - \frac{m(1 + \tau_c)}{w(1 - \tau_\ell)} c_t \tag{19.8}$$

As previously referred, the study to pursue is restricted to the evaluation of the steady-state. Thus, we will resort to Eqs. (19.2), (19.7) and (19.8) to find the steady-state values $a^* := a_{t+1} = a_t$, $c^* := c_{t+1} = c_t$ and $\ell^* := \ell_{t+1} = \ell_t$. Some algebra directs us to the intended expressions:

$$a^* = \left[\frac{\alpha\beta}{1 - \beta} (1 - \tau_s) r \right]^{1/(1-\alpha)} \tag{19.9}$$

$$\ell^* = \frac{1}{1+m} \left\{ 1 - \left(\frac{\alpha\beta}{1 - \beta} \right)^{\alpha/(1-\alpha)} \frac{m \left[(1 - \tau_s) r \right]^{1/(1-\alpha)}}{w(1 - \tau_\ell)} \right\} \tag{19.10}$$

$$c^* = \frac{1}{1+m} \left\{ w \frac{1 - \tau_\ell}{1 + \tau_c} + \left(\frac{\alpha\beta}{1 - \beta} \right)^{\alpha/(1-\alpha)} \frac{\left[(1 - \tau_s) r \right]^{1/(1-\alpha)}}{1 + \tau_c} \right\} \tag{19.11}$$

19.3 Long-Term Policy Implications

Long-term effects of policy changes are straightforward to obtain from the steady-state results.

Proposition 19.1. *Under the proposed setting, an increase on the tax rate on savings has the following long-run impact over the economy: wealth and consumption levels fall and the share of time allocated to working hours suffers a positive change.*

Proof. Just compute first-order derivatives of (19.9), (19.10) and (19.11) with respect to τ_s; we find the results that are evidenced in the proposition, i.e., $\frac{\partial a^*}{\partial \tau_s} < 0$, $\frac{\partial c^*}{\partial \tau_s} < 0$ and $\frac{\partial \ell^*}{\partial \tau_s} > 0$. □

The presented result is intuitive. Whenever a fiscal policy change correspond to an increase on the tax rate over financial wealth earnings, this will lower the accumulated wealth and the resources the agent will have available to consume. If earnings on savings are lowered by a tax increase, then the agent will need to resort to other sources of income, namely she will exchange leisure time by labor time, i.e., share ℓ^* is positively influenced.

For the tax rate on labor income the following outcome is found.

Proposition 19.2. *When the government increases the labor income tax rate, this does not have any impact on the level of wealth. Consumption and the labor share will suffer, in this case, a negative change.*

Proof. Again, compute the derivatives of the steady-state expressions, now with respect to tax rate τ_ℓ; results are $\frac{\partial a^*}{\partial \tau_\ell} = 0$, $\frac{\partial c^*}{\partial \tau_\ell} < 0$ and $\frac{\partial \ell^*}{\partial \tau_\ell} < 0$.   □

Now, the intuition is as follows: by disturbing labor income with a higher tax rate, the willingness to allocate time to working hours falls, i.e., agents will renounce some labor time in favor of leisure time. This reallocation of labor will lead to a lower income that implies lower steady-state consumption levels.

Finally, one can analyze the effect of rising taxes over consumption. A larger tax rate τ_c has the impact that the following proposition describes.

Proposition 19.3. *A positive change in the tax rate on consumption will not influence wealth or the choice concerning the allocation of the household's available time. Relatively to consumption, the long-term value of the variable falls.*

Proof. Proceeding as in the previous occasions, one computes partial derivatives of steady-state values with respect to the tax rate under scrutiny to find the corresponding signs, $\frac{\partial a^*}{\partial \tau_c} = \frac{\partial \ell^*}{\partial \tau_c} = 0$ and $\frac{\partial c^*}{\partial \tau_c} < 0$.   □

A larger tax rate on consumption has impact only over consumption itself. As one should expect, a tax on consumption lowers the amount of available resources that the agent will have available to consume.

Another point deserving evaluation relates to the steady-state level of utility or welfare. For the specified utility function, the steady-state level of utility will be:

$$U(c^*, 1 - \ell^*) =$$

$$\ln\left(\frac{m^m w}{(1+m)^{1+m}} \frac{1 - \tau_\ell}{1 + \tau_c} \left\{ 1 + \left(\frac{\alpha \beta}{1 - \beta} \right)^{\frac{\alpha}{1-\alpha}} \frac{[(1 - \tau_s)r]^{1/(1-\alpha)}}{w(1 - \tau_\ell)} \right\}^{1+m} \right) \quad (19.12)$$

The evaluation of expression (19.12) allows us to understand the impact of fiscal policy changes over long-term utility.

Proposition 19.4. *Long-run utility falls with an increase on savings or consumption tax rates. Utility can vary positively or negatively following a positive change on the labor income tax rate; the positive variation requires the following inequality to hold:*

$$\tau_\ell > 1 - \left(\frac{\alpha \beta}{1 - \beta} \right)^{\frac{\alpha}{1-\alpha}} \frac{m}{w} [(1 - \tau_s)r]^{1/(1-\alpha)}$$

Proof. The signs of the derivatives of steady-state utility with respect to τ_s and τ_c are straightforward to obtain from expression (19.12): $\frac{\partial U^*}{\partial \tau_s} < 0$ and $\frac{\partial U^*}{\partial \tau_c} < 0$.

In what concerns the effect of the labor income tax over utility, the computation of the corresponding derivative gives place to

$$\frac{\partial U^*}{\partial \tau_\ell} = \frac{1}{1 - \tau_\ell} \left[(1 + m) \frac{\phi}{1 + \phi} - 1 \right]$$

with $\phi := \left(\frac{\alpha \beta}{1 - \beta} \right)^{\frac{\alpha}{1 - \alpha}} \frac{[(1 - \tau_s)r]^{1/(1-\alpha)}}{w(1 - \tau_\ell)}$. The above derivative possesses a positive sign if $\phi > 1/m$. Rearranging, the inequality in the proposition is immediately found. □

To improve our understanding of the result in Proposition 19.4, we take a numerical example. The selected values are: $m = 0.2, w = 0.25, r = 0.1, \alpha = 0.25$ and $\beta = 0.96$. In this case, $\tau_\ell = 1 - 0.0675(1 - \tau_s)^{1.3333}$ is the borderline between the two possible outcomes (increasing and decreasing utility following a positive change on τ_ℓ). For the specified values, for instance if $\tau_s = 0, \tau_s = 0.25$, or $\tau_s = 0.5$ then a positive change in utility as the result of $\Delta \tau_\ell > 0$ requires, respectively, $\tau_\ell > 0.9325, \tau_\ell > 0.954$, or $\tau_\ell > 0.9732$.

The information revealed by the example is that an extremely high tax rate on the labor income (relatively to the tax rate on savings) allows for an increase in steady-state utility whenever the tax rate on labor income incurs in a positive change. According to the numbers, though, the most likely result (the only admissible for empirically reasonable tax rates) is the opposite: long-run utility will fall with an increase in τ_ℓ. When the tax on labor earnings increases, the utility will increase through a direct effect (individuals exchange labor by leisure) and an indirect effect will take place as well (diverting labor hours towards leisure time implies lower income levels and therefore consumption will also decline, leading to a fall in utility). According to the proposed example, the second effect tends to dominate, and this becomes increasingly true as the other tax rate (τ_s) assumes larger values.

19.4 A Typology of Diffusion Processes

Based on Young [9], we consider three classes of diffusion processes. Diffusion can occur through contagion, social influence or social learning. Each type of diffusion process has distinctive features, however a common trace exists: in every case, the variable subject to diffusion eventually follows an s-shaped evolution pattern.

Diffusion under contagion implies that people will adopt a given behavior just by being in contact with other individuals that have already adopted such behavior. Diffusion occurs, in this case, much as an epidemic. Models of diffusion by contagion were pioneered by Bass [2, 3] and can be synthesized in a simple logistic dynamic equation of the type:

$$\Delta n_t = \left(\varphi_I^c n_t + \varphi_E^c \right) (1 - n_t), \quad n_0 \text{ given} \tag{19.13}$$

Variable n_t denotes the share of individual agents that have already accessed a given new set of information or that have already adopted a certain behavior; $\Delta n_t := n_{t+1} - n_t$ represents the time change of the assumed endogenous variable. Parameters φ_I^c and φ_E^c are both positive values and they translate the rates at which a current non-adopter 'hears about' the innovation from, respectively, previous adopters and from some external source [in Eq. (19.13), this idea is reflected on the fact that φ_I^c is associated with n_t while parameter φ_E^c is not]. In the extreme case where $\varphi_E^c = 0$, i.e., when contagion is only due to internal sources, we will be in the presence of a conventional s-shaped logistic equation and φ_I^c will represent the rate of contagion (thus, it should be a value lower than 1).

To study the main properties of the contagion equation (19.13), we begin by computing the steady-state value of n_t. Letting $n^* := n_{t+1} = n_t$ we find two solutions: $n^* = -\varphi_E^c/\varphi_I^c$ and $n^* = 1$; only the second solution is admissible, because the parameters are both positive values and the share n_t must lie, at every time moment, between 0 and 1. Thus, the contagion effect implies an equilibrium result where all the universe of individual agents has adopted the behavior or has taken contact with the relevant new information. Furthermore, it is straightforward to realize that the steady-state is a stable point for any $n_0 \in [0, 1]$; to accomplish this result, we just need to notice that, according to (19.13), $\Delta n_t \geq 0$ for all possible values of n_t, i.e., share n_t will increase in time from any admissible initial state until point $n^* = 1$ is reached. Given this global stability result, we can consider a complete diffusion process, that is, we can start at $n_0 = 0$.

The most relevant property concerning Eq. (19.13) is presented in Proposition 19.5.

Proposition 19.5. *Diffusion by contagion allows for an s-shaped adoption process as long as $\varphi_I^c > \varphi_E^c$. The acceleration phase (the phase in which the adoption curve is convex) cannot go beyond the 50 % adoption level.*

Proof. Consider the relation between n_t and Δn_t. One has already verified that if $n_t = 1$ then $\Delta n_t = 0$; this is the steady-state point. We can also compute the instantaneous change in the value of n_t when this share is equal to zero; replacing n_t by zero into (19.13) we obtain $\Delta n_t = \varphi_E^c$. The diffusion process will display an s-shaped form if, starting at $n_t = 0$, the adoption path is, in a first phase, convex and then, following an inflection point, it becomes concave. This inflection point in the time path of n_t translates into a maximum for the function $\Delta n_t = f(n_t)$. If we find a maximum for this function, in the interval $n_t \in [0, 1]$, then the s-shaped diffusion process is confirmed.

The necessary conditions for the existence of a maximum of any twice continuously differentiable function $f(n_t)$ are $f' = 0$ and $f'' \leq 0$. In the case under appreciation, $f' = \varphi_I^c(1 - 2n_t) - \varphi_E^c$ and $f'' = -2\varphi_I^c$. The second-order derivative is, undoubtedly, negative and therefore the maximum exists if condition $f' = 0$ is satisfied; this condition allows to determine the value of share n for which the referred inflection point is accomplished. The respective result is $n^{\max} = \left(\varphi_I^c - \varphi_E^c\right)/(2\varphi_I^c)$. The maximum change, found for $n^{\max}$, will be

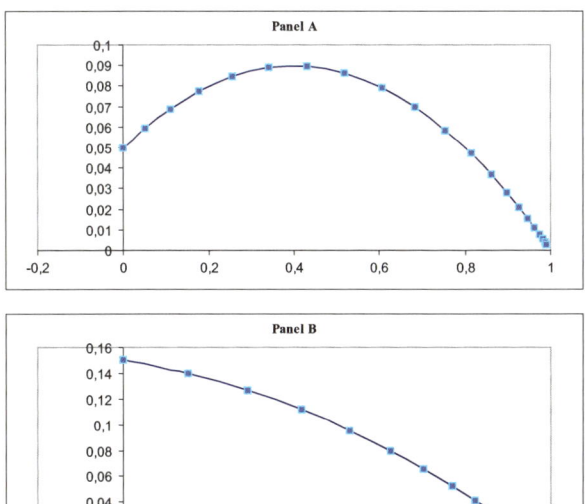

Fig. 19.1 $(n_t, \Delta n_t)$ diagram in the contagion diffusion case. Values of parameters: $\varphi_I^c = 0.25$, $\varphi_E^c = 0.05$ (panel **A**); $\varphi_I^c = 0.1$, $\varphi_E^c = 0.15$ (panel **B**)

$\Delta n^{\max} = f(n^{\max}) = (\varphi_I^c + \varphi_E^c)^2 / (4\varphi_I^c)$; observe that $\Delta n^{\max}$ is necessarily above zero (the value of Δn_t for $n_t = 1$) and it is not an amount inferior to φ_E^c (the value of Δn_t for $n_t = 0$).

The computed value $n^{\max}$ is a quantity below 1, but it can be either positive or negative. If $\varphi_I^c > \varphi_E^c$, then $n^{\max}$ is a positive value and, thus, the inflection point will exist for an admissible value of share n_t: the adoption curve will be convex in a first phase, it reaches a maximum at $n^{\max}$ and then it becomes concave as the convergence towards $n^* = 1$ is completed. In the opposite case, $\varphi_I^c < \varphi_E^c$, the inflection point occurs for a non admissible n_t value and, thus, the function $f(n_t)$ will be decreasing in all the extent of the interval $n_t \in [0, 1]$, i.e., the adoption curve will be concave throughout the diffusion process from $n_0 = 0$ (or any other initial share value) to $n^* = 1$.

The second part of the proposition states that even when $\varphi_I^c > \varphi_E^c$, the acceleration phase is subject to a constraint. In particular, the maximum point $n^{\max}$ is necessarily lower than 0.5, because otherwise φ_E^c would be a non positive value (to confirm this just solve $n^{\max} < 0.5$, in order to obtain an universal condition). ☐

The properties of the diffusion by contagion, referred to in Proposition 19.5, can be illustrated graphically. Figure 19.1 presents two possibilities for the relation between n_t and Δn_t; we depict cases $\varphi_I^c > \varphi_E^c$ and $\varphi_I^c < \varphi_E^c$ (in order to construct the graphics, panel **A** of the figure takes $\varphi_I^c = 0.25$ and $\varphi_E^c = 0.05$; panel **B** considers $\varphi_I^c = 0.1$ and $\varphi_E^c = 0.15$).

In Fig. 19.1, panel **A** shows how n_t varies with its own value for $\varphi_I^c > \varphi_E^c$. The change in n_t is always positive and the system converges towards the stable point $n^* = 1$. An s-shaped evolution for the variable under appreciation is confirmed by the hump-shaped form of the displayed relation. In a first stage, Δn_t increases, then it reaches a maximum and afterwards it begins to fall, meaning that the relation of n_t with time is such that the corresponding path is convex before point $n^{\max}$ is reached and then it becomes concave as the system converges to its resting point $n^* = 1$.

The case $\varphi_I^c < \varphi_E^c$ is represented in panel **B** and, again, it indicates the presence of convergence to $n^* = 1$, for a dynamic process that initiates in some admissible n_0 value. Once more, the contagion effect is stable in the sense that the epidemic will, for sure, spread throughout the entire population. However, in this second case, the maximum of the function does not fall in the interval of possible values of n_t. Therefore, Δn_t decreases along the whole adjustment range. Hence, the s-shaped trajectory for n_t will no longer be observed. This path will have a concave shape indicating that as n_t increases the number of new adopters will be progressively smaller independently of the number of already existing adopters. The border case, where the rates of internal and external current non-adopters are identical is similar in nature with the one in panel **B**, where now the maximum of the function is found at point $n_t = 0$; in this case, $\Delta n^{\max} = \left(\varphi_I^c + \varphi_E^c\right)^2 / (4\varphi_I^c) = \varphi_E^c$. Once again, the adoption curve is strictly concave.

Let us now turn to the setting of diffusion under social influence. In this second scenario, people will adopt the new behavior/absorb the new information when a certain threshold of individuals has already changed their behavior or their degree of attentiveness relatively to the new piece of information.

Individual agents will possess different degrees of responsiveness to social influence (i.e., different adoption thresholds). A low threshold value signifies that agents adopt the innovation even if only a small share of individuals has already adopted; a high threshold will reflect a low responsiveness to the behavior of others. In this type of diffusion process, it is assumed that at a given time t the proportion of individuals whose thresholds have been crossed is given by the cumulative distribution function of n_t, $F(n_t)$. Of this proportion, a subset n_t of individuals have already adopted the innovation and, thus, one can express the share of individuals who have crossed the threshold but have not yet adopted by $F(n_t) - n_t$. We consider $\varphi^{si} > 0$ as the rate of conversion, and therefore the adoption process is presentable under the difference equation

$$\Delta n_t = \varphi^{si} \left[F(n_t) - n_t\right] \quad n_0 \text{ given} \tag{19.14}$$

The specific shape of the diffusion process will depend on the type of distribution one considers, although one can recall some generic properties that help in understanding the main features of the underlying dynamics. A distribution function is, by definition, a non-decreasing function defined in the line of real numbers and with outcomes in the interval $[0, 1]$. Moreover, the derivative F' corresponds to the respective probability density function.

The steady-state is now the point for which $F(n^*) = n^*$; this will not be necessarily accomplished since the cumulative distribution function of n_t will exhibit a value that cannot be lower than n_t. In the examples to explore later, it is true that $F(1)$ is asymptotically equal to 1, and therefore we can consider this to be an asymptotic steady-state; however, if $F(n^*) = n^*$ does not hold for $n_t = 1$, it will not hold for any other, lower than 1, value of the variable. Thus, we just remark that for $n_t = 1$, $\Delta n_t = \varphi^{si} [F(1) - 1]$, which can be zero or a value slightly above zero. We can also compute the change in n_t for $n_t = 0$, which yields $\Delta n_t = \varphi^{si} F(0)$. Finally, it is also a point of interest the one for which the right hand side of (19.14) reaches a maximum. This will occur, as in the contagion case if the mentioned function has a negative second order derivative and it is possible to find the point for which the first derivative is equal to zero. The first derivative is $\varphi^{si} (F' - 1)$ and the second is $\varphi^{si} F''$. Thus, if the derivative of the density function is a negative value, we find a maximum for n_t at the point in which $F' = 1$, i.e., the density function is equal to 1. The diffusion result is expressed in Proposition 19.6.

Proposition 19.6. *Under social influence, the diffusion process is s-shaped for* $F(n^{\max}) - F(0) > n^{\max}$, *with* $n^{\max}$ *the solution of* $F' = 1$.

Proof. Above, one has remarked that the eventual maximum value of n_t is given by the solution of $F' = 1$. Once $n^{\max}$ is found, we observe that $\Delta n^{\max} = \varphi^{si} [F(n^{\max}) - n^{\max}]$. We want to compare this value with the value of Δn_t under $n_t = 0$ and $n_t = 1$. As remarked, the latter case is such that the change in the variable is zero or a value close to zero; thus, it remains to be compared the value of Δn when $n_t = 0$, with $n^{\max}$. If the maximum change lies to the right of $\varphi^{si} F(0)$, we can guarantee, as in the contagion case, the presence of a maximum in the interval $n_t \in (0, 1)$, which implies that the diffusion process will exhibit an inflection point, i.e., a point that guarantees the s-shape form for the trajectory of the share variable. The required condition is the one in the proposition.   □

A common distribution that we can resort to is the normal distribution. Assume an average value $\mu \in (0, 1)$, which represents the expected value of the threshold of adoption for any individual in the population and let σ be the standard deviation of the same process. The corresponding density function is

$$F' = \frac{\exp\left[-\frac{(n_t - \mu)^2}{2\sigma^2}\right]}{\sqrt{2\pi\sigma^2}}.$$

The derivative of this function is straightforward to compute:

$$F'' = -\frac{(n_t - \mu)}{\sigma^2} \frac{\exp\left[-\frac{(n_t - \mu)^2}{2\sigma^2}\right]}{\sqrt{2\pi\sigma^2}}$$

This last expression will correspond to a negative value, for $n_t > \mu$. Therefore, we confirm the existence of a local maximum as long as the value of the variable is above the corresponding average. This maximum is the solution of $F' = 1$, which, in the normal distribution case, corresponds to $n = \mu \pm \sigma \sqrt{-\ln(2\pi\sigma^2)}$. We have two solutions, but only one can correspond to a maximum given the condition $n_t > \mu$; thus, $n^{max} = \mu + \sigma \sqrt{-\ln(2\pi\sigma^2)}$. The diffusion process will be s-shaped (rather than concave throughout) if $n^{max} > 0$, a fact that is straightforward to observe.

Finally, one can look at an environment of social learning. This last setting does not consider that agents adopt a new information set, idea or behavior just because others have done so in the past (this was, in fact, what both the contagion and social influence scenarios implicitly meant). Alternatively, learning implies that agents will weight information about prior outcomes when deciding whether to adopt. Analytically, this translates into the replacement of the distribution function in Eq. (19.14) by the distribution of the sum of all shares n_s from $s = 0$ to $s = t$ (i.e., to the present moment). In the learning case, the relevant variable is the cumulative information generated by all prior adopters from the time they first adopted. Therefore, the dynamic equation translating the diffusion process will be

$$\Delta n_t = \varphi^{sl} \left[F \left(\sum_{s=0}^{t} n_s \right) - n_t \right] \quad n_0 \text{ given} \qquad (19.15)$$

Again, the parameter in the equation, φ^{sl}, represents a positive rate of adoption.

The difference between social influence and social learning relates to what information is assumed relevant in order to formulate a decision. Under social influence, agents choose resorting to the most recent information. Considering social learning, all the payoffs in the past are taken as relevant information. A more realistic approach would be to assume that payoffs received in the past are less valuable in terms of current decisions. Letting $\rho \in (0, 1)$, we could present a more sophisticated version of the social learning dynamics as

$$\Delta n_t = \varphi^{sl} \left[F \left(\sum_{s=0}^{t} \rho^t n_s \right) - n_t \right] \quad n_0 \text{ given} \qquad (19.16)$$

Some basic properties of the social influence case also apply to social learning. More than in the previous case one must expect $F(1)$ to approach 1, and thus $n_t = 1$ will be an asymptotic steady-state. Moreover, the condition underlying the existence of a maximum is of similar nature: $0 < n^{max} < 1$ or, equivalently, $F(0) < F(m^{max}) - n^{max} < F(1) - 1$, with $m_t := \sum_{s=0}^{t} n_s$. A relevant result of the comparison between the social influence and the social learning cases is presented in Proposition 19.7.

Proposition 19.7. *Consider some cumulative distribution function $F(n_t)$. If under both social influence and social learning, the share variable follows an s-shaped evolution, then the inflection point is first crossed for social learning then for social influence.*

Fig. 19.2 $(n_t, \Delta n_t)$ diagram with diffusion under contagion, social influence and social learning

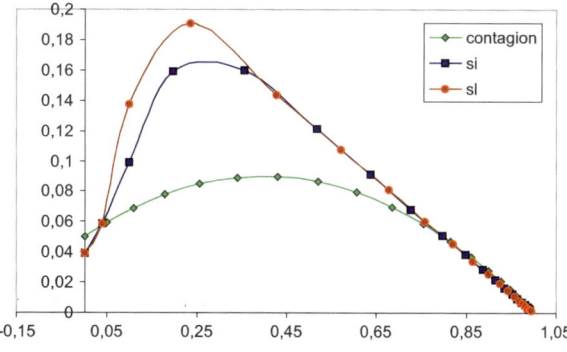

Fig. 19.3 Diffusion trajectories under contagion, social influence and social learning

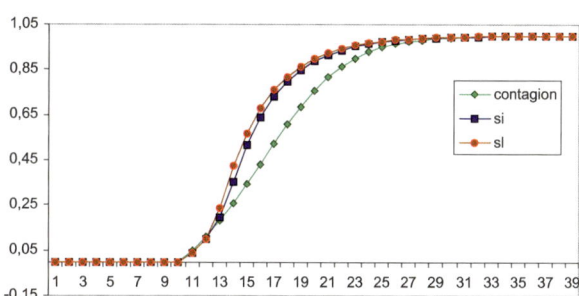

Proof. The result in the proposition is intuitive because in the learning case the argument of the distribution function is a cumulative value, thus a quantity that is larger than the one under social influence. In reality, if $F'\left[(n^{\max})^{si}\right] = 1$ and $F'\left[(m^{\max})^{sl}\right] = 1$, we must have $(n^{\max})^{si} = (m^{\max})^{sl}$, which is equivalent to $(n^{\max})^{si} = \sum\limits_{s=0}^{t-1} n_s + (n^{\max})^{sl}$, i.e., we confirm that $(n^{\max})^{sl}$ must be lower than $(n^{\max})^{si}$. $\square$

We end this section with a numerical example that compares the three types of diffusion processes and that allows to illustrate the described generic results. We assume, for the social influence and for the social learning cases, a normal distribution with $\mu = 0.1$ and $\sigma = 0.1$. The adoption rate parameters will be $\varphi_I^c = \varphi^{si} = \varphi^{sl} = 0.25$ and $\varphi_E^c = 0.05$. Similar adoption rates will allow us to compare outcomes. The results are displayed in Figs. 19.2 and 19.3. Figure 19.2 presents the $(n_t, \Delta n_t)$ diagram for the three kinds of diffusion, while Fig. 19.3 displays the s-shaped trajectories of n_t.

Generic results can be confirmed:

- In the three cases, we have an s-shaped diffusion process, where it is visible the stable nature of the underlying dynamics. After more or less 20 to 25 time periods, the steady-state of full adoption is asymptotically accomplished. In Fig. 19.2, we observe that the relation $(n_t, \Delta n_t)$ is hump-shaped for every possible form of diffusion, what gives place to the diffusion trajectories in Fig. 19.3.

- Simple observation allows to perceive that (for the chosen parameter values) social influence and social learning generate relatively similar outcomes and that the contagion trajectory implies, relatively to the others, a slower adoption rate in a first phase and a recovery after some time periods (around 10).
- For the contagion case, the maximum value of Δn_t occurs at $n^{\max} = \left(\varphi_I^c - \varphi_E^c\right)/(2\varphi_I^c) = 0.4$; this can be confirmed in Fig. 19.3. The statements in Proposition 19.5 are confirmed, i.e., the s-shape form of the adoption path is present because we have chosen values of parameters such that $\varphi_I^c > \varphi_E^c$ and the inflection point $n^{\max}$ lies below the 50 % adoption rate.
- Under social influence, we find the maximum of Δn_t by solving equation $F' = 1$, which, for the selected normal distribution allows for finding the solution $n^{\max} = 0.2663$. Thus, $\Delta n_t = 0.25(0.9519 - 0.2663) = 0.171\,4$. Note that $n^{\max}$ is smaller in the social influence case than in the contagion setup; therefore, the time length in which n_t varies at an increasing rate is larger in the contagion case than under social influence; nevertheless, this second type of diffusion will imply a faster initial increase of the share n_t. Proposition 19.6 is straightforward to apply to this simple example: we confirm that the diffusion process is s-shaped because $F(n^{\max}) - F(0) > n^{\max} \Leftrightarrow 0.9519 - 0.1587 > 0.2663 \Leftrightarrow 0.793\,2 > 0.2663$.
- In what concerns social learning, the first phase of convergence towards full adoption is relatively faster than under social influence. This is an expected result since, as referred, under learning information about past adoptions is also relevant (and not only the number of current adopters). Proposition 19.7 is also confirmed by noticing that the maximum Δn_t is found, in the learning case, for $n^{\max} = 0.2352$, a value that is clearly below the one found in the case of social influence.

19.5 Results on Wealth, Time Allocation and Consumption

Our main purpose is to evaluate the impact of fiscal policy changes over the steady-state values of wealth, time allocation and consumption, in the presence of the assumption on bounded rationality/behavior diffusion. The aggregate response to disturbances is sluggish because only a small share of agents instantly solve the new optimization problem (i.e., the problem involving the new tax rates), while all the other agents will also adopt the behavior compatible with the new fiscal conditions but in posterior time moments, following the diffusion pattern.

Several types of policies can be taken into consideration. For instance, we can distinguish between permanent and temporary policy changes. In the first case, given one of the three types of diffusion processes discussed in the previous section, there will be a gradual departure from the first steady-state set of values in the direction of the post-perturbation equilibrium, which will be accomplished for sure. If the policy change has a temporary nature (e.g., a six-period decrease in the labor income tax rate), some individuals will be able to accommodate their optimal behavior to the tax change, but others, that are slower to react, will continue with their pre-perturbation behavior and when the tax reversal takes place they will not have to change their behavior back. The initial steady-state is recovered as the agents with the faster adoption capacity adjust back to the original tax rate.

The exact mechanism through which the diffusion after a policy shock functions will be the following:

1. The level of wealth is a state variable; as a result, agents in the economy cannot manipulate the value of this variable. If a tax change occurs, this will instantly modify the endowment of wealth held on the aggregate by the economy. The steady-state level of wealth is given in expression (19.9); according to this expression, the only change in a tax rate that modifies the long-run level of wealth is the change in the savings tax rate. Following Proposition 19.1, an increase in τ_s will lead to a fall in a^*. Independently of rationality issues involving the behavior of economic agents, the value of this state variable will always suffer a full instantaneous impact that occurs when a fiscal policy change is triggered. Any change in the tax rates τ_ℓ and τ_c leave the trajectory of a^* unchanged.

2. Sluggishness will basically be the result of the incapacity most agents exhibit in perceiving instantly which is their best decision in terms of labor-leisure choices. As referred, the capacity to immediately re-optimize is restricted to a small share of agents; all the others will just follow the pioneer behavior of the 'innovators' and adopt, sooner or later, a similar kind of labor participation behavior. Thus, when a policy change occurs, agents will gradually (and not instantly) switch from the level of participation in the labor market that allowed to accomplish the pre-shock optimal utility outcome to the labor-leisure option that maximizes the post-perturbation utility level. Recall that the steady-state optimal labor share is given by expression (19.10). From this expression, one has realized that τ_c does not exert any influence over the steady-state time allocation choice; a change in τ_s induces a variation of the same sign in ℓ^*, while $\Delta \tau_\ell$ will lead to a change of the same sign in the leisure share (i.e., a variation of opposite sign in ℓ^*).

    While in Sect. 19.3 we have computed the long-term impact of the policy changes over the allocation of the agents' time, now we are interested in the short-run. The short-run should be interpreted as the time span along which the diffusion process is taking place; therefore, the long-run will be the setting in which the diffusion process is asymptotically terminated. Thus, when a tax change $\Delta \tau$ occurs, this will be perceived, in terms of labor market decisions, only by a share n_t of individuals; all the others will continue acting as if the policy change had not occurred. On the aggregate, when the policy shock takes place, the relevant tax rate involving time allocation decisions will be $\tau' = \tau + n_t \Delta \tau$. That is, at time t, a fraction n_t of the population has already brought the new tax rate (τ') into their decision problem, and a fraction $1 - n_t$ continues to consider as relevant the original tax rate τ. Note that $n_t \Delta \tau$ will coincide with $\Delta \tau$ after some time periods have passed—in the examples of the previous section, around 20 to 25 periods—but there is a transition phase in which the full extent of the tax change is not included in agents' decisions and therefore these depart from the optimal outcome).

3. The other control variable of the problem is consumption. If agents are unable to instantly adapt their time allocation behavior, this inertia will have to be accommodated in terms of consumption levels. If agents work more or less than what is optimal at a given time, this implies that consumption will have to be changed accordingly. From the steady-state conditions we can recover the following expression:

$$c^* = \frac{1}{1 + \tau_c} \left[(1 - \tau_s) r \left(a^* \right)^\alpha + (1 - \tau_\ell) w \ell^* \right] \qquad (19.17)$$

This relation indicates that when a fiscal policy measure is adopted and this implies that ℓ^* does not vary in the exact amount as it should, this non optimal variation must be compensated by consumption, i.e., the consumption level will also change in an amount that differs from the optimal one. Note that these observations make sense only for τ_s and τ_ℓ; tax rate τ_c does not exert any effect over labor-leisure choices and, consequently, consumption can move instantly to the new equilibrium when a change in the consumption tax rate takes place.

The main conclusion to draw is that as long as $n_t < 1$, a change in τ_s or τ_ℓ will imply that the steady-state levels of time allocation and consumption will depart from the corresponding levels that would maximize utility. Thus, the consequence of the assumed kind of bounded rationality will be a loss of long-term welfare, because agents are unable to instantly adapt to new fiscal conditions (this will become clear with the analysis in the following section).

General results are hard to evaluate and typically will not produce informative outcomes. Thus, for an overall assessment of the implications of policy changes, we will consider a numerical example. We will work with the diffusion processes characterized in Sect. 19.4 and displayed in Figs. 19.2 and 19.3. Furthermore, we recover the array of parameter values used in Sect. 19.3, i.e., $m = 0.2$, $w = 0.25$, $r = 0.1$, $\alpha = 0.25$ and $\beta = 0.96$. Finally, we need initial values for the tax rates; let $\tau_s = 0.25$ and $\tau_\ell = \tau_c = 0.2$. The assumed parameter values allow for a straightforward computation of steady-state optimal levels of wealth, time allocation and consumption: $a^* = 0.3448$, $\ell^* = 0.7854$ and $c^* = 0.1788$. The computed time allocation value indicates that in this economy, under the specified settings, around 78.54 % of the available time of the agents is, in the steady-state, allocated to labor hours.

We will consider the following alternative policies:

1. A permanent change in the tax rate on financial income: $\Delta\tau_s = 0.05$;
2. A permanent change in the labor income tax rate: $\Delta\tau_\ell = 0.03$;
3. A temporary change (6 periods) in the labor income tax rate: $\Delta\tau_\ell = -0.05$;
4. A temporary change (6 periods) in the tax rate over consumption: $\Delta\tau_c = 0.02$.

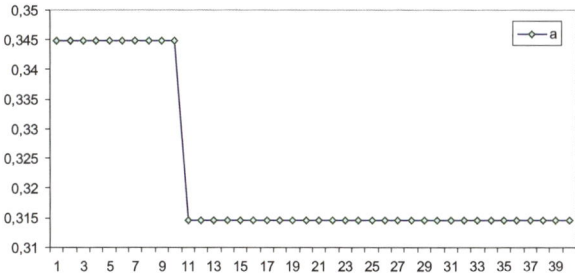

Fig. 19.4 A permanent change in τ_s

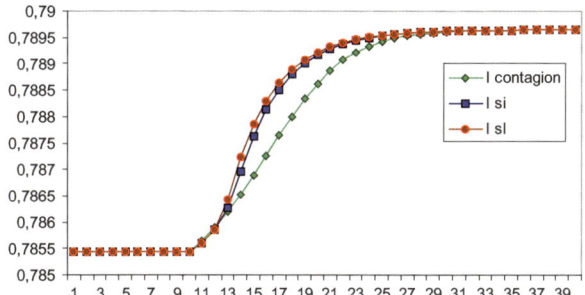

Fig. 19.5 A permanent change in τ_s. Impact on the trajectory of labor time allocation

Many other policy measures could be considered, including fiscal plans combining the simultaneous manipulation of several tax rates. In order to illustrate our arguments, the considered changes are enough.

Consider first the policy measure $\Delta\tau_s = 0.05$. According to Proposition 19.1, this will imply a fall in the level of wealth and in consumption and a reallocation of time in favor of labor hours, if the agents want to continue to maximize utility. The long-run effects are the following: $\Delta a^* = -0.0303$, $\Delta\ell^* = 0.0042$ and $\Delta c^* = -0.0035$. The short-run effects will correspond to the evolution implied by the diffusion process. These effects are depicted in Figs. 19.4, 19.5 and 19.6. The policy change takes place, in the graphics, at time $t = 10$.

The proposed mechanism implies an immediate change in the value of the state variable as the result of the fiscal policy change. In what concerns the control variables, there is a convergence towards the new equilibrium. When the tax rate change takes place, households will start modifying their behavior concerning time allocation at a gradual pace. There is a diffusion process that implies an s-shaped trajectory from the first to the second equilibrium time allocation point, which is accomplished, more or less, in 20 to 25 time periods (see Fig. 19.5). Relatively to consumption, this has to adjust following rule (19.17). Figure 19.6 shows that, with the policy shock, consumption will immediately fall, but because the labor participation does not increase as much as it should in the short-run in order to

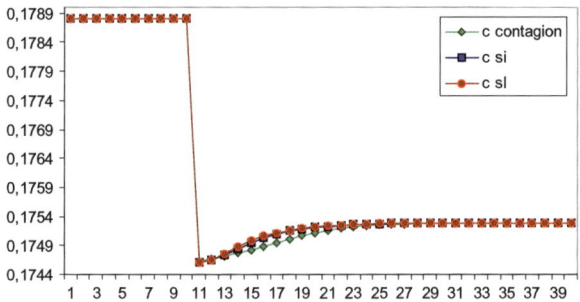

Fig. 19.6 A permanent change in τ_s. Impact on the trajectory of consumption

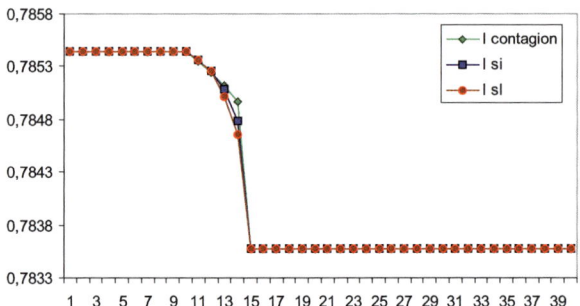

Fig. 19.7 A permanent change in τ_ℓ. Impact on the trajectory of labor time allocation

achieve an optimum, the fall in consumption will initially be stronger than the one observed in the long-run.

Through Figs. 19.4, 19.5, 19.6, besides verifying the evidence on the sluggish adjustment of control variables, we also confirm the different patterns of adjustment for different kinds of diffusion processes: social learning provides the fastest adjustment to the new equilibrium position and contagion the slowest.

Next, assume the permanent positive change in the labor income tax rate. Proposition 19.2 has indicated that this type of policy measure does not have an impact on wealth (this remains at level $a^* = 0.3448$); with respect to the other variables, one should expect a fall in the share of time allocated to labor and also a decrease in consumption. Nevertheless, these are long-term effects that overlook the transitional dynamics phase implied by diffusion. The relevant pictures are now Figs. 19.7 and 19.8.

These figures confirm the fall in the values of ℓ^* and c^* as the result of taking $\Delta\tau_\ell = 0.03$. The changes in the steady-state values are, respectively, $\Delta\ell^* = -0.0019$ and $\Delta c^* = -0.0052$. In the short-run, the labor share will gradually fall to the new steady-state, while consumption decreases, however initially in an extent that is inferior to the long-term change. The differences in the followed trajectories implied by the three different types of diffusion are not too significant, although

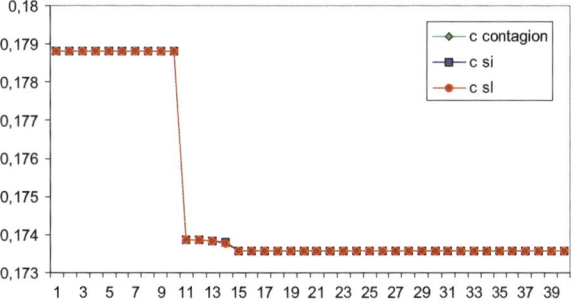

Fig. 19.8 A permanent change in τ_ℓ. Impact on the trajectory of consumption

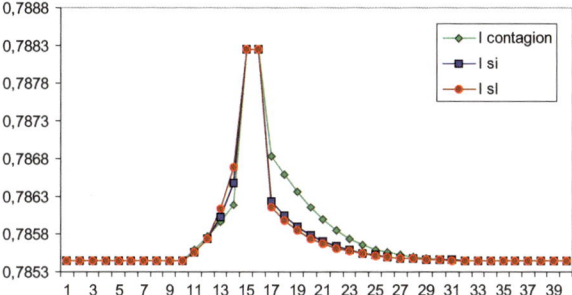

Fig. 19.9 A temporary change in τ_ℓ. Impact on the trajectory of labor time allocation

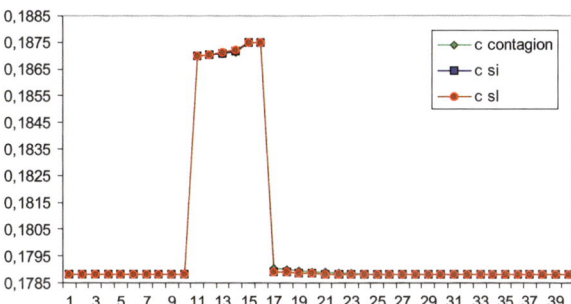

Fig. 19.10 A temporary change in τ_ℓ. Impact on the trajectory of consumption

they are exactly ranked as in the previously illustrated case: social learning provides the fastest convergence towards the long-run locus, followed by the social influence pattern of diffusion, and finally contagion.

Now, consider the temporary policy consisting in a six period fall in the labor income tax. A negative change in τ_ℓ allows for an increase in the steady-state values of ℓ_t and c_t, but now the change in the tax rate is temporary—it will be maintained only for six time periods. Thus, the new equilibrium $\ell^* = 0.7883$ and $c^* = 0.1875$ will never, in fact, be accomplished. There is an inertial convergence to this new equilibrium point following the diffusion process, that is interrupted at $t = 16$ (it was initiated at $t = 10$). When the initial fiscal policy is recovered, a new diffusion process is initiated back to the original steady-state. See Figs. 19.9 and 19.10.

Again, it is social learning that provides a faster initial adjustment to the shock (and also the fastest change back to the original state). Notice that in the case of consumption, the initial change leads to a positive jump and then to a phase of slow positive adjustment towards the new equilibrium that is interrupted when the tax rate changes back to its initial value. The recovery of the initial tax rate produces an immediate jump to a value that is close to the initial steady-state, but a small adjustment process is yet required for the original consumption state to be fully recovered.

Finally, we consider another temporary policy respecting to a change in the consumption tax rate. Proposition 19.3 states that this tax rate has effect solely over consumption, leaving unchanged the level of wealth and the distribution of time between labor and leisure. The assumption underlying diffusion according to which this will occur through a perturbation over the optimal decisions concerning time allocation implies that this kind of policy will not exert any short-run effect. Thus, time trajectories of a^* and ℓ^* are left unchanged, and consumption will be, in a six period time length, lower than the one implied by a tax rate $\tau_c = 0.2$; i.e., $c^* = 0.1788$ for $t = 1, .., 10; 17, \ldots$; and $c^* = 0.1756$ for $t = 11, \ldots, 16$.

19.6 Welfare Effects

If agents are boundedly rational and, as a result, display distinct velocities of reaction when faced with a policy change that has implications over their optimal problem, there is a sluggish adjustment of macro control variables towards a new steady-state, after the policy disturbance takes place. The most relevant consequence of the characterized framework is that we will be faced with temporary departures from maximum utility levels, independently of the nature of the policy (i.e., whether it is temporary or permanent). Long-run utility has been presented in Eq. (19.12) and, according to Proposition 19.4, typically utility falls with a tax increase (the only exception occurs for the labor income tax rate under very strict circumstances).

In this section, we briefly look at the departures from utility maximization given the bounded rational behavior that does not allow for an instant re-optimization.

Take, first, the permanent increase on the savings tax rate. Figure 19.11 displays, for the three types of diffusion processes, the difference between $U(c^*, 1 - \ell^*)$ and the value of utility implied by the observed values of consumption and time allocation under diffusion. Note that for $t = 1, .., 10$, the long-term utility level is the one involving the original tax rate value, while after $t = 10$ the values of c^* and ℓ^* take into consideration the new tax rate.

Although we have found a sluggish adjustment on labor participation following the fiscal shock, we observe an instantaneous departure from the utility level relatively to the benchmark optimal value (note that we are measuring deviations from the optimum, and thus this jump must be interpreted as a loss of utility); afterwards, a sluggish adjustment towards the maximum utility takes place. The optimal outcome is recovered once the diffusion process has ended (20 to 25

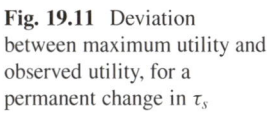

Fig. 19.11 Deviation between maximum utility and observed utility, for a permanent change in τ_s

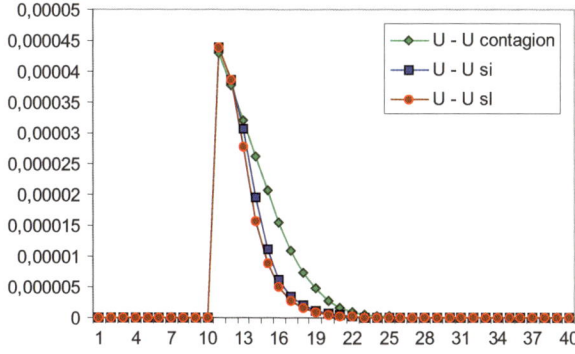

Fig. 19.12 Deviation between maximum utility and observed utility, for a permanent change in τ_ℓ

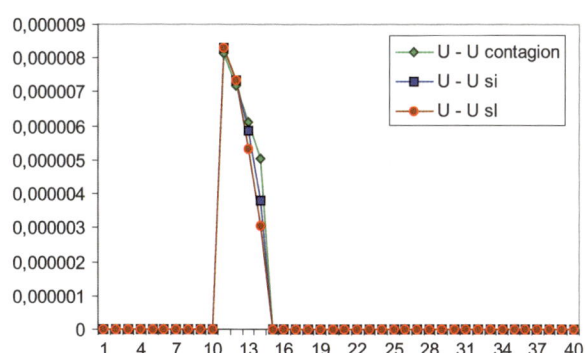

periods). Observe that contagion produces the largest departures relatively to the optimal result of utility maximization; at this level, social learning provides the less penalizing outcome.

The deviation relatively to the maximum utility can also be illustrated for policies $\Delta\tau_\ell = 0.03$ (permanent) and $\Delta\tau_\ell = -0.05$ (temporary); we have seen that for $\Delta\tau_c = 0.02$ there is no deviation relatively to the optimal outcome. Whether the tax rate on labor income falls or goes up, the sluggish adjustment leads to a less then optimal result. This is depicted in Figs. 19.12 and 19.13. The first of these figures presents a pattern of deviation from maximum utility very similar to the one in Fig. 19.11 (in both cases, fiscal policy measures are of a permanent nature). In Fig. 19.13, we have the case of a temporary policy; after the initial change, a relatively large deviation from the optimum is evidenced and, as time goes by, this is attenuated. Six periods later, a new shock occurs, that is, the tax rate changes back to its initial level. As a result, we have a new diffusion process and a new phase of departure relatively to the values of variables that maximize utility. The second deviation is not as strong as the first because at period six not all the agents would have already changed into the optimal behavior imposed by the new tax rate (in the second case there will be less agents adjusting their behavior).

Fig. 19.13 Deviation between maximum utility and observed utility, for a temporary change in τ_ℓ

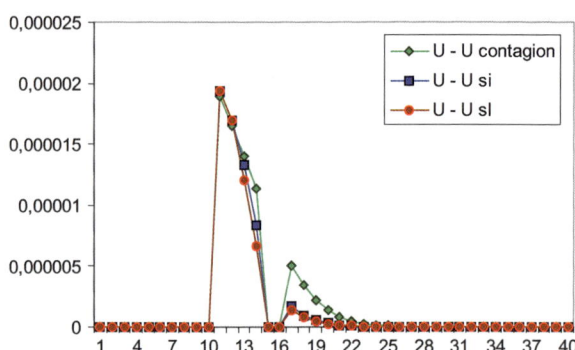

19.7 Conclusion

We have analyzed a standard intertemporal utility maximization model, in a setting where agents make choices concerning the allocation of time and about how much to consume of a given homogeneous good. The only departure from a fully competitive environment involving full rationality relates to the different capacities agents possess in what respects their ability to react to external events (these are basically fiscal policy changes, i.e., tax rate changes). Some agents will eventually have the capacity to re-compute a new optimal set of choices whenever this becomes necessary, but the full majority of the individuals in the population will not 'innovate' at this level; they will 'imitate' the behavior of the pioneer group. Heterogeneity arises because individuals will require different levels of adoption in order to decide whether to adopt themselves.

In our framework, the sluggishness or inertia in perceiving or accepting that economic conditions have changed after a fiscal policy disturbance is felt at the level of labor supply. As agents understand that the new tax rate is no longer compatible with the former optimum, they will gradually change their time allocation choices, following a diffusion process. Given some steady-state endowment of wealth, the departure from an optimal labor-leisure choice has impact over consumption, which is also pushed away from the optimal outcome.

We have considered three forms of diffusion. We can conceive a scenario of pure contagion, where the simple presence in the same geographical space make agents follow, sooner or later, some rule adopted by others; we can also take social influence, a setting where agents adopt the same choices as others because they observe that the behavior is being adopted; and social learning, case in which all the past history concerning adoption is taken into consideration by the individual when deciding about the change in behavior (or about incorporating some set of new information). The developed exercise has indicated that social learning is the more effective type of diffusion process in allowing for a fast aggregate transition from the pre-shock to the post-perturbation equilibrium.

Under diffusion, fiscal policy disturbances have the following specific effects:

1. A positive change in the tax rate on financial wealth earnings leads to an instantaneous fall in the level of wealth, to a modification of labor-leisure choices in favour of labor (this occurs under an s-shaped pattern of evolution over time of the labor share variable) and to a strong fall in consumption (larger than the long-run change) which is followed by a slight increase over time in the direction of the new steady-state consumption level.
2. A positive change in the labor income tax rate does not disturb the level of wealth and leads to an increase in leisure time (relatively to labor time) that occurs, again, under an s-shaped evolution. The aggregate consumption level will instantly decline, in the moment of the policy shock, but for a value slightly above the new steady-state one (the diffusion process will then generate the required adjustment to the long-run outcome).
3. A positive change on the tax rate over consumption does not influence wealth or time allocation decisions. Thus, consumption suffers, immediately, a re-adjustment to the new optimal value.

If the policy measures imply a decline rather than an increase on the tax rates, the effects will be symmetrical to the ones mentioned. Tax changes can also be temporary, and in this case two diffusion processes will occur; the second starts when the tax rate changes back to its initial value and ends when the original optimal point is recovered.

The main consequence of the sluggish reaction of time allocation decisions to tax changes is that there will be a period of less than optimal decisions that produce an aggregate welfare loss in the sense that the level of utility will depart from its maximum value. If fiscal policy is frequently changed, this can be translated in persistent aggregate deviations from the optimal consumption and time allocation decisions. Thus, fiscal stability tends to be welfare enhancing.

Acknowledgements A previous version of this paper was presented in the 24th European Conference on Operational Research (Euro XXIV) [Lisbon University, July 2010]. I would like to thank the organizers and the participants at the conference. I also acknowledge the helpful comments of Alan John Guimarães. The usual disclaimer applies.

References

1. Akerlof, G., Yellen, J.: Can small deviations from rationality make significant differences to economic equilibria? Am. Econ. Rev. **75**, 708–720 (1985)
2. Bass, F.: A new product growth model for consumer durables. Manag. Sci. **15**, 215–227 (1969)
3. Bass, F.: The relationship between diffusion rates, experience curves and demand elasticities for consumer durables and technological innovations. J. Bus. **53**, 551–567 (1980)
4. Beeby, M., Hall, S.G., Henry, B.: Rational expectations and near rational alternatives: how best to form expectations. European Central Bank Working Paper Series, vol. 086. European Central Bank, Frankfurt (2001)

5. Bomfim, A., Diebold, F.: Bounded rationality and strategic complementarity in a macroeconomic model. Econ. J. **107**, 1358–1374 (1997)
6. Gomes, O., Mendes, D.A., Mendes, V.M.: Bounded rational expectations and the stability of interest rate policy. Phys. A **387**, 3882–3890 (2008)
7. Haltiwanger, J., Waldman, M.: Limited rationality and strategic complements: the implications for macroeconomics. Q. J. Econ. **104**, 463–483 (1989)
8. Weder, M.: Near-rational expectations in animal spirits models of aggregate fluctuations. Econ. Model. **21**, 249–265 (2004)
9. Young, H.P.: Innovation diffusion in heterogeneous populations: contagion, social influence, and social learning. Am. Econ. Rev. **99**, 1899–1924 (2009)

Chapter 20
Occupation Times of Exclusion Processes

Patrícia Gonçalves

20.1 Introduction

In these notes we will explore the answer to the following question: given a one-dimensional Markov interaction $\{\eta_t : t \geq 0\}$ with state space Ω and a function $f : \Omega \to \mathbb{R}$, what is the scaling limit of the additive functional:

$$\Gamma_t(f) := \int_0^t f(\eta_s)ds.$$

There is a vast literature on the study of scaling limits of additive functionals of particles systems, but we point out here the seminal work [11], in which the authors give a characterization of the functions f, for which Γ_t has a Brownian motion as scaling limit. There, the study of Γ_t was motivated by the analysis of the motion of a tagged particle. The relation is that, the motion of the tagged particle can be written as a martingale plus an additive functional. Standard theorems for martingales provide the limit of the martingale term, so it remains to characterize the limit of the additive functional, in order to determine the scaling limits of the tagged particle. In [11] they give an abstract condition on f under which the additive functional converges. This condition is described as follows. Suppose that $\{\eta_t : t \geq 0\}$ is a stationary Markov process with state space Ω, with generator $\mathcal{L}$ and that it is reversible with respect to some probability measure ν. For a function f such that $E_\nu[f(\eta)] = 0$, $\theta^{-1/2}\Gamma_{\theta t}(f)$ converges to a Brownian motion, as long as, $\lim_{t \to \infty} t^{-1} E_\nu[(\Gamma_t(f))^2] < \infty$ [11].

P. Gonçalves (✉)
PUC-RIO, Departamento de Matemática, Rua Marquês de São Vicente, no. 225, 22453-900, Gávea, Rio de Janeiro, Brazil

Centro de Matemática da Universidade do Minho, Campus de Gualtar, 4710-057 Braga, Portugal
e-mail: patricia@mat.puc-rio.br

A.A. Pinto and D. Zilberman (eds.), *Modeling, Dynamics, Optimization and Bioeconomics I*, Springer Proceedings in Mathematics & Statistics 73, DOI 10.1007/978-3-319-04849-9_20,
© Springer International Publishing Switzerland 2014

By exploring $t^{-1}E_\nu[(\Gamma_t(f))^2]$, last condition is equivalent to requiring that $f \in \mathcal{H}^{-1}$. The Sobolev space $\mathcal{H}^{-1}$ is the dual, with respect to $L^2(\nu)$, of the space $\mathcal{H}^1$, defined as the set of functions $f \in L^2(\nu)$ such that $\|f\|_1^2 = \int_{\mathcal{E}} -f(\eta)(\mathcal{L}f)(\eta)\nu(d\eta) < \infty$. Usually, is not easy to verify that $f \in \mathcal{H}^{-1}$ and in [16] they came out with a very simple criterium, which gives the Brownian motion limit of $\Gamma_t(f)$. Following the terminology of [16], a local function $f : \Omega \to \mathbb{R}$ whose limiting variance $t^{-1}E_\nu[(\Gamma_t(f))^2]$ is finite (or equivalently $f \in \mathcal{H}^{-1}$) is called an *admissible function*. In that paper the authors prove that for the one-dimensional finite range symmetric exclusion, a function f is admissible for the generator acting on $L^2(\mathbb{P}_\rho)$ (where $\mathbb{P}_\rho$ is the distribution of the exclusion process starting from ν_ρ), if and only if, defining for $\rho \in (0, 1)$ $\varphi_f(\rho) := E_{\nu_\rho}[f(\eta)]$, we have that

$$\varphi_f^j(\tilde{\rho})\Big|_{\tilde{\rho}=\rho} = 0, \quad for \quad j = 0, 1, 2. \tag{20.1}$$

This condition is equivalent to saying that the degree of f is greater or equal to three, namely: f can be written as $f(\eta) := c \prod_{x \in A}(\eta(x) - \rho)$, where c is a constant and $A \subseteq \mathbb{Z}$ with $|A| \geq 3$. In that paper the same result is proved for one-dimensional finite range symmetric zero-range processes, under the condition that, the inverse of the spectral gap for the dynamics restricted to a finite box of size ℓ has second moment bounded from above by $c\ell^4$, where c is a constant.

In [6, 7, 9] we came across with the study of additive functionals, when establishing the equilibrium fluctuations of exclusion type models. In those papers, we establish the limit process governing the fluctuations of particle systems of exclusion type and when characterizing this process as the solution of some stochastic partial differential equation we had the need to derive the Boltzmann-Gibbs Principle. This principle was introduced in [2] and says that:

$$\int_0^t \frac{1}{\sqrt{n}} \sum_{x \in \mathbb{Z}} g(x/n)\{\tau_x f(\eta_s) - \varphi_f(\rho) - \varphi_f'(\rho)(\eta_s(x) - \rho)ds\} \xrightarrow[n \to +\infty]{} 0,$$

in $L^2(\mathbb{P}_\rho)$. Here g is a test function sufficiently smooth and f is a local function defined on Ω—the state space of the Markov process $\{\eta_t : t \geq 0\}$. In [6, 7, 9] we establish a stronger Boltzmann-Gibbs Principle under which, we can identify the limit of the functional above by speeding the processes into longer time scales/stronger asymmetries. Contrarily to our initial purposes, in the previous additive functional, the integrand function is no longer local—since it depends on the process defined on the full lattice. Nevertheless, the proof of the stronger Boltzmann-Gibbs Principle derived in [6, 7, 9] can be formulated in terms of local functions and in [10] we derived the *local Boltzmann-Gibbs Principle*, for exclusion processes satisfying the conditions of Sect. 20.2:

Theorem 20.1 (Local Boltzmann-Gibbs Principle [10]). *Let* $f : \Omega \to \mathbb{E}$ *be a local function, such that* $supp(f) \subseteq \{1, \ldots, k\}$ *and* $\varphi_f(\rho) = 0$. *There exists* $c = c(f, \rho)$ *such that*

(i) if $\varphi'_f(\rho) \neq 0$, *then for any* $t \geq 0$ *and any* $\ell \geq k$:

$$\mathbb{E}_\rho\left[\left(\int_0^t \left\{f(\eta_s) - \varphi'_f(\rho)(\eta_s^\ell - \rho)\right\} ds\right)^2\right] \leq c\left(t\ell + \frac{t^2}{\ell^2}\right),$$

(ii) if $\varphi'_f(\rho) = 0$, *then for any* $t \geq 0$ *and any* $\ell \geq k$:

$$\mathbb{E}_\rho\left[\left(\int_0^t \left\{f(\eta_s) - \frac{\varphi''_f(\rho)}{2}\left((\eta_s^\ell - \rho)^2 - \frac{\rho(1-\rho)}{\ell}\right)\right\} ds\right)^2\right] \leq c\left(t(\log \ell)^2 + \frac{t^2}{\ell^3}\right)$$

where $\eta^\ell := \frac{1}{\ell}\sum_{x=1}^\ell \eta(x)$ *and* $\mathbb{E}_\rho$ *denotes the expectation with respect to* $\mathbb{P}_\rho$.

We will see below that last result allow us to obtain upper bounds on the variance of additive functionals for local functions such that $\varphi_f(\rho) = 0$ and $\varphi'_f(\rho) \neq 0$ or such that $\varphi_f(\rho) = \varphi'_f(\rho) = 0$ and $\varphi''_f(\rho) \neq 0$. Notice that these functions do not satisfy the admissibility condition (20.1) as stated in [16]. Moreover, in the case $\varphi'_f(\rho) \neq 0$, we can also identify the limit of $\Gamma_t(f)$ as a fractional Brownian motion of Hurst exponent $H = 3/4$, for a general class of exclusion processes.

We recall from [16], that in the symmetric finite range exclusion, for an admissible function $f : \Omega \to \mathbb{R}$ as in (20.1) the variance of $\Gamma_t(f)$ is bounded from above by Ct and the invariance principle for $\Gamma_t(f)$ was also established:

$$\frac{1}{\sqrt{\theta}} \int_0^{\theta t} f(\eta_s) ds \xrightarrow[\theta \to +\infty]{} \mathcal{B}(Ct), \tag{20.2}$$

where $\mathcal{B}$ is the standard Brownian motion and C is a constant. In that paper, (20.2) is also proved for the case of symmetric zero-range processes, under the condition on the spectral gap mentioned above. In [15, 17], the previous result was obtained for mean-zero symmetric exclusion processes. In [15] it is also proved that for mean-zero simple exclusion processes, the set of admissible functions are those satisfying condition (20.1). In the finite range symmetric and mean-zero exclusion process, for a local function f such that $\varphi_f(\rho) = 0$ and $\varphi'_f(\rho) \neq 0$, the variance of $\Gamma_t(f)$ is bounded from above by $Ct^{3/2}$ and

$$\frac{1}{\theta^{3/4}} \int_0^{\theta t} f(\eta_s) ds \xrightarrow[\theta \to +\infty]{} \mathcal{B}_{3/4}(Ct), \tag{20.3}$$

where $\mathcal{B}_{3/4}$ is the fractional Brownian motion with Hurst exponent $H = 3/4$ and C is a constant [15]. For symmetric zero-range processes the limit (20.3) was established in [13].

It remains to cover the case of local functions f such that $\varphi_f(\rho) = \varphi'_f(\rho) = 0$ and $\varphi''_f(\rho) \neq 0$. In the case of the symmetric simple exclusion in [15] it is obtained an upper bound for the variance of $\Gamma_t(f)$ and in [13] it is proved that

$$\frac{1}{\sqrt{\theta \log(\theta)}} \int_0^{\theta t} f(\eta_s) ds \xrightarrow[\theta \to +\infty]{} \mathcal{B}(Ct). \tag{20.4}$$

For symmetric zero-range processes last question is open, but (20.4) is conjectured to hold for these processes, see [13].

For non zero mean processes, like for example the asymmetric simple exclusion, much less is known. Obviously that, we can get upper bounds on the variance of $\Gamma_t(f)$ using the symmetric part of the generator and the results presented above. For the asymmetric simple exclusion, it was proved in [14] that for local functions f such that $\varphi_f(\rho) = 0$ and $\varphi'_f(\rho) \neq 0$ or $\varphi_f(\rho) = \varphi'_f(\rho) = 0$ and $\varphi''_f(\rho) \neq 0$, and for $\rho \neq 1/2$, the variance of $\Gamma_t(f)$ is bounded from above by Ct, where C is a constant. In two forthcoming papers [1, 2] we are able to obtain sharp bounds in the remaining cases, namely, for the asymmetric exclusion and also for asymmetric zero-range processes.

Our approach to these problems is completely different from the ones used in the papers mentioned above. We consider general exclusion processes and by using the *local Boltzmann-Gibbs Principle*, we are able to relate additive functionals of local functions f with additive functionals of the density of particles. Then, since for those systems, the Central limit Theorem for the density of particles is very well studied, we obtain upper bounds on the variance of $\Gamma_t(f)$ and we are able to identify its limit.

This paper is organized as follows. On the second section, we define our microscopic dynamics, namely one-dimensional exclusion type models whose dynamics depends on a local function which is assumed to turn the dynamics elliptic and reversible. On the third section, we recall some results on the scaling limit of the density of particles and we state that the additive functional of an Ornstein-Uhlenbeck process evaluated in a proper indicator function, converges and we identify its limit $\mathcal{Z}_t$ as a fractional Brownian motion with Hurst exponent $H = 3/4$. On the fourth section, we state that for local functions f such that $\varphi_f(\rho) = 0$ and $\varphi'_f(\rho) \neq 0$, $n^{-3/2} \Gamma_{tn^2}(f)$ converges as $n \to +\infty$ to $\varphi'_f(\rho) \mathcal{Z}_t$. The fifth section is devoted to the sketch of the proof of the *Local Boltzmann-Gibbs Principle* and on the sixth section we present some examples and we discuss the case of symmetric/asymmetric jump rates.

20.2 Exclusion Processes

In this section we describe our microscopic dynamics. Let $\{\eta_t : t \geq 0\}$ be a Markov process with space state $\Omega := \{0, 1\}^{\mathbb{Z}}$. The occupation variables are defined in such a way that for $x \in \mathbb{Z}$, $\eta(x) = 1$ if the site x is occupied, otherwise $\eta(x) = 0$.

At each site $x \in \mathbb{Z}$, there exists a random time clock, with exponential distribution with parameter 1. If the clock rings at the site x, either there is no particle at that site and one has to wait a new random time, or there is a particle at that site and it jumps according to some rate function that we define as follows. Let $r : \Omega \to \mathbb{R}$ be a local function that satisfies:

(i) There exists $\varepsilon_0 > 0$ such that $\varepsilon_0 < r(\eta) < \varepsilon_0^{-1}$ for any $\eta \in \Omega$. (**Ellipticity**)
(ii) For any $\eta, \xi \in \Omega$, such that $\eta(x) = \xi(x)$ for $x \neq 0, 1$, then $r(\eta) = r(\xi)$. (**Reversibility**)

The dynamics can be formally described by means of a generator, which is given on local functions $f : \Omega \to \mathbb{R}$ by:

$$\mathcal{L}f(\eta) = \sum_{x \in \mathbb{Z}} r(\tau_x \eta)(f(\eta^{x,x+1}) - f(\eta))$$

where

$$\eta^{x,y}(z) = \begin{cases} \eta(y), & z = x \\ \eta(x), & z = y \\ \eta(z), & z \neq x, y \end{cases} \tag{20.5}$$

and τ_x is the space translation by x, namely, for $y \in \mathbb{Z}$ $\tau_x \eta(y) := \eta(x + y)$.

The invariant measures for these processes are $\{\nu_\rho : \rho \in [0, 1]\}$, where for $\rho \in [0, 1]$, ν_ρ denotes the Bernoulli product measure of constant parameter ρ. Under this measure the occupation variables $\{\eta(x) : x \in \mathbb{Z}\}$ are independent and $\nu_\rho(\eta : \eta(x) = 1) = \rho$. Here and in the sequel, for $T > 0$, we denote by $\mathcal{D}([0, T], \Omega)$ ($\mathcal{C}([0, T], \Omega)$) the space of càdlàg (continuous) trajectories from $[0, T)$ to Ω. We denote by $\mathbb{E}_\rho$ the expectation with respect to $\mathbb{P}_\rho$—the distribution of $\{\eta_t : t \geq 0\}$ in the space $\mathcal{D}([0, T], \Omega)$ starting from ν_ρ.

20.3 Scaling Limits of the Density of Particles

As mentioned in the introduction, our approach is to related the additive functional of local functions f such that $\varphi_f(\rho) = 0$ and $\varphi'_f(\rho) \neq 0$, with the additive functional of the density of particles. Then, by using the known results on the scaling limits of the density of particles we are able to deduce the corresponding scaling limits for the additive functional of f. We start by recalling the results that concern the density of particles for the exclusion processes that we have defined above.

20.3.1 Hydrodynamic Limit

For each configuration η we denote by $\pi^n(\eta; du)$ the empirical measure given by:

$$\pi^n(\eta; du) = \frac{1}{n} \sum_{x \in \mathbb{Z}} \eta(x) \delta_{x/n}$$

where $\delta_{x/n}$ is the Dirac measure at x/n and $\pi_t^n(\eta, du) := \pi^n(\eta_t, du)$.

Under a diffusive scaling of time tn^2 and for a set of initial measures associated to a sufficiently smooth profile, the hydrodynamic limit for $\{\eta_t : t \geq 0\}$ was obtained by [5]. The hydrodynamic limit is a Law of Large Numbers for the empirical measure in the following sense. Fix an initial profile sufficiently smooth $\gamma : \mathbb{R} \to [0, 1]$. If for an initial distribution $\{\mu_n : n \geq 1\}$ $(\eta_0 \sim \mu_n)$ associated to the profile $\gamma(\cdot)$, the empirical measure at time $t = 0$ converges to the deterministic measure $\gamma(u) du$, then for any time $t > 0$, the empirical measure at time t converges to the deterministic measure $\rho(t, u) du$, where $\rho(t, u)$ is the unique weak solution of the corresponding hydrodynamic equation with initial condition $\gamma(\cdot)$.

20.3.2 Equilibrium Fluctuations

Now we recall the Central Limit Theorem for the empirical measure for the exclusion processes described above and starting from the invariant state ν_ρ. Let $\mathcal{S}(\mathbb{R})$ denote the Schwarz space of test functions and let $\mathcal{S}'(\mathbb{R})$ be its dual. For $g \in \mathcal{S}(\mathbb{R})$, the density fluctuation field is defined as

$$\mathcal{Y}_t^n(g) := \frac{1}{\sqrt{n}} \sum_{x \in \mathbb{Z}} g\left(\frac{x}{n}\right) \{\eta_{tn^2}(x) - \rho\}. \tag{20.6}$$

It was proved in [3] that $\{\mathcal{Y}_t^n : t \in [0, T]\}$ converges in distribution with respect to the Skorohod topology of $\mathcal{D}([0, T], \mathcal{S}'(\mathbb{R}))$ to the stationary solution of the Ornstein-Uhlenbeck equation

$$d\mathcal{Y}_t = D(\rho)\Delta \mathcal{Y}_t dt + \sqrt{2D(\rho)\rho(1-\rho)} \nabla d\mathcal{B}_t, \tag{20.7}$$

where $\mathcal{B}_t$ is a $\mathcal{S}'(\mathbb{R})$-valued Brownian motion and $D(\rho)$ is the diffusion coefficient. In particular, this means that the trajectories of the limit field $\mathcal{Y}_t$ are in $\mathcal{C}([0, T], \mathcal{S}'(\mathbb{R}))$ and that $\mathcal{Y}_0$ is a white noise of variance $\rho(1-\rho)$—namely for any $g \in \mathcal{S}(\mathbb{R})$, the real-valued random variable $\mathcal{Y}_0(g)$ has a normal distribution of mean zero and variance $\rho(1-\rho) \int (g(x))^2 dx$.

Now, we state a fundamental result in which we state the convergence of the additive functional of $\mathcal{Y}_t$ solution of (20.7):

Theorem 20.2. *Fix a stationary solution $\{\mathcal{Y}_t : t \in [0, T]\}$ of (20.7). For $x \in \mathbb{R}$, let $i_\varepsilon(x) : y \mapsto \varepsilon^{-1} 1_{(0,1]}((y - x)\varepsilon^{-1})$. For each $\varepsilon \in (0, 1)$, let $\{\mathcal{Z}_t^\varepsilon : t \in [0, T]\}$ be defined as*

$$\mathcal{Z}_t^\varepsilon = \int_0^t \mathcal{Y}_s(i_\varepsilon) ds.$$

Then, the process $\{\mathcal{Z}_t^\varepsilon : t \in [0, T]\}$ converges in distribution with respect to the uniform topology of $\mathcal{C}([0, T], \mathbb{R})$, as $\varepsilon \to 0$, to a fractional Brownian motion $\{\mathcal{Z}_t : t \in [0, T]\}$ of Hurst exponent $H = 3/4$.

20.4   Additive Functionals

As mentioned in the introduction, our goal consists in obtaining functional limit theorems for observables of the processes $\{\eta_t : t \geq 0\}$ as defined in Sect. 20.2. For these processes it holds that:

Theorem 20.3. *For a local function $f : \Omega \to \mathbb{R}$, the process $\{\Gamma_{tn^2}(f) : t \in [0, T]\}$ defined as*

$$\Gamma_{tn^2}(f) = \frac{1}{n^{3/2}} \int_0^{tn^2} \left(f(\eta_s) - \varphi_f(\rho) \right) ds \qquad (20.8)$$

converges in distribution with respect to the uniform topology of $\mathcal{C}([0, T], \mathbb{R})$ to $\{\varphi'_f(\rho)\mathcal{Z}_t : t \in [0, T]\}$, where $\{\mathcal{Z}_t : t \in [0, T]\}$ is the same as in Theorem 20.2.

The proof of this result is a consequence of the local Boltzmann-Gibbs Principle whose proof is sketched in the next section.

20.5   Proof of the *Local Boltzmann-Gibbs Principle*

The proof of the local Boltzmann-Gibbs Principle as stated in Theorem 20.1 is divided into four steps. The main ingredients that we use are the Kipnis-Varadhan inequality (see [11]) and the spectral gap inequality (see [12]).

1. Firstly, we compare the additive functional of f with the additive functional of $\psi_f(\ell) := E_\rho[f | \sum_{x=1}^\ell \eta(x)]$, using the:

Lemma 20.1 (One-block estimate). *Let $f : \Omega \to \mathbb{R}$ be a local function such that $\varphi_f(\rho) = 0$. Then, there exists $c = c(f, \rho)$ such that for any $\ell \geq k$ and any $t \geq 0$:*

$$\mathbb{E}_\rho\left[\left(\int_0^t \{f(\eta_s) - \psi_f(\ell; \eta_s)\}ds\right)^2\right] \leq ct\ell^2 \mathbf{Var}(f; \nu_\rho),$$

where $\mathbf{Var}(f; \nu_\rho)$ denotes the variance of f with respect to ν_ρ.

2. Secondly, we compare the additive functional of $\psi_f(\ell)$ with the additive functional of $\psi_f(2\ell)$, using the:

Lemma 20.2 (Renormalization step). *Let* $f : \Omega \to \mathbb{R}$ *be a local function such that* $\varphi_f(\rho) = 0$. *There exists* $c = c(f, \rho)$ *such that for any* $\ell \geq k$ *and any* $t \geq 0$:

$$\mathbb{E}_\rho\left[\left(\int_0^t \{\psi_f(\ell; \eta_s) - \psi_f(2\ell; \eta_s)\}ds\right)^2\right] \leq \begin{cases} ct\ell, \text{if } \varphi'_f(\rho) \neq 0, \\ ct, \text{if } \varphi'_f(\rho) = 0. \end{cases}$$

3. Thirdly, we compare the additive functional of $\psi_f(\ell)$ with the additive functional of $\psi_f(2^m\ell)$, using the renormalization step m times.

Lemma 20.3 (Two-blocks estimate). *Let* $f : \Omega \to \mathbb{R}$ *be a local function such that* $\varphi_f(\rho) = 0$. *Then, there exists* $c = c(f, \rho)$ *such that for any* $\ell \geq k$ *and any* $t \geq 0$:

$$\mathbb{E}_\rho\left[\left(\int_0^t \psi_f(k; \eta_s) - \psi_f(\ell; \eta_s)ds\right)^2\right] \leq \begin{cases} ct\ell, \text{if } \varphi'_f(\rho) \neq 0, \\ ct(\log \ell)^2, \text{if } \varphi'_f(\rho) = 0. \end{cases}$$

4. Finally, we replace $\psi_f(2^m\ell)$ by the corresponding function of $\eta^{2^m\ell}$ using the:

Proposition 20.1 (Equivalence of Ensembles). *Let* $f : \Omega \to \mathbb{R}$ *be a local function. Then there exists a constant* $c = c(f, \rho)$ *such that for any* $\ell \geq k$:

$$\int \left(\psi_f(\ell, \eta) - \varphi'_f(\rho)(\eta^\ell - \rho) - \frac{\varphi''_f(\rho)}{2}\left((\eta_s^\ell - \rho)^2 - \frac{\rho(1-\rho)}{\ell}\right)\right)^2 d\nu_\rho \leq \frac{c}{\ell^3}.$$

20.6 Examples

In this section we present some examples for which we can derive the precise statement of the theorems given above. We start by the mean-zero exclusion process.

20.6.1 Mean-Zero Exclusion

The mean-zero exclusion process is defined as in Sect. 20.2 , but in this case after an exponential time of parameter 1, a particle at the site x jumps to the site $x + y$ with probability $p(y)$. We assume the following conditions on the probability measure $p : \mathbb{Z} \setminus \{0\} \to [0, 1]$:

(1) $p(\cdot)$ has finite range, that is, there exists $M > 0$ such that $p(z) = 0$ whenever $|z| > M$;
(2) $p(\cdot)$ is irreducible, i.e. $\mathbb{Z} = \text{span}\{z \in \mathbb{Z}; p(z) > 0\}$;
(3) $p(\cdot)$ has *mean-zero*: $\sum_{z \in \mathbb{Z}} z p(z) = 0$.

Example. If we take $p(1) = 2/3$, $p(-2) = 1/3$ and $p(z) = 0$ if $z \neq -2, 1$, then the process is an example of an *asymmetric mean-zero* exclusion.

We define the Markov process $\{\eta_t^{\text{ex}} : t \geq 0\}$, whose generator acts over local functions $f : \Omega \to \mathbb{R}$ as

$$\mathcal{L}_{\text{ex}} f(\eta) = \sum_{x,y \in \mathbb{Z}} p(y)\eta(x)(1 - \eta(x + y))(f(\eta^{x,x+y}) - f(\eta)),$$

with $p(\cdot)$ satisfying (1), (2) and (3) and $\eta^{x,x+y}$ as in (20.5). The measures $\{\nu_\rho : \rho \in [0, 1]\}$ are invariant, but they are not necessarily reversible (that is true if and only if $p(\cdot)$ is symmetric). Thus, asymmetric mean-zero exclusion processes are diffusive and non-reversible systems. We can define the density fluctuation field $\{\mathcal{Y}_t^n : t \in [0, T]\}$ as in (20.6) and we have that:

Proposition 20.2. *The process $\{\mathcal{Y}_t^n : t \in [0, T]\}$ converges in distribution with respect to the Skorohod topology of $\mathcal{D}([0, T], \mathcal{S}'(\mathbb{R}))$ to the stationary solution of the Ornstein-Uhlenbeck equation*

$$d\mathcal{Y}_t = D(\rho)\Delta\mathcal{Y}_t dt + \sqrt{2D(\rho)\rho(1 - \rho)(\rho)}\nabla d\mathcal{B}_t,$$

where $D(\rho)$ is the diffusion coefficient.

The results presented above allow us to get the scaling limits of additive functionals as in Theorem 20.3. We notice that in spite of having stated the theorem for reversible systems (see condition (ii) on r), we can prove the *local Boltzmann-Gibbs Principle* for non-reversible systems, since the Kipnis-Varadhan inequality also fits these systems, see [4] for details.

20.6.2 Symmetric Simple Exclusion

Consider $\mathcal{L}_{ex}$ as above with $p(\cdot)$ such that $p(1) = p(-1) = 1/2$ and $p(z) = 0$ for $z \neq -1, 1$. We notice that for this process the measures $\{\nu_\rho : \rho \in [0, 1]\}$ are invariant and reversible. In this case we have that:

Proposition 20.3. *The process $\{\mathcal{Y}_t^n : t \in [0, T]\}$ converges in distribution with respect to the Skorohod topology of $\mathcal{D}([0, T], \mathcal{S}'(\mathbb{R}))$ to the stationary solution of* *the Ornstein-Uhlenbeck equation*

$$d\mathcal{Y}_t = \frac{1}{2}\Delta\mathcal{Y}_t dt + \sqrt{\rho(1-\rho)}\nabla d\mathcal{B}_t.$$

The results presented above allow us to get the scaling limits of additive functionals as stated in Theorem 20.3.

20.6.3 The Weakly Asymmetric Simple Exclusion

Now, we introduce an exclusion type process which has a drift towards the right. For that purpose, take $\mathcal{L}_{ex}$ as above with $p(\cdot)$ given by $p_n(1) = \frac{1}{2} + \frac{a_n}{2}$, $p_n(-1) = \frac{1}{2} - \frac{a_n}{2}$ and $p_n(z) = 0$ if $z \neq -1, 1$. The measures $\{v_\rho : \rho \in [0,1]\}$ are invariant but not reversible.

20.6.3.1 The Hydrodynamic Scaling

If $a_n := \frac{1}{n}$, then we have that

Proposition 20.4. *The process $\{\mathcal{Y}_t^n : t \in [0,T]\}$ converges in distribution with respect to the Skorohod topology of $\mathcal{D}([0,T], \mathcal{S}'(\mathbb{R}))$ to the stationary solution of the Ornstein-Uhlenbeck equation*

$$d\mathcal{Y}_t = \frac{1}{2}\Delta\mathcal{Y}_t dt + (1 - 2\rho)\nabla\mathcal{Y}_t dt + \sqrt{\rho(1-\rho)}\nabla d\mathcal{B}_t.$$

In this case the Ornstein-Uhlenbeck process has a drift, nevertheless one can get the same result as stated in Theorem 20.3.

20.6.3.2 The KPZ Scaling

Fix a density $\rho = 1/2$. Then, inserting this in the previous stochastic partial differential equation, we can see that the limit field is the same as in the symmetric simple exclusion (so a weak asymmetry does not have influence!), see [9]. In this case the "correct" *strength asymmetry* is $a_n = 1/\sqrt{n}$. In this case we have

Proposition 20.5. *The process $\{\mathcal{Y}_t^n : t \in [0,T]\}$ converges in distribution with respect to the Skorohod topology of $\mathcal{D}([0,T], \mathcal{S}'(\mathbb{R}))$ to the stationary solution of the stochastic Burgers equation:*

$$d\mathcal{Y}_t = \frac{1}{2}\Delta\mathcal{Y}_t dt + \left(\nabla\mathcal{Y}_t\right)^2 dt + \sqrt{\rho(1-\rho)}\nabla d\mathcal{B}_t. \tag{20.9}$$

In this case, the limit density field is no longer an Ornstein-Uhlenbeck process, so Theorem 20.2 is not useful in this case. Nevertheless, for $\mathcal{Y}_t$ solution of (20.9) we can also prove that:

Theorem 20.4. *Let $\{\mathcal{Y}_t : t \in [0, T]\}$ be a stationary solution of (20.9). For $\varepsilon > 0$, let $\tilde{\mathcal{Z}}_t^\varepsilon = \int_0^t \mathcal{Y}_s(i_\varepsilon)ds$. Then there exists $\{\tilde{\mathcal{Z}}_t : t \in [0, T]\}$ such that, $\{\tilde{\mathcal{Z}}_t^\varepsilon : t \in [0, T]\}$ converges in distribution with respect to the uniform topology of $\mathcal{C}([0, T], \mathbb{R})$, as $\varepsilon \to 0$, to $\{\tilde{\mathcal{Z}}_t : t \in [0, T]\}$.*

And as a consequence we have that

Theorem 20.5. *Let $f : \Omega \to \mathbb{R}$ be a local function such that $\varphi_f(1/2) = 0$. Then, $\{\Gamma_{tn^2}(f) : t \in [0, T]\}$ as defined in (20.8) converges in distribution with respect to the uniform topology of $\mathcal{C}([0, T], \mathbb{R})$ to $\{\cong'_f(1/2)\tilde{\mathcal{Z}}_t : t \in [0, T]\}$, where $\tilde{\mathcal{Z}}_t$ is the same as in Theorem 20.4.*

20.6.4 Symmetric Simple Exclusion/Asymmetric Simple Exclusion

Here we discuss the differences between the bounds on the variance of additive functionals of the symmetric simple exclusion (ssep) and the asymmetric simple exclusion process (asep), both defined on $\mathbb{Z}$. The latter process is defined through $\mathcal{L}_{ex}$ as above, but with $p(1) := p$, $p(-1) := 1 - p$ with $p \neq 1/2$ and $p(z) = 0$ for $z \neq -1, 1$.

Let f be a local function.

(1) If $\varphi_f(\rho) = 0$ and $\varphi'_f(\rho) \neq 0$, then:

$$\mathbf{Var}(\Gamma_t(f); \nu_\rho) \leq Ct^{3/2} \ in \ ssep$$

and

$$\mathbf{Var}(\Gamma_t(f); \nu_\rho) \leq \begin{cases} Ct^{4/3}, & \rho = \frac{1}{2} \\ Ct, & \rho \neq \frac{1}{2} \end{cases} \ in \ asep.$$

With the results presented above one gets the correct upper bound for the ssep. The method presented above does not give the correct upper bound in the asep. In [1, 14] it is proved the correct bound in the asep when $\rho \neq 1/2$ and in [1] the upper bound $t^{3/2}$ is obtained when $\rho = 1/2$. The correct upper bound when $\rho = 1/2$ is still out of reach.

(2) If $\varphi_f(\rho) = \varphi'_f(\rho) = 0$, $\varphi''_f(\rho) \neq 0$, then:

$$\mathbf{Var}(\Gamma_t(f); \nu_\rho) \leq Ct \log(t) \ in \ ssep$$

and

$$\mathbf{Var}(\Gamma_t(f); \nu_\rho) \le Ct \quad in \quad asep.$$

With the results presented above one gets the upper bound $Ct(log(t))^2$ in the ssep. The correct upper bound was obtained in [13]. In the asep, in [1, 14] the correct upper bound was obtained for $\rho \ne 1/2$ ($\rho = 1/2$).

(3) If $\varphi_f(\rho) = \varphi'_f(\rho) = \varphi''_f(\rho) = 0$, $\varphi'''_f(\rho) \ne 0$, then $\mathbf{Var}(\Gamma_t(f); \nu_\rho) \le Ct$ for both ssep and asep.

This bound was firstly obtained in [16] and with the results presented above we can also get the correct upper bound in these cases. Above C is a constant.

Acknowledgements The author thanks FCT (Portugal) for support through the research project "Non-Equilibrium Statistical Physics" PTDC/MAT/109844/2009 and the Research Centre of Mathematics of the University of Minho, for the financial support provided by "FEDER" through the "Programa Operacional Factores de Competitividade COMPETE" and by FCT through the research project PEst-C/MAT/UI0013/2011.

References

1. Bernardin, C., Gonçalves, P., Sethuraman, S.: Equilibrium fluctuations of additive functionals of zero-range models. In: Conference Proceedings of Particle Systems and Partial Differential Equations (in press)
2. Brox, T, Rost, H.: Equilibrium fluctuations of stochastic particle systems: the role of conserved quantities. Ann. Probab. **12**(3), 742–759 (1984)
3. Chang, C.C.: Equilibrium fluctuations of gradient reversible particle systems. Probab. Theory Relat. Fields **100**(3), 269–283 (1994)
4. Chang, C., Landim, C., Olla, S.: Equilibrium fluctuations of asymmetric simple exclusion processes in dimension $d \ge 3$. Probab. Theory Relat. Fields **119**(3), 381–409 (2001)
5. Funaki, T., Uchiyama, K., Yau, Y.: Hydrodynamic limit for lattice gas reversible under Bernoulli measures. Nonlinear Stochastic PDE's. Springer, New York (1996)
6. Gonçalves, P.: Central limit theorem for a tagged particle in asymmetric simple exclusion. Stoch. Process Appl. **118**, 474–502 (2008)
7. Gonçalves, P., Jara, M.: Universality of KPZ equation. Available online at arXiv:1003.4478 (2011)
8. Gonçalves, P., Jara, M.: Nonlinear fluctuations of weakly asymmetric interacting particle systems. Arch. Rational Mech. Anal. **212**(2), 597–644 (2014)
9. Gonçalves, P., Jara, M.: Crossover to the KPZ equation. Ann. Henri Poincaré **13**(4), 813–826 (2012)
10. Gonçalves, P., Jara, M.: Scaling limits of additive functionals of interacting particle systems. Commun. Pure Appl. Math. **66**(5), 649–677 (2013)
11. Kipnis, C., Varadhan, S.: Central limit theorem for additive functionals of reversible Markov processes and applications to simple exclusions. Commun. Math. Phys. **104**(1), 1–19 (1986)
12. Quastel, J.: Diffusion of color in the simple exclusion process. Commun. Pure Appl. Math. **45**(6), 623–679 (1992)
13. Quastel, J., Jankowski, H., Sheriff, J.: Central limit theorem for zero-range processes. Special issue dedicated to Daniel W. Stroock and Srinivasa S.R. Varadhan on the occasion of their 60th birthday. Methods Appl. Anal. **9**(3), 393–406 (2002)

14. Seppäläinen, T., Sethuraman S.: Transience of second-class particles and diffusive bounds for additive functionals in one-dimensional asymmetric and exclusion processes. Ann. Probab. **31**(1), 148–169 (2003)
15. Sethuraman, S.: Central limit theorems for additive functionals of the simple exclusion process. Ann. Probab. **28**, 277–302 (2000)
16. Sethuraman, S., Xu, L.: A central limit theorem for reversible exclusion and zero-range particle systems. Ann. Probab. **24**(4), 1842–1870 (1996)
17. Varadhan, S.: Self-diffusion of a tagged particle in equilibrium for asymmetric mean zero random walk with simple exclusion. Ann. de l'Institut Henri Poincaré **31**, 273–285 (1995)

Chapter 21
Error Estimates for a Coupled Continuous-Discontinuous FEM for the Two-Layer Shallow Water Equations

Pedro S. Gonçalves, Bruno M. Pereira, and Juha H. Videman

21.1 Introduction

The viscous shallow-water equations form a set of hyperbolic/parabolic partial differential equations that govern the flow when the vertical length scale of the fluid motion is much smaller than the typical horizontal scale. They are derived from the incompressible Navier-Stokes equations via elimination of the vertical velocity by depth-integration. The shallow-water equations, by constituting a much simplified form of the primitive equations, are often used in modeling large-scale atmospheric and oceanic circulation when the horizontal wave length is large compared to the depth, cf. [20]. Moreover, they are widely applied in studying fluid motion in shallow seas, coastal regions, estuaries, rivers, wetlands and salt marshes, see [21, 22] for an overview of shallow-water hydrodynamics. Shallow-water equations, possibly coupled with other models, can be used for simulation of complex processes such as storm surges, freshwater-saltwater interactions, contaminant transport, tidal fluctuations and overland flow, see, e.g., [7, 8, 17].

The so called primitive form of the shallow-water equations consists of a hyperbolic continuity equation for the surface elevation (or, equivalently, water depth) and of parabolic equations for the (depth-averaged) horizontal velocity components. The viscous terms arise from a eddy viscosity closure model. In the conservative form, the unknown quantities are the linear momentum and the surface elevation.

P.S. Gonçalves • B.M. Pereira • J.H. Videman (✉)
CAMGSD/Department of Mathematics, Instituto Superior Técnico, University of Lisbon, Av. Rovisco Pais 1, 1049-001 Lisboa, Portugal
e-mail: pegoncal@math.ist.utl.pt; bmpereira@adm.isel.pt; videman@math.ist.utl.pt

A.A. Pinto and D. Zilberman (eds.), *Modeling, Dynamics, Optimization and Bioeconomics I*, Springer Proceedings in Mathematics & Statistics 73, DOI 10.1007/978-3-319-04849-9_21,
© Springer International Publishing Switzerland 2014

Various numerical approximation schemes have been developed for the shallow-water equations since late 1970s, cf. [17, 21]. Often, to avoid spurious spatial oscillations under complicated flow regimes (complex geometries, fronts, shocks), the continuity equation has been replaced with a second-order wave continuity equation, see, e.g., [14, 17], and the error analysis in [3, 5].

More recently, there has been an attempt to go back to the primitive form of the equations by resorting to discontinuous approximating spaces, cf. [1, 4, 9, 10]. Discontinuous Galerkin methods can be formulated elementwise and lead to globally discontinuous piecewise polynomial approximations, the connection between neighboring elements arising from numerical traces. They seem to be well suited for handling advection-dominated flows and for the use of non-conforming meshes. Besides, they support polynomial approximations of different orders in different elements and are locally conservative, i.e. the rate of change of water elevation is balanced by advective fluxes element by element.

The discontinuous Galerkin finite element method was first introduced by Reed and Hill [18] to solve a hyperbolic neutron transport equation, see also [16] and [12] for the mathematical analysis of the method. Later, the method was successfully extended to hyperbolic systems, elliptic equations as well as to convection-diffusion and Navier-Stokes equations, see the review article [6] and the references therein.

One of the main simplifications of the shallow-water equations is the assumption that the density is constant in the vertical. However, in many situations mentioned above, there can be small but distinctive density differences along the depth, think for example of fresh/warm water flowing above salty/cold water or the spreading of oil sheets. This type of stratified structure of the flow is very common in coastal regions, estuaries, inlets and channels.

In this article, we develop a coupled continuous/discontinuous Galerkin method for the two-layer (multi-layer) shallow-water equations and derive a priori error estimates. Our scheme is based on one of methods proposed and analysed by Dawson and Proft for the single layer shallow-water equations, cf. [9], i.e. we apply a standard continuous Galerkin method for the discretization of the momentum equations and a discontinuous Galerkin method for the discretization of the continuity equations. The discontinuous approach is advantageous when local conservation is important. Besides, the non-conservative form of the equations is more suitable for coupling with other e.g., transport) schemes. Most of the existing numerical schemes for the two-layer shallow-water equations are either based on the conservative form, cf. [13, 15], or deal with the inviscid case, see, e.g., [19].

The plan of the paper is as follows. In Sect. 21.2, we derive the multi-layer shallow water equations from the incompressible Navier-Stokes equations by depth-integration. In Sect. 21.3, we introduce our notation and present the semi-discrete Galerkin formulation. Then we state our main result and outline the principal steps of its proof. We leave the numerical examples and the validation of our scheme with real-world data to a forthcoming work.

21.2 Multi-Layer Shallow Water Model

Let us start by recalling the derivation of the multi-layer viscous shallow-water equations, see, e.g., Vallis [20]. Consider n immiscible viscous fluid layers of different constant densities $\rho_1, \rho_2, \ldots, \rho_n$ placed on top of one another and numbered from top to bottom, see Fig. 21.1. Assume, for gravitational stability, that the density increases with depth, i.e. $\rho_1 < \rho_2 < \ldots < \rho_n$, and that the vertical scale of the fluid motion, say the sum of the mean (constant) heights of the fluid layers, is much smaller than its horizontal scale. Hence, the hydrostatic approximation is valid in each fluid layer, i.e.

$$\frac{\partial p_k}{\partial z} = -\rho_k g, \qquad k = 1, 2, \ldots, n, \tag{21.1}$$

where p_k is the (hydrostatic) pressure in layer k and g is the acceleration due to gravity.

Assume moreover that the bottom is flat, thus the sum of the equilibrium depths, denoted by $H = \sum_{k=1}^{n} H_k$, of the fluid layers is constant, and place the origin of Cartesian coordinates $(\mathbf{x}, z) = (x_1, x_2, z)$ at the equilibrium position of the free surface at $z = 0$. Denoting by η_1 the perturbed position of the free surface and by $\eta_k, k = 2, \ldots, n$, the perturbed positions of the interfaces between the fluid layers, the layer heights $\zeta_k, k = 1, \ldots, n$, can be written as

$$\zeta_k(\mathbf{x}, t) = \eta_k(\mathbf{x}, t) - \eta_{k+1}(\mathbf{x}, t), \qquad k = 1, \ldots, n - 1,$$

$$\zeta_n(\mathbf{x}, t) = H + \eta_n(\mathbf{x}, t).$$

Integrating (21.1) with respect to z and assuming constant atmospheric pressure at the free surface, we obtain within the top fluid layer

$$p_1(\mathbf{x}, t) - p_{\text{atm}} = \rho_1 g(\eta_1(\mathbf{x}, t) - z) =$$

$$= \rho_1 g \left(\sum_{k=1}^{n} \zeta_k(\mathbf{x}, t) - H - z \right), \qquad \eta_2(\mathbf{x}, t) < z < \eta_1(\mathbf{x}, t). \tag{21.2}$$

The linear momentum balance in the top layer is therefore

$$\partial_t \mathbf{u}_1 + (\mathbf{u}_1 \cdot \nabla) \mathbf{u}_1 - \nu_1 \Delta \mathbf{u}_1 = -g \nabla \eta_1, \tag{21.3}$$

where $\nu_1 = \mu / \rho_1$ is the kinematic viscosity in the top layer (μ stands for the horizontal eddy viscosity, assumed to be constant throughout the fluid) and $\mathbf{u}_1 = \mathbf{u}_1(\mathbf{x}, t)$ is the fluid velocity in the top layer. The mass balance within the uppermost layer takes the form

Fig. 21.1 A multi-layer
model with free surface

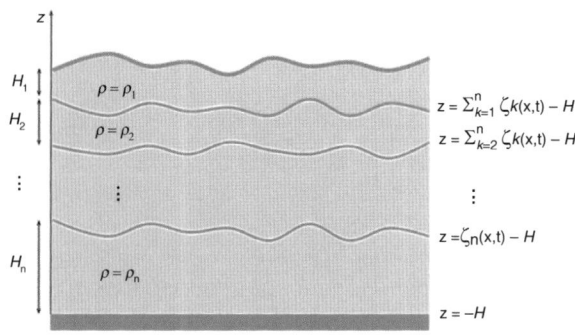

$$\partial_t \zeta_1 + \nabla \cdot (\zeta_1 \mathbf{u}_1) = 0,$$

where $\partial_t = \partial/\partial t$.

In the kth layer, for $k = 2, \ldots, n$, we integrate the hydrostatic equation (21.1)
from z to η_k and use the continuity of the pressure to obtain

$$p_k - p_{\text{atm}} = \sum_{j=1}^{k-1} \rho_j \, g(\eta_j - \eta_{j+1}) + \rho_k \, g \, (\eta_k - z), \qquad \eta_{k+1} < z < \eta_k \,.$$

Consequently, the momentum equation becomes

$$\partial_t \mathbf{u}_k + (\mathbf{u}_k \cdot \nabla) \, \mathbf{u}_k - \nu_k \varDelta \mathbf{u}_k = -\frac{g}{\rho_k} \left(\sum_{j=1}^{k-1} \rho_j (\nabla \eta_j - \nabla \eta_{j+1}) + \rho_k \nabla \eta_k \right)$$

$$= -\frac{g}{\rho_k} \left(\rho_1 \nabla \eta_1 + \sum_{j=2}^{k} (\rho_j - \rho_{j-1}) \nabla \eta_j \right). \tag{21.4}$$

Making the usual Boussinesq approximation, we may replace ρ_k by ρ_1 except in the
gravity terms arising from the density differences. Equation (21.4) thus simplifies to

$$\partial_t \mathbf{u}_k + (\mathbf{u}_k \cdot \nabla) \, \mathbf{u}_k - \nu \varDelta \mathbf{u}_k + g \nabla \eta_1 = -\sum_{j=2}^{k} g'_j \nabla \eta_j, \tag{21.5}$$

where $\nu = \mu/\rho_1$ and $g'_k = g \, (\rho_k - \rho_{k-1})/\rho_1$ is the reduced gravity in layer k. The
continuity equation is given by

$$\partial_t \zeta_k + \nabla \cdot (\zeta_k \mathbf{u}_k) = 0 \,. \tag{21.6}$$

Collecting Eqs. (21.3), (21.5) and (21.6), we arrive at the following coupled system for the viscous multi-layer shallow water equations

$$
\begin{cases}
\partial_t \mathbf{u}_1 + (\mathbf{u}_1 \cdot \nabla)\mathbf{u}_1 - \nu \Delta \mathbf{u}_1 + g \sum_{j=1}^{n} \nabla \zeta_j = 0, \\[2mm]
\partial_t \mathbf{u}_k + (\mathbf{u}_k \cdot \nabla)\mathbf{u}_k - \nu \Delta \mathbf{u}_k + g \sum_{j=1}^{n} \nabla \zeta_j = -\sum_{j=2}^{k} g'_j \sum_{i=j}^{n} \nabla \zeta_i, \quad k = 2, \dots, n \\[2mm]
\partial_t \zeta_k + \nabla \cdot (\zeta_k \mathbf{u}_k) = 0, \quad\quad\quad\quad\quad\quad\quad\quad\quad\quad\quad\quad k = 1, \dots, n
\end{cases}
\tag{21.7}
$$

where we have also taken into account that

$$
\eta_k = \sum_{j=k}^{n} \zeta_j - H .
$$

Let $\Omega \subset \mathbb{R}^2$ be a bounded domain with its piecewise smooth boundary $\partial \Omega$ divided, in each layer and at each time instant $t \in (0, T)$, into inflow and outflow parts $\partial \Omega_I$ and $\partial \Omega_O$ defined by

$$
\partial \Omega_I = \{ \mathbf{x} \in \partial \Omega \mid \mathbf{u}_k \cdot \mathbf{n} < 0 \}, \qquad \partial \Omega_O = \{ \mathbf{x} \in \partial \Omega \mid \mathbf{u}_k \cdot \mathbf{n} \geq 0 \},
$$

where n is the outward unit vector normal to $\partial \Omega$. In each layer $k = 1, \dots, n$, we prescribe the initial conditions

$$
\zeta_k(\mathbf{x}, 0) = \zeta_0^{(k)}(\mathbf{x}), \qquad \mathbf{u}_k(\mathbf{x}, 0) = \mathbf{u}_0^{(k)}(\mathbf{x}), \qquad \mathbf{x} \in \Omega ,
\tag{21.8}
$$

the inflow boundary condition (e.g., river inflow)

$$
\zeta_k(\mathbf{x}, t) = \hat{\zeta}_k(\mathbf{x}, t), \qquad (\mathbf{x}, t) \in \partial \Omega_I \times (0, T),
\tag{21.9}
$$

and the no-slip boundary condition for the velocity field

$$
\mathbf{u}_k(\mathbf{x}, t) = \hat{\mathbf{u}}_k(\mathbf{x}, t), \qquad (\mathbf{x}, t) \in \partial \Omega \times (0, T),
\tag{21.10}
$$

($\mathbf{u}_k \cdot \mathbf{n} = 0$ at land boundaries).

21.2.1   Two-Layer Shallow Water Equations

Taking $n = 2$ in (21.7), we obtain the following coupled set of nonlinear partial differential equations for the unknown functions $\mathbf{u}_1, \zeta_1, \mathbf{u}_2$ and ζ_2

$$
\begin{cases}
\partial_t \mathbf{u}_1 + (\mathbf{u}_1 \cdot \nabla) \mathbf{u}_1 - \nu \Delta \mathbf{u}_1 + g \, \nabla \zeta_1 = -g \, \nabla \zeta_2 \\
\partial_t \zeta_1 + \nabla \cdot (\zeta_1 \, \mathbf{u}_1) = 0 \\
\partial_t \mathbf{u}_2 + (\mathbf{u}_2 \cdot \nabla) \mathbf{u}_2 - \nu \Delta \mathbf{u}_2 + (g + g') \nabla \zeta_2 = -g \nabla \zeta_1 \\
\partial_t \zeta_2 + \nabla \cdot (\zeta_2 \, \mathbf{u}_2) = 0 \, ,
\end{cases}
\tag{21.11}
$$

to be solved in $\Omega \times (0, T)$ subject to the initial and boundary conditions (21.8), (21.9) and (21.10) (with $k = 1, 2$). We assume that this problem admits a unique solution, see below for the regularity assumptions.

Remark 21.1. It is easier to analyse the equations when they are written in terms of the layer heights as in (21.11). The surface/interface elevations are recovered from the formulae

$$
\zeta_1 = \eta_1 - \eta_2 , \qquad \zeta_2 = \eta_2 + H .
$$

21.3   A Priori Error Estimate for the Two-Layer Model

We shall adopt standard notation, i.e., $(\cdot, \cdot)_\Omega$ is the inner product in $L_2(\Omega)$, $H^k(\Omega)$ denote the usual Sobolev spaces and $H_0^1(\Omega)$ is the space of functions in $H^1(\Omega)$ with zero trace at $\partial\Omega$. When there is no danger of confusion, the norm in $L_2(\Omega)$ is denoted simply by $\|\cdot\|$. The $L_2(\gamma)$-inner product over one-dimensional surface γ is denoted by $\langle \cdot, \cdot \rangle_\gamma$ and the corresponding norm by $\|\cdot\|_\gamma$.

Let $\{\mathscr{T}_h\}_{h>0}$ be a family of conforming, quasi-uniform partitions of Ω into affine-equivalent, shape regular finite elements (triangles or quadrilaterals) Ω_i, for $i = 1, 2, \ldots, N_h$. Moreover, let $h_i = \text{diam } \Omega_i$, $h = \max_i h_i$ and denote by γ_j, for $j = 1, \ldots, M_h$, the interior edges of the partition $\mathscr{T}_h$.

For each $t > 0$, we approximate $\mathbf{u}_j(\cdot, t)$ by continuous, piecewise polynomial functions $\mathbf{u}_h^{(j)}(\cdot, t) \in [V_h]^2$ and $\zeta_j(\cdot, t)$ by a (possibly) discontinuous polynomials $\zeta_h^{(j)}(\cdot, t) \in W_h$. This means that V_h and W_h are finite-dimensional subspaces of $H^1(\Omega)$ defined by

$$
V_h = \{v \in C(\overline{\Omega}) \mid v \in \mathscr{S}_k(\Omega_i) \; \forall \, \Omega_i \in \mathscr{T}_h\} ,
$$

$$
W_h = \{\xi : \overline{\Omega} \to \mathbb{R} \mid \xi \in \mathscr{S}_k(\Omega_i) \; \forall \, \Omega_i \in \mathscr{T}_h\},
$$

where $\mathscr{S}_k(\Omega_i) \supset \mathscr{P}_k(\Omega_i)$ with $\mathscr{P}_k(\Omega_i)$ denoting the space of all polynomials of degree $k \geq 1$ on Ω_i. We also define $V_{h,0} = V_h \cap H_0^1(\Omega)$ and, for any $\xi \in W_h$, let

$$
\xi^+(\mathbf{x}) = \lim_{\varepsilon \to 0^+} \xi(\mathbf{x} + \varepsilon \, \mathbf{n}_j) , \qquad \xi^-(\mathbf{x}) = \lim_{\varepsilon \to 0^-} \xi(\mathbf{x} + \varepsilon \, \mathbf{n}_j) , \qquad \mathbf{x} \in \gamma_j ,
$$

where $\mathbf{n}_j$ is a unit normal at γ_j and denote the jump of ξ over γ_j by $[\xi] = \xi^- - \xi^+$.

21.3.1   Semi-Discrete Galerkin Formulation

The semi-discrete Galerkin formulation of problem (21.11) reads as follows:
Find $\mathbf{u}_h^{(1)}, \mathbf{u}_h^{(2)} : [0, T] \rightarrow [V_h]^2$ and $\zeta_h^{(1)}, \zeta_h^{(2)} : [0, T] \rightarrow W_h$ satisfying

$$\left(\partial_t \mathbf{u}_h^{(1)}(t), \mathbf{w}_h^{(1)}\right)_\Omega + \left((\mathbf{u}_h^{(1)}(t) \cdot \nabla) \mathbf{u}_h^{(1)}(t), \mathbf{w}_h^{(1)}\right)_\Omega + \nu \left(\nabla \mathbf{u}_h^{(1)}(t), \nabla \mathbf{w}_h^{(1)}\right)_\Omega -$$

$$- g \sum_j \left\langle [\zeta_h^{(1)}(t)], \mathbf{w}_h^{(1)} \cdot \mathbf{n}_j \right\rangle_{\gamma_j} + g \sum_i \left(\nabla \zeta_h^{(1)}(t), \mathbf{w}_h^{(1)}\right)_{\Omega_i} -$$

$$- g \sum_j \left\langle [\zeta_h^{(2)}(t)], \mathbf{w}_h^{(1)} \cdot \mathbf{n}_j \right\rangle_{\gamma_j}$$

$$+ g \sum_i \left(\nabla \zeta_h^{(2)}(t), \mathbf{w}_h^{(1)}\right)_{\Omega_i} = 0, \qquad \forall \, \mathbf{w}_h^{(1)} \in [V_{h,0}]^2$$

$$\sum_i \left(\partial_t \zeta_h^{(1)}(t), \xi_h^{(1)}\right)_{\Omega_i} - \sum_i \left(\zeta_h^{(1)}(t) \, \mathbf{u}_h^{(1)}(t), \nabla \xi_h^{(1)}\right)_{\Omega_i} +$$

$$+ \sum_j \left\langle \zeta_h^{(1)}(t)^\uparrow \mathbf{u}_h^{(1)}(t) \cdot \mathbf{n_j}, [\xi_h^{(1)}] \right\rangle_{\gamma_j} +$$

$$+ \left\langle \zeta_h^{(1)}(t) \, \hat{\mathbf{u}}_1(t) \cdot \mathbf{n}, \xi_h^{(1)} \right\rangle_{\partial\Omega_O} = - \left\langle \hat\zeta_1(t) \, \hat{\mathbf{u}}_1(t) \cdot \mathbf{n}, \xi_h^{(1)} \right\rangle_{\partial\Omega_I}, \qquad \forall \, \xi_h^{(1)} \in W_h$$

$$\left(\partial_t \mathbf{u}_h^{(2)}(t), \mathbf{w}_h^{(2)}\right)_\Omega + \left((\mathbf{u}_h^{(2)}(t) \cdot \nabla) \mathbf{u}_h^{(2)}(t), \mathbf{w}_h^{(2)}\right)_\Omega + \nu \left(\nabla \mathbf{u}_h^{(2)}(t), \nabla \mathbf{w_h}^{(2)}\right)_\Omega -$$

$$- (g + g') \sum_j \left\langle [\zeta_h^{(2)}(t)], \mathbf{w}_h^{(2)} \cdot \mathbf{n}_j \right\rangle_{\gamma_j} + (g + g') \sum_i \left(\nabla \zeta_h^{(2)}(t), \mathbf{w}_h^{(2)}\right)_{\Omega_i} -$$

$$- g \sum_j \left\langle [\zeta_h^{(1)}(t)], \mathbf{w}_h^{(2)} \cdot \mathbf{n}_j \right\rangle_{\gamma_j}$$

$$+ g \sum_i \left(\nabla \zeta_h^{(1)}(t), \mathbf{w}_h^{(2)}\right)_{\Omega_i} = 0, \qquad \forall \, \mathbf{w}_h^{(2)} \in [V_{h,0}]^2$$

$$\sum_i \left(\partial_t \zeta_h^{(2)}(t), \xi_h^{(2)}\right)_{\Omega_i} - \sum_i \left(\zeta_h^{(2)}(t) \, \mathbf{u}_h^{(2)}(t), \nabla \xi_h^{(2)}\right)_{\Omega_i} +$$

$$+ \sum_j \left\langle \zeta_h^{(2)}(t)^\uparrow \mathbf{u}_h^{(2)}(t) \cdot \mathbf{n_j}, [\xi_h^{(2)}] \right\rangle_{\gamma_j} +$$

$$+ \left\langle \zeta_h^{(2)}(t) \, \hat{\mathbf{u}}_2(t) \cdot \mathbf{n}, \xi_h^{(2)} \right\rangle_{\partial\Omega_O} = - \left\langle \hat\zeta_2(t) \, \hat{\mathbf{u}}_2(t) \cdot \mathbf{n}, \xi_h^{(2)} \right\rangle_{\partial\Omega_I} \quad \forall \, \xi_h^{(2)} \in W_h$$

$$\tag{21.12}$$

$$\left(\mathbf{u}_h^{(1)}(0) - \mathbf{u}_0^{(1)}, \mathbf{w}_h^{(1)}\right)_\Omega = 0, \qquad \forall \mathbf{w}_h^{(1)} \in [V_{h,0}]^2$$

$$\left(\zeta_h^{(1)}(0) - \zeta_0^{(1)}, \xi_h^{(1)}\right)_\Omega = 0, \qquad \forall \xi_h^{(1)} \in W_h$$

$$\left(\mathbf{u}_h^{(2)}(0) - \mathbf{u}_0^{(2)}, \mathbf{w}_h^{(2)}\right)_\Omega = 0, \qquad \forall \mathbf{w}_h^{(2)} \in [V_{h,0}]^2$$

$$\left(\zeta_h^{(2)}(0) - \zeta_0^{(2)}, \xi_h^{(2)}\right)_\Omega = 0, \qquad \forall \xi_h^{(2)} \in W_h$$

The summations in i above are over all elements of the partition and the sums in j over all interior edges γ_j. Moreover, $\zeta_h^\uparrow$ denotes the upwind value of ζ_h across the interior edges, i.e.

$$\zeta_h^\uparrow = \begin{cases} \zeta_h^-, & \text{if } \mathbf{u}_h \cdot \mathbf{n}_j > 0 \text{ at } \gamma_j \\ \zeta_h^+, & \text{if } \mathbf{u}_h \cdot \mathbf{n}_j \le 0 \text{ at } \gamma_j \end{cases}$$

Note that the jump terms in (21.12)$_{1-4}$ vanish for (continuous) exact solution. Therefore, multiplying the four equations in (21.11) by test functions $\mathbf{w}_1 \in H_0^1(\Omega)$, $\xi_1 \in H^1(\Omega)$, $\mathbf{w}_2 \in H_0^1(\Omega)$ and $\xi_2 \in H^1(\Omega)$, respectively, performing some integrations by parts and using the boundary conditions (21.9) and (21.10), one easily sees that the method is consistent. The jump (stabilization) terms are added to the momentum equations in order to handle the integrations by parts in the coupling terms.

Equation (21.12) define a system of nonlinear ordinary differential equations that can be solved by any standard method (e.g., Runge-Kutta, finite differences) and we can thus assume the existence of a discrete solution, for each $t > 0$.

21.3.2 A Priori Error Estimate

Let us assume that the exact solution possesses the regularity

$$\mathbf{u}_1, \mathbf{u}_2 \in L^2\left(0, T; [H^{k+1}(\Omega)]^2\right) \cap L^\infty\left(0, T; [W^{1,\infty}(\Omega)]^2\right),$$

$$\partial_t \mathbf{u}_1, \partial_t \mathbf{u}_2 \in L^2\left(0, T; [H^k(\Omega)]^2\right),$$

$$\zeta_1, \zeta_2 \in L^2\left(0, T; H^{k+1}(\Omega)\right),$$

$$\partial_t \zeta_1, \partial_t \zeta_2 \in L^2\left(0, T; H^k(\Omega)\right),$$

and that

$$\mathbf{u}_0^{(1)}, \mathbf{u}_0^{(2)} \in [H^k(\Omega)]^2, \qquad \zeta_0^{(1)}, \zeta_0^{(2)} \in H^k(\Omega),$$

with the corresponding norms all bounded by some constant

$$K = K\left(\mathbf{u}_1, \mathbf{u}_2, \zeta_1, \zeta_2, \mathbf{u}_0^{(1)}, \mathbf{u}_0^{(2)}, \zeta_0^{(1)}, \zeta_0^{(2)}\right) > 0.$$

Given that the error estimate is bound to be suboptimal, it suffices to define $\Pi \mathbf{u}_j(\cdot, t) \in [V_h]^2$ and $\Pi \mathbf{u}_0^{(j)} \in [V_h]^2$ as the $L^2(\Omega)$-projections of $\mathbf{u}_j(\cdot, t)$ and $\mathbf{u}_0^{(j)}$, $j = 1, 2$, respectively. Moreover, we let $\zeta_I^{(j)}(\cdot, t) \in V_h$ be the continuous interpolant of $\zeta_j(\cdot, t)$, denote the approximation (projection or interpolation) errors as

$$\boldsymbol{\theta}_{\mathbf{u}}^{(j)} = \mathbf{u}_j - \Pi \mathbf{u}^{(j)}, \qquad \theta_\zeta^{(j)} = \zeta_j - \zeta_I^{(j)}$$

and introduce the (affine) errors

$$\boldsymbol{\psi}_{\mathbf{u}}^{(j)} = \mathbf{u}_h^{(j)} - \Pi \mathbf{u}^{(j)}, \qquad \psi_\zeta^{(j)} = \zeta_h^{(j)} - \zeta_I^{(j)}.$$

The finite element errors are thus $\mathbf{u}_j - \mathbf{u}_h^{(j)} = \boldsymbol{\theta}_{\mathbf{u}}^{(j)} - \boldsymbol{\psi}_{\mathbf{u}}^{(j)}$ and $\zeta_j - \zeta_h^{(j)} = \theta_\zeta^{(j)} - \psi_\zeta^{(j)}$.

We recall that, see [2, Chap. 4], there exists a constant $K_I = K(\zeta_1, \zeta_2) > 0$ such that

$$\|\zeta_I^{(1)}\|_{L^\infty(0,T;W^{1,\infty}(\Omega))} + \|\zeta_I^{(2)}\|_{L^\infty(0,T;W^{1,\infty}(\Omega))} \leq K_I ,$$

and *assume* that there exists a constant $K_* > 0$ such that

$$\sum_{j=1}^{2} \left(\|\zeta_h^{(j)}\|_{L^\infty(0,T;L^\infty(\Omega))} + \|\mathbf{u}_h^{(j)}\|_{L^\infty(0,T;L^\infty(\Omega))} \right) \leq K_* , \tag{21.13}$$

where $K_* \geq 2 \max\{K_I, K\}$; recall that the constant $K > 0$ bounds, in particular, the $L^\infty(0, T; L^\infty(\Omega))$-norms of $\mathbf{u}_1$ and $\mathbf{u}_2$. We will see in the end that, for $k > 1$ and for h small enough, the bound (21.13) is always satisfied, i.e., it is not an extra assumption on the discrete solution, see e.g., [4, 11] or [9] for similar reasoning.

We will now state our main result

Theorem 21.1. *For $\mathbf{u}_1$, $\mathbf{u}_2$, ζ_1, ζ_2 smooth enough, we have the following error estimate*

$$\sum_{j=1}^{2} \left(\|(\zeta_j - \zeta_h^{(j)})(T)\| + \|(\mathbf{u}_j - \mathbf{u}_h^{(j)})(T)\| \right) \leq C h^k , \tag{21.14}$$

where the constant $C = C(g, g', \nu, T, K, K_I, K_*) > 0$ does not depend on h nor k.

Proof. We will sketch only the main ideas of the proof and, to simplify the notation, omit the functions' dependence on t whenever possible.

First, for each of the four equations in $(21.12)_{1-4}$, we consider the corresponding continuous counterpart and, taking as test functions $\boldsymbol{\psi}_{\mathbf{u}}^{(j)} \in [V_{h,0}]^2 \subset [H_0^1(\Omega)]^2$ and $\psi_\zeta^{(j)} \in W_h \subset H^1(\Omega)$, subtract the equations pairwise, integrate in t over the interval $[0, T]$ and sum all the resulting equations.

Aiming at bounding the affine errors, we collect to the left-hand side the (non-negative) terms

$$|||(\boldsymbol{\psi}_{\mathbf{u}}^{(1)}, \psi_{\zeta}^{(1)}, \boldsymbol{\psi}_{\mathbf{u}}^{(2)}, \psi_{\zeta}^{(2)})|||^2 :=$$

$$= \|\psi_{\zeta}^{(1)}(T)\|^2 + \|\psi_{\zeta}^{(2)}(T)\|^2 + \|\boldsymbol{\psi}_{\mathbf{u}}^{(1)}(T)\|^2 + \|\boldsymbol{\psi}_{\mathbf{u}}^{(2)}(T)\|^2$$

$$+ \int_0^T \sum_j \langle |\mathbf{u}_h^{(1)} \cdot \mathbf{n}_j|, [\psi_{\zeta}^{(1)}]^2 \rangle_{\gamma_j} dt + \int_0^T \sum_j \langle |\mathbf{u}_h^{(2)} \cdot \mathbf{n}_j|, [\psi_{\zeta}^{(2)}]^2 \rangle_{\gamma_j} dt$$

$$+ \int_0^T \langle |\mathbf{u}_h^{(1)} \cdot \mathbf{n}|, (\psi_{\zeta}^{(1)})^2 \rangle_{\partial\Omega_1} dt + \int_0^T \langle |\mathbf{u}_h^{(2)} \cdot \mathbf{n}|, (\psi_{\zeta}^{(2)})^2 \rangle_{\partial\Omega_1} dt$$

$$+ \int_0^T \langle |\hat{\mathbf{u}}_1 \cdot \mathbf{n}|, (\psi_{\zeta}^{(1)})^2 \rangle_{\partial\Omega_1} dt + \int_0^T \langle |\hat{\mathbf{u}}_2 \cdot \mathbf{n}|, (\psi_{\zeta}^{(2)})^2 \rangle_{\partial\Omega_0} dt$$

$$+ \nu \int_0^T \|\nabla\boldsymbol{\psi}_{\mathbf{u}}^{(1)}\|^2 dt + \nu \int_0^T \|\nabla\boldsymbol{\psi}_{\mathbf{u}}^{(2)}\|^2 dt .$$

As for the terms left on the right-hand side, the linear terms (bilinear in the variational form) are easily bounded by the approximation error and the terms

$$\frac{\nu}{4} \int_0^T \|\nabla\boldsymbol{\psi}_{\mathbf{u}}^{(1)}\|^2 dt + \frac{\nu}{4} \int_0^T \|\nabla\boldsymbol{\psi}_{\mathbf{u}}^{(2)}\|^2 + C_1 \left(\int_0^T \|\psi_{\zeta}^{(1)}\|^2 dt + \int_0^T \|\psi_{\zeta}^{(2)}\|^2 dt \right),$$

where $C_1 = C_1(\nu, g, g') > 0$. To estimate the nonlinear terms, we write them as

$$(\mathbf{u}_j \cdot \nabla)\mathbf{u}_j - (\mathbf{u}_h^{(j)} \cdot \nabla)\mathbf{u}_h^{(j)} = (\boldsymbol{\theta}_{\mathbf{u}}^{(j)} - \boldsymbol{\psi}_{\mathbf{u}}^{(j)}) \cdot \nabla \mathbf{u}_j + \mathbf{u}_h^{(j)} \cdot \nabla (\boldsymbol{\theta}_{\mathbf{u}}^{(j)} - \boldsymbol{\psi}_{\mathbf{u}}^{(j)})$$

$$\zeta_j \mathbf{u}_j - \zeta_h^{(j)}\mathbf{u}_h^{(j)} = \zeta_j \mathbf{u}_j - \zeta_I^{(j)}\mathbf{u}_h^{(j)} + \psi_{\zeta}^{(j)}\mathbf{u}_h^{(j)} .$$

and integrate the terms

$$\left(\zeta_j \mathbf{u}_j - \zeta_I^{(j)}\mathbf{u}_h^{(j)} + \psi_{\zeta}^{(j)}\mathbf{u}_h^{(j)}, \nabla\psi_{\zeta}^{(j)} \right)_{\Omega_i}$$

by parts. Using the regularity of the exact solution and the bounds K_I and K_* of the interpolant $\zeta_I^{(j)}$ and the discrete solution, one readily obtains an estimate for the integrals over the domain Ω in terms of the approximation error and

$$\frac{\nu}{4} \int_0^T \|\nabla\boldsymbol{\psi}_{\mathbf{u}}^{(1)}\|^2 dt + \frac{\nu}{4} \int_0^T \|\nabla\boldsymbol{\psi}_{\mathbf{u}}^{(2)}\|^2 + C_2 \sum_{j=1}^2 \left(\int_0^T \|\psi_{\zeta}^{(j)}\|^2 dt + \int_0^T \|\boldsymbol{\psi}_{\mathbf{u}}^{(j)}\|^2 dt \right),$$

where $C_2 = C_2(\nu, K, K_I, K_*) > 0$.

The nonlinear boundary terms (the linear boundary terms in the momentum equations vanish after partial integration) can be handled with the trace inequality, cf. [2, Theorem 1.6.6],

$$\|u\|_{L^2(\partial\Omega)} \leq c \, \|u\|_{L^2(\Omega)}^{1/2} \|u\|_{H^1(\Omega)}^{1/2} \qquad \forall u \in H^1(\Omega),$$

and the inverse inequality (e.g., [2])

$$\|u_h\|_{H^1(\Omega)} \leq C h^{-1} \|u_h\|_{L^2(\Omega)} \qquad \forall u_h \in W_h .$$

The final upper bound for the nonlinear boundary terms is of same form as for the other nonlinear terms.

Recalling the standard approximation estimates (cf. [2]) for the projection and interpolation errors

$$\int_0^T \left(\|\partial_t \theta_\zeta^{(j)}\|^2 + \|\nabla\theta_\zeta^{(j)}\|^2 + h^{-2}\|\theta_\zeta^{(j)}\|^2 \right) dt + \|\theta_\zeta^{(j)}(0)\|^2 \leq C_3 \, h^{2k} ,$$

$$\int_0^T \left(\|\nabla\boldsymbol{\theta}_{\mathbf{u}}^{(j)}\|^2 + h^{-2}\|\boldsymbol{\theta}_{\mathbf{u}}^{(j)}\|^2 \right) dt \leq C_4 \, h^{2k} ,$$

and collecting all the estimates, we obtain

$$\left\| \left(\boldsymbol{\psi}_{\mathbf{u}}^{(1)}, \psi_\zeta^{(1)}, \boldsymbol{\psi}_{\mathbf{u}}^{(2)}, \psi_\zeta^{(2)} \right) \right\|^2$$

$$\leq C_5 \, h^{2k} + C_6 \sum_{j=1}^2 \left(\int_0^T \|\psi_\zeta^{(j)}\|^2 dt + \int_0^T \|\boldsymbol{\psi}_{\mathbf{u}}^{(j)}\|^2 dt \right).$$

Now, a simple application of Gronwall's lemma yields

$$\|\psi_\zeta^{(1)}(T)\|^2 + \|\psi_\zeta^{(2)}(T)\|^2 + \|\boldsymbol{\psi}_{\mathbf{u}}^{(1)}(T)\|^2 + \|\boldsymbol{\psi}_{\mathbf{u}}^{(2)}(T)\|^2 \leq C_7 \, h^{2k}$$

and, given that

$$\sum_{j=1}^2 \left(\|\theta_\zeta^{(j)}(T)\|^2 + \|\boldsymbol{\theta}_{\mathbf{u}}^{(j)}(T)\|^2 \right) \leq C_8 \, h^{2k} ,$$

the a priori error estimate (21.14) finally follows from the triangle inequality.   □

Remark 21.2. If $k > 1$ and $h \ll 1$ is sufficiently small, we can argue as follows, see also [4, 9], and remove the dependence on K_* in estimate (21.14). In fact, the inverse inequality, see [2, Theorem 4.5.11],

$$\|u_h(\cdot, t)\|_{L^\infty(\Omega)} \leq C_9 h^{-1} \|u_h(\cdot, t)\|_{L^2(\Omega)} \qquad \forall u_h(\cdot, t) \in W_h$$

shows that

$$\|\zeta_h^{(j)}\|_{L^\infty(0,T;L^\infty(\Omega))} \leq \|\zeta_I^{(j)}\|_{L^\infty(0,T;L^\infty(\Omega))} + \|\psi_\zeta^{(j)}\|_{L^\infty(0,T;L^\infty(\Omega))}$$

$$\leq K_I + C_9 h^{-1} C h^k \ll 2 K_I \leq K_*$$

and

$$\|\mathbf{u}_h^{(j)}\|_{L^\infty(0,T;L^\infty(\Omega))} \leq \|\mathbf{u}_j\|_{L^\infty(0,T,L^\infty(\Omega))} + \|\boldsymbol{\theta}_{\mathbf{u}}^{(j)}\|_{L^\infty(0,T,L^\infty(\Omega))}$$

$$+ \|\boldsymbol{\psi}_{\mathbf{u}}^{(j)}\|_{L^\infty(0,T,L^\infty(\Omega))}$$

$$\leq K + C_{10} h^k + C_{11} h^{k-1} \ll 2K \leq K_* .$$

Remark 21.3. It is straightforward to take into account Earth's rotation by adding the Coriolis acceleration terms $f \mathbf{k} \times \mathbf{u}_1$ and $f \mathbf{k} \times \mathbf{u}_2$ to the left-hand side of the momentum equations $(21.11)_1$ and $(21.11)_3$, with f denoting the Coriolis parameter, positive on the Northern Hemisphere and in general depending on the latitude, and $\mathbf{k}$ standing for the upward-pointing Cartesian unit vector. Equally simple is to consider a not-flat bottom. In this case, it suffices to substitute ζ_2 by $\zeta_2 + \eta_b$ in the momentum equations, with $\eta_b = \eta_b(x, y)$ denoting a (given) bottom profile. The momentum equations may also include linear or nonlinear bottom friction terms (in the bottom layer) and forcing terms, such as wind stress or atmospheric pressure gradient in the top layer. All these terms can be treated without further complications and have only been omitted to clarify the presentation.

Acknowledgements The first author was supported by a FCT (Fundação para a Ciência e a Tecnologia) fellowship SFRH/BD/70749/2010. The second author was partially supported by a FCT fellowship SFRH/BD/36583/2007. The last author was partially supported by the project UTAustin/MAT/0035/2008.

References

1. Aizinger, V., Dawson, C.N.: Discontinuous Galerkin methods for two-dimensional flow and transport in shallow water. Adv. Water Res. **25**, 67–84 (2002)
2. Brenner, S.C., Scott, L.R.: The Mathematical Theory of Finite Element Methods. 3rd edn. Springer, New York (2008)
3. Chippada, S., Dawson, C.N., Martinez, M.L., Wheeler, M.F.: Finite element approximations to the system of shallow-water equations I: continuous-time a priori estimates. SIAM J. Numer. Anal. **35**, 692–711 (1998)
4. Chippada, S., Dawson, C.N., Martinez, M.L., Wheeler, M.F.: A Godunov-type finite volume method for the system of shallow water equations. Comput. Methods Appl. Mech. Eng. **151**, 105–29 (1998)
5. Chippada, S., Dawson, C.N., Martinez, M.L., Wheeler, M.F.: Finite element approximations to the system of shallow-water equations II: discrete-time a priori estimates. SIAM J. Numer. Anal. **36**, 226–250 (1999)

6. Cockburn, B.: Discontinuous Galerkin methods for computational fluid mechanics. In: Stein, E., de Borst, R., Hughes, T.J.R. (eds.) Encyclopedia of Computational Mechanics. Fluids, vol. 3, pp. 91–127. Wiley, Chichester (2004)

7. Dawson, C.N.: Analysis of discontinuous finite element methods for ground water/surface water coupling. SIAM J. Numer. Anal. **44**, 1375–1404 (2006)

8. Dawson, C.N.: A continuous/discontinuous Galerkin framework for modeling coupled subsurface and surface water flow. Comput. Geosciences **12**, 451–472 (2008)

9. Dawson, C., Proft, J.: Discontinuous and coupled continuous/discontinuous Galerkin methods for the shallow water equations. Comput. Methods Appl. Mech. Eng. **191**, 4721–4746 (2002)

10. Dawson, C., Proft, J.: Coupled discontinuous and continuous Galerkin finite element methods for the depth-integrated shallow water equations. Comput. Methods Appl. Mech. Eng. **193**, 289–318 (2004)

11. Ewing, R.E., Wheeler, M.F.: Galerkin methods for miscible displacement problems in porous media. SIAM J. Numer. Anal. **17**, 351–365 (1980)

12. Johnson, C., Pitkäranta, J.: An analysis of the discontinuous Galerkin method for a scalar hyperbolic equation. Math. Comput. **46**, 1–26 (1986)

13. Kanayama, H., Dan, H.: A finite element scheme for two-layer viscous shallow-water equations. Jpn. J. Ind. Appl. Math. **23**, 163–191 (2006)

14. Kolar, R., Westerink, J., Cantekin, M., Blain, C.: Aspects of nonlinear simulations using shallow water models based on the wave continuity equation. Comput. Fluids **23**, 523–538 (1994)

15. Kurganov, A., Petrova, G.: Central-upwind schemes for two-layer shallow water equations. SIAM J. Sci. Comput. **31**, 1742–1773 (2009)

16. LeSaint, P., Raviart, P.A.: On a finite element method for solving the neutron transport equation. In: de Boor, C. (ed.) Mathematical Aspects of Finite Elements in Partial Differential Equations, pp. 89–145. Academic, New York (1974)

17. Lynch D.R., Gray, W.: A wave equation model for finite element tidal computations. Comput. Fluids **7**, 207–228 (1979)

18. Reed, W.H., Hill, T.R.: Triangular mesh methods for the neutron transport equation. Technical Report LA-UR-73-479, Los Alamos Scientific Laboratory, Los Alamos (1973)

19. Salmon, R.: Numerical solution of the two-layer shallow-water equations with bottom topography. J. Mar. Res. **60**, 605–638 (2002)

20. Vallis, G.K.: Atmospheric and Oceanic Fluid Dynamics: Fundamentals and Large-Scale Circulation. Cambridge University Press, Cambridge (2006)

21. Vreugdenhil, C.B.: Numerical Methods for Shallow-Water Flow. Kluwer, Boston (1994)

22. Weiyan, T.: Shallow Water Hydrodynamics. Elsevier Oceanography Series, vol. 55. Elsevier, Amsterdam (1992)

Chapter 22
Bankruptcy Triggering Asset Value: Continuous Time Finance Approach

Karel Janda and Jakub Rojcek

22.1 Introduction

The aim of this paper is to provide an answer to a simple question: "At what share price shall an investor expect a company goes bankrupt?" Answering this question will be provided be means of the Game Theory Analysis of Options introduced in [10]. This useful method consists of two basic legs: game theory and option pricing. In the second part of this work we will attempt to provide an example of a listed company going bankrupt. In this example we will apply the theory to a real-life on the financial markets. This work could be further considered when building algorithmic trading models seeking for final warning values of the stocks traded.

The basic insight of the technique which we use in this chapter is separating a given problem into valuation of the payoffs and the analysis of strategic interactions between parties of a contract. We will handle the two separated parts by means of option pricing[1] and game theory, respectively. Ziegler in [10] presents this method

---

[1] Referring to [2] and [6].

K. Janda (✉)
Faculty of Social Sciences, Institute of Economic Studies, Charles University in Prague, Prague, Czech Republic

Faculty of Finance and Accounting, Department of Banking and Insurance, University of Economics, Prague, Czech Republic
e-mail: karel-janda@seznam.cz

J. Rojcek
Department of Banking and Finance, University of Zurich, Zurich, Switzerland
e-mail: jakub.rojcek@gmail.com

A.A. Pinto and D. Zilberman (eds.), *Modeling, Dynamics, Optimization and Bioeconomics I*, Springer Proceedings in Mathematics & Statistics 73, DOI 10.1007/978-3-319-04849-9__22,
© Springer International Publishing Switzerland 2014

as an attempt to integrate game theory and option pricing. However, we see that this apparatus could be further enhanced by magnifying the amount of game theory tools integrated. According to Ziegler, this game theory analysis of options is applicable to the following:

- *pricing* of contingent claims and corporate securities when economic agents can behave strategically,
- analysis of *incentive effects* of some common contractual financial arrangements, and
- design of *incentive contracts* aiming at resolving conflicts of interest between economic agents.

Bankruptcy prediction has been studied since [1] using the z-score. Wilson and Sharda [9] argue that Bankruptcy predictors using neural networks out-perform this previous discriminate type models. Most of the predictors, however, represent an econometric model, trying to fit the default data using whole range of explanatory variables as e.g. in logit model of [8]. These statistical models usually do not capture effects of different law procedures as in [4], who theoretically studied the impact of the U.S. bankruptcy procedure with renegotiation possibility on the valuation of corporate securities and capital structure decisions. Moreover, the various incentives of stakeholders in the company also play an important and often omitted role in bankruptcy decisions. In this paper, we build a game theoretic model, which produces a bankruptcy triggering asset value in closed form based only on a small set of parameters.

22.2   The Method of Game Theory Analysis of Options

In the book *The Game Theory Analysis of Options: Corporate Finance and Financial Intermediation* [10] Alexandre Ziegler attempts to combine game theory and option pricing. He argues that when he uses option pricing for obtaining players' future payoffs, these payoffs are arbitrage-free, discounted to the present and at the same time the price of risk is taken into account. Afterwards, he moves on with inserting these values into strategic games between the agents.

More tangibly, in our case, when we will try to find out *the bankruptcy triggering asset value*, we will set up a game of three parties all having a stake in a company. First of all, there will be a manager equity holder who will possess the shares of the company and run the firm. Secondly, there will be debt holders, who will buy the liability issued by the company, and at last, but not least there will be an investor interested in the company's dividend and trying to quit earlier than the business goes under.

22.2.1   Three-Step Methodology

Taking step back to introducing the method, it can be digested into the three main steps:

1. At first, the game is defined. The players' action sets, the sequence of their choices and the consequent payoffs are specified.
2. The second step is to value the players' future uncertain payoffs using option pricing theory. Players' possible actions enter the valuation formulas as parameters (for example the risk of the asset chosen by an agent).
3. The ultimate stage is utilizing the backward induction or sub-game perfection starting with the last period. Here the last period refers to the last decision to be made (for example computing the bankruptcy triggering asset value and then move backwards).

In classical game theory[2] the game is solved by considering expected utilities of all players. Here, Ziegler [10] tries to replace expected utility maximization with the maximization of the value of an option. Furthermore, he claims that the main advantage over the classical approach lies in separating the valuation problem (step 2) from the strategic interactions analysis (step 3). By and large, this means that the analysis and solving the game can be frequently collapsed into finding a first-order condition for a maximum or minimum in the option's price at each decision node of the game, where we are handling uncertain payoffs.

To better grasp the practicalities of the method, let's consider a two players game, each choosing his or her optimal strategy: *Player*1 chooses strategy A and immediately afterwards *Player*2 chooses his strategy B. These strategies are mutually best responses and determine the ultimate current arbitrage-free payoffs given by $G(A, B, S, t)$ for *Player*1 and $H(A, B, S, t)$ for *Player*2. The players' strategies mean choosing one of the parameters of the differential equation

$$\frac{1}{2}\sigma^2 S^2 F_{SS} + (rS - a)F_S + F_t - rF_1 + b = 0, \qquad (22.1)$$

as well as its boundary conditions.

In the last stage of the game, Player 2 chooses that strategy B, which maximizes his expected payoff:

$$\frac{\partial H(A, B, S, t)}{\partial B} = 0, \qquad (22.2)$$

provided that B is not a boundary solution. The solution yields an optimal strategy $\overline{B} = \overline{B}(A, S, t)$, which might depend on *Player*1's strategy A, underlying asset's value S and time t.

---

[2]See for example [7].

*Player*1 must anticipate *Player*2's consequent strategy choice $\overline{B}$ and maximize value of his own payoff, G, using first-order condition:

$$\frac{dG(A, \overline{B}, S, t)}{dA} = \frac{\partial G(A, \overline{B}, S, t)}{\partial A} + \frac{\partial G(A, \overline{B}, S, t)}{\partial \overline{B}} \frac{d\overline{B}(A, S, t)}{dA} = 0, \quad (22.3)$$

which yields an optimal strategy $\overline{A} = \overline{A}(S, t)$, that may again depend on the value of the underlying asset and time. The term

$$\frac{\partial G(A, \overline{B}, S, t)}{\partial \overline{B}} \frac{d\overline{B}(A, S, t)}{dA}, \quad (22.4)$$

in the first-order condition reflects the gist of backward induction, i.e. the indirect effect of *Player*1's strategy on his expected payoff that results from the influence his own choice has on *Player*2's optimal strategy $\overline{B}$.

22.2.2 When Is the Method Appropriate?

However appealing the visage of the method may be, it still has to satisfy some theoretical requirements to be appropriate.

At first, we must answer the already mentioned issue: is an option value a good proxy for expected utility? First we should note that it is nicely translating *uncertain future payoffs* and adjusting them for risk to the present value, then we should realize that here is a *monotonic increasing*, however not necessarily linear, relationship between the option's value and agent's utility. This implies that any utility maximization choice by the agent is also optimizing the value of option and vice versa. The answer is then 'yes', it is a good proxy.

All of this is true under condition that the option's value is correct. And when it is correct? It is correct only in the case that the option pricing implicit adjustments on time and uncertainty are consistent with the *underlying information structure* and with agent's *preferences*. This statement needs a couple of words more for explanation. The main requirement is that the underlying asset's state-price density is lognormal. This will be the case for example, if the underlying asset's price follows a geometric Brownian motion and the risk-free rate is constant, or if the aggregate endowment in the economy follows a geometric Brownian motion and investors have constant relative risk aversion (CRRA) utility.

Ziegler moreover notes that the method does not require that the underlying asset S be traded to be applicable. We will essentially require only that there exists a traded security whose price is driven by the same Wiener process dB_t as the underlying asset's value and investors can trade these securities continuously with zero transaction costs. This is because under these conditions, as shown by Brennan and Schwartz [3] a continuously rebalanced self-financing portfolio can be constructed that replicates the underlying asset. In our case, we will consider only companies which were traded in high volumes on the stock exchange, e.g. Bear Stearns.

22.2.3 For What Problems Is It Suitable?

The game theory analysis of options could be considered for the investigation of strategic interactions in which a direct evaluation of players' expected utilities is cumbersome.

At first, as mentioned above, it can well handle problems in which uncertainty takes place.

Secondly, it is suitable for problems in which the payoff time is not precisely specified and thus payoffs cannot be easily discounted. The time of the payoff may be driven by exogenous uncertainty alone, or may even depend endogenously on decisions made by the players.[3] The technique is suited to value payoffs that occur at random times and to analyze timing or optimal stopping problems. This all is again due to option pricing.

The third type of problems is the presence of option value in players' payoffs, i.e. when these payoffs are a nonlinear function of the underlying asset's value. This makes the method valuable in the field of real option analysis.

22.3 Bankruptcy Triggering Asset Value

22.3.1 Introduction

In this section, we will draw our attention to determining a theoretical asset value, or price of a share, at which a bankruptcy is triggered. In other words, at the bankruptcy triggering asset value, the shareholders should switch from a long position and sell the shares before the company goes under. This is due to bankruptcy costs and the subordination of their claims in the company to the holders of company's debt.

22.3.2 The Model

Our model draws its resemblance to the model of Junior debt presented in [10].

---

[3]This will be precisely the case of our analysis, when the bankruptcy will occur upon the managing shareholder's impulse.

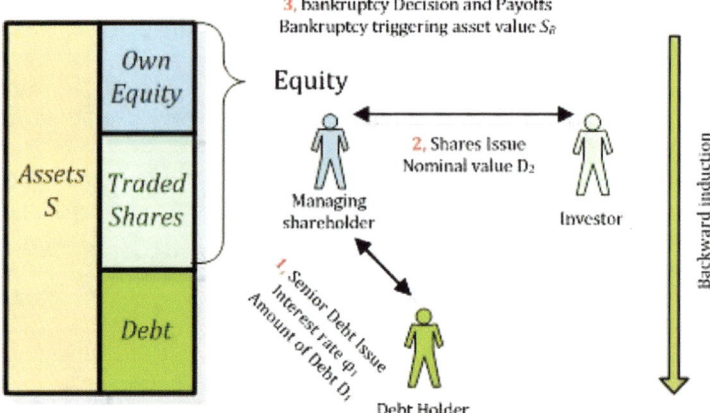

Fig. 22.1 Structure of the game

Ziegler examined a situation of three parties having a stake in a company. At first, there were shareholders maximizing the value of equity in the company. Secondly, they issued a senior bond, which bore priority of being paid-off first in the case of bankruptcy. At third, the company issued a junior debt which was subordinated to the senior one meaning that the payments to the junior bond holders could be made only if the full promised payment to the senior debt holders has been made.

We have modified the model and replaced the junior bond by publicly offered shares. Thus the new setting is shaped by the decision of the following three stakeholders:

- Managing shareholders of the company, whose main interest is to maximize the value of the own equity of the company. In case of bankruptcy, paying off their stakes hold the least priority.
- Debt holders, who acquired the debt issued by the company. They are concerned about the regular interest payments and have absolute priority in case of bankruptcy.
- Investors, who are interested in buying the listed shares of the company. Their payoffs in case of bankruptcy enjoy priority over those of managing shareholders.

The sequence of events is following:

At first, the company management issues a debt of the face value of D_1 at interest rate ϕ_1. This debt is perpetual. Secondly, an investor into the company buys shares equivalent to the δ times the value of the firm's total assets after this purchase S, together δS. Then managing shareholder chooses his optimal bankruptcy triggering asset value S_B^*, which maximizes the total value of the firm's own equity. Figure 22.1 may be helpful for better visualization of the setting.

The value of the firm's assets, S, is assumed to follow the geometric Brownian motion

$$dS_t = \mu S_t dt + \sigma S_t dB_t. \tag{22.5}$$

Asset substitution is not possible, so that the parameters μ and σ are known to all parties. Asset sales are prohibited. Hence, any net cash outflows associated with interest payments must be financed by selling additional equity. The rate of return on the riskless asset is r.

We further assume that in order to finance a project, an agent borrows from a lender with whom he reaches the following agreement: in exchange of a loan F_1, the borrower is to pay an instantaneous interest of $\phi_1 D_1 dt$ to the lender, where D_1 and ϕ_1 denote the face value of debt and the interest rate, respectively. Debt is assumed to be perpetual. Ziegler argues that perpetuality of the debt is a more realistic setting as the company does not usually finish its activities after the debt matures, rather acquires a new debt for its on-going business. Moreover, the perpetuity of the debt makes it easier to value the debt, as the partial differential equation turns into ordinary differential equation not depending on time t.

Finally, assume that the borrower is able to acquire additional funds on an exchange through public offering of its shares. These shares are naturally subordinated to the debt. The company attains δ proportion of total assets S, $\delta \in (0, 1)$ in this way. Let denote the value of this claim F_2. However, in the event of bankruptcy, investor is eligible to require only the nominal value of the shares representing his stake in the company, i.e. D_2, which equals number of the shares times nominal value of a share.

22.3.3 The Value of the Company and Its Securities

After we have specified the game, we need to value each player's payoffs using option pricing theory, treating all the players' decision variables as parameters. We can now compute the value of lender's claim F_1.

Proposition 22.1 (The Value of Debt). *If the current value of assets, S, follows geometric brownian motion and the contract is perpetual satisfying boundary conditions $F_1(S_B) = \min[(1 - \alpha)S_B, D_1]$ and $F_1(\infty) = \frac{\phi_1 D_1}{r}$, then the value of the company's debt, F_1, is given by*

$$F_1(S) = \begin{cases} \frac{\phi_1 D_1}{r}\left(1 - \left(\frac{S}{S_B}\right)^{-\gamma}\right) + (1 - \alpha)S_B\left(\frac{S}{S_B}\right)^{-\gamma} & \text{if}(1 - \alpha)S_B < D_1 \\ \frac{\phi_1 D_1}{r}\left(1 - \left(\frac{S}{S_B}\right)^{-\gamma}\right) + D_1\left(\frac{S}{S_B}\right)^{-\gamma} & \text{if}(1 - \alpha)S_B \geq D_1 \end{cases}$$

$$(22.6)$$

where $\gamma \equiv \frac{2r}{\sigma^2}$.

For computation of this proposition please see the appendix to this article "The Value of Debt" in Appendix 1. The meaning is that values of the senior debt equals the value of the perpetuity, $\frac{\phi_1 D_1}{r}$ times the risk-neutral probability that the bankruptcy will not occur, $1 - (S/S_B)^{-\gamma}$, plus the payoff to the lender in the case that the bankruptcy takes place. This differs case by case, depending on the value disposable, $(1 - \alpha)S_B$ being lower or bigger than D_1.

Proposition 22.2 (The Value of Listed Shares). *If the current value of assets, S, follows geometric brownian motion and the contract is perpetual satisfying boundary conditions* $F_2(S_B) = (1 - \alpha)S_B - \min[(1 - \alpha)S_B, D_1] = \max[0, \min[(1 - \alpha)S_B - D_1, D_2]]$, $F_2(\infty) = \delta S$ *and* $F_2(0) = 0$, *then the value of listed shares is given by*

$$
F_2(S) = \begin{cases}
\delta S + S_B \left(\dfrac{S}{S_B}\right)^{-\gamma} & \text{if}(1 - \alpha)S_B \leq D_1 \\[2ex]
\delta S + \left((1 - \alpha - \delta)S_B - D_1\right)\left(\dfrac{S}{S_B}\right)^{-\gamma} & \text{if} D_1 < (1 - \alpha)S_B \leq D_1 + D_2 \\[2ex]
\delta S + \left(D_2 - \delta S_B\right)\left(\dfrac{S}{S_B}\right)^{-\gamma} & \text{if} D_1 + D_2 \leq (1 - \alpha)S_B
\end{cases}
$$

(22.7)

where $\gamma \equiv \dfrac{2r}{\sigma^2}$.

Derivation of this result is available in the appendix to this article "The Value of Listed Shares" in Appendix 1. The value of the listed equity can be interpreted as the value of the portion of the company's value less the value that could be lost for investor in case of bankruptcy times the probability of bankruptcy

In eliciting the total value of the firm W, Ziegler draws on [5] and from there we know that the total value of the firm reflects three terms: the firm's asset value, S, the value of the tax deduction of interest payments, TB, less the value of bankruptcy costs, K. We can summarize the total value of the company in a proposition.

Proposition 22.3 (The Value of the Company). *If the current value of assets, S follows geometric brownian motion, the current value of bankruptcy costs, K, satisfy the boundary conditions* $K(S_B) = \alpha S_B$ *and* $K(\infty) = 0$, *and moreover the current value of the tax benefits, TB, satisfy the boundaries* $TB(S_B) = 0$ *and* $TB(\infty) = \theta \frac{\phi_1 D_1}{r}$, *then the total value of the company is given by*

$$
W(S) = S + TB(S) - K(S) = S + \theta\frac{\phi_1 D_1}{r}\left(1 - \left(\frac{S}{S_B}\right)^{-\gamma}\right) - \alpha S_B\left(\frac{S}{S_B}\right)^{-\gamma}.
$$

(22.8)

where $\gamma \equiv \dfrac{2r}{\sigma^2}$.

You can see the computation of this result in the appendix to this article "The Value of Company" in Appendix 1. The above equation means that, the whole value of the company is given by the value of its assets S, plus the present value of tax shield, $\theta\frac{\phi_1 D_1}{r}$, times the risk-neutral probability that bankruptcy does not occur, minus the value lost in the case of bankruptcy, αS_B, times risk-neutral probability of the company going under, $(S/S_B)^{-\gamma}$.

Provided that now we have calculated the total value of the firm, we may easily compute the value of equity, which will be maximized by the managing shareholder.

22.3.4 The Value of Equity

Hereby, we will give value of equity for both, an own equity and equity as a whole. The whole equity consists of own equity plus listed shares. The difference becomes apparent when the managing shareholders will start to maximize the value of each of them in separate cases. In the first case, they will maximize the value of the whole equity as they should have been supposed to. In the second case, they will only maximize the value of the equity in which they take the greatest stake—in the own equity. The distinction in the two options is that in the event of bankruptcy, the holders of own equity will participate on the break-up value only after the holders of listed shares have been satisfied in their claims on the company.[4]

22.3.4.1 Total Value of Equity

Firstly, we will give the value of whole equity. The value of equity is naturally a difference between the value of the firm, W, and the value of outstanding debt, F_1:

$$E(S) = W(S) - F_1(S). \qquad (22.9)$$

Because we still do not know what will be the bankruptcy triggering asset value, S_B, we have to compute the value of equity for two cases, please see the appendix "Total value of Equity" in Appendix 2 for the details of the computation. Here, we will state the final result:

$$E(S) = \begin{cases} S - (1-\theta)\frac{\phi_1 D_1}{r}\left(1 - \left(\frac{S}{S_B}\right)^{-\gamma}\right) - S_B\left(\frac{S}{S_B}\right)^{-\gamma} & \text{if } (1-\alpha)S_B < D_1 \\ S - (1-\theta)\frac{\phi_1 D_1}{r}\left(1 - \left(\frac{S}{S_B}\right)^{-\gamma}\right) - (\alpha S_B + D_1)\left(\frac{S}{S_B}\right)^{-\gamma} & \text{if } (1-\alpha)S_B \geq D_1 \end{cases}$$

$$(22.10)$$

The intuition behind what we have computed as the value of equity is very similar to the previous equations. It means that the value of equity is always given by a value of the asset S, minus the present value of the tax-adjusted cost of interest payments to the firm times the risk-neutral probability that bankruptcy does not occur, minus the value of assets lost in the event of bankruptcy times the risk-neutral probability of bankruptcy.

---

[4]However, companies tend to be liquidated when the value of equity is near zero. So we are here mainly interested in the "warnings" of bankruptcy, hinting at the most favorable situation for managing shareholder.

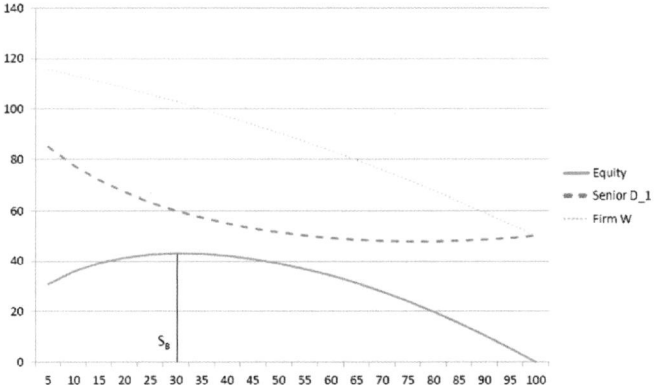

Fig. 22.2 Firm, debt and equity values as functions of bankruptcy triggering asset value. $\theta = 0.19, \alpha = 0.5, S = 100, r = 0.05, \phi_1 = 0.1, D_1 = 40, \phi_2 = 0.12, \delta = 0.15 \, \sigma^2 = 0.16$

22.3.4.2 The Value of Own Equity

In this part of the paper, we will compute the value of own equity. The value of own equity is apparently a difference between the value of the firm, W, the value of outstanding debt, F_1, and the value of listed shares, F_2:

$$E(S) = W(S) - F_1(S) - F_2(S). \tag{22.11}$$

The computation is again available in the appendix "The Value of own equity" in Appendix 2. The value of own equity then is:

$$E(S) = W(S) - F_1(S) - F_2(S) =$$
$$= (1 - \delta)\left(S - S_B\left(\frac{S}{S_B}\right)^{-\gamma}\right) - (1 - \theta)\frac{\phi_1 D_1}{r}\left(1 - \left(\frac{S}{S_B}\right)^{-\gamma}\right) +$$
$$+ \left((1 - \alpha)S_B - D_1 - D_2\right)\left(\frac{S}{S_B}\right)^{-\gamma}. \tag{22.12}$$

In the Fig. 22.2 are plotted values of Own Equity, Senior Debt, Listed Shares and the Firm's value for parameters $\theta = 0.19, \alpha = 0.5, S = 100, r = 0.05, \phi_1 = 0.1, D_1 = 40, \phi_2 = 0.12, \delta = 0.15 \, \sigma^2 = 0.16$. For the simple case, when the equity is intact we may estimate the optimal bankruptcy triggering asset value graphically.

Nevertheless, we are now adequately equipped to solve the optimization problem for the managing shareholder.

22.3.5   The Equity Holders' Optimal Bankruptcy Choice

In this chapter of the paper, we will compute the optimal bankruptcy strategies for managing shareholders (1) when they are optimizing the whole value of equity and (2) when they maximize only the own equity. Consequently, we will compare the values and find the possible risks for the potential investors into the company.

22.3.5.1   The Equity Holders' Optimal Bankruptcy Choice for Non-Listed Company

We will now determine the optimal bankruptcy strategy for equity holders. They are trying to set S_B so as to maximize the current value of equity. This will be done by finding first-order conditions and solving for S_B. We again distinguish two cases as we do not know what will be $(1 - \alpha)S_B$ compared to D_1:

Proposition 22.4 (The Equity Holders' Optimal Bankruptcy Choice for Non-Listed Company). *If the current value of assets, S, follows geometric brownian motion. The of value of equity is given by (22.10), and moreover if $S_B > 0$, then the optimal bankruptcy choice for the owners of the company maximizing the value of the equity is*

$$S_B^* = \begin{cases} (1 - \theta)\frac{\phi_1 D_1}{\sigma^2/2 + r} & \text{if}(1 - \alpha)S_B < D_1 \\ (1 - \theta)\frac{D_1(\phi_1 - r)}{\alpha(\sigma^2/2 + r)} & \text{if}(1 - \alpha)S_B \geq D_1. \end{cases} \tag{22.13}$$

For computation of this result see the appendix "The Equity Holders' Optimal Bankruptcy Choice for Non-Listed Company" in Appendix 3. These are the bankruptcy triggering asset values when we assume that managing shareholders of the company maximize the value of the whole equity. Let's now proceed to computation of bankruptcy triggering asset value for the case when they optimize only the own equity part of the whole equity.

22.3.5.2   The Equity Holders' Optimal Bankruptcy Choice for Listed Company

Let us now compute the bankruptcy triggering asset value S_B for the managing shareholder optimizing only the own equity part of the whole equity. This proceeds similarly to previous part and involves finding first order conditions and solving for S_B. We can again distinguish two cases in which $(1 - \alpha)S_B$ is compared with the residual claims, $D_1 + D_2$.

Proposition 22.5 (The Equity Holders' Optimal Bankruptcy Choice for Listed Company). *If the current value of assets, S, follows geometric brownian motion. The of value of the own equity is given by (22.12), and moreover if $S_B > 0$, then* *the optimal bankruptcy choice for the owners of the company maximizing the value of the equity is*

$$S_B^* = \begin{cases} \dfrac{1-\theta}{1-\delta}\dfrac{\phi_1 D_1}{\sigma^2/2+r} & \text{if}(1-\alpha)S_B < D_1 + D_2 \\[3ex] \dfrac{\left((1-\theta)\dfrac{\phi_1 D_1}{r}-D_1-D_2\right)\gamma}{(1-\delta-(1-\alpha)(\gamma+1))} & \text{if}(1-\alpha)S_B \geq D_1 + D_2. \end{cases} \tag{22.14}$$

where $\gamma \equiv \frac{2r}{\sigma^2}$.

Computation of this result can be found in the appendix "The Equity Holders' Optimal Bankruptcy Choice for Listed Company" in Appendix 3. In comparison with the first branch of (22.13) we may see that the value is multiplied by term $\frac{1}{1-\delta}$ which means that the more the company is leveraged, the lower is the bankruptcy triggering asset value. The value in the case of fully leveraged company coincides with the value for a non-listed company.

Now we are sufficiently equipped to address the question of when the company goes bankrupt in different motivation schemes. In the next section we will apply this theoretical results to the case of investment bank Bear Stearns.

22.4 Case Study: Bear Stearns

Credit crunch, Financial Crisis, Recession... These have been only some of the most frequently used vocabulary throughout 2008–2009. In the times when the trust is lost, the financial markets and financial institution suffer hard because the capital moves extremely fast nowadays. The Wall Street investment bank established in 1923 and made public in 1985, Bear Stearns & Co. Inc., went bare and down due to its extreme exposure to CMOs, CDOs and another types of assets backed securities and structured products. After all, Bear seemed to Fed definitely "too big to fail" and bailed it out with the help of its fiduciary JP Morgan which then proposed an acquisition contract to which Bear Stearns agreed on 29th of May 2008.

Throughout 2007 the stock Bear Stearns had been traded on the levels over USD 140 up to USD 170. The last weeks before the final Fed's decision on Bear Stearns you could see in the graph (Fig. 22.3) the stock on the levels around USD 80 which was mainly in the line with the other companies from financial sector experiencing the economic downturn. After the Fed's decision, the stock immediately plunged into level of USD 2 from springboard of USD 60, which should have been the offer from JP Morgan. And a few days later adjusted to USD 10 which was the reconciled version of the acquisition prospect.

Nonetheless, from the point of academic researchers, it is a good opportunity, how to ascertain the usefulness of our models. Therefore, we are now about to compute the bankruptcy triggering asset value in case of Bear Stearns and compare it to the reality. The fact the company did not actually bankrupt is not important

Fig. 22.3 Optimal
bankruptcy strategy and the
price development

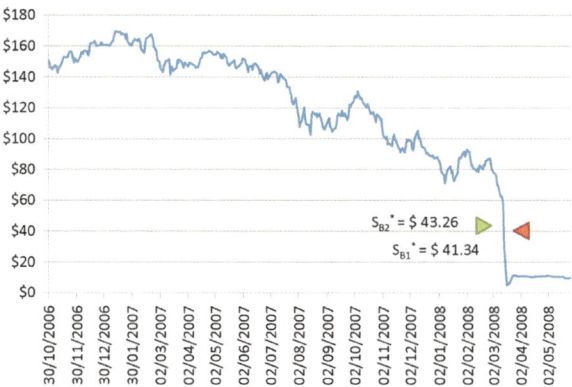

for our analysis, because in the end of the day the firm has been acquired by its
peer after an eminent decrease in its value. Thus, this example is valuable for our
analysis. Let us first have a look at the way of gathering the data.

22.4.1   Data Extraction for the Model

Equations for computing the bankruptcy triggering asset value (22.13) and (22.14)
which we will use, have five and seven variables as their parameters, respectively.

1. tax rate θ
2. debt D_1
3. ϕ_1 as an interest rate on debt D_1
4. nominal value of the investors' shares D_2
5. $\delta \in [0, 1)$ as a current value of a stake in the company which investors hold
6. risk-free interest rate r, and
7. volatility of the stock σ.

We will use the **Case 1** equations

$$S_B^* = (1 - \theta)\frac{\phi_1 D_1}{\sigma^2/2 + r},$$

(22.15)

$$S_B^* = \frac{1 - \theta}{1 - \delta}\frac{\phi_1 D_1}{\sigma^2/2 + r}.$$

(22.16)

as the value $(1 - \alpha)S_B$ is less than $D_1 + D_2$. It follows from our assumption
that the final offered price of USD 10 per share was not higher than the value
shareholders of Bear Stearns expected to receive upon bankruptcy. We also assume
that "bankruptcy costs" are higher than 0. Now, we will calculate the parameters
of S_B. Let's start with tax rate.

The tax rate θ is computed as an average percentage tax shield of the company during the last 3 years 2005–2007 from its profit and loss statements. We came to number of 32.4 % what is very much in line with the corporate tax rate in the U.S. which is 35 %.[5]

The amount of debt outstanding, D_1, is taken directly from the company's 2007 balance sheet and amounts up to USD 383,569 million.

The interest rate on the debt was computed as interest expenses taken from the Bear's P&L 2007 statement, USD 10,206 million, divided by the debt D_1. This equals to 2.66 % which is a number achievable by bigger companies on nowadays financial markets.

Risk-free interest rate r is an interest rate on three years U.S. government bonds and its value is 4 %.[6]

The last, but certainly not insignificant parameter to our equation is the volatility or standard deviation σ, which we calculated over the 2007 stock's performance and rebased into percentage equivalent. The value is 17.8 % and $\sigma^2 = 0.03$.

We are now ready to give the bankruptcy asset triggering value for the case of Bear Stearns.

22.4.2 Computation of the Bankruptcy Triggering Asset Value

In this part of the paper, we apply our findings on a real-life case, on Bear Stearns case. We utilize the parameters we computed on the previous pages and insert them into Eq. (22.71). We get

$$S_B^* = (1 - 0.324)\frac{0.0266 * 383,569,000,000}{0.03/2 + 0.04} = 123,258,201,000. \quad (22.17)$$

which we would like to compare with the actual stock price development.

The re-basement will be done in the following manner:

We have to compare comparable, so we would like to put on one side S_B^* per share and the share price on the other. Therefore, we multiply S_B^* by δ. δ stands here for a proportion of the whole market value of traded stocks in the assets on the balance sheet. We arrive to $\delta = 4.45$ %. Now, we multiply S_B^* by δ, we get 908,879,000. Finally, we divide this number by the number of common shares, 132,738,565, and obtain a per share bankruptcy triggering asset value $S_B^* = $ **USD 41.34**. We may now insert this into Fig. 22.3 of the share price development.

In case when managing shareholders maximize only their portion of the equity upon bankruptcy we basically multiply the first equation by the term

---

[5]On tax details in the U.S., please use visit www.irs.gov Internal *Revenue Service U.S. Department of Treasury.*

[6]Please visit http://www.treasurydirect.gov *Treasury Direct.*

$\frac{1}{1-\delta} = \frac{1}{1-0.0445}$ which shifts the bankruptcy triggering asset value up a bit. In our case by roughly 4.7 %. Altogether the bankruptcy signal in this case is $S_B^* = \mathbf{41.34 * 1.0445 = USD\ 43.26}$.

In this situation, the payoff for holders of traded equity would be naturally negative, as neither of the values USD 41.34 or USD 43.26 do not reach the average stock's price over the year 2007, which had been USD 132.6.

From market data, we may observe that the warning call according to the optimal bankruptcy which we have computed, would occur during Friday 14th of March 2008, where the high price was USD 54.79 and low USD 26.85. The situation going on was described by Stephen Labaton in New York Times (4.4.2008) by following: *"The testimony disclosed that Treasury Secretary Henry M. Paulson Jr. had insisted that Bear be paid a very low price for its stock by JP Morgan Chase. The testimony also offered more details about the pressures on Bear. The firm's chief executive, Alan D. Schwartz, said that he thought on the morning of Friday, March 14, that he had engineered a loan, backed by the Federal Reserve Bank of New York, that bought him 28 days to find a solution. But he said he realized that he had misunderstood the terms of the loan when the Fed decided later that day that the loan would last only through the weekend and that he had only until Sunday afternoon to find a buyer for the 85-year-old firm. The testimony also disclosed that regulators were unaware of Bear's precarious health and did not know until the afternoon of Thursday, March 13, that the firm was planning to file for bankruptcy protection the next morning."*

Drawing only to this one case study, we may see that the logic of our model could be applied on real-life on financial markets. If the investor sold his shares for proposed USD 41.34, he would not suffer the following loss of another 75 % of the stock's price when it came down to USD 2 and stabilized on USD 10 after a bit more generous offer was made by JPMorgan.

However, there is a need for more thorough case studies and surveys to be undergone in order to precisely estimate the behavior of the model according to its parameters and, accordingly, to find ways how to calculate the variables as precisely as possible.

22.5   Conclusion

The combination of modern financial economics and microeconomics can produce very interesting insights into the real-life on financial markets. The game theory serves here to define strategic interactions between the players. The option pricing is used to translate uncertain future payoffs to the present value with a variety of use of its boundary conditions.

In this paper, we have utilized the approach proposed by Ziegler [10] and combined the means of game theory and option pricing to compute the optimal bankruptcy strategy for owners of a listed company and non-listed company.

Moreover, this value is given as an easily computable closed formula. This fact makes the method appealing when one considers programmable solution to the answer "At what share price is the company likely to go under?", producing a valuable warning signal.

In the last part of the paper, we have surveyed an authentic situation of the investment bank Bear Stearns. We have come to conclusion that the model derived is applicable on the real-life data and thus it is worth to examine its future applications and case studies in a thorough survey. The managing owners indeed tend to file for bankruptcy on higher stock price than they would do in the case they optimized the value for all shareholders. On the other hand, some other effects may take role as for example the hope that the company would be rescued and did not have to go under. This could create another incentive for waiting while the stock price falls further down.

Acknowledgements The work on this paper was supported by the Grant Agency of the Czech Republic (grants 403/10/1235 and 402/11/0948) and by institutional support VSE IP100040. Karel Janda acknowledges research support provided during his long-term visits at University of California, Berkeley and Australian National University and the support he obtains as an Affiliate Fellow at CERGE-EI. The views expressed here are those of the authors and not necessarily those of our institutions. All remaining errors are solely our responsibility.

Appendix 1: The Value of the Company and Its Securities

The Value of Debt

F_1 must satisfy the ordinary differential equation

$$\frac{1}{2}\sigma^2 S^2 F_1'' + rSF_1' - rF_1 + \phi_1 D_1 = 0, \tag{22.18}$$

which does not depend on time t as the contract is perpetual. Equation (22.18) has general solution

$$F_1 = \alpha_0 + \alpha_1 S + \alpha_2 S^{-\gamma}, \quad \gamma \equiv \frac{2r}{\sigma^2}. \tag{22.19}$$

This solution is subject to boundary conditions

$$F_1(S_B) = \min[(1-\alpha)S_B, D_1] \tag{22.20}$$

and

$$F_1(\infty) = \frac{\phi_1 D_1}{r}. \tag{22.21}$$

Boundary condition (22.20) means that what remains in the jar of assets after bankruptcy, $(1 - \alpha)S_B$, is poured into the jar of the debt up to the level of D_1. Condition (22.21) states that as asset value S becomes very large, bankruptcy becomes an irrelevant option and the value of the lender's claim equals the value of a risk less bond, $\phi_1 D_1/r$. From (22.21), $\alpha_1 = 0$ and

$$\alpha_0 = \frac{\phi_1 D_1}{r}. \tag{22.22}$$

Hence, we can write

$$F_1(S) = \frac{\phi_1 D_1}{r} + \alpha_2 S^{-\gamma}. \tag{22.23}$$

Due to the fact that we do not know whether $(1 - \alpha)S_B$ is bigger or lower than D_1, we do not know which of the two values for boundary condition (22.20) to utilize. The result is that, we have to analyze two separate cases: $(1 - \alpha)S_B < D_1$ and $(1 - \alpha)S_B \geq D_1$.

Case 1: $(1 - \alpha)S_B < D_1$ In this case, using condition (22.20) yields

$$F_1(S_B) = (1 - \alpha)S_B = \frac{\phi_1 D_1}{r} + \alpha_2 S_B^{-\gamma}. \tag{22.24}$$

By extracting α_2 we obtain

$$\alpha_2 = \left((1 - \alpha)S_B - \frac{\phi_1 D_1}{r}\right)S_B^{\gamma}. \tag{22.25}$$

Therefore, the value of senior debt, F_1, is given by

$$F_1(S) = \frac{\phi_1 D_1}{r} + \left((1 - \alpha)S_B - \frac{\phi_1 D_1}{r}\right)\left(\frac{S}{S_B}\right)^{-\gamma}. \tag{22.26}$$

Case 2: $(1 - \alpha)S_B \geq D_1$ In this case, using condition (22.20) yields

$$F_1(S_B) = D_1 = \frac{\phi_1 D_1}{r} + \alpha_2 S_B^{-\gamma}. \tag{22.27}$$

Solving for α_2 yields

$$\alpha_2 = D_1 + \left(1 - \frac{\phi_1}{r}\right)\left(\frac{S}{S_B}\right)^{-\gamma}. \tag{22.28}$$

Therefore, the value of senior debt, F_1, is given by

$$F_1(S) = \frac{\phi_1 D_1}{r} + D_1\left(1 - \frac{\phi_1}{r}\right)\left(\frac{S}{S_B}\right)^{-\gamma}. \tag{22.29}$$

Putting (22.26) and (22.29) together, gains the value of senior debt, F_1, as

$$F_1(S) = \begin{cases} \frac{\phi_1 D_1}{r} + \left((1-\alpha)S_B - \frac{\phi_1 D_1}{r} \right) \left(\frac{S}{S_B} \right)^{-\gamma} & \text{if} (1-\alpha)S_B < D_1 \\ \frac{\phi_1 D_1}{r} + D_1 \left(1 - \frac{\phi_1}{r} \right) \left(\frac{S}{S_B} \right)^{-\gamma} & \text{if} (1-\alpha)S_B \geq D_1 \end{cases}$$

$$(22.30)$$

The Value of Listed Shares

The way to determine the value of the listed shares, F_2, of the company would be very similar to what we have been doing on the previous pages while computing F_1. It also has to satisfy the differential equation (22.18). Thus, F_2 must be of form

$$F_2 = \alpha_0 + \alpha_1 S + \alpha_2 S^{-\gamma}, \quad \gamma \equiv \frac{2r}{\sigma^2}. \tag{22.31}$$

The difference lies in the boundary conditions applied. They are

$$F_2(S_B) = (1-\alpha)S_B - \min[(1-\alpha)S_B, D_1] = \max[0, \min[(1-\alpha)S_B - D_1, D_2]], \tag{22.32}$$

$$F_2(\infty) = \delta S \tag{22.33}$$

and

$$F_2(0) = 0. \tag{22.34}$$

In the case of the shares listed we need to utilize all three boundary conditions instead of only two applied in the case of debt. The reason is that condition (22.33) states, that also value F_2 approaches infinity with value of assets going beyond any boundaries, $\delta S \rightarrow \infty$ as $S \rightarrow \infty$. The condition (22.32) means that if we take the same jar of what remained from assets after the bankruptcy, $(1-\alpha)S_B$, as in the case of debt, we shall take a look into it and see if there remains something to satiate shareholders thirst. If yes, we will start pouring it into the pot of shareholders claims up to the level of D_2 or until we have anything to pour in there. The additional condition (22.34) is nothing less than when the value of firm is zero, also the value of the listed equity is zero. Let us now employ the boundaries to obtain the value F_2.

From (22.33) we know that $\alpha_1 = \delta$, because $\alpha_2 S^{-2r/\sigma^2}$ approaches zero as $S \rightarrow \infty$. Furthermore, using (22.34) we obtain that $\alpha_0 = 0$ what implies that

$$F_2(S) = \delta S + \alpha_2 S^{-\gamma}. \tag{22.35}$$

In the sequel, we need to distinguish three cases stemming from the condition (22.32).

Case 1: $(1 - \alpha)S_B < D_1$
This transforms into condition

$$F_2(S_B) = 0. \tag{22.36}$$

and following equation

$$F_2(S_B) = \delta S_B + \alpha_2 S_B^{-\gamma} = 0. \tag{22.37}$$

with the following solution for α_2

$$\alpha_2 = -\delta S_B^{1+\gamma}. \tag{22.38}$$

Case 2: $D_1 + D_2 > (1 - \alpha)S_B \geq D_1$
With the following condition

$$F_2(S_B) = (1 - \alpha)S_B - D_1. \tag{22.39}$$

and following equation

$$F_2(S_B) = \delta S_B + \alpha_2 S_B^{-\gamma} = (1 - \alpha)S_B - D_1. \tag{22.40}$$

with the following solution for α_2

$$\alpha_2 = (1 - \alpha - \delta)S_B^{1+\gamma} - D_1 S_B^{\gamma}. \tag{22.41}$$

Case 3: $D_2 + D_1 \leq (1 - \alpha)S_B$
The condition in this case is

$$F_2(S_B) = D_2. \tag{22.42}$$

and following equation

$$F_2(S_B) = \delta S_B + \alpha_2 S_B^{-\gamma} = D_2. \tag{22.43}$$

with the following solution for α_2

$$\alpha_2 = (D_2 - \delta S_B)\left(\frac{S}{S_B}\right)^{-\gamma}. \tag{22.44}$$

When we gather all the results for each of the cases (22.36), (22.39) and (22.42) with different α_2 (22.38), (22.41) and (22.44), we obtain the general solution for F_2, depending on the volume of $(1 - \alpha)S_B$.

$$
F_2(S) = \begin{cases} \delta S + S_B \left(\frac{S}{S_B}\right)^{-\gamma} & \text{if}(1-\alpha)S_B \le D_1 \\ \delta S + \left((1-\alpha-\delta)S_B - D_1\right)\left(\frac{S}{S_B}\right)^{-\gamma} & \text{if}\, D_1 < (1-\alpha)S_B \le D_1 + D_2 \\ \delta S + \left(D_2 - \delta S_B\right)\left(\frac{S}{S_B}\right)^{-\gamma} & \text{if}\, D_1 + D_2 \le (1-\alpha)S_B \end{cases}
$$

$$(22.45)$$

The Value of Listed Company

The current value of bankruptcy costs K must satisfy

$$
K = \alpha_0 + \alpha_1 S + \alpha_2 S^{-\gamma}, \qquad \gamma \equiv \frac{2r}{\sigma^2} \tag{22.46}
$$

with boundary conditions

$$
K(S_B) = \alpha S_B \tag{22.47}
$$

and

$$
K(\infty) = 0. \tag{22.48}
$$

Condition (22.47) says that at the time of bankruptcy, the value of bankruptcy costs equals αS_B. Boundary (22.48) reflects that when the value of asset is very large, bankruptcy becomes irrelevant an thus the current value of bankruptcy costs is zero.

From (22.47), $\alpha_0 = \alpha_1 = 0$, and from (22.48), we get

$$
K(S_B) = \alpha S_B = \alpha_2 S_B^{-\gamma}. \tag{22.49}
$$

From here,

$$
\alpha_2 = \alpha S_B^{1+\gamma}. \tag{22.50}
$$

Therefore, the current value of bankruptcy costs is given by

$$
K(S) = \alpha_2 S^{-\gamma} = \alpha S_B \left(\frac{S}{S_B}\right)^{-\gamma}. \tag{22.51}
$$

For computing the tax benefits, we will make a use of the tax rate θ. Then the current value of the tax benefits, TB, must again satisfy

$$
TB = \alpha_0 + \alpha_1 S + \alpha_2 S^{-\gamma}, \qquad \gamma \equiv \frac{2r}{\sigma^2} \tag{22.52}
$$

with boundary conditions

$$TB(S_B) = 0 \tag{22.53}$$

and

$$TB(\infty) = \theta \frac{\phi_1 D_1}{r}. \tag{22.54}$$

Boundary condition (22.53) says that the tax benefits are lost if bankruptcy occurs and condition (22.54) says that, as asset value becomes very large, the bankruptcy turns out as an irrelevant option and the value of the tax benefits is θ times the value of risk-free senior debt.[7]

From (22.53), $\alpha_1 = 0$ and $\alpha_0 = \theta(\frac{\phi_1 D_1}{r})$. Substituting into (22.46) and using (22.53) yields the condition

$$TB(S_B) = \theta \frac{\phi_1 D_1}{r} + \alpha_2 S_B^{-\gamma} = 0. \tag{22.55}$$

Solving for α_2, we obtain

$$\alpha_2 = -\theta \frac{\phi_1 D_1}{r} S_B^{\gamma}. \tag{22.56}$$

Hence, the current value of the tax benefits equals

$$TB(S) = \theta \frac{\phi_1 D_1}{r} \left(1 - \left(\frac{S}{S_B} \right)^{-\gamma} \right). \tag{22.57}$$

Using (22.51) and (22.57), the total value of the firm, W, is given by

$$W(S) = S + TB(S) - K(S) = S + \theta \frac{\phi_1 D_1}{r} \left(1 - \left(\frac{S}{S_B} \right)^{-\gamma} \right) - \alpha S_B \left(\frac{S}{S_B} \right)^{-\gamma}. \tag{22.58}$$

Appendix 2: The Value of Equity

Total Value of Equity

Case 1: $(1 - \alpha) S_B < D_1$. In this case, senior debt is worth

$$F_1(S) = \frac{\phi_1 D_1}{r} \left(1 - \left(\frac{S}{S_B} \right)^{-\gamma} \right) + (1 - \alpha) S_B \left(\frac{S}{S_B} \right)^{-\gamma}. \tag{22.59}$$

---

[7]We assume that there are no tax benefits from the listed equity part of liabilities. Either because we assume that they are not a subject to double taxation or because the company pays-out no dividend.

Hence, with subtracting F_1 from W, we get

$$E(S) = W(S) - F_1(S) = S + \theta \frac{\phi_1 D_1}{r}\left(1 - \left(\frac{S}{S_B}\right)^{-\gamma}\right) -$$

$$- \alpha S_B\left(\frac{S}{S_B}\right)^{-\gamma} - \frac{\phi_1 D_1}{r}\left(1 - \left(\frac{S}{S_B}\right)^{-\gamma}\right) - (1-\alpha)S_B\left(\frac{S}{S_B}\right)^{-\gamma} =$$

$$= S - (1-\theta)\frac{\phi_1 D_1}{r}\left(1 - \left(\frac{S}{S_B}\right)^{-\gamma}\right) - S_B\left(\frac{S}{S_B}\right)^{-\gamma}.$$

$$(22.60)$$

Case 2: $D_1 \leq (1-\alpha)S_B$. The senior debt in this case is worth

$$F_1(S) = \frac{\phi_1 D_1}{r}\left(1 - \left(\frac{S}{S_B}\right)^{-\gamma}\right) + D_1\left(\frac{S}{S_B}\right)^{-\gamma}. \qquad (22.61)$$

The equity value now turns into

$$E(S) = W(S) - F_1(S) = S + \theta\frac{\phi_1 D_1}{r}\left(1 - \left(\frac{S}{S_B}\right)^{-\gamma}\right) -$$

$$- \alpha S_B\left(\frac{S}{S_B}\right)^{-\gamma} - \frac{\phi_1 D_1}{r}\left(1 - \left(\frac{S}{S_B}\right)^{-\gamma}\right) - D_1\left(\frac{S}{S_B}\right)^{-\gamma} =$$

$$= S - (1-\theta)\frac{\phi_1 D_1}{r}\left(1 - \left(\frac{S}{S_B}\right)^{-\gamma}\right) - (\alpha S_B + D_1)\left(\frac{S}{S_B}\right)^{-\gamma}. \qquad (22.62)$$

Putting (22.60) and (22.62) together, we obtain a summary for value $E(S)$ of equity

$$E(S) = \begin{cases} S - (1-\theta)\frac{\phi_1 D_1}{r}\left(1 - \left(\frac{S}{S_B}\right)^{-\gamma}\right) - S_B\left(\frac{S}{S_B}\right)^{-\gamma} & \text{if}(1-\alpha)S_B < D_1 \\ S - (1-\theta)\frac{\phi_1 D_1}{r}\left(1 - \left(\frac{S}{S_B}\right)^{-\gamma}\right) - (\alpha S_B + D_1)\left(\frac{S}{S_B}\right)^{-\gamma} & \text{if}(1-\alpha)S_B \geq D_1 \end{cases}$$

$$(22.63)$$

The Value of Own Equity

Case 1: $(1-\alpha)S_B < D_1 + D_2$. In this case, senior debt is worth

$$F_1(S) = \frac{\phi_1 D_1}{r}\left(1 - \left(\frac{S}{S_B}\right)^{-\gamma}\right) + (1-\alpha)S_B\left(\frac{S}{S_B}\right)^{-\gamma}, \qquad (22.64)$$

and shares listed

$$F_2(S) = \delta S + S_B \left(\frac{S}{S_B} \right)^{-\gamma}.$$ (22.65)

Hence, with subtracting F_1 and F_2 from W, we get

$$E(S) = W(S) - F_1(S) - F_2(S) = (1 - \delta) \left(S - S_B \left(\frac{S}{S_B} \right)^{-\gamma} \right) -$$
$$- (1 - \theta) \frac{\phi_1 D_1}{r} \left(1 - \left(\frac{S}{S_B} \right)^{-\gamma} \right).$$ (22.66)

Case 2: $(1 - \alpha) S_B \geq D_1 + D_2$
The senior debt in this case is worth

$$F_1(S) = \frac{\phi_1 D_1}{r} \left(1 - \left(\frac{S}{S_B} \right)^{-\gamma} \right) + D_1 \left(\frac{S}{S_B} \right)^{-\gamma},$$ (22.67)

and shares listed

$$F_2(S) = \delta S + \left(D_2 - \delta S_B \right) \left(\frac{S}{S_B} \right)^{-\gamma}$$ (22.68)

The equity value now turns into

$$E(S) = W(S) - F_1(S) - F_2(S) =$$
$$= (1 - \delta) \left(S - S_B \left(\frac{S}{S_B} \right)^{-\gamma} \right) - (1 - \theta) \frac{\phi_1 D_1}{r} \left(1 - \left(\frac{S}{S_B} \right)^{-\gamma} \right) +$$
$$+ \left((1 - \alpha) S_B - D_1 - D_2 \right) \left(\frac{S}{S_B} \right)^{-\gamma}.$$ (22.69)

Appendix 3: The Equity Holders' Optimal Bankruptcy Choice

The Equity Holders' Optimal Bankruptcy Choice for Non-Listed Company

Case 1: $(1 - \alpha) S_B < D_1$. In this case, using the upper branch of (22.63) we arrive to the first-order condition

$$\frac{\partial E(S)}{\partial S_B} = (1 - \theta) \frac{\phi_1 D_1}{r} \frac{\gamma}{S_B} \left(\frac{S}{S_B} \right)^{-\gamma} - (1 + \gamma) \left(\frac{S}{S_B} \right)^{-\gamma} = 0.$$ (22.70)

Extracting S_B and simplifying yields the optimal bankruptcy strategy[8]

$$S_B^* = (1 - \theta)\frac{\phi_1 D_1}{\sigma^2/2 + r}. \tag{22.71}$$

The result is that the optimal bankruptcy strategy S_B^* does not depend on the current asset value S is a quite interesting finding.

Case 2: $D_1 \leq (1 - \alpha)S_B$. The first-order condition now turns into

$$\frac{\partial E(S)}{\partial S_B} = \left((1-\theta)\frac{\phi_1 D_1}{r} - D_1\right)\frac{\gamma}{S_B}\left(\frac{S}{S_B}\right)^{-\gamma} - \alpha(1+\gamma)\left(\frac{S}{S_B}\right)^{-\gamma} = 0 \tag{22.72}$$

from which we can extract the optimal bankruptcy strategy[9]

$$S_B^* = \frac{r}{\alpha(\sigma^2/2 + r)}\left((1-\theta)\frac{\phi_1 D_1}{r} - D_1\right) = (1-\theta)\frac{D_1(\phi_1 - r)}{\alpha(\sigma^2/2 + r)}. \tag{22.73}$$

Putting (22.71) and (22.73) together, gains the bankruptcy triggering asset value S_B^* which we are looking for

$$S_B^* = \begin{cases} (1 - \theta)\frac{\phi_1 D_1}{\sigma^2/2 + r} & \text{if}(1 - \alpha)S_B < D_1 \\ (1 - \theta)\frac{D_1(\phi_1 - r)}{\alpha(\sigma^2/2 + r)} & \text{if}(1 - \alpha)S_B \geq D_1. \end{cases} \tag{22.74}$$

The Equity Holders' Optimal Bankruptcy Choice for Listed Company

Case 1: $(1 - \alpha)S_B < D_1 + D_2$. In this case, using (22.69) we arrive to the first-order condition

$$\frac{\partial E(S)}{\partial S_B} = -(1-\delta)\left(\frac{S}{S_B}\right)^{-\gamma}(\gamma + 1) + (1 - \theta)\frac{\phi_1 D_1}{r}\frac{\gamma}{S_B}\left(\frac{S}{S_B}\right)^{-\gamma} = 0. \tag{22.75}$$

---

[8]It is a maximum and not a minimum since $\frac{\partial^2 E(S)}{\partial S_B^2} = -S_B^{\gamma-2}S^{-\gamma}\gamma(1 - \theta)\frac{\phi_1 D_1}{r} < 0$.

[9]It is again maximum, since $\frac{\partial^2 E(S)}{\partial S_B^2} = -\gamma S_B^{\gamma-2}S^{-\gamma}\left((1-\theta)\frac{\phi_1 D_1}{r} - D_1\right) < 0$, where the inequality stems from (22.73) and from that $S_B > 0$.

Extracting S_B and simplifying yields the optimal bankruptcy strategy[10]

$$S_B^* = \frac{1-\theta}{1-\delta} \frac{\phi_1 D_1}{\sigma^2/2 + r}. \tag{22.76}$$

The result is that the optimal bankruptcy strategy S_B^* does not depend on the current asset value S is a quite interesting finding.

Case 2: $D_1 + D_2 \leq (1-\alpha)S_B$. The first-order condition now turns into

$$\frac{\partial E(S)}{\partial S_B} = -(1-\delta)\left(\frac{S}{S_B}\right)^{-\gamma}(\gamma+1) +$$

$$+ \left((1-\theta)\frac{\phi_1 D_1}{r} - (D_1 + D_2)\right)\frac{\gamma}{S_B}\left(\frac{S}{S_B}\right)^{-\gamma} + (1-\alpha)(1+\gamma)\left(\frac{S}{S_B}\right)^{-\gamma} = 0 \tag{22.77}$$

from which we can extract the optimal bankruptcy strategy

$$S_B^* = \frac{\left((1-\theta)\frac{\phi_1 D_1}{r} - D_1 - D_2\right)\gamma}{(1-\delta-(1-\alpha)(\gamma+1))}. \tag{22.78}$$

Putting (22.76) and (22.78) together, gains the bankruptcy triggering asset value S_B^* which we are looking for

$$S_B^* = \begin{cases} \frac{(1-\theta)}{1-\delta}\frac{\phi_1 D_1}{\sigma^2/2+r} & \text{if}(1-\alpha)S_B < D_1 + D_2 \\ \frac{\left((1-\theta)\frac{\phi_1 D_1}{r} - D_1 - D_2\right)\gamma}{(1-\delta-(1-\alpha)(\gamma+1))} & \text{if}(1-\alpha)S_B \geq D_1 + D_2. \end{cases} \tag{22.79}$$

References

1. Altman, E.I.: Financial ratios, discriminant analysis and the prediction of corporate bankruptcy. J. Finance **23**(4), 589–609 (1968)
2. Black, F., Scholes, M.: The pricing of options and corporate liabilities. J. Polit. Econ. **81**, 659–683 (1973)
3. Brennan, M.J., Schwartz, E.S.: Evaluating natural resource investments. J. Bus. **58**, 135–157 (1995)

---

[10]It is a maximum and not a minimum since $\frac{\partial^2 E(S)}{\partial S_B^2} = -(1-\delta)(\gamma+1)\gamma\frac{1}{S_B}\left(\frac{S}{S_B}\right)^{-\gamma} +$

$(1-\theta)\frac{\phi_1 D_1}{r}\gamma(\gamma-1)\frac{1}{S_B}\left(\frac{S}{S_B}\right)^{-\gamma} < 0$, where we use (22.76) and the fact that $S_B > 0$.

4. François, P., Morellec, E.: Capital structure and asset prices: some effects of bankruptcy procedures. J. Bus. **77**(2), 387–411 (2004)
5. Leland, H.E.: Corporate debt value, ond covenants, and optimal capital structure. J. Finance **49**, 1213–1252 (1994)
6. Merton, R.C.: Theory of rational option pricing. Bell J. Econ. Manag. Sci. **4**(s.l.), 141–183 (1973)
7. Osborne, M.J., Rubinstein, A.: A Course in Game Theory. MIT Press, Cambridge (1994). ISBN 0-262-15041-7
8. Tseng, F.M., Lin, L.: Quadratic interval logit model for forecasting bankruptcy. Omega **33**(1), 85–91 (2005)
9. Wilson, R.L., Sharda, R.: Bankruptcy prediction using neural networks. Decis. Support Syst. **11**, 545–557 (1994). (North-Holland)
10. Ziegler, A.: A Game Theory Analysis of Options. Springer, Berlin (2004). ISBN:354020668X

Chapter 23
Economic Impact Analysis of Potential Trade Restrictions on Biotech Maize in Latin American Countries

Nicholas Kalaitzandonakes, James Kaufman, and Douglas Miller

23.1 Introduction

Since their initial commercialization in the mid-1990s, the introduction and adoption of new biotech crops has continued unabated. In 2010, 29 countries cultivated biotech crops on 148 million hectares [13]. The broad adoption has been driven by farm-level yield and efficiency gains which have translated into billions of dollars of economic gains every year [2, 3, 8, 14, 17, 19].

Because major exporting countries have led their adoption, biotech crops represent a substantial share of key agricultural commodities (maize, soybeans, cotton and canola) which are broadly traded in international markets. Yet, as the biotech pipeline has accelerated and the trade of biotech crops has become more extensive, alerts about the chance for "regulatory asynchronicity" and ensuing trade disruptions have become more frequent [1, 6, 15, 20].

Biotech crops are strictly regulated for food and environmental safety at a national level. To date, more than 120 biotech events and 24 biotech crops have been approved for use or cultivation in various countries. As the biotech pipeline has expanded, however, regulatory approvals of new biotech crops across different countries have become less synchronized. Under such conditions, a large and increasing number of new biotech crops has received regulatory approval for use and cultivation in one or more countries but is still unauthorized in others [20]. Since asynchronicity in regulatory approvals between producing and importing countries

---

N. Kalaitzandonakes (✉) • D. Miller
Department of Agricultural and Applied Economics, University of Missouri,
Columbia, MO, USA
e-mail: kalaitzandonakesN@missouri.edu

J. Kaufman
Economics and Management of Agrobiotechnology Center, University of Missouri,
Columbia, MO, USA

A.A. Pinto and D. Zilberman (eds.), *Modeling, Dynamics, Optimization and Bioeconomics I*, Springer Proceedings in Mathematics & Statistics 73, DOI 10.1007/978-3-319-04849-9__23,
© Springer International Publishing Switzerland 2014

implies that some agricultural commodity trade flows may contain unauthorized material, it could lead to costly trade disruptions [6, 16].

Latin America is home to major agricultural commodity exporters that have led in the adoption of biotech crops. It is also home to significant importers who have not been significant adopters of biotechnology. In a few occasions, trade in the continent has been interrupted as a result of regulatory asynchronicity of biotech crops among importers and exporters.[1] Yet, such events have, generally, been infrequent. In this paper, we are interested in the chance for sustained regulatory asynchronicity among importing and exporting countries in Latin America, the potential for ongoing trade disruptions and the relevant economic implications.

23.2 Volume and Importance of Agricultural Commodity Trade in Latin America

Because of geographic proximity and inherent market integration trade patterns of agricultural commodities in the Latin America (LA) region are best understood when considering LA and North America (NA) together. The trade of maize is an instructive example of agricultural commodity exchanges in these regions.[2]

The United States is by far the largest producer and exporter of maize in the world while Argentina and Brazil are the second and third largest exporters and contribute significant amounts in the global trade of maize every year (Table 23.1). Paraguay is a smaller but meaningful exporter of maize as well. Together these top four exporters accounted, on average, for 81 % of global maize exports in the last 5 years. The United States and Brazil use a large share of their maize production to feed livestock (and in the case of the US to feed an expanding ethanol industry). Accordingly, exports comprised a relatively small portion of their total maize crop in the last 5 years-approximately 15 %. In contrast, Argentina and Paraguay produce maize largely for international markets and as a result their exports accounted, respectively, for 69 and 80 % of their annual crop over the same period (Table 23.1).

Most LA countries are net maize importers accounting together for 23 % of global imports over the 2007–2011 period. The largest maize importer is Mexico which, despite its significant domestic maize production, imported more than 8.5 MMT of maize per year in the last 5 years. Colombia, Venezuela, Peru and the Dominican Republic are also relatively large importers of maize. Most LA countries depend heavily on imports for their domestic needs as their share of imports to

---

[1]For instance, imports of soybean meal and soybean oil to Ecuador were stopped for several weeks by the Ministry of Agriculture in May 2005 causing great difficulties in local poultry, animal feed and tuna canning industries. These led the government of Ecuador to reverse its prior decision and permit the restart of the imports [24].

[2]We review the trade of maize here in some detail but we note that the trade of other major agricultural commodities follows similar patterns.

Table 23.1 Maize production, use and trade, avg. 2007/2008–2011/2012 (1,000 MT)

	Domestic use	Exports	Imports	production	Exports/ production (%)	Imports/ domestic use (%)
United States	275,007	49,539	442	320,895	15	0
Brazil	47,300	8,705	645	56,840	15	1
Argentina	6,960	15,121	7	22,063	69	0
Paraguay	300	1,460	12	1,834	80	4
% of world exports		82 %				
Mexico	31,040	223	8,564	22,560	1	28
Canada	11,872	749	1,865	10,703	7	16
Colombia	5,140	1	3,497	1,580	0	68
Venezuela	3,250	—	1,487	1,749	0	46
Peru	3,180	9	1,576	1,627	1	50
Guatemala	1,840	9	654	1,190	1	36
El Salvador	1,350	3	496	851	0	37
Ecuador	1,310	16	414	946	2	32
Cuba	1,120	—	771	360	0	69
Dominican Republic	1,090	—	1,044	35	0	96
Honduras	990	1	383	603	0	39
Uruguay	720	117	405	440	27	56
Bolivia	700	13	15	699	2	2
Costa Rica	685	—	657	18	0	96
Nicaragua	570	5	129	440	1	23
Panama	455	—	365	82	0	80
Haiti	250	—	0	250	0	0
Jamaica	245	—	242	2	0	99
Trinidad and Tobago	110	—	103	5	0	93
Guyana	37	—	32	5	0	86
% of world imports			23 %			
World	813,670	92,524	90,373	817,255	11	11

Source: USDA, PS&D Database

domestic use is rather high (Table 23.1). For instance, the Dominican Republic, Costa Rica and Jamaica depend almost entirely on imports while Colombia, Venezuela, Peru and El Salvador have significant domestic production but still depend on trade to satisfy 1/3–2/3 of their domestic consumption. Haiti and Bolivia have small domestic markets and are self-sufficient.

23.3   Patterns of Trade Flows

With major exporters and significant importers in close proximity, much of the trade of maize in the Americas occurs through a dense network of exchanges crisscrossing the continent. The US is by far the largest supplier of maize to LA, with a relatively large share shipped to Central America, Caribbean, and South America countries. For instance, over a typical 4 year period (2005–2008), the US exported 37 % of its maize to LA countries (Table 23.2). A large share of these exports went to Mexico and Canada. However, Colombia, the Dominican Republic and other countries in Central America were meaningful markets. In fact, only a few countries (e.g. Chile and Peru) received more maize from other Latin American sources (Argentina) than from the US.

Argentina and Brazil typically trade a larger share of their maize crop to Europe, South East Asia and the Middle East. Because of such orientation, Argentina exported only 18 % of its maize to LA countries over the 2005–2008 period while Brazil exported only 2 %. The majority of maize exports from Argentina were sent to neighbors in South America including Chile, Peru and Colombia. Other LA importers (e.g. Paraguay, Mexico and Canada) similarly trade mostly with adjacent countries (Table 23.2).

23.4   Biotech Use and Regulatory Approvals

Given the importance of the maize trade for most countries in LA, trade disruptions could have significant impacts on supplies and prices. Since the presence of unapproved biotech events in the supplies of an exporter could trigger bilateral trade disruptions with one or more importers, it is of interest to examine the potential economic implications of such events for both importers and exporters in the region. In this context, the use of biotechnology in NA and LA countries is of interest.

All major exporters in the Americas have adopted biotech maize varieties quickly and extensively (Table 23.3). Among the major maize exporters, the US and Argentina have led in the adoption of biotech maize with levels exceeding 80 % of their harvested hectareage in 2009. Brazil approved and commercialized biotech maize for the first time in 2008 but it has since increased its adoption at a rapid rate and is expected to reach adoption levels similar to those in the US and Argentina in the next 2 years. Paraguay is the only maize exporter that has not approved and, officially, does not grow biotech maize. Nevertheless, the USDA has estimated that 90 % of Paraguay's maize harvested hectares in 2010 were planted with biotech hybrids imported illegally from neighboring countries [25]. Among net maize importers, only Canada and Uruguay have adopted biotech maize extensively (over 80 % of hectares). A small share of the maize hectares in Honduras were

Table 23.2 Maize trade flows in the Americas: 2005–2008 (1,000 MT)

Exports (→) Imports (↓)	Argentina	Brazil	Canada	Mexico	Paraguay	USA	ROW	Am. % of imports
Argentina	—	1	—	0	44	4	21	70 %
Bahamas	0	—	—	—	—	1	0	100 %
Barbados	—	—	0	—	—	36	0	100 %
Belize	—	—	—	0	—	3	0	99 %
Bermuda	—	—	0	—	—	0	—	100 %
Bolivia	2	1	—	—	—	0	0	97 %
Brazil	25	—	—	0	1,049	1	0	100 %
Canada	0	—	—	0	0	2,623	16	99 %
Cayman Isds	—	48	—	—	40	0	0	100 %
Chile	1,335	20	0	0	4	325	4	100 %
Colombia	177	89	—	0	—	3,170	18	99 %
Costa Rica	0	—	0	32	—	702	2	100 %
Cuba	5	—	—	0	—	778	0	100 %
Dominica	—	—	0	—	—	1	0	99 %
Dominica Rep.	13	—	—	—	—	1,205	0	100 %
Ecuador	29	12	—	0	—	427	0	100 %
El Salvador	0	—	—	89	—	521	12	98 %
Grenada	—	—	0	—	—	12	0	100 %
Guatemala	0	—	—	6	—	737	0	100 %
Guyana	—	—	—	—	—	26	3	90 %
Honduras	0	—	—	0	—	350	1	100 %
Jamaica	—	—	0	—	—	259	0	100 %
Mexico	2	1	—	—	—	9,535	27	100 %
Nicaragua	—	—	—	0	—	137	1	100 %
Panama	0	0	—	0	—	424	0	100 %
Paraguay	2	11	—	—	—	0	0	100 %
Peru	1,101	19	—	0	15	377	17	99 %
Suriname	—	—	0	—	—	13	5	71 %
Trinidad & Tobago	0	—	0	—	—	116	0	100 %
USA	39	2	223	64	1	—	83	80 %
Uruguay	69	0	—	—	347	0	0	100 %
Venezuela	9	4	—	4	—	898	1	100 %
ROW	13,140	9,588	395	0	196	39,138	—	
Am. % of exports	18 %	2 %	36 %	100 %	88 %	37 %		

Source: GTIS, Global Trade Atlas

also planted with biotech hybrids in 2009 while a limited amount of hectares used biotech maize under a "controlled planting program" in Colombia during the same year [12].[3]

---

[3]Costa Rica and Chile also have a small number of hectares that grow biotech maize seeds for export markets.

Table 23.3 Maize
hectareage and biotech
adoption in the Americas in
2009

Country	Maize	
	Harvested hectares	% Biotech
USA	32,169,000	85 %
Brazil	12,925,000	36 %
Argentina	2,750,000	83 %
Canada	1,142,000	84 %
Paraguay	600,000	0 %[a]
Uruguay	96,000	82 %
Bolivia	310,000	0 %
Mexico	6,280,000	0 %
Honduras	362,000	4 %
Colombia	550,000	0 %[b]

[a] Biotech maize has not been approved in Paraguay;
however, USDA FAS estimated 90 % adoption
[b] ISAAA reports that 35,000 hectares of maize were
grown under a "controlled planting program"
Sources: ISAAA, USDA

The adoption levels of biotech maize in Table 23.3 illustrate the divide in the use of biotechnology among exporting and importing countries in the Americas: Exporters have adopted the technology while importers have not. Yet, aggregate adoption levels, such as those reported in Table 23.3, often obscure important differences in the varietal plantings among biotech adopters as well. As the biotech pipeline has continued to evolve and grow over time, newer biotech traits and "stacks" of multiple traits have been introduced in some countries but not in others. As a result, adopters often use "different generations" of biotech crop technologies. Regulatory approvals across countries are typically good indicators of such potential differences.

Table 23.4 lists the regulatory approvals of new biotech maize events for use and planting which have been granted over time by countries in NA and LA.[4] From these tables, it is readily apparent that even among the countries that have broadly adopted biotechnology, there are significant differences. The US and Canada have reviewed and approved new biotech events on an ongoing basis and as a result they have had access to the newest biotech crops available. Argentina and Brazil have reviewed and approved significantly fewer new biotech events, though Brazil has greatly accelerated its deregulations in the last 3 years. Mexico, and to a lesser extent

---

[4]Approvals for new biotech events are generally granted for food use, feed use, processing, importation and planting. Individual countries may use particular approvals and not others. For example, some countries grant food and feed approvals, while others require an importation approval when food and feed safety for imports is necessary. Countries grant planting approvals when permission to grow a new biotech event is sought. In Table 23.4 food, feed, processing and importation approvals are summarily presented as "Use" approvals, implying that a new biotech event (imported or otherwise) can be placed in the market for consumption. Planting approvals (placed in parenthesis) are separately indicated.

Table 23.4 Approvals of biotech maize events for use and (planting) in the Americas[a]

	1995	1996	1997	1998	1999	2000	2001	2002	2003	2004	2005	2006	2007	2008	2009	2010	2011	Total
US	3(7)	10(3)	1(3)	2(2)	(1)	2(1)	3(2)	0	1(1)	5(2)	1(1)	4(2)	7(4)	2(1)	5(4)	7(1)	(2)	53(37)
Canada	1	4(7)	6(3)	1(1)	0	0	2(2)	1(1)	2(1)	3(1)	3(3)	6(6)	3(2)	2(2)	4(2)	6(2)	0	44(33)
Argentina	0	(1)	0	4(3)	0	0	1(1)	0	0	1(1)	4(3)	0	1(1)	1	1	5(4)	(2)	18(16)
Brazil	0	0	0	0	0	0	0	0	0	0	0	0	2(1)	5(6)	8(7)	2(3)	1(1)	18(18)
Colombia	0	0	0	0	0	0	0	0	1	1	1	1(1)	1	5(3)	1	4	0	15(4)
Mexico	0	0	0	0	0	0	0	4	1	4	0	4	8	5	1	9	0	36
Paraguay	0	0	0	0	0	0	0	0	0	0	0	0	0	0	0	0	0	0
Uruguay	0	0	0	0	0	0	0	0	1(1)	1(1)	0	0	0	0	0	0	(4)	2(6)
Costa Rica	0	0	0	0	0	0	0	0	0	0	0	0	0	0	0	0	0	0
Honduras	0	0	0	0	0	0	0	1(1)	0	0	0	0	0	1(1)	1(1)	0	0	3(3)
Bolivia	0	0	0	0	0	0	0	0	0	0	0	0	0	0	0	0	0	0
Chile	0	0	0	0	0	0	0	0	0	0	0	0	(1)	0	0	0	0	0(1)
El Salvador	0	0	0	0	0	0	0	0	0	0	0	0	0	0	3	0	0	3

[a]Use approvals consist of the earliest Food, Feed, Import or Processing approval for an individual event *Source*: ISAAA, GM Approval Database. http://www.isaaa.org/gmapprovaldatabase/default.asp

Colombia, has approved a large number of new biotech events but mostly for use and not for planting. As a result, Mexico and Colombia have continued to import and consume biotech maize but not grow them much. All other adopters have approved and used a limited part of the biotech maize pipeline.

While there are significant differences in the regulatory approvals and use of new biotechnologies among adopters, there are even larger discrepancies among the countries that are not listed in Table 23.4. Indeed, some LA countries have no biosafety regulatory framework at all (e.g. Dominican Republic and Panama). Others have been developing their biosafety regulations (e.g. El Salvador, Venezuela, and Chile) but it is unclear when they might be able to complete them. And there are other LA countries that use ministerial decisions, decrees and other executive decisions to manage regulatory approvals for field trials or importation on an "as needed" basis (e.g. Guatemala and Paraguay).

A number of factors contribute to this wide diversity in the stage of development of regulatory approvals and use of biotechnology, chief among them however are: (a) an accelerating pipeline; (b) diverse national priorities, government policies and demands by various stakeholder groups; and (c) national differences in technical capacity and financial resources needed to support a robust regulatory system. The factors can explain, at least in part, the significant differences in the portion of the biotech pipeline that has been deregulated in the different NA and LA countries. But they also serve to highlight an emerging problem: As new events are brought to market at an increasing rate, the divergent regulatory capacities of individual countries imply the chance for ongoing (structural) asynchronicities in the regulatory approvals of new biotech crops across the Americas. As such, the possibility of exporting biotech commodities that have not been authorized in one or more LA importing countries will be high in the future.

23.5   Asynchronicity, LLP and Potential Economic Impacts of Trade Disruption: An Empirical Investigation

If asynchronicity in the regulatory approvals of new biotech crops among exporting and importing countries in the Americas were to lead to disruptions in bilateral trade flows, importers would need to identify alternative supplies which exporters would need to find alternative markets for their products. We are then interested in the economic consequences of such potential disruptions on importers and exporters alike. Because only the trade flows between specific importers and exporters may be affected, a model that explains bilateral trade flows must be used for quantitative analysis. Here, we use a spatial equilibrium framework to examine the economic impact of disruptions in bilateral trade flows of maize among importers and exporters in NA and LA.

Modern spatial equilibrium concepts were first developed by Enke [7] and Samuelson [18] and were later formalized by Takayama and Judge [21] as quadratic programming problems. Spatial equilibrium models have been used in studies of commodity markets to examine the potential price impacts from changes in trade due to improvements in infrastructure (roads, canals, lock and dam systems, ports, etc.); country-specific tariffs and/or quotas; country-specific trade bans or embargoes; trade restrictions associated with disease outbreaks or epidemics, phytosanitary concerns; and other related issues.

Although spatial equilibrium models have been applied to a wide range of economic problems, the segment of the spatial equilibrium literature that examines the economic impacts of bilateral trade disruptions is relatively small. Much of the modern empirical and theoretical literature on such trade disruptions was initially motivated by several trade embargoes in the world markets for petroleum and other commodities in the 1970s as well as the 1980 grain embargo imposed by the US on the Soviet Union (e.g. see [23]).

Since the group of grain embargo studies was published in the 1980s, there have been some more recent applications of the Takayama-Judge spatial equilibrium models to studies of trade disruptions in international agricultural markets. Recent examples include the study of the world rice market by Chen et al. [5] and the study of bans on genetically modified products in the international soy complex by Sobolevsky et al. [19]. Although much of the most recent empirical research on international trade issues is now based on other types of models [e.g., computable general equilibrium (CGE) or Armington-type models], the Takayama-Judge spatial equilibrium models have certain advantages in analyzing the impact of trade disruptions. For instance, Armington-type models are based on constant elasticity of substitution (CES) specifications, which imply that the trade flows of countries that do not normally trade will remain zero under any type of disruption and change in global trade. This is limiting when the potential redistribution of trade in response to disruptions of traditional trade flows is of interest.

23.6   Conceptual Model

The spatial equilibrium model developed for this analysis represents the global maize sector, and the basic model structure is based on the general approach for multi-region trade that is described by Takayama and Judge [21]. The objective of the spatial equilibrium model is to maximize net social welfare (consumer and producer surplus minus transportation costs and import and export tariffs) by choosing the bilateral trade flows subject to relevant behavioral constraints (i.e., market equilibrium and no-arbitrage constraints).

23.6.1   Notation

The market quantities and prices for maize are the market equilibrium levels that are derived from the excess supply-demand functions for maize in each country. We use linear quantity-dependent supply and demand specifications

Supply: $Q_i^S = \delta_{1i} + \delta_{2i} p_i^S$ $\qquad$ Demand: $Q_i^D = \gamma_{1i} - \gamma_{2i} p_i^D$, $\qquad$ (23.1)

where p_i^S is the supply price of maize in country i, p_i^D is the demand price in country i, Q_j^S is the aggregate quantity of maize supplied in country j, Q_i^D is the aggregate quantity of maize demanded in county i, and all of the model parameters are positive values. We allow the aggregate quantity demanded to differ from the aggregate quantity supplied so that individual countries may be net importers or exporters of maize, but the market equilibrium is imposed in each market by restricting the supply and demand prices to be equal ($p_i^S = p_i^D$). Accordingly, we can form the linear aggregate excess demand function

$$Q_i^D - Q_i^S = (\gamma_{1i} - \delta_{1i}) - (\gamma_{2i} + \delta_{2i}) p_i = \beta_{1i} - \beta_{2i} p_i \qquad (23.2)$$

for country i. To determine the market equilibrium outcomes, we calibrate the excess demand functions by using observed price and import or export quantity data for each country. For example, the slope of the excess demand equation in (23.2) is computed from excess demand elasticity estimates taken from prior studies, and the intercept coefficient is calibrated from the observed price and quantity data.

Given these linear excess demand equations, the net social welfare values may be computed for country i as

$$\sum_i \beta_{1i} p_i - \sum_i \beta_{2i} p_i^2 / 2. \qquad (23.3)$$

The objective of this formulation for the spatial equilibrium problem is to choose the trade flows (imports or exports) that maximize the net social welfare expression in (23.3) subject to the behavioral constraints. The market equilibrium relationships for each country are expressed in (23.2), and we also require that international price differences may not be subject to arbitrage. In particular, the price differences between any two markets should not exceed the per-unit transport costs between the markets. Further, the no-arbitrage conditions should account for the possibility that countries may impose import or export tariffs on their trade volumes. If a country imposes a specific (per-unit) import tariff (t_{ij}^M) or export tariff (t_{ij}^X) on maize, the tariff is added to or subtracted from the associated transportation rate for maize flowing to or from the country. To represent a spatial equilibrium such that no arbitrage between the countries is possible, we also impose the following constraints on the prices

$$|p_i - p_j| \leq w_{ij} + (t_{ij}^X + t_{ij}^M) p_i. \qquad (23.4)$$

Here, the absolute price differences between countries i and j are less than or equal to the transportation and per-unit import and export tariff rates. The solution values from the spatial equilibrium models include the aggregate excess supply and demand quantities and the market-clearing prices for maize in each country. These values are used as the baseline for our trade scenario analysis.

23.6.2   Imposing Potential Bilateral Trade Restrictions

If trade between countries h and k is restricted, then the no-arbitrage constraint on the price difference between these markets (e.g., (23.4)) does not have to hold at the spatial equilibrium. To remove these constraints from the optimization problem, we may increase the constraint bound in (23.4), which may be interpreted as imposing a prohibitively large per-unit cost of trade between the parties. However, the other no-arbitrage constraints between these countries and all third parties are still imposed, so the price difference between countries h and k under these trade restrictions cannot become implausibly large as long as there are other willing trading partners with available stocks to exchange (i.e., an interior solution exists for each country). In such cases, the price changes in countries h and k largely reflect the additional costs of reallocating the displaced quantities to other markets. Under extreme trade disruptions in which alternative trading partners may have limited ability to make up for the displaced quantities, the no-arbitrage constraints may no longer hold between countries h and k and all third parties due to corner solutions. In such cases, the no-arbitrage constraints are removed for all interactions involving countries h or k, but the no-arbitrage constraints are retained for all relationships among the third parties. Thus, the price changes resulting from the trade disruption may be large in countries h and k.

23.7   Empirical Analysis

23.7.1   Data

In order to operationalize the spatial equilibrium model set out in the previous section, detailed data on production, consumption, trade flows, prices and freight rates, as well as tariffs and quotas for maize in selected countries and groups of countries were collected or constructed. Trade flow data was derived from the United Nations Comtrade database [22] and this data set was validated and augmented by additional information on trade flows taken from Global Trade Atlas of Global Trade Information Services (GTIS) [11] as well as from FAOSTAT-TRADESTAT [9]. Each of these data sets is based on national customs data collected by origin and/or destination countries. In particular, the maize trade flows were collected for HS Code 100500.

Detailed annual bilateral trade flow data for maize in various countries was collected for individual countries and some selected country groupings. Initially, the data was collected for all available trading partners and where one dataset omitted a potential trading partner it was complemented with data from the other data sources to ensure that relevant trade flows were not excluded. Trade flow data was aggregated into 38 countries and country groupings yielding a symmetric 38×38 matrix of bilateral trade flows (Table 23.5).

Table 23.5 Countries and country groupings

Acronym	Description
EU25	European Union 25
BRA	Brazil
ARG	Argentina
USA	United States of America
CHN	China
PAR	Paraguay
CAN	Canada
MEX	Mexico
BUR	Bulgaria & Romania
WBA	Western Balkans
REU	Rest of Europe
RUB	Russia & Belarus
UKR	Ukraine
CAM	Central America
VEN	Venezuela
CHL	Chile
URU	Uruguay
BOL	Bolivia
NWSA	NW countries in South America (Colombia, Ecuador and Peru)
IND	India
JAP	Japan
TLD	Thailand
SKR	South Korea
INDO	Indonesia
MLAY	Malaysia
PHIL	Philippines
ANZ	Australia & New Zealand
MOR	Morocco
TUN	Tunisia
ALG	Algeria
EGY	Egypt
TUR	Turkey
ISR	Israel
LDC	Least Developed Countries
AFR	Non-LDC African Countries in the ACP
C&P	Non-LDC Caribbean & Pacific Island countries in ACP
MIDE	Middle East (Syria, Iran, Iraq, Saudi Arabia, UAE)
ROW	Rest of World

As the data reported by origin countries and destination countries did not always match, the maximum value of the two reporting countries was taken for the final trade flow. Trade flow data was available in both volume and value. Although volume was of primary interest for the analysis, value data allowed the calculation of implied per-unit costs for various trades which were, in turn, used in the validation of global trade prices (discussed below).

Domestic supply and demand data came from FAOSTAT and was validated with USDA (PS&D) [26] data. FAOSTAT data is reported on a calendar year basis and was used to indicate each country's excess supply and demand conditions. The composition of domestic demand (feed, food, industrial demand) was used in applying and weighting the appropriate elasticities to the trade model. Demand and supply elasticities were obtained from the CAPRI [4] model. Where data was unavailable, comparable elasticities were taken from FAPRI [10] and WATSIM [27].

Freight rates for all possible routes implied in the constructed $38 \times x38$ trade matrix used in the analysis were estimated through regression analysis. Actual freight rates reported for maize were obtained from Maritime Research. These rates were regressed against the distance covered in each individual trade as well as against selected indexes of bunker and fixtures for panamax and handy size vessels typically used in dry bulk commodity trade. The regression equation was then used to estimate freight rates for all routes and years in the analysis.

Cash port prices reported by USA, Brazil, Argentina and the EU for the dominant trading ports were the basis of the global trade prices used in the model. Using each country's share of trade with these four countries an average FOB price was constructed for the port of origin. A per-unit weighted average of transportation cost and tariff was added to the FOB price to derive a CIF price for each importer. These prices were validated by the implied per-unit import costs calculated from GTIS trade data to ensure consistency.

Annual import tariff data was collected from the WTO tariff database [28] using the country's average applied tariff. This data was validated with tariff rates maintained by FAPRI. All export tariffs used were from FAPRI. Tariffs were calculated for country aggregates by weighting the volume of imports (or exports) and average tariff paid for each country.

23.7.2   Baseline Development and Model Validation

The spatial equilibrium model developed here is a simplified representation of world trade in maize. The model is not a forecasting tool and should not be understood as such. Still, it must effectively represent the direction and magnitude of changes that might occur in response to a given disruption of bilateral trade. In this context, an effective representation of observed supply, demand and trade by the model for the countries of interest is important. Hence, the model must be validated for its effectiveness to approximate observed demand, supply and trade conditions in any particular year.

The model was calibrated with GTIS trade data from the respective calendar years. Solving the model provides estimates of supply and demand as well as imports and exports for all 38 countries/country groups in the analysis. These baseline estimates can be compared with observed data in order to evaluate the adequacy of the empirical model. Deviations of model-derived baseline estimates from actual excess demand and supply figures for maize—expressed as (baseline-actual)/actual—ranged from −29 to 12 % for all countries and years in the analysis. The average deviations were much smaller than the extreme values—under 5 %—with the relevant deviations for the world market at 3.6 % for 2005 and 2.3 % for 2007, the 2 years for which the model was calibrated.[5] Similarly, calculated trade flows for all large importers and exporters closely matched observed trade flows. Small volume trades were not represented as effectively, which is typical in spatial equilibrium models using annual data. Such trades typically represent opportunistic transactions within a year and are difficult to represent through annual averages. In all, the baseline model runs were considered effective.

23.8 Empirical Results

With an effective baseline in hand, the impact of potential bilateral trade disruptions for selected countries or regions in LA could be examined through scenario analysis as outlined in the modeling section above. Two scenarios were considered and each scenario has two components. Scenario 1 involves the disruption of bilateral trade between a single large importer (Mexico) and (1) its main supplier (US) or (2) its major supplier and another large exporter (US and Argentina). Scenario 2 involves the disruption of trade for a block of smaller importers (Colombia, Ecuador, and Peru) located in northwest South American (NWSA) and (1) their main supplier (US) and (2) their main supplier and another large exporter (US and Argentina).[6] Note that the importer and its major supplier are neighboring countries in Scenario 1, and alternative supplies have to be procured from farther away when bilateral trade is disrupted. In contrast, the importer and the primary suppliers are more distant from one another in Scenario 2 and alternative suppliers to NWSA are closer. Finally, both scenarios are examined under two distinct types of markets situations as represented by 2007 and 2005 (i.e. the baseline years) reflecting above and below average global supply conditions.

23.8.1 Scenario 1

Mexico normally gets nearly all of its maize imports from the United States. For example, Mexico imported more than 8 million metric tons of maize in 2007, and

[5]2005 was chosen as a baseline year because it was representative of a short global maize crop and hence of tight market conditions. 2007 was chosen as an alternative baseline year since global production was above average and hence it was representative of a well-supplied global maize market. The two alternative baselines are used to envelop "normal" market conditions.

[6]The importing countries in scenarios 1 and 2 were chosen in order to provide effective examples of differential size in imports and relative proximity to suppliers.

the US supplied all but roughly 80,000 metric tons of this amount. The reasons for this strong trade relationship include proximity as well as the size and dependability of the US market and the convenient and inexpensive freight access through various modes of transport (i.e., rail and ocean vessels).

If regulatory asynchronicity could lead to finding material in the US maize supply chain that were unauthorized in Mexico for consumption, trade disruptions could ensue. As such, Mexico would seek maize imports from alternative suppliers. As well, US maize exports would be consumed domestically or delivered to other importers. Due to the larger distances involved in finding alternative markets for both the US and Mexico, we expect that freight costs would significantly increase for both countries. In particular, the domestic price of maize in Mexico should rise, and the domestic price of maize in the US should fall as the trade volumes shift to other partners. The maize price should rise in the countries that increase exports to Mexico (e.g., Argentina), and the price should decline in those countries that attract new imports from the US. If the scenario is extended to include a bilateral trade disruptions in maize trade between Mexico and one of the alternative suppliers (Argentina), we should expect the maize price in Mexico to increase even more, and the prices of maize in the US and Argentina to decline.

The Scenario 1 results for 2005 are presented in Tables 23.6 and 23.7 for 2007. In both tables, the leftmost columns present the results for the US-only trade disruption, and the impact of a disruption in the bilateral trade between Mexico the US and Argentina appears in the rightmost columns of the tables. For 2005, the unavailability of US imports increase the price of maize in Mexico by about 15.3 %. As expected, the maize price in alternative suppliers to Mexico (e.g., Argentina, Brazil, Paraguay, and Venezuela) increase, and the price in countries that import maize from the US (e.g., Canada and Central America) decline. The price of the US declines by 1.2 % as a result of the trade disruption. The same general pattern is observed under the 2007 data, though the relative price impacts are smaller than in 2005 due to the presence of more plentiful global maize supplies in 2007. In particular, the increase in the Mexico maize price is about 9.3 % under the 2007 situation.

If maize trade between Mexico and the US and Argentina was interrupted due to regulatory asynchronicity and the presence of unapproved biotech events in the supplies of these exporters, the price of maize in Mexico would increase by 20.3 % under the 2005 situation and by 12.6 % under the 2007 situation. Maize prices both in the US and in Argentina would decline by 0.5–1.4 % in the US and 0.2–0.5 % in Argentina, respectively. As before, the relative magnitudes of the price changes are more pronounced under the short-supply situation in 2005 than in 2007. Overall, we find that the maize price in Mexico would increase by about 9–20 % under the various situations in this study, and the relative impacts on maize prices in the relevant trading partners are considerably smaller (i.e., less than 2 % across all cases).

Table 23.6 Price and supply impacts (% changes against baseline)

Scenario 1–2005 baseline						
	No imports from the US			No imports from the US & Argentina		
	Price	Exports	Imports	Price	Exports	Imports
	%change			%change		
EU27	−0.9	0.0	0.0	−1.0	0.0	0.0
BRA	0.4	0.0	0.0	3.3	1.3	0.0
ARG	0.4	0.3	0.0	−0.5	−1.6	0.0
USA	−1.2	−0.3	0.0	−1.4	−0.5	0.0
CHN	−0.2	0.0	0.0	−1.0	0.0	0.0
PAR	2.8	1.1	0.0	7.7	13.5	0.0
CAN	−1.0	0.0	0.1	−1.2	0.0	0.1
MEX	15.3	0.0	−3.4	20.3	0.0	−6.5
BUR	0.3	0.0	0.0	−1.2	0.0	0.1
WBA	2.0	0.1	0.0	1.0	0.1	0.0
REU	2.5	3.5	0.0	13.4	23.0	0.0
RUB	2.2	0.0	−1.1	2.4	0.0	−1.3
UKR	0.3	0.1	0.0	−1.3	−0.9	0.0
CAM	−1.0	0.0	0.7	−1.2	0.0	0.9
VEN	8.7	0.0	0.0	6.4	0.0	0.0
CHL	0.3	0.0	−0.3	−0.4	0.0	0.4
URU	0.4	0.0	0.0	−0.4	0.0	0.1
BOL	−0.4	0.0	0.0	1.1	0.0	0.0
NWSA	−1.0	0.0	0.9	−0.4	0.0	0.4
IND	−1.0	0.0	0.0	−0.7	0.0	0.0
JAP	−0.7	0.0	0.7	−0.9	0.0	0.8
TLD	2.1	0.0	0.0	−3.3	−0.1	0.0
SKR	−0.8	0.0	2.3	−0.9	0.0	2.1
INDO	0.3	0.0	0.0	−0.3	0.0	0.0
MLAY	0.3	0.0	−0.9	−0.3	0.0	0.9
PHIL	5.8	0.2	0.0	−0.3	0.0	0.0
ANZ	0.3	0.0	0.0	−0.3	0.0	0.0
MOR	0.3	0.0	−1.1	−1.1	0.0	3.1
TUN	0.3	0.0	−2.4	−1.0	0.0	5.3
ALG	0.3	0.0	−1.3	−1.0	0.0	3.4
EGY	0.3	0.0	−0.4	−1.2	0.0	2.0
TUR	0.3	0.0	0.0	−1.2	0.0	0.0
ISR	0.3	0.0	−1.8	−1.2	0.0	5.0
LDC	0.3	0.0	−0.1	−0.3	0.0	0.1
AFR	0.3	0.0	0.0	−0.4	0.0	0.0
C&P	−1.0	0.0	2.2	−1.1	0.0	2.3
MIDE	0.3	0.0	−0.4	−0.3	0.0	0.5
ROW	0.3	0.0	0.0	−0.3	0.0	0.0

Table 23.7 Price and supply impacts (% changes against baseline)

Scenario 1–2007 baseline						
	No imports from the US			No imports from the US & Argentina		
	Price	Exports	Imports	Price	Exports	Imports
	%change			%change		
EU27	−0.45	0.00	0.40	−0.32	0.00	0.14
BRA	0.20	0.13	0.00	0.68	0.29	0.00
ARG	0.05	0.05	0.00	−0.20	−0.20	0.00
USA	−0.54	−0.14	0.00	−0.59	−0.20	0.00
CHN	−0.46	−0.02	0.00	−0.39	−0.02	0.00
PAR	0.16	0.76	0.00	0.83	1.05	0.00
CAN	−0.50	0.00	0.04	−0.17	0.00	0.02
MEX	9.28	0.00	−2.12	12.62	0.00	−2.88
BUR	0.04	0.00	−0.05	−0.16	0.00	0.23
WBA	0.04	0.00	−0.01	−0.16	0.00	0.05
REU	0.04	0.00	−0.05	−0.16	0.00	0.21
RUB	−0.04	0.00	0.01	−0.15	0.00	0.02
UKR	0.04	0.01	0.00	−0.16	−0.03	0.00
CAM	−0.48	0.00	0.36	−0.17	0.00	0.12
VEN	−0.49	0.00	0.16	−0.17	0.00	0.06
CHL	0.04	0.00	−0.06	−0.17	0.00	0.27
URU	0.04	0.00	−0.01	−0.18	0.00	0.04
BOL	1.57	0.03	0.00	2.44	0.05	0.00
NWSA	0.04	0.00	−0.05	0.59	0.00	−0.69
IND	0.14	0.02	0.00	0.13	0.02	0.00
JAP	−0.42	0.00	0.33	−0.36	0.00	0.28
TLD	0.65	0.09	0.00	0.31	0.04	0.00
SKR	−0.43	0.00	0.35	−0.36	0.00	0.30
INDO	0.04	0.00	0.00	−0.13	0.00	0.01
MLAY	0.03	0.00	−0.10	−0.15	0.00	0.43
PHIL	−0.21	−0.03	0.00	−0.39	−0.06	0.00
ANZ	0.04	0.00	0.00	−0.16	0.00	0.01
MOR	0.04	0.00	−0.09	−0.16	0.00	0.39
TUN	0.04	0.00	−0.12	−0.16	0.00	0.53
ALG	0.04	0.00	−0.11	−0.16	0.00	0.49
EGY	0.04	0.00	−0.05	−0.16	0.00	0.22
TUR	0.04	0.00	−0.02	−0.15	0.00	0.08
ISR	0.04	0.00	−0.15	−0.15	0.00	0.62
LDC	0.03	0.00	0.00	−0.15	0.00	0.02
AFR	0.04	0.00	0.00	−0.16	0.00	0.02
C&P	−0.48	0.00	0.98	−0.17	0.00	0.34
MIDE	0.03	0.00	−0.08	−0.15	0.00	0.33
ROW	0.03	0.00	−0.01	−0.15	0.00	0.05

23.8.2   Scenario 2

Like Mexico, the small countries in NWSA (Colombia, Ecuador, and Peru) import much of their maize from the US. However, these countries are more distant from the US than Mexico, and they are closer to other sources of maize imports (e.g., Argentina, Brazil, and Paraguay). As well, these countries are smaller trading partners, and they should find it easier to replace the disrupted trade quantities. Thus, we expect that the price impacts of the trade disruptions in these markets should be smaller than in Scenario 1.

The Scenario 2 results for 2005 are presented in Table 23.8. Here, the disruption of US imports increases the maize price by 2.4 % in the importing countries and decreases the US price by less than 1 %. As well, the potential suppliers to NWSA (Argentina, Brazil, Paraguay, and other South American countries) see slightly higher maize prices, and the maize prices in other markets that buy US exports (e.g., Canada and Mexico) decline by small amounts. When both US and Argentine maize imports are interrupted, the price in NWSA increases by 8.2 %, the prices in the US and Argentina decline by small amounts, and the prices in other maize-producing nations of South America increase slightly. Finally, the 2007 outcomes are uniformly smaller in magnitude (as illustrated in the Scenario 1 results). Overall, we find that bilateral trade disruptions between smaller countries like those in NWSA and the US and Argentina cause modest (2–8 %) increases in the domestic maize price of the importing countries.

23.9   Summary and Concluding Comments

LA is home to large number of agricultural commodity importers that depend heavily on such imports for their consumption. Almost all of these agricultural commodity imports are procured from major exporters in the Americas. All exporting countries in the Americas have adopted biotech crops extensively while Latin America importers, generally, have not. The degree to which biotech crops have been regulated in these countries has also been quite variable.

Some exporting countries have been approving new biotech crops for use and planting on an ongoing basis. As a result, these exporting countries have had access to the full biotech pipeline over time. Other exporting countries have approved and planted fewer new biotech crops. Importers have also deregulated new biotech crops at variable speeds. A few importers have approved a large number of new biotech crops for consumption but not for planting but most importing countries in LA have approved few if any. As a result, structural asynchronicity in the regulatory approvals of biotech crops among importing countries in LA and their suppliers has been ongoing and it is likely to worsen as the biotech pipeline continues to expand.

In the past, regulatory asynchronicity has led to few trade disruptions but that could change in the near future as new biosafety laws take effect in many LA

Table 23.8 Price and supply impacts (% changes against baseline)

	No imports from the US			No imports from the US & Argentina		
	Price	Exports	Imports	Price	Exports	Imports
	%change			%change		
EU27	−0.15	0.00	0.01	0.49	0.00	−0.01
BRA	0.41	0.06	0.00	2.32	2.43	0.00
ARG	1.25	1.66	0.00	−0.94	−1.27	0.00
USA	−0.46	−0.18	0.00	−0.05	−0.02	0.00
CHN	0.35	0.00	0.00	−0.59	0.15	0.00
PAR	0.62	0.47	0.00	8.97	8.83	0.00
CAN	−0.41	0.00	0.04	−0.59	0.00	0.06
MEX	−0.41	0.00	0.13	0.37	0.00	−0.12
BUR	−0.35	0.00	0.02	−0.97	0.00	0.06
WBA	−0.38	−0.03	0.00	−1.05	−0.08	0.00
REU	0.58	1.00	0.00	1.88	3.22	0.00
RUB	0.50	0.00	−0.28	1.62	0.00	−0.89
UKR	−0.38	−0.26	0.00	−1.06	−0.72	0.00
CAM	−0.39	0.00	0.30	0.43	0.00	0.01
VEN	−0.39	0.00	0.02	−0.60	0.00	0.03
CHL	0.39	0.00	−0.45	−0.29	0.00	0.34
URU	0.42	0.00	−0.08	−0.32	0.00	0.06
BOL	0.52	0.00	0.00	−0.17	0.00	0.00
NWSA	2.38	0.00	−0.81	8.15	0.00	−9.07
IND	0.48	0.01	0.00	−0.19	0.00	0.00
JAP	−0.31	0.00	0.26	−0.52	0.00	0.43
TLD	0.48	0.02	0.00	0.65	0.13	0.00
SKR	−0.32	0.00	0.72	−0.54	0.00	1.24
INDO	0.32	0.00	−0.01	0.26	0.00	−0.61
MLAY	0.32	0.00	−0.92	0.25	0.00	−0.16
PHIL	0.37	0.02	0.00	−0.28	−0.01	0.00
ANZ	0.35	0.00	−0.01	−0.24	0.00	0.00
MOR	−0.36	0.00	1.05	−0.56	0.00	1.65
TUN	−0.34	0.00	1.80	−0.57	0.00	2.97
ALG	−0.35	0.00	1.16	−0.57	0.00	1.91
EGY	−0.34	0.00	0.56	−0.94	0.00	1.57
TUR	−0.34	0.00	0.01	−0.96	0.00	0.03
ISR	−0.33	0.00	1.46	−0.93	0.00	4.06
LDC	0.33	0.00	−0.09	−0.25	0.00	0.07
AFR	0.37	0.00	−0.03	−0.28	0.00	0.02
C&P	−0.39	0.00	0.79	−0.63	0.00	1.28
MIDE	0.32	0.00	−0.56	−0.25	0.00	0.42
ROW	0.33	0.00	0.00	−0.25	0.00	0.00

Scenario 2–2005 baseline

importing countries. Other countries with existing biosafety laws may also see their regulatory capacity tested by the increasing flow of new biotech crops and face, at least temporary, asynchronicity with their trade partners. Under these conditions, LA countries that import large amounts of agricultural commodities may benefit from the adoption of national LLP policies, including the use of the CODEX Annex, which could resolve short term pressures from asynchronicity and low level presence of unauthorized material in their imports. A more significant and long-lasting problem, however, for many importing LA countries could be structural asynchronicity from limited regulatory capacity.

Many Latin American importing countries with limited technical and scientific regulatory capacity and financial resources will confront a difficult reality as they implement more fully their biosafety laws. They will have to find ways to effectively evaluate the safety of new biotech crops at a fast pace or risk costly trade disruptions with trade partners. Pulling regulatory resources through regional partnerships and leveraging reviews and assessments from countries with well-developed regulatory capacity may be some options among others.

Trade disruptions are an unpalatable alternative as they can be costly. In our empirical analysis of maize trade in LA we estimated that for smaller importing countries whose trade can be more easily shifted across alternative suppliers, prices would increase 2–8 % depending on the supply conditions in the global market. Even such modest price increases translate into significant outlays, however. Applying, for instance, these increases on the 2005–2009 maize import spending of Colombia, Ecuador and Peru would result, on average, in 20–80 million higher outlays per year. For larger importers, like Mexico, the estimated price increases were higher, 9–20 % depending on the supply conditions of the global market. Applying, these price increases on the 2005–2009 maize import spending of Mexico would result, on average, in 130–294 million higher outlays per year.

These increased outlays do not reflect additional costs from foregone value added activity, potential short term supply disruptions, quality differences in alternative supplies, and added transaction costs. They also do not reflect the tighter supply conditions experienced in international markets in more recent years. For instance, in the last 2 years, China has turned from a modest net exporter of maize to a meaningful net importer, a trend expected to strengthen in the future. Similarly, the US has continued to increase the amount of maize retained for its growing ethanol industry. Tighter supply conditions would only worsen potential price increases from trade disruptions. The number of alternative exporters could also dwindle. In our scenario analysis, trade from up to two major exporters was restricted leaving ample alternative supplies for trade substitution. In the face of structural asynchronicity, however, this may not be the case leading to larger and more abrupt price changes. Finally, adding such potential economic costs across all relevant commodities provides a full scope of the problem in hand.

Amid renewed interest in food security in the face of escalating commodity prices and tight global supplies of agricultural commodities, trade will continue to play an important moderating role. A predictable and effective regulatory environment that minimizes the structural asynchronicity of regulatory approvals in LA importing countries and their suppliers is desirable in order to keep trade options open and agricultural commodity prices in check.

References

1. Backus, G.B.C., Berkhout, P., de Kleijn, A.J., Eaton, D.J.F., Franke, L., Lotz, B., van Mil, E.M., Roza, P., Uffelmann, W.: EU Policy on GMOs: A Quick Scan of the Economic Consequences. LEI, Report 2008-070 (2008)
2. Brookes, G., Yu, T.H., Tokgoz, S., Elobeid, A.: The production and price impact of biotech corn, canola, and soybean crops. AgBioForum **13**(1), 25–52 (2000)
3. Carpenter, J.E.: Peer-reviewed surveys indicate positive impact of commercialized GM crops. Nat. Biotechnol. **28**(4), 319–321 (2010)
4. Common Agricultural Policy Regionalized Impact Modeling System (CAPRI). http://www.capri-model.org. Accessed December 2012
5. Chen, C., McCarl, B., Chang, C.: Estimating the impacts of government interventions in the international rice market. Can. J. Agr. Econ. **54**, 81–100 (2006)
6. EC DG Agriculture Report: Economic Impact of Unapproved GMOs on EU Feed Imports and Livestock production, European Union Directorate General of Agriculture and Rural Development, Brussels (2007)
7. Enke, S.: Equilibrium among spatially separated markets: solution by electrical analogue. Econometrica **19**, 40–47 (1951)
8. Falck-Zepeda, J.B., Traxler, G., Nelson, R.G.: Surplus distribution from the introduction of a biotechnology innovation. Am. J. Agr. Econ. **82**, 360–369 (2000)
9. Food and Agriculture Organization of the United Nations, FAOSTAT, TRADESTAT. http://faostat.fao.org/site/406/default.aspx. Accessed March 2013
10. Food and Agricultural Policy Research Institute "Elasticity Database". http://www.fapri.iastate.edu/tools/elasticity.aspx. Accessed March 2013
11. Global Trade Information Services "Global Trade Atlas". http://www.gtis.com/english/GTIS_revisit.html. Accessed March 2013
12. James, C.: Global Status of Commercialized Biotech/GM Crops ISAAA Brief No. 41-2009. ISAAA, Ithaca (2009)
13. James, C.: Global Status of Commercialized Biotech/GM Crops: 2010 ISAAA Brief 42-2010. ISAAA Ithaca (2010)
14. Konduru, S., Kruse, J., Kalaitzandonakes, N.G.: The Global Economic Impacts of Roundup Ready Soybeans, in Genetics and Genomics of Soybeans. Gary Stacey, Springer, New York (2008)
15. Krueger, R., Buanec, B.L.: Action needed to harmonize regulation of low-level presence of biotech traits. Nat. Biotechnol. **26**, 161–162 (2008)
16. Philippidis, G.: EU import restrictions on genetically modified feeds: impacts on Spanish, EU and global livestock sectors. Span. J. Agr. Res. **8**(1), 3–17 (2010)
17. Qaim, M.: The economics of genetically modified crops. Ann. Rev. Res. Econ. **1**, 665–694 (2009)
18. Samuelson, P.: Spatial price equilibrium and linear programming. Am. Econ. Rev. **42**, 283–303 (1952)
19. Sobolevsky, A., Moschini, G., Lapan, H.: Genetically modified crops and product differentiation: trade and welfare effects in the soybean complex. Am. J. Agr. Econ. **87**, 621–644 (2005)
20. Stein, A.J., Rodríguez-Cerezo, E.: International trade and the global pipeline of new GM crops. Nat. Biotechnol. **28**, 23–25 (2010)

21. Takayama, T., Judge, G.: Spatial and Temporal Price and Allocation Models. North-Holland Publishing, Amsterdam (1971)
22. United Nations "Comtrade". http://comtrade.un.org/db/default.aspx. Accessed March 2013
23. United States Department of Agriculture, Embargoes, Surplus Disposal, and U.S. Agriculture, Agricultural Economic Report 564, Economic Research Service, Washington, DC (1986)
24. U.S. Department of Agriculture, Ecuador: Agricultural Biotechnology. Annual GAIN Report. Foreign Agricultural Service (FAS), Washington, DC. http://gain.fas.usda.gov/Recent%20GAIN%20Publications/Agricultural%20Biotechnology%20Annual_Quito_Ecuador_7-6-2011.pdf (July 6, 2011)
25. U.S. Department of Agriculture, Paraguay: Agricultural Biotechnology. Annual GAIN Report. Foreign Agricultural Service (FAS), Washington, DC. http://gain.fas.usda.gov/Recent%20GAIN%20Publications/Agricultural%20Biotechnology%20Annual_Buenos%20Aires_Paraguay_7-18-2011.pdf (July 15, 2011)
26. U.S. Department of Agriculture, Foreign Agriculture Service, Production Supply and Distribution Database. http://www.fas.usda.gov/psdonline/. Accessed March 2013
27. World Agriculture Trade Simulation Model (WATSIM). http://www.ilr1.uni-bonn.de/agpo/rsrch/watsim/databasewatsim02.xls
28. World Trade Organization. WTO Tariff Database. http://tariffdata.wto.org/. Accessed December 2012

Chapter 24
How Venture Capital Creates Externalities in the Bioeconomy: A Geographical Perspective

Christos Kolympiris and Nicholas Kalaitzandonakes

24.1 Introduction

Venture capital is an engine for economic growth because it promotes innovation, entrepreneurship and wealth creation [38,55,83,94]. One way venture capitalists act as a conduit to economic growth is by providing financing and managerial services to firms with innovation potential but limited business expertise and sources of finance [21, 61, 76]. Alongside, venture capital also tends to bring about positive externalities to the overall economy (the externality impact). Such externalities often emanate from the enhancement of the knowledge base of a given region associated with venture capital activity and result to improvements in innovation, increases in firm births as well as to the creation of new industries [32, 57, 82, 83, 96].

Some of the studies that analyze the externality impact of venture capital are not occupied with how physically far this impact goes. Other studies, largely due to the aim of their research, examine whether venture capital affects economic outcomes in large administrative regional units such as Metropolitan Statistical Areas (MSAs). This methodological approach implicitly specifies a priori the (broad) MSA as the unit that represents the spatial scope of the externality impact of venture capital.

An earlier version of this work was published as "The geographic extent of venture capital externalities on innovation, Kolympiris C. and Kalaitzandonakes N., 2013, Venture Capital: An International Journal of Entrepreneurial Finance 15, 199–236."

C. Kolympiris
Wageningen University, Hollandseweg 1, 6706 KN Wageningen (Building 201),
De Leeuwenborch, Room 5059, Wageningen, The Netherlands
e-mail: christos.kolympiris@wur.nl

N. Kalaitzandonakes (✉)
Department of Agricultural and Applied Economics, University of Missouri, 125 A Mumford
Hall, Columbia, MO 65211, USA
e-mail: kalaitzandonakesn@missouri.edu

A.A. Pinto and D. Zilberman (eds.), *Modeling, Dynamics, Optimization
and Bioeconomics I*, Springer Proceedings in Mathematics & Statistics 73,
DOI 10.1007/978-3-319-04849-9_24,
© Springer International Publishing Switzerland 2014

Nevertheless, venture capitalists show a general tendency to invest locally as a means to mitigate agency problems with target firms and benefit from network externalities [22, 32, 41, 87].[1] Indeed, the spatial extent of such relationships has been found to be in the order of a few miles, [56, 59, 93]. Insofar as the narrow geographic scope that VCs operate mirrors the range they create externalities via the improvement of the regional knowledge base, there is a divergence between the use of broad administrative units in relevant prior studies and the implied narrow extent of the externality impact of venture capital.[2] By extension, this divergence invites research that examines the question of just how physically far does that externality impact go. A question that we address in this paper.

For the analysis we focus on the externality impact of venture capital on innovation. Operationally, we approximate innovative performance with patents [15,18,31,57][3] which is a methodological choice that sides with the generally strong relationship between patents and innovations in life sciences which is the industry we focus on [7]. Then, we associate the patenting rate of a focal life sciences firm (LSF) with the amount of venture capital investments towards life sciences that have taken place at increasingly distant narrow spatial units from the LSF. To strengthen the robustness of our findings we employ an instrumental variables approach that addresses the potential attraction of venture capitalist to already innovative regions and firms. Further, we account for a number of sources that can complement venture capital in generating local externalities and these include the research intensity of local universities and the agglomeration of similar firms. As well, we consult with previous literature [78] and examine whether the externality impact of venture capital is sensitive to the stage of firm growth that venture capital funds are directed to. To do so, we adopt a novel methodological approach as we partition venture capital funds to the stage of firm growth they correspond.

The reason we focus on LSFs is twofold. First, venture capitalists invest heavily in life sciences [81] which, by extension, makes the industry at hand a suitable setting for the present study. Second, life sciences is a knowledge-based industry which should then allow any externalities from venture capital that associate with improvements in the knowledge base of a given region to show. To distinguish the externality impact of venture capital from improvements that are brought about from the involvement of venture capitalists on target firms, we focus on small LSFs

---

[1]Using Germany as their case study, Fritsch and Schilder [36] demonstrate that country-specific features can undermine the significance of spatial proximity between venture capitalists and target firms.

[2]To illustrate the geographic scope of an MSA, Stuart and Sorenson [90] report that in their study the average area of an MSA was 10,515 square miles.

[3]Despite their wide use, patents have certain shortcomings as a measure of innovation. For instance, innovative firms may not patent for strategic reasons [92] or maybe patents relate more to invention rather than to innovation [67]. Albeit a less than perfect proxy for innovation, patents are generally still a reliable measure of innovation [3].

that have won Small Business Innovation Research (SBIR)[4] grants. The majority of the SBIR firms are relatively small and at their very early stages of research, which then partly explains why most of them are not recipients of venture capital funds. Further, because small firms generally reap the most benefits from spatial externalities [66], we expect our focus on firms of that size to allow us to measure their strength. From a policy perspective, the present study is expected to shed more light on the effectiveness of ongoing efforts from governments around the world to attract venture capital to a region based on the rationale that it is a core element of successful regional innovation systems partly because it creates externalities [44,60]. If the strength, geographic extent and underlying cause of such externalities is not well understood, the effectiveness of the relevant policies can be hampered. Further, given the aims of the SBIR program to promote innovation [25] and the increasing popularity of like programs around the world (i.e. UK's SIRI program, Australia's IIF program and programs in Sweden, Russia, Canada, UK etc.) [27,98] an understanding of potential sources of complementarity between publicly-funded firms and private investments could induce greater returns on the public funds invested through these programs. We proceed with the rest of the paper as follows: In Sect. 24.2 we review the relevant literature and form our theoretical expectations. In Sects. 24.3, 24.4 and 24.5 we present our empirical methods and explain our data sources. In Sect. 24.6 we present the empirical results of our study and we summarize and conclude in Sect. 24.7.

24.2 What Venture Capitalists Do and How They Generate Local Externalities

Venture Capitalists (VCs) tend to invest in high risk young firms with a potential to yield returns in the range of 25–35 % or above within a short period of time [101]. On the downside of the high returns, target firms do not typically have an established record which then makes their evaluation a thorny task. Insofar as the lack of an established record also reflects entrepreneurs with more information about the firm compared to the VCs, information asymmetries are expected to grow [37, 85]. To confront these asymmetries and meet the targeted financial returns, VCs usually tie their compensation with the performance of the target firm, which then creates incentives for them to nurture the firm in question to financial success.

To reduce agency costs and improve their expected compensation via an increase in the potential performance of the target firm, VCs tend to employ two main mechanisms. First, they spend a considerable amount of their time in monitoring and

---

[4]The SBIR program is the largest federal program in the US and it provides seed and early stage funding to promising small firms in cutting edge research areas such as life sciences, electronics, materials and energy conversion. Promoting innovation is among the stated goals of the program and it has generally been found successful in achieving that goal [11].

providing managerial advice and direction to the firms they invest in [39,54,78,94]. Second, in order to cope with the limited publicly available information about target firms, VCs use their local networks primarily with other VCs to gain more information [50,65]. These two mechanisms are expected to be the main drivers of the externality impact of venture capital on innovation.

Largely because of their hands-on investment approach, venture capitalists often improve the performance of the target firms [16, 57, 73]. In turn, better firms can advance the knowledge base of a given region through a number of ways. For instance, the quality of the relevant actors tends to boost the quality of knowledge that circulates within informal knowledge feedback loops among proximate firms and evidence suggests that the performance of a given firm hinges on the performance of nearby firms [14, 56]. Further, better firms can equip the local labor pool with highly trained professionals who, via labor mobility, can also increase the performance of their new employer [34]. As well, professional networks maintained by research alliances, contractual agreements or other forms of formal partnerships can also benefit from the participation of better firms in such networks especially because knowledge tends to diffuse among members of the networks [12, 51, 69]. It is important to note that the participation of firms in professional networks is not a given and the same holds for the participation of local firms in those networks. However, the realities of knowledge intensive industries such as life sciences prompt firms to increasingly engage in such networks [77] mainly with other local firms [87, 90]. For instance, mounting costs, long operating cycles, complex regulatory environment and scientific uncertainties are potent in life sciences [28, 46]. Such industry characteristics often render research efforts by a sole firm insufficient and firms turn to collaborative schemes in order to confront them. These collaborative schemes tend to materialize among nearby firms [70] in large part because spatial proximity typically eases the flow of knowledge as joint experiments, visits and informal encounters are less costly and more likely to occur.

The networks that VCs form with other VCs[5] in order to facilitate the flow of information among them with regard to target firms, constitute the second main mechanism under which VCs can generate positive externalities.[6] The knowledge that circulates in networks of VCs relies heavily on interpersonal contacts and other forms of human capital. As such, this knowledge often leaks in local circles and then forms the basis for improvement in the knowledge base of a given region. In a way, VCs become what [102] terms "knowledge brokers" and a number of authors have noted this role of VCs [32,37,47,84]. For example, Florida and Kenney [32] stress the importance of VC networks as a means to mobilize valuable resources, Shane and Cable [84] present evidence how investment-specific information transfers

---

[5]Note that the composition of those networks is not necessarily confined only to VCs but it can also include other industry professionals such as lawyers and accountants [75,103].

[6]This is not to say that positive externalities cannot arise from a single VC. But, because externalities typically emanate from the flow of knowledge, networks of VCs are expected to be stronger in that respect.

often through social networks and [37] explains how important is the knowledge generated by VCs. Collectively, the aforementioned mechanisms initiated by VCs should work towards developing the externality impact of venture capital.

The next relevant inquiry is just to which geographic distance do venture capitalists generate externalities. To form our theoretical expectations for this question we consult two related strands of literature. First we look into literature that analyzes the spatial extent of relationships between VCs and between VCs and target firms. Using this literature we approximate how far VCs can generate externalities. The second strand of literature we employ, identifies the geographical range of spatial externalities among firms. The contributions in this literature assist us in approximating how far do externalities from the improvement in the performance of a focal firm go.

On theoretical grounds, spatial proximity between VCs and target firm is crucial mainly because it allows VCs to monitor and coach their target firms more intensively as regular visits are easier and transaction and opportunity costs from travelling are minimized [33, 87, 102].[7] Indeed, empirical studies have corroborated the importance of spatial proximity for the relationship between VCs and target firms and have thus indirectly provided an estimate of the range in which VCs can improve the knowledge base of a focal region. Lerner [59] finds that the probability of a VC becoming a board member on a target firm is close to 50 % when the target firm is within 5 miles; a probability that is twice the magnitude of the comparable probability of target firms located more than 500 miles away. Reporting similar estimates in magnitude, Kolympiris et al. [56] discover that the magnitude of venture capital funds accumulated by a given biotechnology firm moves in the same direction with funds raised by similar firms located within 10 miles; they also report that the density of non-funding VCs in the same radius is instrumental for the growth of venture capital funds for a particular firm. Finally, Tian [93] observes substantially more intense involvement of VCs, in the form of more financing rounds, to target firms when they both locate within a 25 miles radius.[8] A common feature of the aforementioned studies is that they have discovered spatial relationships to be potent at the spatial unit that is the narrowest and the closest to the focal actor of all the spatial units considered in each study (5 miles for [59], 10 miles for [56] and 25 miles for [93]). The fact that in all these studies the narrowest and the most proximate unit was the one with the most pronounced effect, suggests that these estimates may present an upper bound of the spatial extent of the externalities that emanate from VCs.

---

[7]In this section we concentrate on the impact of spatial proximity on post-investment activities; proximity is relevant for pre-investment activities as well but such discussion is beyond our scope.

[8]In related empirical studies, that do not focus on the spatial extent of VC relationships but estimate what drives the investment decision of VCs ex ante, Lutz et al. [63], Cumming and Dai [26] and [87] also highlight how significant geographic proximity between VCs and target firms is in shaping that decision.

This proposition is supported by evidence presented in the second stream of research we examine here; research that analyzes the geographic scope of spatial externalities among firms. Seminal early contributions in this literature indicated that spatial externalities are confined in space but did not estimate neither an upper geographical bound on these relationships nor whether the strength of the externalities varies with different degrees of proximity (e.g. [10, 52]). Leveraging developments in geographical information systems and computing power, more recent contributions have picked up where the early works left off. For instance, Aharonson et al. [4] discovered that a significant form of spatial externalities, knowledge spillovers, within one third of a mile were instrumental in explaining the location of Canadian biotechnology firms. In a similar vein, Wallsten [97] estimated that the probability of a focal firm to win an SBIR grant was significantly affected by the density of previous SBIR winners in a 0.1 miles radius and attributed these finding, in part, to informal conversations among employees of different firms. As well, Rosenthal and Strange [79] discovered that spatial externalities within a 1 miles radius had significant explanatory power on the location of new firms. Along these lines, Graham [40] estimated localization externalities to exhaust themselves within a roughly 6.2 miles radius and depending on industry to be stronger at 0.6 miles. In sum, these estimates are in line with the notion that spatial externalities tend to be confined at very short distances perhaps because the marginal cost of knowledge transmission is an increasing function of distance as it relies on spatially bounded mechanisms such as labor mobility and human interactions [9]. For the case at hand, the implication of those findings is that externalities that arise from venture capital are expected to hold largely at immediate proximity and be spatially bounded.

24.3   Methods and Procedures

Consistent with our theoretical arguments, the empirical model needs to associate the innovative performance of a given life-sciences firm (LSF) with venture capital activity that generates externalities whose strength is potent at close proximity and decays with distance. We approximate innovative performance with patents and employ the amount of venture capital funds to capture venture capital activity. Accordingly, we specify a model in which the dependent variable y_i is the total number of patents granted to LSF i from its birth and in the period of time it was active as an independent firm from 1983 to 2006. Relatedly, we specify independent variables that measure venture capital investments towards LSFs situated at increasing distances from the origin LSF.[9] Given that the dependent

---

[9]Alternatively we could use a dependent variable that reflects the number of patents per year. Nevertheless, we did not opt for this approach because differential and often unobserved lags in the dates of discovery, patent submission and patent issuance could make the allocation of innovative

variable is an observed count (patents), the general form of the expected count is formulated as follows[10]:

$$E(y_i | X_i) = m(X_i \beta + \varepsilon), \qquad (24.1)$$

where function $m()$ is a link function that maps the linear combination of the explanatory variables into an expected count that is non-negative and X_i is a vector of explanatory variables described below.

As previously introduced, we test our main theoretical expectations with independent variables that measure the inflation-adjusted amount of funds from venture capitalists allocated to LSFs situated at increasingly distant narrow spatial units from the focal LSF. We partition these amounts to the stage of firm growth they are directed to. We do so in order to account for the specialization of venture capitalists in investing at firms that are in the stage of growth in which they try to narrow the gap between already existing prototypes or/and ideas and commercial success [30, 35, 101]. By extension, funds towards the firm growth stage that VCs specialize maybe more effective in improving the knowledge base of a given region. Along these lines, there is evidence suggesting that the usefulness of venture capital involvement in target firms is sensitive to the stage of firm growth that financing is directed to [78].

To define the geographic scope of the units we measure where venture capital investments occur we follow [97] who employed radii that reflect immediate proximity; in particular, we measure the venture capital disbursements to LSFs located within the following miles rings from the focal LSF: 0–0.1, 0.1–0.5, 0.5–1 and so on. To allocate venture capital funds to firm growth stages we follow the classification of [72]: disbursements up to $25K correspond to the seed stage where the main task is typically to develop a working prototype, disbursements from $25[11] to $500K correspond to the startup stage where the main task tends to be the completion of beta testing, disbursements from $500 to $2,000K correspond to the first stage (where VCs specialize) in which firms likely try to achieve market penetration and disbursements larger than $2,000K correspond to later stages that aim to increased sales and growth.

---

performance by year an exceedingly difficult task. Relatedly, we performed robustness checks for potential temporal lag effects between the timing of the venture capital investments and the strength of the externality impact. In these tests, the venture capital investments were limited to those that occurred only 1, 2, 3 or 4 years from the birth of the focal LSF. The tests yielded qualitatively similar results to those presented in Figs. 24.5, 24.6, 24.7, 24.8 and 24.9.

[10]We use the Negative Binomial maximum likelihood estimator. The Negative Binomial estimator was chosen to address observed overdispersion and to overcome the standard assumption of the Poisson model of equal conditional means and variances, which was not met for the dependent variable in our sample.

[11]In fact, Parhankangas [72] specified $100K as the lower bound of that stage. In order to include venture capital disbursement between $25 and $100K in the analysis we included such amounts in the start-up stage. Robustness checks in which these amounts were either included in the seed stage or were excluded from the analysis yielded qualitatively similar results to the results presented here.

After the model presented in (1) is estimated, we approximate the cutoff point above which the externality impact of venture capital ceases to exist as the most distant unit from the origin LSF where the association between the venture capital funds allocated to firms in that unit and the patenting rate of the focal firm is economically small and statistically insignificant. For instance, if the estimated coefficient of the variable that measures the funds disbursed to LSFs 0.1–0.5 miles from the origin LSF is economically small and statistically insignificant, and the same holds for the variables that measure the venture capital disbursements at more distant units (0.5–1, 1–1.5 and so on), then we would conclude that the externality impact of venture capital is exhausted at 0.5 miles. To infer whether that effect attenuates with distance we compare the magnitude and statistical significance of the estimates that correspond to that unit with the estimates that correspond to units that are closer to the origin firm. Continuing from the previous example, if the estimate of the spatial ring that reflects venture capital funds within 0–0.1 miles from the origin firm was economically larger and statistically stronger from the estimate that reflects venture capital funds within 0.1–0.5 miles, we would conclude that the externalities from venture capital are dying off with distance since they are potent at 0.1 miles and weaker at more distant spatial rings.

To measure whether the externality impact of venture capital investments is indeed sensitive to the stage of funds provided, we compare the magnitude and the statistical significance of the estimates that correspond to different stages. For instance if, as expected, the estimates that measure the effect of first stage funds are economically stronger and exhibit higher statistical significance compared to the estimates that represent seed stage funds, then we would conclude that the specialization of venture capital investments on first stage is reflected on the externality impact of venture capital on innovation.

Moving to remaining control variables that describe the regional environment, LSFs that are close to universities may benefit from potential knowledge spillovers and other forms of agglomeration economies [2, 74]. Previous studies in this research have shown that such benefits from universities can extend up to the Metropolitan Statistical Area (MSA) level [1, 5, 95]. Accordingly, we include a variable (NIH) that measures the inflation adjusted sum of life-sciences funds from the National Institutes of Health awarded to universities in the MSA of the focal LSF as a proxy for the intensity of related university research in the region and the associated spillover effects.[12] We expect a positive sign for the coefficient of this variable.

Besides expected benefits from the increased performance of VC-backed LSFs, which are part of the venture capital variables, local non VC-backed LSFs can also have a positive influence on the knowledge base of the focal firm. This

---

[12]Note that besides research intensity, we expect the NIH variable to also be capturing the underlying quality and subsequent reputation of the local universities because NIH funds are awarded on a very competitive basis. As we explain in detail in Sect. 24.4, this observation is instrumental for the robustness of the instrumental variable we use in the empirical analysis.

holds because LSFs may exploit spatial externalities when they locate close to similar firms and a number of studies have documented such effects (e.g. [3, 53, 79]).[13] For instance, the so called, "local buzz" allows valuable knowledge transfers, such as failures in scientific experiments, to diffuse, mainly, locally [8, 13, 89]. Agglomerations of proximate LSFs may also increase the availability to specialized labor pools and other service providers (lawyers, consultants) [29] because labor and service providers can be attracted to potential customers and agglomerations of similar firms can also enhance labor pools through employee turnover. While such spatial externalities are expected to increase the pace of innovation for a particular firm, their effect is largely confined in space. That is, the plausible benefits associated with a rich regional environment are expected to be potent among close proximity of the relevant actors [4, 40, 79, 97]. Against this background, we include in the analysis a set of variables (SBIR_0.1, SBIR_0.5 etc.) that measures the number of non VC-backed SBIR winners that have been awarded at least one patent after the birth of the focal LSF and are located in the same geographic units we considered for the case of venture capital investments (0–0.1 miles, 0.1–0.5 miles and so on). We expect positive coefficients for these variables and their magnitude to decrease as we move from closer to more distant firms, reflecting thus an impact that attenuates with distance.

To complement regional characteristics that can enhance the knowledge base of a particular LSF, characteristics of the local environment that relate more to the general business environment should also be considered in the empirical analysis. For example, some states provide support services to LSFs through their Small Business Administration offices but also through private organizations.[14] Other regional characteristics may include the monetary costs of doing business in the state which are partly determined by tax rates and zoning ordinances. Collectively, highly effective consulting organizations and favorable business climate could enhance the capacity of LSFs to improve their performance and consequently become more innovative. Therefore, the relative effectiveness of these features must be considered in the empirical analysis. As we explain in detail in Sect. 24.6 we employ appropriate estimation techniques to account for such considerations.[15]

---

[13]Among others, contributions from [19] and [91] take a critical stand towards positive spatial externalities.

[14]Examples of private organizations that offer consulting services on securing SBIR grants include Foresight S&T in Rhode Island and the Larta Institute in California and the District of Columbia.

[15]Because our sample covers a lengthy period (23 years) in which relevant state characteristics are expected to change, we do not include associated independent variables, whose historical availability is also limited, in the empirical specification. Further, some of these state characteristics may be difficult to observe (e.g. business climate), and hence to approximate with associated variables which adds to our methodological approach.

Alongside the effects of the regional environment, we expect features of the focal LSF to affect the rate that it produces patents. As such, we include in (1) a set of relevant control variables. In line with previous arguments, firms that receive financial and managerial support from venture capitalists are expected to be more innovative largely due to the coaching and monitoring services of VCs. Accordingly, we include a dummy control variable that takes the value of 1 for VC-backed LSFs and takes the value of 0 otherwise (VC) and expect a positive sign.

A number of studies have pointed to operational efficiencies including scale and scope economies that strengthen the link between firm size and innovation [24]. Specifically to patent production, Hall and Ziedonis [43] narrowed the economies of scale advantage to the fixed costs related to maintaining legal departments that deal with intellectual properties issues and a number of empirical studies have supported the hypothesized positive relationship between increased firm size and patent production [14, 43]. To incorporate such considerations in the analysis we include a control variable that measures the number of employees of the origin LSF[16] (Size) and expect a positive sign for the coefficient at hand.

We also analyze the potential effects of time on an LSF's development process with a variable that measures the age of the LSF. In particular, we include a variable that measures the number of years in our sample (1983–2006) that a firm was active as an independent firm and before potential mergers, acquisitions and bankruptcies (Age). Older firms may develop skills and routines that could make them more prolific in producing patents but the case might also be that as firms age, they focus their efforts not so much on developing innovations and protecting them (i.e. patenting) but more so towards increased sales and the like. Therefore, due to such potentially conflicting effects we do not form priors for the effect of the control variable at hand on patent production.

As well, firms with increased research and development (R&D) expenditures are expected to produce more patents because expenditures of this sort tend to improve the knowledge base of a given firm which in turn can result to increased innovative measures like patents. Corroborating such theoretical expectations a large body of empirical work has reported a positive relationship between patents and R&D intensity (e.g. [17, 45, 71]). The SBIR program provides research funds in phases where Phase 1 grants are used to explore the scientific and commercial feasibility of an idea/technology and typically do not exceed $100,000 and Phase 2 grants are considerably larger and are given to the most meriting Phase 1 winners.[17] Along these lines , we include two variables that measure the total sum of funds from Phase 1 (Phase1) and Phase 2 (Phase2) SBIR grants received by the focal LSF and expect positive signs.

---

[16]We, however, followed the codification scheme described in Figs. 24.3 and 24.4 because the number of employees is typically reported by firms in discrete categories.

[17]There is also a Phase 3 but federal agencies do not provide funds during it.

24.4   Identification Strategy and Endogeneity

Before we present a detailed discussion of the data used to estimate the empirical model we explicitly discuss two important considerations that relate to the robustness of our empirical estimates. Based on theoretical arguments and extant empirical evidence presented previously, we expect venture capital investments to assist nearby LSFs in becoming more innovative. Nevertheless, the potential for a reinforcing process in which venture capital activity and innovative firms grow together [64] suggests that we cannot rule out the case that venture capital investments follow rather than spur innovation [49, 58, 86]. Accordingly, VCs will tend to invest in regions that host already innovative firms. If this is the case, our estimates can be plagued by such endogeneity.

Endogeneity is a recurring issue in research on geographic agglomeration and a number of studies have recognized that it can be present in cases similar to ours where the interplay between firm performance and regional attributes can be dynamic [23, 42, 48, 80]. The general evidence in these studies is that if endogeneity is present, then its impact on empirical estimates is relatively small. Similar conclusions have been drawn from empirical estimates that address endogeneity concerns specifically on venture capital [16, 82, 83].

Instrumental variables are a common way to test and correct for endogeneity that can impact estimated coefficients [100]. In our application, an instrumental variable needs to address the possibility that venture capital investments promote local innovation but they can also be attracted to innovative firms that typically share the same location. Accordingly, such an instrumental variable needs to meet two main criteria: it should be correlated with local venture capital activity but uncorrelated with the potential attraction of VCs to closely located innovative firms. Starting from the premise that local institutional investors tend to direct their funds towards local VCs, we follow [83] in using the level of state university endowments as the basis of our instrumental variable. University endowments are a large component of institutional investors whose funds are managed by VCs; hence the magnitude of these endowments should correlate with local venture capital investments. At the same time, institutional investors rarely invest in small firms directly, which then suggests that they do not have a direct impact on the innovative local environment that VCs can be attracted to.[18] These characteristics make state university endowments a suitable candidate for an instrumental variable.

---

[18]Notice that while direct investments from university endowments to young firms are rare, the local density of young innovative firms may be correlated with university endowments through an alternative process. Typically, the most reputable academic institutions realize the largest endowments. At the same time, these kinds of institutions tend to be research-intensive, which often prompts young innovative firms to locate close to them mainly in order to reap spatial externalities and other sorts of proximity effects. By extension, university endowments maybe correlated with the innovative character of regions that attract VCs. We account for this potential relationship by including the NIH variable in the analysis, which can capture the research intensity and underlying quality/reputation of the local universities as NIH funds are awarded on a very competitive basis.

However, before we construct the variable we need to tackle two practical issues. The first issue is whether the level of university endowments correlates with total venture capital funds or whether it correlates only with funds that are directed to a particular firm growth stage. On theoretical grounds, we expect university endowments to correlate with total venture capital funds because institutional investors are typically "passive investors" and the allocation of funds to different firm growth stages is a task of the venture capitalists. To support our expectations, the correlation between university endowments and total funds in our sample was significantly stronger than correlations of university endowments with early and later stage funds. The second issue is to determine the regional unit of the venture capital variable to be instrumented. In order for the instrumental variable to have validity we need to find the venture capital disbursements directed to firms at a certain region that have a high correlation with endowments of state universities. Consistent with studies that have approximated the spatial extent of venture capital relationships [56, 59, 93] we found that such correlation is stronger between endowments of state universities and venture capital funds allocated to firms within a 10 miles radius from a focal LSF (0.65).[19]

Along these lines, we construct a variable that represents the sum of the average inflation adjusted endowment amount of each university located in the same state with the focal LSF over time. We use this variable as an instrument for the total venture capital disbursements directed to firms within a 10 miles radius from the focal firm. Note that the variable we form is not specified at the narrow units we consider in the empirical model and does not partition venture capital funds to firm growth stages. However, we do expect it to capture the essential goal of our exercise, which is to check whether in our sample endogeneity hampers estimates that measure the externality impact of venture capital investments on local innovative performance. Perhaps more importantly, having in mind the potential of biased estimates when instrumental variables are misspecified [100], we opt for a variable that we expect it to have desired econometric properties. The second issue we discuss here also relates to endogeneity. As introduced previously, venture capital funds directed towards different firm growth stages may have a different impact on innovation. However, VC's typically provide funds in a stepwise fashion: target firms receive funds of a particular stage conditional on having received funds (and have met certain research milestones) of a previous stage. This funding structure suggests that the funds of each stage are not determined independently but rather jointly. Because of this joint process, an econometric specification which includes simultaneously all the variables that measure the level of VC disbursements of different stages, is expected to yield estimates that could make measuring the separate impact of each type of funds on innovation an exercise with mounting difficulties. As a remedy to the issue at hand, we build four models where the funds of each stage are included separately while the remaining variables remain the same. That is, the Seed_01, Seed_05, Seed_1 variables are included in one model, the

---

[19]See the correlation table in the Appendix for more details.

Startup_01, Startup_05, Startup_1 are separately included in a second model and so on. Common to all the models are the Age, Size, VC, Phase1, Phase2, SBIR_01, SBIR_05 and SBIR_1 variables.

24.5 Data Sources and Presentation

InKnowVation, Inc. provided a dataset on the SBIR grants awarded to LSFs from the first year that the SBIR program provided an award in 1983 up to 2006. The dataset included specific information about the dollar amount and nature of each grant as well as about the LSF that won each individual grant. This information was used to construct the dependent variable and the Age, Size, VC, Phase1, and Phase2 variables. For the Phase1 and Phase2 variables we converted the nominal amounts of each grant to real amounts ($2,006) using the CPI. The data from InKnowVation, Inc. included the address of each SBIR winner which we converted to geographic coordinates using the ArcGis® software to develop the SBIR_0.1, SBIR_0.5,and SBIR_1 variables. Thomson's Financial SDC Platinum Database (SDC) provided the dollar amount, round date and recipient firm information—including name and address—of all venture capital investments towards life-sciences firms from 1983 to 2006. After the dollar amounts were allocated to firm growth stages and were then adjusted for inflation using the CPI, the addresses of the recipient firms were converted to coordinates. Using that information we constructed the variables that describe the venture capital disbursements around each focal SBIR winner that occurred after firm birth and while the given LSF was active as an independent firm in the period between 1983 and 2006 (Seed_01, Seed_05, Seed_1, Startup_01, Startup_05, Startup_1, First_01, First_05, First_1, Later_01, Later_05, Later_1).[20]

The NIH variable was built with data collected from NIH and reflected life science grants[21] from the National Institutes of Health. Similar to the Phase1 and Phase2 variables, the nominal amounts of the NIH grants were converted to $2,006 using the CPI. Finally, the online directory on university endowments maintained

---

[20]In the first stage of the instrumental variable approach presented in Figs. 24.10 and 24.11 we construct variables that employ the number of patents that each of the proximate firms that eventually received venture capital investments was granted before such investments took place. To collect the number of patents per firm we searched in the online patent search engine maintained the United States Patent and Trademark Office for patents issued before the first venture capital investment where the focal firm was listed as the applicant/assignee. To ensure that our search was not prone to different name recordings of the same firm we run the searches with different versions of the name of each firm (e.g. instead of *inc.* we tried *inc*).

[21]To sort out life-sciences grants from the total population of grants from the National Institutes of Health we consulted with life-sciences researchers employed at the authors' institution. The list was composed of more than 400 terms, including the following: enzyme, peptide, antigen, mutation, clone, immunoassay, coli, hormone, neuron, PCR, cytokines, gene, collagen, bioreactor, elisa, nucleotide, plasmid, biomass, bacillus, bioassay, embryo and genetic.

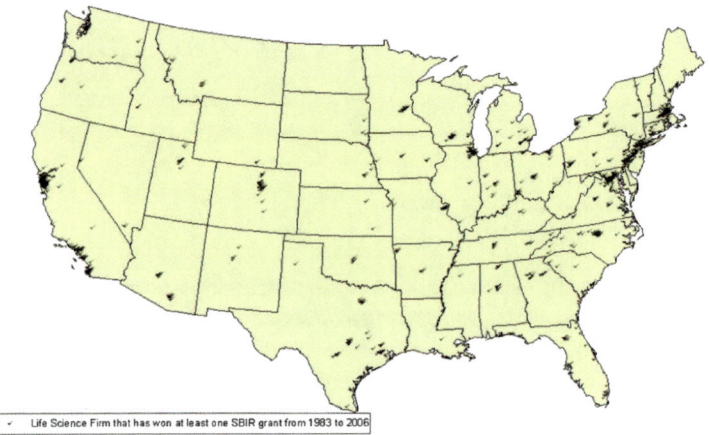

Fig. 24.1 Life science firms include in the dataset (1,671 firms)

by the Chronicles of Higher Education was used to construct the instrumental variable.[22]

The map presented in Fig. 24.1 indicates the location of the life-sciences firms included in the dataset. Most of the firms are located at the East and West Coast but approximately one third of the firms are located in urban and rural interior regions of the US. The wide geographical distribution of the firms in the dataset reinforces the use of narrow geographical units in the empirical models: for instance, 10 or 20 miles in crowded La Jolla, CA might not have the same meaning of (immediate) proximity when compared to 10 or 20 miles in typical wide-space rural college town Lincoln, NE. On the other hand, the degree of proximity of 0.1 miles in La Jolla, CA and Lincoln, NE should be roughly more comparable in the two regions.[23] Particularly for firms in our dataset we expect narrow units of this kind to depict comparable measures of proximity because, regardless of location, most of these firms are young and as such they tend to locate in very close physical proximity to each other often in business incubators and similar facilities that regularly occupy analogous spaces both in dense and sparse regions.

As seen in Fig. 24.2, the dependent variable is left skewed with slightly more than half of the LSFs in the sample not having any patents and approximately one out of three LSFs having between one and ten patents. Further, as also depicted by the four times larger standard deviation of the dependent variable when compared

---

[22]The first year in which the Chronicles of Higher Education report data on university endowments is 1999. Therefore, the average endowment of each state university is calculated with the corresponding value for 1999 as the starting point.

[23]The use of narrow units in the analysis offers as an additional methodological advantage in that we escape estimation issues that relate to required spatial corrections in the data when the unit of analysis is administrative units such as states [6].

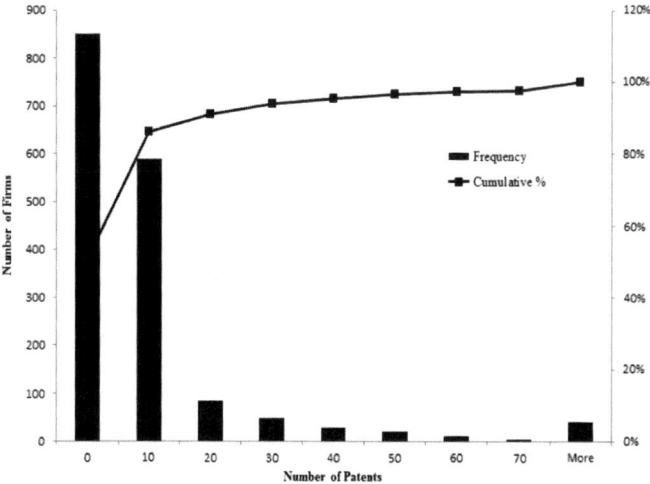

Fig. 24.2 Total patents awarded from 1983 to 2006 to the 1,671 life sciences firms in the sample

Table 1. Descriptive Statistics of the Variables Used in the Empirical Models

Variable / Statistic	Code	Number of Observations[1]	Mean	Median	Mode	Standard Deviation
Total number of patents awarded to Life Sciences Firm (LSF) from 1983 to 2006	Patents	1671	7.978	0.000	0.000	33.189
Number of years the LSF is active in the period from 1983 to 2006	Age	1671	5.929	5.000	0.000	5.508
Variable that is increasing with the number of the employees at the origin LSF	Size	1671	3.338	2.000	1.000	3.002
Dummy variable that takes the value of 1 for LSFs that have received VC funds	VC	545				
Total amount from Phase1 funds awarded to an LSF from 1983 to 2006 (2006 M. $)	Phase1	1671	0.470	0.250	0.091	0.757
Total amount from Phase2 funds awarded to an LSF from 1983 to 2006 (2006 M. $)	Phase2	1671	1.411	0.337	0.000	6.361
Total amount from life-sciences funds awarded from the NIH to universities in the MSA of the LSF from 1992 to 2006 (2006 M. $)[2]	NIH	1671	818.907	670.639	1,279.807	759.277
Number of non VC-backed LSFs located within 0.1 miles from the origin LSF that had won an SBIR grant and had been awarded a patent	SBIR_01	1671	0.066	0.000	0.000	0.327
Number of non VC-backed LSFs located within 0.1 to 0.5 miles from the origin LSF that had won an SBIR grant and had been awarded a patent	SBIR_05	1671	0.129	0.000	0.000	0.507
Number of non VC-backed LSFs located within 0.5 to 1 miles from the origin LSF that had won an SBIR grant and had been awarded a patent	SBIR_1	1671	0.229	0.000	0.000	0.804
Number of non VC-backed LSFs located within 10 miles from the origin LSF that had won an SBIR grant and had been awarded a patent	SBIR_10	1671	2.431	0.000	0.000	0.000
Number of non VC-backed LSFs located within 10 to 20 miles from the origin LSF that had won an SBIR grant and had been awarded a patent	SBIR_20	1671	1.412	0.000	0.000	0.000
Total amount of venture capital seed stage financing awarded (after LSF birth) to LSFs located 0 to 0.1 miles from the origin LSF (2006 M.$)	Seed_01	1671	0.005	0.000	0.000	0.033
Total amount of venture capital seed stage financing awarded (after LSF birth) to LSFs located 0.10001 to 0.5 miles from the origin LSF (2006 M.$)	Seed_05	1671	0.009	0.000	0.000	0.047
Total amount of venture capital seed stage financing awarded (after LSF birth) to LSFs located 0.50001 to 1 miles from the origin LSF (2006 M.$)	Seed_1	1671	0.017	0.000	0.000	0.071

[1] In the case of the VC variable the figure measures the number of firms that have received venture capital funds

[2] The variable is constructed with 1992, and not 1983 which is the beginning point of our analysis, as the first year because the NIH data at hand includes values that start from 1992

For the size variable we followed the following classification scheme: 1 for LSFs with 1 to 4 employees, 2 for LSFs with 5 to 9 employees, 3 for LSFs with 10 to 14 employees, 4 for LSFs with 15 to 19 employees, 5 for LSFs with 20 to 24 employees, 6 for LSFs with 25 to 49 employees, 7 for LSFs with 50 to 74 employees, 8 for LSFs with 75 to 99 employees, 9 for LSFs with 100 to 149 employees, 10 for LSFs with 150 to 249 employees, 11 for

Fig. 24.3 Descriptive statistics of the variables used in the empirical models

with the mean (Figs. 24.3 and 24.4), the sample includes LSFs with varying degrees of patent production. In sum, these features of the dependent variable illustrate why we used a count model to study the relationship between patents and venture capital

Table 1 continued. Descriptive Statistics of the Variables Used in the Empirical Models

Variable / Statistic	Code	Number of Observations[1]	Mean	Median	Mode	Standard Deviation
Total amount of venture capital startup stage financing awarded (after LSF birth) to LSFs located 0 to 0.1 miles from the origin LSF (2006 M.$)	Startup_01	1671	0.045	0.000	0.000	0.185
Total amount of venture capital startup stage financing awarded (after LSF birth) to LSFs located 0.10001 to 0.5 miles from the origin LSF (2006 M.$)	Startup_05	1671	0.132	0.000	0.000	0.436
Total amount of venture capital startup stage financing awarded (after LSF birth) to LSFs located 0.50001 to 1 miles from the origin LSF (2006 M.$)	Startup_1	1671	0.187	0.000	0.000	0.611
Total amount of venture capital first stage financing awarded (after LSF birth) to LSFs located 0 to 0.1 miles from the origin LSF (2006 M.$)	First_01	1671	0.373	0.000	0.000	1.240
Total amount of venture capital first stage financing awarded (after LSF birth) to LSFs located 0.10001 to 0.5 miles from the origin LSF (2006 M.$)	First_05	1671	1.287	0.000	0.000	4.312
Total amount of venture capital first stage financing awarded (after LSF birth) to LSFs located 0.50001 to 1 miles from the origin LSF (2006 M.$)	First_1	1671	1.771	0.000	0.000	5.141
Total amount of venture capital later stage financing awarded (after LSF birth) to LSFs located 0 to 0.1 miles from the origin LSF (2006 M.$)	Later_01	1671	15.320	0.000	0.000	55.284
Total amount of venture capital later stage financing awarded (after LSF birth) to LSFs located 0.10001 to 0.5 miles from the origin LSF (2006 M.$)	Later_05	1671	66.404	0.000	0.000	215.851
Total amount of venture capital later stage financing awarded (after LSF birth) to LSFs located 0.50001 to 1 miles from the origin LSF (2006 M.$)	Later_1	1671	99.644	0.000	0.000	292.249
Total amount of venture capital funds (including all stages) awarded (after LSF birth) to LSFs located 0 to 0.1 miles from the origin LSF (2006 M.$)	AllFunds_01	1671	15.743	0.000	0.000	55.972
Total amount of venture capital funds (including all stages) awarded (after LSF birth) to LSFs located 0.10001 to 0.5 miles from the origin LSF (200	AllFunds_05	1671	67.831	0.000	0.000	219.977
Total amount of venture capital funds (including all stages) awarded (after LSF birth) to LSFs located 0.50001 to 1 miles from the origin LSF (2006	AllFunds_1	1671	101.619	0.000	0.000	297.245
Total amount of venture capital funds (including all stages) awarded (after LSF birth) to LSFs located 0 to 10 miles from the origin LSF (2006 M.$)	AllFunds_10	1671	1,157.294	118.392	0.000	1,924.476
Total number of patents granted to VC-backed LSFs that are located 0 to 10 miles from the orign firm before they received venture capital funds	VCFirmsPatents_10	1671	30.229	0.000	0.000	73.669
Total number of patents granted to VC-backed LSFs that are located 10 to 20 miles from the orign firm before they received venture capital funds	VCFirmsPatents_20	1671	24.965	0.000	0.000	78.263
Sum of average endowments (1999 - 2006) of each university in the same state as the focal LSF (2006 M.$)	Endowments	1671	17,898.326	15,595.805	30,815.761	14,147.769

[1] In the case of the VC variable the figure measures the number of firms that have received venture capital funds

[2] The variable is constructed with 1992, and not 1983 which is the beggining point of our analysis, as the first year because the NIH data at hand includes values that start from 1992

For the size variable we followed the following classification scheme: 1 for LSFs with 1 to 4 employees, 2 for LSFs with 5 to 9 employees, 3 for LSFs with 10 to 14 employees, 4 for LSFs with 15 to 19 employees, 5 for LSFs with 20 to 24 employees, 6 for LSFs with 25 to 49 employees, 7 for LSFs with 50 to 74 employees, 8 for LSFs with 75 to 99 employees, 9 for LSFs with 100 to 149 employees, 10 for LSFs with 150 to 249 employees, 11 for LSFs with 250 to 500

Fig. 24.4 Continued: descriptive statistics of the variables used in the empirical models

activity and indicate why the assumption of equality between the variance and the mean of the standard Poisson model was not met for the case at hand (hence we opted for the negative binomial estimator).

As presented in Figs. 24.3 and 24.4, our sample consists of relatively homogeneous LSFs in terms of age and size. Most of the LSFs were 5 years old by 2006 and were employing about 13 employees. One third of them had received venture capital funds and over time had accumulated just short of half a million from Phase 1 SBIR grants and more than 1.4 million from Phase 2 SBIR grants.

With regard to features that describe the knowledge base of the regional environment, half of the LSFs in our sample did not have any non VC-backed SBIR winners with a patent located in the spatial units considered. Nevertheless, the standard deviation of all the variables that measure the density of SBIR LSFs in the spatial units at hand is larger than the arithmetic mean; this observation ties with Fig. 24.1 and indicates that our sample consists of LSFs located in both densely and sparsely populated regions in the US. Similar trends are observed for the level of venture capital disbursements towards LSFs located at increasingly distant spatial units from the focal LSF. That is, most LSFs in our sample were not in immediate proximity to venture capital investments but some LSFs were close to substantial levels of such investments. For example, most of the firms had no first stage venture capital funds invested in 0.1 miles and the average corresponding amount for our sample was $430,687; at the same time, but not shown in Figs. 24.3 and 24.4, the maximum observed amounts were in the range of $12 million and occurred in San Francisco, CA and Boston, MA.

24.6 Estimation Results

Figures 24.5, 24.6, 24.7, 24.8 and 24.9 presents the count estimates and the marginal effects of the model described in Sect. 3.1 The fitted Negative Binomial model was based on an exponential specification for the conditional mean and the parameter estimates were computed by maximum likelihood (ML).[24] Given the fitted model, we computed the estimated marginal effect for each of the continuous explanatory variables in the model, $E_x \left[\frac{\partial E(y|x)}{\partial x_j} \right] = \beta_j E[exp(x'\beta)]$ [99].[25] This expression indicates that the marginal effects are potentially different for each observation, and we compute the sample average of the estimated marginal effects for all observations and report these values in Figs. 24.5, 24.6, 24.7, 24.8 and 24.9. In Sect. 3.2 we presented characteristics of the economic environment of the state where LSFs reside such as the business climate and the efficacy of private consulting organization that can induce LSFs of the same state to overperform or underperform jointly. These potential relationships may lead to violations in the assumption of independence across observations [68, 88]. To evaluate whether such potential violations are present and correct for them, Figs. 24.5, 24.6, 24.7, 24.8 and 24.9 exhibits estimates with standard errors clustered at the state level. As a robustness check we also report heteroskedasticity-robust standard errors and standard errors without a correction neither for heteroskedasticity nor for the potential violation of the independence across observations assumption. To illustrate whether allocating the VC investments to funds directed towards different firm growth stages is a fruitful exercise when estimating the externality impact of VC on innovation, the last model of Figs. 24.5, 24.6, 24.7, 24.8 and 24.9 presents estimates that do not partition VC investments to different firm growth stages. The most distant spatial unit we consider in the analysis is the one that measures venture capital disbursements and the density of SBIR winners with patents within 0.5–1 miles from the origin LSFs. We stop at this unit because it is already two spatial units apart from the last unit we find statistically significant relationships and because variables defined at even more spatially distant units (1–1.5 miles, 1.5–2 miles and so on) proved to be statistically insignificant and economically small in unreported results.

The Likelihood Ratio statistic for all the models is statistically significant at the 1 % level, which indicates that the models presented in Figs. 24.5, 24.6, 24.7, 24.8

---

[24]The estimation with standard errors clustered at the state level is carried out with generalized estimating equations which is a method to estimate the standard errors which first estimates the variability within the defined cluster and then sums across all clusters [104].

[25]The marginal effects for dummy variables were computed as the change in the expected number of counts (patents) when the value of the variable goes from 0 to 1 and keeping the remaining variables at their mean value.

and 24.9 have explanatory power. Also, McFadden's pseudo-R^2 statistic[26] is about 0.13 for all the models. Finally, the condition number for the set of explanatory variables (between 5.96 and 6.20) is within acceptable levels and alleviates potential inference issues associated with multicollinearity.

The externality impact of venture capital appears sensitive to the firm growth stage that venture capital funds are directed to. Regardless of the type of standard errors used, the estimates that measure the impact of seed and startup stage funding on the innovative performance of nearby LSFs are statistically insignificant. VCs do not specialize on seed and startup stage financing, and the externalities that they generate from investments of this kind, as measured in our estimates, appears to mirror that fact. On the other hand, the statistical significance and the magnitude of the estimated marginal effects indicate that the amount of venture capital first stage funds invested locally is instrumental in promoting local innovation. Importantly, such impact is confined within 0.1 miles from the origin LSF. More specifically, one additional million of first stage VC funds invested in LSFs in 0.1 miles from the focal LSF is associated with approximately 0.17 more patents for the focal LSF. To put the magnitude of that effect in perspective, note that most LSFs in the dataset did not have any patents and compare this 0 modal value with the 2.24[27] more patents that LSFs located at regions with the highest VC first stage investments are expected to produce).

With regard to the estimates that pertain to the effect of later stage VC funds on innovation, only the coefficients with the standard errors clustered at the state level indicate a positive effect, which is relatively small in magnitude and also confined within a 0.1 miles radius. Comparing the count estimates and the marginal effects of the First_01 and the Later_01 variables reinforces the importance of first stage funds; the First_01 estimates are 69 times larger than the Later_01 estimates. Even when evaluated at the maximum value of the Later_01 variable (560 M. $), the expected number of patents associated with nearby later stage investing ($560 \times 0.0025 = 1.40$) is close to one patent less than the expected additional number of patents from first stage funding (2.24). In sum, because of the small economic magnitude and the statistical significance in only one of the models, the evidence towards the beneficial effect of venture capital later stage funding on innovation is not well established. All in all, the estimates that measure whether venture capital spurs innovation via an externality process strongly suggest that perhaps because VCs tend to specialize on first stage funds, investments of that stage have a significantly larger positive externality impact on local innovation than earlier and later stage funds. As well, consistent with theoretical expectations, the externality impact of venture capital on innovation is geographically confined and exhausted within very narrow spatial units—spanning just 0.1 miles in our sample.

---

[26]McFadden's R2 is analogous to the OLS R2 where the log likelihood value for the null model replaces the total sum of squares and the log likelihood value for the unrestricted model replaces the residual sum of squares. An increase in the McFadden statistic indicates better model fit [62].

[27]0.1738×12.876 (the maximum value of the First_01 variable) $= 2.24$.

Table 2. Marginal Effects and Estimated Counts of Negative Binomial Model . The Dependent Variable is the Total Number of Patents Awarded to a Given Life Sciences Firm from 1983 to 2006.

Variables	Code	Marginal Effect	Estimated Count	Standard Errors Clustered at the State Level	Heteroskedasticity-Robust Standard Errors	Unadjusted Standard Errors
				Model with seed stage funds		
Intercept	Intercept		-1.3345	0.1162 ***	0.1280 ***	0.1062 ***
Number of years the LSF is active in the period from 1983 to 2006	Age	0.2888	0.1081	0.0074 ***	0.0097 ***	0.0093 ***
Variable that is increasing with the number of the employees at the origin LSF	Size	0.8635	0.3232	0.0169 ***	0.0195 ***	0.0175 ***
Dummy variable that takes the value of 1 for LSFs that have received VC funds	VC	2.8457	0.9131	0.1738 ***	0.1238 ***	0.1051 ***
Total amount from Phase1 funds awarded to origin LSF from 1983 to 2006 (2006 M. $)	Phase1	0.0370	0.0138	0.0557	0.0732	0.0861
Total amount from Phase2 funds awarded to origin LSF from 1983 to 2006 (2006 M. $)	Phase2	-0.0065	-0.0024	0.0050	0.0065	0.0104
Total amount from life-sciences funds awarded from the NIH to universities in the MSA of the LSF from 1992 to 2006 (2006 M. $)	NIH	-0.0001	0.0000	0.0001	0.0001	0.0001
Number of non VC-backed LSFs located within 0.1 miles from the origin LSF that had won an SBIR grant and had been awarded a patent	SBIR_01	-0.0126	-0.0047	0.1256	0.1170	0.1199
Number of non VC-backed LSFs located within 0.1 to 0.5 miles from the origin LSF that had won an SBIR grant and had been awarded a patent	SBIR_05	-0.1690	-0.0632	0.0420	0.0607	0.0837
Number of non VC-backed LSFs located within 0.5 to 1 miles from the origin LSF that had won an SBIR grant and had been awarded a patent	SBIR_1	0.1616	0.0605	0.0275 **	0.0551	0.0589
Total amount of venture capital seed stage financing awarded (after LSF birth) to LSFs located 0 to 0.1 miles from the origin LSF (2006 M.$)	Seed_01	-1.3223	-0.4949	1.5101	0.9072	1.2606
Total amount of venture capital seed stage financing awarded (after LSF birth) to LSFs located 0.10001 to 0.5 miles from the origin LSF (2006 M.$)	Seed_05	-0.6816	-0.2551	0.6638	0.7712	0.8856
Total amount of venture capital seed stage financing awarded (after LSF birth) to LSFs located 0.50001 to 1 miles from the origin LSF (2006 M.$)	Seed_1	1.1989	0.4487	0.3054	0.5642	0.6296
McFadden's Adjusted Pseudo R²		0.1300				
Likelihood Ratio test of overall model fit		916.96	***			
Multicollinearity Condition Number		5.96				
Number of Observations		1,333				

*** .01 significance, ** .05 significance, * .10 significance

Fig. 24.5 Marginal effects and estimated counts of negative binomial model. The dependent variable is the total number of patents awarded to a given life sciences from 1983 to 2006

Table 2 continued. Marginal Effects and Estimated Counts of Negative Binomial Model . The Dependent Variable is the Total Number of Patents Awarded to a Given Life Sciences Firm from 1983 to 2006.

Variables	Code	Marginal Effect	Estimated Count	Standard Errors Clustered at the State Level	Heteroskedasticity-Robust Standard Errors	Unadjusted Standard Errors
				Model with startup stage funds		
Intercept	Intercept		-1.3447	0.1188 ***	0.1278 ***	0.1061 ***
Number of years the LSF is active in the period from 1983 to 2006	Age	0.2861	0.1073	0.0071 ***	0.0097 ***	0.0094 ***
Variable that is increasing with the number of the employees at the origin LSF	Size	0.8666	0.3249	0.0168 ***	0.0197 ***	0.0176 ***
Dummy variable that takes the value of 1 for LSFs that have received VC funds	VC	2.7658	0.8926	0.1695 ***	0.1233 ***	0.1060 ***
Total amount from Phase1 funds awarded to origin LSF from 1983 to 2006 (2006 M. $)	Phase1	0.0384	0.0144	0.0540	0.0730	0.0855
Total amount from Phase2 funds awarded to origin LSF from 1983 to 2006 (2006 M. $)	Phase2	-0.0048	-0.0018	0.0055	0.0066	0.0104
Total amount from life-sciences funds awarded from the NIH to universities in the MSA of the LSF from 1992 to 2006 (2006 M. $)	NIH	-0.0001	0.0000	0.0001	0.0001	0.0001
Number of non VC-backed LSFs located within 0.1 miles from the origin LSF that had won an SBIR grant and had been awarded a patent	SBIR_01	-0.0553	-0.0207	0.1178	0.1148	0.1241
Number of non VC-backed LSFs located within 0.1 to 0.5 miles from the origin LSF that had won an SBIR grant and had been awarded a patent	SBIR_05	-0.1963	-0.0736	0.0437 *	0.0632	0.0862
Number of non VC-backed LSFs located within 0.5 to 1 miles from the origin LSF that had won an SBIR grant and had been awarded a patent	SBIR_1	0.2829	0.1061	0.0314 ***	0.0706	0.0714
Total amount of venture capital startup stage financing awarded (after LSF birth) to LSFs located 0 to 0.1 miles from the origin LSF (2006 M.$)	Startup_01	0.4803	0.1801	0.1383	0.1560	0.2174
Total amount of venture capital startup stage financing awarded (after LSF birth) to LSFs located 0.10001 to 0.5 miles from the origin LSF (2006 M.$)	Startup_05	0.2656	0.0996	0.1404	0.0911	0.1067
Total amount of venture capital startup stage financing awarded (after LSF birth) to LSFs located 0.50001 to 1 miles from the origin LSF (2006 M.$)	Startup_1	-0.2313	-0.0867	0.0912	0.0798	0.0873
McFadden's Adjusted Pseudo R²		0.1300				
Likelihood Ratio test of overall model fit		518.33	***			
Multicollinearity Condition Number		6.12				
Number of Observations		1,333				

*** .01 significance, ** .05 significance, * .10 significance

Fig. 24.6 Continued: marginal effects and estimated counts of negative binomial model. The dependent variable is the total number of patents awarded to a given life sciences from 1983 to 2006

Table 2 continued. Marginal Effects and Estimated Counts of Negative Binomial Model . The Dependent Variable is the Total Number of Patents Awarded to a Given Life Sciences Firm from 1983 to 2006.

Variables	Code	Marginal Effect	Estimated Count	Standard Errors Clustered at the State Level	Heteroskedasticity - Robust Standard Errors	Unadjusted Standard Errors
		Model with first stage funds				
Intercept	Intercept		-1.3491	0.1117 ***	0.1267 ***	0.1062 ***
Number of years the LSF is active in the period from 1983 to 2006	Age	0.2893	0.1087	0.0072 ***	0.0097 ***	0.0094 ***
Variable that is increasing with the number of the employees at the origin LSF	Size	0.8548	0.3211	0.0166 ***	0.0195 ***	0.0176 ***
Dummy variable that takes the value of 1 for LSFs that have received VC funds	VC	2.7751	0.8965	0.1706 ***	0.1237 ***	0.1053 ***
Total amount from Phase1 funds awarded to origin LSF from 1983 to 2006 (2006 M. $)	Phase1	0.0685	0.0257	0.0571	0.0736	0.0851
Total amount from Phase2 funds awarded to origin LSF from 1983 to 2006 (2006 M. $)	Phase2	-0.0076	-0.0028	0.0053	0.0063	0.0102
Total amount from life-sciences funds awarded from the NIH to universities in the MSA of the LSF from 1992 to 2006 (2006 M. $)	NIH	-0.0001	0.0000	0.0001	0.0001	0.0001
Number of non VC-backed LSFs located within 0.1 miles from the origin LSF that had won an SBIR grant and had been awarded a patent	SBIR_01	-0.1772	-0.0665	0.1284	0.1203	0.1232
Number of non VC-backed LSFs located within 0.1 to 0.5 miles from the origin LSF that had won an SBIR grant and had been awarded a patent	SBIR_05	-0.2034	-0.0764	0.0465	0.0635	0.0843
Number of non VC-backed LSFs located within 0.5 to 1 miles from the origin LSF that had won an SBIR grant and had been awarded a patent	SBIR_1	0.2136	0.0802	0.0632	0.0747	0.0723
Total amount of venture capital first stage financing awarded (after LSF birth) to LSFs located 0 to 0.1 miles from the origin LSF (2006 M.$)	First_01	0.1738	0.0653	0.0347 *	0.0368 *	0.0356 *
Total amount of venture capital first stage financing awarded (after LSF birth) to LSFs located 0.10001 to 0.5 miles from the origin LSF (2006 M.$)	First_05	0.0203	0.0076	0.0163	0.0100	0.0117
Total amount of venture capital first stage financing awarded (after LSF birth) to LSFs located 0.50001 to 1 miles from the origin LSF (2006 M.$)	First_1	-0.0144	-0.0054	0.0196	0.0114	0.0125
McFadden's Adjusted Pseudo R²		0.1350				
Likelihood Ratio test of overall model fit		920.66	***			
Multicollinearity Condition Number		6.20				
Number of Observations		1,333				

*** .01 significance, ** .05 significance, * .10 significance

Fig. 24.7 Continued: marginal effects and estimated counts of negative binomial model. The dependent variable is the total number of patents awarded to a given life sciences from 1983 to 2006

Table 2 continued. Marginal Effects and Estimated Counts of Negative Binomial Model . The Dependent Variable is the Total Number of Patents Awarded to a Given Life Sciences Firm from 1983 to 2006.

Variables	Code	Marginal Effect	Estimated Count	Standard Errors Clustered at the State Level	Heteroskedasticity - Robust Standard Errors	Unadjusted Standard Errors
		Model with later stage funds				
Intercept	Intercept		-1.3379	0.1140 ***	0.1280 ***	0.1058 ***
Number of years the LSF is active in the period from 1983 to 2006	Age	0.2904	0.1088	0.0075 ***	0.0098 ***	0.0095 ***
Variable that is increasing with the number of the employees at the origin LSF	Size	0.8549	0.3204	0.0166 ***	0.0195 ***	0.0178 ***
Dummy variable that takes the value of 1 for LSFs that have received VC funds	VC	2.7645	0.8920	0.1678 ***	0.1252 ***	0.1055 ***
Total amount from Phase1 funds awarded to origin LSF from 1983 to 2006 (2006 M. $)	Phase1	0.0488	0.0183	0.0558	0.0738	0.0852
Total amount from Phase2 funds awarded to origin LSF from 1983 to 2006 (2006 M. $)	Phase2	-0.0060	-0.0022	0.0054	0.0065	0.0103
Total amount from life-sciences funds awarded from the NIH to universities in the MSA of the LSF from 1992 to 2006 (2006 M. $)	NIH	-0.0001	0.0000	0.0001	0.0001	0.0001
Number of non VC-backed LSFs located within 0.1 miles from the origin LSF that had won an SBIR grant and had been awarded a patent	SBIR_01	-0.0614	-0.0230	0.1263	0.1155	0.1204
Number of non VC-backed LSFs located within 0.1 to 0.5 miles from the origin LSF that had won an SBIR grant and had been awarded a patent	SBIR_05	-0.1581	-0.0593	0.0384	0.0605	0.0829
Number of non VC-backed LSFs located within 0.5 to 1 miles from the origin LSF that had won an SBIR grant and had been awarded a patent	SBIR_1	0.1869	0.0701	0.0414 *	0.0652	0.0647
Total amount of venture capital later stage financing awarded (after LSF birth) to LSFs located 0 to 0.1 miles from the origin LSF (2006 M.$)	Later_01	0.0025	0.0009	0.0005 *	0.0007	0.0009
Total amount of venture capital later stage financing awarded (after LSF birth) to LSFs located 0.10001 to 0.5 miles from the origin LSF (2006 M.$)	Later_05	0.0004	0.0002	0.0004	0.0002	0.0002
Total amount of venture capital later stage financing awarded (after LSF birth) to LSFs located 0.50001 to 1 miles from the origin LSF (2006 M.$)	Later_1	-0.0003	-0.0001	0.0003	0.0002	0.0002
McFadden's Adjusted Pseudo R²		0.1346				
Likelihood Ratio test of overall model fit		918.46	***			
Multicollinearity Condition Number		6.19				
Number of Observations		1,333				

*** .01 significance, ** .05 significance, * .10 significance

Fig. 24.8 Continued: marginal effects and estimated counts of negative binomial model. The dependent variable is the total number of patents awarded to a given life sciences from 1983 to 2006

Table 2 continued. Marginal Effects and Estimated Counts of Negative Binomial Model . The Dependent Variable is the Total Number of Patents Awarded to a Given Life Sciences Firm from 1983 to 2006.

Variables	Code	Marginal Effect	Estimated Count	Standard Errors Clustered at the State Level	Heteroskedasticity-Robust Standard Errors	Unadjusted Standard Errors
Intercept	Intercept		-1.3381	0.1139 ***	0.1279 ***	0.1058 ***
Number of years the LSF is active in the period from 1983 to 2006	Age	0.2904	0.1088	0.0075 ***	0.0098 ***	0.0095 ***
Variable that is increasing with the number of the employees at the origin LSF	Size	0.8547	0.3204	0.0166 ***	0.0195 ***	0.0178 ***
Dummy variable that takes the value of 1 for LSFs that have received VC funds	VC	2.7628	0.8916	0.1676 ***	0.1252 ***	0.1055 ***
Total amount from Phase1 funds awarded to origin LSF from 1983 to 2006 (2006 M. $)	Phase1	0.0491	0.0184	0.0559	0.0738	0.0852
Total amount from Phase2 funds awarded to origin LSF from 1983 to 2006 (2006 M. $)	Phase2	-0.0060	-0.0022	0.0054	0.0065	0.0103
Total amount from life-sciences funds awarded from the NIH to universities in the MSA of the LSF from 1992 to 2006 (2006 M. $)	NIH	-0.0001	0.0000	0.0001	0.0001	0.0001
Number of non VC-backed LSFs located within 0.1 miles from the origin LSF that had won an SBIR grant and had been awarded a patent	SBIR_01	-0.0654	-0.0245	0.1266	0.1156	0.1205
Number of non VC-backed LSFs located within 0.1 to 0.5 miles from the origin LSF that had won an SBIR grant and had been awarded a patent	SBIR_05	-0.1584	-0.0594	0.0386	0.0605	0.0829
Number of non VC-backed LSFs located within 0.5 to 1 miles from the origin LSF that had won an SBIR grant and had been awarded a patent	SBIR_1	0.1877	0.0703	0.0420 *	0.0654	0.0649
Total amount of venture capital of all stages awarded (after LSF birth) to LSFs located 0 to 0.1 miles from the origin LSF (2006 M.$)	AllFunds_01	0.0026	0.0010	0.0005 **	0.0007	0.0009
Total amount of venture capital of all stages awarded (after LSF birth) to LSFs located 0.10001 to 0.5 miles from the origin LSF (2006 M.$)	AllFunds_05	0.0004	0.0001	0.0004	0.0002	0.0002
Total amount of venture capital of all stages awarded (after LSF birth) to LSFs located 0.50001 to 1 miles from the origin LSF (2006 M.$)	AllFunds_1	-0.0003	-0.0001	0.0003	0.0002	0.0002
McFadden's Adjusted Pseudo R^2			0.1346			
Likelihood Ratio test of overall model fit			918.55 ***			
Multicollinearity Condition Number			6.19			
Number of Observations			1,333			

*** .01 significance, ** .05 significance, * .10 significance

Fig. 24.9 Continued: marginal effects and estimated counts of negative binomial model. The dependent variable is the total number of patents awarded to a given life sciences from 1983 to 2006

Further, it is important to point at the similarities between the estimated coefficients of the models that include later stage funding and total venture capital funding respectively (the last two models in Figs. 24.5, 24.6, 24.7, 24.8 and 24.9). Because later stage funds are considerably larger than funds from earlier stages, the vector that measures the later stage VC disbursements closely resembles the vector of the total VC funding (see Figs. 24.3 and 24.4). Therefore, the similarity in the estimates is expected. What we find particularly interesting is that if the VC funds were not partitioned to stages towards firm growth, the externality impact of venture capital on innovation could have been masked. If we were basing our conclusions on the model with the total VC funding, we would (erroneously) have inferred that the externality impact of venture capital on innovation is limited: the marginal effect of first stage funds alone is 0.17 and the marginal effect of total funds is 68 times smaller −0.0026 while total funds reach statistical significance in only one model. All in all, these findings side with [78] early observation that the stage of funding needs to be accounted for when measuring the effectiveness of venture capital on promoting innovation, wealth creation and the like. Moreover, the findings are in line with the stream of studies that find knowledge transmission to span within immediate proximity [4, 40, 79, 97] and imply that existing estimates on the spatial extent of venture capital relationships may indicate an upper bound of such geographic scope [56, 59, 93].

Pertaining to control variables that describe the knowledge base of the regional environment we report mixed results. We find that life-sciences funds directed towards universities located in the same MSA with the focal LSF have limited explanatory power in the patenting rate of a focal LSF. On the other hand, the density of non VC–backed SBIR winners located 0.5–1 miles from the focal LSF is statistically significant and positive in three models where the errors are clustered at the state level.[28] Given the statistical insignificance of these variables in the rest of the models, the evidence towards the beneficial effects of colocation with non VC-backed SBIR firms is weak.

Contrary to the results with regard to the density of SBIR firms, the variables that describe LSF-specific features seem to dominate the propensity of a given LSF to patent. For instance, one additional year of age for the origin LSF corresponds to roughly 0.29 more patents and suggests that the benefits associated with potential experience gained over time allow LSFs to produce more patents. Further, our results corroborate the findings of previous literature that VCs tend to improve the innovative performance of the target firms (see for instance [61] for evidence specific on SBIR winners). LSFs that received venture capital funds had approximately 2.7 more patents than non-VC backed LSFs. The magnitude of this coefficient reiterates the impact of VC on innovation and highlights the significance of LSF-specific characteristics in determining the rate that a firm innovates. As a case in point, the aforementioned 2.7 marginal effect has the largest magnitude of all marginal effects included in the empirical models. The size of the LSF was also found to be a significant predictor of the propensity to patent. As LSFs grow, scale and scope efficiencies appear to make them more prolific in patenting. Finally, contrary to expectations, our estimates strongly reject the proposition that the total amount from SBIR Phase 1 and Phase 2 funds has an effect on the patenting rate of the focal LSF. A potential driver behind this finding is the long lags between initial discoveries and innovation. SBIR grants are largely used to set up operations. Hence, their contribution might not lie on bridging the gap between prototypes and innovations but rather on assisting LSFs towards the development of prototypes.

We performed the instrumental variable analysis following the two-step methodology outlined in [20, p. 593] and the results are presented in Figs. 24.10 and 24.11. In the first stage, the total amount of venture capital funds invested to LSFs within 10 miles from the origin LSF was regressed on characteristics that describe the innovative character of a given region such as the research intensity of local universities and the characteristics of already innovative local firms. Because the geographic scope of the regional unit of analysis for the dependent variable was 10 miles (see relevant discussion in Sect. 24.4), the variables that describe the regional environment were defined using the same spatial scope. To better approximate the regional environment we also include variables that describe it up

---

[28]Note that in one model with standard errors clustered at the state level, the SBIR_05 estimate is negative and statistically significant. However, the statistical significance is marginal (p-value 0.095). Accordingly, this estimate should be interpreted with caution.

Table 3. Negative Binomial model with state university endowments as an instrumental variable for venture capital funds from all stages invested to Life Sciences Firms Located within 10 miles from the origin Life Sciences Firm

Variables	First stage : Dependent Variable is the total amount of venture capital funds from all stages invested to Life Sciences Firms located within 10 miles from the origin Life Sciences Firm		Second Stage : The Dependent Variable is the Total Number of Patents Awarded to a Given Life Sciences Firm from 1983 to 2006.				
	Estimates	Heteroskedasticity - Robust Standard Errors	Marginal Effect	Count Estimates	Standard Errors Clustered at the State level	Heteroskedasticity - Robust Standard Errors	Unadjusted Standard Errors
Intercept	-211.1498	38.9105 ***		-1.2589	0.1140 ***	0.1293 ***	0.1066 ***
Sum of average university endowments from universities located in the same state with the origin LSF (2006 M. $)	0.0837	0.0045 ***					
Number of years the LSF is active in the period from 1983 to 2006			0.2682	0.1007	0.0088 ***	0.0103 ***	0.0103 ***
Variable that is increasing with the number of the employees at the origin LSF			0.8442	0.3171	0.0163 ***	0.0200 ***	0.0174 ***
Dummy variable that takes the value of 1 for LSFs that have received VC funds			2.6254	0.8559	0.1852 ***	0.1317 ***	0.1044 ***
Total amount from Phase1 funds awarded to origin LSF from 1983 to 2006 (2006 M. $)			-0.0291	-0.0109	0.0554	0.0742	0.0857
Total amount from Phase2 funds awarded to origin LSF from 1983 to 2006 (2006 M. $)			-0.0051	-0.0019	0.0056	0.0063	0.0102
Total amount from life-sciences funds awarded from the NIH to universities in the MSA of the LSF from 1992 to 2006 (2006 M. $)	-0.2163	0.0404 ***	-0.0001	0.0000	0.0001	0.0001	0.0001
Number of non VC-backed LSFs located within 10 miles from the origin LSF that had won an SBIR grant and had been awarded a patent			-0.0038	-0.0014	0.0107	0.0114	0.0123
Number of non VC-backed LSFs located within 10 to 20 miles from the origin LSF that had won an SBIR grant and had been awarded a patent			0.0860	0.0323	0.0117 **	0.0163 **	0.0155 **
Total number of patents granted to VC-backed LSFs that are located 0 to 10 miles from the origin firm before they received venture capital funds	5.6835	0.7344 ***					
Total number of patents granted to VC backed LSFs that are located 10 to 20 miles from the origin firm before they received venture capital funds	1.2902	0.5676 **					
Instrumental variable			0.0002	0.0001	0.0000 ***	0.0000 *	0.0000 **
Venture capital funds from all stages invested to Life Sciences Firms Located within 10 miles from the origin Life Sciences Firm							
Adjusted R²	0.469						
F-test for overall model significance	231.990	***					
McFadden's Adjusted Pseudo R²			0.1320				
Likelihood Ratio test of overall model fit			924.01	***			
Multicollinearity Condition Number	3.96		6.17				
Number of Observations	1,404		1,330				

*** .01 significance, ** .05 significance, * .10 significance

Fig. 24.10 Negative binomial model with state university endowments as an instrumental variable for venture capital funds from all stages invested to life sciences firms located within 10 miles from the origin life science firm

to 20 miles from the origin LSF, which is the most distant radius we discovered statistically significant relationships. In the second stage, the residuals of the first stage were used as an independent variable in a count model that measures the impact of LSF-specific and regional characteristics on the number of patents awarded to a given LSF from 1983 to 2006.

Three general conclusions can be drawn from Figs. 24.10 and 24.11. First, potential endogeneity does not appear to have a substantial impact on the magnitude of the estimated coefficients. The count estimates and the marginal effects of the instrumental variable and the instrumented variable (venture capital funds in a 10 miles radius from focal LSF) are well within the same range (the marginal effects were 0.000212 and 0.000225 respectively). The estimate of the instrumented variable appears slightly larger than then estimate of the instrumental variable but the difference is relatively small and suggests that endogeneity most likely inflates the coefficients only slightly. These results afford us significant comfort in positing that in our sample estimates that measure the local externality impact of venture capital on innovation are not seriously plagued by endogeneity. The second general conclusion is that LSF-specific characteristics remain robust predictors of the patenting rate of a focal LSF. The magnitude and statistical significance of the estimates that represent LSF-specific characteristics in Figs. 24.5, 24.6, 24.7, 24.8 and 24.9 and Figs. 24.10 and 24.11 are very similar. The third general conclusion

Table 3 continued. Negative Binomial model with state university endowments as an instrumental variable for venture capital funds from all stages invested to Life Sciences Firms Located within 10 miles from the origin Life Sciences Firm

| | | Comparable Model to Second Stage: The Dependent Variable is the Total Number of Patents Awarded to a Given Life Sciences Firm from 1983 to 2006. | | |
Variables	Marginal Effect	Count Estimates	Standard Errors Clustered at the State level	Heteroskedasticity - Robust Standard Errors	Unadjusted Standard Errors
Intercept		-1.3555	0.1118 ***	0.1305 ***	0.1083 ***
Sum of average university endowments from universities located in the same state with the origin LSF (2006 M. $)					
Number of years the LSF is active in the period from 1983 to 2006	0.2886	0.1090	0.0093 ***	0.0109 ***	0.0111 ***
Variable that is increasing with the number of the employees at the origin LSF	0.8322	0.3143	0.0163 ***	0.0195 ***	0.0175 ***
Dummy variable that takes the value of 1 for LSFs that have received VC funds	2.6588	0.8688	0.1740 ***	0.1259 ***	0.1040 ***
Total amount from Phase1 funds awarded to origin LSF from 1983 to 2006 (2006 M. $)	-0.0345	-0.0130	0.0594	0.0757	0.0863
Total amount from Phase2 funds awarded to origin LSF from 1983 to 2006 (2006 M. $)	-0.0032	-0.0012	0.0057	0.0065	0.0104
Total amount from life-sciences funds awarded from the NIH to universities in the MSA of the LSF from 1992 to 2006 (2006 M. $)	-0.0002	-0.0001	0.0001	0.0001	0.0001
Number of non VC-backed LSFs located within 10 miles from the origin LSF that had won an SBIR grant and had been awarded a patent	-0.0306	-0.0116	0.0080	0.0125	0.0137
Number of non VC-backed LSFs located within 10 to 20 miles from the origin LSF that had won an SBIR grant and had been awarded a patent	0.0566	0.0214	0.0105 **	0.0166	0.0154
Total number of patents granted to VC-backed LSFs that are located 0 to 10 miles from the orign firm before they received venture capital funds					
Total number of patents granted to VC-backed LSFs that are located 10 to 20 miles from the orign firm before they received venture capital funds					
Instrumental variable					
Venture capital funds from all stages invested to Life Sciences Firms Located within 10 miles from the origin Life Sciences Firm	0.0002	0.0001	0.0000 ***	0.0000 ***	0.0000 ***
Adjusted R²					
F-test for overall model significance					
McFadden's Adjusted Pseudo R²	0.1333				
Likelihood Ratio test of overall model fit	928.26	***			
Multicollinearity Condition Number	7.00				
Number of Observations	1,333				

*** .01 significance, ** .05 significance, * .10 significance

Fig. 24.11 Continued: negative binomial model with state university endowments as an instrumental variable for venture capital funds from all stages invested to life sciences firms located within 10 miles from the origin life science firm

is that externalities from venture capital on innovation can be masked if small spatial units are aggregated in larger units. The count estimate of the total amount of funds disbursed within 0.1 miles from the focal LSF presented in Figs. 24.5, 24.6, 24.7, 24.8 and 24.9 is 11.3 times larger than the corresponding estimate for the 10 miles radius presented in Figs. 24.10 and 24.11 (0.000962 and 0.000085 respectively). To be clear, in Figs. 24.10 and 24.11 the 0.1 miles radius is included in the 10 miles radius. If these findings can be generalized to different samples that refer to firms in industries where effective knowledge transfer takes place across larger spaces, then the common methodological approach of the existing literature to encompass narrow geographic areas in larger spatial units might have yielded estimates that present a lower bound of the impact of venture capital on innovation.

24.7 Discussion and Concluding Comments

Potentially prompted by the central role of venture capital as a means to promote innovation, a number of studies have demonstrated that venture capital activity tends to create externalities that can assist firms especially in knowledge-intensive industries to improve their innovative performance. These externalities typically emanate from two main sources: first from the networks of venture capitalists that create and circulate relevant knowledge and second from the improvements venture capitalists bring about to target firms. In turn, better firms tend to equip the local knowledge feedback loop among close by similar firms with superior knowledge, improve the quality of the local labor pool and allow participating firms of local professional networks to source superior knowledge. What has gone largely unexamined in these studies is the geographic extent of the externality impact of venture capital on innovation. This holds largely because the few studies that are occupied with geographical aspects of venture capital externalities implicitly treat the geographic scope of those externalities as given. That is, they examine the effects of venture capital activity on economic outcomes that occur in spatially large, predefined, administrative units such as MSAs. This methodological approach diverges from theoretical and empirical contributions which imply that VCs transfer knowledge that initiates the aforementioned externality impact in physical space that is significantly narrower than MSAs or other sorts of administrative regions.

The present study was partly initiated by such considerations. We found that for the case of life sciences winners of SBIR grants, venture capital funds that mirror the specialization of VCs in assisting firms in narrowing the gap between existing prototypes and commercial outcomes (so-called first stage funds) generate externalities that significantly improve the patenting rate of small life sciences firms. Importantly, these externalities appear to be realized only by firms situated within 0.1 miles from the location of venture capital disbursements. Venture capital investments at more distant spatial units or of different stages of financing were not found to generate externalities. Using an instrumental variables approach we found support that for the sample at hand estimates that measure the externality impact of venture capital on innovation are not significantly plagued by endogeneity that can arise from the attraction of VCs to already innovative regions and firms. Finally, despite the significant externality impact of venture capital on innovation, firm specific characteristics including age and VC support, had the strongest explanatory power in explaining the patenting rate of a focal LSF.

The findings of this study can be extended in a number of ways and here we offer some specific suggestions. For instance, mainly because we focused on life-sciences young firms with a general lack of commercial outcomes we approximated innovativeness with patent counts. Given the acknowledged difficulties of patents in capturing innovation, future work can analyze the robustness or our findings with alternative measures of innovation or with patent counts that are adjusted by the closeness to final commercial outcomes. As well, we discussed how various forms of knowledge transfer (labor mobility, network externalities, knowledge spillovers and others) are expected to shape the externality impact of venture capital. Future research can scrutinize the relative weight of each mechanism in forming the externality impact of venture capital.

To complement the aforementioned specific improvements that can be brought by future work, this study has a number of more long term scholarly and policy implications. From a methodological standpoint, classifying venture capital funds to firm growth stages seems a promising avenue for research on venture capital. The extant relevant research has largely, so far, amalgamated all types of venture capital funds to a common measure. This methodological approach might have masked the externality contribution of venture capital to innovation. Different types of venture capital funds appear to associate with varying degrees of the externality impact on innovation. However, in the existing literature these separate contributions are typically aggregated in a single measurement in which weaker and stronger effects are presented as one.

Further, our work adds evidence that pertains to venture capital to a relatively small but increasing stream of studies that highlights the importance of human capital on transferring knowledge and reports findings that indicate the narrow geographic extent of knowledge transmission (e.g. [4, 40, 79, 97]). However, we focused on small life-sciences firms that do not need large spaces to operate and accordingly tend to locate "across halls" often in the same building. These characteristics along with the knowledge-based nature of the life-sciences industry, present a fertile ground for locally confined knowledge transmission. As a result, more research that uses narrow units in the analysis and focuses on firms from industries that need more physical space could help delineate how strong are industry features in determining the spatial extent of externalities.

Relatedly, the use of narrow spatial units in the analysis can inform the discussion of how venture capital can help local economies. Evidence on how far the externality impact of venture capital extends is, at best, scarce and perhaps more importantly might have estimated a lower bound of that effect because typically narrow spatial units have been aggregated to larger units. But, keeping in mind the geographical features of the small life-sciences firms in the dataset, our evidence suggests that the externality impact of venture capital on innovation is more pronounced in short distances. Therefore, when the weak effect of large distances is mixed with the strong effect of short distances, the overall effect might appear to be smaller than what it really is. More work is needed to evaluate this proposition and the present study can be a novel step towards that end.

From a policy perspective, this work corroborates previous findings (e.g. [57, 82, 83]) that venture capital is a conduit to innovation and this role extends beyond the direct beneficial effects of the involvement of VCs on target firms. Therefore, popular efforts to attract venture capital appear grounded insofar as they target the relevant geographical unit, which seems to be considerably smaller than MSAs, cities, zip codes and other administrative units. Consequently, the challenge for policy makers and other practitioners is to devise and implement measures that attract venture capital in units that reflect immediate proximity such as incubators, office parks and similar facilities. If our findings can be generalized and if, as expected, increased innovation measures correspond to increases in wealth and economic growth, they would suggest a rationale for public intervention in such structures. Nevertheless, firm-specific characteristics seem to carry the most weight

Appendix Table. Correlations Among Variables Used in the Empirical Models

	1	2	3	4	5	6	7	8	9	10	11	12	13	14	15	16	17	18	19	20	21	22	23	24	25	26	27	28	29	30	31
(1) Patents	1																														
(2) Age	0.240	1																													
(3) Size	0.409	0.369	1																												
(4) VC	0.259	0.122	0.538	1																											
(5) Phase1	0.116	0.324	0.159	0.072	1																										
(6) Phase2	0.081	0.230	0.131	0.061	0.781	1																									
(7) NIH	0.044	0.027	0.093	0.123	0.015	0.034	1																								
(8) SBIR_01	0.041	0.068	0.006	0.037	0.079	0.079	-0.018	1																							
(9) SBIR_05	0.127	0.197	0.200	0.149	0.068	0.033	-0.024	0.090	1																						
(10) SBIR_1	0.129	0.179	0.145	0.110	0.073	0.023	-0.019	0.161	0.391	1																					
(11) SBIR_10	0.197	0.436	0.259	0.179	0.227	0.166	0.007	0.215	0.459	0.606	1																				
(12) SBIR_20	0.243	0.370	0.307	0.195	0.274	0.228	0.143	0.001	0.050	0.111	0.265	1																			
(13) Seed_01	0.009	0.017	0.021	0.047	0.016	-0.013	-0.005	0.084	0.028	0.101	0.085	0.005	1																		
(14) Seed_05	0.068	0.088	0.126	0.115	-0.007	-0.001	0.027	0.068	0.264	0.150	0.196	0.011	0.134	1																	
(15) Seed_1	0.139	0.091	0.127	0.087	0.084	0.023	0.018	0.161	0.322	0.411	0.379	0.041	0.099	0.210	1																
(16) Startup_01	0.032	0.035	0.069	0.139	0.000	-0.019	0.061	0.176	0.140	0.125	0.182	0.028	0.189	0.078	0.193	1															
(17) Startup_05	0.116	0.153	0.199	0.179	0.036	0.003	0.046	0.116	0.382	0.402	0.413	0.099	0.061	0.477	0.394	0.075	1														
(18) Startup_1	0.133	0.106	0.169	0.142	0.044	0.028	0.037	0.158	0.379	0.675	0.488	0.065	0.184	0.214	0.498	0.177	0.417	1													
(19) First_01	0.085	0.015	0.110	0.155	-0.002	-0.003	0.072	0.243	0.167	0.164	0.224	0.050	0.133	0.108	0.205	0.493	0.171	0.247	1												
(20) First_05	0.153	0.127	0.225	0.214	0.022	0.001	0.057	0.095	0.350	0.421	0.421	0.126	0.023	0.354	0.408	0.116	0.857	0.496	0.200	1											
(21) First_1	0.128	0.090	0.185	0.161	0.048	0.011	0.044	0.136	0.379	0.693	0.509	0.095	0.136	0.227	0.469	0.181	0.501	0.879	0.261	0.592	1										
(22) Later_01	0.121	0.001	0.154	0.202	0.027	-0.004	0.076	0.099	0.096	0.209	0.207	0.064	0.104	0.121	0.278	0.320	0.281	0.327	0.486	0.360	0.357	1									
(23) Later_05	0.145	0.048	0.218	0.219	0.016	-0.003	0.075	0.092	0.245	0.361	0.347	0.103	0.070	0.324	0.410	0.130	0.730	0.430	0.239	0.875	0.564	0.445	1								
(24) Later_1	0.136	0.022	0.183	0.165	0.060	0.031	0.075	0.113	0.312	0.541	0.424	0.082	0.099	0.220	0.471	0.140	0.742	0.238	0.587	0.872	0.381	0.619	1								
(25) AllFunds_01	0.121	0.001	0.155	0.204	0.026	-0.004	0.077	0.103	0.099	0.211	0.210	0.065	0.107	0.123	0.280	0.331	0.282	0.329	0.504	0.361	0.360	0.964	0.446	0.382	1						
(26) AllFunds_05	0.146	0.050	0.219	0.220	0.017	-0.003	0.075	0.092	0.248	0.363	0.350	0.104	0.069	0.326	0.411	0.130	0.735	0.432	0.239	0.880	0.566	0.445	0.982	0.620	0.445	1					
(27) AllFunds_1	0.136	0.024	0.183	0.165	0.060	0.031	0.075	0.113	0.314	0.545	0.426	0.082	0.100	0.221	0.472	0.141	0.449	0.747	0.239	0.589	0.876	0.381	0.620	0.975	0.382	0.621	1				
(28) AllFunds_10	0.142	0.080	0.209	0.224	0.090	0.049	0.149	0.072	0.260	0.452	0.656	0.249	0.093	0.142	0.323	0.198	0.376	0.477	0.283	0.479	0.594	0.372	0.503	0.621	0.375	0.504	0.622	1			
(29) VCFirmsPatents_10	-0.049	-0.238	-0.076	0.035	-0.052	-0.043	0.016	0.004	0.051	0.059	0.069	-0.042	0.010	0.011	0.082	0.059	0.072	0.059	0.177	0.096	0.126	0.190	0.171	0.226	0.191	0.170	0.225	0.412	1		
(30) VCFirmsPatents_20	-0.015	-0.176	0.025	0.116	-0.020	-0.009	0.146	-0.030	-0.048	-0.029	-0.074	0.192	0.009	-0.005	0.006	-0.018	0.068	0.005	0.056	0.146	0.096	0.162	0.273	0.207	0.161	0.271	0.206	0.268	0.180	1	
(31) Endowments	0.129	0.040	0.169	0.157	0.089	0.069	0.336	-0.015	0.088	0.235	0.337	0.359	0.003	0.075	0.141	0.099	0.205	0.270	0.166	0.277	0.345	0.250	0.308	0.385	0.251	0.308	0.385	0.646	0.321	0.312	1

Fig. 24.12 Correlations among variables used in the empirical models

in determining how innovative a firm can be, which suggests that policies should target firms of certain cohorts. Along these lines, our work indicates that the externality impact of venture capital can increase the social returns in the form of innovation from federally funded SBIR-type programs.

Appendix

See Fig. 24.12.

Acknowledgements Research funding provided by the Ewing Marion Kauffman Foundation Strategic Grant #20050176 is gratefully acknowledged.

References

1. Abel, J., Deitz, R.: Do colleges and universities increase their region's human capital? J. Econ. Geography **12**, 667–691 (2012)
2. Acosta, M., Coronado, D., Flores, E. University spillovers and new business location in high-technology sectors: Spanish evidence. Small Bus. Econ. **36**(3), 365–376 (2011)

3. Acs, Z.J., Anselin, L., Varga, A.: Patents and innovation counts as measures of regional production of new knowledge. Res. Pol. **31**(7), 1069–1085 (2002)
4. Aharonson, B.S., Baum, J.a.C., Feldman, M.P.: Desperately seeking spillovers? Increasing returns, industrial organization and the location of new entrants in geographic and technological space. Ind. Corporate Change **16**(1), 89–130 (2007)
5. Anselin, L., Varga, A., Acs, Z.J.: Geographic and sectoral characteristics of academic knowledge externalities. Pap. Reg. Sci. **79**(4), 435–443 (2000)
6. Arauzo-Carod, J.M., Manjón-Antolín, M.: (Optimal) spatial aggregation in the determinants of industrial location. Small Bus. Econ. **39**, 1–14 (2009)
7. Arundel, A., Kabla, I.: What percentage of innovations are patented? empirical estimates for European firms. Res. Pol. **27**(2), 127–41 (1998)
8. Asheim, B., Gertler, M.: The geography of innovation. In: Fagerberg, J., Mowery, D.C., Nelson, R.R. (eds.) The Oxford Handbook of Innovation, pp. 291–317. Oxford University Press, Oxford (2005)
9. Audretsch, D.B.: Agglomeration and the location of innovative activity. Oxford Rev. Econ. Pol. **14**(2), 18 (1998)
10. Audretsch, D.B., Feldman, M.P.: R&d spillovers and the geography of innovation and production. Am. Econ. Rev. **86**(3), 630–640 (1996)
11. Audretsch, D.B., Link, A., Scott, J.: Public/private technology partnerships: evaluating sbir-supported research. Res. Pol. **31**(1), 145 (2002)
12. Autant-Bernard, C., Mairesse, J., Massard, N.: Spatial knowledge diffusion through collaborative networks. Pap. Reg. Sci. **86**(3), 341–350 (2007)
13. Bathelt, H., Malmberg, A., Maskell, P. Clusters and knowledge: local buzz, global pipelines and the process of knowledge creation. Prog. Hum. Geography **28**, 31–56 (2004)
14. Beaudry, C., Breschi, S.: Are firms in clusters really more innovative? Econ. Innov. New Technol. **12**(4), 325–342 (2003)
15. Bertoni, F., Croce, A., D'adda, D.: Venture capital investments and patenting activity of high-tech startups: a micro-econometric firm-level analysis. Venture Capital Int. J. Entrepreneurial Finance **12**(4), 307–326 (2010)
16. Bertoni, F., Colombo, M.G., Grilli, L.: Venture capital financing and the growth of high-tech start-ups: disentangling treatment from selection effects. Res. Pol. **40**(7), 1028–1043 (2011)
17. Blundell, R., Griffith, R., Windmeijer, F.: Individual effects and dynamics in count data models. J. Econometrics **108**(1), 113–131 (2002)
18. Boix, R., Galletto, V.: Innovation and industrial districts: a first approach to the measurement and determinants of the i-district effect. Reg. Stud. **43**(9), 1117–1133 (2009)
19. Breschi, S., Lissoni, F.: Localised knowledge spillovers vs. innovative milieux: knowledge *tacitness* reconsidered. Pap. Reg. Sci. **80**(3), 255–273 (2001)
20. Cameron, A.D., Trivedi, K.P.: Microeconometrics Using Stata. Stata Press, College Station (2009)
21. Chemmanur, J.T., Krishnan, K., Nandy, K.D.: How does venture capital financing improve efficiency in private firms? a look beneath the surface. Rev. Financ. Stud. **24**(12), 4037–4090 (2011)
22. Chen, H., Gompers, P., Kovner, A., Lerner, J.: Buy local? the geography of venture capital. J. Urban Econ. **67**(1), 90–102 (2010)
23. Ciccone, A.: Agglomeration effects in europe. Eur. Econ. Rev. **46**(2), 213–227 (2002)
24. Cohen, W.M., Levin, R.C.: Empirical studies of innovation and market structure. In: Handbook of Industrial Organization, pp. 1059–1107. Elsevier, Amsterdam (1989)
25. Cooper, R.S.: Purpose and performance of the small business innovation research (sbir) program. Small Bus. Econ. **20**(2), 137–151 (2003)
26. Cumming, D., Dai, N.: Local bias in venture capital investments. J. Empirical Finance **17**(3), 362–380 (2010)
27. Cumming, D., Fleming, G., Schwienbacher, A.: The structure of venture capital funds. In: Landström, H. (ed.) Handbook of Research on Venture Capital, vol. 18. Edward Elgar, Cheltenham/Nortampton (2007)

28. Dimasi, J.A., Grabowski, H.G.: The cost of biopharmaceutical r&d: is biotech different? Managerial Decis. Econ. **28**(4–5), 469–479 (2007)
29. Duranton, G., Puga, D.: Micro-foundations of urban agglomeration economies. In: Henderson, J.V., Thisse, J.-F. (eds.) Handbook of Urban and Regional Economics, pp. 2063–2117. Elsevier/North-Holland, New York (2004)
30. Elango, B., Fried, V.H., Hisrich, R.D., Polonchek, A.: How venture capital firms differ. J. Bus. Venturing **10**(2), 157–179 (1995)
31. Engel, D., Keilbach, M.: Firm-level implications of early stage venture capital investment–an empirical investigation. J. Empirical Finance **14**(2), 150–167 (2007)
32. Florida, R.L., Kenney, M.: Venture capital, high technology and regional development. Reg. Stud. **22**(1), 33–48 (1988)
33. Florida, R.L., Smith, D.F.: Venture capital formation, investment, and regional industrialization. Ann. Assoc. Am. Geographers **83**(3), 434–451 (1993)
34. Franco, A.M., Filson, D.: Spin outs: knowledge diffusion through employee mobility. RAND J. Econ. **37**(4), 841–860 (2006)
35. Fried, V.H., Hisrich, R.D. Toward a model of venture capital investment decision making. Financ. Manag. **23**, 28–37 (1994)
36. Fritsch, M., Schilder, D.: Does venture capital investment really require spatial proximity? an empirical investigation. Environ. Plann. **40**(9), 2114–2131 (2008)
37. Gompers, P.A.: Optimal investment, monitoring, and the staging of venture capital. J. Finance **50**, 1461–1489 (1995)
38. Gompers, P.A., Lerner, J.: The Money of Invention: How Venture Capital Creates New Wealth. Harvard Business Press, Boston (2001)
39. Gorman, M., Sahlman, W.A.: What do venture capitalists do? J. Bus. Venturing **4**(4), 231–248 (1989)
40. Graham, D.J.: Identifying urbanisation and localisation externalities in manufacturing and service industries. Pap. Reg. Sci. **88**(1), 63–84 (2009)
41. Gupta, A.K., Sapienza, H.J.: Determinants of venture capital firms' preferences regarding the industry diversity and geographic scope of their investments. J. Bus. Venturing **7**(5), 347–362 (1992)
42. Hall, R., Ciccone, A.: Productivity and the density of economic activity. Am. Econ. Rev. **86**(1), 54–70 (1996)
43. Hall, H.B., Ziedonis, H.R.: The patent paradox revisited: an empirical study of patenting in the us semiconductor industry, 1979–1995. RAND J. Econ. **32**(1), 101–128 (2001)
44. Harding, R.: Plugging the knowledge gap: an international comparison of the role for policy in the venture capital market. Venture Capital Int. J. Entrepreneurial Finance **4**(1), 59–76 (2002)
45. Hausman, J., Hall, B.H., Griliches, Z.: Econometric models for count data with an application to the patents-r & d relationship. Econometrica J. Econometric Soc. **52**(4), 909–938 (1984)
46. Haussler, C., Zademach, H.-M.: Cluster performance reconsidered: structure, linkages and paths in the german biotechnology industry, 1996–2003. Schmalenbach Bus. Rev. **59**(3), 261–281 (2007)
47. Hellmann, T.: Venture capitalists: the coaches of silicon valley. In: Lee, C.-M., Miller, W.F., Hancock, K.G., Rowen, H.S. (eds.) The Silicon Valley Edge, pp. 276–294. Stanford University Press, Stanford (2000)
48. Henderson, J.V.: Marshall's scale economies. J. Urban Econ. **53**(1), 1–28 (2003)
49. Hirukawa, M., Ueda, M.: Venture capital and innovation: which is first? Pac. Econ. Rev. **16**(4), 421–465 (2011)
50. Hochberg, Y.V., Ljungqvist, A., Lu, Y.: Whom you know matters: venture capital networks and investment performance. J. Finance **62**(1), 251–301 (2007)
51. Huggins, R., Johnston, A.: Knowledge flow and inter-firm networks: the influence of network resources, spatial proximity and firm size. Entrepreneurship Reg. Dev. **22**(5), 457–484 (2010)
52. Jaffe, A.B.: Real effects of academic research. Am. Econ. Rev. **79**, 957–970 (1989)
53. Jaffe, A.B., Trajtenberg, M., Henderson, R.: Geographic localization of knowledge spillovers as evidenced by patent citations. Q. J. Econ. **108**(3), 577 (1993)

54. Kaplan, S.N., Strömberg, P.: Financial contracting theory meets the real world: an empirical analysis of venture capital contracts. Rev. Econ. Stud. **70**(2), 281–315 (2003)
55. Kenney, M.: How venture capital became a component of the us national system of innovation. Ind. Corporate Change **20**(6), 1677–723 (2011)
56. Kolympiris, C., Kalaitzandonakes, N., Miller, D.: Spatial collocation and venture capital in the us biotechnology industry. Res. Pol. **40**, 1188–1199 (2011)
57. Kortum, S., Lerner, J.: Assessing the contribution of venture capital to innovation. RAND J. Econ. **31**, 674–692 (2000)
58. Kreft, S.F., Sobel, R.S.: Public policy, entrepreneurship, and economic freedom. Cato J. **25**(3), 595 (2005)
59. Lerner, J.: Venture capitalists and the oversight of private firms. J. Finance **50**, 301–318 (1995)
60. Lerner, J.: Boulevard of Broken Dreams: Why Public Efforts to Boost Entrepreneurship and Venture Capital Have Failed–and What to Do About It. Princeton University Press, Princeton (2009)
61. Link, A.N., Ruhm, C.J.: Bringing science to market: commercializing from nih sbir awards. Econ. Innov New Technol. **18**(4), 381–402 (2009)
62. Long, J.S.: Regression Models for Categorical and Limited Dependent Variables, vol. 7. Sage Publications, Thousand Oaks (1997)
63. Lutz, E., Bender, M., Achleitner, A.K., Kaserer, C.: Importance of spatial proximity between venture capital investors and investees in Germany. J. Bus. Res. **66**(11), 2346–2354 (2013)
64. Martin, R., Sunley, P., Turner, D.: Taking risks in regions: the geographical anatomy of europe's emerging venture capital market. J. Econ. Geography **2**(2), 121 (2002)
65. Mason, C.: Venture capital: a geographical perspective. In: Landström, H. (ed.) Handbook of Research on Venture Capital, pp. 86–112. Edward Elgar Publishing, Northampton (2007)
66. Mccann, B.T., Folta, T.B.: Performance differentials within geographic clusters. J. Bus. Venturing **26**(1), 104–123 (2011)
67. Moser, P.: How do patent laws influence innovation? Evidence from nineteenth-century world's fairs. Am. Econ. Rev. **95**(4), 1214–1236 (2005)
68. Nichols, A., Schaffer, M.: Clustered Standard Errors in Stata: Stata Users Group (2007)
69. Owen-Smith, J., Powell, W.: Knowledge networks as channels and conduits: the effects of spillovers in the boston biotechnology community. Organ. Sci. **15**(1), 5–21 (2004)
70. Paier, M., Scherngell, T.: Determinants of collaboration in european r&d networks: empirical evidence from a discrete choice model. Industry Innov. **18**(1), 89–104 (2011)
71. Pakes, A., Griliches, Z.: Patents and r&d at the firm level: a first report. Econ. Lett. **5**(4), 377–381 (1980)
72. Parhankangas, A.: An overview of research on early stage venture capital: current status and future directions. In: Landström, H. (ed.) Handbook of Research on Venture Capital, pp. 253–280. Edward Elgar Publishing, Northampton (2007)
73. Peneder, M.: The impact of venture capital on innovation behaviour and firm growth. Venture Capital Int. J. Entrepreneurial Finance **12**(2), 83–107 (2010)
74. Ponds, R., Van Oort, F., Frenken, K.: The geographical and institutional proximity of research collaboration. Pap. Reg. Sci. **86**(3), 423–443 (2007)
75. Powell, W.W., Koput, K.W., Bowie, J.I., Smith-Doerr, L.: The spatial clustering of science and capital: accounting for biotech firm-venture capital relationships. Reg. Stud. **36**(3), 291–305 (2002)
76. Puri, M., Zarutskie, R.: On the life cycle dynamics of venture-capital and non-venture-capital-financed firms. J. Financ. **67**(6), 2247–2293 (2012)
77. Roijakkers, N., Hagedoorn, J.: Inter-firm r&d partnering in pharmaceutical biotechnology since 1975: trends, patterns, and networks. Res. Pol. **35**(3), 431–446 (2006)
78. Rosenstein, J., Bruno, A.V., Bygrave, W.D., Taylor, N.T.: The ceo, venture capitalists, and the board. J. Bus. Venturing **8**(2), 99–113 (1993)
79. Rosenthal, S.S., Strange, W.C.: Geography, industrial organization, and agglomeration. Rev. Econ. Stat. **85**(2), 377–393 (2003)

80. Rosenthal, S.S., Strange, W.C.: The attenuation of human capital spillovers. J. Urban Econ. **64**(2), 373–389 (2008)
81. Rosiello, A., Parris, S.: The patterns of venture capital investment in the uk bio-healthcare sector: the role of proximity, cumulative learning and specialisation. Venture Capital Int. J. Entrepreneurial Finance **11**(3), 185–211 (2009)
82. Samila, S., Sorenson, O.: Venture capital as a catalyst to commercialization. Res. Pol. **39**(10), 1348–1360 (2010)
83. Samila, S., Sorenson, O.: Venture capital, entrepreneurship, and economic growth. Rev. Econ. Stat. **93**(1), 338–349 (2011)
84. Shane, S., Cable, D.: Network ties, reputation, and the financing of new ventures. Manag. Sci. **48**(3), 364–381 (2002)
85. Shepherd, D.A., Zacharakis, A.: The venture capitalist-entrepreneur relationship: control, trust and confidence in co-operative behaviour. Venture Capital Int. J. Entrepreneurial Finance **3**(2), 129–149 (2001)
86. Sørensen, M.: How smart is smart money? A two-sided matching model of venture capital. J. Finance **62**(6), 2725–2762 (2007)
87. Sorenson, O., Stuart, T.E.: Syndication networks and the spatial distribution of venture capital investments. Am. J. Sociol. **106**(6), 1546–1588 (2001)
88. Stimson, J.A.: Regression in space and time: a statistical essay. Am. J. Polit. Sci. **29**, 914–947 (1985)
89. Storper, M.: The Regional World: Territorial Development in a Global Economy. The Guilford Press, New York (1997)
90. Stuart, T.E., Sorenson, O.: The geography of opportunity: spatial heterogeneity in founding rates and the performance of biotechnology firms. Res. Pol. **32**(2), 229–253 (2003)
91. Tappeiner, G., Hauser, C., Walde, J.: Regional knowledge spillovers: fact or artifact? Res. Pol. **37**(5), 861–874 (2008)
92. Teece, D.J.: Profiting from technological innovation: Implications for integration, collaboration, licensing and public policy. Res. Pol. **15**(6), 285–305 (1986)
93. Tian, X.: The causes and consequences of venture capital stage financing. J. Financ. Econ. **101**, 132–159 (2011)
94. Timmons, J.A., Bygrave, W.D.: Venture capital's role in financing innovation for economic growth. J. Bus. Venturing **1**(2), 161–176 (1986)
95. Varga, A.: Local academic knowledge transfers and the concentration of economic activity. J. Reg. Sci. **40**(2), 289–309 (2000)
96. Von Burg, U., Kenney, M.: Venture capital and the birth of the local area networking industry. Res. Pol. **29**(9), 1135–1155 (2000)
97. Wallsten, S.J.: An empirical test of geographic knowledge spillovers using geographic information systems and firm-level data. Reg. Sci. Urban Econ. **31**(5), 571–599 (2001)
98. Wessner, W.C.: Venture Funding and the Nih Sbir Program. The National Academies Press, Washington, DC (2009)
99. Winkelmann, R.: Econometric Analysis of Count Data. Springer, Berlin (2008)
100. Wooldridge, J.: Econometric Analysis of Cross Section and Panel Data. MIT Press, Cambridge (2002)
101. Zider, B.: How venture capital works. Harv. Bus. Rev. **76**(6), 131–139 (1998)
102. Zook, M.A.: The knowledge brokers: venture capitalists, tacit knowledge and regional development. Int. J. Urban Reg. Res. **28**(3), 621–641 (2004)
103. Zook, M.A.: The Geography of the Internet Industry. Venture Capital, Dot-Coms, and Local Knowledge. Wiley-Blackwell, Malden (2005)
104. Zorn, C.: Comparing gee and robust standard errors for conditionally dependent data. Polit. Res. Q. **59**(3), 329 (2006)

Chapter 25
Inverse Problems in Complex Multi-Modal Regulatory Networks Based on Uncertain Clustered Data

Erik Kropat, Gerhard-Wilhelm Weber, Sırma Zeynep Alparslan-Gök, and Ayşe Özmen

25.1 Introduction

The modeling and prediction of *regulatory networks* is of considerable importance in many disciplines such as finance, biology, medicine and life sciences. The identification of the underlying network topology and the regulating effects allows to gain deeper insights in the hidden relationships between the entities under consideration. This is even more promising as the technical developments of the last decades have produced a huge amount of data that is still waiting for a deeper analysis. Although many theoretical contributions from various disciplines have focussed on the analysis of such systems, the identification of regulatory networks

E. Kropat
Department of Informatics, Institute of Applied Computer Science, Universität der Bundeswehr München, Werner-Heisenberg-Weg 39, 85577 Neubiberg, Germany
e-mail: Erik.Kropat@unibw.de

G.-W. Weber
Departments of Financial Mathematics, Scientific Computing and Actuarial Sciences, Institute of Applied Mathematics, Middle East Technical University, 06531 Ankara, Turkey
e-mail: gweber@metu.edu.tr

S.Z. Alparslan-Gök
Faculty of Arts and Sciences, Department of Mathematics, Süleyman Demirel University, 32 260 Isparta, Turkey
e-mail: zeynepalparslan@yahoo.com; sirmagok@sdu.edu.tr

A. Özmen
Institute of Applied Mathematics, Middle East Technical University, 06531 Ankara, Turkey
e-mail: ayseozmen19@gmail.com

A.A. Pinto and D. Zilberman (eds.), *Modeling, Dynamics, Optimization and Bioeconomics I*, Springer Proceedings in Mathematics & Statistics 73, DOI 10.1007/978-3-319-04849-9__25,
© Springer International Publishing Switzerland 2014

from real-world data is still challenging mathematics. In particular, the presence of noise and data uncertainty raises serious problems to be dealt with on both the theoretical and computational side. There are many sources of uncertainty in the real world. We refer here to technological and market uncertainty, noise in observation and experimental design, incomplete information and vagueness in decision making. Beside this, the regulatory system has often to be further extended and improved with regard to the unknown effects of additional parameters and factors which may exert a disturbing influence on the key variables (target variables) under consideration. All these dynamical networks and multi-modal systems are affected

- by uncertainty in the data, both in their input and their output parts, or, in other words,
- by uncertainty in the scenarios and by random fluctuation,
- by the necessity to reduce the model complexity, i.e., to regularize, rarefy and stabilize.

In this regard, we are on the way between complete determinism in processes and the rich randomness as it can be investigated by stochastic calculus and, especially, Lévy processes. In 2002, we started our modelling of processes related with *genetic networks* in the deterministic case where, then, in the following years, we included the role of additional environmental factors which yielded us our multi-modal systems such as *gene-environment* and *eco-finance networks*. Since in these kinds of dynamics, the impact of the environmental items became implied as additive "shift" terms which can also be called as *perturbations*, we arrived at our first implication of noise. Having once entered the domain of uncertainty, we went on working it out, firstly, by *interval uncertainty* where, however, the dependencies and correlations between the various items from biology, medicine, these sectors of ecology, education and finance were not taken into account yet [48, 51]. We treated those modelling tasks by the help of *Chebychev approximation* and *Generalized Semi-Infinite Optimization* [46, 47, 49, 50, 52, 54, 55, 58]. By turning to the case of *ellipsoidal uncertainty* [25–27] and, as far as splines were used for approximation, by applying *Multivariate Adaptive Regression Spline* instead of *Generalized Additive Models* [43–45], we could overcome that drawback and we included stochastic dependencies and interactions into our model. Here, the dimensions of the ellipsoids are motivated by additional information related to the model items and their similarities, i.e., on how much they are close to each other and how the distribution of such clusters expresses itself geometrically in ellipsoidal forms. In game theoretical contexts, we called these clusters *(sub) coalitions*. In this way, we arrived at a family or, in particular, sequence of ellipsoids which can be regarded as the bodies which contain our target or environmental variables at the corresponding times, i.e., the processes studied, by confidence levels of, e.g., 95 %.

As it is clearly understood today, environmental factors constitute an essential group of regulating components and by including these additional variables the models performance can be significantly improved. The advantage of such an refinement has been demonstrated for example in [29], where it is shown that prediction and classification performances of supervised learning methods for the most complex genome-wide human disease classification can be greatly improved by considering environmental aspects. Many other examples from biology and life sciences refer to regulatory systems where environmental effects are strongly involved. Among them are, e.g., *metabolic networks* [9, 35, 51], *immunological networks* [18], *social-* and *ecological networks* [17]. We refer to [1, 12, 13, 15, 16, 19, 21, 22, 32, 40–42, 53, 61, 62] for applications, practical examples and numerical calculations.

Whenever we want to particularly address items to the financial sector among the target variables or the environmental variables which, in fact, maybe be regarded in a dual relationship mutually, then we arrive at *eco-finance networks* [24, 59]. This interpretation and variety of our studies also represents that the identification of dynamics related with the Kyoto Protocol, where financial expenditures and emissions reduced interact in time (TEM model) [21,23,28,36–39]. Financial negotiation processes, represented in the way of collaborative game theory [57, 58], and the identification and dynamics of financial processes given by stochastic differential equations and their time-discretized versions [56, 60], are an important part our research. Incorporating uncertainty in cooperative game theory is motivated by the need to handle uncertain outcomes in collaborative situations. Interval uncertainty is a natural instance of uncertainty which influences cooperation. A broader overview on recent developments on interval solutions and their applications can be found in [2, 11]. Cooperative games are the games whose characteristic functions are interval valued, i.e., the worth of a coalition is not a real number, but a compact interval of real numbers. This means that one observes a lower and an upper bound of the considered coalitions. This is very important from computational and algorithmic viewpoint. We notice that the approach is general, since the characteristic function interval-values may result from solving general optimization problems. Cooperative interval games and interval solution concepts are useful tools for modeling various economic and Operations Research situations where payoffs are affected by interval uncertainty. The interval Baker-Thompson rule for solving the aircraft fee problem of an airport with one runway when there is uncertainty regarding the costs of the pieces of the runway is presented and identified in [5] and an axiomatic characterization of the interval Baker-Thompson rule is given in [3]. Further, one-machine sequencing situations with interval data are considered in [4] for which they present different possible scenarios and extend to the interval setting classical results regarding well known rules and sequencing games. Two classical bankruptcy rules, namely the proportional rule and the rights-egalitarian rule, are extended in [10] using a cooperative interval game approach. They show that interval allocations generated by such rules belong to the interval core of related cooperative interval games. Finally, Moretti et al. [30] deal with cost allocation problems arising from connection situations where edge costs are closed intervals of real numbers, and to solve such problems, they extend classical solutions from the theory of minimum cost spanning tree games.

Recent studies on target-environment and gene-environment networks focussed on systems with functionally related groups of target and environmental factors. These groups are identified in a preprocessing step of clustering and classification and the corresponding uncertain multivariate states are represented by ellipsoids [25–27]. The interaction of clusters is determined by affine-linear equations based on ellipsoidal calculus. Various regression problems are introduced for an identification of unknown system parameters from (ellipsoidal) measurement data. In addition, problems of network rarefication and the corresponding mixed-integer regression problems as well as a further relaxation by means of continuous optimization have been addressed in [25]. For further details on the underlying set-theoretic regression theory and the solvability by semi-definite programming we refer to [25–27].

In this paper, we further extend this approach and offer a new perspective where potentially overlapping clusters of targets and groups of environmental factors take influence on the states and values of *single targets* and *single environmental variables*.

The comparison of measurements and predictions of the model leads to a regression model for parameter estimation. Since clusters can be affected by noise and errors, the uncertain multivariate states are represented by ellipsoids what refers to the concept of *robustness* for mathematical programming problems. This approach complements and further extends the framework developed in [25–27] for multi-modal systems under ellipsoidal uncertainty.

The chapter is organized as follows: In Sect. 25.2 some basic facts and notation about target and environmental variables as well as the partitioning of data in possibly overlapping clusters are provided. Then, in Sect. 25.3, a time-discrete linear model is introduced that relates the single variables and the multivariate states of groups of target and environmental factors. The corresponding regression model for parameter estimation is addressed in Sect. 25.4. In a further step, data uncertainty becomes included into our modelling in Sects. 25.5 and 25.6, where the multivariate states of clusters are represented in terms of ellipsoids. Hereby, the corresponding regression models can be reformulated in terms of robust counterpart programs.

25.2   Target-Environment Networks

The time-discrete *target-environment regulatory systems* under consideration consist of n targets and m environmental factors and, thus, constitutes a *two-modal system*. The expression values of the target variables are given by the vector $\mathbb{X} = [\mathbb{X}_1, \ldots, \mathbb{X}_n]'$ and the vector $\mathbb{E} = [\mathbb{E}_1, \ldots, \mathbb{E}_m]'$ denotes the states of the environmental variables, where $[\cdot]'$ stands for the transposition of a matrix or vector. Data mining methods like clustering and classification as well as statistical data analysis can be used for an identification of functionally related groups of targets and environmental factors. These groups can show a direct interaction, but they can also have a regulating effect on single targets or environmental factors. In this paper, we focus on the interactions between clusters and single targets or environmental factors. For a deeper analysis of inter-cluster regulatory networks under ellipsoidal uncertainty we refer to [25–27].

When a cluster partition is established, the set of targets can be divided in R clusters $C_r \subset \{1, \dots, n\}$, $r = 1, \dots, R$. Similarly, the set of all environmental items is divided in S clusters $D_s \subset \{1, \dots, m\}$, $s = 1, \dots, S$. Depending on the data structure and the data mining method used, the clusters might be disjoint or overlapping [20]. We note that in case of a strict sub-division of variables, the relations $C_{r_1} \cap C_{r_2} = \emptyset$ for all $r_1 \neq r_2$ and $D_{s_1} \cap D_{s_2} = \emptyset$ for all $s_1 \neq s_2$ are fulfilled. However, in many applications a single entity might be involved in more than one regulating cycle and for this reason we do not explicitly impose such restrictions and refer to the more general situation of *overlapping clusters*. According to the cluster structure, we introduce the sub-vector $X_r \in \mathbb{R}^{|C_r|}$ of X as the restriction of X given by elements of C_r. In the same way, the sub-vector $E_s \in \mathbb{R}^{|D_s|}$ is defined as the restriction of $\mathbb{E}$ given by elements of D_s.

25.3 The Time-Discrete Model

In this section, we introduce a time-discrete model for the states of the targets $\mathbb{X}_j$, $j = 1, \dots, n$, and environmental factors $\mathbb{E}_i$, $i = 1, \dots, m$. Four types of interactions and regulating effects are involved:

(TT) target cluster → target variable,
(ET) environmental cluster → target variable,
(TE) target cluster → environment variable,
(EE) environmental cluster → environment variable.

When we refer to cluster partitions with potentially overlapping clusters, single entities can refer to more than one group of data items. In such a situation, the *target-environment regulatory model* can be formulated as follows:

$$
\left.
\begin{aligned}
\mathbb{X}_j^{(\kappa+1)} &= \zeta_{j0}^T + \sum_{r=1}^{R} [X_r^{(\kappa)}]' \Theta_{jr}^{TT} + \sum_{s=1}^{S} [E_s^{(\kappa)}]' \Theta_{js}^{ET}, \\
\mathbb{E}_i^{(\kappa+1)} &= \zeta_{i0}^E + \sum_{r=1}^{R} [X_r^{(\kappa)}]' \Theta_{ir}^{TE} + \sum_{s=1}^{S} [E_s^{(\kappa)}]' \Theta_{is}^{EE}
\end{aligned}
\right\}
\tag{CM}
$$

with $\kappa \geq 0$, where (CM) stands for *cluster model*. The initial values $\mathbb{X}^{(0)}$ and $\mathbb{E}^{(0)}$ can be given by the first measurements of targets and environmental factors, i.e.,

$\mathbb{X}^{(0)} := \overline{\mathbb{X}}^{(0)}$ and $\mathbb{E}^{(0)} := \overline{\mathbb{E}}^{(0)}$. The vectors Θ_{jr}^{TT} and Θ_{ir}^{TE} are $|C_r|$-subvectors of the parameter vectors $\Theta_j^{TT} \in \mathbb{R}^n$ and $\Theta_i^{TE} \in \mathbb{R}^n$, respectively. These subvectors are given by the indices of cluster C_r. Similarly, the vectors Θ_{js}^{ET} and Θ_{is}^{EE} are $|D_s|$-subvectors of the parameter vectors $\Theta_j^{ET} \in \mathbb{R}^m$ and $\Theta_i^{EE} \in \mathbb{R}^m$. The additional parameters $\zeta_{j0}^T, \zeta_{i0}^E \in \mathbb{R}$ are intercepts. We note that if all clusters are disjoint, the aforementioned subvectors correspond to distinct parts of the parameter vectors, but we do not make this restriction here.

25.4 The Regression Problem

We now turn to an estimation of parameters of the cluster model (CM). For a regression analysis, the predictions of (CM) have to be compared with the states of targets $\overline{\mathbb{X}}^{(\kappa)} = \left[\overline{\mathbb{X}}_1^{(\kappa)}, \ldots, \overline{\mathbb{X}}_n^{(\kappa)} \right]' \in \mathbb{R}^n$ and environmental observations $\overline{\mathbb{E}}^{(\kappa)} = \left[\overline{\mathbb{E}}_1^{(\kappa)}, \ldots, \overline{\mathbb{E}}_m^{(\kappa)} \right]' \in \mathbb{R}^m, \kappa = 0, 1, \ldots, T$, which are obtained from measurements taken at sampling times $t_0 < t_1 < \ldots < t_T$. By inserting these *measurements* in model (CM) we obtain the following *predictions*:

$$\hat{\mathbb{X}}_j^{(\kappa+1)} = \zeta_{j0}^T + \sum_{r=1}^{R} \left[\overline{X}_r^{(\kappa)} \right]' \Theta_{jr}^{TT} + \sum_{s=1}^{S} \left[\overline{E}_s^{(\kappa)} \right]' \Theta_{js}^{ET},$$

$$\hat{\mathbb{E}}_i^{(\kappa+1)} = \zeta_{i0}^E + \sum_{r=1}^{R} \left[\overline{X}_r^{(\kappa)} \right]' \Theta_{ir}^{TE} + \sum_{s=1}^{S} \left[\overline{E}_s^{(\kappa)} \right]' \Theta_{is}^{EE},$$

where $\kappa = 0, 1, \ldots, T - 1$. We use the initial values $\hat{\mathbb{X}}_j^{(0)} := \overline{\mathbb{X}}_j^{(0)}$, $\hat{\mathbb{E}}_i^{(0)} := \overline{\mathbb{E}}_i^{(0)}$ and define the vectors $\hat{\mathbb{X}}^{(\kappa)} = \left[\hat{\mathbb{X}}_1^{(\kappa)}, \ldots, \hat{\mathbb{X}}_n^{(\kappa)} \right]'$ and $\hat{\mathbb{E}}^{(\kappa)} = \left[\hat{\mathbb{E}}_1^{(\kappa)}, \ldots, \hat{\mathbb{E}}_m^{(\kappa)} \right]'$, where $\kappa = 0, 1, \ldots, T; i = 1, \ldots, n; j = 1, \ldots, m$.

When we compare measurements and predictions, we obtain the following regression problem:

$$\text{Minimize} \quad \sum_{\kappa=1}^{T} \left\{ \sum_{j=1}^{n} \left| \hat{\mathbb{X}}_j^{(\kappa)} - \overline{\mathbb{X}}_j^{(\kappa)} \right| + \sum_{i=1}^{m} \left| \hat{\mathbb{E}}_i^{(\kappa)} - \overline{\mathbb{E}}_i^{(\kappa)} \right| \right\}. \tag{RP}$$

25.5 Ellipsoidal Uncertainty

Ben-Tal and Nemirovski introduced the concept of robustness for programming problems where data is subject to ellipsoidal uncertainty [6, 7]. In general, an *ellipsoid* in $\mathbb{R}^p$ will be parameterized in terms of its center $c \in \mathbb{R}^p$ and a symmetric non-negative definite *configuration (or shape) matrix* $\Sigma \in \mathbb{R}^{p \times p}$ as

$$\mathcal{E}(c, \Sigma) = \{\Sigma u + c \mid \|u\|_2 \leq 1\}.$$

In order to include data uncertainty into our model, we now assume that the states of the clusters of target variables and environmental factors are subject to ellipsoidal uncertainty. That means, our regression analysis will be based on set-valued data

$$X_r^{(\kappa)} \in \mathcal{E}\big(\overline{X}_r^{(\kappa)}, \overline{\Sigma}_r^{(\kappa)}\big) \subset \mathbb{R}^{|C_r|},$$

$$E_s^{(\kappa)} \in \mathcal{E}\big(\overline{E}_s^{(\kappa)}, \overline{\Pi}_s^{(\kappa)}\big) \subset \mathbb{R}^{|D_s|},$$

with $\kappa = 0, 1, \ldots, T$. The measurements $\overline{X}_r^{(\kappa)}$ and $\overline{E}_s^{(\kappa)}$ determine the centers of the ellipsoids and the corresponding symmetric shape matrices $\overline{\Sigma}_r^{(\kappa)}$ and $\overline{\Pi}_s^{(\kappa)}$ are given by the variance-covariance matrices of cluster data what also refers to partial correlations and partial variances of cluster elements.

25.6   Robust Regression Under Ellipsoidal Uncertainty

Measurements and observations of targets and environmental factors are usually affected by uncertainty. The regression problem (RP) depends on crisp (numerical) measurements and does not reflect the disturbing influence of unprecise data. For this reason, we now turn to robust regression models with regard to data sets with (overlapping) cluster partition. There are several ways to describe data uncertainty from a set-theoretic perspective. When an individual error can be assigned to each target and environmental factor, the corresponding states of variables are given by intervals, whereas the states of clusters are represented by hyperrectangles. When errors of clusters elements are correlated, non-paraxial sets have to be considered and the polyhedral uncertainty sets can be replaced by error ellipsoids.

In order to include data uncertainty in the regression problem (RP) it is convenient to reformulate this model as follows:

$$\text{Minimize } \sum_{\kappa=1}^{T} \left\{ \sum_{j=1}^{n} p_j^{(\kappa)} + \sum_{i=1}^{m} q_i^{(\kappa)} \right\}$$

$$\text{such that } \left| \hat{\mathbb{X}}_j^{(\kappa)} - \overline{\mathbb{X}}_j^{(\kappa)} \right| \leq p_j^{(\kappa)} \ (\kappa = 1, \ldots, T; \ j = 1, \ldots, n),$$

$$\left| \hat{\mathbb{E}}_i^{(\kappa)} - \overline{\mathbb{E}}_i^{(\kappa)} \right| \leq q_i^{(\kappa)} \ (\kappa = 1, \ldots, T; \ i = 1, \ldots, m).$$

This problem can be equivalently written as

$$\text{Minimize} \sum_{\kappa=1}^{T} \left\{ \sum_{j=1}^{n} p_j^{(\kappa)} + \sum_{i=1}^{m} q_i^{(\kappa)} \right\}$$

$$\text{such that} \left| \zeta_{j0}^{T} + \sum_{r=1}^{R} [\overline{X}_r^{(\kappa-1)}]' \Theta_{jr}^{TT} + \sum_{s=1}^{S} [\overline{E}_s^{(\kappa-1)}]' \Theta_{js}^{ET} - \overline{\mathbb{X}}_j^{(\kappa)} \right| \le p_j^{(\kappa)},$$

$$\left| \zeta_{i0}^{E} + \sum_{r=1}^{R} [\overline{X}_r^{(\kappa-1)}]' \Theta_{ir}^{TE} + \sum_{s=1}^{S} [\overline{E}_s^{(\kappa-1)}]' \Theta_{is}^{EE} - \overline{\mathbb{E}}_i^{(\kappa)} \right| \le q_i^{(\kappa)}$$

$$(\kappa = 1, \ldots, T; \ j = 1, \ldots, n; \ i = 1, \ldots, m).$$

We assume that the constraints are satisfied for all realizations of the states $X_r^{(\kappa)} \in \mathscr{E}(\overline{X}_r^{(\kappa)}, \overline{\Sigma}_r^{(\kappa)})$ and $E_s^{(\kappa)} \in \mathscr{E}(\overline{E}_s^{(\kappa)}, \overline{\Pi}_s^{(\kappa)})$ and in this way we obtain the following robust regression problem with uncertain ellipsoidal states:

$$\text{Minimize} \sum_{\kappa=1}^{T} \left\{ \sum_{j=1}^{n} p_j^{(\kappa)} + \sum_{i=1}^{m} q_i^{(\kappa)} \right\}$$

$$\text{such that} \left| \zeta_{j0}^{T} + \sum_{r=1}^{R} [X_r^{(\kappa-1)}]' \Theta_{jr}^{TT} + \sum_{s=1}^{S} [E_s^{(\kappa-1)}]' \Theta_{js}^{ET} - \overline{\mathbb{X}}_j^{(\kappa)} \right| \le p_j^{(\kappa)},$$

$$\left| \zeta_{i0}^{E} + \sum_{r=1}^{R} [X_r^{(\kappa-1)}]' \Theta_{ir}^{TE} + \sum_{s=1}^{S} [E_s^{(\kappa-1)}]' \Theta_{is}^{EE} - \overline{\mathbb{E}}_i^{(\kappa)} \right| \le q_i^{(\kappa)}$$

$$(\kappa = 1, \ldots, T; \ j = 1, \ldots, n; \ i = 1, \ldots, m)$$

$$\forall X_r^{(\kappa)} \in \mathscr{E}(\overline{X}_r^{(\kappa)}, \overline{\Sigma}_r^{(\kappa)}) \ (\kappa = 0, \ldots, T-1; \ r = 1, \ldots, R),$$

$$\forall E_s^{(\kappa)} \in \mathscr{E}(\overline{E}_s^{(\kappa)}, \overline{\Pi}_s^{(\kappa)}) \ (\kappa = 0, \ldots, T-1; \ s = 1, \ldots, S).$$

The above problem can be rewritten:

$$\text{Minimize} \sum_{\kappa=1}^{T} \left\{ \sum_{j=1}^{n} p_j^{(\kappa)} + \sum_{i=1}^{m} q_i^{(\kappa)} \right\}$$

$$\text{such that} \quad \zeta_{j0}^{T} + \sum_{r=1}^{R} [X_r^{(\kappa-1)}]' \Theta_{jr}^{TT} + \sum_{s=1}^{S} [E_s^{(\kappa-1)}]' \Theta_{js}^{ET} - \overline{\mathbb{X}}_j^{(\kappa)} \le p_j^{(\kappa)},$$

$$-\zeta_{j0}^{T} - \sum_{r=1}^{R} [X_r^{(\kappa-1)}]' \Theta_{jr}^{TT} - \sum_{s=1}^{S} [E_s^{(\kappa-1)}]' \Theta_{js}^{ET} + \overline{\mathbb{X}}_j^{(\kappa)} \le p_j^{(\kappa)},$$

$$\zeta_{i0}^E + \sum_{r=1}^{R} [X_r^{(\kappa-1)}]' \Theta_{ir}^{TE} + \sum_{s=1}^{S} [E_s^{(\kappa-1)}]' \Theta_{is}^{EE} - \overline{\mathbb{E}}_i^{(\kappa)} \leq q_i^{(\kappa)},$$

$$-\zeta_{i0}^E - \sum_{r=1}^{R} [X_r^{(\kappa-1)}]' \Theta_{ir}^{TE} - \sum_{s=1}^{S} [E_s^{(\kappa-1)}]' \Theta_{is}^{EE} + \overline{\mathbb{E}}_i^{(\kappa)} \leq q_i^{(\kappa)}$$

$$(\kappa = 1, \ldots, T; \; j = 1, \ldots, n; \; i = 1, \ldots, m)$$

$$\forall X_r^{(\kappa)} \in \mathscr{E}\big(\overline{X}_r^{(\kappa)}, \overline{\Sigma}_r^{(\kappa)}\big) \; (\kappa = 0, \ldots, T-1; \; r = 1, \ldots, R),$$

$$\forall E_s^{(\kappa)} \in \mathscr{E}\big(\overline{E}_s^{(\kappa)}, \overline{\Pi}_s^{(\kappa)}\big) \; (\kappa = 0, \ldots, T-1; \; s = 1, \ldots, S).$$

This problem has an infinite number of constraints as it depends on all possible realizations of ellipsoidal states of targets and environmental factors. Another reformulation of this problem can be obtained when the ellipsoids are represented as follows:

$$\mathscr{E}\big(\overline{X}_r^{(\kappa)}, \overline{\Sigma}_r^{(\kappa)}\big) = \big\{\overline{X}_r^{(\kappa)} + \overline{\Sigma}_r^{(\kappa)} u_r \;\big|\; \|u_r\|_2 \leq 1\big\},$$

$$\mathscr{E}\big(\overline{E}_s^{(\kappa)}, \overline{\Pi}_s^{(\kappa)}\big) = \big\{\overline{E}_s^{(\kappa)} + \overline{\Pi}_s^{(\kappa)} v_s \;\big|\; \|v_s\|_2 \leq 1\big\}.$$

With

$$U_r := \big\{u_r \in \mathbb{R}^{|C_r|} \;\big|\; \|u_r\|_2 \leq 1\big\}, \; r = 1, \ldots, R,$$

$$V_s := \big\{v_s \in \mathbb{R}^{|D_s|} \;\big|\; \|v_s\|_2 \leq 1\big\}, \; s = 1, \ldots, S$$

we then obtain the equivalent problem

Minimize
$$\sum_{\kappa=1}^{T} \left\{ \sum_{j=1}^{n} p_j^{(\kappa)} + \sum_{i=1}^{m} q_i^{(\kappa)} \right\}$$

such that
$$\zeta_{j0}^T + \sum_{r=1}^{R} [\overline{X}_r^{(\kappa-1)}]' \Theta_{jr}^{TT} + \sum_{r=1}^{R} \max_{u_r \in U_r} \big\{u_r' \, \overline{\Sigma}_r^{(\kappa-1)} \, \Theta_{jr}^{TT}\big\}$$

$$+ \sum_{s=1}^{S} [\overline{E}_s^{(\kappa-1)}]' \Theta_{js}^{ET} + \sum_{s=1}^{S} \max_{v_s \in V_s} \big\{v_s' \, \overline{\Pi}_s^{(\kappa-1)} \, \Theta_{js}^{ET}\big\}$$

$$- \overline{\mathbb{X}}_j^{(\kappa)} \leq p_j^{(\kappa)} \; (\kappa = 1, \ldots, T; \; j = 1, \ldots, n),$$

$$-\zeta_{j0}^T - \sum_{r=1}^{R} [\overline{X}_r^{(\kappa-1)}]' \Theta_{jr}^{TT} - \sum_{r=1}^{R} \max_{u_r \in U_r} \big\{u_r' \, \overline{\Sigma}_r^{(\kappa-1)} \, \Theta_{jr}^{TT}\big\}$$

$$- \sum_{s=1}^{S} \left[\overline{E}_s^{(\kappa-1)} \right]' \Theta_{js}^{ET} - \sum_{s=1}^{S} \max_{v_s \in V_s} \left\{ v_s' \, \overline{\Pi}_s^{(\kappa-1)} \, \Theta_{js}^{ET} \right\}$$

$$+ \overline{\mathbb{X}}_j^{(\kappa)} \leq p_j^{(\kappa)} \quad (\kappa = 1, \ldots, T; \; j = 1, \ldots, n),$$

$$\zeta_{i0}^{E} + \sum_{r=1}^{R} \left[\overline{X}_r^{(\kappa-1)} \right]' \Theta_{ir}^{TE} + \sum_{r=1}^{R} \max_{u_r \in U_r} \left\{ u_r' \, \overline{\Sigma}_r^{(\kappa-1)} \, \Theta_{ir}^{TE} \right\}$$

$$+ \sum_{s=1}^{S} \left[\overline{E}_s^{(\kappa-1)} \right]' \Theta_{is}^{EE} + \sum_{s=1}^{S} \max_{v_s \in V_s} \left\{ v_s' \, \overline{\Pi}_s^{(\kappa-1)} \, \Theta_{is}^{EE} \right\}$$

$$- \overline{\mathbb{E}}_i^{(\kappa)} \leq q_i^{(\kappa)} \quad (\kappa = 1, \ldots, T; \; i = 1, \ldots, m),$$

$$-\zeta_{i0}^{E} - \sum_{r=1}^{R} \left[\overline{X}_r^{(\kappa-1)} \right]' \Theta_{ir}^{TE} - \sum_{r=1}^{R} \max_{u_r \in U_r} \left\{ u_r' \, \overline{\Sigma}_r^{(\kappa-1)} \, \Theta_{ir}^{TE} \right\}$$

$$- \sum_{s=1}^{S} \left[\overline{E}_s^{(\kappa-1)} \right]' \Theta_{is}^{EE} - \sum_{s=1}^{S} \max_{v_s \in V_s} \left\{ v_s' \, \overline{\Pi}_s^{(\kappa-1)} \, \Theta_{is}^{EE} \right\}$$

$$+ \overline{\mathbb{E}}_i^{(\kappa)} \leq q_i^{(\kappa)} \quad (\kappa = 1, \ldots, T; \; i = 1, \ldots, m).$$

The equations

$$\max_{u_r \in U_r} \left\{ u_r' \, \overline{\Sigma}_r^{(\kappa)} \, \Theta_{jr}^{TT} \right\} = \max_{u_r \in U_r} \left\{ -u_r' \, \overline{\Sigma}_r^{(\kappa)} \, \Theta_{jr}^{TT} \right\} = \left\| \overline{\Sigma}_r^{(\kappa)} \Theta_{jr}^{TT} \right\|_2,$$

$$\max_{u_r \in U_r} \left\{ u_r' \, \overline{\Sigma}_r^{(\kappa)} \, \Theta_{ir}^{TE} \right\} = \max_{u_r \in U_r} \left\{ -u_r' \, \overline{\Sigma}_r^{(\kappa)} \, \Theta_{ir}^{TE} \right\} = \left\| \overline{\Sigma}_r^{(\kappa)} \Theta_{ir}^{TE} \right\|_2,$$

$$\max_{v_s \in V_s} \left\{ v_s' \, \overline{\Pi}_s^{(\kappa)} \, \Theta_{is}^{ET} \right\} = \max_{v_s \in V_s} \left\{ -v_s' \, \overline{\Pi}_s^{(\kappa)} \, \Theta_{is}^{ET} \right\} = \left\| \overline{\Pi}_s^{(\kappa)} \Theta_{is}^{ET} \right\|_2,$$

$$\max_{v_s \in V_s} \left\{ v_s' \, \overline{\Pi}_s^{(\kappa)} \, \Theta_{is}^{EE} \right\} = \max_{v_s \in V_s} \left\{ -v_s' \, \overline{\Pi}_s^{(\kappa)} \, \Theta_{is}^{EE} \right\} = \left\| \overline{\Pi}_s^{(\kappa)} \Theta_{is}^{EE} \right\|_2$$

lead to a further description of the regression problem:

Minimize
$$\sum_{\kappa=1}^{T} \left\{ \sum_{j=1}^{n} p_j^{(\kappa)} + \sum_{i=1}^{m} q_i^{(\kappa)} \right\}$$

such that
$$\left| \zeta_{j0}^{T} + \sum_{r=1}^{R} \left[\overline{X}_r^{(\kappa-1)} \right]' \Theta_{jr}^{TT} + \sum_{s=1}^{S} \left[\overline{E}_s^{(\kappa-1)} \right]' \Theta_{js}^{ET} - \overline{\mathbb{X}}_j^{(\kappa)} \right|$$

$$+ \sum_{r=1}^{R} \left\| \overline{\Sigma}_r^{(\kappa-1)} \Theta_{jr}^{ET} \right\|_2 + \sum_{s=1}^{S} \left\| \overline{\Pi}_s^{(\kappa-1)} \Theta_{js}^{ET} \right\|_2 \leq p_j^{(\kappa)},$$

$$\left| \zeta_{i0}^E + \sum_{r=1}^R [\,\overline{X}_r^{(\kappa-1)}\,]' \Theta_{ir}^{TE} + \sum_{s=1}^S [\,\overline{E}_s^{(\kappa-1)}\,]' \Theta_{is}^{EE} - \overline{\mathbb{E}}_i^{(\kappa)} \right|$$

$$+ \sum_{r=1}^R \|\,\overline{\Sigma}_r^{(\kappa-1)} \Theta_{ir}^{TE}\,\|_2 + \sum_{s=1}^S \|\,\overline{\Pi}_s^{(\kappa-1)} \Theta_{is}^{EE}\,\|_2 \;\le\; q_i^{(\kappa)}$$

$$(\kappa = 1, \ldots, T;\; j = 1, \ldots, n;\; i = 1, \ldots, m).$$

Finally, with the vectors

$$\Theta_j^T = \left[\zeta_{j0}^T, \Theta_{j1}^{TT}, \ldots, \Theta_{jR}^{TT}, \Theta_{j1}^{ET}, \ldots, \Theta_{jS}^{ET} \right]^T,$$

$$\Theta_i^E = \left[\zeta_{i0}^E, \Theta_{i1}^{TE}, \ldots, \Theta_{iR}^{TE}, \Theta_{i1}^{EE}, \ldots, \Theta_{iS}^{EE} \right]^T,$$

$$c^{(\kappa)} = \left[1, \overline{X}_1^{(\kappa)}, \ldots, \overline{X}_R^{(\kappa)}, \overline{E}_1^{(\kappa)}, \ldots, \overline{E}_S^{(\kappa)} \right]^T$$

we obtain the following program for an estimation of the parameters of the cluster model (CM) based on ellipsoidal uncertainty:

Minimize $\displaystyle \sum_{\kappa=1}^T \left\{ \sum_{j=1}^n p_j^{(\kappa)} + \sum_{i=1}^m q_i^{(\kappa)} \right\}$

such that $\displaystyle \left| [c^{(\kappa-1)}]' \Theta_j^T - \overline{\mathbb{X}}_j^{(\kappa)} \right| + \sum_{r=1}^R \|\,\overline{\Sigma}_r^{(\kappa-1)} \Theta_{jr}^{TT}\,\|_2 + \sum_{s=1}^S \|\,\overline{\Pi}_s^{(\kappa-1)} \Theta_{js}^{ET}\,\|_2 \;\le\; p_j^{(\kappa)},$

$$\left| [c^{(\kappa-1)}]' \Theta_i^E - \overline{\mathbb{E}}_i^{(\kappa)} \right| + \sum_{r=1}^R \|\,\overline{\Sigma}_r^{(\kappa-1)} \Theta_{ir}^{TE}\,\|_2 + \sum_{s=1}^S \|\,\overline{\Pi}_s^{(\kappa-1)} \Theta_{is}^{EE}\,\|_2 \;\le\; q_i^{(\kappa)}$$

$$(\kappa = 1, \ldots, T;\; j = 1, \ldots, n;\; i = 1, \ldots, m). \tag{RCPE}$$

To be able to solve the problems like RCPE, stochastic programming, dynamic programming and robust optimization methods are principle methods which cope with uncertainty. Although it seems that the areas of them overlap, they are developed freely of one another. Stochastic programming methods present the uncertain data by scenarios which are created in advance while dynamic programming methods handle stochastic uncertain systems over multiple stages. As an alternative to stochastic and dynamic programming methods, robust optimization methods deal with uncertainty as deterministic, but do not limit parameter values to point estimates [14]. The purpose of robust optimization is to find an optimal or near optimal solution which is feasible for any values of the uncertain parameters in prespecified uncertainty sets that have special shape such as polyhedral and ellipsoidal. For further details on robust optimization and the numerical treatment of the corresponding uncertainty-affected programming problems with polyhedral and ellipsoidal uncertainty we refer to [8, 31, 33, 34].

25.7 Conclusion

In this chapter, we analyzed inverse problems for target-environment networks under ellipsoidal uncertainty. This theoretical framework is particularly suited for parameter identification of gene-environment networks in system genetics and computational biology as well as eco-finance networks of OR-applications. This approach constitutes a further extension of our analysis of target-environment networks in OR that are based on interval arithmetics where Chebychev approximation and generalized semi-infinite optimization are considered. In this paper, we focused on time-discrete two-modal models that determine the response of single target variables and environmental factors to the actual states of potentially overlapping clusters or coalitions of system variables. This complements our recently introduced concept of target-environment networks for an analysis of the intrinsic interactions and synergetic connections between clusters. The underlying regression models are based on ellipsoidal calculus and in future work, combinations of both approaches have to considered such that clusters may take influence on target and environmental clusters as well as single genes and single environmental factors simultaneously.

References

1. Akhmet, M.U., Gebert, J., Öktem, H., Pickl, S.W., Weber, G.-W.: An improved algorithm for analytical modeling and anticipation of gene expression patterns. J. Comput. Technol. **10**(4), 3–20 (2005)
2. Alparslan Gök, S.Z.: Cooperative Interval Games: Theory and Applications. LAP-Lambert Academic Publishing House, Saarbrücken (2010) (PROJECT-ID: 5124, ISBN: 978-3-8383-3430-1)
3. Alparslan Gök, S.Z.: An axiomatic characterization of interval Baker Thompson rule. J. Appl. Math. (2012). doi:10.1155/2012/218792
4. Alparslan Gök, S.Z., Branzei, R., Fragnelli, V., Tijs, S.: Sequencing interval situations and related games. Cent. Eur. J. Operations Res. **63** (2008, to appear). Tilburg University, Center for Economic Research, The Netherlands, CentER DP. doi:10.1007/s10100-011-0226-3
5. Alparslan Gök, S.Z., Branzei, R., Tijs, S.H.: Airport interval games and their shapley value. Oper. Res. Decis. **2**, 9–18 (2009). ISSN: 1230-1868
6. Ben-Tal, A.: Conic and robust optimization. Lecture Notes. Available at http://iew3.technion. ac.il/Home/Users/morbt.phtml (2002)
7. Ben-Tal, A., Nemirovski, A.: Convex Optimization in Engineering - Modeling, Analysis, Algorithms. Faculty of Industrial Engineering and Management, Technion-City. Available at www.isa.ewi.tudelft.nl/roos/courses/WI4218/tud00r.pdf
8. Ben-Tal, A., El Ghaoui, L., Nemirovski, A.: Robust Optimization. Princeton University Press, Princeton (2009)
9. Borenstein, E., Feldman, M.W.: Topological signatures of species interactions in metabolic networks. J. Comput. Biol. **16**(2), 191–200 (2009). doi:10.1089/cmb.2008.06TT

10. Branzei, R., Alparslan Gök, S.Z.: Bankruptcy problems with interval uncertainity. Econ. Bull. **3**(56), 1–10 (2008)
11. Branzei, R., Branzei, O., Alparslan Gök, S.Z., Tijs, S.: Cooperative Interval Games: A Survey. Cent. Eur. J. Oper. Res. (CEJOR) **18**(3), 397–411 (2010). doi:10.1007/s10100-009-0116-0
12. Chen, T., He, H.L., Church, G.M.: Modeling gene expression with differential equations. Proc. Pac. Symp. Biocomput. **4**, 29–40 (1999)
13. Ergenç, T., Weber, G.-W.: Modeling and prediction of gene-expression patterns reconsidered with Runge-Kutta discretization. In the special issue at the occasion of seventieth birthday of Prof. Dr. Karl Roesner, TU Darmstadt. J. Comput. Technol. **9**(6), 40–48 (2004)
14. Fabozzi, F.J., Kolm, P.N., Pachamanova, D.A., Focardi, S.M.: Robust Portfolio Optimization and Management. Wiley Finance, New Jersey (2007)
15. Gebert, J., Lätsch, M., Quek, E.M.P., Weber, G.-W.: Analyzing and optimizing genetic network structure via path-finding. J. Comput. Technol. **9**(3), 3–12 (2004)
16. Gebert, J., Lätsch, M., Pickl, S.W., Weber, G.-W., Wünschiers, R.: An algorithm to analyze stability of gene-expression pattern. In: Anthony, M., Boros, E., Hammer, P.L., Kogan, A. (guest eds.) special issue discrete mathematics and data mining II. Discrete Appl. Math. **154**(7), 1140–1156 (2006)
17. Gökmen, A., Kayalgil, S., Weber, G.-W., Gökmen, I., Ecevit, M., Sürmeli, A., Bali, T., Ecevit, Y., Gökmen, H., DeTombe, D.J.: Balaban valley project: improving the quality of life in rural area in Turkey. Int. Sci. J. Methods Models Complex. **7**(1), 24 pp. (2004)
18. Harris, J.R., Nystad, W., Magnus, P.: Using genes and environments to define asthma and related phenotypes: applications to multivariate data. Clin. Exp. Allergy **28**(1), 43–45 (1998)
19. Hoon, M.D., Imoto, S., Kobayashi, K., Ogasawara, N., Miyano, S.: Inferring gene regulatory networks from time-ordered gene expression data of Bacillus subtilis using differential equations. Proc. Pac. Symp. Biocomput. **8**, 17–28 (2003)
20. Höppner, F., Klawonn, F., Kruse, R., Runkler, T.: Fuzzy Cluster Analysis: Methods for Classification, Data Analysis and Image Recognition. Wiley, New York (1999)
21. Işcanoğlu, A., Weber, G.-W., Taylan, P.: Predicting Default Probabilities with Generalized Additive Models for Emerging Markets. Invited lecture. Graduate Summer School on New Advances in Statistics, METU (2007)
22. Jong, H.D.: Modeling and simulation of genetic regulatory systems: a literature review. J. Comput. Biol. **9**, 103–129 (2002)
23. Krabs, W., Pickl, S.: A game-theoretic treatment of a time-discrete emission reduction model. Int. Game Theory Rev. **6**(1), 21–34 (2004)
24. Kropat, E., Weber, G.-W., Akteke-Öztürk, B.: Eco-finance networks under uncertainty. In: Herskovits, J., Canelas, A., Cortes, H., Aroztegui, M. (eds.) Proceedings of the International Conference on Engineering Optimization, EngOpt 2008, Rio de Janeiro (2008). ISBN: 978857650156-5 (CD)
25. Kropat, E., Weber, G.-W., Rückmann, J.-J.: Regression analysis for clusters in gene-environment networks based on ellipsoidal calculus and optimization. In the special issue in honour of Professor Alexander Rubinov of Dynamics of Continuous. Discrete Impulsive Syst. Ser. B Appl. Algorithms **17**(5), 639–657 (2010)
26. Kropat, E., Weber, G.-W., Belen, S.: Dynamical gene-environment networks under ellipsoidal uncertainty - Set-theoretic regression analysis based on ellipsoidal OR. In: Peixoto, M.M., Pinto, A.A., Rand, D.A. (eds.) Dynamics, Games and Science I. Springer Proceedings in Mathematics, vol. 1, pp. 545–571. Springer, Berlin/Heidelberg (2011). ISBN: 978-3-642-11455-7
27. Kropat, E., Weber, G.-W., Pedamallu, C.S.: Regulatory networks under ellipsoidal uncertainty - Data analysis and prediction by optimization theory and dynamical systems. In: Holmes, D.E., Jain, L.S. (eds.) Data Mining: Foundations and Intelligent Paradigms: Volume 2: Statistical, Bayesian, Time Series and Other Theoretical Aspects. ISRL, vol. 24, pp. 27–56. Springer, Berlin (2012). ISBN: 978-3642232404
28. Li, Y.F., Venkatesh, S., Li, D.: Modeling global emissions and residues of pesticided. Environ. Model. Assess. **9**, 237–243 (2004)

29. Liu, Q., Yang, J., Chen, Z., Yang, M.Q., Sung, A.H., Hunag, X.: Supervised learning-based tagSNP selection for genome-wide disease classifications. BMC Genom. **9**(1), S6 (2007)
30. Moretti, S., Alparslan Gök, S.Z., Branzei, R., Tijs, S.: Connection situations under uncertainty and cost monotonic solutions. Comput. Oper. Res. **38**(11), 1638–1645 (2011). http://dx.doi.org/10.1016/j.cor.2011.02.004
31. Nemirovski, A.: Modern Convex Optimization. Lecture at PASCAL Workshop, Thurnau, 16–18 March (2005)
32. Özöğür, S.: Mathematical modelling of enzymatic reactions, simulation and parameter estimation. MSc. thesis at Institute of Applied Mathematics, METU, Ankara (2005)
33. Özmen, A.: Robust conic quadratic programming applied to quality improvement- A robustification of CMARS. Master Thesis, METU, Ankara (2010)
34. Özmen, A., Weber, G.-W., Batmaz, I., Kropat, E.: RCMARS: robustification of CMARS with different scenarios under polyhedral uncertainty set. Commun. Nonlin. Sci. Numer. Simulat. (CNSNS) (2011). doi:10.1016/j.cnsns.2011.04.001
35. Partner, M., Kashtan, N., Alon, U.: Environmental variability and modularity of bacterial metabolic network. BMC Evol. Biol. **7**, 169 (2007). doi:10.1186/1471-2148-7-169
36. Pickl, S.: Der τ-value als Kontrollparameter - Modellierung und Analyse eines Joint-Implementation Programmes mithilfe der dynamischen kooperativen Spieltheorie und der diskreten Optimierung. Thesis, Darmstadt University of Technology, Department of Mathematics (1998)
37. Pickl, S.: An iterative solution to the nonlinear time-discrete TEM model - the occurence of chaos and a control theoretic algorithmic approach. AIP Conf. Proc. **627**(1), 196–205 (2002)
38. Pickl, S., Weber, G.-W.: Optimization of a time-discrete nonlinear dynamical system from a problem of ecology - an analytical and numerical approach. J. Comput. Technol. **6**(1), 43–52 (2001)
39. Pickl, S., Kropat, E., Hahn, H.: The impact of uncertain emission trading markets on interactive resource planning processes and international emission trading experiments. Climatic Change, Special Issue Benefits of Dealing with Uncertainty in Greenhouse Gas Inventories, vol. 103, no. 1–2, pp. 327–338 (2010). Also in: T. White, M. Jonas, Z. Nahorski, S. Nilsson (eds.): Greenhouse Gas Inventories - Dealing With Uncertainty. Springer, Dordrecht (2011) (ISBN: 978-94-007-1669-8)
40. Taştan, M.: Analysis and prediction of gene expression patterns by dynamical systems, and by a combinatorial algorithm. MSc Thesis, Institute of Applied Mathematics, METU (2005)
41. Taştan, M., Pickl, S.W., Weber, G.-W.: Mathematical modeling and stability analysis of gene-expression patterns in an extended space and with Runge-Kutta discretization. In: Proceedings of Operations Research 2005, Bremen, Springer, pp. 443–450 (2005)
42. Taştan, M., Ergenç, T., Pickl, S.W., Weber, G.-W.: Stability analysis of gene expression patterns by dynamical systems and a combinatorial algorithm. In: HIBIT – Proceedings of International Symposium on Health Informatics and Bioinformatics, Turkey '05, Antalya, pp. 67–75 (2005)
43. Taylan, P., Weber, G.-W., Beck, A.: New approaches to regression by generalized additive models and continuous optimization for modern applications in finance, science and technology. In the special issue in honour of Prof. Dr. Alexander Rubinov, B. Burachik and X. Yang (guest eds.) Optimization **56**(5–6), 675–698 (2007)
44. Taylan, P., Weber, G.-W., Yerlikaya, F.: A new approach to multivariate adaptive regression spline by using Tikhonov regularization and continuous optimization. Selected Papers at the Occasion of 20th EURO Mini Conference Continuous Optimization and Knowledge-Based Technologies, Neringa, Lithuania, 20–23 May 2008. TOP **18**(2), 377–395 (2010)
45. Taylan, P., Weber, G.-W., Liu, L., Yerlikaya, F.: On foundations of parameter estimation for generalized partial linear models with B-splines and continuous optimization. Comput. Math. Appl. **60**(1), 134–143 (2010)
46. Uğur, Ö., Weber, G.-W.: Optimization and dynamics of gene-environment networks with intervals. In the Special Issue at the occasion of the 5th Ballarat Workshop on Global and Non-Smooth Optimization: Theory, Methods and Applications, 28–30 November 2006. J. Ind. Manag. Optim. **3**(2), 357–379 (2007)

47. Uğur, Ö., Pickl, S.W., Weber, G.-W., Wünschiers, R.: Operational research meets biology: an algorithmic approach to analyze genetic networks and biological energy production. Optimization **58**(1), 1–22 (2009)
48. Weber, G.-W., Tezel, A.: On generalized semi-infinite optimization of genetic networks. TOP **15**(1), 65–77 (2007)
49. Weber, G.-W., Alparslan-Gök, S.Z., Dikmen, N.: Environmental and life sciences: gene-environment networks - optimization, games and control - a survey on recent achievements. Invited paper. J. Organisational Transformation Soc. Change (Special Issue) **5**(3), 197–233 (2008)
50. Weber, G.-W., Tezel, A., Taylan, P., Soyler, A., Çetin, M.: Mathematical contributions to dynamics and optimization of gene-environment networks. In the Special Issue: In Celebration of Prof. Dr. Dr. Hubertus Th. Jongen's 60th Birthday, Pallaschke, D., Stein, O. (guest eds.). Optimization **57**(2), 353–377 (2008)
51. Weber, G.-W., Taylan, P., Alparslan-Gök, S.-Z., Özöğür, S., Akteke-Öztürk, B.: Optimization of gene-environment networks in the presence of errors and uncertainty with Chebychev approximation. TOP **16**(2), 284–318 (2008)
52. Weber, G.-W., Alparslan-Gök, S.Z., Söyler, B.: A new mathematical approach in environmental and life sciences: gene-environment networks and their dynamics. Environ. Model. Assess. **14**(2), 267–288 (2009)
53. Weber, G.-W., Özögür-Akyüz, S., Kropat, E.: A review on data mining and continuous optimization applications in computational biology and medicine. Embryo Today Birth Defects Res. C **87**, 165–181 (2009)
54. Weber, G.-W., Kropat, E., Akteke-Öztürk, B., Görgülü, Z.-K.: A survey on OR and mathematical methods applied on gene-environment networks. In the Special Issue on Innovative Approaches for Decision Analysis in Energy, Health, and Life Sciences. Cent. Eur. J. Operations Res. (CEJOR) at the occasion of EURO XXII 2007 (Prague, 8–11 July 2007) **17**(3), 315–341 (2009)
55. Weber, G.-W., Uğur, Ö., Taylan, P., Tezel, A.: On optimization, dynamics and uncertainty: a tutorial for gene-environment networks. In the Special Issue Networks in Computational Biology. Discrete Appl. Math. **157**(10), 2494–2513 (2009)
56. Weber, G.-W., Taylan, P., Yıldırak, K., Görgülü, Z.K.: Financial regression and organization. In the Special Issue on Optimization in Finance, of Dynamics of Continuous. Discrete Impulsive Syst. Ser. B (DCDIS-B) **17**(1b), 149–174 (2010)
57. Weber, G.-W., Branzei, R., Alparslan Gök, S.Z.: On cooperative ellipsoidal games. In: 24th Mini EURO Conference - On Continuous Optimization and Information-Based Technologies in the Financial Sector, MEC EurOPT 2010, Selected Papers, ISI Proceedings, Izmir, pp. 369–372, 23–26 June 2010
58. Weber, G.-W., Branzei, R., Alparslan Gök, S.Z.: On the ellipsoidal core for cooperative games under ellipsoidal uncertainty. In: Proceedings of 2nd International Conference on Engineering Optimization, Lisbon, 6–9 September (2010) (on a CD-Rom)
59. Weber, G.-W., Kropat, E., Tezel, A., Belen, S.: Optimization applied on regulatory and eco-finance networks - survey and new developments. Pac. J. Optim. **6**(2), 319–340 (2010)
60. Weber, G.-W., Taylan, P., Görgülü, Z.-K., Abd. Rahman, H., Bahar, A.: Parameter estimation in stochastic differential equations. In: Peixoto, M., Rand, D., Pinto, A. (eds.) Dynamics, Games and Science II (in Honour of Mauricio Peixoto and David Rand). Springer Proceedings in Mathematics, vol. 2, pp. 703–733. Springer, New York (2011)
61. Yılmaz, F.B.: A mathematical modeling and approximation of gene expression patterns by linear and quadratic regulatory relations and analysis of gene networks. MSc thesis, Institute of Applied Mathematics, METU, Ankara (2004)
62. Yılmaz, F.B., Öktem, H., Weber, G.-W.: Mathematical modeling and approximation of gene expression patterns and gene networks. In: Fleuren, F., den Hertog, D., Kort, P. (eds.) Operations Research Proceedings, pp. 280–287. Springer, Berlin (2005)

Chapter 26
Financial Bubbles

Efsun Kürüm, Gerhard-Wilhelm Weber, and Cem İyigün

26.1 Introduction

Basically, financial bubbles are artificial increases in price which are created by high expectations of individuals; in other words, artificial price augmentation originated from greediness of human beings. Since some investors do not know how to manage their risk when the bubbles burst, this causes big disappointment resulting even in committing suicide. Moreover, their results affect the whole world like prompting domino stones.

For instance, because of the recent real-estate bubble and the subsequent subprime crises, 4 trillion dollar global cumulative losses of financial institutions have occurred and more than 6 million job losses happened in America up to 2009 [25]. It is still very difficult to characterize, to estimate and to prevent them from further inflating in advance.

26.1.1 What Are Bubbles?

A bubble is said to occur if an asset price goes beyond its *fundamental value*. Here, the adversity is to determine the fundamental value of an asset. This value of an asset is arranged in equilibrium, endogenously. It is generally not given, i.e., not

E. Kürüm (✉) • G.-W. Weber
Institute of Applied Mathematics, METU, 06531 Ankara, Turkey
e-mail: efsun.kurum@gmail.com; gweber@metu.edu.tr

C. İyigün
Department of Industrial Engineering, METU, 06531 Ankara, Turkey
e-mail: iyigun@metu.edu.tr

A.A. Pinto and D. Zilberman (eds.), *Modeling, Dynamics, Optimization and Bioeconomics I*, Springer Proceedings in Mathematics & Statistics 73, DOI 10.1007/978-3-319-04849-9_26,
© Springer International Publishing Switzerland 2014

exogenously. The fundamental value determines whether a bubble took place at all and which component of the price is due to a bubble [3].

An investor is willing to buy or hold an asset at a price which is above its fundamental value if he thinks that the asset can be resold at an even higher price in a later trading round [3]. This trading behaviour is called "speculation" by *Harrison* and *Kreps* [10]. As a result, the fundamental value can be thought as the price which an investor is eager to pay if he is compelled to hold the asset forever, i.e., if he is not allowed to retrade [3].

However, the fundamental value is generally insufficient constrained and it is not possible to distinguish between an exponentially growing fundamental price and an exponentially growing bubble price. Therefore, a more precise definition of bubbles is required to detect them [25]. In the Conclusion and Outlook part, Sect. 26.5, alternative approaches will be introduced for that purpose.

26.1.2 Why Bubbles Matter?

When any bubble goes burst, some—in fact, often many—people lose. Sometimes, experienced speculators can realize the end of a bubble and cut their positions in time, since they generally know how to limit their risk to a bearable level. The investors with very little experience of how to manage risks hold the asset at the late phase of a bubble; therefore, they are damaged by the bubbles excessively [4].

If a bubble affected only a few investors, it would be of a limited importance because it has little effect on the overall economy. Bubbles can cause major problems when they originate in an asset that is commonly held. Then, when the bubble bursts, not only a large number of people suffer directly, but it also interacts with the economy. Therefore, dangerous bubbles are generally the ones in a stock market [4].

26.2 Famous Historical Bubbles

26.2.1 Tulip Mania

The tulip mania is considered as the first recorded financial bubble; occurred in the 1630s in Holland. In 1593, tulips were introduced by a botanist *Carolus Clusius* who brought them from Ottoman Empire. He planted his garden to examine them for medicinal purposes. Then, his neighbours stole some of his bulbs in order to make some quick money. Hence, they gave rise to start the process of Dutch bulb trade [24].

In 1634, the madness of having tulips became too much and daily work was neglected. Even the lowest members of the society joined the tulip trade. Until 1635,

Table 26.1 Items and their values which were traded for a single Viceroy bulb [22]

Item	Value (florins)
Two lasts of wheat	448
Four lasts of rye	558
Four fat oxen	480
Eight fat swine	240
Twelve fat sheep	120
Two hogsheads of wine	70
Four casks of beer	32
Two tons of butter	192
A complete bed	100
A suit of clothes	80
A silver drinking cup	60
Total	**2,500**

the mania ascended, the prices increased and many people invested fortunes to own tulip bulbs. Then, to sell tulips by their weight, in *perits* became necessity. Perit was a small weight which is less than a grain. In fact, 480 grains equalled 1 ounce. Prices for different varieties were as follows:

Admiral Liefken, weighing 400 perits = 4,400 florins; *Admiral Von der Eyk*, weighing 446 perits = 1,260 florins; *Shilder* of 106 perits = 1,615 florins; *Viceroy* of 400 perits = 3,000 florins; *Semper Augustus*, weighing 200 perits = 5,500 florins [22].

To perceive, the value of a single tulip, Table 26.1 can be useful. It was recorded in terms of items and their values by one of the authors of that time, *Munting*.

As with all gambling mania, at the beginning, confidence was high because everybody was gaining. Tulip stocks were speculated in the rise and fall by the tulip-brokers. They earned much money, buying the stocks when prices fell, and selling when they rose. Many people suddenly became rich. Everyone imagined that this process would last forever. Eventually, wise investors began to recognize that this imaginary world could not last forever [22]. After that, as it happens in many cases of speculative bubbles, some prudent people started to sell and freezed their profits. As everyone tried to sell while not many were buying, a domino effect became realized and prices became lower and lower. This caused people to panic and sell regardless of losses [2]. Confidence was destroyed and the tulip market collapsed in February 1637, abruptly as shown in Fig. 26.1.

26.2.2 South Sea Bubble

The South Sea Company was established in 1711 by Earl of Oxford *Harley*. As a result of the war between Spain and Britain, British Government left 10 million pounds in debt. Harley offered the government paying back 10 million pounds of debt to ameliorate the government's financial condition. In turn, the

Fig. 26.1 A standardized price index for tulip bulb contracts [23]

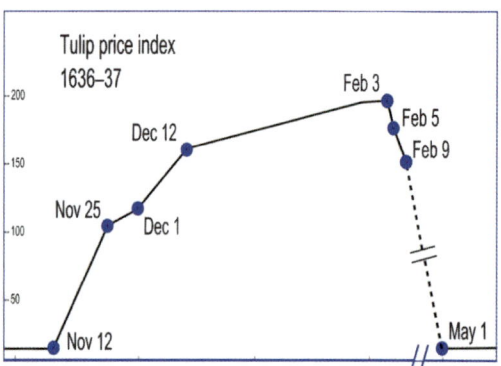

government proposed 6 % interest to the company [21]. In addition to 6 % interest, the government suggested privileged trading rights with Spain's colonies in the South Seas, today known as South America [5].

To finance its operations, the company issued stock to investors and the company's shares were snatched by investors, rapidly. After their first share issue success, the company issued more shares. Although the company had an inexperienced administration team, investors gathered this stock, competitively [5]. It was thought that this company "could never fail".

The war between Britain and Spain began again in 1718; therefore all trading occasions stopped. However, this event did not deter the investors for buying stocks. After a while, management of the South Sea Company recognized that they were not generating much profit from company's operations. For this reason, to generate revenues, they decided to place more importance on issuing stock instead of making actual trade. Meanwhile, the leaders of the company decided to sell their shares since they noticed that the company's stocks were fabulously overvalued relative to its profit. At that time, the other investors did not realize that, actually, the firm was scarcely profitable [5].

Eventually, rumors spread among investors about the firm's management team had sold all of their shares in the company. Then, panic selling immediately started and soon the South Sea Company's shares became worthless as is evident from the Fig. 26.2. *Isaac Newton* lost over 20,000 pounds. He stated "I can calculate the motion of heavenly bodies, but not the madness of people" [5]. After the South Sea Bubble busted, Britain's economy collapsed in spite of the government's endeavors and to heal completely almost took a century.

26.2.3 1929 Great Depression

The end of World War I brought a new era to America. An era of confidence, optimism and welfare were being experienced in the United States during 1920s. After World War I, industrialization and evolution of new technologies such as

Fig. 26.2 Log scale price for south sea company stocks [7]

radio, automobile, air flight bolstered the economic and cultural boom. The Dow Jones Industrial Average, DJIA, increased throughout the 1920s and because of the country's strong economic conditions, most of the economists thought shares were the most confident investment. These caused many investors to buy stocks, greedily [6].

In a little while, investors purchased stocks *on margin* which means that the buyer would invest some of his own money, but the rest compensated by the broker. In those years, only 10–20 % of the stock price had to be paid by the buyer and, hence, 80–90 % of the cost of the stock price would be paid by the broker. If the price of the shares declined lower than the amount of the loan, the broker would probably issue a *margin call*, i.e., the buyer must pay back his loan as cash immediately. Therefore, to buy shares on margin could be very risky. However, in the 1920s, many people who expected to make a lot of money on the stock market easily, called speculators, purchase stocks on margin. They supposed that this rising process in prices never end; so they could not recognize how serious the risk was which they were taking [17].

The Dow Jones bounced from 60 to 400 from 1921 to 1929. This generated a lot of new millionaires. Many people mortgaged their homes and invested their savings into stock market. However, few people really had knowledge about the companies in which they invested [6].

In 1929, from June through August, stocks prices reached their highest levels. Economist *Irving Fisher* stated that "Stock prices have reached what looks like a permanently high plateau", this was the statement many speculators wanted to hear and believe. On 3rd of September 1929, the DJIA closed at 381.17; hence, the stock

Fig. 26.3 Dow Jones Industrial Average index from 1915 to 1942 [20]

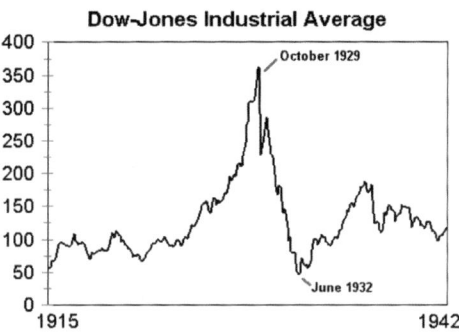

market had reached its peak. Two days later, the market commenced falling. Stock prices fluctuated throughout September and into October [17].

As indicated in Fig. 26.3, a strong bear market, a market condition in which the prices of shares were decreasing, had started by October 1929. On 24th of October 1929 which is known as *Black Thursday*, panic selling started since investors distinguished the stock market boom was actually an over inflated speculative bubble [6]. Rumors spread that people commit suicide [17]. Although the Federal Reserve Bank increased interest rates several times to relieve stock market and overheated economy in 1929, this could not prevent sad end.

When the stock market crashed on 28th and 29th of October, millionaire margin investors went bankrupt instantly. In November 1929, DJIA sharply declined from 400 to 145. Over $5 billion worth of market capitalization had been vanished from stocks that were trading on the New York Stock Exchange in just three days. The stock market crash of 1929 caused a great economic crisis known as the *Great Depression* [6].

As we observe from the famous bubble examples above, irrational expectations always trigger destructive financial crashes.

Indeed, this big and yet vaguely understood phenomenon of "bubbles" asks for a high academic excellence. Academicians and central bankers still try to find a method to prevent from bubbles and try to develop a model which estimates bubbles, previously. One of the successful models which detects bubbles was developed by *Anders Johansen, Olivier Ledoit* and *Didier Sornette*, called *Johansen-Ledoit-Sornette (JLS) model*.

26.3 The Johansen-Ledoit-Sornette Model

At a symposium sponsored by the Federal Reserve Bank of Kansas City, in 2002, the former Federal Reserve chairman *Alan Greenspan* said: "... we recognized that, despite our suspicions, it was very difficult to definitively identify a bubble until after the fact—that is, when its bursting confirmed its existence". In his speech, he declared that "It seems reasonable to generalize from our recent experience that no low-risk, low-cost, incremental monetary tightening exists that can reliably deflate a bubble. But is there some policy that can at least limit the size of a bubble and, hence, its destructive fallout? From the evidence to date, the answer appears to be *no*. But we do need to know more about the behavior of equity premiums and bubbles and their impact on economic activity" [9]. As we understood from his speech, there is no exact theory which can estimate the bubbles previously, and as he underlined that *how it* important to identify them before they blow up.

Anders Johansen, Olivier Ledoit and *Didier Sornette* have proposed a theoretical model, known as the *Johansen-Ledoit-Sornette (JLS)* model, to designate bubbles in advance [25]. By analyzing the cumulative human behaviors, the price dynamics is defined by the model during a time interval related with a bubble process (also called a *regime*). Also, after a bubble regime, the most probable crash time can be estimated by the JLS model previously. This model actually does not describe bubbles by exponential prices but rather by *faster-than-exponential growth* of price. The reason for this arises from imitation and herding behavior of noise traders, called as *irrational investors* [25].

26.3.1 Evolvement of the Price

In the JLS model, only a speculative asset is considered that does not pay dividends. The interest rate, risk aversion, information asymmetry, and the market-clearing condition are ignored for simplicity. Rational expectations are equivalent to the martingale hypothesis [19]:

$$E_t\left[p(t')\right] = p(t) \quad \forall t' > t. \tag{26.1}$$

Here, the price of the asset at time t and the expectation conditional on information up to time t are showed by $p(t)$ and $E_t\left[\cdot\right]$, respectively. The solution of Eq. (26.1) is a constant, if the asset price is not permitted to fluctuate under the effect of noise as follows:

$$p(t) = p(t_0).$$

Here, t_0 displays some initial time and the meaning of this equality is that there is no change in the price. Since the asset price pays no dividend, without loss of generality, for simplicity, its fundamental value is $p(t) = 0$. Thus, a positive value of $p(t)$ generates a speculative bubble [13]. In the model, the crash is the jump, each jump is defined as a bubble and small jumps are not considered. A jump process can be indicated by j and the value zero is attained before the crash, and the value one after the crash, as representing in the following:

$$j = \begin{cases} 0, & \text{before crash,} \\ 1, & \text{after crash.} \end{cases}$$

Fig. 26.4 The cumulative distribution function (cdf) of the time of the crash

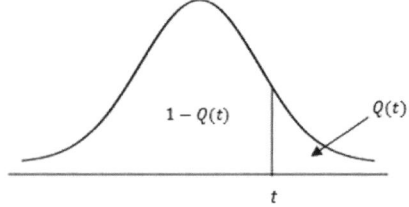

As depicted in Fig. 26.4, the cumulative distribution function (cdf) of the time of the crash, t_c, is called $1 - Q(t)$ and the probability density function (pdf) is

$$q(t) := \frac{dQ}{dt}(t) \approx \frac{\Delta Q}{\Delta t},$$

The hazard rate of the crash is defined as [13]:

$$h(t) = q(t)/[1 - Q(t)],$$

where $Q(t) := Q(t_c > t)$. The hazard rate $h(t)$ compares that change of probability with the likelihood $1 - Q(t) = Q(t_c \le t)$.

According to the JLS model, the asset price is accepted to drop by a certain percentage $\kappa \in (0, 1)$ in case of a crash, and the dynamics of the asset price before the crash is given by [19]:

$$dp_t = \mu(t)p(t)dt - \kappa p(t). \tag{26.2}$$

In order to satisfy the martingale condition for the price process, in Eq. (26.2), the time dependent drift, $\mu(t)$, is chosen, namely

$$E_t[dp_t] = \mu(t)p(t)dt - \kappa p(t)h(t)dt = 0$$

$$\Rightarrow \mu(t) = \kappa h(t). \tag{26.3}$$

Substituting Eq. (26.3) into Eq. (26.2) will give us Eq. (26.4) below:

$$dp = \kappa h(t)p(t)dt - \kappa p(t)dj. \tag{26.4}$$

While $j = 0$ being the case before the crash, we get an ordinary differential equation, ODE:

$$dp = \kappa h(t)p(t)dt \Rightarrow \frac{dp}{dt} = \kappa h(t)p(t) \Rightarrow p'(t) = \kappa h(t)p(t)$$

$$\Rightarrow \frac{p'(t)}{p(t)} = \kappa h(t). \tag{26.5}$$

Integrating the Eq. (26.5) from both sides gives

$$\int_{t_0}^{t} \frac{p'(s)}{p(s)} ds = \int_{t_0}^{t} \kappa h(s) ds.$$

Thus, the following expression is obtained

$$\log\left[\frac{p(t)}{p(t_0)}\right] = \kappa \int_{t_0}^{t} h(s) ds \quad \text{before the crash.} \quad (26.6)$$

According to this new statement, under the condition that there is no crash, any increase of the probability of a crash implies an increase of the price speed in order to fulfill the martingale property. With [13] we note that the hazard rate $h(t)$ of the crash drives the price; that rate is not restricted by any bound so far.

To clarify the investors' behaviours, the model considers a *network* of agents. Each of the *agents* is indexed by an integer $i = 1, 2, \ldots, I$, and the set of agents connected with i is indicated by $N(i)$. According to some graph like in Fig. 26.5, the set of agents, $N(i)$, is directly linked with agent i. An assumption in the model is that agent i can have only one of two possible states, $s_i \in \{-1, +1\}$. Here, the values -1 and $+1$ can be interpreted as buy or sell, bullish or bearish or, in technical or Boolean terms, "on or off". The state of investor i is determined by the model as follows [19]:

$$s_i = \text{sign}\left(K \sum_{j \in N(i)} s_j + \sigma \epsilon_i\right). \quad (26.7)$$

Here, K is a positive constant and ϵ_i is independently distributed with the standard normal distribution. The parameter K, called the *coupling strength*, controls the tendency towards imitation and σ governs the tendency towards *idiosyncratic* behavior. The parameter σ can be interpreted as an *environmental* factor. Bigger values of K reflect *strong imitation* [13]. If $K = 0$, only environmental effects determine the decision of the trader.

The ratio K/σ represents the outcome of the order and disorder decision; in other words, it symbolizes the probability of a crash. Also, as long as the average of all values K was strictly positive, the model allows some of the values of K even to be negative. The meaning of the negative K is that the investor does not accept other investors' decisions.

As for the *susceptibility* of the system, it can be explained by adding a global influence term G to the Eq. (26.7):

$$s_i = \text{sign}\left(K \sum_{j \in N(i)} s_j + \sigma \epsilon_i + G\right). \quad (26.8)$$

In Eq. (26.8), $K \sum_{j \in N(i)} s_j + \sigma \epsilon_i$ corresponds to individual decision and G can be explained as an average group effect to the investor's decision.

The parameter G will converge to the state $+1$ if $G > 0$; G will approximate the favour state -1 if $G < 0$ and, if $G = 0$, no global influence will exist.

The average state is defined as $M = (1/I) \sum_{i=1}^{I} s_i$. When there is no global influence, expectation of the average state is zero, $E(M) = 0$, i.e., agents are equally divided between the two states.

Provided that a positive (negative) global influence is given, agents in the positive (negative) state will outnumber the others, $E(M) \cdot G \geq 0$. With this notation, the *susceptibility* of the system is defined as [13]:

$$\chi = \frac{d(E(M))}{dG}\Big|_{G=0}. \tag{26.9}$$

The sensitivity of the average state to a small global influence is measured by this susceptibility. It also has a second interpretation, namely, as the expected squared deviance (variance) of the average state M around its mean of zero, caused by the random idiosyncratic shocks ϵ_i.

The susceptibility is affected by the structure of the network. At the basis of the JLS model, there are two kinds of network structures: The two-*dimensional grid* and the *hierarchical diamond lattice*.

26.3.2 Two-Dimensional Grid

In this network structure, the agents are placed on a two-dimensional grid in the Euclidean plane. Each agent has four nearest neighbors: one to the North, one to the South, one to the East and one to the West. The related parameter K/σ evaluates the propensity towards imitation relative to the tendency towards idiosyncratic behavior [13].

The properties of the system are arranged by a critical point K_c. While $K < K_c$, *disorder* decision wins. The meaning of this, the sensitivity is small for a small global influence, imitation only propagates, like in a *chain reaction* or even *cascading*, between close neighbors and the susceptibility χ of the system is finite.

When K rises and approaches K_c, *order* decision reigns, i.e., the system is extremely sensitive to a small global perturbation, imitation propagates over long distances. In this situation, the susceptibility χ of the system goes to infinity with respect to the power law. This behaviour of the χ can be explained mathematically as follows [19] :

$$\chi \approx A(K_c - K)^{-\gamma}, \quad \gamma > 1. \tag{26.10}$$

Here, the parameter K does not even need to be deterministic, but it could also be a stochastic process, as long as it proceeds slowly enough. The value t_c is called the *first time* such that $K(t_c) = K_c$. Then, before the critical date t_c, the following approximation is obtained from Taylor expansion:

$$K_c - K(t) \approx \text{constant} \cdot (t_c - t).$$

Since the hazard rate of the crash behaves similar to the susceptibility around the critical point, this approximation yields the following statement:

$$h(t) \approx B \cdot (t_c - t)^{-\alpha}. \tag{26.11}$$

Here, B is a positive constant, and if the bubble has not busted yet, to prevent the price from going to infinity when approaching t_c, α should be between 0 and 1. Since the crash could happen at some time prior to t_c, although the crash is not most probably, as stated in [13], t_c is not necessarily the time of the crash. Finally, substituting Eq. (26.6) into Eq. (26.6) gives us the following relation:

$$\log p(t) = \log p(t_c) + \kappa B \int_{t_c}^{t} (t_c - s)^{-\alpha} ds,$$

i.e.,

$$\log p(t) = \log p(t_c) - \frac{\kappa B}{1 - \alpha}(t_c - t)^{1-\alpha}.$$

Hence,

$$\log p(t) = \log p(t_c) - \frac{\kappa B}{\beta}(t_c - t)^{\beta} \quad \text{before the crash.} \tag{26.12}$$

Since $\alpha \in (0, 1)$, $\beta := 1 - \alpha \in (0, 1)$ and $p(t_c)$ is the price at the critical time t_c. The logarithm of the price before the crash also obeys the power law. As the price process approaches the critical date, the slope of the log price (the expected return per unit of time) becomes limitless. This helps to compensate boundless probability of the crash in the next instant [19].

26.3.3 Hierarchical Diamond Lattice

Another network structure is the hierarchical diamond lattice. To obtain this lattice, the process starts with a pair of investors who are connected to each other. Then, these connections are changed according to a *diamond* structure. On the diamond, two original investors reside two opposite vertices and where the two new investors occupy on the other two vertices. There are four links included by this diamond. For each one of these four links, it is altered by a diamond in the aforementioned way,

Fig. 26.5 The relationship
between traders and links

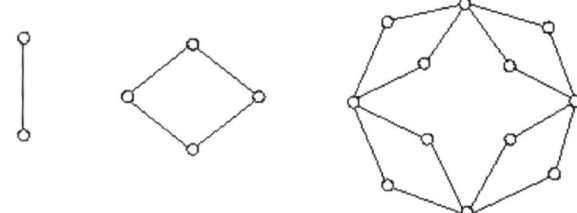

and this operation is iterated [18]. Then, a diamond lattice is formed, as illustrated
in Fig. 26.5.

In Fig. 26.5, circles and edges show traders and links, respectively. After p
iterations, $(2/3)(2 + 4^p)$ investors and 4^p connections between them are obtained
[18]. This can be proven by induction principle.

The basic properties of the hierarchical diamond lattice are similar to the two-
dimensional grid. In this network structure, there is also a critical point K_c. The
susceptibility is restricted while $K < K_c$, and it goes to infinity as K increases
towards K_c. Unlike from two-dimensional grid, here, in the hierarchical diamond
lattice, the critical exponent can be a complex number. The general solution for the
susceptibility is a sum of terms like the one in $\chi \approx A(K_c - K)^{-\gamma}$, but now with
complex exponents $-\gamma + iw$, etc. [13]. This approximation is evaluated gradually
as follows:

$$\chi \approx \mathrm{Re}\left[A_0(K_c - K)^{-\gamma} + A_1(K_c - K)^{-\gamma+iw} + \ldots\right], \tag{26.13}$$

$$\chi \approx \mathrm{Re}\left[A_0(K_c - K)^{-\gamma}\right] + \mathrm{Re}\left[A_1(K_c - K)^{-\gamma+iw}\right] + \ldots,$$

$$\chi \approx \mathrm{Re}\left[A_0(K_c - K)^{-\gamma}\right] + \mathrm{Re}\left[A_1(K_c - K)^{-\gamma}(K_c - K)^{iw}\right] + \ldots,$$

$$\chi \approx \mathrm{Re}\left[A_0(K_c - K)^{-\gamma}\right] + \mathrm{Re}\left[A_1(K_c - K)^{-\gamma}e^{iw\log(K_c-K)}\right] + \ldots,$$

$$\chi \approx A_0(K_c - K)^{-\gamma} + A_1(K_c - K)^{-\gamma}\mathrm{Re}\left[e^{iw\log(K_c-K)}\right] + \ldots,$$

$$\chi \approx A_0(K_c - K)^{-\gamma} +$$
$$A_1(K_c - K)^{-\gamma}\mathrm{Re}\left[\cos(w\log(K_c - K)) + i\sin(w\log(K_c - K))\right] + \ldots,$$

$$\chi \approx A_0(K_c - K)^{-\gamma} + A_1(K_c - K)^{-\gamma}\cos\left[w\log(K_c - K) + \psi\right] + \ldots.$$

Here, $\mathrm{Re}[\cdot]$ indicates the real part of a complex number and γ, A_0, A_1, w, ψ are real
numbers. Hence, the *power law* is corrected by oscillations whose frequency busts
as the price process approach the critical time. These accelerating oscillations are
called log-periodic, and their frequency are $\lambda = w/2\pi$ is called the log-frequency
[13]. If the same steps are applied, Eq. (26.13) is obtained for the hazard rate of a
crash:

$$h(t) \approx B_0(t_c - t)^{-\alpha} + B_1(t_c - t)^{-\alpha} \cos\left[w\log(t_c - t) + \psi'\right]. \qquad (26.14)$$

Substituting Eq. (26.14) into Eq. (26.12), we get:

$$\log[p(t)] = \log[p(t_c)] - \frac{\kappa}{\beta}\left\{B_0(t_c - t)^{\beta} + B_1(t_c - t)^{\beta} \cos\left[w\log(t_c - t) + \phi\right]\right\},$$
$$(26.15)$$

and (26.15) can be rewritten as

$$y(t) \approx A + B(t_c - t)^{\beta} + C(t_c - t)^{\beta} \cos\left[w\log(t_c - t) + \phi\right], \qquad (26.16)$$

where $A = \log[p(t_c)]$, $B = -\kappa B_0/\beta$, $C = -\kappa B_1/\beta$ and ϕ is another phase constant.

26.4 The Fitting Process

In the fitting process, so as to diminish the number of unknown parameters, three linearly embedded variables, A, B, C, have been optimally adjusted and calculated from the obtained values of the nonlinear parameters. For this reason, approximate Eq. (26.16) is rewritten as [11]

$$y(t) \approx A + Bf(t) + Cg(t), \qquad (26.17)$$

where

$$f(t) = \begin{cases} (t_c - t)^{\beta}, & \text{for a speculative bubble,} \\ (t - t_c)^{\beta}, & \text{for an antibubble.} \end{cases}$$

and

$$g(t) = \begin{cases} (t_c - t)^{\beta} \cos(w\log(t_c - t) + \phi), & \text{for a speculative bubble,} \\ (t - t_c)^{\beta} \cos(w\log(t_c - t) + \phi), & \text{for an antibubble,} \end{cases}$$

An *antibubble* is the opposite of the speculative bubble. Like a speculative bubble, it also follows a log-periodic power law (LPPL) but, of course, with decelerating oscillations and generally being bearish inclined instead of bullish. The term antibubble comes from the term *antiparticle* in physics, since an antiparticle is similar to its sibling particle, except that it carries opposite charges and demolishes its sibling particle when it comes across with it [12].

The best values of the linear parameters are calculated for each choice of the nonlinear parameters by applying ordinary least-squares (OLS) method as follows by Gaussian normal equations [11]:

$$Xb = y \implies (X^T X)b = X^T y.$$

$$
\begin{pmatrix}
N & \sum_{i=1}^{N} f(t_i) & \sum_{i=1}^{N} g(t_i) \\
\sum_{i=1}^{N} f(t_i) & \sum_{i=1}^{N} f(t_i)^2 & \sum_{i=1}^{N} f(t_i)g(t_i) \\
\sum_{i=1}^{N} g(t_i) & \sum_{i=1}^{N} f(t_i)g(t_i) & \sum_{i=1}^{N} g(t_i)^2
\end{pmatrix}
\begin{pmatrix} A \\ B \\ C \end{pmatrix}
=
\begin{pmatrix}
\sum_{i=1}^{N} \log p_i \\
\sum_{i=1}^{N} \log p_i \, f(t_i) \\
\sum_{i=1}^{N} \log p_i \, g(t_i)
\end{pmatrix},
$$

where

$$
X := N
\begin{pmatrix}
1 & f(t_1) & g(t_1) \\
\vdots & \vdots & \vdots \\
1 & f(t_N) & g(t_N)
\end{pmatrix},
\; b :=
\begin{pmatrix} A \\ B \\ C \end{pmatrix}
\text{ and } y :=
\begin{pmatrix} \log p_1 \\ \vdots \\ \log p_N \end{pmatrix}.
$$

Here, t_i ($i = 1, 2, \ldots, N$) are the times of the price sampling (t_i, p_i). The general solution of this equation is given by

$$\hat{b} = (X^T X)^{-1} X^T y.$$

Of course, we may assume that the number of data is greater than to the number of unknown, p, i.e., $N \geq p = 3$, now the solution A, B and C become inserted into problem (26.17) and that the design matrix X has full rank. In case of ill-conditionedness, we may apply a regularization technique [1].

Fitting a function to data is a nonlinear estimation problem of the residuals sum of squares, RSS, where the objective function is given by [11]

$$\min_{\theta} \; F(\theta) := \sum_{i=1}^{N} (y_\theta(t_i) - y_i)^2 \, ;$$

Here, $\theta = (t_c, \phi, w, \beta)^T$ is the vector of unknown parameters, and $y_\theta(\cdot) := y(\cdot)$, depending on θ.

The function of the residual sum of squares, F, is an analytically complicated or a strongly nonconvex function, and it can comprise multiple local minima with quite similar values. Here, the aim is to find the global minimum and to optimize the objective function with methods like the *downhill-simplex method* or the *quasi-Newton method*, this could be hazardous since these methods can be tricky by directing a local minimum rather than the global minimum [11]. Therefore, only more model-free *global optimization* solution methods were considered, such as *Simulated Annealing*, *Taboo Search* and *Genetic Algorithm*, etc.. Here, we used the Genetic Algorithm [14–16], and then as an example of a speculative bubble, the fitted Log-index for the crisis of the year 1987 in US stock market was obtained according to *JLS model* as shown in Fig. 26.6. In the figure, t_c demonstrates the estimated bubble-burst time by *JLS* model and t_R stands for the real bubble-burst time.

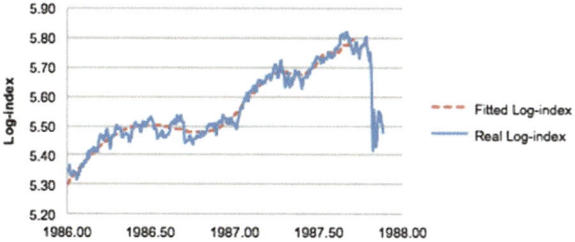

Fig. 26.6 Fitted Log-index process for the year 1987 at S&P 500 index.
Parameters: $A = 328.63$, $B = -85.26$, $C = -17.47$, $t_c = 1987.7055$, $\beta = 0.82$, $w = 13.64$, $\phi = 1.40$ and $t_R = 1987.8222$. $R^2 = 0.9694$

26.5 Conclusion and Outlook

As the president of the Federal Reserve Bank of New York, *William C. Dudley* emphasized in 2010 [8]: "... what I am proposing is that we try—try to identify bubbles in real time, try to develop tools to address those bubbles, try to use those tools when appropriate to limit the size of those bubbles and, therefore, try to limit the damage when those bubbles burst". To prevent from the devastating results of bubbles, we should develop new tools, methods or models. Namely, we should understand their dynamics. For this purpose, bubbles could be represented as a mathematical object, so advantages of *optimization, pattern recognition tools* and *machine learning* can be used as different solution approaches to detect, forecast and control their size, shape and position.

Here, we suggest a new and pioneering mathematical approach where the logarithmic price process is represented with ellipsoids. In fact, the process is divided into intervals and each interval is described with an ellipsoid. On the basis of this approach, a bubble is dealt with geometrically. By observing the volume of the ellipsoids, a bubble can be detected or the behaviour of the volume change can be used as an *early warning system* against destructive outcomes of the bubbles. After identifying and constructing ellipsoids throughout the price process, they can be clustered; each cluster may depict a bubble structure with fitting a covering ellipsoid.

Acknowledgements This research has been supported by TUBITAK (The Scientific and Techno-logical Research Council of Turkey) under the grant number 112T744.

References

1. Aster, R.C., Borchers, B., Thurber, C.H.: Parameter Estimation and Inverse Problems, 2nd edn. Academic (2012)
2. Beattie, A.: Market Crashes: The Tulip and Bulb Craze. http://www.investopedia.com/features/crashes/crashes2.asp#axzz2ED0bIWnb (2012). Cited 10 Dec 2012

3. Brunnermeier, M.K.: Asset Pricing Under Asymetric Information, Bubbles, Crashes, Technical Analysis, and Herding. Oxford University Press, New York (2001)
4. Calverley, J.: Bubbles and How to Survive Them. N. Brealey Publishing, Boston (2004)
5. Colombo, J.: Stock Market Crash, The South Sea Bubble (1719–1720). http://www.stock-market-crash.net/south-sea-bubble/ (2012). Cited 10 Dec 2012
6. Colombo, J.: Stock Market Crash, The Stock Market Crash of 1929. http://www.stock-market-crash.net/1929-crash/ (2012). Cited 15 Dec 2012
7. Colombo, J.: The Bubble Bubble, We May Be On the Verge of the Next Major Bubble Boom. http://www.thebubblebubble.com/next-bubble/ (2012). Cited 1 Jan 2013
8. Goldstein, J.: New York Fed chief: We should "try to identify bubbles", National Public Radio, www.npr.org/blogs/money/2010/04/the_friday_podcast.html (2010).
9. Greenspan, A.: Economic volatility. Remarks at a symposiums sponsored by the Federal Reserve Bank of Kansas City, Jackson Hole, Wyoming, www.federalreserve.gov/boarddocs/Speeches/2002/20020830/ (2002)
10. Harrison, J.M., Kreps, D.M.: Speculative investor behavior in a stock market with heterogeneous expectations. Q. J. Econ. **92**(2), 323–336 (1978)
11. Jacobsson, E.: How to predict crashes in financial markets with the Log-Periodic Power Law. Technical Report (2009)
12. Johansen, A., Sornette, D.: Financial "anti-bubbles": Log-periodicity in Gold and Nikkei collapses. Int. J. Mod. Phys. C **10**, 563–575 (1999)
13. Johansen, A., Ledoit, O., Sornette, D.: Crashes as critical points. Int. J. Theor. Appl. Finance **3**(2), 219–255 (2000)
14. Matlab User's guide: Genetic Algorithm and Direct Search Toolbox, www.mathworks.com//help/releases/R13sp2/pdf_doc/gads/gads_tb.pdf (2012)
15. Nougues, J.M., Grau, M.D., Puigjaner, L.: Parameter estimation with genetic algorithm in control of fed-batch reactors. Chem. Eng. Process. **41**, 303–309 (2012)
16. Özöğür, S.: Mathematical Modelling of Enzymatic Reactions, Simulation and Parameter Estimation. MSc thesis, Institute of Applied Mathematics of METU Ankara, Turkey (2005)
17. Rosenberg, J.: The Stock Market Crash of 1929. http://history1900s.about.com/od/1920s/a/stockcrash1929.htm (2012). Cited 10 Dec 2012
18. Sornette, D.: Why Stock Markets Crash: Critical Events in Complex Financial Systems. Princeton University Press, New Jersey (2003)
19. Sornette, D.: Critical market crashes. Phys. Rep. **378**, 1–98 (2003)
20. The Econ Review: Stock Market Crash Causes Depression. http://www.econreview.com/events/stocks1929b.htm. Cited 1 Jan 2013
21. The South Sea Bubble: A Stock Market Bubble. http://www.thesouthseabubble.com/. Cited 10 Dec 2012
22. The tulipomania: An Investing Bubble. http://www.thetulipomania.com/. Cited 10 Dec 2012
23. Tulip mania: http://en.wikipedia.org/wiki/Tulip_mania. Cited 1 Jan 2013
24. Wood, C.: The Dutch Tulip Bubble of 1637. http://www.damninteresting.com/the-dutch-tulip-bubble-of-1637/ (2006) . Cited 10 Dec 2012
25. Yan, W.: Identification and Forecasts of Financial Bubbles. Ph.D thesis, ETH Zurich, Switzerland (2011)

Chapter 27
Modern Applied Mathematics for Alternative Modeling of the Atmospheric Effects on Satellite Images

Semih Kuter, Gerhard Wilhelm Weber, Ayşe Özmen, and Zuhal Akyürek

27.1 Introduction

To estimate unknown (predictor, response) dependency (or model) from training data (consisting of a finite number of observations) with good prediction (generalization) capabilities for future (test) data is the basic aim of the learning task [10,25]. While regression is to learn a mapping from the input space, X, to the output space, Y, where this mapping, f, is called an *estimator*, which is used for predicting *quantitative* outputs (i.e., $X = \mathbb{R}^d$, $Y = \mathbb{R}$); the aim of classification is to learn a mapping from the feature space, X, to a label space, Y, where this mapping, f, is called a *classifier*, which is used for predicting *qualitative* outputs (i.e., $X = \mathbb{R}^d$, $Y = \{0, 1\}$). Although the naming convention for the learning task depends on the output type, both have common characteristics and can be regarded as a task in function approximation [21].

As in almost all branches of science and engineering, regression and classification also play a crucial role in *geographic information systems* (GIS) and *remote sensing* (RS) fields, where images taken by earth-observing satellites are used to extract various kinds of information concerning the earth environment.

S. Kuter (✉)
Faculty of Forestry, Department of Forest Engineering, Çankırı Karatekin University, 18200, Çankırı, Turkey
e-mail: semihkuter@yahoo.com

G.W. Weber • A. Özmen
Institute of Applied Mathematics, Middle East Technical University, 06800, Ankara, Turkey
e-mail: gweber@metu.edu.tr; ayseozmen19@gmail.com

Z. Akyürek
Faculty of Engineering, Department of Civil Engineering, Middle East Technical University, 06800, Ankara, Turkey
e-mail: zakyurek@metu.edu.tr

A.A. Pinto and D. Zilberman (eds.), *Modeling, Dynamics, Optimization and Bioeconomics I*, Springer Proceedings in Mathematics & Statistics 73, DOI 10.1007/978-3-319-04849-9__27,
© Springer International Publishing Switzerland 2014

Many kinds of parametric and nonparametric regression and classification methods have been applied for information extraction from remotely sensed earth data [32, 42, 43, 55, 65, 67, 71]. However, enumerating them all is not the goal of this chapter. Instead, the main motivation is to revisit the two non-parametric regression techniques, well-known *Multivariate Adaptive Regression Splines* (MARS) and recently introduced *Conic Multivariate Adaptive Regression Splines* (CMARS), together with preliminary results of their applications on satellite image data as a real-life earth science example.

According to its modern usage in earth sciences, RS is briefly defined as the use of aerial sensor technologies in order to acquire data about earth's surface and atmosphere through propagation of electromagnetic radiation, and the extraction of information by processing, analyzing, and applying those data [50, 54]. Since RS makes it possible for users to collect, interpret, and manipulate data sets over large areas, which are often dangerous or inaccessible, it has become a powerful and preferred tool for many scientific disciplines.

When examining earth's surface from a space-based RS platform, the atmosphere is a large factor in the uncertainty associated with surface reflectance measurements [46]. A fundamental problem within RS of earth is to correct land surface reflectance data gathered by space-based sensors for the perturbations introduced by the passage of radiation through earth's atmosphere [24, 45]. Failure to correct for the effects of the atmosphere can lead to substantially different values for the surface reflectances to those expected, and will therefore have a significant effect on any conclusions drawn from such data, particularly when examining long time series over which the atmospheric parameters do not remain constant [45].

There are two main approaches in atmospheric correction methodology: relative and absolute radiometric correction of atmospheric attenuation. According to the former, the image histograms of different bands are shifted to the left, assuming that the offset of the histogram from zero brightness value is due to the atmospheric scattering. In the latter, the whole process of atmospheric attenuation is numerically modeled. The parameters of atmospheric temperature, pressure, relative humidity, visibility, solar zenith angle, solar distance from the earth and band wavelength are input values, which have to be a priori determined. The finally calculated value is the target reflectance, which is supposed to be equal to the in-situ spectroradiometrically measured reflectance value, as long as the numerical model is accurate and reliable [38].

Numerous absolute radiometric correction methods, which are based on rigorous *radiative transfer* (RT) modeling, have been proposed in order to remove atmospheric effects from images taken by sensors with different technical specs and operational objectives, installed either on aircraft or satellite platforms such as *Simulation of a Satellite Signal in the Solar Spectrum* (5S) [49, 56], *Second Simulation of a Satellite Signal in the Solar Spectrum* (6S) [26, 27, 60, 61], *MODerate resolution atmospheric TRANsmission* (MODTRAN) [7, 68], *Fast Line-of-sight Atmospheric Analysis of Spectral Hypercubes* (FLAASH) [3, 19, 34], *Atmospheric CORrection Now* (ACORN) [1], *Atmospheric and Topographic CORrection* (ATCOR) [51, 52], and *QUick Atmospheric Correction* (QUAC) [8]. However, it has to be noted that the high accuracy available with RT models can not actually be reached when working on large areas, due to unknown atmospheric parameters. Besides, models based on a rigorous treatment of the RT problem, like 5S or 6S, are too expensive and time consuming to be used on an operational basis for the atmospheric correction of large numbers of satellite images, especially those acquired with *large field of view* (LFOV) sensors, since in these images each pixel has different observational geometry and each atmospheric parameter has to be calculated for each pixel [45, 49].

Introduced by physicist and statistician Friedman [17, 18], MARS is an innovative nonparametric regression procedure that does not make any specific assumption about the underlying functional relationship between response and predictor variables [13, 59]. The algorithm estimates the model function at two stages; forward pass and backward pass. In the first stage, similar to forward stepwise linear regression, the basis functions and their products are used to generate a maximal model that overfits data. Then in backward pass stage, the model is pruned stepwise by eliminating the terms which result in smallest increase in the residual squared error until a predefined threshold is reached. Although MARS seems similar to its predecessor *Classification and Regression Trees* (CART) in that it partitions the data into two or more parts, it differs from decision tree techniques since nonlinearity of the models is approximated by different regression slopes in the corresponding intervals of each predictor variable in MARS [13, 64].

CMARS ('C' stands not only for *convex*, but also *continuous* and *conic*) has recently been developed as an alternative method to MARS. In CMARS, instead of applying backward pass algorithm, a *Penalized Residual Sum of Squares* (PRSS) is introduced to MARS as a *Tikhonov regularization* (TR) problem, and this two-objective optimization problem is treated using the continuous optimization technique called *Conic Quadratic Programming* (CQP) [41, 59, 64]. Within this context, CMARS is more model-based and employs continuous, well-structured convex optimization which uses *Interior Point Methods* and their codes.

A large literature is available for MARS, which has many successful applications in broad range of fields such as operational research, marketing and finance [2, 13, 31, 39, 62, 70]; biology [4, 29]; ecology and forestry [16, 20, 22, 23, 30, 37, 44, 63]; simulation and computation [5, 6]; geophysics [28]; engineering [9, 36, 53, 72]; medical and biomedical sciences [11, 12, 47, 66, 69]; as well as in GIS and RS [14, 15, 48]. Although not as large as its predecessor, a growing literature can be found on CMARS [2, 41, 57–59, 64].

The main scope of this chapter is to introduce the preliminary results of the efforts being made for building models through MARS and CMARS that can be applied for the atmospheric correction of *MODerate-resolution Imaging Spectroradiometer* (MODIS) images on local scale. The results obtained by MARS and CMARS methods are compared with the results obtained by the *Simplified Method for Atmospheric Correction* (SMAC) [49], which is an atmospheric correction technique for satellite measurements in the solar spectrum developed by Rahman and Dedieu in the early 1990s and several hundred times faster than the more detailed RT models like 5S and 6S. The remainder of this chapter is organized as follows. The next section gives basic overview of MARS and CMARS. In Sect. 27.3, a brief insight into RT equation and SMAC method is given. The methodology and the results are represented in Sect. 27.4. And finally, Sect. 27.5 concludes this chapter.

27.2   Overview of MARS and CMARS Methods

In this section, the basics of MARS and CMARS algorithms are given based on [2, 17, 21, 40, 41, 57, 59, 64].

Based on a modified recursive partitioning methodology, MARS algorithm is an extension of CART and both are similar in the partitioning of intervals, where two symmetric basis functions (BFs) are constructed at the knot location. However, continuous piecewise linear functions are used in MARS algorithm and a continuous model that provides a more effective way to model nonlinearities is produced, whereas CART uses indicator functions that leads to lack of continuity, which severely affects the model accuracy. Selection of BFs is data-based and specific to the problem in MARS, which makes it an adaptive regression procedure suitable for solving high dimensional problems. In MARS model building, piecewise linear BFs are fitted in such a way that additive and interactive effects of the predictors are taken into account to determine the response variable.

MARS uses expansions of the truncated piecewise linear basis functions of the form:

$$(x - \tau)_+ = \begin{cases} x - \tau, & \text{if } x > \tau, \\ 0, & \text{otherwise,} \end{cases}$$

$$(\tau - x)_+ = \begin{cases} \tau - x, & \text{if } x < \tau, \\ 0, & \text{otherwise.} \end{cases}$$

where $x, \tau \in \mathbb{R}$. These two functions are called as a reflected pair and the symbol '+' indicates that only the positive parts are used, and otherwise zero.

General model on the relation between input and response is:

$$Y = f(\boldsymbol{x}) + \varepsilon, \tag{27.1}$$

where Y is a response variable, $\boldsymbol{x} = (x_1, x_2, \ldots, x_p)^T$ is a vector of predictors and ε is an additive stochastic component with zero mean and finite variance. The logic in MARS is to generate reflected pairs for each input x_j $(j = 1, 2, \ldots, p)$ with p-dimensional knots $\boldsymbol{\tau}_i = (\tau_{i,1}, \tau_{i,2}, \ldots, \tau_{i,p})^T$ at each input data vectors $\overline{\boldsymbol{x}}_i = (\overline{x}_{i,1}, \overline{x}_{i,2}, \ldots, \overline{x}_{i,p})^T$ of that input $i = (1, 2, \ldots, N)$. Therefore, the collection of BFs is:

$$C := \{(x_j - \tau)_+, (\tau - x_j)_+ \mid \tau \in \{x_{1,j}, x_{2,j}, \ldots, x_{N,j}\}, \ j \in \{1, 2, \ldots, p\}\}, \tag{27.2}$$

where N denotes the number of observations, p is the dimension of the input space. The tensor products of univariate spline functions are applied for the generalization of spline fitting in higher dimensions, which leads to multivariate spline BFs with the following form:

$$B_m(\boldsymbol{x}) =: \prod_{j=1}^{K_m} [s_{\kappa_j^m}.(x_{\kappa_j^m} - \tau_{\kappa_j^m})], \tag{27.3}$$

where the total number of truncated linear functions multiplied in the mth BF is denoted by K_m, $x_{\kappa_j^m}$ is the corresponding input variable of the kth truncated linear function in the mth BF, $\tau_{\kappa_j^m}$ is the corresponding knot value for the variable $x_{\kappa_j^m}$ and, finally, $s_{\kappa_j^m} \in \{\pm 1\}$.

Estimation of model function $f(\boldsymbol{x})$ is carried out by forward and backward pass algorithms in MARS. Although the former is analogous to forward stepwise linear regression, use of functions from the set C and their products are allowed, generating the maximal model that overfits the data. The initial model has the following form:

$$f(\boldsymbol{x}) = \beta_0 + \sum_{m=1}^{M} \beta_m B_m(\boldsymbol{x}), \tag{27.4}$$

where B_m is either a function or product of two or more functions from the set C, M is the number of BFs in the current model, and β_0 is the intercept term. To compare the possible BFs, a *lack-of-fit* (LOF) criterion is used.

In MARS algorithm, each BF must have different input variables. At each step, with one of the reflected pair in the set C, all products of a function $B_m(\boldsymbol{x})$ in the model set are considered as a new function pair. The product that results in the largest decrease in residual error is added into the current model and has the form:

$$\beta_{M+1} B_k(\boldsymbol{x}).(x_j - \tau)_+ + \beta_{M+2} B_k(\boldsymbol{x}).(\tau - x_j)_+, \tag{27.5}$$

where β_{M+1} and β_{M+2} are coefficients estimated by least squares. The products satisfying the above mentioned condition are successively added to the model until a user-defined value M_{max} is reached.

At the end, a large model that typically overfits the initial data is obtained. Then, backward pass is applied in order to prevent the model obtained in forward pass from over-fitting by decreasing the complexity of the model without degrading the fit to the data. It removes the BFs that give the smallest increase in the residual sum of squares at each step, producing an estimated best model $\hat{f}_\alpha$ with the optimal value α. Here, α expresses some complexity of estimation, and also note that M_{max} is reduced to M. In order to find the optimal value of α, cross-validation could be used; however, for the sake of decreasing computational burden, MARS uses *generalized cross-validation* (GCV) that shows the LOF. GCV is defined as:

$$\text{LOF}(\hat{f}_\alpha) = \text{GCV}(\alpha) := \frac{\sum_{i=1}^{N}(y_i - \hat{f}_\alpha(x_i))^2}{(1 - Q(\alpha)/N)^2}, \tag{27.6}$$

where N is the number of sample observations, $Q(\alpha) = u + dK$ with K representing the number of knots which are selected in forward pass and u is the number of linearly independent functions in the model, d denotes a cost for each BF optimization and, usually, $d = 3$ ($d = 2$ is used when the model is additive). The numerator is the conventional residual sum of squares, which is penalized by the denominator that accounts for the increasing variance in case of increasing model complexity, i.e., while larger $Q(\alpha)$ creates a smaller model with less number of BFs, smaller $Q(\alpha)$ generates a larger model with more BFs. Using the LOF criteria, the best model is chosen according to backward pass that minimizes GCV.

As briefly discussed so far, MARS uses two algorithms (forward pass and backward pass), through which it tries to accomplish two tasks simultaneously: a good fit to the data, yet a simple model. In CMARS, backward pass of MARS is not employed; instead, PRSS with BFs is collected during forward pass and penalty terms are added to the least squares estimation in order to control the LOF, introducing an alternative point of view of the complexity and the stability of the estimation. Therefore, PRSS summed up in forward pass stage of the MARS method is in the following form:

$$\text{PRSS} := \sum_{i=1}^{N}(y_i - f(\overline{x}_i))^2 + \sum_{m=1}^{M_{\max}}\phi_m \sum_{\substack{|\alpha|=1 \\ \alpha=(\alpha_1,\alpha_2)^T}}^{2} \sum_{\substack{r<s \\ r,s\in V_m}} \int \beta_m^2 \left[D_{r,s}^\alpha B_m(t^m)\right]^2 dt^m, \tag{27.7}$$

where $V_m := \{\kappa_j^m \mid j = 1, 2, \ldots, K_m\}$ is the variable set associated with the mth BF B_m; $t_m = (t_{m_1}, t_{m_2}, \ldots, t_{m_{K_m}})^T$ denotes the vector of variables contributing to the mth BF B_m; the parameters $\phi_m \geq 0$ serve as penalty parameters ($m = 1, 2, \ldots, M_{\max}$); and, finally, $D_{r,s}^\alpha B_m(t^m) := \frac{\partial^\alpha B_m}{\partial^{\alpha_1} t_r^m \partial^{\alpha_2} t_s^m}(t^m)$ for $\alpha = (\alpha_1, \alpha_2)^T$, $|\alpha| := \alpha_1 + \alpha_2$, where $\alpha_1, \alpha_2 \in \{0, 1\}$. Then, the trade-off between accuracy and complexity in this optimization problem is established through the penalty parameters ϕ_m, and PRSS can be approximated as:

$$\text{PRSS} \approx \left\| y - B(\tilde{d})\beta \right\|_2^2 + \sum_{m=1}^{M_{\max}}\phi_m \sum_{i=1}^{(N+1)^{K_m}} L_{im}^2 \beta_m^2, \tag{27.8}$$

where $y := (y_1, y_2, \ldots, y_n)^T$ is the vector of responses, $B(\tilde{d}) = (B(\tilde{d}_1), B(\tilde{d}_2), \ldots, B(\tilde{d}_N))^T$ is an $(N \times (M_{\max} + 1))$ matrix, $\| \cdot \|_2$ is the Euclidean norm.

Rather than using distinct penalty parameters for each derivative in (27.8), solely one penalty parameter ($\phi = \phi_m := \varphi^2$) can be used, and then PRSS becomes:

$$\text{PRSS} \approx \left\| y - B(\widetilde{d})\beta \right\|_2^2 + \phi \left\| L\beta \right\|_2^2, \tag{27.9}$$

where L is a diagonal $((M_{\max}+1) \times (M_{\max}+1))$ matrix and β is an $((M_{\max}+1) \times 1)$ parameter vector that is estimated through the data points. Then the PRSS problem turns into a classical TR problem with $\phi > 0, \phi = \varphi^2$, for some $\varphi \in \mathbb{R}$. The TR problem stated in (27.9) can be handled by using CQP with an appropriate choice of a bound $\widetilde{Z} \in \mathbb{R}$:

$$\begin{aligned} \min_{t,\beta} \quad & t, \\ \text{subject to} \quad & \left\| y - B(\widetilde{d})\beta \right\|_2 \le t, \\ & \left\| L\beta \right\|_2 \le \sqrt{\widetilde{Z}}. \end{aligned} \tag{27.10}$$

It is important to point out that the values of $\widetilde{Z}$ must be obtained through a careful learning process. When the modern methods of *continuous optimization techniques* are applied, the CQP can be expressed by the following basic notation:

$$\begin{aligned} \text{minimize}_{x} \quad & c^T x, \\ \text{subject to} \quad & \left\| D_i x - d_i \right\|_2 \le p_i^T x - q_i, \quad (i = 1, 2, \ldots, k). \end{aligned} \tag{27.11}$$

The parametrical upper bound $\widetilde{Z}$ in a constraint of the CQP and the penalty parameter ϕ in the PRSS are associated, and the value for ϕ can be found via TR. This method of regularization employs a logarithmically-scaled efficiency curve obtained by plotting the optimal solutions to problem (27.9) according to a large (finite) number of parameter values, in the form of points in a coordinate scheme with two axes, where one axis indicates the complexity and the other one is used for the length of the residual vector. As a result of this regularization, an L-curve with some 'kink' (corner) kind of a point on the efficiency boundary, which is regarded to be the closest to the origin, is obtained. Then, this point is often chosen together with the corresponding penalty parameter.

27.3 A Brief Insight into RT Equation and SMAC Algorithm

There are various numerical RT models that have been proposed during the last three decades; however, this section does not aim to treat all these models in a detailed fashion. Instead, RT equation, within the context of 6S model, and SMAC algorithm are briefly introduced based on [45, 46, 49, 60, 61].

The *top of atmospheric* (TOA) signal measured by the sensor is seriously affected by gaseous absorption, molecular scattering and aerosols. In addition, view zenith

Fig. 27.1 Geometry of the problem. θ_s: solar zenith angle, θ_v: viewing zenith angle, Δ_φ: relative azimuth between the Sun and satellite direction

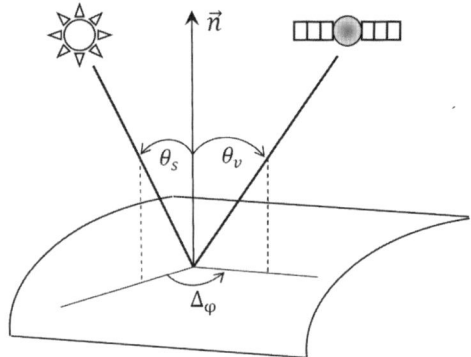

angle (VZA) and solar zenith angle (SZA) also play a major role in determining the effects of the atmosphere. If the zenith angle is far from nadir then the photon must travel through a much larger portion of the atmosphere, and thus the chance of an absorption or scattering event greatly increases.

6S model is the basic code underlying the MODIS atmospheric correction algorithm, in which the surface/atmosphere bidirectional reflectance distribution function is used in order to compute the surface reflectance. If ρ_s is the spectral surface reflectance of the Lambertian and uniform target, the TAO spectral reflectance ρ_{TOA} at the satellite level is expressed as

$$\rho_{TOA}(\theta_s, \theta_v, \Delta\varphi) = T_g(O_3, O_2, CO_2)\left[\rho_R + (\rho_{R+A} - \rho_R)T_g(U_{H_2O}/2)\right.$$

$$\left. + T_{R+A}\frac{\rho_s}{1 - \rho_s S_{R+A}}T_g(U_{H_2O})\right],$$

(27.12)

where θ_s and θ_v are SZA and VZA, respectively, $\Delta\varphi$ is relative azimuth between the Sun and satellite direction, T_g refers to total gaseous transmission (for the solar radiation, H_2O, CO_2, O_2 and O_3 are the principle absorbing gases), ρ_R is the molecular scattering intrinsic reflectance, ρ_{R+A} is the intrinsic reflectance of the molecules and aerosols, T_{R+A} is the transmission of the molecules and aerosols and S_{R+A} is the spherical albedo (the term $1 - \rho_s S_{R+A}$ takes the multiple scattering between the surface and the atmosphere into account). The geometry of the problem is illustrated in Fig. 27.1.

SMAC is a computationally fast and accurate technique for the atmospheric correction of satellite measurements in the solar spectrum and it is based on the parametrization of the RT equations. A separate equation for each of the atmospheric interaction processes is defined and the coefficients of these equations are adjusted to match an accurate RT model. For the minimization of the atmospheric effects and the calculation of surface reflectance, relatively simple set of equations are used in SMAC. The algorithm needs seven input variables: TOA reflectance

(ρ_{TOA}), SZA (θ_s), VZA (θ_v), relative azimuth angle $(\Delta\varphi)$, aerosol optical depth at 550 nm (τ_{550}), water vapor content of the atmosphere (uWV) and the ozone content of the atmosphere (uO$_3$). Then the surface reflectance ρ_s is calculated by the following equation:

$$\rho_s(\theta_s, \theta_v, \Delta\varphi) = \frac{\rho_s'}{T_G T(\theta_s) T(\theta_v) + S\rho_s'},\tag{27.13}$$

where

$$\rho_s' = \rho_{TOA}(\theta_s, \theta_v, \Delta\varphi) - T_G \rho_a(\theta_s, \theta_v, \Delta\varphi).$$

Here, T_G is the gaseous transmittance of the atmosphere, $T(\theta_s)$ and $T(\theta_v)$ are the total downward and upward atmospheric transmittances respectively, S refers to the albedo of the land surface, and ρ_a is the component of measured reflectance produced by the atmosphere. Each of these new variables is calculated internally by SMAC based upon other simple equations and a series of predefined coefficients that are dependent upon the instrument used to acquire the reflectance data. SMAC works on a pixel-by-pixel basis, computing the surface reflectance for each before moving on to the next.

Despite its age, the primary reason why SMAC approach is still used by many groups and projects lies in the simple fact that it can perform an atmospheric correction in a short time. Real-time atmospheric correction of large datasets collected by sensors with high-temporal resolution via detailed RT codes like 6S is operationally impractical. To give an example, processing of a single MSG (Meteosat Second Generation) subset containing 4.4 million pixels for each of the three visible and near-infrared channels takes approximately 25 seconds of runtime for SMAC, but more than 20 minutes for 6S. Although other correction techniques that provide accurate results and are still significantly faster than 6S exist, like QUAC, FLAASH or ATCOR, these are often commercial and provide little access to the inner structure of the algorithm, whereas SMAC is free and open-source, which means that anyone can use and modify SMAC to their own needs.

27.4 MARS and CMARS for the Local Atmospheric Correction of MODIS Images

MODIS is the name of the two scientific instruments operated by NASA: one has been on board to Terra satellite since 1999, and the other to Aqua satellite since 2002. The former passes from north to south across the equator in the morning, whereas the latter passes south to north over the equator in the afternoon, and both capture data in 36 spectral bands ranging in wavelength from 0.4 to 14.4 μm at varying spatial resolutions (detailed info on MODIS is available at http://modis.gsfc.nasa.gov/).

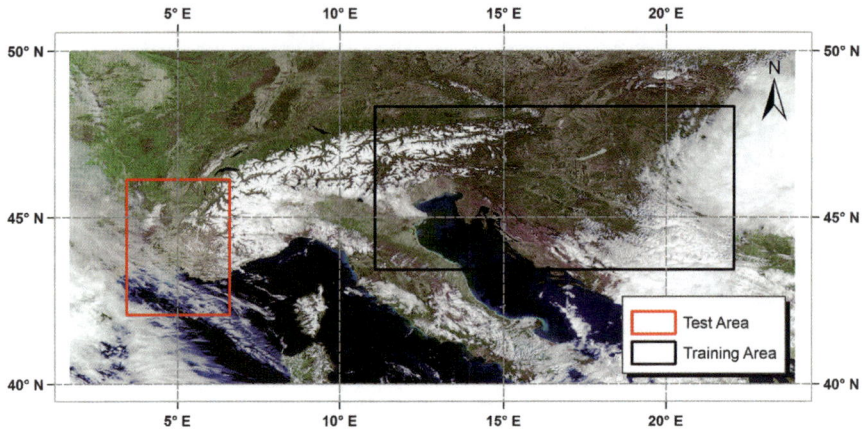

Fig. 27.2 RGB color composite MODIS image of the study area (22.04.2005-S4)

Five sets of images taken by Terra on 11.10.2001 (S1), 10.03.2002 (S2), 06.12.2003 (S3), 22.04.2005 (S4) and 13.01.2006 (S5) over European Alps are used as data set (Fig. 27.2) and the calculations are performed for the 4th reflective solar band (0.545–0.565 μm).

Model training is carried out on S4, and model testing is performed on each set. For model training, a training polygon (including cloud, snow, water, vegetation and bare land cover types) is drawn and 10,000 random points are generated in this polygon. Then, for each point, TOA reflectance value is extracted from Terra Level 1B product (MOD02HKM), while atmospherically corrected surface reflectance value is obtained from Terra Level 2G product (MOD09GA), together with its geographic latitude and longitude. For MARS modelling, the observation equation has the following form:

$$
\begin{bmatrix} y_1 \\ y_2 \\ \vdots \\ y_k \\ \vdots \\ y_N \end{bmatrix} = \begin{bmatrix} ACSref(\boldsymbol{x}_1) \\ ACSref(\boldsymbol{x}_2) \\ \vdots \\ ACSref(\boldsymbol{x}_k) \\ \vdots \\ ACSref(\boldsymbol{x}_N) \end{bmatrix} + \begin{bmatrix} \varepsilon_1 \\ \varepsilon_2 \\ \vdots \\ \varepsilon_k \\ \vdots \\ \varepsilon_N \end{bmatrix}, \tag{27.14}
$$

where $ACSref(\boldsymbol{x})$ is the model function estimated by MARS that gives the atmospherically corrected surface reflectance with $\boldsymbol{x} = (\lambda, \phi, \kappa)^T$, where λ and ϕ refer to the geographic longitude and latitude, respectively, κ is the TOA reflectance value given by MOD02HKM product, N is the total number of observations, y_k represents the atmospherically corrected surface reflectance value given by MOD09GA product at location $\boldsymbol{x}_k = (\lambda_k, \phi_k, \kappa_k)^T$, and ε_k is the measurement error.

For the MARS model building, Salford Systems MARS® Ver. 3 is used [33], and the maximal model obtained in the forward pass stage has the following form:

$$BF1 = \max\{0, \ \kappa - 0.460874\},$$

$$BF2 = \max\{0, \ 0.460874 - \kappa\},$$

$$BF3 = \max\{0, \ \phi - 47.0316\} \times \max\{0, \ \kappa - 0.460874\},$$

$$BF4 = \max\{0, \ 47.0316 - \phi\} \times \max\{0, \ \kappa - 0.460874\},$$

$$BF5 = \max\{0, \ \phi - 47.2438\},$$

$$BF6 = \max\{0, \ 47.2438 - \phi\},$$

$$BF7 = \max\{0, \ \lambda - 22.0669\} \times \max\{0, \ 47.2438 - \phi\},$$

$$BF8 = \max\{0, \ 22.0669 - \lambda\} \times \max\{0, \ 47.2438 - \phi\},$$

$$BF9 = \max\{0, \ \lambda - 19.1247\} \times \max\{0, \ 0.460874 - \kappa\},$$

$$BF10 = \max\{0, \ 19.1247 - \lambda\} \times \max\{0, \ 0.460874 - \kappa\},$$

$$BF11 = \max\{0, \ \kappa - 0.172034\},$$

$$BF12 = \max\{0, \ 0.172034 - \kappa\},$$

$$BF13 = \max\{0, \ \lambda - 22.0642\} \times \max\{0, \ \kappa - 0.172034\},$$

$$BF14 = \max\{0, \ 22.0642 - \lambda\} \times \max\{0, \ \kappa - 0.172034\},$$

$$BF15 = \max\{0, \ \phi - 43.4389\} \times \max\{0, \ 0.172034 - \kappa\}.$$

After the backward pass, the final form of the MARS model is obtained with the GCV value of 0.0049 and the adjusted R^2 of 0.8913, and given as

$$
\begin{aligned}
Y = {} & 0.681299 + 1.39677 \times BF1 - 1.56786 \times BF2 - 2.85519 \times BF3 \\
& - 5.85391 \times BF7 - 0.00163948 \times BF8 + 0.0109712 \times BF10 \\
& - 0.14844 \times BF11 + 16.8514 \times BF13 - 0.00966538 \times BF14 \\
& - 0.0495191 \times BF15.
\end{aligned}
$$

For CMARS, all of the BFs obtained in the forward pass are used for the formulation of PRSS in the form of a CQP problem, and then it is solved by using MOSEK™ software [35]. The obtained CMARS model has the following form:

$$
\begin{aligned}
Y = {} & 0.5043 + 0.6372 \times BF1 - 0.4603 \times BF2 - 2.7299 \times BF3 - 0.0776 \times BF4 \\
& + 0.0038 \times BF5 + 0.0053 \times BF6 - 5.5764 \times BF7 - 0.0015 \times BF8 \\
& - 0.0176 \times BF9 - 0.104 \times BF10 + 0.4940 \times BF11 - 0.6916 \times BF12 \\
& + 14.4283 \times BF13 - 0.0093 \times BF14 - 0.0257 \times BF15.
\end{aligned}
$$

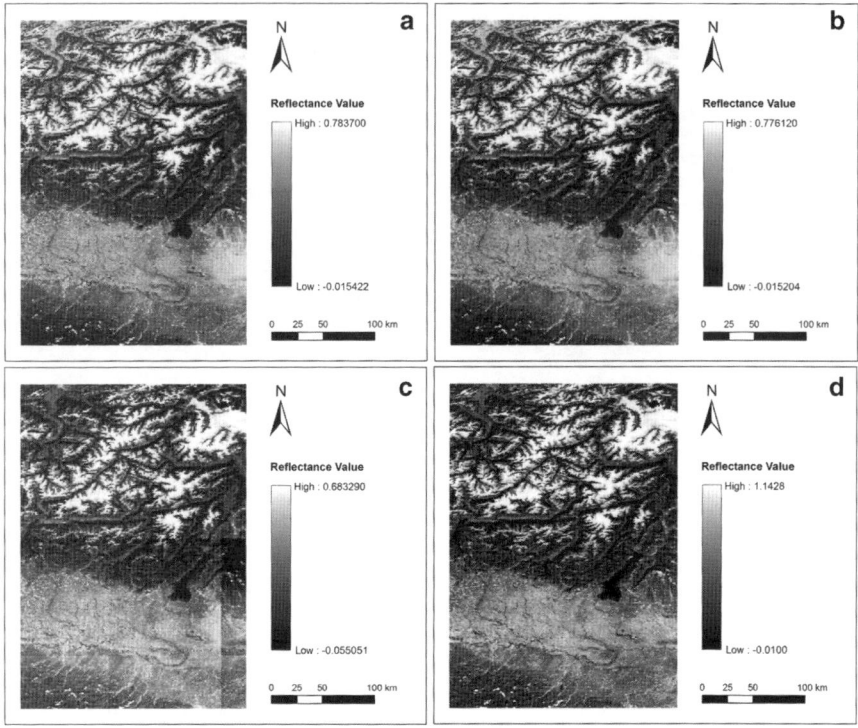

Fig. 27.3 Images obtained by each model for S2 (10.03.2002): (**a**) MARS, (**b**) CMARS, (**c**) SMAC, (**d**) MOD09GA

Table 27.1 RMSE values for MARS, CMARS and SMAC

Test set	MARS	CMARS	SMAC
S1	0.08296	0.08168	0.08629
S2	0.04618	0.04569	0.07150
S3	0.06505	0.06465	0.11834
S4	0.24085	0.24011	0.27556
S5	0.17004	0.16844	0.18680

TOA reflectance values of each test area are also corrected for atmospheric effects by using SMAC algorithm. The necessary atmospheric parameters for SMAC are retrieved from Terra Level 3 daily atmospheric product (MOD08D3). Images constructed by each method for S2 can be seen in Fig. 27.3.

Performance of MARS, CMARS and SMAC methods is tested on the predefined test polygon on each set against surface reflectance values of MOD09GA product in terms of RMSE, and the results are given in Table 27.1.

27.5 Concluding Remarks

In this chapter, MARS and CMARS methods are briefly discussed from an earth science point of view, together with their novel application for the removal of atmospheric effects on MODIS images. The main advantages of MARS and CMARS are that they are nonparametric methods and can be applied without prior assumptions about the underlying distribution. In addition, due to their space partitioning and adaptive spline fitting characteristics, they do not require gridding and large amounts of data with large gaps can be handled.

The results obtained by both methods are also compared with the results of SMAC algorithm, simplified version of a conventional RT model, 6S. Even though CMARS seems to perform slightly better than MARS, it is more model-based and uses the advantages of modern techniques of continuous optimization, especially CQP. On the other hand, according to the test results, both MARS and CMARS significantly outperform SMAC. One interesting finding of the preliminary results is that although the model is trained on a single image (S4) and then applied on each test area, the obtained MARS and CMARS models have better performances than SMAC, which should be investigated further.

Within the light of the preliminary results, MARS and CMARS seem to be alternative tools for atmospheric correction and can also be utilized for other problems related with different research areas in earth sciences. The future work will focus on the application of MARS and CMARS, as well as the new RCMARS [41] (Robust CMARS: the refined version of CMARS by applying robust optimization technique in order to cope with data uncertainty), on larger data sets with different wavelength bands in order to analyze their performances in a more detailed manner.

References

1. ACORN 4.0: User's Guide Stand-alone Version, Version 4.0, Analytical Imaging and Geophysics LLC (Jan 2002)
2. Alp, Ö.S., Büyükbebeci, E., Çekiç, A.İ., Özkurt, F.Y., Taylan, P., Weber G.-W.: CMARS and GAM & CQP-Modern optimization methods applied to international credit default prediction. J. Comput. Appl. Math. **235**, 4639–4651 (2011)
3. Anderson, G.P., Pukall, B., Allred, C.L., Jeong, L.S., Hoke, M., Chetwynd, J.H., Adler-Golden, S.M., Berk, A., Bernstein, L.S., Richtsmeier, S.C., Acharya, P.K., Matthew, M.W.: FLAASH and MODTRAN4: state-of-the-art atmospheric correction for hyperspectral data. IEEE Aerosp. Conf. Proc. **4**, 177–181 (1999)
4. Balshi, M.S., McGuire, A.D., Duffis, P., Flannigan, M., Walsh, J., and Melillo, J.: Assessing the response of area burned to changing climate in western boreal North America using a multivariate adaptive regression splines (MARS) approach. Glob. Chang. Biol. **15**, 578–600 (2009)
5. Banks, D.L., Olszewski, R.T., Maxion, R.A.: Comparing methods for multivariate nonparametric regression. Commun. Stat. Simul. Comput. **32**, 541–571 (2003)

6. Ben-Ari, E.N., Steinberg, D.M.: Modeling data from computer experiments: an empirical comparison of kriging with MARS and projection pursuit regression. Qual. Eng. **19**, 327–338 (2007)
7. Berk, A., Bernstein, L.S., Anderson, G.P., Acharya, P.K., Robertson, D.C., Chetwynd, J.H., Adler-Golden, S.M.: MODTRAN cloud and multiple scattering upgrades with application to AVIRIS. Remote Sens. Environ. **65**, 367–375 (1998)
8. Bernstein, L., Adler-Golden, S., Sundberg, R., Levine, R., Perkins, T., Berk, A., Ratkowski, A.J., Felde, F., Hoke, M.L.: A new method for atmospheric correction and aerosol optical property retrieval for VISSWIR multi-and hyperspectral imaging sensors: QUAC (QUick Atmospheric Correction). In: Proceedings of IGARSS 2005, 25th International Geoscience and Remote Sensing Symposium, Seoul, 25–29 July 2005. ISBN: 0-7803-9050-4
9. Briand, L.C., Freimut, B., Vollei, F.: Using multiple adaptive regression splines to support decision making in code inspections. J. Syst. Softw. **73**, 205–217 (2004)
10. Cherkassky, V., Mulier, F.: Learning from Data: Concepts, Theory, and Methods, 2nd ed. Wiley, Hoboken (2007)
11. Chou, S.M., Lee, T.S., Shao, Y.E., Chen, I.F.: Mining the breast cancer pattern using artificial neural networks and multivariate adaptive regression splines. Expert Syst. Appl. **27**, 133–142 (2004)
12. Cook, N.R., Zee, R.Y.L., Ridker, P.M.: Tree and spline based association analysis of gene-gene interaction models for ischemic stroke. Stat. Med. **23**, 1439–1453 (2004)
13. Deichmann, J., Esghi, A., Haughton, D., Sayek, S., Teebagy, N.: Application of multiple adaptive regression splines (MARS) in direct response modelling. J. Interact. Mark. **16**, 15–27 (2002)
14. Durmaz, M., Karslıoğlu, M.O.: Non-parametric regional VTEC modeling with multivariate adaptive regression B-Splines. Adv. Space Res. **48**, 1523–1530 (2011)
15. Durmaz, M., Karslıoğlu, M.O., Nohutcu, M.: Regional VTEC modeling with multivariate adaptive regression splines. Adv. Space Res. **46**, 180–189 (2010)
16. Felicísimo, A.M., Gomez, A., Muñoz, J., Schnabel, S., Gonçalves, A.: Potential distribution of forest species in dehesas of extremadura (Spain). In: Schnabel, S., Gonçalves, S. (eds.) Sustainability of Agrosilvopastoral Systems: Dehesa Montados, vol. 37, pp. 231–246. Catena Verlag, Reiskirchen (2005)
17. Friedman, J.H.: Multivariate adaptive regression splines. Ann. Stat. **19**, 1–141 (1991)
18. Friedman, J.H.: Fast MARS. Technical Report No. 110, Laboratuary for Computational Statistics, Department of Statistics, Stanford University (May 1993)
19. Gao, B., Montes, M.J., Davis, C.O., Goetz, A.F.H.: Atmospheric correction algorithms for hyperspectral remote sensing data of land and ocean. Remote Sens. Environ. **113**, S17–S24 (2009)
20. Guitérrez, A.G., Schnabel, S., Contador, J.F.L.: Using and comparing two nonparametric methods (CART and MARS) to model the potential distribution of gullies. Ecol. Model. **220**, 3630–3637 (2009)
21. Hastie, T., Tibshirani, R., Friedman, J.: The Elements of Statistical Learning: Data Mining, Inference, and Prediction, 2nd edn. Springer, New York (2009)
22. Heikkinen, R.K., Luoto, M., Kuussaari, M., Toivonen, T.: Modelling the spatial distribution of a threatened butterfly: impacts of scale and statistical technique. Landsc. Urban Plan. **79**, 347–357 (2007)
23. Henne, P.D., Hu, F.S., Cleland, D.T.: Lake-effect snow as the dominant control of mesic-forest distribution in Michigan, USA. J. Ecol. **95**, 517–529 (2007)
24. Herman, B.M., Browning, S.R.: The effect of aerosols on the Earth-atmosphere albedo. J. Atmos. Sci. **32**, 1430–1445 (1975)
25. Jekabsons, G.: Adaptive basis function construction: an approach for adaptive building of sparse polynomial regression models. In: Zhang, Y. (ed.) Machine Learning. In-Tech, Croatia (2010)
26. Kotchenova, S.Y., Vermote, E.F.: Validation of a vector version of the 6S radiative transfer code for atmospheric correction of satellite data: part II: homogeneous lambertian and anisotropic surfaces. Appl. Opt. **46**, 4455–4464 (2007)

27. Kotchenova, S.Y., Vermote, E.F., Matarrese, R., Klemm, F.M.: Validation of a vector version of the 6S radiative transfer code for atmospheric correction of satellite data: part I: path radiance. Appl. Opt. **45**, 6762–6774 (2006)
28. Krzyścin, J.W., Eerme, K., Janouch, M.: Long-term variations of the UV-B radiation over Central Europe as derived from the reconstructed UV time series. Ann. Geophysicae **22**, 1473–1485 (2004)
29. Leathwick, J.R., Rowe, D., Richardson, J., Elith, J., Hastie, T.: Using multivariate adaptive regression splines to predict the distributions of New Zealand's freshwater diadromous fish. Freshw. Biol. **50**, 2034–2052 (2005)
30. Leathwick, J.R., Elith, J., Hastie, T.: Comparative performance of generalized additive models and multivariate adaptive regression splines for statistical modelling of species distributions. Ecol. Model. **199**, 188–196 (2006)
31. Lee, T., Chiu, C., Chou, Y., Lu, C.: Mining the customer credit using classification and regression tree and multivariate adaptive regression splines. Comput. Stat. Data Anal. **50**, 1113–1130 (2006)
32. Lhermitte, S., Verbesselt, J., Verstraeten, W.W., Coppin, P.: A comparison of time series similarity measures for classification and change detection of ecosystem dynamics. Remote Sens. Environ. **115**, 3129–3152 (2011)
33. MARS®: Salford Systems (software available at http://www.salfordsystems.com)
34. Matthew, M., Adler-Golden, S., Berk, A., Felde, G., Anderson, G., Gorodetzky, D., Paswaters, S., Shippert, M.: Atmospheric correction of spectral imagery: evaluation of the FLAASH algorithm with AVIRIS data. Proc. SPIE **5093**, 474–482 (2003)
35. MOSEK™: (software available at http://www.mosek.com)
36. Mukhopadhyay, A., Iqbal, A.: Prediction of mechanical property of steel strips using multivariate adaptive regression splines. J. Appl. Stat. **36**, 1–9 (2009)
37. Muñoz, J., Felicísimo, A.M.: Comparison of statistical methods commonly used in predictive modeling. J. Veg. Sci. **15**, 285–292 (2004)
38. Nikolakopoulos, K.G., Vaiopoulos, D.A., Skianis, G.A.: A comparative study of different atmospheric correction algorithms over an area with complex geomorphology in Western Peloponnese, Greece. In: Proceedings of International Geoscience and Remote Sensing Symposium, 2002 (IGARSS'02), vol. 4, Toronto, 24–28 June 2002, pp. 2492–2494. IEEE (2002)
39. Osei-Bryson, K.M., Ko, M.: Exploring the relationship between information technology investments and firm performance using regression splines analysis. Inform. Manag. **42**, 1–13 (2004)
40. Özmen, A.: Robust conic quadratic programming applied to quality improvement: a robustification of CMARS, M.Sc. thesis, Middle East Technical University (2010)
41. Özmen, A., Weber, G.-W., Batmaz, İ., Kropat, E.: RCMARS: Robustification of CMARS with different scenarios under polyhedral uncertainty set. Commun. Nonlin. Sci. Numer. Simulat. **16**, 4780–4787 (2011)
42. Permuter, H., Francos, J., Jermyn, I.: A study of Gaussian mixture models of color and texture features for image classification and segmentation. Pattern Recognit. **39**, 695–706 (2006)
43. Pouteau, R., Rambal, S., Ratte, J.P., Gogé F., Joffre, R., Winkel, T.: Downscaling MODIS-derived maps using GIS and boosted regression trees: the case of frost occurrence over the arid Andean highlands of Bolivia. Remote Sens. Environ. **115**, 117–129 (2011)
44. Prasad, A.M., Iverson, L.R., Liaw, A.: Newer classification and regression tree techniques: bagging and random forests for ecological prediction. Ecosystems **9**, 181–199 (2006)
45. Proud, S.R., Fensholt, R., Rasmussen, M.O., Sandholt, I.: A comparison of the effectiveness of 6S and SMAC in correcting for atmospheric interference of meteosat second generation images. J. Geophysical Res. **115** (2010). doi:10.1029/2009JD013693
46. Proud, S.R., Rasmussen, M.O., Fensholt, R., Sandholt, I., Shisanya, C., Mutero, W., Mbow, C., Anyamba, A.: Improving the SMAC atmospheric correction code by analysis of meteosat second generation NDVI and surface reflectance data. Remote Sens. Environ. **114**, 1687–1698 (2010)

47. Put, R., Xua, Q.S., Massart, D.L., Vander Heyden, Y.: Multivariate adaptive regression splines (MARS) in chromatographic quantitative structure-retention relationship studies. J. Chromatogr. A **1055**, 11–19 (2004)
48. Quirós, E., Felicísimo A.M., Cuartero, A.: Testing multivariate adaptive regression splines (MARS) as a method of land cover classification of TERRA-ASTER satellite images. Sensors **9**, 9011–9028 (2009)
49. Rahman, H., Dedieu, G.: SMAC: a simplified method for the atmospheric correction of satellite measurements in the solar spectrum. Int. J. Remote Sens. **15**, 123–143 (1994)
50. Richards, J.A., Jia, X.: Remote Sensing Digital Image Analysis: An Introduction, 4th edn. Springer, Berlin (2006)
51. Richter, R.: ATCOR-2/3 User Guide (Version 7.1), DLR-German Aerospace Center, Remote Sensing Data Center (2010)
52. Richter, R., Schlaepfer, D.: Geo-atmospheric processing of airborne imaging spectrometry data: part 2: atmospheric/topographic correction. Int. J. Remote Sens. **23**, 2631–2649 (2002)
53. Samui, P., Das, S., Kim, D.: Uplift capacity of suction caisson in clay using multivariate adaptive regression spline. Ocean Eng. **38**, 2123–2127 (2011)
54. Schowengerdt, R.A.: Remote Sensing: Models and Methods for Image Processing, 3rd edn. Academic Press, Burlington (2007)
55. Steele, B.M.: Combining multiple classifiers: an application using spatial and remotely sensed information for land cover type mapping. Remote Sens. Environ. **74**, 545–556 (2000)
56. Tanre, D., Deroo, C., Duhaut, P., Herman, M. , Morcrette, J.J.: Description of a computer code to simulate the satellite signal in the solar spectrum: the 5S code. Int. J. Remote Sens. **11**, 659–668 (1990)
57. Taylan, P., Weber, G.-W., Beck, A.: New approaches to regression by generalized additive models and continuous optimization for modern applications in finance, science and technology. Optim. J. Math. Program. Oper. Res. **56**(5–6), 675–698 (2007)
58. Taylan, P., Weber, G.-W., Liu, L., Yerlikaya-Özkurt, F.: On the foundations of parameter estimation for generalized partial linear models with B-splines and continuous optimization. Comput. Math. Appl. **60**, 134–143 (2010)
59. Taylan, P., Weber, G.-W., Yerlikaya-Özkurt, F.: A new approach to multivariate adaptive regression splines by using Tikhonov regularization and continuous optimization. Top **18**(2), 377–395 (2010)
60. Vermote, E.F., Tanré, D., Deuzé, J.L., Herman, M., Morcrette, J.: Second simulation of the satellite signal in the solar spectrum, 6S: an overview. IEEE Trans. Geosci. Remote Sens. **35**, 675–686 (1997)
61. Vermote, E.F., El Saleous, N.Z., Justice, J.O.: Atmospheric correction of MODIS data in the visible to middle infrared: first results. Remote Sens. Environ. **83**, 97–111 (2002)
62. Vidoli, F.: Evaluating the water sector in Italy through a two stage method using the conditional robust nonparametric frontier and multivariate adaptive regression splines. Eur. J. Oper. Res. **212**, 583–595 (2011)
63. Viscarra Rossel, R.A., Behrens, T.: Using data mining to model and interpret soil diffuse reflectance spectra. Geoderma **158**, 46–54 (2010)
64. Weber, G.-W., Batmaz, İ., Köksal, G., Taylan, P., Yerlikaya-Özkurt, F.: CMARS: a new contribution to nonparametric regression with multivariate adaptive regression splines supported by continuous optimization. Inverse Probl. Sci. Eng. (2011). doi:10.1080/17415977.2011.624770
65. Xu, M., Watanachaturaporn, P., Varshney, P.K., Arora, M.K.: Decision tree regression for soft classification of remote sensing data. Remote Sens. Environ. **97**, 322–336 (2005)
66. York, T.P., Eaves, L.J., van den Oord, E.J.C.G.: Multivariate adaptive regression splines: a powerful method for detecting disease-risk relationship differences among subgroups. Stat. Med. **25**, 1355–1367 (2006)
67. Yu, J., Ekström, M.: Multispectral image classication using wavelets: a simulation study. Pattern Recognit. **36**, 889–898 (2003)

68. Yuanliu, X., Runsheng, W., Shengwei, L., Suming, Y., Bokun, Y.: Atmospheric correction of hyperspectral data using MODTRAN model. In: Remote Sensing of the Environment: 16th National Symposium on Remote Sensing of China: Proceedings of SPIE, pp. 7123, 2008. doi: 10.1117/12.815552
69. Zakeri, I.F., Adolph, A.L., Puyau, M.R., Vohra, F.A., Butte, N.F.: Multivariate adaptive regression splines models for the prediction of energy expenditure in children and adolescents. J. Appl. Physiol. **108**, 128–136 (2010)
70. Zareipour, H., Bhattacharya, K., Cañizares, C.A.: Forecasting the hourly Ontario energy price by multivariate adaptive regression splines. In: Power Engineering Society General Meeting, Montreal. IEEE (2006)
71. Zhang, W., Wang, W., Wu, F.: The application of multi-variable optimum regression analysis to remote sensing imageries in monitoring landslide disaster. Energy Procedia **16**, 190–196 (2012)
72. Zhou, Y., Leung, H.: Predicting object-oriented software maintainability using multivariate adaptive regression splines. J. Syst. Softw. **80**, 1349–1361 (2007)

Chapter 28
Modeling Ethanol Investment Decisions

C.-Y. Cynthia Lin

28.1 Introduction

Recently the support of biofuel production has been a politically sensitive topic. Politicians have pushed for support for fuel ethanol production as an environmentally friendly alternative to imported oil, as well as a way to boost farm profits and improve rural livelihoods. Several government policies actively promote ethanol production via tax incentives and mandates, and these policies are blamed for rising food prices around the world [14]. It is important to understand the factors that have motivated the significant local investments in the ethanol industry that have been made since the mid-1990s both in the U.S. and worldwide.

Fuel ethanol has been in use in the United States since the time of the Model T Ford (the original flex-fuel vehicle), and while the United States passed Brazil in ethanol production in 2005, today ethanol is mostly relegated to status as a gasoline additive. The first US ethanol boom began as a result of the oil embargoes in 1973 and 1979. The desire for more energy self-sufficiency, the resulting legislation (in the form of federal income tax credits and blender's credits that continue today), and the phase out of leaded gasoline led to the construction of 153 new plants by 1985 [4]. These plants were tiny by today's standards, with an average capacity of eight million gallons per year, and by 1991 only 35 were still operational due to poor business judgment and bad engineering [4, 18].

The second US ethanol boom began in the mid-1990s and hit full-stride by the early 2000s. Several factors contributed to this most recent boom. The Clean Air Act of 1990 mandated use of oxygenates in gasoline, of which ethanol is one, and the subsequent phase out and ban of MTBE as additive beginning in the late 1990s

C.-Y.C. Lin (✉)
University of California at Davis, Davis, CA, USA
e-mail: cclin@primal.ucdavis.edu

A.A. Pinto and D. Zilberman (eds.), *Modeling, Dynamics, Optimization and Bioeconomics I*, Springer Proceedings in Mathematics & Statistics 73, DOI 10.1007/978-3-319-04849-9__28, further increased demand for ethanol. Additionally the Renewable Fuel Standards of the Energy Policy Act of 2005 mandated ethanol production floors beginning in 2007, which rise to 36 billion gallons per year in 2033. Over this time period, the number of ethanol plants rose from 35 plants in 1991, to 50 in 1999, to 192 in September of 2010 for a total capacity of 13 billion gallons per year.

In addition to the policy and demand-side contributors to the recent ethanol boom, this new industry growth has been accompanied by changes in plant management and technology. Most significantly, the average capacity of plants in our focus region was 62 million gallons per year in 2008 up from 8 million gallons per year in 1985. In the mid-1990s the industry began designing more efficient plants, which use natural gas instead of coal as fuel [4]. Ownership is also shifting to streamlined corporate owners with multiple plants. Historically, farmer-owned plants had a large share of the market, though by 2007 only 11 % of new capacity was farmer owned, while the largest five corporations had 42 % of capacity in 2008 [6].

This recent boom, in addition to industry changes in technology and ownership structure, beg an analysis of investment decisions. Most ethanol plants use corn as a feedstock, and thus are located in the Midwestern United States, where the majority of the corn in the US is grown; these plants are the focus of our study. Since biofuels have been touted as a way to enhance profits in rural areas, where grain prices have remained stagnant over time, it is important to determine what factors affect decisions about when and where to invest in building new ethanol plants. In [11], we model this decision in this paper using both reduced-form and structural models.

Even when excluding the U.S., which was the country with the largest fuel-ethanol production in 2009, the fuel-ethanol industry has been growing rapidly in the rest of the world (ROW). Ethanol-producing countries in the ROW include Brazil, Canada, China, and Thailand, as well as countries in Europe. There are 191 fuel-ethanol plants in the ROW, which is a little more than in the U.S., and 82 % of them were built after 2005. In [12,13], we estimate a model of the investment timing game in ethanol plants worldwide that allows for the choice among different feedstocks.

In Europe, 20 countries have fuel ethanol production and most of the fuel ethanol plants were built after 2000. The development of European biofuel is based on two Directives: the Renewable Energy Directive (RED) of 2003/30/EC sets indicative targets of 2 % renewable fuels in transport by 2005 and 5.75 % by 2010 but is not legally binding, and the RED of 2009/28/EC is made mandatory and therefore legally binding. The main policies fuel ethanol policies in Europe include a tax credit, a blending mandate and R&D support. Most of the policies were implemented after 2003. Empirical research shows that the effects of policies for the U.S. fuel ethanol production are positive [9, 11, 17], however, whether the stimulation effects of the government policies play the same role in Europe is not yet clear, especially for the different varieties of feedstocks.

There is a related literature on food manufacturing location decisions which begins with a basic model of determinants of manufacturing establishment growth. One example is Goetz's [7] analysis of the determinants of rural food manufacturing establishment growth. He considers the effects of the following factors: access to output markets, labor force composition and quality, transportation infrastructure, government intervention, and availability of raw materials. In his model, location decisions involve a two-step process where regions are first chosen for broader consideration, and then the choice is narrowed within each region.

The decision to build a fuel ethanol plant involves a dynamic decision-making process. Investors have to consider the entire stream of per-period payoffs from now into the future. Because values of the state variables can change over time, for example if ethanol and feedstock prices change, or if there is a chance that neighboring plants might be constructed and start production, there is an option value to waiting that makes the decision dynamic rather than static.

When the decision of an investor is affected by neighbors' investment decisions, it becomes a more complicated decision-making problem. There are two types of effects that add a strategic (or non-cooperative) dimension to the potential entrants' investment timing decisions. The first type of effect is a competition effect, which arises if there is more than one ethanol plant located in one region so that these plants compete in feedstock supply when they choose the same feedstock or compete in local fuel ethanol market given limited demand. The competition effect deters ethanol plants for entering in regions where there are other ethanol plants already present. The second type of effect is an agglomeration effect: if there are several ethanol plants located in the same region, the existing plants may have developed transportation and marketing infrastructure and/or an educated work force that new plants can benefit from. The agglomeration effect induces an ethanol plant to locate near other plants, since a fuel ethanol plant benefits from the existence of other plants. Owing to both competition and agglomeration effects, the dynamic decision-making problem faced by the potential ethanol plants is not merely a single-agent problem, but rather can be viewed as a non-cooperative game in which plants behave strategically and base decisions on their neighbors' strategies.

There are few studies that specifically address the location determinants of ethanol plants in the United States and this is the first that econometrically models the entry location decision. Sarmiento and Wilson [17] use a cross-sectional discrete choice model to analyze agricultural characteristics and spatial dimensions that determine plant location. Similarly, Lambert et al. [9] use a cross-sectional discrete choice model spatial clustering to look at factors that affect the presence of ethanol plants and proposed plants in a given county, and also isolate clusters that may attract investment. Finally, Haddad et al. [8] model state-by-state spatial determinants of plant location, and find significant differences across states.

While these studies analyze similar factors as Goetz [7], they are cross-sectional and so cannot address investment timing decisions and thus only focus on location determinants, providing a starting point for our analysis as far as identifying potentially important exogenous factors. However, the results of these studies are not always qualitatively similar because of the different empirical specifications and regional foci. For example, the first two studies find that access to corn is an important location determinant. However, Haddad et al. [8] do not find access to corn significant, though they note that following location theory (see [7]), firms might choose a region with a lot of corn, and in the second stage will make their location decision based on other factors; this two-stage decision-making process is why they find differences in location determinants across states.

None of the studies adequately addresses the potential competition between plants in the location decision. Haddad et al. [8] do not include potential spatial competition between plants as an explanatory variable, perhaps since most competing plants are not located in the same county. Lambert et al. [9] include plants established before 2000 as an explanatory variable and find a negative impact on new plants and announcements between 2000 and 2007, though there is no analysis of spatial relationship or relative timing on those (potential) entrants. Sarmiento and Wilson [17] employ a cross-sectional model of plant location with a spatially lagged dependent variable in order to estimate the competitive effect between plants. They find a large negative effect of a nearby plant on the probability of another plant locating nearby, and furthermore, that this effect decreases with distance. It is important to note however that these competitive effects only describe the relationship between existing plants; neither of these models has a time element and without panel data it is not possible to analyze the effect of competition on entry.

This article reviews some of the papers my co-authors and I have written analyzing what factors affect the decision to invest in building new ethanol plants using a dynamic structural econometric model of the investment timing game. This work improves upon the previous literature by estimating a dynamic model with panel data, and by directly estimating the effect of covariates on the investment decision itself. The results of my research will help determine which policies and factors can promote fuel-ethanol industry development.

28.2   Theoretical Model

We model whether or not there is an investment in an ethanol plant i in county k in year t. Investment in an ethanol plant is irreversible and, in each year t, all investment decisions are made simultaneously. I_{ikt} is an indicator of whether there is an investment in a new ethanol plant i in county k in year t.

Because the profits from investing in building a new ethanol plant depend on market conditions such as feedstock price that vary stochastically over time, a potential entrant that hopes to make a dynamically optimal decision would need to account for the option value to waiting before making this irreversible investment [3]. The dynamic decision-making problem faced by a potential entrant is even more complicated when its profits are affected not only by exogenous market conditions, but also by the existence or potential entry of nearby plants.

The covariates X_{kt} describe the state of the input and output markets. The state variable a_{kt} is the number of other plants in the county. The investment decision in each county k in year t depends on the state of the county $\Omega_{kt} = (a_{kt}, X_{kt})$ through its effect on the profits from investing. The state variables a_{kt} and X_{kt} evolve according to a first-order Markov process and summarize the direct effect of the past on the current environment.

The profit from investing in an ethanol plant in year t is denoted by $\pi(a_{kt}, X_{kt})$, which is the expected revenue from the plant minus the expected costs. The value function for a potential entrant i in county k in period t can be written as:

$$V(a_{kt}, X_{kt}) = \max\{\pi(a_{kt}, X_{kt}), \beta V^c(a_{kt}, X_{kt}). \tag{28.1}$$

The payoff will depend on whether the potential entrant decides to wait and not build an ethanol plant in year t, indicated by $I_{ikt} = 0$, or to build an ethanol plant, indicated by $I_{ikt} = 1$. If the potential entrant chooses to build an ethanol plant in year τ, he will receive the payoff $\pi(a_{kt}, X_{kt})$. If the potential entrant chooses not to invest at time t, then he receives the discounted continuation value $\beta V^c(a_{kt}, X_{kt})$, where β is a discount rate, and $V^c(a_{kt}, X_{kt})$ is the continuation value to waiting. Whether or not there is a new investment depends on which of these options yield the highest payoff in that particular period.

The continuation value $V^c(a_{kt}, X_{kt})$ is the expected value of the next period's value function, conditional on not building an ethanol plant in the current period, and is given by:

$$V^c(a_{kt}, X_{kt}) = E\left[V(a_{k,t+1}, X_{k,t+1})|a_{kt}, X_{kt}, I_{ikt} = 0\right]. \tag{28.2}$$

There will be a new investment in an ethanol plant i in county k in year t if the profits from the investment are greater than the continuation value from waiting.

The state variable a_{kt} represents the strategic interactions among plants. Plants located nearby have the potential to create positive and negative externalities for entering plants. In terms of positive externalities, such as agglomeration effects, there could be benefits for a new plant from taking advantage of the transportation or marketing infrastructure or the educated work force already developed by an existing plant (see e.g., [5, 7, 9]). In terms of negative (pecuniary) externalities, plants could compete in both the output and input markets. For example, Sarmiento and Wilson [17] explain that their estimated negative competitive effect is due to competition in feedstock procurement. If corn markets are localized a shift in demand from a new plant could cause increased prices.

The state variables in X_{kt} model the expected profits from the sale of ethanol, which include variables describing prices in the output market as well as the costs of production.

28.3 Econometric Methodology

In [11], we first estimate a reduced-form discrete choice model by regressing the probability of investment in an ethanol plant on the covariates using a fixed-effects logit model. It is logical to begin with the fixed-effects model since unobservable county characteristics might explain their ability to attract investment in an ethanol plant. The major advantage of a fixed-effects model is the addition of the county-specific effect. This allows us to control for unobservable county traits, such as openness to or promotion of business, that remain fixed over time. This is particularly important since the resolution of our data is not ideal and some variables are not observed at the county level. We next specify the strategic variable, other plants, as capacity instead of count. We hypothesize that larger or smaller competing capacities could have different effects. Third, we group the competing plants by type: singlets (plants that have no sister plant with the same owner), ethanol-only firms, and conglomerates. We hypothesize that different types of operators may produce different externalities (either positive and negative) towards potential entrants. We estimate a fourth model using entry of different type of plant (singlet, ethanol-only, and conglomerate) as the dependent variable in a pooled multinomial logit model.

In [11], we follow up the reduced-form models of investment in an ethanol plant with a structural model. We use a structural model for several reasons. First, it is interesting to estimate the effect of the state variable the expected payoff from investing in an ethanol plant. In the reduced-form model, we estimate the effect of these variables on the probability of investment. Second, the structural model makes it possible to estimate the strategic interaction between different regions that produce corn and could invest in ethanol plants. Thirdly, the structural model explicitly models the dynamic investment decision, including the continuation value to waiting. In contrast, the reduced-form model only estimates the per-period probability of investment.

As explained by Reiss and Wolak [16], a structural econometric model is one that combines economic theory with a statistical model, enabling us to estimate structural parameters. Incorporating firm dynamics into structural econometric models enhances our understanding of behavior and also enables us to estimate structural parameters which have a transparent interpretation within the theoretical model that frames the empirical investigation [1].

Dynamic discrete choice structural models are useful tools in the analysis of economic and social phenomena whenever strategic interactions are an important aspect of individual behavior. In the ethanol market, because a firm's costs and market demand hinge on the structure of market, a firm's decision depends on its conjecture about competitors' behavior. This type of model assumes agents are forward looking and maximize the expected discounted value of the entire stream of payoffs. Agents are assumed to make decisions based only on historic information directly related to current payoffs, and history only influences current decisions insofar as it impacts a state variable that summarize the direct influence of the past on current payoffs.

Recent papers have developed techniques for estimating dynamic games between multiple agents. Pakes et al. [15] illustrated a dynamic entry/exit game where the structural parameters could be estimated semi-parametrically. Bajari et al. [2] added to this literature by describing the estimation of the parameters in a dynamic game with continuous control variables.

In [11], we estimate a model of the investment timing game in corn ethanol plants in the United States. This model follows my previous work estimating a structural econometric model of the multi-stage dynamic investment timing game in offshore petroleum production [10], which is based on an econometric model developed by Pakes et al. [15]. In Lin [10], I build on the work of Pakes et al. [15] on discrete games of entry and exit by examining sequential investments with a finite horizon. The econometric estimation technique takes place in two steps. In the first step, the continuation value is estimated nonparametrically and used to form the model's estimate of the investment probabilities. In the second step, the investment probabilities predicted by the model are matched with the empirical investment probabilities in the data using generalized method of moments.

In [12, 13], we estimate a model of the investment timing game in ethanol plants worldwide that allows for the choice among different feedstocks. This research differs from previous studies of the investment and location of ethanol plants because it models the decision as a dynamic one rather than a static one, because it allows for the choice among multiple feedstocks rather than just one feedstock such as corn, because its strategic framework allows the estimation of strategic interactions among plants, and because it uses international data rather than data from the U.S.

28.4 Conclusions

The results of both the reduced-form and structural estimation in [11] indicate that there is an important strategic component to investment in ethanol plants. This net-negative effect may be due to localized competition, though we also find that plant identity matters as far as the strength and sign of this effect. We cannot explore the plant identity question (or exit) more in our current framework, but this is subject of ongoing work.

In [11], we also find that intensity of corn production is important in determining local investment in both models. Corn is bulky and transportation is not cheap, so it is beneficial for plants to locate where they have good access to feedstock.

In [11], we find mixed results of the effects of input and output prices across the different models and specifications. This inconsistency is potentially due to the data resolution, which, at state and national level, is not ideal. That said, we were still able to find some effect of prices indicating that (1) they do matter, and (2) even with less than ideal data source, our model is strong enough to tease out some of these effects.

In [11], we also find mixed results of the policy variables. In the reduced-form specification, we find that the state producer tax breaks are important, and the MTBE ban is significant in the structural model. The differences here may be due to the models themselves. The MTBE ban affects the market via increased ethanol demand and higher expected prices; thus in the structural model it leads to increased expected profits from investment. The producer support/tax break only applies to some plants, and only to some of the production, so we may not have identified this effect in the structural model, where the profit function is the same for each county. Further analysis of the effects of policies on investment in ethanol plants is subject of ongoing work.

In [12, 13], we first construct a dynamic discrete choice model for a potential fuel ethanol plant in which the investor maximizes its present discounted value of its entire stream of payoffs, and in which the decisions of other plants in the same local market affect an investor's decision.

The innovative features of our model are the consideration of interactions between fuel ethanol plants and the dynamic decision making framework. Once the dynamic decision making process has been fully described, the effects of economic, policy and strategic variables on profit can be estimated through a semiparametric approach.

Our results in [13] show that the potential investor considers various exogenous conditions: higher ethanol prices, gasoline prices, plenty of feedstocks and government support are very helpful to improve the profits, while a high natural gas price would harm the profits although some of the coefficients on natural gas are not significant. The negative competition effects of livestock vary across the choices of feedstocks. There is a negative effect on profits for all types of fuel ethanol plants which come from existing plants. In addition, the strategic interactions resulting from competition and agglomeration effects from other potential entrants are not obvious.

Thus, according to our results, we find that, in the United States, competition between plants is enough to deter local investments. We also find that availability of feedstock is important in determining plant location. We also find that in the United States state producer tax credits and the federal MTBE ban have a positive effect on ethanol investment [11]. In Europe, competition between plants deters local investments and ethanol support policies encourage investments [13]. In Canada, competition between plants is enough to deter local investments, the availability of feedstock is important in determining plant location, and the effects of policy support for wheat-based plants are significant [12].

References

1. Aguirregabiria, V., Mira, P.: Dynamic discrete choice structural models: a survey. J. Econ. **156**(1), 38–67 (2010)
2. Bajari, P., Benkard, C.L., Levin, J.: Estimating dynamic models of imperfect competition. Econometrica **75**(5), 1331–1370 (2007)
3. Dixit, A.K., Pindyck, R.S.: Investment under Uncertainty. Princeton: Princeton University Press, Princeton (1994)
4. DOE [Department of Energy]: Energy timelines: ethanol (2008). Revised June 2008. Accessed Jan 2009. http://www.eia.gov/kids/energy.cfm?page=tl_ethanol
5. Ellison, G., Glaeser, E.L.: The geographic concentration of industry: does natural advantage explain agglomeration? Am. Econ. Rev. **89**(2), 311–316 (1999)

6. FTC [Federal Trade Commission]: Report on Ethanol Market Concentration (2008)
7. Goetz, S.: State- and county-level determinants of food manufacturing establishment growth: 1987–1993. Am. J. Agr. Econ. **79**, 838–850 (1997)
8. Haddad, M.A., Taylor, G., Owusu, F.: Locational choices in of the ethanol industry in the Midwest corn best. Econ. Dev. Q. **24**(1), 74–86 (2010)
9. Lambert, D.M., Wilcox, M., English, A., Stewart, L.: Ethanol plant location determinants and county comparative advantage. J. Agr. Appl. Econ. **40**, 117–135 (2008)
10. Lin, C.-Y.C.: Strategic decision-making with information and extraction externalities: a structural model of the multi-stage investment timing game in offshore petroleum production. Rev. Econ. Stat. **95**(5), 1601–1621 (2013)
11. Lin, C.-Y.C., Thome, K.: Investment in corn-ethanol plants in the Midwestern United States: an analysis using reduced-form and structural models. Working paper, University of California at Davis (2013)
12. Lin, C.-Y.C., Yi, F.: Ethanol plant investment in Canada: a structural model. Working paper. University of California at Davis (2013)
13. Lin, C.-Y.C., Yi, F.: What factors affect the decision to invest in a fuel ethanol plant? a structural model of the ethanol investment timing game. Working paper, University of California at Davis (2013)
14. Mitchell, D.: A note on rising food prices. Policy research. Working Paper no. 4862, The World Bank Development Prospects Group (July 2008)
15. Pakes, A., Ostrovsky, M., Berry, S.: Simple estimators for the parameters of discrete dynamic games (with entry and exit examples). RAND J. Econ. **38**(2), 373 (2007)
16. Peter C.R., Wolak, F.A.: Structural econometric modeling: rationales and examples from industrial organization. In: Heckman, J.J., Leamer, E.E. (eds.) Handbook of Econometrics, vol. 6A, pp. 4277–4415. Elsevier, Amsterdam (2007)
17. Sarmiento, C., Wilson, W.W.: Spatial competition and ethanol plant location decisions. In: Annual Meeting of the American Agricultural Economics Association, Orlando, 27–29 July 2008 (2007)
18. Urbanchuk, J.M.: Economic Impacts on the Farm Community of Cooperative Ownership of Ethanol Production. National Corn Growers Association Report (2006)

Chapter 29
Ethanol and Distiller's Grain: Implications of the Multiproduct Firm on United States Bioenergy Policy

Charles B. Moss, Andrew Schmitz, and Troy G. Schmitz

29.1 Forward

The use of corn for ethanol production has spawned considerable debate. There is little agreement among economists as to whether U.S. ethanol policy generates a welfare net loss or welfare net gain to society [16]. In modeling the impact of ethanol policy, there are several key components, including, for example, the impact of increased ethanol output on gasoline prices. If the impact is positive, the benefits from ethanol policy exceed the costs, but the reverse can be true if the impact is small or nonexistent [15]. More recently, the potential effect of an ethanol byproduct (distiller's grain) has become increasingly important. According to Dennis Conley "We're set on ethanol as the product and distiller's grain as the byproduct, but that could shift. Distiller's grain may become the product and ethanol the byproduct" [2]. Given that the ethanol process removes nutritional content from the corn in the process of creating ethanol, the primary profit center for distillers will likely continue to be ethanol. However, the potential value of distiller's grain undoubtedly affects the profitability of ethanol producers. This chapter incorporates some of the effects of distiller's grain into the benefit-cost of ethanol policy. It is possible that some of these considerations may yield a benefit-cost ratio greater than one if that is not already the case in the absence of distiller's grain.

The evolution of bioenergy policy over the past two decades has led to a variety of studies into the effects of ethanol on U.S. agriculture, food supply, and energy markets [17]. The growing consensus is that the Volumetric Ethanol Excise Tax

C.B. Moss (✉) • A. Schmitz
University of Florida, Gainesville, FL, USA
e-mail: cbmoss@ifas.ufl.edu; aschmitz@ufl.edu

T.G. Schmitz
Arizona State University, Tempe, AZ, USA
e-mail: troy.schmitz@asu.edu

A.A. Pinto and D. Zilberman (eds.), *Modeling, Dynamics, Optimization and Bioeconomics I*, Springer Proceedings in Mathematics & Statistics 73, DOI 10.1007/978-3-319-04849-9_29,
© Springer International Publishing Switzerland 2014

Table 29.1 Prices relative to the price of corn and quantity of dry distiller's grain

| Market year | Relative prices (to cwt of corn) | | | | | DDGS Produced (mil metric ton) |
	DDGS	Sorghum	Barley	Ethanol	Gasoline	
1981/82	1.6311	0.8982	1.1573	1.9052	1.1237	
1982/83	1.5938	0.9685	0.9974	1.8758	1.0294	
1983/84	1.4836	0.8531	0.8977	1.3622	0.7552	
1984/85	1.0016	0.8837	1.0158	1.6972	0.8952	
1985/86	1.3954	0.8664	1.0359	1.5957	0.7628	
1986/87	2.1630	0.9147	1.2522	2.0938	1.0236	
1987/88	2.0026	0.8775	1.0885	1.6273	0.7878	
1988/89	1.5542	0.8929	1.2861	1.3614	0.6531	
1989/90	1.4543	0.8898	1.1963	1.5088	0.8008	
1990/91	1.5518	0.9309	1.0950	1.6108	0.9139	
1991/92	1.4583	0.9475	1.0338	1.5575	0.7551	
1992/93	1.6493	0.9144	1.1498	1.6243	0.8308	1.2
1993/94	1.3864	0.9251	0.9287	1.3169	0.6272	1.6
1994/95	1.3130	0.9416	1.0479	1.4423	0.7083	1.0
1995/96	1.3490	0.9835	1.0406	1.1249	0.5891	0.2
1996/97	1.4442	0.8617	1.1796	1.2200	0.6010	0.5
1997/98	1.2186	0.9103	1.1427	1.2579	0.5339	0.8
1998/99	1.2290	0.8573	1.1907	1.4036	0.7373	1.0
1999/00	1.2292	0.8615	1.3654	1.8603	1.3577	0.9
2000/01	1.2202	1.0201	1.3306	2.4128	1.4366	1.6
2001/02	1.1398	0.9836	1.3147	1.5741	1.0648	2.0
2002/03	1.0631	0.9993	1.3678	1.5207	1.1667	4.3
2003/04	1.3358	0.9858	1.3643	1.8725	1.3431	6.1
2004/05	1.0324	0.8672	1.4045	2.3379	2.1294	7.3
2005/06	1.1983	0.9324	1.4758	3.6027	2.7253	9.5
2006/07	1.0046	1.0832	1.0938	2.1023	1.8920	13.2
2007/08	1.0435	0.9707	1.1167	1.6411	1.8644	20.5
2008/09	0.8068	0.7890	1.5431	1.2322	1.1161	25.6
2009/10	0.9136	0.9070	1.5315	1.4651	1.6649	32.0
2010/11	1.0099	0.9686	0.8694	1.4144	1.5086	33.2
2011/12	0.9129	0.9630	1.0035	1.0954	1.3242	

Source: Author's computations using data from the USDA's Feed Grains Yearbook [20], http://www.ers.usda.gov/data-products/feed-grains-database/feed-grains-yearbook-tables.aspx

Credit (VEETC) which expired in 2012 led to significant distortions; however, these distortions were complicated by agricultural policies that were slated to expire at the beginning of 2012. Thus, several studies suggested that the VEETC be replaced with a mandated minimum blend ratio [4–6]. In 2012, a drought in the United States further complicated the policy debate by increasing corn prices to historic levels. The resulting pressure on food prices, primarily through increases in livestock prices, spurred a debate pitting bioenergy goals against food security. Table 29.1 presents the prices for dried distiller's grains (DDGS), grain sorghum,

Fig. 29.1 Price and quantity of dry distiller's grains

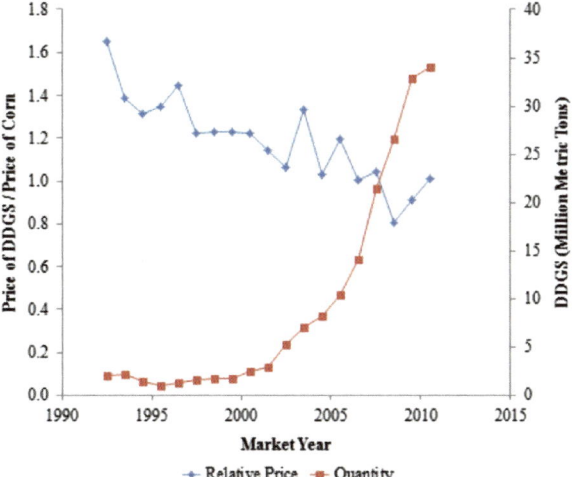

barley, ethanol, and gasoline relative to the price of corn along with the quantity of DDGS marketed in each crop year. These data indicate that 9.5 million metric tons of distiller's grain was sold in the United States in the 2005/2006 market year. This quantity increased almost 3.5 times, to 33.2 million metric tons in the 2010/2011 market period. During this period of expansion, the relative price of distiller's grains was close to one. These results are graphically depicted in Fig. 29.1.

29.2   Market for Distiller's Grain

In earlier work by Schmitz et al. [15] on the benefits and costs of U.S. ethanol policy, the price of distiller's grain, along with the quantity used, was assumed to be zero. However, this is no longer the case. "Distiller's grain, once regarded with about as much excitement as leftover mashed potatoes, is becoming an increasingly important part of sales from ethanol production" [10]. Ethanol plants are selling distiller's grain, a byproduct of ethanol production, to domestic and international customers. Every one bushel of corn processed yields about 2.8 gallons of ethanol and 17 pounds of distiller's grain [19]. In 2011, ethanol accounted for 27.3 % of corn usage; in 2012, 5 billion bushels of corn were converted to ethanol, with 1.547 billion bushels of ethanol byproducts used as livestock feed [12].

As an important ethanol byproduct, distiller's grain is used for different products, such as high-protein, high-energy feedstock (dairy, beef, swine, poultry, and aquaculture industries); fertilizers; and weed inhibitors. Distiller's grain is sold as dried distiller's grains (DDG), wet distiller's grains (WDG), and distiller's grains with solubles (WDGS or DDGS). The use of DDGS for livestock feed has become more important over time (see Table 29.1). In general, DDGS provides both a source of energy and proteins.

Table 29.2 Livestock ration data

Feed	Dry Matter	Crude Protein	Crude Fiber	Ca	P	Poultry ME	Swine Digestible Protein	DE (kcal/kg)	ME (kcal/kg)	TDN (%)
Alfalfa hay	92.2	16.7	25.8							
Barley grain	89.0	11.0	5.5	0.03	0.36	2,640	8.2	3,120	3,040	70
Corn grain	89.0	8.5	2.2	0.02	0.28	3,350	7.0	3,488	3,275	79
Sorghum grain	89.0	11.0	2.0	0.04	0.29	3,250	7.8	3,453	3,229	78
Wheat grain	89.0	12.7	3.0	0.05	0.36	3,071	11.7	3,520	3,277	80
Corn distiller's grain	92.0	26.9	4.0	0.35	1.37	2,932	16.1	3,300	2,976	75
Cottonseed meal	92.5	50.0	8.5	0.16	1.01	2,150	45.0	3,018	2,569	68
Soybean meal	89.0	44.0	7.0	0.29	0.65	2,230	41.7	3,300	2,825	75

Source: Kellems and Church [13]

The use of distiller's grains continues to expand and partially replace corn and soybean meal in the U.S. feed market. DDGS [distiller's grain] can supply both energy and protein in livestock and poultry rations, but use may be limited due to nutritional or price considerations [8, p. 2]. The usefulness of DDGS is limited by the removal of starches for the production of ethanol and by the fact that the proteins created are basically identical to those found in raw corn:

DDGS is considerably lower in carbohydrates than corn (because the starch has been fermented into ethanol); however, the other components (fat, protein, fiber, ash) are increased three- to four-fold compared with the original grain. The higher fat content of DDGS (9–10 %) tends to offset its higher fiber content, resulting in DE [digestible energy] and ME [metabolizable energy] values that are comparable to those of corn. The protein is poor in quality because the amino acid makeup is not greatly different from that of corn. Growing-finishing pigs can handle diets containing 20–30 % DDGS without any depression in growth performance, provided that the diets are formulated properly [13, pp. 259–260].

The demand for DDGS for livestock production is dependent on the nutritional characteristics of the byproduct, compared to alternatives, including corn. As depicted in Table 29.2, the crude protein in DDGS is over twice that in corn, but the digestible energy for swine is 5 % lower. While the process reduces the digestible energy, the energy available from DDGS is still higher than that available from barley, but somewhat less than that available from grain sorghum.

The livestock ration model represented an early mathematical programming formulation developed in agricultural economics [14]. In these formulations producers (i.e., feedlots, dairy producers) select that set of feed stuffs (x) given a set of prices for each feed stuff (w) to minimize the cost of feed subject to a set linear technical coefficients (A)

Fig. 29.2 Change in demand
from change in quality

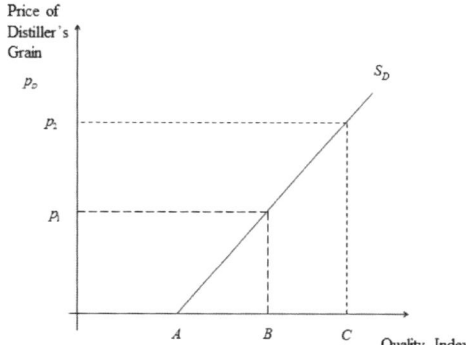

$$\left.\begin{array}{c} \min\limits_{x} w'x \\ Ax \geq b \end{array}\right\} \implies \left\{x_k^*(w, A, b)\right. \tag{29.1}$$

which results in a derived demand curve for each feed stuff $(x_k^*(\cdot))$. As one feed stuff
(i.e., DDGS) becomes cheaper relative to another feed stuff (i.e., corn), producers
use more of the cheaper feed stuff subject to the technical coefficients.

While distiller's grain is a good bypass protein (protein which is not degraded in
the rumen also referred to as undegradable intake protein) source for animal feed, it
can be a challenge due to feeding and storage management issues [11]. For instance,
feed rations may only require from 8 to 15 pounds of wet distiller's grain per cow
on a daily basis, so for a herd of 50 cows, that is 400–750 pounds daily. Also, most
ethanol plants will only sell distiller's grain in 50,000-pound truck loads.

Distiller's grain with solubles (WDGS or DDGS) is the most economically
valuable byproduct of distiller's grain, with DDGS being the more prominent.
WDGS is usually limited to large farms because it has a short shelf life (less
than 1 week) due to its moisture content, and is only shipped by truck in large
quantities. DDGS, on the other hand, has a long shelf life, which allows it to be
stored for months in storage facilities, including silos. Distiller's grain prices tend
to follow corn and soybean meal prices and are quoted in dollars per ton [7]. A
2011 USDA/ERS report found that in the United States, 1 metric ton of DDGS can
replace 1.22 metric tons of corn and soybean meal, with DDGS now surpassing
soybean meal as the No. 2 feed stuff [3, 9].

Consider Fig. 29.2 where the price of distiller's grain is plotted against a quality
index. At time t, the quantity of corn x produces y of distiller's grain at a quality
index of A. This generates a zero price in the distiller's grain market. For a quality
index of B, the price of distiller's grain at time $t + 1$ the price of distiller's grain
becomes p_1. For a quality index of C at time $t + 2$ the price of distiller's grain
becomes p_2. The supply curve S_D is upward sloping. The total revenue for x amount
of corn producing y of distiller's grain is $p_2 y$ given a quality index of C. For a
quality index of B the total revenue for the sale of y of distiller's grain at price P_1
is $p_1 y$.

Fig. 29.3 Market for
distiller's grain

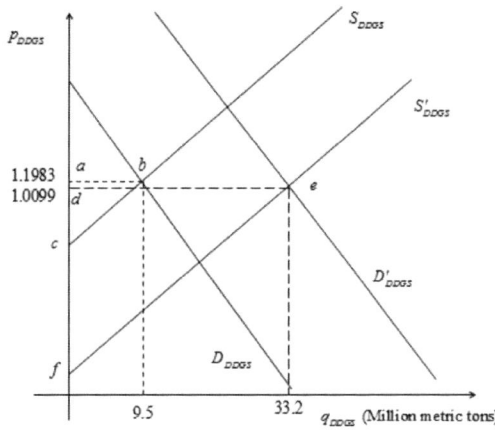

The market for DDGS is depicted graphically in Fig. 29.3. The initial supply of DDGS is depicted as S_{DDGS} while the initial demand for DDGS is depicted as D_{DDGS} (i.e., the market clearing price and quantity in the 2005/2006 market year). As ethanol production increased, the supply of DDGS shifts outward to S'_{DDGS}. This outward shift in the supply of distiller's grain results in a reduction in the quantity of corn usable for livestock feed. This results in an increase in the price of corn and contributes to an outward shift in the demand for distiller's grain to D'_{DDGS}. Together these shifts imply a slight reduction in the relative price of distiller's grain (i.e., relative to corn prices) and an increased quantity demanded of distiller's grain. From a benefit-cost ratio perspective, the producer surplus from the production of distiller's grain increases from abc to def. This increase in producer surplus is affected by improvements in the quality of distiller's grain.

29.3 Production, Byproducts, and the Multiproduct Firm

The change in the markets for distiller's grain has significant implications for the supply of ethanol. Several studies have recognized the possible consequences of byproducts, namely distiller's grain, on the production of ethanol.

As one bushel of corn produces 2.8 gallons of ethanol and 31 % of the value of corn is returned to the market in the form of byproducts, a tax credit of 51 ¢/gal. translates into approximately a \$2.04/bu. increase in the corn price [5, p. 477]. The notion implicit in this quote is that if the VEETC is set so as to make ethanol production economically profitable, then any profits to the sale of the byproducts is pure profit. Attempting to interpret this quotation we assume that profit can be computed as

$$\pi = p_1 q_1 + p_2 q_2 - p_c q_c - w' x \qquad (29.2)$$

where π is the profit, p_1 is the price of ethanol, q_1 is the quantity of ethanol, p_2 is the price of distiller's grain, q_2 is the quantity of distiller's grain, p_c is the price of corn, q_c is the quantity of corn, w is the price of other inputs (i.e., energy and labor), and x is the level of those inputs. Following the quotation, we assume that each bushel of corn yields 2.8 gallons of ethanol. In addition, we assume that $p_2q_2 = 0.31p_cq_c$. Thus, Eq. (29.1) becomes

$$\pi = 2.8p_1q_c + 0.31p_cq_c - p_cq_c - w'x$$

$$\Rightarrow \frac{\pi}{q_c} = 2.8p_1 + 0.31p_c - p_c - w'x^* \tag{29.3}$$

where x^* denotes the normalized input use (i.e., the level of other inputs used per bushel of corn processed into ethanol). Imposing a zero-profit condition implies a break-even price of corn (p_c^*) of

$$p_c^* = \frac{2.8p_1 - w'x^*}{1 - 0,31} = \frac{2.8p_1 - w'x^*}{0.69}. \tag{29.4}$$

Following de Gorter and Just

$$p_c^* - p_c = \frac{2.8(p_1 + \tau) - w'x^*}{0.69} = 0.24 \tag{29.5}$$

where p_c is the market price for corn and τ is the VEETC.

This study reformulates the simple profit function presented in Eq. (29.1) as

$$\pi(p, w) = \max_{x,y} - w'x$$

$$\text{s.t} \quad h(y, x) = 0 \tag{29.6}$$

where $\pi(p, w)$ is the profit function, $p = (p_1 \ p_2)'$ is the vector of output prices, $q = (q_1 \ q_2)'$ is the vector of outputs, $w = (w_1(= p_c) \ w_2 \ \cdots)$ is a vector of input prices with the first input price being the price of corn, $x = (x_1(= q_c) \ x_2 \ \cdots)'$ is a vector of inputs with the first input being the level of corn purchased, and $h(y, x)$ is the multiproduct production function. The tradeoff between output levels is typically represented in the production possibility frontier ($h(y, \bar{x})$), constructed by fixing the input levels ($\bar{x}$) as depicted in Fig. 29.4. Maximizing profits holding the input level constant yields

$$d\pi = p_1dy_1 + p_2dy_2 = 0 \Rightarrow \begin{cases} \frac{dy_1}{dy_2} = -\frac{p_2}{p_1} \\ \\ \frac{dy_1}{dy_2} = \frac{\frac{\partial h(y,\bar{x})}{dy_2}}{\frac{\partial h(y,\bar{x})}{dy_1}} \end{cases} \tag{29.7}$$

Raising the price of one of the inputs (in this case the price of ethanol by the level of the VEETC (τ)) rotates the price line, yielding an increase in the production of

Fig. 29.4 Production
possibility frontier

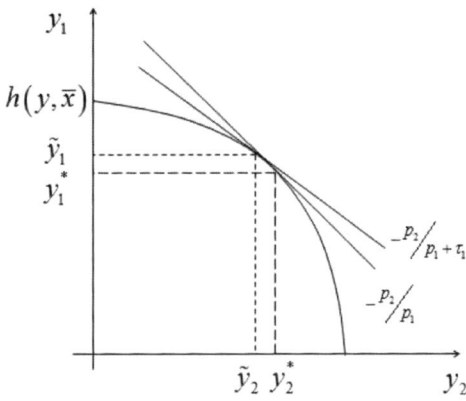

Fig. 29.5 Production
possibility frontier with
increased inputs

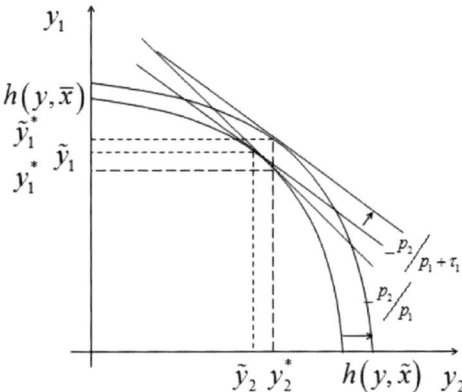

ethanol from y_1^* to $\tilde{y}_1$ and decreasing the level of distiller's grain from y_2^* to $\tilde{y}_2$. Of course, this analysis assumes that the level of inputs is held constant. Actually, as the price of ethanol increases, additional resources will be drawn into production. Figure 29.4 depicts the scenario where the production process exhibits constant returns to scale for all inputs. In this case the level of inputs used increases from $\bar{x}$ to $\tilde{x}$.

A key point in our discussion is related to the specification of the multiproduct production function and the relationship between the specification of the production function and the shape of the production possibility frontier. Specifically, the shape of the production possibility frontier depicted in Fig. 29.5 implies that the relative output level changes as output prices change

$$\frac{\tilde{y}_1^*}{y_2^*} > \frac{y_1^*}{y_2^*}. \tag{29.8}$$

Figure 29.6 presents a slightly different specification, such that the combination of outputs is determined by a fixed-ratio technology. This technology more closely

Fig. 29.6 Fixed proportion production possibilities frontier

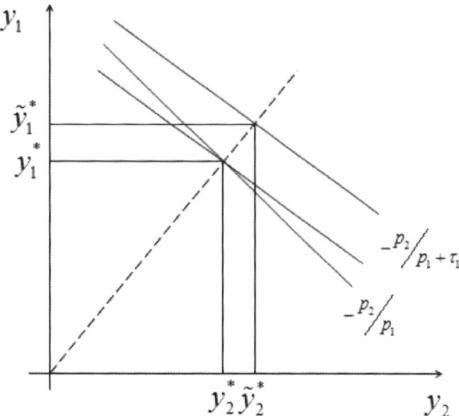

resembles the production of distiller's grain in combination with ethanol. Once the optimal level of ethanol has been determined, the optimal level of distiller's grain to be produced is fixed

$$\frac{y_1}{y_2} = \theta \implies y_2 = \frac{y_1}{\theta}. \tag{29.9}$$

The firm's profit function presented in Eq. (29.6) can then be rewritten as

$$\pi(p_1, p_2, w) = \max_{y_1, x} \left(p_1 + \frac{p_2}{\theta} \right) y_1 - w'x \left. \right\} \implies \max_x \left(p_1 + \frac{p_2}{\theta} \right) h_1(x) - w'x$$
$$\text{s.t } y_1 - h_1(x) = 0 \tag{29.10}$$

or the economic decision is based on the total sales which is proportional to the level of ethanol produced.

The mathematical formulation in Eq. (29.10) allows the analysis of a variety of scenarios that may be relevant for ethanol policy. As a first example, we assume that distiller's grain has a zero price. In this case, the first-order conditions for profit maximization are

$$x_j \left(p_1 \frac{\partial h_1(x)}{\partial x_j} - w_j \right) = 0$$

$$p_1 \frac{\partial h_1(x)}{\partial x_j} - w_j \leq 0 \tag{29.11}$$

for all inputs. The basic assumption here is that without some form of tax credit or minimum blend ratio, the result would be a corner solution

$$x_j = 0$$

$$p_1 \frac{\partial h_1(x)}{\partial x_j} - w_j \leq 0 \qquad (29.12)$$

so that the marginal value product for the first (or any) increment in input is less than the price of the input. Thus, taking $j = 1$, the level of corn demanded by ethanol producers is zero. In this case, the VEETC could be added to yield a positive level of production

$$(p_1 + \tau) \frac{\partial h_1(x)}{\partial x_j} - w_j = 0 \quad \Leftrightarrow \quad x_j > 0. \qquad (29.13)$$

Thus, some level of VEETC is sufficient to generate a positive level of ethanol production and demand for corn to produce ethanol.

Next, consider two possibilities for a non-zero price of distiller's grain. First, consider the scenario where the production of distiller's grain is a nuisance (i.e., something has to be paid to dispose of the output). Building on Eq. (29.10)

$$x_j \left[\left(p_1 + \frac{p_2}{\theta} + \tau \right) \frac{\partial h_1(x)}{\partial x_j} - w_j \right] = 0$$

$$\left(p_1 + \frac{p_2}{\theta} + \tau \right) \frac{\partial h_1(x)}{\partial x_j} - w_j \leq 0. \qquad (29.14)$$

Comparing the result in Eq. (29.14) with Eq. (29.13), if $p_2 < 0$, then τ must increase to offset the cost of nuisance output. Alternatively, if we assume that $p_2 < 0$, the level of the VEETC required to generate a positive level of ethanol production declines. Thus, the supply of ethanol can be written as $q_1^*(p_1, p_2, w, \theta)$. In general, this supply function is increasing in the price of ethanol and the price of distiller's grain (considering the fact that the price of ethanol may actually be negative), decreasing in input prices, and increasing in θ as long as the price of distiller's grain is positive but decreasing in θ if the price of distiller's grain is negative. Similarly, the demand for corn by distillers can be written as $x_1^*(p_1, p_2, w, \theta)$. This demand function is similarly increasing in the price of ethanol and distiller's grain and increasing in θ as long as the price of distiller's grain is positive but decreasing in θ if the price of distiller's grain is negative. However, the relationship between other inputs depends on whether those inputs are complements or substitutes.

29.4 Ethanol Processor Profitability

Because distiller's grain has positive value, the profitability of producing ethanol not only depends on the price of corn and gasoline, but also on the degree of correlation between the price of distiller's grain and the price of corn and soybeans (the main ingredients for livestock feed). Corn and distiller's grain prices are positively

Table 29.3 Corn and distiller's grain prices (Iowa)

November 1-5, 2012	Corn $/bushel	Distiller's grain (10 % moisture) $/ton
November 1	7.50	269
November 2	7.48	268
November 3	7.33	260
November 4	7.48	254
November 5	7.63	256
Average	7.48	261
November 1-4, 2011	Corn $/bushel	Distiller's grain (10 % moisture) $/ton
November 1	5.47	151
November 2	5.42	152
November 3	5.06	152
November 4	4.99	155
Average	5.24	153

Source: Agricultural Marketing Resource Center [1]; USDA Livestock and Grain Market News, State Ethanol Plant Reports [18]

Fig. 29.7 Profitability of ethanol production

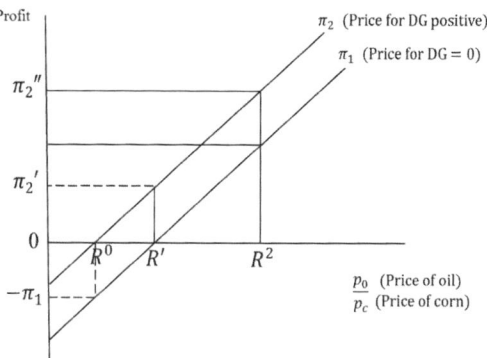

correlated (Table 29.3). For example, for an average corn price of $7.48 per bushel, the average distiller's grain price was $261 per ton. For a lower average corn price of $5.24 per bushel, the average distiller's grain price was $153 per ton. The fact that the price of distiller's grain is positively correlated to the price of corn means that, even though the price of corn may increase which lowers the profitability from the production of ethanol, there is a partial offset due to the increase in the price of the byproduct, distiller's grain.

Figure 29.7 illustrates geometrically the impact of taking into account the market for distiller's grain on ethanol processor profitability. Consider π_1 where the price of distiller's grain equals zero (DG = 0). For a $\frac{p_0 \text{ (price of oil)}}{p_c \text{ (price of corn)}}$ of R', the processor makes zero profit. However, for a ratio of R_2, profits are π_1'.

Now consider π_2 where the distiller's grain is positively priced. At a rate of R', unlike before, profits are positive (i.e., π_2''). At a ratio of R^0, profits are zero but before profits are negative (i.e., $-\pi_1$).

29.5 Conclusions

In all likelihood, the United States will continue to use large portions of its corn crop for ethanol production. The production of ethanol in the United States is highly dependent on the U.S. energy policy. The profitability of using ethanol varies from year to year, and in part depends on the size of the ethanol distillery plant. In addition, as we show, added revenue for ethanol distillers is obtainable from the sale of a major byproduct, distiller's grain. Through technological improvements, distiller's grain is now an integral part of livestock rations.

In the future, it will be interesting to see what will happen with the production of ethanol relative to the blend rate mandate. If ethanol production greatly exceeds the minimum mandated requirement, which it well might, then the production of distiller's grain will increase over the minimum mandated levels for ethanol. Thus, the production of distiller's grain could well increase. The price of distiller's grain could fall, as could the price of ethanol.

References

1. Agricultural Marketing Resource Center: Corn and Distiller's Grain Prices in Iowa. Agricultural Marketing Resource Center, Iowa State University, Ames (2013)
2. Anderson, T.: Economist: distiller's grain could replace ethanol as main product. Midwestern Producer, 29 January 2013
3. Bevill, K.: USDA finds distiller's grains offer greater feed value than corn. EthanolProducer.com. http://www.ethanolproducer.com/articles/8272 (2011)
4. de Gorter, H., Just, D.R.: "Water" in the U.S. ethanol tax credit and mandate: implications for rectangular deadweight costs and the corn-oil price relationship. Rev. Agr. Econ. **30**(3), 397–410 (2008)
5. de Gorter, H., Just, D.R.: The welfare economics of a biofuel tax credit and the interaction effects with price contingent farm subsidies. Am. J. Agr. Econ. **91**(2), 477–488 (2009)
6. de Gorter, H., Just, D.R.: The economics of a blend mandate for biofuels. Am. J. Agr. Econ. **91**(3), 738–750 (2009)
7. Dooley, F., Cox, M., Cox, L.: Distillers Grain Handbook: A Guide for Indiana Producers to Using DDGS for Animal Feed. Report for Indiana Corn Marketing Council, West Lafayette (2008)
8. Hoffman, L., Baker, A.: Market Issues and Prospects for U.S. Distiller's Grains: Supply, Use, and Price Relationships. FDS-10k–01. United States Department of Agriculture, Washington, DC (2010)
9. Hoffman, L., Baker, A.: Estimating the Substitution of Distillers Grain for Corn and Soybean Meal in the U.S. Feed Complex. Report for the Economic Research Service, United States Department of Agriculture, Washington, DC (2011)
10. Hovey, A.: Distillers grain gaining importance as byproduct of ethanol. Lincoln J. Star, 28 January 2013
11. Iowa Beef Center: Using distillers grains in alternative cow-calf production systems. IBC 43. Iowa Beef Center, Iowa State University, Ames (2010)
12. Jessen, H.: World of corn report breaks down corn used for ethanol, DDGS. EthanolProducer.com. http://www.ethanolproducer.com/articles/8611 (2012)

13. Kellems, R.O., Church, D.C.: Livestock Feeds and Feeding, 6th edn. Prentice-Hall, Upper Saddle River (2010)
14. Moss, C.B.: Applied optimization in agriculture. In: Pardalos, P.M., Resende, M.G.C. (eds.) Handbook of Applied Optimization, pp. 957–966. Oxford University Press, London (2002)
15. Schmitz, A., Moss, C., Schmitz, T.: Ethanol: no free lunch. J. Agr. Food Industrial Organization 5(2, Article 3) (2007). http://www.bepress.com/jafio/vol5/iss2/art3
16. Schmitz, A., Moss, C.B., Schmitz, T.G., Furtan, H.W., Schmitz, H.C.: Agricultural Policy, Agribusiness, and Rent-Seeking Behaviour, 2nd edn. University of Toronto Press, Toronto (2010)
17. Schmitz, A., Wilson, N., Moss, C., Zilberman, D. (eds.): The Economics of Alternative Energy Sources and Globalization. Bentham Science Publishers, Oak Park (2011)
18. United States Department of Agriculture: USDA Livestock and Grain Market News, State Ethanol Plant Reports. United States Department of Agriculture, Washington, DC (2013)
19. United States Department of Agriculture, Economics Research Service: Corn: Background. United States Department of Agriculture, Economic Research Service, Washington, DC. http://www.ers.usda/gov/topics/crops/corn/background.aspx (2013)
20. United States Department of Agriculture, Economic Research Service: Feed Grains Yearbook. United States Department of Agriculture, Economic Research Service, Washington, DC. http://www.ers.usda.gov/data-products/feed-grains-database/feed-grains-yearbook-tables.aspx (2013)

Chapter 30
A Review on Protein-Protein Interaction Network Databases

Chandra Sekhar Pedamallu and Linet Ozdamar

30.1 Introduction

Cells are the structural and functional units of all known living organisms. These carry out numerous functions, from DNA replication, cell replication, protein synthesis, and energy production to molecule transport, to various inter- and intracellular signaling. Many of these fundamental processes require cascades of biochemical reactions that are catalyzed by possibly interacting protein enzymes. Other proteins provide structural support for the cells, form scaffolds for intracellular localization, or serve as chaperones or as transporters. The large-scale study of all cellular proteins is known as Proteomics [2,5]. One of the main goals of proteomics is to map the interactions of proteins. Interactomes, study of interaction networks, for dozens of model organisms have been established experimentally. Functions of proteins can be defined by their complex interactions and by their positions in interaction networks. Protein-protein interaction information plays a vital role in basic biological research; it also helps in the discovery of novel drug targets for the treatment of various chronic diseases. Experimental probing of protein-protein interactions requires labor-intensive techniques, such as co-immunoprecipitation, or affinity chromatography [26]. High-throughput experimental techniques, such as yeast two-hybrid [34] and mass spectrometry [11] are also available for large-scale detection of protein-protein interactions, and for the exploration of protein sequences, structures, and relationships in complete genomes [26].

C.S. Pedamallu (✉)
Department of Medical Oncology, Dana-Farber Cancer Institute, Boston, MA, USA
The Broad Institute of MIT and Harvard, Boston, MA, USA
e-mail: pcs.murali@gmail.com

L. Ozdamar
Department of Systems Engineering, Yeditepe University, Istanbul, Turkey
e-mail: linetozdamar@lycos.com

A.A. Pinto and D. Zilberman (eds.), *Modeling, Dynamics, Optimization and Bioeconomics I*, Springer Proceedings in Mathematics & Statistics 73, DOI 10.1007/978-3-319-04849-9__30,
© Springer International Publishing Switzerland 2014

Following these advances, numerous computational methods have been developed to predict protein-protein interaction networks based on sequence or (and) structural features of the proteins. These computational approaches use phylogenetic profiling [21, 28], homologous interacting partner analysis [1], structural pattern comparisons [4, 15, 19], Bayesian network modeling [13], data mining techniques [20, 36] and so on.

There are several surveys on computational methods that predict protein-protein interaction networks (e.g., [23, 26, 31]). Complementing efforts have been made to centralize protein-protein interaction data through the construction of databases, such as STRING [29], MINT [7], BioGRID [27], FPPI [35], HAPPI [8], PIP [18], DIP [25], POINeT [16] and IntAct [3]. These databases can be classified as general databases and specialized databases. In this paper, we attempt to provide a summary of available databases.

30.2   Protein-Protein Interaction Databases

PPI databases can be grouped into two categories, (1) General databases that contain interactions networks from a wide variety of organisms; (2) Specialized databases that contain interaction networks from specific organisms. These databases are used during efforts of protein-protein interaction predictions.

30.2.1   General Databases

30.2.1.1   Search Tool for the Retrieval of Interacting Genes [29]

Search Tool for the Retrieval of Interacting Genes (STRING) is a comprehensive database that provides both experimental as well as predicted interaction information. Each of interactions in STRING are provided with a confidence score, and accessory information such as protein domains and 3D structures are made available, all within a stable and consistent identifier space. Other features that are included in STRING are interactive network viewer that can cluster networks on demand, updated on-screen previews of structural information including homology models, extensive data updates and strongly improved connectivity and integration with third-party resources. The current version of STRING covers more than 5,214,234 proteins from 1,133 organisms range from Bacteria, Archaea to Homo sapiens.

The resource can be reached at http://string-db.org.

30.2.1.2 Molecular INTeration Database [7]

Molecular INTeration database (MINT) is a public repository for molecular interactions reported in peer-reviewed journals. It mainly focuses on experimentally verified protein-protein interactions mined from the scientific literature by expert curators. The interactions curated and validated in MINT are automatically imported, according to its properties, by one or more sister databases that include human interactions database—HomoMINT (http://mint.bio.uniroma2.it/domino), domain-peptide interactions DOMINO (http://mint.bio.uniroma2.it/domino), all virus-virus and virus-host interactions databases, VirusMINT (http://mint.bio.uniroma2.it/virusmint).

The scoring function used for the MINT is based on size of experiment, type of experiment, evidence of direct interaction (i.e. two-hybrid) with respect to experimental support, number of interaction partners detected in single purification, sequence similarity of ortholog proteins (in case of human proteome in HomoMINT), and the number of publications supporting the interaction. The resulting score ranges between 0 to 1 (well supported evidence). This database contains interaction networks from Homo sapiens, C. elegans, Bacteria, and various Viruses.

The resources can be reached at http://mint.bio.uniroma2.it/mint/Welcome.do.

30.2.1.3 Biological General Repository for Interaction Datasets [27]

Biological General Repository for Interaction Datasets (BioGRID) is a public database that archives and disseminates genetic and protein interaction data from model organisms and humans (http://www.thebiogrid.org). It currently holds 347,966 interactions (170,162 genetic, 177,804 protein) curated from both high-throughput data sets and individual focused studies, as derived from over 23,000 publications in the primary literature. All interaction data are freely provided through the search index and available via download in a wide variety of standardized formats. This database contains interaction networks from Homo sapiens, C. elegans, Plant, Mouse, and different bacterial species.

The resources can be reached at http://thebiogrid.org/.

30.2.1.4 IntAct [3]

IntAct is an open data molecular interaction database abstracted from the literature or from direct data depositions by expert curators following a deep annotation model providing a high level of detail. It contains over 268,920 binary interactions, 57,741 proteins and 13,802 experiments. The search interface allows the user to iteratively develop complex queries, exploiting the detailed annotation with hierarchically controlled vocabularies. This database contains interaction information from a wide variety of organisms that includes but not limited to Homo sapiens, Mus musculus,

Drosophila melanogaster, Caenorhabditis elegans, Escherichia coli and Arabidosis thaliana.

The resources can be reached at http://www.ebi.ac.uk/intact/.

30.2.1.5   POINeT [16]

POINeT is an integrated web service that processes protein-protein interaction searching, analysis and visualization. It merges protein-protein interaction and tissue-specific expression data from multiple resources including DIP (http://dip. doe-mbi.ucla.edu/), MINT [7], BIND (http://www.bind.ca), HPRD (http://www. hprd.org), MIPS (http://mips.gsf.de/proj/ppi/), CYGD, BioGRID [27] and NCBI interaction (ftp://ftp.ncbi.nlm.nih.gov/gene/GeneRIF/interactions.gz). The tissue-specific PPIs and the number of research papers supporting the PPIs can be filtered with user-adjustable threshold values and are dynamically updated in the viewer. The network constructed in POINeT can be readily analyzed with, for example, the built-in centrality calculation module and an integrated network viewer. Nodes in global networks can also be ranked and filtered using various network analysis formulas, i.e., centralities. To prioritize the sub-network, a ranking filtered method (S3) is developed to uncover potential novel mediators in the midbody network. This database contains interaction information from wide variety of organisms.

The resources can be reached at http://poinet.bioinformatics.tw/.

30.2.1.6   Reactome [9, 12]

Reactome is a database of pathways and reactions (pathway steps) in human biology that have been curated by expert biologist researchers which is extensively cross-referenced to other resources e.g. NCBI, Ensembl, UniProt, UCSC Genome Browser, HapMap, KEGG (Gene and Compound), ChEBI, PubMed and GO. It includes many events in biology that involve changes in state, such as binding, activation, translocation and degradation, in addition to classical biochemical reactions. Reactome contains inferred orthologous reactions for over 20 non-human species including mouse, rat, chicken, puffer fish, worm, fly, yeast, rice, Arabidopsis and E.coli.

The resources can be reached at http://www.reactome.org/ReactomeGWT/ entrypoint.html.

30.2.1.7   iRefWeb [30]

iRefWeb provides a web interface to protein interaction data consolidated from ten public databases that includes BIND, BioGRID, CORUM, DIP, IntAct, HPRD, MINT, MPact, MPPI and OPHID. It provides an overview of the consolidated protein-protein interaction landscape and shows how it can be automatically cropped to help generate meaningful organism-specific interactomes. iRefWeb presents aggregated interactions for a protein of interest, and various statistical summaries of the data across databases, such as the number of organism-specific interactions, proteins and cited publications.

The resources can be reached at http://wodaklab.org/iRefWeb and http://wodaklab.org/iRefWeb/.

30.2.1.8   Database of Interacting Proteins [25]

Database of interacting proteins (DIP) is a database that catalogs experimentally determined protein-protein. It combines information from a variety of sources to create a single, consistent set of protein-protein interactions. The data stored within the DIP database were curated, both, manually by expert curators and also automatically using computational approaches that utilize the knowledge about the protein-protein interaction networks extracted from the most reliable, core subset of the DIP data. DIP contains 23,201 proteins from 372 organisms including but not limited to Homo sapiens, Mouse, E. Coli, Rat, Bakers Yeast, and 71,276 interactions.

The resources can be reached at http://dip.doe-mbi.ucla.edu/dip/Main.cgi.

30.2.1.9   CORUM [24]

CORUM is a collection of experimentally verified mammalian protein complexes. All information presented in the database is obtained from individual experiments published in scientific data; however, data from high-throughput experiments are excluded. The majority of protein complexes in CORUM originate from human. The resources can be reached at http://mips.helmholtz-muenchen.de/genre/proj/corum/index.html.

30.2.2   Specialized Databases

30.2.2.1   Arabidopsis Thaliana Protein Interactome Database [22]

The AtPID (Arabidopsis thaliana Protein Interactome Database) represents protein-protein interaction networks, domain architecture, ortholog information and GO annotation in the Arabidopsis thaliana proteome. The protein-protein interaction pairs are predicted by integrating several methods with the Naive Bayesian Classifier. AtPID contains 28,062 putative PPIs.

The resources can be reached at http://www.megabionet.org/atpid/webfile/

30.2.2.2 Human Protein Reference Database [14]

Human Protein Reference Database (HPRD) is a database of curated proteomic information pertaining to human proteins. It integrates information pertaining to domain architecture, post-translational modifications, interaction networks and disease association for each protein in the human proteome. All the information in HPRD has been manually extracted from the literature by expert biologists. HPRD contains 39,194 protein-protein interactions from 30,047 protein entries.

The resources can be reached at http://www.hprd.org/.

30.2.2.3 Fusarium Graminearum Protein-Protein Interaction Database [35]

Fusarium graminearum protein-protein interaction (FPPI) database provides comprehensive information of protein-protein interactions (PPIs) of Fusarium graminearum. The PPIs are predicted based on both interologs from several PPI databases of seven species and domain-domain interactions experimentally determined based on protein structures. FPPI database contains 223,166 interactions among 7,406 proteins and 27,102 interactions among 3,745 proteins in the core PPI set.

The resources can be reached at http://csb.shu.edu.cn/fppi/.

30.2.2.4 Human Annotated and Predicted Protein Interaction Database [8]

The Human Annotated and Predicted Protein Interaction (HAPPI) database is developed to integrate publicly available human protein interaction data from BIND, OPHID, MINT, IntAct, HPRD, and STRING databases into a data warehouse. In the data warehouse, various types of sequence, structure, pathway, and literature annotation data from established bioinformatics resources such as NCBI, PubMed, UniProt, HUGO, EBI, PDB were also integrated. HAPPI contains 601,757 protein-protein interactions and associations from 70,829 curated human proteins.

The resources can be reached at http://discern.uits.iu.edu:8340/HAPPI/index.html.

30.2.2.5 Human Protein-Protein Interaction Prediction Database [18]

The Human protein-protein interaction prediction database (PIPs) is a resource for studying protein-protein interactions in human. It contains predictions of more than 37,000 high probability interactions of which more than 34,000 are not reported in the interaction databases HPRD, BIND, DIP or OPHID. The interactions in PIPs were calculated by a Bayesian method that combines information from expression, orthology, domain co-occurrence, post-translational modifications and sub-cellular location. The predictions also take into account the topology of the predicted interaction network.

The resources can be reached at http://www.compbio.dundee.ac.uk/www-pips.

30.3   Brief Note on Application of Protein-Protein Interaction Networks

Study of protein-protein interaction networks is fundamental to understanding the key biological systems across different organisms. Protein-protein interaction networks can be used to protein function prediction (e.g. [17, 33]), drug discovery in cancer, autoimmune diseases, etc. [10, 32], understanding of successful plant defense mechanisms against pathogens [6] and etc.

30.4   Conclusion

Protein-protein interaction networks helps in understanding biological processes in living cells. There are several experimental and computational approaches developed to identify and predict these interaction networks experimentally and *in silico* respectively. Moreover, there are several complementing efforts made to centralize protein-protein interaction data through the construction of databases from experimental and computational protein-protein interactions networks. In this paper, we attempt to provide a summary of most widely used protein-protein interactions databases. These databases will serve as a platform for researchers to mine the data in a systematic fashion and employ them to predict the protein function, identifying important proteins in diseases and so on.

References

1. Aloy, P., Russell, R.B.: InterPreTS: protein interaction prediction through tertiary structure. Bioinformatics **19**(1), 161–162 (2003)
2. Anderson, N.L., Anderson, N.G.: Proteome and proteomics: new technologies, new concepts, and new words. Electrophoresis **19**(11), 1853–1861 (1998)
3. Aranda, B., Achuthan P., Alam-Faruque, Y., Armean, I., Bridge, A., Derow, C., Feuermann, M., Ghanbarian, A.T., Kerrien, S., Khadake, J., Kerssemakers, J., Leroy, C., Menden, M., Michaut, M., Montecchi-Palazzi, L., Neuhauser, S.N., Orchard, S., Perreau, V., Roechert, B., van Eijk, K., Hermjakob, H.: The IntAct molecular interaction database in 2010. Nucleic Acids Res. **38**(Database issue), D525–D531 (2010)
4. Aytuna, A.S., Keskin, O., Gursoy, A.: Prediction of protein-protein interactions by combining structure and sequence conservation in protein interfaces. Bioinformatics **21**(12), 2850–2855 (2005)

5. Blackstock, W.P., Weir, M.P.: Proteomics: quantitative and physical mapping of cellular proteins. Trends Biotechnol. **17**(3), 121–127 (1999)
6. Bogdanove, A.J.: Protein-protein interactions in pathogen recognition by plants. Plant Mol. Biol. **50**(6), 981–989 (2002)
7. Ceol, A., Chatr Aryamontri, A., Licata, L., Peluso, D., Briganti, L., Perfetto, L., Castagnoli, L., Cesareni, G.: MINT, the molecular interaction database: 2009 update. Nucleic Acids Res. **38**(Database issue), D532–D539 (2010)
8. Chen, J.Y., Mamidipalli, S., Huan, T.: HAPPI: an online database of comprehensive human annotated and predicted protein interactions. BMC Genomics **10**(Suppl 1), S16 (2009)
9. Croft, D., O'Kelly, G., Wu, G., Haw, R., Gillespie, M., Matthews, L., Caudy, M., Garapati, P., Gopinath, G., Jassal, B., Jupe, S., Kalatskaya, I., Mahajan, S., May, B., Ndegwa, N., Schmidt, E., Shamovsky, V., Yung, C., Birney, E., Hermjakob, H., D'Eustachio, P., Stein, L.: Reactome: a database of reactions, pathways and biological processes. Nucleic Acids Res. **39**(Database issue), D691–D697 (2011)
10. Drews, J.: Drug discovery: a historical perspective. Science **287**(5460), 1960–1964 (2000)
11. Figeys, D., McBroom, L.D., Moran, M.F.: Mass spectrometry for the study of protein-protein interactions. Methods **24**(3), 230–239 (2001)
12. Hermjakob, H., D'Eustachio, P., Stein, L.: Reactome: a database of reactions, pathways and biological processes. Nucleic Acids Res. **39**(Database issue), D691–D697 (2011)
13. Jansen, R., Yu, H., Greenbaum, D., Kluger, Y., Krogan, N.J., Chung, S., Emili, A., Snyder, M., Greenblatt, J.F., Gerstein, M.: A Bayesian networks approach for predicting protein-protein interactions from genomic data. Science **302**(5644), 449–453 (2003)
14. Keshava Prasad, T.S., Goel, R., Kandasamy, K., Keerthikumar, S., Kumar, S., Mathivanan, S., Telikicherla, D., Raju, R., Shafreen, B., Venugopal, A., Balakrishnan, L., Marimuthu, A., Banerjee, S., Somanathan, D.S., Sebastian, A., Rani, S., Ray, S., Harrys Kishore, C.J., Kanth, S., Ahmed, M., Kashyap, M.K., Mohmood, R., Ramachandra, Y.L., Krishna, V., Rahiman, B.A., Mohan, S., Ranganathan, P., Ramabadran, S., Chaerkady, R., Pandey, A.: Human protein reference database–2009 update. Nucleic Acids Res. **37**(Database issue), D767–D772 (2009)
15. Keskin, O., Ma, B., Nussinov, R.: Hot regions int protein-protein interactions: The organization and contribution of structurally conserved hot spot residues. J. Mol. Biol. **345** 1281–1294 (2004)
16. Lee, S.A., Chan, C.H., Chen, T.C., Yang, C.Y., Huangm K.C., Tsai, C.H., Lai, J.M., Wang, F.S., Kao, C.Y., Huang, C.Y.: POINeT: protein interactome with sub-network analysis and hub prioritization. BMC Bioinformatics **21**(10), 114 (2009)
17. Marcotte, E.M., Pellegrini, M., Ng H.,-L., Rice, D.W., Yeates, T.O., Eisenberg, D.: Detecting protein function and protein-protein interactions from genome sequences. Science **285**(5428), 751–753 (1999)
18. McDowall, M.D., Scott, M.S., Barton, G.J.: PIPs: human protein-protein interaction prediction database. Nucleic Acids Res. **37**(Database issue), D651–D656 (2009)
19. Ogmen, U., Keskin, O., Aytuna, A.S., Nussinov, R., Gursoy, A.: PRISM: protein interactions by structural matching. Nucleic Acids Res. **33** (Web Server issue), W331–W336 (2005)
20. Paradesi, M.S.R., Caragea, D., Hsu, W.H.: Incorporating graph features for predicting protein-protein interactions. In: Li, X.-L., Ng, S.-K. (eds.) Biological Data Mining in Protein Interaction Networks. IGI Publishers, USA (2008)
21. Pellegrini, M., Marcotte, E.M., Thompson, M.J., Eisenberg, D., Yeates, T.O.: Assigning protein functions by comparative genome analysis: protein phylogenetic profiles. Proc. Natl. Acad. Sci. USA **96**, 4285–4288 (1999)
22. Peng, L., Weidong, Z., Yuhua, L., Feng, X., Jigang ,W., Tieliu, S.: AtPID: the overall hierarchical functional protein interaction network interface and analytic platform for Arabidopsis. Nucleic Acids Res. **39**(suppl 1), D1130–D1133 (2011)
23. Pitre, S., Alamgir, M., Green, J.R., Dumontier, M., Dehne, F., Golshani, A.: Computational methods for predicting protein-protein interactions. Adv. Biochem. Eng. Biotechnol. **110**, 247–267 (2008)

24. Ruepp, A., Waegele, B., Lechner, M., Brauner, B., Dunger-Kaltenbach, I., Fobo, G., Frishman, G., Montrone, C., Mewes, H.W.: CORUM: the comprehensive resource of mammalian protein complexes–2009. Nucleic Acids Res. **38**(Database issue), D497–D501 (2010)
25. Salwinski, L., Miller, C.S., Smith, A.J., Pettit, F.K., Bowie, J.U., Eisenberg, D.: The database of interacting proteins: 2004 update. Nucleic Acids Res. **32**(Database issue), D449–D451 (2004)
26. Skrabanek, L., Saini, H.K., Bader, G.D., Enright, A.J.: Computational prediction of protein-protein interactions. Mol. Biotechnol. **38**(1), 1–17 (2008)
27. Stark, C., Breitkreutz, B.J., Chatr-Aryamontri, A., Boucher, L., Oughtred, R., Livstone, M.S., Nixon, J., Van Auken, K., Wang, X., Shi, X., Reguly, T., Rust, J.M., Winter, A., Dolinski, K., Tyers, M.: The BioGRID interaction database: 2011 update. Nucleic Acids Res. **39**(Database issue), D698–D704 (2011)
28. Sun, J., Xu J., Liu, Z., Liu, Q., Zhao, A., Shi, T., Li, Y.: Refined phylogenetic profiles method for predicting protein-protein interactions. Bioinformatics **21**(16),3409–3415 (2005)
29. Szklarczyk, D., Franceschini, A., Kuhn, M., Simonovic, M., Roth, A., Minguez, P., Doerks, T., Stark, M., Muller, J., Bork, P., Jensen, L.J., von Mering, C.: The STRING database in 2011: functional interaction networks of proteins, globally integrated and scored. Nucleic Acids Res. **39**(Database issue), D561–D568 (2011)
30. Turner, B., Razick, S., Turinsky, A.L., Vlasblom, J., Crowdy, E.K., Cho, E., Morrison, K., Donaldson, I.M.,Wodak, S.J.: iRefWeb: Interactive analysis of consolidated protein interaction data and their supporting evidence. Database. 2010: baq023 (2010)
31. Valencia, A., Pazos, F.: Computational methods for the prediction of protein interactions. Curr. Opin. Struct. Biol. **12**(3), 368–373 (2002)
32. Valkov E., Sharpe T., Marsh M., Greive S., Hyvönen M.: Targeting protein-protein interactions and fragment-based drug discovery. Top Curr. Chem. **317**, 145–179 (2012)
33. Vazquez , A., Flammini, A., Maritan, A., Vespignani, A.: Global protein function prediction from protein-protein interaction networks. Nat. Biotechnol. **21**, 697–700 (2003)
34. Young, K.H.: Yeast two-hybrid: so many interactions, (in) so little time . . .. Biol. Reprod. **58**, 302–311 (1998)
35. Zhao, X.M., Zhang, X.W., Tang, W.H., Chen, L.: FPPI: fusarium graminearum protein-protein interaction database. J. Proteome Res. **8**(10), 4714–4721 (2009)
36. Zhou, D., He, Y.: Extracting interactions between proteins from the literature. J. Biomed. Inform. **14**(2), 393–407 (2008)

Chapter 31
On the Use of Cross Impact Analysis for Enhancing Performance in Primary School Education

Chandra Sekhar Pedamallu, Linet Ozdamar, and Gerhard-Wilhelm Weber

31.1 Introduction

The first stage of compulsory education is primary or elementary education. In most countries, it is compulsory for children to receive primary education, though in many jurisdictions it is permissible for parents to provide it. The transition to secondary school or high school is somewhat arbitrary, but it generally occurs at about 11 or 12 years of age. Some educational systems have separate middle schools with the transition to the final stage of education taking place at around the age of fourteen.

The major goals of primary education are achieving basic literacy and numeracy amongst all pupils, as well as establishing foundations in science, geography, history and other social sciences. The relative priority of various areas, and the methods used to teach them, are areas of considerable political debate. Some of the expected benefits from primary education are the reduction of infant mortality rate, population growth rate, crude birth and death rate, increasing skills of the future work force and so on.

Keeping all school age children enrolled at a school is especially important in developing countries where government policies toward industrialization lead to massive migration from rural areas to cities [14, 16]. Security related internal

C.S. Pedamallu (✉)
Department of Medical Oncology, Dana-Farber Cancer Institute, Boston, MA, USA
The Broad Institute of MIT and Harvard, Cambridge, MA, USA
e-mail: pcs.murali@gmail.com

L. Ozdamar
Department of Systems Engineering, Yeditepe University, Kayisdagi, 34755 Istanbul, Turkey
e-mail: linetozdamar@lycos.com; lozdamar@hotmail.com

G.-W. Weber
Institute of Applied Mathematics, Middle East Technical University, 06531 Ankara, Turkey
e-mail: gweber@metu.edu.tr

A.A. Pinto and D. Zilberman (eds.), *Modeling, Dynamics, Optimization and Bioeconomics I*, Springer Proceedings in Mathematics & Statistics 73, DOI 10.1007/978-3-319-04849-9__31,
© Springer International Publishing Switzerland 2014

displacement also takes place resulting in very high migration rates (e.g., the armed conflict in Turkey resulted in a net migration rate of about 50 % [9]. Migrant populations usually develop their own squatter areas where crime rates and illiteracy are high. Goksen and Cemalciler [9] suggest that structural risk factors such as child labour, illiterate mother, unstable employment of the father are potential reasons for boys dropping out more than girls. Previous weak academic achievement and social alienation in an urban environment create a lot of psychological and academic difficulties for students who are transferred to urban schools where teachers are unprepared for classroom diversity [2, 4]. However, economic difficulty is still regarded as the main factor for dropout, malnutrition and child labor [6, 27]. Other impact factors for poor academic performance are overcrowded, unattractive learning environments (World Bank 2005), lack of infrastructure, and the factor of not speaking the formal country language [22].

Trying to improve enrolment rates and student performance in squatter areas certainly deserves serious attention for the general welfare of the society. Therefore, there are several models proposed to study the factors influencing the primary school enrollment and progression. These are logistic regression models [1], poisson regression models [1], system models [5, 15, 19, 24], behavioral models [7, 11] constructed for the context of different countries. Several factors which influence the school enrollment and drop outs are identified in various studies. Some of the vital factors at the macro level are social, economic and logistics factors [7], and at the micro level there are parental education, household wealth/income, distance to school, financial assistance to students and quality of school [1, 7, 21]. An early system dynamics model to investigate the low efficiency of primary education in Latin America is introduced by [24]. This model investigates the progression through primary school and includes causal chains leading to progression, dropout and repetition of students. Karadeli et al. [15] develop a model to analyze the future quality of the Turkish educational system based on the budget of the Ministry of National education. In this model, quality of education and progression of students is influenced by the student to teacher ratio and student to class ratio. Altamirano and van Daalen [5] propose a system dynamics model to analyze the educational system of Nicaragua and helps in identifying and analyzing the consequences of policies that are aimed at improving the coverage of the different educational programs, reducing illiteracy and increasing the average number of schooling years of the population. This study shows that implementing literacy programs and introducing a program in which families in extreme poverty receive a subsidy has an effect on school coverage as well as on the number of illiterate people. More recently, [11] shows that school quality and grade completion by students are directly linked. The World Bank has published several reports on achieving universal primary education [8, 23]. In particular, Serge [23] focuses on the infrastructure challenge in Sub-Saharan Africa and the constraints to scale up at an affordable cost.

In this study, we present the Cross Impact Analysis (CIA) as a system dynamics method that enables the construction of a model relating entities and attributes relevant to the primary education system. Then, the CIA simulates the model construct and observes the changing system status. Furthermore, the CIA permits the integration of proposed policies into the model construct and enables the comparison of the simulated system with the one that is augmented by such policies. Thereby, it is possible to observe the effectiveness of policies on school enrollment rates and student performance. Here, we briefly describe how CIA method is utilized to achieve these goals in two developing countries.

31.2   The Model

The model proposed here is developed by using the cross impact analysis method (CIA). The CIA method is one of the most popular systems thinking approach developed for identifying the relationships among the variables defining the systems [10, 13, 28]. This method first was developed by Theodore Gordon and Olaf Helmer in 1966 in an attempt to answer a question whether perceptions of how future events may interact with each other can be used in forecasting. As it is well known, most events and trends are interdependent in some ways. The CIA method provides an analytical approach to the probabilities of an element in a forecast set, and it helps to assess probabilities in view of judgments about potential interactions between those elements. (We refer to [17] and [18] for more detailed information on system dynamics modeling.) CIA has been used to model and simulate several real-time problems (for example: [12, 20, 25, 26, 29]). Here, we briefly describe the steps of the CIA method through a block diagram given in Fig. 31.1.

31.2.1   Definition of the System

Systems defined based on entities, which interact with each other and produce some outputs that are either designed or natural. A system receives inputs and converts them through a process and produces outputs. All the outputs of a system need not be desirable. In the present context, the system represents the primary education system.

a. Environment

Every system functions in an Environment, which provides inputs to the system and receives outputs from the system. In our context, the Environment is the society.

b. Structure

All systems have a Structure. The 'body' of a system's structure is represented by the entities of the system and their interrelationships or linkages or connections. The entities in our system are defined as follows.

1. student,
2. teacher,

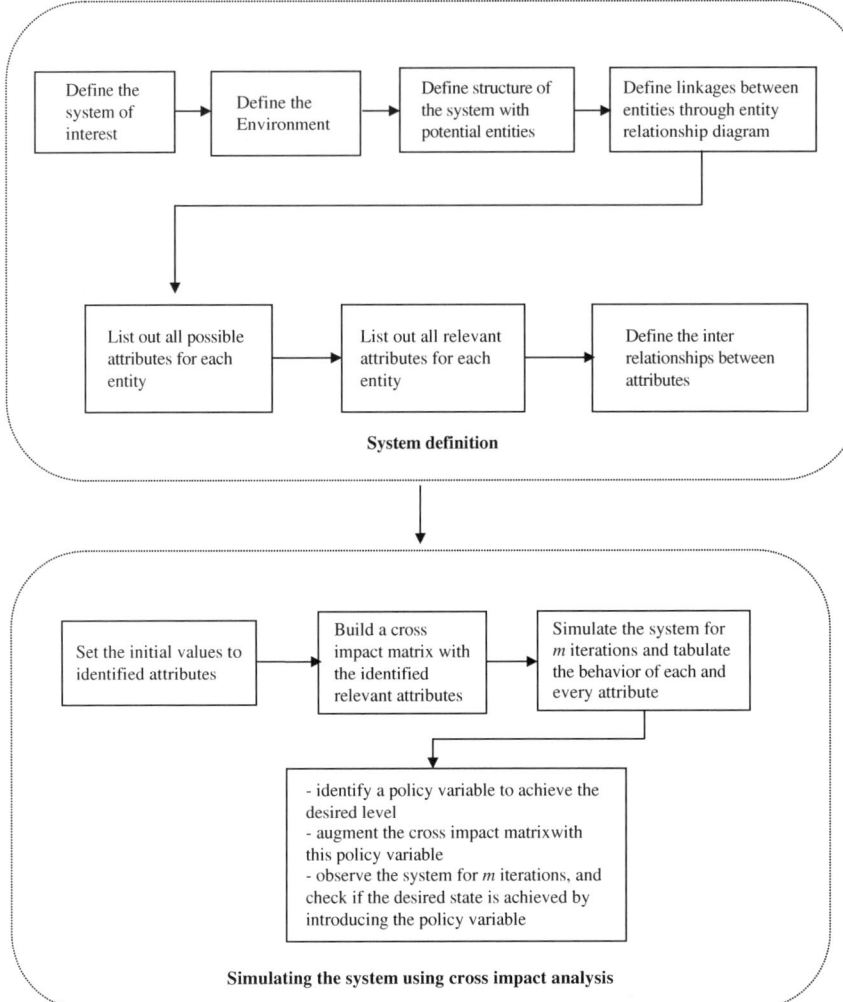

Fig. 31.1 Block diagram for the steps of the CIA method

3. parents,
4. educational officials,
5. infrastructure and
6. local community.

c. Linkages

The linkages among entities may be physical (e.g., facilitates), electro-magnetic (e.g., electrical, electronic and communications systems, and so on), and information-based (e.g., influence, and so on). It is important to try and understand,

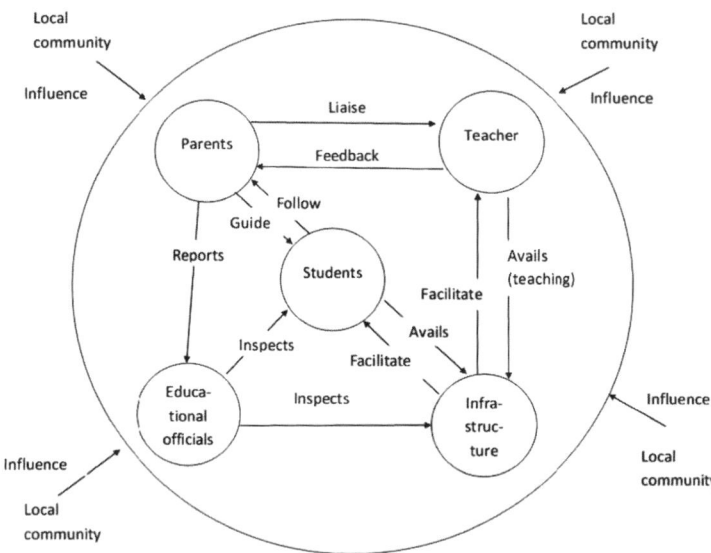

Fig. 31.2 Entity relationship diagram for the primary education system

what linkages exist in the system's structure, which entities are linked with each other, and the implications of these linkages on the behavior of the entities in particular. The entity relationship diagram of the system is illustrated in Fig. 31.2. Exchange of matter, information and/or spirit between two entities causes a change in the state of both entities. This is reflected as system behavior.

31.2.2   System Entities and Relationships Equations

The dynamic change of the system state is referred to as *system behavior*. The state of a system is an instantaneous snapshot of levels (or, amounts) of the relevant attributes (or, characteristics) possessed by the entities that constitute the system. In all systems, every entity possesses many attributes, but only a few attributes are "relevant" with respect to the problem at hand. Some attributes are of immediate or short-term relevance while others may be of relevance in the long run. The choice of relevant attributes has to be made carefully, keeping in mind both the short-term and long-term consequences of solutions (decisions). All attributes can be associated with given levels that may indicate quantitative or qualitative possession.

When entities interact through their attributes, the levels of the attributes might change, i.e., the system behaves in certain directions. Some changes in attribute levels may be desirable while others may not be so. Each attribute influences several others, thus creating a web of complex interactions which eventually determine system behavior. In other terms, attributes are variables that vary from time to time. They can vary in the system in an unsupervised way. However, variables can be controlled directly or indirectly, and partially by introducing new intervention policies. The interrelationships among variables should be analyzed carefully before introducing new policies.

The following **conjectures** are valid in the systems approach (the following is motivated by [13]).

(a) *Modeling and forecasting the behavior of complex systems are necessary if we are to exert some degree of control over them.*

(b) *Properties of variables and interactions in large scale system variables are bounded such that:*

(i) *System variables are bounded.* It is now widely recognized that any variable of human significance cannot increase indefinitely. There must be distinct limits. In an appropriate set of units these can always be set to a value between one and zero:

(ii) *A variable increases or decreases according to whether the net impact of the other variables is positive or negative.*

To preserve boundedness, $x_i(t + \Delta t)$ is calculated by the transformation

$$x_i(t + \Delta t) = x_i(t)^{P_i}, \tag{31.1}$$

where the exponent $P_i(t)$ is given by

$$P_i(t) = \frac{(1 + \Delta t |sum\ of\ negative\ impacts\ on\ x_i|)}{(1 + \Delta t |sum\ of\ positive\ impacts\ on\ x_i|)}. \tag{31.2}$$

(iii) *A variables response to a given impact decreases to zero as that variable approaches its upper or lower bound. It is generally found that bounded growth and decay processes exhibit this sigmoidal character.*

(iv) *All other things being kept fixed (constant), a variable (attribute) will produce a greater impact on the system as it grows larger* (ceteris paribus*).*

(v) *Complex interactions are described by a looped network of binary interactions (this is the basis of the* cross impact analysis*).*

31.2.3 Simulating the System Using Cross Impact Analysis

There are four steps to follow while implementing the cross impact analysis in our case. First, we conduct the simulation by considering the primary education system without human intervention. Then, we run the same analysis after implementing some selected policy variables such as infrastructure improvement and observe the change in system dynamics.

We now describe how we construct the model in the following four steps.

Step 1.   Set the initial values for attributes.

Step 2.   Build a cross impact matrix with the identified relevant attributes. Summing the effects of column attributes on rows shows the effect of each attribute in the matrix. The parameters α_{ij} can be determined by creating a pairwise correlation matrix after collecting the data, and these can be adjusted by subjective assessment.

Step 3.   Simulate the system for a number of 50 iterations (m iterations) and tabulate the behavior of each and every attribute in each every iteration. Plot the results on a worksheet.

Step 4.   Identify a policy variable to achieve the desired level or state and augment the cross impact matrix with this policy variable with the qualitative assessment of pairwise attribute interactions. Re-simulate the model.

31.3   Implementation

The method is implemented on two applications, the first of which is based on survey data collected from Gujarat state of India [19]. The goal of the model is to measure the impact of infrastructure on primary school enrolment rates. The second application involves primary education performance data collected from the slum districts of 24 provinces in Turkey [3] where variables that influence student learning and teacher's classroom practices are identified.

31.3.1   Gujarat Application

The set of attributes for the Gujarat application are given below.

Entity 1: Student:

1.1   Level of Enrollment (*loe*).
1.2   Level of boys dropouts in a school (*lbd*).
1.3   Level of girls dropouts in a school (*lgd*).
1.4   Level of repeaters in a school (*lr*).

Entity 2: Teacher:

2.1   Level of perceived quality of teaching by the Students (*lts*).
2.2   Level of perceived quality of teaching by the Parents (*ltp*).

Entity 3: Parents:

3.1   Educational level of parents (*elp*).
3.2   Income level of parents (*ilp*).
3.3   Level of expectations from school by the parents (*leps*).

Table 31.1 Impact rates of variables (attributes)

Representation of impact	Value	Description
+ + ++	0.8	Very strong positive effect
+ + +	0.6	Strong positive effect
++	0.4	Moderate positive effect
+	0.2	Mild positive effect
0	0	Neutral
-	−0.2	Mild negative effect
–	−0.4	Moderate negative effect
——	−0.6	Strong negative effect
——	−0.8	Very strong negative effect

Entity 4: Educational officials:

4.1 Level of perceived quality of teaching by the District educational officer (DEO) (*ltd*).

Entity 5: Infrastructure:

5.1 Level of Space and ventilation available in a Classroom (*lsv*).
5.2 Level of cleanliness and other facilities such as board, mats, table/chair, educational aids (maps, toys, charts, etc.) (*lc*).
5.3 Level of sanitation facilities for general purpose (for both boys and girls) (*ls_g*).
5.4 Level of separate sanitation facilities for girls (*ls_s*).
5.5 Level of drinking water facility available (*ldw*).
5.6 Level of availability of Playground area and other equipment for children used in playing (*lpa*).
5.7 Level of bad organisation in the classrooms (*lbo*):

   (a) Number of cases in which more than one class is conducted in a single instructional classroom.
   (b) Number of cases in which more than 40 people are accommodating in a single instructional classroom.

Entity 6: Local community:

6.1 Level of participation of local community (*llc*).
6.2 Level of awareness of local community about educational benefits (*lale*).

The initial attribute values are obtained from the survey data reported in [19]. We quantify qualitative attribute impacts in Table 31.1. A cross-impact matrix for the attributes listed above is illustrated in Table 31.2. We apply Step 3 and illustrate, in Fig. 31.3, the simulation of the system for 50 iterations without any policy related variables. It is observed that there is sharp increase in enrollment rate at the beginning phase of the simulation (i.e., for the first 12 iterations). However, there is a steady decrease in the enrollment rate after a certain period of time. A similar kind of trend is observed in the level of boy dropouts in the first four iterations and in the

Table 31.2 Cross impact matrix for primary education system

	loe	lsv	lc	elp	ilp	leps	lts	ltp	ltd	ls_g	ls_s	lpa	llc	lale	lr	lbd	lgd	lbo	ldw
loe	o	+	+	+	+	+	+	+	o	+	+	+	o	o	o	−	o	−	+
lsv	−	o	o	o	o	o	o	o	+	o	o	o	+	+	o	o	o	−	+
lc	−	o	o	o	o	o	o	o	+	o	o	o	+	+	o	o	o	−	+
elp	o	o	o	o	o	o	o	o	o	o	o	o	o	o	o	o	o	o	o
ilp	o	o	o	o	o	o	o	o	o	o	o	o	o	o	o	o	o	o	o
leps	o	o	o	o	o	o	o	o	o	o	o	o	o	o	o	o	o	o	o
lts	o	+	+	o	o	o	o	+	+	o	o	+	o	o	o	o	o	−	o
ltp	o	o	o	o	o	o	+	o	+	o	o	+	o	o	o	o	o	−	o
ltd	o	o	o	o	o	o	o	o	o	o	o	o	o	o	o	o	o	−	o
ls_g	−	o	o	o	o	o	o	o	+	o	o	o	+	+	o	o	o	o	o
ls_s	−	o	o	o	o	o	o	o	+	o	o	o	+	+	o	o	o	o	o
lpa	−	o	o	o	o	o	o	o	o	o	o	o	+	+	o	o	o	o	o
llc	−	o	o	o	o	o	o	o	o	o	o	o	+	+	o	o	o	o	o
lale	o	o	o	o	o	o	o	o	o	o	o	o	o	o	o	+	o	o	o
lr	+	o	o	o	o	o	o	o	o	o	o	o	−	−	o	+	o	+	o
lbd	o	−	−	−	−	−	−	−	o	−	−	−	−	−	+	o	o	+	−
lgd	o	−	−	−	−	−	−	−	o	−	−	−	−	−	+	o	o	+	−
lbo	+	o	o	o	o	o	o	o	−	o	o	o	o	o	o	o	o	o	o
ldw	−	o	o	o	o	o	o	o	o	o	o	o	+	+	o	o	o	o	o

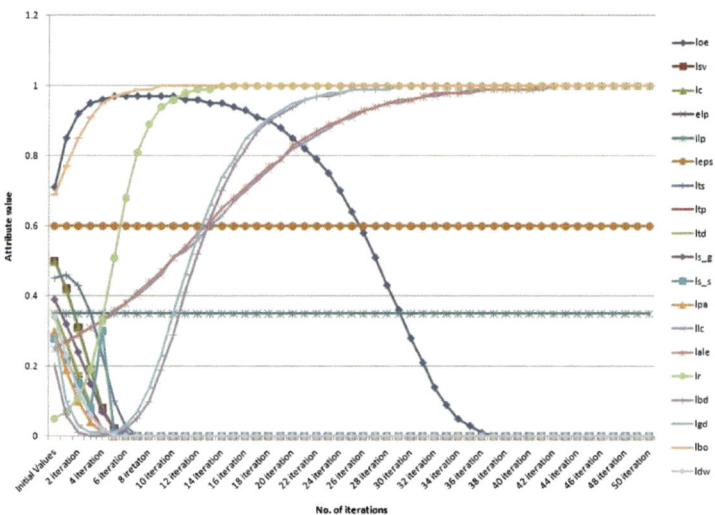

Fig. 31.3 Behavior of primary educational system before adding the policy variable

level of girl dropouts in the first five iterations. This early amelioration in the dropout rates is short lived, and both boy and girl dropouts increase steadily thereafter. We validate the simulation results by comparing them with observed levels of enrolment, dropouts and repeaters published by Directorate of Primary Education, Gandhinagar (http://gujarat-education.gov.in/primary/mahiti/shada-chodvano-dar-eng.htm).

In order to observe the effect of infrastructure attributes, we include them as policy variables in our next step. The policy variable that is selected involves additional investment in the infrastructure related attributes and elements which we call it as "policy variable". In Step 4, we augment the cross impact table with the policy variable. We re-simulate the system and observe it for 50 iterations, and check if the desired state is achieved by introducing the policy variable. We then compare the results obtained in the two simulation runs. Figure 31.4 illustrates the results of the simulated system after adding the identified policy variable in Step 4. Here, it is observed that the policy variable is effective on improving the enrollment and dropout and repeater rates.

After a policy variable related to infrastructure improvements is introduced, a positive impact is observed on the level of space and ventilation available in classrooms, level of cleanliness and other facilities such as board, mats, table/chair, educational aids (maps, toys, charts, etc.), level of separate sanitation facilities for girls, level of general sanitation facilities, level of available drinking water facilities, and class organization. These impacts are discussed with education officials, parents, students, and other local community people. By introducing this policy variable, the enrollment rate has improved steadily from an initial value of 0.71 to unity in a few iterations. Further, the level of repeaters increased to a value of

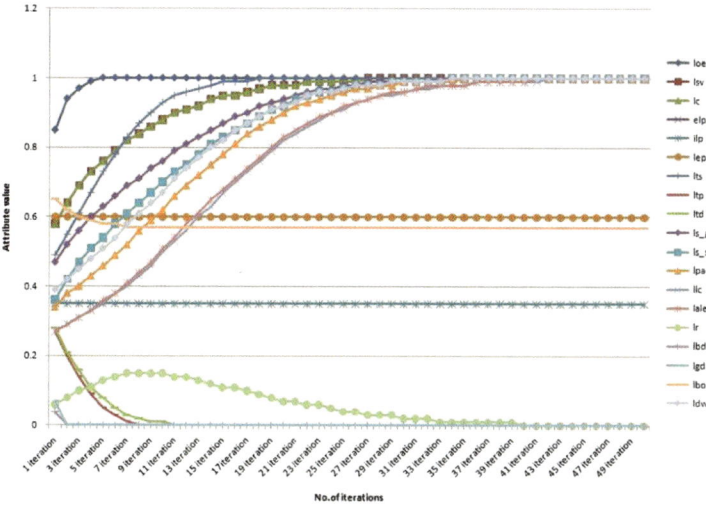

Fig. 31.4 Behavior of primary educational system after adding the policy variable

0.12 from an initial value of 0.05 in first 14 iterations, and then declined thereafter. This is logical in the sense that an improvement in the infrastructure does not have an instant impact on the level repeaters, but it would have an instant impact on the enrollment rate because students and parents are more eager to have the children attend a nice looking healthy school. The level of bad organization in the classroom is not greatly affected by the improvement in infrastructure facilities because there are several other attributes that influence this variable such as the level of perceived quality of teaching by the district educational officer and the number of teachers available for teaching. Consequently, the level of bad classroom organization is reduced from 0.69 to 0.57 in the second simulation run. In previous studies found in the literature, it is observed that the quality and the number of teachers have significant impacts on the enrollment, dropouts and repeaters. The design of our proposed model is sufficiently flexible to accommodate those impacts in future studies.

To summarize, in this study, we find that infrastructural facilities have significant impacts on the enrollment, dropout and repeater rates.

31.3.2   Turkey Application

The set of attributes for the Turkish case are listed below. These are suggested by [3].
   Entity 1: Student:

F1.1   Not believing in education
F1.2   Dislike of school books
F1.3   Lack of good Turkish language skills

F1.4   Dislike of school
F1.5   Weak academic self confidence
F1.6   Frequent disciplinary problem

Entity 2: Teacher:

F2.1   Relating to guidance teacher

Entity 3: Family:

F3.1   Economic difficulty
F3.2   Child labor
F3.3   Large families
F3.4   Malnutrition
F3.5   Homes with poor infrastructure (e.g., lack of heating, lack of room)
F3.6   Different home language
F3.7   Parent interest

Entity 4: School environment:

F4.1   Violence at school
F4.2   Alien school environment
F4.3   Lecture not meeting student needs

As in the previous application, we conduct the simulation by considering the primary education system without any new policy intervention. Then, we run the same analysis after implementing some selected policy variables such as daily distribution of milk, financial aid, and similar measures in order to observe the change in system state (student academic performance).

In Step 1, the initial values are obtained by considering some statistical measures on teacher's perceptions of the major challenges faced by migrant communities [3]. For example. 77.4 % of the teachers report in the survey that large family size appear to impact student's schooling in a negative manner; therefore, we assign the initial value of 0.774 to attribute code F3.3, "large family".

In Table 31.3, we provide the correlation matrix between the variables considered in the survey data [3]. The cross impact matrix is needed for calculating $P_i(0)$ and we build this matrix based on the strength of the relationships reflected by the correlation matrix given in Table 31.3 and based on the literature survey. The variables in Akar's survey deal with household, social milieu and academic issues that may challenge and affect student's academic lives the most. Those variables are listed under four entities as student, teacher, parent, and school environment. The correlation matrix implies that the strongest correlations (significant at the 0.05 level) are related to household variables and human capital as can be seen in Table 31.3. The family speaking in another language than the formal language is found to correlate highly with child labor, crowded family structure, and low belief in schooling. Therefore, there are strong correlations between economic difficulty and malnutrition; large family and child labor; large family and malnutrition; home language and large family; poor home infrastructure and large family. Different home language is also positively correlated to disbelief in education and dislike

Table 31.3 Correlation matrix for pairs of attributes

	F3.1	F3.2	F3.3	F4.2	F3.4	F1.1	F1.2	F1.3	F1.4	F4.3	F3.5	F3.6	F4.1	F1.5	F1.6	F2.1	F3.7
F3.1	–																
F3.2	0.296	–															
F3.3	0.397	0.486	–														
F4.2	0.187	0.202	0.311	–													
F3.4	0.421	0.282	0.453	0.307	–												
F1.1	0.103	0.270	0.288	0.264	0.238	–											
F1.2	0.140	0.101	0.140	0.137	0.204	0.254	–										
F1.3	0.199	0.134	0.260	0.119	0.349	0.297	0.298	–									
F1.4	0.054	0.204	0.222	0.211	0.171	0.457	0.226	0.227	–								
F4.3	0.131	0.098	0.167	0.157	0.210	0.200	0.462	0.238	0.298	–							
F3.5	0.385	0.258	0.491	0.235	0.502	0.236	0.194	0.346	0.212	0.279	–						
F3.6	0.232	0.452	0.488	0.221	0.355	0.263	0.178	0.341	0.204	0.164	0.442	–					
F4.1	0.051	0.235	0.137	0.065	0.146	0.122	0.215	0.084	0.133	0.161	0.097	0.248	–				
F1.5	0.261	0.187	0.338	0.229	0.290	0.311	0.205	0.329	0.312	0.289	0.438	0.298	0.079	–			
F1.6	0.023	0.152	0.179	0.165	0.088	0.260	0.051	0.063	0.262	0.048	0.075	0.169	0.095	0.157	–		
F2.1	0.040	0.044	0.048	0.077	0.001	0.022	−0.035	−0.002	0.034	−0.049	0.011	0.008	−0.014	0.037	0.209	–	
F3.7	−0.166	−0.316	−0.294	−0.075	−0.175	−0.235	−0.071	−0.177	−0.223	−0.077	−0.221	−0.251	−0.036	−0.255	−0.240	−0.023	–

of school. These students do not speak the formal language well, are in economic difficulty and have weak academic confidence. The last three attributes are the ones having the most positive correlations with the rest of the attributes. Parental interest is negatively correlated with other attributes, the reason being obvious: parental support is high when the family is not in economic difficulty, since parents have more time to take better care of their children.

After simulating the system as is, six different policy variables are selected, and a seventh policy that is a combination four policies. The proposed policies are listed below.

P_1. Daily distribution of milk.

P_2. Distribution of coal, food etc. to below poverty line families.

P_3. The three-children per family aspiration by government.

P_4. School infrastructure renovation.

P_5. Providing adult education for poor parents in migrant communities.

P_6. Providing 1 year of Turkish language class before migrant student attends urban school (17 % of the Turkish population has a mother tongue different from Turkish (http://betam.bahcesehir.edu.tr).

P_7. Combined policy: $P_2 + P_4 + P_5 + P_6$: The combined policy is proposed because each policy listed above addresses a different aspect of the situation. While the mere basics of survival in the city are maintained by P2, policy P5 would enable the parents to find better employment and improve the migrant family's economic conditions. On the other hand, policies P4 and P6 target the academic performance of students directly.

Before the simulation augmented with policies starts, the desirable state of each attribute is stored: an attribute that has a negative impact on academic performance should stabilize at a level of 0.0 at the end of the simulation. On the other hand, an attribute with a positive impact should ideally reach the level of 1.0.

During the simulation, the level of each attribute, $x_i(t)$, is observed in each iteration and recorded. At the end of the simulation, the specific iteration at which each attribute stabilizes at the desirable level of 0.0 or 1.0 is reported, if it ever reaches the desirable state. If, on the contrary, the attribute converges to the undesirable state (1.0 for negative impact attributes and 0.0 for positive impact attributes), then, the iteration at which the attribute reaches its undesirable state is reported. It is possible that under some policies the system will converge to an undesirable state while in others it will converge to the desirable state.

When we compare two policies that both end up in undesirable states, we take a look at the iterations at which the attributes converge to their undesirable states. The policy that causes the latest convergence to the undesirable state is considered more successful. We calculate the average number of iterations in which the system stabilizes at its undesirable state. In a similar fashion, when two policies are able to stabilize the system at its desirable state, we prefer the policy that renders this stabilization in the minimum average number of iterations. That is, the policy that moves the system to the desirable state sooner is preferable. We can also compare the effects of the policies by comparing the average number of iterations that the

Table 31.4 The number of iterations where each variable reaches its desirable and undesirable level under every policy

Variables	Initial value	Undesirable value	Desired value	NO_P	P1	P2	P3	P4	P5	P6	P7
F1.1	0.693	**1**	0	**15**	**19**	**21**	**15**	**31**	**39**	**23**	23
F1.2	0.405	**1**	0	**18**	**18**	**20**	**18**	**50**	**21**	**50**	20
F1.3	0.509	**1**	0	**14**	**15**	**16**	**14**	**17**	**19**	**31**	28
F1.4	0.261	**1**	0	**16**	**24**	**21**	**16**	**36**	**22**	**30**	22
F1.5	0.707	**1**	0	**19**	**20**	**28**	**19**	**50**	**50**	**25**	14
F1.6	0.5	**1**	0	**25**	**25**	**25**	**25**	**26**	**25**	**25**	28
F2.1	0.5	**0**	1	**12**	**12**	**12**	**12**	**23**	**12**	**14**	43
F3.1	0.815	**1**	0	**13**	**16**	**24**	**11**	**15**	**35**	**15**	36
F3.2	0.5	**1**	0	**20**	**27**	**50**	**16**	**25**	**50**	**20**	29
F3.3	0.774	**1**	0	**17**	**18**	**21**	**12**	**21**	**50**	**17**	33
F3.4	0.729	**1**	0	**16**	**41**	**35**	**13**	**21**	**47**	**17**	30
F3.5	0.799	**1**	0	**14**	**18**	**30**	**12**	**18**	**29**	**15**	32
F3.6	0.658	**1**	0	**16**	**17**	**19**	**16**	**23**	**27**	**43**	25
F3.7	0.5	**0**	1	**10**	**16**	**37**	**6**	**44**	**50**	**12**	23
F4.1	0.5	**1**	0	**16**	**17**	**18**	**16**	**25**	**19**	**20**	44
F4.2	0.619	**1**	0	**15**	**19**	**17**	**15**	**32**	**17**	**23**	33
F4.3	0.5	**1**	0	**19**	**20**	**21**	**19**	**50**	**23**	**36**	12
Average				16	20	24	15	30	31	24	28
Standard deviation				3	7	9	4	12	14	11	9
Maximum				25	41	50	25	50	50	50	44

Bold indicates the iteration where a variable stabilizes at its undesirable value; normal value indicates the iteration where a variable stabilizes at its undesirable value

system stabilizes under a given policy to the average number of iterations where the system achieves the same state under no policy at all. This comparison gives an indication of how a given policy slows down system degradation.

After simulating the system for 50 iterations using Eq. (31.1), we summarize the results according to the performance indicator. We provide the summary of results in Table 31.3. In Table 31.3, the first column refers to the attribute codes listed previously. The second column illustrates the initial values of the attributes, the third and fourth columns show the desired (ideal) and undesirable (worst case) stabilization levels of attributes at system convergence. All columns between the sixth and the twelfth columns stand for the simulation runs that have one policy added to the system as an attribute. The header for each column indicates which policy has been added. The fifth column stands for the simulation run where the 'do nothing' option is used, that is, no policy is added to the system. All columns between the fifth and the eleventh columns indicate the earliest iterations at which each attribute converges to its undesirable state.

In the twelfth column where policy P_7 is added to the system, we can see that all attributes converge to their desirable states. The results in Table 31.3 show us that none of the policy variables except for the combined policy P_7 lead to a desirable system state convergence. The policy variables P_1 to P_6 are not sufficiently effective when they are implemented singularly (Table 31.4).

Under the current conditions (the "do nothing" option), we can see that due to worsening economic conditions, parental interest in education will be reduced to null very soon. Indeed, given the high jobless rate in Turkey (around 15 % if stay-home people are not counted, and, nearly 56 % jobless rate if stay-home women are included, http://nkg.tuik.gov.tr), we do not think the economic conditions of poor people will improve in the foreseeable future. We also observe that student disciplinary issue is not affected by any of the policies except for the combined policy, implying that psychological problems are the hardest to solve. Policy P_1, delays the deterioration of the attribute levels of child labor, malnutrition and school dislike. Policy P_2, distribution of aid, involves more extensive economic help, and impacts the deterioration of economic difficulty, child labor, malnutrition and home environment. However, it is noted that feeding students with milk is more effective in preventing malnutrition, because parents tend to utilize government aid for other purposes as well and their number one priority is not feeding their children properly. However, government aid does tend to reduce child labor and it helps in keeping migrant children at school. Policy P_3 (the three children per family aspiration) deteriorates all system state, and obviously, the rate of population increase has to slow down if the society wants to improve the job market. Policy P_4 (improving school infrastructure) ameliorates the school environment, and students tend to like education and books more. Policy P_4 also improves the academic confidence of students, and in our opinion, in the long run, it reduces the drop out rate. Policy P_5 (adult education for migrant parents) improves economic conditions, and has some positive impact on prevent child labor. Policy P_5 also have positive impact on the family size by teaching parents about birth control. Educating the parents has a positive impact on the attitude of their offspring towards schooling. This policy has also an interesting effect on academic self-confidence of students, it seems that when parents learn about the rewards of education, they become enthusiastic about the schooling of their children, and hence, the children become more proud of being students. Policy P_6 (Turkish language class) has a positive effect on school dislike, naturally, when a student can actually read his/her books, he or she becomes more interested in school work. Finally, the combined policy P_7 is the sole policy to drive the system to the desired state for all the attributes. Obviously, P_7 tries to attack all major causes of student's low academic performance; however, it is also the most expensive option.

31.4 Conclusion

A cross-impact model is developed here to build system dynamics models and study the influence of various policies on primary education enrollment and academic performance. The cross-impact matrix illustrates the influence of one variable over the others and it also has a provision to identify the impact variables (i.e., policy variables). Here, we present two implementations of this method in two different developing countries, India and Turkey. We illustrate which policies might lead the primary education system to its desired state of excellence.

Acknowledgements The authors thank Dr. Hanife Akar (Dept of Educational Sciences, Middle East Technical University), Prof. L. S. Ganesh (Indian Institute of Technology Madras, Chennai, India) for their fruitful discussions. We also wish to thank Mrs. Anupama Pedamallu for helping us with data entry.

References

1. Admassu, K.: Primary school enrollment and progression in ethiopia: family and school factors. In: American Sociological Association Annual Meeting, Boston, 31st July 2008
2. Akar, H.: Poverty, and schooling in Turkey: a needs assessment study. In: Presentation at Workshop on Complex Societal Problems, Sustainable Living and Development, IAM, METU, Ankara, 13–16 May 2008
3. Akar, H.: Challenges for schools in communities with internal migration flows. Int. J. Educ. Dev. **30**, 263–276 (2010)
4. Aksel, S., Gun, Z., YlmazIrmak, T., Celenci, B.: Migration and psychological status of adolescents in Turkey. Adolescence **42**(167), 589–602 (2007)
5. Altamirano, M.A., van Daalen, C.E.: A system dynamics model of primary and secondary education in Nicaragua. In: 22nd International Conference of the System Dynamics Society, Oxford, 25–29 July 2004
6. Baslevent, C., Dayioglu, M.: The effect of squatter housing on income distribution in Urban Turkey. Urban Stud. **42**(1), 32–45 (2005)
7. Benson, H.: Household demand for primary schooling in ethiopia: preliminary findings. In: Annual Meeting of the American Educational Research Association, San Francisco, 18–22 April 1995
8. Bruns, B., Mingat, A., Ramahatra, R.: Achieving Universal Primary Education by 2015, a Chance for Every Child. The World Bank, Washington, DC (2003)
9. Goksen, F., Cemalciler, F.: Social capital and cultural distance as predictors of early school dropout : implications for community action for Turkish internal migrants. Int. J. Intercult. Relat. **34**,163–175 (2010)
10. Gordon, T.J., Hayward, H.: Initial experiments with the cross-impact matrix method of forecasting. Futures **1**, 100–116 (1968)
11. Hanushek, E.A., Lavy, V., Kohtaro, H.: Do students care about school quality? Determinants of dropout behavior in developing countries. J. Hum. Cap. **2**(1), 69–105 (2008)
12. Hayashi, A., Tokimatsu, K., Yamamoto, H., Mori, S.: Narrative scenario development based on cross-impact analysis for the evaluation of global-warming mitigation options. Appl. Energy **83**(10), 1062–1075 (2006)
13. Julius, K.: A primer for a new cross-impact language: KSIM. In: Harold, A.L., Murray, T. (eds.) The Delphi Method: Techniques and Applications. Addison-Wesley, Reading (2002)
14. Kagitcibasi, C., Cemalcilar, Z., Baydar, N.: Children of rural to urban migration: an integrative intervention for adaptation to social change. ISSBD Bull. **1**(55),10–14 (2009)
15. Karadeli, N., Kaya, O., Keskin, B.B.: Dynamic Modeling of Basic Education in Turkey. Senior graduation project, Bogazici university, Turkey (2001)
16. Kulu, H., Billari, F.: Multilevel analysis of internal migration in a transitional country: the case of Estonia. Reg. Stud. **38**(6), 679–696 (2004)
17. Lane, D.C.: Social theory and system dynamics practice. Eur. J. Oper. Res. **113**(3), 501–527 (1999)
18. Mohapatra, P.K.J., Mandal, P., Bora, M.C.: Introduction to System Dynamics Modeling. Universities Press (India) Limited, Hyderabad (1994)
19. Pedamallu, C.S.: Externally aided construction of school rooms for primary classes- preparation of project report. Master's dissertation, Indian Institute of Technology Madras, Chennai (2001)

20. Pedamallu, C.S., Ozdamar, L., Kropat, E., Weber, G-W.: A system dynamics model for intentional transmission of HIV/AIDS using cross impact analysis. Cent. Eur. J. Oper. Res. (2010). doi: http://dx.doi.org/10.1007/s10100-010-0183-2
21. Rena, R.: Factors affecting the enrolment and the retention of students at primary education. Essays Educ. **22**, 102–112 (2007)
22. Sahin, I.: Cultural responsiveness of school curriculum and students' failure in Turkey. Interchange **34**(4), 383–420 (2003)
23. Serg, T.: School Construction Strategies for Universal Primary Education in Africa. The World Bank, Washington, DC (2009)
24. Terlou, B., van Kuijk, E., Vennix, J.A.M.: A system dynamics model of efficiency of primary education in Latin America. In: Proceedings of the International Conference of the System Dynamics Society, pp. 578–587 (1991)
25. Thorleuchter, D., Van den Poel, D., Prinzie, A.: A compared R&D-based and patent-based cross impact analysis for identifying relationships between technologies. Technol. Forecast. Soc. Change **77**(7),1037–1050 (2010)
26. Torres, N., Olaya, C.: Tackling the mess: causal-loop conceptualization of solid waste management systems through cross-impact. In: 28th International Conference of the System Dynamics Society, Seoul, 25–29 July 2010
27. Tsui, M.: Family income, home environment, parenting, and mathematics achievement of children in China and the United States. Educ. Urban Soc. **37**(3), 336–355 (2005)
28. Weimer-Jehle, W.: Cross-impact balances: a system-theoretical approach to cross-impact analysis. Technol. Forecast. Soc. Change **73**(4), 334–361 (2006)
29. Wu, G.: Application of cross-impact analysis to the relationship between aldehyde dehydrogenase 2 allele and the flusing syndrome. Alcohol Alcoholism **35**(1), 55–59 (2000)

Chapter 32
The Bioeconomics of Migration: A Selective Review Towards a Modelling Perspective

E.V. Petracou, A. Xepapadeas, and A.N. Yannacopoulos

32.1 Introduction

Migration develops different spatio-temporal patterns depending on specific historical social and economic conditions. Furthermore, it is an issue of paramount importance in social, political and economic theory, that infiltrates many aspects of everyday life.

The modelling of migration is a very challenging issue from both the theoretical and practical point of view and nowadays its study involves a variety of disciplines ranging from the social sciences and economics to science, mathematics and statistics and extending to the arts and philosophy, since migration is a multi-dimensional and multi-faceted complex phenomenon and the mobilities of people have intensified and have become a 'permanent' feature of social reality.

It is the purpose of this expository chapter to (a) present a brief account of the theoretical study on migration as has been developed since the 1970s placing the emphasis on the economic aspects of migration (b) develop a bioeconomic type framework for the monitoring of mobility within different geographical regions, incorporating various socio-economic factors in the decision to migrate using a discrete choice model and (c) coupling the population dynamics to a simple

E.V. Petracou (✉)
Department of Geography, University of the Aegean, Lesvos, Greece
e-mail: ipetr@geo.aegean.gr

A. Xepapadeas
Department of International and European Economic Studies, Athens University of Economics and Business, Athens, Greece
e-mail: xepapad@aueb.gr

A.N. Yannacopoulos
Department of Statistics, Athens University of Economics and Business, Athens, Greece
e-mail: ayannaco@aueb.gr

A.A. Pinto and D. Zilberman (eds.), *Modeling, Dynamics, Optimization and Bioeconomics I*, Springer Proceedings in Mathematics & Statistics 73, DOI 10.1007/978-3-319-04849-9__32,
© Springer International Publishing Switzerland 2014

growth model and thus creating a spatio-temporal dynamical system that can monitor the complex interaction between the economy and migration and discuss the mathematical, economic and simulation challenges of such models. It must be admitted that there exists a huge literature on the field ranging from econometric studies to studies in economic growth theory, therefore our references to the literature are very selective.

32.2   Theories and Models of Migration

After the nineteenth century the emergence of modern state with territorial borders turned population into an important concept made it a fundamental issue for political interest and introduced migration alongside with fertility and mortality as one of the main subjects of the newly developed science of demography. As a result, the first attempts to construct a migration theory, or better not only to find out the rules and laws under which migration occurs but also changes in time, were in demographic terms and were influenced by the empirical observations of that time. Simultaneously, the emergence and establishment of the modern state and its national borders meant that movements of people were divided into different types i.e., internal-international/migrants-refugees etc. Specific importance was attached in migration studies to the so called immigration or settler countries such as the USA, Canada and Australia.

Later on, due to the fact that the Second World War triggered 'huge' migration labour movements within Europe and also to Europe to assist rebuilding of Western Europe after the War, scholars' interest in migration expanded considerably, turning their attention from just the traditional immigration states to the European states and to almost all states either as immigration or emigration ones.

An important group of theories of migration are called 'equilibrium' (or functional or orthodox theories) and have as their unit of analysis the individual and their explanation for the migration movements is based on the grounds of free rational choice. Migration is seen as caused by wage or income differentials between geographical areas. The motivation and decision making to migrate is seen exclusively from the point of migrant's perceptions and their interests. The presupposition is that the individual is relatively free and well-informed about wage differentials or generally about the labour market situation between countries and decides to migrate, on the basis of cost-benefit analysis. In these theories migration is seen as a means for individuals' ends.

Equilibrium theories are based on elements of neoclassical economic, modernisation theories and microeconomic perspective. In these theories, the perspective and the logic of the push-pull factors are predominant, focusing on the starting, continuing and stopping points of migration, (rural-urban. international migration) while its essential assumption is that remittances and the return of skilled migrants to the source region will stimulate economic growth.

According to a general theory developed by Lee [22] both origin and destination places are characterised by sets of plus and minus factors while between them there are sets of intervening obstacles such as distance, actual physical barriers and immigration laws, and all the above factors are causes for migration. Migration is thus seen as an almost rational individual decision in the terms of balancing costs and benefits.

In a more precise way than Lee's obstacles and sets of plus and minus factors, equilibrium theories especially in the neoclassical context, explain causes for migration on the basis of market disequilibrium between geographical areas. Traditional agricultural areas are characterised by low productivity, a supply of labour and low wages, while modern industrial urban areas represent a high productivity, a demand of labour and high wages. Thus, there is a rational transference of labour from the rural sector to the urban sector. In this perspective, Todaro's model [17] is the most developed one, because it takes into account probabilistic factors in explaining the continuing migration in the urban areas, even in the case of high unemployment. He points out that the decision to migrate depends on 'expected' rather than actual urban-rural real wage differentials and does not assume urban full employment.

Generally, equilibrium theories argue that the equilibrium between wage differences is achieved by the aggregation of individuals' decisions to migrate and focus on geographic differences in wages and unemployment, underlying the importance of push factors in migration decision making. It can be said that these theories describe the ways that actual migration is considered and functions. Both their units of analysis, the individual, and the social context in which migration takes place are taken for granted and are seen as natural and stable. The market rationality is identified with individual rationality so, the exclusive definition of individual interest is maximisation of income.

A last point is that in the logic of cost-benefit analysis for individuals and markets, other aspects of social life, such as political, ideological, social, cultural, which constitute migration are ignored. Although, it can be argued that Lee's model incorporates these factors, they are concealed in sets of plus and minus factors and intervening obstacles, in which political and legal frameworks are described as distance or actual physical barriers or just immigration laws.

A second group of theories are the so called structural or historical-structural or conflict theories. These theories emphasise structural factors which force individuals, mainly as members of the working class, to migrate according to the needs of capital and as a result labour migration intensifies the existing exploitative relations and uneven development among countries. Migration is explained in the terms of social forces and constraints posed on migrants, that is migration is seen as a structural response and a manipulated behaviour.

The historical-structuralist perspective has been founded on different versions of theories such as dependency, world system and articulation or modes of production developed by Amin, Wallerstein and the neo-Marxists (e.g. Meillassoux). Each theorist relates migration—mainly labour migration—with other broader social phenomena. For example, Piore [28] pays attention to the dual labour market at the global level, Nikolinakos [26] focuses on dependency between periphery and centre,

Castells [7] on uneven development inherent in the capitalist mode of production. and finally, Portes [29] shows that migration is a link between spatially separate modes of production due to the coexistence of modes of production.

Although structuralist theories include interesting elements of migration. they also have inconsistencies in a sense similar to those of individualistic ones, since they view capitalism as merely an economic system which develops according to some objective laws and they perceive migrants as functional elements, or mere economic units, whose actions and practice are responses to the requirements of capital such as the need for a reserve army or to split indigenous labour movement. Later research on migration, in order to avoid the above dichotomy and fragmentation in the explanation of migration, attempted to integrate both approaches in a migration theory. In order to connect the two opposite approaches researchers introduce additional units such as the 'household', 'social systems' and 'social networks'.

The third perspective of migration which attempts to combine both individual and structural approaches in the explanation of migration includes a variety of theories. In this logic, Wood argues that the unit of analysis should shift to 'household', because a migration decision may not be made merely by an individual but made by a group of individuals. This approach is similar to that adopted in the world-system approach, in which groups adopt strategies in order to overcome the limitations imposed by their socio-economic and physical environment. He argues that household behaviour is based on a series of 'sustainance dynamic strategies including geographical mobility by which household actively strives to achieve a fit between its consumption necessities, the labour power at its disposal and the alternatives for generating monetary and non-monetary income'. The 'household' must adopt dynamic, flexible and innovative strategies in order to respond to the structural factors. Both seasonal or permanent migration are seen as such strategies, permitting the household to achieve its goal which is attaining a desired level of consumption and investment.

Moreover, in the process of integrating individual and structural migration perspectives and of explaining the continuation of migration, some researchers apply the notion of networks in migration theories. In these terms, Kearney [18] introduces the concept of the 'Articulator Migrant Network' (AMN) in order to capture the complex processes of micro differentiation that occur in 'traditional' communities (characterised by the non capitalist mode of production) as they become articulated with the developed world' (characterised by the capitalist mode of production). Kearney's theory links household and migrant networks in anthropological research as efficient units of analysis of migration in connection with economic development. Owing to the fact that return migration has a negative or neutral influence on modernisation of underdevelopment countries, he identified 'household' as the appropriate unit of analysis for migration, since it reveals the role of women in production and marketing activities in the households that derive income for migrants. Moreover, the concept of the AMN can address the movements of migrants into various 'spaces' 'not only geographic but also economic and social niches and also the flow of surplus and goods within the migrant community'.

Another theory is developed by Portes and Kelly [29] who adopt the 'social network' perspective in order to explain the stability of migrant flows after the original causes have disappeared. Social networks are links between the domestic unit, and the global economic system, which includes household and extending to the family and community levels. Social networks construct and are also constructed by collective relationships across time and space. They argue that migration flows do not respond automatically to economic or political changes due to the mediation of social networks which become key structures and stabilise migration flows.

Massey and his collaborators in their exploration of migration within the context of network theory, argue that migration can be seen as 'a self-sustaining diffusion process: in which migration networks are sets of interpersonal ties between migrants, former migrants and non migrants in origin and destination areas' and these networks regulate migration behaviour. Migration networks, when expanding, assist to reduce the risks and costs of migration and 'once the number of network connections in country of origin reaches a critical level migration becomes self-perpetuating because migration itself creates the social structure to sustain it'.

The main proposition of network theories is that migration is continued almost constantly, even though not with the same volume, after the constitution of migration networks among coethnics and consequently migration becomes 'a self perpetuating social structure', independent from social and economic causes from which the migration movement started. This happened because the development of migration networks reduces costs and risks for new migrant movements and eventually migration spreads to all socioeconomic segments of the sending society. Due to well developed networks, governments have great difficulty in controlling migration while migration continues irrespectively of the kind of migration policies that governments apply.

Another theory which refers to the causes for the continuation of migrant flows is the so called institutional theory. Its initial assumption is that there is not a correspondence between the numbers of people who seek entry into rich countries and the number of visas that these countries issue. Due to this fact, an underground market for migration is developed exploiting migrants. A similar argument to those of network theories is developed by proponents of institutional theory who assume that international migration movements continue and liberate themselves from the initial causes of migration while governments are unable to curb further migration due to the pressures which both underground market and humanitarian groups put on the governments.

Fawcett [10] associates 'migration systems' with the notion of social networks in order to stress linkages between people (i.e. personal or family networks) and the economic and political linkages among sending and receiving countries and the relation between personal and non personal links. Under the same perspective, Kritz and Zlotnik [21] argue that the systems approach includes a group of countries linked by migration flows which can capture structural conditions in these countries and also economic and social ties between them. Moreover, with the incorporation of networks in the systems approach, it is possible for a migration theory to explain who is likely to become a migrant and who actually migrates, that is, to connect structures with potential migrants.

In fact, the attempts to integrate the individual and structural migration theories seem to be based on the assumption that these two elements of migration, that is, individuals and structures exist as independent but they acquire a cause-effect relationship in the case of migration. On these grounds, the above segmentation in migration research is taken for granted and these theories search for an intermediate unit of analysis in order to connect these independent but related units of analysis in migration, and consequently to explain previous inconsistencies of theories related to stability of migration flows but not in the terms of labour migration. Thus, the household and networks have been the concepts which replace the individual or migrant workers as units of analysis.

As [3] in their study of migration emphasise, migration decision making is organised by a set of dynamic as well as pre-established social relations. Another attempt to reconcile competing migration theories and evaluate them on empirical grounds is made by Massey and his collaborators [25]. The first part of their work focuses on examining assumptions and propositions of the leading but competing migration theories, dividing them into theories which explain origins of migration and theories which provide support for the continuation of international migration movements. Their stated goal is the usage of some propositions from each migration theory because each one can offer useful insights into the understanding of the multi-dimensionality of migration and in analysing various levels of migration. In the second part of their work they review empirical migration studies as the basis for the evaluation of existing theoretical propositions in order to show deficiencies and inconsistencies of migration theories and choose the correct assumptions of theoretical elaborations which are connected with dimensions of migrations.

In order to overcome the polarisation of individual and structural perspectives in migration theory. other researchers introduce Gidden's 'structuration theory' into migration analysis. In this framework. Goss and Lindquist [15] explain international migration as 'the result of knowledgeable individuals undertaking strategic action within institutions—specifically the institution of migration—which operate according to recognisable rules and which attribute resources accordingly'.

In this theory, structures are defined as rules and resources which both constraint and enable individuals' action. Then, individuals' actions and practices produce. reproduce and change the social structures. In the case of international labour migration, international migrant institutions have emerged by practices of knowledgeable individuals—potential migrants, return migrants—and the agents of organisations (from migrant associations to multinational corporations) and other institutions (from kinship to the state). Both individuals and agents draw upon sets of rules in order to increase access to resources.

The preceding selective review of migration theories shows that theoretical elaborations of migration are expanding and including various perspectives influenced by both broader theoretical issues, and contemporary issues. In fact, it cannot be said that there is a theory which can explain the multi-dimensionality and complexity of migration, but each theory reveals important aspects which can lead to a fertile and ongoing discussion of the subject.

It can be said that existing migration theories give the general directions for a new theoretical construction, in these terms, migration is an ongoing social phenomenon, whose understanding depends on the exploration of the social historical context globally. Migration incorporates social, economic, political, cultural, ideological and ethical dimensions and a theory of migration cannot exhaust its understanding focusing exclusively on states, economics and migrants but it also has to take into account the general social framework. Therefore, migration should be situated in a global context and a theory of migration must study the broader ways that societies are organised in order to explain migration as a contemporary social phenomenon.

32.3   Modelling the Population Dynamics

Consider a set of countries or geographical regions $\mathcal{G} = \{1, \cdots, N\}$. Each of these regions i supports a population u_i. For the time being we do not specify the type of the population. Depending on the use of the model, the population may refer to labour force of a particular type (e.g. specialized or unspecialized labor), migrant population of specific characteristics or the general population.

The population may change with respect to time thus leading to a real valued function $u_i : [0, T] \to \mathbb{R}$; $u_i(t)$ being the population of region i at time $t \in [0, T]$. We will focus on migrant population and consider $u_i(t)$ as the total population of migrants in a region i at time t. For the time being we will consider the migrant community as homogeneous, i.e., we will not distinguish between migrants of various characteristics, such as nationality, gender, occupation, high or low skilled workers etc.

We next consider a connectivity structure on $\mathcal{G}$. Let $\mathcal{E} \subset \mathcal{G} \times \mathcal{G}$ be the set of edges of $\mathcal{G}$. The pair $(i, j) \in \mathcal{E}$ if and only if there exists a migration flow from i to j. The pair $(\mathcal{G}, \mathcal{E})$ constitutes a graph, which will be called the *migration graph* $\mathcal{M}$. This graph is considered as a directed graph, the pairs (i, j) and (j, i) being considered as different; it is not necessary that $(i, j) \in \mathcal{E}$ implies that also $(j, i) \in \mathcal{E}$ unless there is also a possibility of return migration from region j to region i.

Through the connectivity structure of the graph, there may be secondary connections between different regions. For instance even if $(i, j) \notin \mathcal{E}$, therefore, not being a direct connection between regions $i = i_1$ and $j = i_n$ one may find a path to go from i_1 to i_n *indirectly*. That means that there may exist a sequence $i = i_1 \to i_2 \to \cdots i_{n-1} \to i_n = j$, with the property $(i_r, i_{r+1}) \in \mathcal{E}, r = 1, \cdots n-1$. This is a path, connecting i to j in n moves. This indirect connectivity is of particular importance to our model, since it implies that regions what are apparently not connected directly may be indirectly connected; therefore what happens in one of them may have an effect (even in long term) to what happens to the other. A graph such that for any $i, j \in \mathcal{G}$ there exists a path connecting i with j is called a connected graph.

Clearly, the direct connectivity between two vertices of the graph i and j is not enough to provide a complete description of the migration flow between i and j. Some notion of intensity of migration between these two vertices must be defined, which will quantify whether there is a pronounced migration movement from i to j or not. We will then consider a positive weight $w(i, j)$, quantifying this. It is convenient to scale this weight so that $w(i, j) \in [0, 1]$. Then, this positive weight can be interpreted as the probability of migrating from region i to region j. The larger $w(i, j)$ the more likely is it for an agent to migrate from i to j. First of all we emphasize that in general $w(i, j) \neq w(j, i)$. Furthermore, because of the adopted scaling it must hold that

$$\sum_{(i,j)\in\mathscr{E}} w(i, j) = 1, \quad \forall i \in \mathscr{G},$$

meaning that an agent originating from region i will end up to one of the regions of $\mathscr{G}$, directly connected with i with probability 1, allowing of course for the possibility of the agent staying in i. This is the probability of a single migration (single transition) so only the set of edges $\mathscr{E}$ is considered.

From the geographical point of view some comments are in order. The set of edges, which describes the "direct" connectivity of the migration graph, in some sense defines the geography of the regions. One could dare to use the word topology here, in the sense that $\mathscr{E}$ defines a system of neighbourhoods on the migration graph $\mathscr{M} = (\mathscr{G}, \mathscr{E})$, describing affinity of certain regions with others. Geographical distance (physical distance) on the globe plays no particular role in $\mathscr{E}$. Regions that may be miles apart (such as India and UK for example) may present very strong connectivity properties (indicated both by $\mathscr{E}$ as well as the corresponding weight structure $w(i, j)$) because of bilateral agreements etc.

The introduction of the weights may now simplify the exposition. Define a generalized weight function $w : \mathscr{G} \times \mathscr{G} \to [0, 1]$, with the property $w(i, j) = 0$ if $(i, j) \neq \mathscr{E}$. Therefore, the knowledge of the weight function w automatically provides knowledge of the edge structure $\mathscr{E}$, clearly

$$\mathscr{E} := \{(i, j) \in \mathscr{G} \times \mathscr{G} : w(i, j) \neq 0\}.$$

We will then build our model on the directed weighted graph $\mathscr{M} = (\mathscr{G}, w)$. We may further define the space of all functions u on $\mathscr{G}$, $u(i) =: u_i$, $i \in \mathscr{G}$, such that $\sum_{i \in \mathscr{G}} |u(i)|^2 < \infty$ which is a Hilbert space, $L^2(\mathscr{G})$.

Suppose that the population on $\mathscr{G}$ is given by the function $u : \mathscr{G} \to \mathbb{R}^N$. We will try to compare the population at a site $i \in \mathscr{G}$ between two times t and $t + 1$ (time is considered discrete without loss of generality). If $w(i, j)$ is considered as the probability of migrating from $i \in \mathscr{G}$ to $j \in \mathscr{G}$ then the total number of agents to move from i to j from t to $t + h$ is $w(i, j)u_i(t)h$. Since they may choose to move to any of the possible destinations the total outflow of agents from i in $[t, t + h]$ is

$$\mathscr{O}_i = \sum_{\{j:(i,j)\in\mathscr{E}\},j\neq i} w(i,j)u_i(t) = \left(\sum_{\{j:(i,j)\in\mathscr{E}\}} w(i,j)\right)u_i(t)h,$$

where the sum is taken over all the edges $\mathscr{E}$. Similarly, agents from region $j \in \mathscr{G}$ may choose to migrate to region $i \in \mathscr{G}$ with probability $w(j,i)$ in the time interval $[t, t+h]$, so that the total number of agents to move from j to i in this time interval is $w(j,i)u_j(t)h$. Then the total inflow of migrant population to i in $[t, t+h]$ is

$$\mathscr{I}_i = \sum_{\{j:(i,j)\in\mathscr{E}\},j\neq i} w(j,i)u_j(t)h. \tag{32.1}$$

Of course there is local changes of the population at i (birth-death) with the total rate of change being $a(i)u_i(t)$, where $a(i)$ will be assumed to be either positive or negative depending on whether local population increases or decays.

The total balance of the population at $i \in \mathscr{G}$ then yields

$$u_i(t+h) = u_i(t) - (\alpha_i u_i(t) + \mathscr{I}_i - \mathscr{O}_i)h$$

which is conveniently written as

$$\frac{u_i(t+h) - u_i(t)}{h} = -a_i u_i(t) + \sum_{\{j:(i,j)\in\mathscr{E}\},j\neq i} w(j,i)(u_j(t) - u_i(t)),$$

where

$$a_i = \alpha_i - \sum_{\{j:(i,j)\in\mathscr{E}\}} (w(j,i) - w(i,j)).$$

This equation is in the general form of a master equation (see e.g. [16] for an example of the use of a master equation in the study of interregional migration).

Define the operators

$$(Au)_i = -a_i u_i + \sum_j w(j,i)(u_i - u_j)$$

and its adjoint operator

$$(A^*v)_i = -a_i^* v_i + \sum_j w(i,j)(v_i - v_j)$$

where $a_i^* = a_i + \sum_j (w(j,i) - w(i,j))$. The adjoint operator A^* is defined through the standard property $\langle Au, v \rangle = \langle u, A^*v \rangle$, for all $u, v \in L^2(\mathscr{G})$. The operator

$A : L^2(\mathscr{G}) \rightarrow L^2(\mathscr{G})$ is called the weighted graph Laplacian corresponding to the weight function w. The graph Laplacian is a symmetric operator if the weights are symmetric, i.e., if $w(i, j) = w(j, i)$ for all $i, j \in \mathscr{G}$.

The total balance equation is then written in compact operator form as

$$\frac{u(t+h) - u(t)}{h} = Au(t), \tag{32.2}$$

or equivalently as

$$u(t + h) = (I + hA)u(t).$$

One may either assume that $h \rightarrow 0$ and obtain an ordinary differential equation on the graph for the evolution of the population, or assume that the time unit of interest is finite and set without loss of generality $h = 1$ and obtain a difference equation for the evolution of the population. We will choose this option here for simplicity.

An alternative form of writing the model is the following. As before, let $w(i, j)$ be the probability that somebody moves from location i to location j, so that the inflow of immigrants into country i is given by $\mathscr{I}_i$ as in Eq. (32.1). Also let $w(i, i)$ be the joint probability that an individual at i survives from t to $t + 1$ and decides to stay at location i. Then the agents from location i that remain in this location (between the periods t and $t + 1$) are $w(i, i)u_i(t)$. To the new comers we must include the newborns which are $f_i u_i(t)$ where f_i is a local growth rate. Then the population balance equation is

$$u_i(t+1) = (f_i + w(i,i))u_i + \sum_{\{j : (i,j) \in \mathscr{E}\}, j \neq i} w(j,i)u_j(t),$$

or in matrix form as

$$u(t+1) = Tu(t)$$

where

$$T = \begin{pmatrix} f_1 + w(1,1) & w(2,1) & \cdots & w(N,1) \\ w(1,2) & f_2 + w(2,2) & \cdots & w(N,2) \\ \cdots & \cdots & \cdots & \cdots \\ w(1,N) & w(2,N) & \cdots & f_N + w(N,N) \end{pmatrix}$$

which clearly is a positive matrix.

To make the above discussion useful as a model for migration we need to complete the following important steps:

1. Construct the relevant migration graph $\mathscr{M} = (\mathscr{G}, \mathscr{E})$ by identifying the edge structure $\mathscr{E}$.
2. Find the relevant weight function w. This will in principle depend on a number of factors and parameters, which when modelled properly will lead to a well versed decision theoretic tool to assist policy making.
3. Study the quantitative and qualitative properties of the time dependent and steady state equations (32.2) and how these depend on the connectivity properties of the migration graph. Then, one can monitor how changes in the connectivity of the migration graph, that may be influenced by policies, bilateral agreements etc. may affect the long term behaviour of the population distribution on $\mathscr{G}$. To this point this is a linear equation, but in the next section will be turned into a nonlinear equation through coupling with an model for economic growth (a fact that will internalize the migration dynamics and make the model more interesting).

The above steps require also the employment of statistical techniques in order to infer the important quantities of the model, topological (e.g. $\mathscr{E}$) or quantitative (e.g., w) from available data. Furthermore, our study will highlight directions that may be needed in data collection so as to acquire data that are sufficient so as to reveal the desired quantities.

32.4 A Fine Tuning of the Model: Calculation of w

An important feature of the model, which so far is a generic book counting model of population movement into different spatial locations (countries) is the matrix of migration probabilities (w_{ij}). The model will only be of use in the understanding of migration if the migration probabilities from location to location are specified in a realistic fashion. Our modelling of the migration probabilities will be a combination of the neoclassical (equilibrium) theories, which assume that the migration decision is based on economic grounds as well as on theories based on individual and structuralist approaches (which argue for the importance of effects such as networks) as presented in Sect. 32.1.

For example, let us consider the probability $w(i, j)$ of an agent at location i to migrate to location j. This can be modelled in terms of a discrete choice model. Discrete choice models have been very useful tools in modelling a decision maker's choice between mutually exclusive and collectively exhaustive alternatives (for an introduction to discrete choice models see e.g., [34]). These models are related to the maximization of random utility functions [24]. Discrete choice models have been used by various authors and in various contexts to model the migration decision (see e.g. [19, 27] and references therein).

According to this model consider a representative agent residing at time t in region i and facing the decision (lottery) to migrate to another region j (the case where $j = i$ is included and corresponds to staying in the region where the agent is originally situated). Let $V_{ij}(t)$ be the utility of this agent that decides to realize a move from region i to region j between the times instances t and $t + 1$. This utility is subject to a random term ϵ_{ij}. This random term models either uncertainty with respect to the utility that an agent is likely to face, on account of incomplete knowledge of the situation she is likely to face or stochastic changes in the overall underlying situation. Alternatively, this term may model possible deviations of the agent from rationality (it can easily be argued that migrants may not always act rationally, for a variety of reasons). Therefore, the overall utility of an agent that decides to migrate from region i to region r will be a random function $V_{ir}(t) + \epsilon_{ir}$. The probability that this agent initially located at region i prefers to move to location r rather than to location k is

$$p_{i,r>k} := P(V_{ir}(t) + \epsilon_{ir} > V_{ik}(t) + \epsilon_{ik}) = P(V_{ir}(t) - V_{ik}(t) > \epsilon_{ik} - \epsilon_{ir})$$

The probability that an agent decides to move from site i to site j is the probability that $p_{i,j>k}$ for all $k \neq j$. This can be expressed in terms of utilities as

$$w(i, j) = P(V_{ij}(t) - V_{ik}(t) > \epsilon_{ik} - \epsilon_{ij}, \ \forall j \neq k)$$
$$= P(\epsilon_{ik} < V_{ij}(t) - V_{ik}(t) + \epsilon_{ij}, \ \forall k \neq j)$$
$$= \prod_{k=1}^{N-1} P(\epsilon_{ik} < V_{ij}(t) - V_{ik}(t) + \epsilon_{ij})$$

where in the last expression the independence of the ϵ_{ik} for different values of k has been used.

This probability can be calculated numerically, or even analytically for certain choices of functional form for the distribution of the error terms ϵ_{ij}. For example, under the generic choice of Gumbel type distributions for this error term, one can complete the calculation and derive a multinomial type model for the probability of a representative agent to migrate from i to j of the form

$$w(i, j) = \frac{e^{V_{ij}}}{\sum_{k=1}^{N} e^{V_{ik}}}$$

where clearly the denominator in the above expression serves as a normalization factor. The choice of the Gumbel family of distributions for the error term is not accidental or simply in order to simplify the algebra. It derives from a deep result in probability theory related to the asymptotic behaviour of independent errors, the celebrated Fisher-Tippet-Gnedenko theorem (see e.g., [20] or [9]). According to this theorem, if the error terms are drawn from a wide variety of distributions (including the normal, lognormal, exponential, gamma and the logistic) the distribution of their maximum follows the Gumbel distribution.

To complete the model, and fully specify the probability of moving from country i to country j, it remains to specify the utility levels V_{ik} which in some sense corresponds to the average (mean) utility and is the deterministic part of the random utility. This captures the average 'rational' behaviour of the agent and should depend on both the characteristics of the agents as well as on 'external' characteristics of the economy. For example the average utility level V_{ij} may depend on a number of economic factors, such as the wage difference between the two countries, the difference in the cost of living etc, as well as on a number of qualitative variables, modeled in statistics in terms of categorical variables. Such variables may model structural issues, for example, are there good provisions for the welcome of migrants in the host country, are there migrant networks that may facilitate the settlement of migrants in the host country etc. A particularly convenient class of models are linear models for V_{ij} of the form

$$V_{ij} = \sum_{\ell=1}^{m} \beta_\ell Y_\ell^{ij}$$

where $Y^{ij} = (Y_1^{ij}, \cdots, Y_m^{ij})$ is a vector of m variables (continuous valued or categorical) that models several characteristics of regions i and j that may influence the migrants decision to move from region i to region j. For example Y_1^{ij} may correspond to the ratio or difference of the wage rates in the two regions, Y_2^{ij} may correspond to the ratio or difference of the unemployment rate in the two regions etc. whereas other variables may be categorical, i.e. may take the value 1 is there is a network of migrants from region i operating in region j that facilitates settlement and 0 otherwise, or 1 is there are bilateral agreements between the governments of i and j and 0 otherwise etc. The vector of coefficients $\beta = (\beta_1, \cdots, \beta_m)$ (assumed for simplicity to be common for all utility pairs V_{ij}) provide an idea of the sensitivity of the utility levels on the various possible external factors. This choice leads to the multinomial logit model for the probability of migration between sites,

$$w(i, j) = \frac{\exp(\sum_{\ell=1}^{m} \beta_\ell Y_\ell^{ij})}{\sum_{k=1}^{N} \exp(\sum_{\ell=1}^{m} \beta_\ell Y_\ell^{ik})}$$

This model can be estimated using standard statistical techniques e.g. maximum likelihood methods that will allow us to estimate the vector β, and thus calibrate the model into real life data.

One way to understand the above model is as $w(i, j) = w(i, j \mid Y^{ij})$, that is that the above formula provides the probability of an agent to migrate from i to j is

$$w(i, j) = \frac{\exp(\sum_{\ell=1}^{m} \beta_\ell y_\ell^{ij})}{\sum_{k=1}^{N} \exp(\sum_{\ell=1}^{m} \beta_\ell y_\ell^{ik})}$$

given that the macroeconomic or behavioural or structural variables Y^{ij} take the value $Y^{ij} = y^{ij}$. Clearly, these macroeconomic variables are subject to periodic or stochastic variability, that may be accounted for to shocks that arise in the economy.

32.5 Including the Migration Model into an Economic Growth Model

Migration is at the same time affected by economic factors but also affects the economy, since labour (including migrant labour) is a very important factor of production. The question of how migration affects global economic growth (i.e., the economic growth of the various regions in the world economy, modelled here as the graph $\mathscr{G}$) is an interesting one that deserves our attention.

Obviously, this is a very intriguing and deep question, (see for instance Chap. 9 in [4] and the references therein to get an idea of the considerable activity in the field; see also e.g., [11,12,33,35] and references therein for alternative or related models) the answer of which would require a full treatise rather than simply an expository chapter, whose main objective is to outline a fundamental modelling framework. In this vein, we propose a simple economic growth model which when coupled with the population Eq. (32.7) can model and predict growth and population patterns on the graph $\mathscr{G}$.

Our starting point in the modelling of economic growth is the celebrated Solow model. Let K_i be the capital stock of region $i \in \mathscr{G}$. This capital stock is subject to temporal change through production and consumption. The production of output Y_i at site $i \in \mathscr{G}$ is modelled through a neoclassical production function $Y_i = f(A_i, K_i, L_i)$ where L_i is the labour population at site i and A_i is a region specific productivity parameter. The temporal change of capital stock at site $i \in \mathscr{G}$ is given by

$$K_i(t+1) = s_i F(A_i, K_i(t), L_i(t)) + (1-\delta)K_i(t)$$

where s_i is the average propensity to save of region i, and δ is the depreciation rate of capital. For simplicity, we may assume that the production function and the depreciation rate of capital are common for all regions. Clearly, this assumption may (and will) be relaxed. Let us now assume full employment and that u_i represents the total population of labour force in region i therefore, $L_i = u_i$. We may therefore express the capital stock change in i by

$$K_i(t+1) = s_i F(A_i, K_i(t), u_i(t)) + (1-\delta)K_i(t), \quad i \in \mathscr{G}.$$

A convenient choice for the production function is a Cobb-Douglas type production function of the form

$$F(A, K, L) = K^\alpha (AL)^{1-\alpha}, \quad \alpha < 1,$$

where A is the index of the technology (Harrod neutral, see, e.g., [4, p. 52]). This capital stock update equation is complemented with the population monitoring equation

$$u(t+1) = Tu_t \tag{32.3}$$

However, one should note that the migration decision depends on the macroeconomic variables, and in the present model this is shown by the dependence of the
transition probabilities $w(i, j)$ on macroeconomic variables.

The model may also be expressed in terms of per capita variables, i.e. in terms
of $k_i = K_i/(A_i u_i)$. Then noting that $F(K, Au) = Au f(k)$ where $f(k) := F(k, 1)$
we see that

$$k_i(t+1) = \frac{A_i(t)u_i(t)}{A_i(t+1)u_i(t+1)}(f(k_i(t)) + (1-\delta)k_i(t)). \qquad (32.4)$$

Note that in a single region setting and under the standard assumption of steady
growth of the population, i.e. that $u_i(t+1) = (1+n)u_i(t)$ and constant $A_i(t)$ this
equation would reduce to the standard form of the Solow model, involving only the
per capita capital stock. An alternative approach is to assume technological change
of the form $A_i(t+1) = (1 + g_{i,A})A_i(t)$, leading essentially to the same model
but with a modified parameter on account of the technological change. The effect
of the technological change may be included in the local growth rate of labour,
modifying the factor f_i. In the following, we assume, that the factor f_i accounts also
for the effect of technological change and, without loss of generality, set $A_i = 1$.[1]
On the other hand, concerning labour, here we assume that the population evolves
according to the population monitoring equation (32.3) which is in fact a coupled
system of n equations and this assumption is not in general valid, except perhaps in
an asymptotic sense. Another way to express (32.4) is to express

$$\frac{u_i(t+1)}{u_i(t)} = f_i + w(i, i) + \sum_{\{j:(i,j)\in\mathscr{E}\}, j \neq i} w(j, i)\frac{u_j(t)}{u_i(t)}$$

and then rewrite the per capita capital stock evolution law as

$$k_i(t+1) \qquad (32.5)$$

$$= \left(f_i + w(i, i) + \sum_{\{j:(i,j)\in\mathscr{E}\}, j \neq i} w(j, i)\frac{u_j(t)}{u_i(t)} \right)^{-1} (Af(k_i(t)) + (1-\delta)k_i(t)).$$

The modelling of the transition probabilities can be done using the discrete
choice model of Sect. 32.4, by proper interpretation of the "decision variables" Y.
For example a decision variable Y_1^{ij} may be the difference between the per capita
output in each region

$$Y_1^{ij} = \frac{Y_j}{u_j} - \frac{Y_i}{u_i} = f(k_j) - f(k_i).$$

---

[1]Alternatively we simply measure all quantities in efficiency units AL.

This implies that agents will turn to migrate from region i towards region i with more propensity, the higher the difference between the per capita output in the corresponding regions is. Within the context of the Solow model, under constant returns to scale, this quantity is related to the wage difference between the regions. On the other hand if $Y_1^{ij} < 0$ then an agent originating at region i has no incentive to migrate to region j therefore, $w(i, j) = 0$. Another possible choice for the decision variable can be the difference between the capital stock between the various regions, with the same interpretation as above.

Another factor we can introduce is a qualitative factor (a caterogical variable) modelling the existence or not of migrant networks facilitating the settlement of migrants. This can be modelled by $Y_2^{ij} = \epsilon_{ij}$ a matrix consisting of 1 and 0, the element ij is 1 if there exists a network of migrants from region i that facilitates the settlement in region j and 0 otherwise. Empirical studies of migration decision imply that such qualitative factors (e.g. networks, herding effects as well as cultural factors) may play a very important role and in some cases even outweight economic factors.

One may introduce other factors, e.g., the effect of the supply and demand of labour in the employment rate etc. Furthermore, we may introduce capital flows and in addition to labour flows, different sectors in the economy and separate labour force into skilled and unskilled (dual markets), time lags in the migration decision process etc. In some sense the sky is the limit when it comes to modelling, however, one should be careful to keep a model down to its bare essentials in order to keep it functional and useful. Since the main objective of this expository chapter is to present a simple working model as an illustration of the modelling strategy we will limit ourselves to this rather simplistic model.

Collecting all the above, we end up with the following "bioeconomic" type model for the interaction of migration and the macroeconomy

$$K_i(t+1) = s_i F(K_i(t), A_i u_i(t)) + (1-\delta)K_i(t), \quad i \in \mathcal{G},$$

$$u_i(t+1) = (f_i + w(i,i))u_i + \sum_{\{j:(i,j)\in\mathcal{E}\}, j\neq i} w(j,i)u_j(t)$$

$$w(i,j) = \frac{\exp(\sum_{\ell=1}^{2} \beta_\ell y_\ell^{ij})}{\sum_{k=1}^{N} \exp(\sum_{\ell=1}^{2} \beta_\ell y_\ell^{ik})}$$

or its scaled form

$$k_i(t+1) = \left(f_i + w(i,i) + \sum_{\{j:(i,j)\in\mathcal{E}\}, j\neq i} w(j,i)\frac{u_j(t)}{u_i(t)} \right)^{-1} (Af(k_i(t))$$

$$+(1-\delta)k_i(t)), \quad i \in \mathcal{G},$$

$$u_i(t+1) = (f_i + w(i,i))u_i + \sum_{\{j:(i,j)\in\mathscr{E}\}, j\neq i} w(j,i)u_j(t)$$

$$w(i,j) = \frac{\exp(\sum_{\ell=1}^{2} \beta_\ell y_\ell^{ij})}{\sum_{k=1}^{N} \exp(\sum_{\ell=1}^{2} \beta_\ell y_\ell^{ik})}$$

where

$$Y_1^{ij} = \frac{Y_j}{u_j} - \frac{Y_i}{u_i} = f(k_j) - f(k_i)$$

$$Y_2^{ij} = \epsilon_{ij} = \begin{cases} 1 & \text{network exists} \\ 0 & \text{otherwise.} \end{cases}$$

This is a dynamical system, the evolution of which may provide some intuition concerning the transitory and asymptotic spatio-temporal behaviour of the migration process as well as its effects and dependences on the economic variables. As there are no constraints to the movement of migrants from region to region, and as we assume that labour force that has arrived from region j to region i is treated in the same terms as local labour force, our model may be a reasonable model for e.g. the Eurozone.[2] The model may be augmented with such features for as to include sanctions, difference in wages between local labour and migrant labour etc.

The modeling proposed here is very schematic and of course may be improved along various directions. For example, a more elaborate model for economic growth than that of Solow may be used, which may internalize employment rates, care for more than one sectors in the economy, consider the effects of dual markets etc. Also the decision to migrate can be modelled in a more elaborate fashion, including more factors, time lags etc. Empirical evidence based on econometric studies of migration seems to show that qualitative effects such as culture, network formation etc. may in some cases outweight economic effects (see e.g. [5, 14] or [30] and references therein). The modelling of migration networks included here is rather schematic, more elaborate models can be investigated using ideas from [1] or [6]. At any rate, even this simple model serves well within the limited scope of this expository chapter to capture the intricate interdependence structure of migration and the economy and designate the salient features of the bioeconomics of migration.

---

[2]Since the Treaty of Rome which aimed at the establishment of a common market the free movement of capital, goods and people has been established. In the beginning, the right of free movement referred to workers and later on (the Maastrich, Amsterdam and Shengen treaties) referred to the general free mobility of nationals of the EU member states.

32.6   Economic, Mathematical and Simulation Challenges of the Model

The long term behaviour of the proposed dynamical system is of interest from both the point of view of demographic and migration issues as well as from the point of view of economic dynamics. A good understanding of such issues will be important from the point of view of economic and migration policy as well. Eventhough our model is simple enough, it is still too complicated to deal with analytically, therefore one should resort to extended simulations and scenario based studies. However, there are still some interesting qualitative results we may obtain from this simple model proposed here, which is one of the main reasons for introducing our various simplifying assumptions.

32.6.1   Long Term Behaviour

To get an idea of the type of a priori information we can obtain from our model, let us recall one of the fundamental features of the Solow growth model. In the standard (single region) model, assuming a steady growth rate for the population $(1 + n_i)$ it is well known (see e.g. [4]) that the economy reaches a steady state in per capita output $k_i^* = A_i^{1/(1-\alpha_i)} (n + \delta)^{1/(\alpha_i - 1)}$. This steady state depends on the characteristics of the economy (i.e., the production function) as well as the population growth rate n_i. Of course the coupled nature of the system does not allow us to assume a constant growth rate for each economy. However, it is seen that if we manage to find an upper bound for the quantities $u_i(t + 1)/u_i(t)$, $i \in \mathscr{G}$, this will provide an upper bound for the quantity n_i and therefore, a lower bound for k_i^*. If we can show that an upper bound exists for the growth rate of the population and if the capital evolution equation satisfy monotonicity properties that guarantee the validity of a comparison principle, then we can provide a lower bound for the limiting state for the per capita capital k_i^*.

Suppose that for some reason the transition matrix is constant in time and given. This means that somehow the economies have approached an equilibrium state so that the per capita ratios $K_i/u_i = k_i$ that provide the incentives for migration[3] are either constant or are varying but at a very slow rate, so that they may well be approximated as constant. This approximation is also true if we assume that other effects except economic effects play an important role in the decision to migrate, a fact supported sometimes by empirical studies. The population equation is then effectively a constant coefficient difference equation of the form $u(t + 1) = \mathsf{T}u(t)$.

---

[3]Recall that in our model the incentive for migration between two regions is the difference in $\frac{Y_i}{u_i}$ which for the Cobb-Douglas function is equal to the difference between $f_i(k_i)$. Note also that the functions f_i are monotone.

Assuming that all eigenvalues of the matrix T are real and different ordered as $\sigma(T) = \{\lambda_0, \lambda_1, \cdots, \lambda_{N-1}\}$ with corresponding eigenvectors $\{v_0, v_1, \cdots, v_{N-1}\}$ the general solution may be expressed as

$$u(t) = \sum_{j=0}^{N-1} C_j v_j \lambda_j^t,$$

where C_j are arbitrary constants to be specified by the initial condition and in particular are such that $u(0) = \sum_{j=0}^{N-1} C_j v_j$. This formula will hold for complex valued eigenvalues and has to be modified accordingly in case where there are multiple eigenvalues. This general solution gives us a general idea about the possible growth rates of the various populations. For example if v_0 does not contain any zero elements then for a "generic" initial condition $u(0)$ we expect all elements of the vector $u(t)$ to grow at the rate λ_0^t. At the positions where v_0 has a zero element we expect a slower growth rate which will coincide with one of the other eigenvalues etc. Crude as this information may sound it is probably the most we can say for the behaviour of the population asymptotically.

The non negativity of the matrix T ensures that there is a great deal of structure on the eigenvalues and eigenvectors of T, given by the celebrated Perron-Frobenius theorem. Furthermore, this theorem links the topology and connectivity of the migration graph with the long term behaviour of the system. For the sake of the reader we briefly recall (for a full account of this beautiful theory see e.g. [32]) that according to the Perron-Frobenius theory, If the matrix T is irreducible and primitive then it has a non-negative algebraically simple eigenvalue λ_0 that strictly dominates all the others, and the corresponding left and right eigenvectors related to this eigenvalue are positive.

This, importantly involves the geometric and topological properties of the migration graph. This comes through the condition of irreducibility and primitivity that is needed for the validity of the Perron-Frobenius theorem. Recall, that a positive matrix $T \geq 0$ is called irreducible and primitive if there exists a $m \in \mathbb{N}$ such that $T^m > 0$. This definition, implies that the irreducibility of the matrix T is related to the connectivity properties of the graph related to the migration process. In particular T is irreducible if the graph $\mathscr{G}$ is strongly connected. That means that there is a path taking an agent from any location i to any location j as long as we wait long enough (at least as long as m time units where m is least integer such that $T^m > 0$). Furthermore the Perron-Frobenius eigenvalue (which in fact coincides with the spectral radius of the matrix T) can be a priori estimated by

$$\min_i \sum_{j=1}^{N} t_{ij} \leq r \leq \max_i \sum_{j=1}^{N} t_{ij}$$

i.e. lies between the minimum and the maximum of the row sums of the matrix T.

The long term behaviour of the system $u(t + 1) = \mathsf{T}u(t)$ is given by the Perron-Frobenius theorem. In fact $u(t) = \mathsf{T}^t u(0)$ where $u(0)$ is the initial state of the population, and as a consequence of the Perron-Frobenius theorem

$$\lim_{t \to \infty} \frac{\mathsf{T}^t}{\lambda_0} = v w^{tr}$$

where v and w are the left and right eigenvectors corresponding to λ_0, normalized so that $v^{tr} w = 1$. This result leads us to the approximate asymptotic result that the fastest growth rate of the populations in $\mathscr{G}$ will be related to λ_0. This implies that $u_i(t + 1)/u_i(t)$ will tend to λ_0 asymptotically as $t \to \infty$. The action of the matrix $v w^{tr}$ on the initial population distribution (i.e., in effect the structure of the left eigenvector v) will reveal which regions are expected to have a growth rate equal to the Perron-Frobenius eigenvector. When inserted into the Solow equations this provides an idea of the growth rates of these economies, in the sense that in the long run the per capita capital for these regions should follow

$$k_i(t + 1) \simeq \frac{1}{\lambda_0}(A_i f(k_i) - (1 - \delta)k_i).$$

This may give us a rough idea of the growth rate allowed for these economies. The economies (regions) which correspond to zeroes of the left Perron-Frobenius will necessarily have population growth rates which are lower that λ_0 and that will correspond to a different eigenvalue in the spectrum of T. Knowledge of this eigenvalue (and assuming that it is simple) in conjunction with the Solow equation for this region (where the relevant population growth rate is inserted) will give an estimate of the relevant long run steady state for the per capita capital.

This procedure has a slight catch. The matrix T depends on the steady state of the economies, and so in turn the eigenvalues of T which give the long run behaviour of the population in the various regions depend on this steady state. On the other hand, these eigenvalues characterize the steady state of the economies since the population growth rate enters the Solow equation. This is a nonlinear problem and the procedure we have just described may be turned into a fixed point argument so that we may obtain approximations to the relevant long run behaviour of the system. The implementation of such a fixed point argument will require certain results on the behaviour of the eigenvalues of T as the matrix varies.

There is an alternative way to view the above arguments. Let us assume that we wish to drive the system into a prescribed steady state $\bar{k}^* = (k_1^*, \cdots, k_N^*)$. This is equivalent to prescribing a given capital growth rate for each economy $\bar{g} = (g_1, \cdots, g_N)$, since a simple calculation using the Solow equation and the Cobb-Douglas production function implies that a prescribed growth rate g_i for country i (in terms of the capital stock K_i) specifies the asymptotic behaviour of the per capita capital, in terms of

$$\rho_i := \frac{K_i}{u_i} = k_i = \left(\frac{g_i + \delta}{s_i A_i} \right)^{\frac{1}{1 - \alpha_i}} .$$

This ratio determines the migration matrix and through that the matrix T. Therefore, given a growth rate vector $\bar{g}$, the vector $\bar{\rho} = (\rho_1, \cdots, \rho_N)$ is specified and that determines the matrix T which we will denote as $\mathsf{T}(\bar{\rho})$ for the same reason as above. The above asymptotic behaviour in turn prescribes an asymptotic growth rate for the population in each region since by the definition of k_i we have that asymptotically in time

$$u_i(t) \sim \frac{1}{\rho_i} K_i(t) \sim \frac{1}{\rho_i} (1 + g_i)^t .$$

Since the asymptotic behaviour of the population dynamics equation is specified by the spectral decomposition of the matrix $\mathsf{T}(\bar{\rho})$, this provides a link between the vector $\bar{g}$ (or equivalently $\bar{\rho}$) and the eigenvalues of the matrix $\mathsf{T}(\bar{\rho})$. This is a nonlinear, inverse eigenvalue problem that will allow us to specify the allowed values of the vector $\bar{g}$ for which such a prescribed behaviour may be asymptotically supported by the system. These problems can be treated numerically and in general are difficult problems to handle. However, if treated it may provide interesting information on the "spatial" structure of the vector $\bar{g}$ (or equivalently $\bar{k}$) and allow us to infer on question of whether migration of labour may contribute to phenomena like unconditional convergence, conditional convergence or club convergence of the economies (for a definition of the relevant notions in a single country model see [13]; these notions may be extended to our model of N coupled economies and provide important insight on the spatial distribution of growth). The particular case where the problem is solvable for a spatially homogeneous $\bar{g}$ (i.e., in the case where $g_i = g$ for all $i \in \mathcal{G}$) corresponds to the phenomenon of convergence for the economies.

32.6.2   Non Homogeneous and Random Versions of the Model

It is a very naive assumption that the probabilities an agent from location i will migrate to location j, as well as the local growth rates of labour are constant for every time period. These depend on the economic and social conditions and clearly change with time. Therefore, a better model for the population would be a temporally non homogeneous model of the form

$$u(t + 1) = \mathsf{T}(t)u(t)$$

therefore, given the state $u(s)$ we may find the state $u(t)$, $s < t$ by the forward iteration scheme,

$$u(t) = \mathsf{T}(t - 1)\mathsf{T}(t - 2) \cdots \mathsf{T}(s)u(s)$$

or in more compact form in terms of the "propagator"

$$u(t) = U(t, s)u(s), \quad s < t,$$
$$U(t, s) := \mathsf{T}(t - 1)\mathsf{T}(t - 2) \cdots \mathsf{T}(s).$$

Each of the matrices $T(k)$, $k = s, \cdots, t - 1$ is a positive matrix, therefore, we are now dealing with products of positive matrices. Ergodic theory may provide some help in understanding the long term behaviour of the system.

As also argued in Sect. 32.2 the migration decision often depends on contingencies (the general economic and social framework) which can be modelled as random variables using a properly selected probability space. Generalization of the model to $u(t + 1) = \mathsf{T}(t, \omega)u(t)$ where in general T is a stochastic process are also clearly feasible. This is interesting since it allows us to introduce into the model random effects depending on uncertainty and fluctuations in the environment, as well as to introduce effects related to the agents decision making, which may be rational or irrational.

A very convenient way to introduce these random effects into our model is by assuming that the macroeconomic variables Y^{ij}, that play the role of indicating variables in the discrete choice model of Sect. 32.4 that determine the migration probabilities, depend on a set of hidden random variables, H. These take values on a metric space $\mathbb{X}$, which without loss of generality assume that $\mathbb{X} = \mathbb{R}^d$. The random variables H are in some sense the factors that drive the economy. Then, the transition matrix structure for the system $u(t + 1) = \mathsf{T}u(t)$ can be expressed simply by observations of the factors H that drive the economy. If $H(t)$ is the value of the factors at time t then $\mathsf{T}_t(\omega) = \mathsf{T}(H(t))$. A simpler structure can be given to this model if we assume that each of the values that the random variable $H(t)$ can take is obtained by choosing random variables from a probability space $(\Omega, \mathscr{F}, P) := (\mathbb{X}, \mathscr{F}, P)$ in the following way: Consider a transformation $\theta : \Omega \to \Omega$ that preserves the probability measure P, i.e. $P(\theta(A)) = P(A)$ for every $A \subset \Omega$. Let us choose a sequence of matrix valued random variables $\{\mathsf{T}_t\}$. Each of these T_t is assumed to contain the transition probabilities that may describe migration tendencies between the different regions from time t to time $t + 1$. We now make the further assumption that all these matrices may be generated through a single (random) matrix variable $\mathsf{T} : \Omega \to \mathbb{R}_+^{N \times N}$ using the transformation θ by

$$\mathsf{T}_t(\omega) = \mathsf{T}(\theta^t \omega). \tag{32.6}$$

This assumption introduces a stationarity assumption in the migration probabilities.

The state of the population at time t at each region is then understood as a random variable, $u(t, omega)$ which can be obtained by the solution of the random dynamical system

$$u(t + 1, \omega) = \mathsf{T}_t(\omega)u(t, \omega). \tag{32.7}$$

The treatment of this system is greatly simplified by the assumption that $T_t(\omega)$ is of the special form (32.6), as this allows the use of the powerful techniques of ergodic theory in the study of the long term dynamics of (32.7).

This stationarity assumption allows us to state and understand the long term behaviour of the system, using ergodic theory. A random version of the Perron-Frobenius theory (see e.g. [2]) is useful in that. This theory guarantees that, under certain technical assumptions, that there exist a unique positive random unit vector u and a positive random scalar r with $\ln^+ r \in L^1(\Omega, \mathscr{F}, P)$ such that $T(\omega)u(\omega) = r(\omega)u(\omega)$. This captures the asymptotic long term behaviour of the system in the sense that there exists an invariant splitting of $\mathbb{R}^N$ as $\mathbb{R}^N = W(\omega) \oplus \mathbb{R}\,u(\omega)$, such that $T(\omega)W(\omega) \subset W(\theta\omega)$, and if we define $\phi_t(\omega) = T(\theta^{t-1}\omega) \cdots T(\omega)$, then, if $x \neq W(\omega)$

$$\lim_{t \to \infty} \frac{1}{t} \ln \left(|\phi_t(\omega)x| \right) = \lambda = \mathbb{E}[r]$$

where λ is the top Lyapunov exponent (in case $x \in W(\omega)$ then this limit is strictly less that λ). Therefore, even for the random case we have that in some sense the long term temporal behaviour of the system is captured by the random vector u.

When viewed in full coupling with the Solow model, we obtain a fully coupled nonlinear random dynamical system, the long term behaviour of which will provide us with information on the asymptotic distribution of population and capital in the economy. The steady state of the random dynamical system is now not a vector but a vector valued random variable whose law is invariant under the action of the random dynamical system. The calculation of this random fixed point is not a very easy task and in most cases requires detailed numerical work. The question of random effects in growth theory has been studied within the standard Solow model (including one region and without the introduction of migration) in [31] where using the theory of random dynamical systems and a random version of the Banach fixed point theorem the existence of a random fixed point was proved. The multi-region case coupled with the migration dynamics presents a considerably more complicated problem, which is worth of further investigation. The theory of monotone random dynamical systems (see e.g. [8]) is expected to play an important role in this study.

32.6.3   Simulation

It must be clear from the above discussion that analytic techniques will not take us very far with our model, which albeit simple is still very complex for analytic treatment. One must resort to simulation, which can help us generate multiple scenarios concerning the evolution of the population dynamics and the economy which may assist in policy making. The simulation of the model as such is not too demanding, when the transition probabilities are not assumed to be random. What seems to be more demanding is the calibration of the model to realistic parameters

Table 32.1 Initial conditions for the simulation (*Source*: Micro Time Series, available on line at http://econ.worldbank.org/WBSITE/EXTERNAL/EXTDEC/EXTRESEARCH/0,, contentMDK:20701055~pagePK:64214825~piPK:64214943~theSitePK:469382,00.html

Country	Output per worker (y)	Capital stock per worker (k)	Workers (u)
Greece GR (1)	17,717 USD	39,423 USD	3,867,047
Portugal PT (2)	16,637 USD	28,973 USD	4,435,469
Germany DE (3)	30,099 USD	79,049 USD	30,126,905

that may fit the real world. For example, the calibration of the discrete choice model that determines the location in which a migrant chooses to migrate to has to be based on questionnaires, or micro-econometric studies (see e.g. [23]). While there are standard techniques for dealing with such problems, the collection of data can be a problem, especially when it comes to undocumented migration.

We now present the results of simulations of the model, for three regions.[4] As initial conditions for the simulation we have used data for the output per worker (y), capital stock per worker (k) and worker population (u) taken from the Macro Time Series (available online from the World Bank). The initial condition is chosen to be the 1990 data for two reasons. The first is technical, this is the last date where capital per worker data has been published and without this data available the estimation of the production function requires more sophisticated econometric techniques outside the scope of the present paper. The second is that this date is sufficiently far from the present day, so that (a) we may test for the plausibility of our results using our historical experience and (b) be sufficiently remote from the present day Eurozone crisis. For the sake of argument we have used data from Greece, Portugal and Germany and therefore we name the relevant regions after these countries for simplicity. The data are presented in Table 32.1. Using the available data and assuming the coefficient α_i in the range 0.2–0.4 we estimate the production function and then use this production function to simulate the model. For the particular run presented here we have chosen $\alpha_1 = \alpha_2 = \alpha_3 = 0.3$. The f_i are taken to be $f_1 = f_2 = 0.015$ and $f_3 = 0.02$, where in this we have also included the rate of technical change. The propensity to save is taken to be $s_1 = s_2 = 0.1$ and $s_3 = 0.2$ which is a reasonable estimate and in accordance with available data.

In Figs. 32.1 and 32.2 we present the results of a simulation of the model for these initial conditions and these parameters. In Fig. 32.1 we present on the top panel the evolution of the capital stock per worker (k), on the mid panel the evolution of the output per worker (y) and the ratio of workers divided by the number of workers in the absence of migration $u_i(t)/(1 + f_i)^t$ as a function of time for the three regions. The time scale is deliberately taken to be long in order to show that the model is well behaves globally in time. The results show a migratory movement from regions 1 and 2 to region 3 as expected, and the log run macro economic

---

[4]The number of regions is chosen to be so low only for simplicity and economy in the presentation of the results, there is no limitation as to the number of regions.

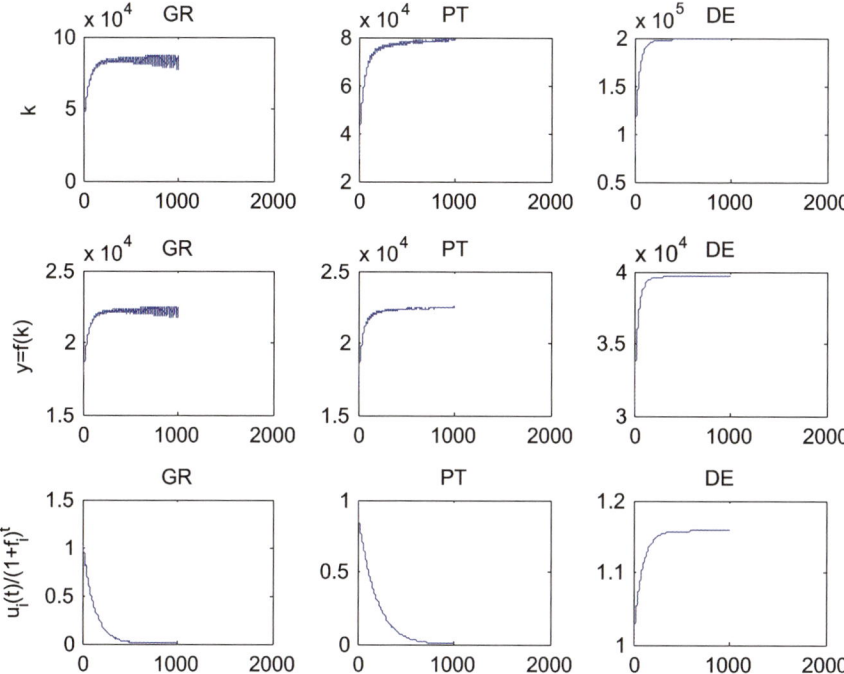

Fig. 32.1 Simulation of the model for three countries. Country number 3 (Germany) has an advantage in production (as shown in the choice of the relevant Cobb-Douglas parameter). The long run macroeconomic variables for the three regions and the evolution of the worker population

quantities k and y show a lead for region 3, also as expected. The long run behaviour of the population as predicted by the simulation matches the pattern predicted by the Perron-Frobenius theory. In Fig. 32.2 we present the effects of migration on the capital stock K_i and the overall production Y_i for region 3 (the receiving country). On the first panel we plot the temporal evolution of the quantity $\frac{K_i - K_{i,0}}{K_{i,0}}$ where K_i is the prediction for the capital stock in region i from the model presented here and $K_{i,0}$ is the relevant quantity as predicted by the standard Solow model without migration. On the second panel we plot the quantity $\frac{Y_i - Y_{i,0}}{Y_{i,0}}$ where Y_i is the prediction for the output in region i from the model presented here and $K_{i,0}$ is the relevant quantity as predicted by the standard Solow model without migration. It is seen that region 3 benefits as an effect of the migrant worker influx. A similar calculation for the sending countries shows that regions 1 and 2 suffer a loss as an effect of labour outflow.

It has to be stressed that the simulation presented here is for the sake of illustration of the general behaviour of the model only and as it is cannot be used for quantitative predictive purposes. For that, one has to employ more sophisticated econometric methods for the estimation of the parameters of the growth model than the simple estimation performed here. Furthermore, detailed micro-econometric

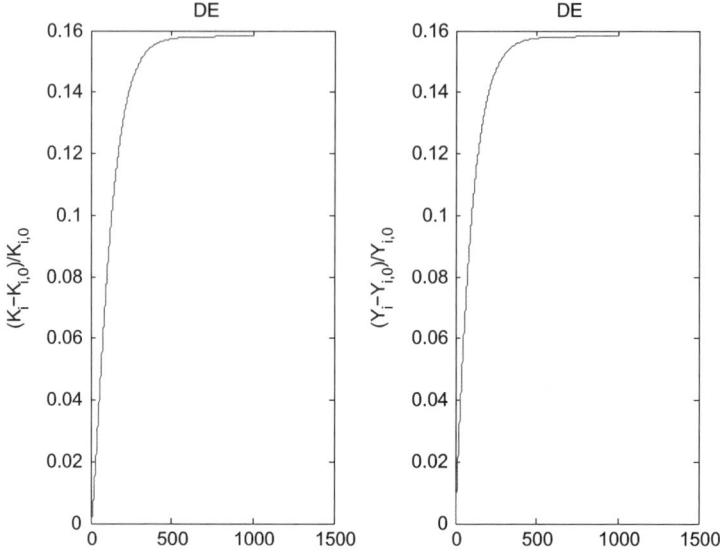

Fig. 32.2 Effect of migration on the macroeconomic variables of the receiving country

modelling of the migration probabilities should be made, based on quantitative and qualitative methods. This is important as the migration probabilities play an important role in the model. The proper calibration of the model is clearly beyond the scope of the present work. However, when properly calibrated one may use this model to run a number of different scenarios, including any number of regions, and trying different types of network effects. This may provide a fairly good understanding of the dynamics of this complex system, and the relative importance of the various factors and parameters in its evolution. The inclusion of randomness complicates things a bit further as it requires detailed modelling of the random effects as well as Monte-Carlo simulations in order to assess the effects of randomness in the model.

32.7 Conclusion

We have attempted a selective review of migration theories with a focus towards a modelling framework for the bioeconomics of migration. We also present a simple model, based on a set of difference equations monitoring the motion of agents between N regions, where the migration probabilities depend on economic as well as on network effects. This system is coupled with a simple multi-region Solow model that monitors economic growth and the migration decision is based partly on the evolution of the economy in these regions. The final model is a nonlinear dynamical system that provides information on the evolution of the economies and the population. Certain qualitative features of the model are addressed and some comments on simulation are made. Extensions are numerous and quite interesting and are expected to provide useful insight for the decision making process in migration policy.

Acknowledgements This research has been co-financed by the European Union (European Social Fund—ESF) and Greek national funds through the Operational Program "Education and Lifelong Learning" of the National Strategic Reference Framework (NSRF)—Research Funding Program: "ARISTEIA—Athens University of Economics and Business—Spatiotemporal Dynamics in Economics".

References

1. Akerlof, G.A.: Social distance and social decisions. Econometrica J. Econometric Soc. **65**, 1005–1027 (1997)
2. Arnold, L.: Random Dynamical Systems. Springer, New York(1998)
3. Bach, R.L., Schraml, L.A.: Migration, crisis and theoretical conflict. Int. Migration Rev. **16** 320–341 (1982)
4. Barro, R.J., Sala-I-Martin, X.: Economic Growth. MIT Press, Cambridge (2004)
5. Bodvarsson, Ö.B., Van den Berg, H.: The Economics of Immigration: Theory and Policy. Springer, New York(2009)
6. Brock, W.A., Durlauf,S.N.: Discrete choice with social interactions. Rev. Econ. Stud. **68**(2), 235–260 (2001)
7. Castells, M.: Immigrant workers and class struggles in advanced capitalism: the western european experience. Polit. Soc. **5**(1), 33–66 (1975)
8. Chueshov, I.: Monotone Random Systems: Theory and Applications, Vol. 1779. Springer, New York (2002)
9. De Haan, L., Ferreira, A.: Extreme Value Theory: An Introduction. Springer, New York (2006)
10. Fawcett, J.T.: Networks, linkages, and migration systems. Int. Migr. Rev. **23**, 671–680 (1989)
11. Fischer, M.M.: A spatial mankiw–romer–weil model: theory and evidence. Ann. Reg. Sci. **47**(2), 419–436 (2011)
12. Fukuchi, T.: Long-run development of a multi-regional economy. Pap. Reg. Sci. **79**(1), 1–31 (2000)
13. Galor, O.: Convergence? inferences from theoretical models. Econ. J. **106**, 1056–1069 (1996)
14. Geis, W., Uebelmesser, S., Werding, M.: How Do Migrants Choose Their Destination Country?: an Analysis of Institutional Determinants. CESifo (Center for Economic Studies & Ifo Institute for Economic Research), Munich (2008)
15. Goss, J., Lindquist, B.: Conceptualizing international labor migration: a structuration perspective. Int. Migr. Rev. **29**, 317–351 (1995)
16. Haag, G., Weidlich, W.: A stochastic theory of interregional migration. Geographical Anal. **16**(4), 331–357 (1984)
17. Harris, J.R., Todaro, M.P.: Migration, unemployment and development: a two-sector analysis. Am. Econ. Rev. **60**(1), 126–142 (1970)
18. Kearney, M.: From the invisible hand to visible feet: anthropological studies of migration and development. Annu. Rev. Anthropol. **15**, 331–361 (1986)
19. Kennan, J., Walker, J.R.: The effect of expected income on individual migration decisions. Econometrica **79**(1), 211–251 (2011)
20. Kotz, S., Nadarajah, S.: Extreme Value Distributions: Theory and Applications. World Scientific, River Edge (2000)

21. Kritz, M.M. Lim, L.L., Zlotnik, H.: International Migration Systems: A Global Approach. Oxford University Press, New York (1992)
22. Lee, E.S.: A theory of migration. Demography 3(1), 47–57 (1966)
23. LeSage, J.P., Pace, R.K.: Spatial econometric modeling of origin-destination flows. J. Reg. Sci. 48(5), 941–967 (2008)
24. Manski, C.F.: The structure of random utility models. Theory Decis. 8(3), 229–254 (1977)
25. Massey, D.S., Arango, J., Hugo, G., Kouaouci, A., Pellegrino, A., Taylor, J.E.: Theories of international migration: a review and appraisal. Popul. Dev. Rev. 19, 431–466 (1993)
26. Nikolinakos, M.: Notes towards a general theory of migration in late capitalism. Race Class 17(1), 5–17 (1975)
27. Pellegrini, P.A., Fotheringham, A.S.: Modelling spatial choice: a review and synthesis in a migration context. Prog. Hum. Geography 26(4), 487–510 (2002)
28. Piore, M.J.: Birds of Passage. Cambridge Books, UK (1980)
29. Portes, A., Kelly, M.P.F.: Images of movement in a changing world: a review of current theories of international migration. Int. Rev. Comp. Public Pol. 1, 15–33 (1989)
30. Radu, D.: Social interactions in economic models of migration: a review and appraisal. J. Ethnic Migration Stud. 34(4), 531–548 (2008)
31. Schenk-Hoppé, K.R., Schmalfuss, B.: Random fixed points in a stochastic solow growth model. J. Math. Econ. 36(1), 19–30 (2001)
32. Seneta, E.: Non-negative Matrices and Markov Chains. Springer, New York (2006)
33. Sheppard, E.: Urban system population dynamics: incorporating nonlinearities. Geographical Anal. 17(1), 47–73 (1985)
34. Train, K.: Discrete Choice Methods with Simulation. Cambridge University Press, New York (2003)
35. Weber, L.: Demographic Change and Economic Growth: Simulations on Growth Models. Springer, New York (2010)

Chapter 33
Complete versus Incomplete Information in the Hotelling Model

Alberto Adrego Pinto and Telmo Parreira

33.1 Introduction

Since the seminal work of Hotelling [14], the model of spatial competition has been seen by many researchers as an attractive framework for analyzing oligopoly markets. The Hotelling model became one of the most important methods of analyzing product differentiation.

In his model, Hotelling present a city represented by a line segment where a uniformly distributed continuum of consumers have to buy a homogeneous good. Consumers have to support linear transportation costs when buying the good in one of the two firms of the city. The firms compete in a two-staged location-price game, where simultaneously choose their location and afterwards set their prices in order to maximize their profits. After this seminal work, a lot of research has been done in related models (see [1–4, 13, 15–21, 24–26]).

In this work, we do not study the Hotelling models in which the location choice by the firms plays a major rule, but models where the location of firms are fixed, this is, models of price competition under spatial nature. We assume that the firms are located at the extremes of the line and so we do not study the first subgame in location strategies. Our main goal is to compare the price formation in the Hotelling

A.A. Pinto
LIAAD-INESC TEC, Porto, Portugal

Faculty of Sciences, Department of Mathematics, University of Porto, Rua do Campo Alegre, 687, 4169-007, Porto, Portugal
e-mail: aapinto@fc.up.pt

T. Parreira
Department of Mathematics, University of Minho, Campus de Gualtar, 4710-057 Braga, Portugal

LIAAD-INESC TEC, Porto, Portugal
e-mail: telmoparreira@hotmail.com

A.A. Pinto and D. Zilberman (eds.), *Modeling, Dynamics, Optimization and Bioeconomics I*, Springer Proceedings in Mathematics & Statistics 73, DOI 10.1007/978-3-319-04849-9_33,
© Springer International Publishing Switzerland 2014

model with complete and incomplete information in the production costs of both firms. The incomplete information consists in each firm to know its production cost but to be uncertain about the competitor cost as usual in oligopoly theory (see [5–12, 22, 23]).

We introduce the bounded costs BC condition that defines a bound for the production costs in terms only of the exogenous variables that are the transportation cost and the road length of the segment line (see Sect. 33.5). Hence, the prices strategies do not appear in the BC condition. Under the bounded costs BC condition, the Nash price strategy for the firms exists. We explicitly determine for the profit, consumer surplus and welfare, the quantitative economical advantages and disadvantages between having complete or incomplete information in the production costs. We prove that, in expected value, the consumer surplus and the welfare are greater with incomplete information than with the complete information and the difference is determined by the variances of the probability distributions (see Eqs. (33.17) and (33.19)). In expected value, the profit it is greater for the firm with higher variance for the probability distribution of its productions costs with incomplete information than with the complete information. However, in expected value, the profit can be smaller for the firm with lower variance for the probability distribution of its productions costs with incomplete information than with the complete information (see Eqs. (33.14) and (33.15)). We conclude by observing that if the probability distribution of the costs is the same for both firms then, in expected value, the profit it is greater for the firms with incomplete information than with the complete information and the welfare is also greater with incomplete information than with the complete information.

Our results are universal, in the incomplete information scenario, in the sense that they apply to all probability distributions in the production costs. All the results presented in this chapter are proved in [18].

33.2   Hotelling Model

The buyers of a commodity will be supposed uniformly distributed along a line with length l. In the two ends of the line there are two firms A and B, located at positions 0 and l respectively, selling the same commodity with unitary *production costs* c_A and c_B. No customer has any preference for either seller except on the ground of price plus *transportation cost* t. We will assume that each consumer buys a single unit of the commodity, in each unit of time and in each unit of length of the line. Denote A's *price* by p_A and B's *price* by p_B. The point of division $x = x(p_A, p_B) \in]0, l[$ between the regions served by the two entrepreneurs is determined by the condition that at this place it is a matter of indifference whether one buys from A or from B.

The point x is the location of the *indifferent consumer* to buy from firm A or firm B, if

$$p_A + tx = p_B + t(l - x).$$

Solving for x, we obtain

$$x = \frac{p_B - p_A + tl}{2t}.$$

Both firms have a non-empty market share if and only if $x \in]0, l[$. Hence, both firms have a non-empty market share if and only if the prices satisfy

$$|p_A - p_B| < tl. \tag{33.1}$$

Assuming inequality (33.1), both firms A and B have a non-empty demand (x and $l - x$) and the *profits* of the two firms are defined respectively by

$$\pi_A = (p_A - c_A) x = (p_A - c_A) \frac{p_B - p_A + tl}{2t}; \tag{33.2}$$

and

$$\pi_B = (p_B - c_B)(l - x) = (p_B - c_B) \frac{p_A - p_B + tl}{2t}. \tag{33.3}$$

Two of the fundamental economic quantities in oligopoly theory are the consumer surplus CS and the welfare W that are computed as follows. Let us denote by v_T be the total amount that consumers are willing to pay for the commodity. The total amount $v(y)$ that a consumer located at y pays for the commodity is given by

$$v(y) = \begin{cases} p_A + ty & \text{if } 0 < y < x; \\ p_B + t(l - y) & \text{if } x < y < l. \end{cases}$$

The *consumer surplus* CS is the difference between the total amount that a consumer is willing to pay v_T and the total amount that the consumer pays $v(y)$

$$CS = \int_0^l v_T - v(y) \mathrm{d}y. \tag{33.4}$$

The *welfare* W is given by adding the profits of firms A and B with the consumer surplus

$$W = CS + \pi_A + \pi_B. \tag{33.5}$$

A price strategy $(\underline{p}_A, \underline{p}_B)$ for both firms is a *Nash* price strategy, if for every deviation of the price $\underline{p}_A$ the profit π_A of firm A decreases, and for every deviation of the price $\underline{p}_B$ the profit π_B of firm B decreases.

Let us compute the Nash price strategy $(\underline{p}_A, \underline{p}_B)$. Differentiating π_A with respect to p_A and π_B with respect to p_B and equalizing to zero, we obtain the first order conditions (FOC). The FOC implies that

$$\underline{p}_A = tl + \frac{1}{3}(2c_A + c_B) \tag{33.6}$$

and

$$\underline{p}_B = tl + \frac{1}{3}(c_A + 2c_B). \tag{33.7}$$

We note that the first order conditions refer to jointly optimizing the profit function (33.2) with respect to the price p_A and the profit function (33.3) with respect to the price p_B.

Since the profit functions (33.2) and (33.3) are concave, the second-order conditions for this maximization problem are satisfied and so the prices (33.6) and (33.7) are indeed maxima for the functions (33.2) and (33.3), respectively. The corresponding equilibrium profits are given by

$$\underline{\pi}_A = \frac{(3tl + c_B - c_A)^2}{18t} \tag{33.8}$$

and

$$\underline{\pi}_B = \frac{(3tl + c_A - c_B)^2}{18t}. \tag{33.9}$$

Furthermore, the consumer indifference location corresponding to the maximizers $\underline{p}_A$ and $\underline{p}_B$ of the profit functions π_A and π_B is

$$\underline{x} = \frac{l}{2} + \frac{c_B - c_A}{6t}.$$

Finally, for the pair of prices $(\underline{p}_A, \underline{p}_B)$ to be a Nash price strategy, we need assumption (33.1) to be satisfied with respect to these pair of prices. We observe that assumption (33.1) is satisfied with respect to the pair of prices $(\underline{p}_A, \underline{p}_B)$ if and only if the following condition with respect to the production costs is satisfied.

Definition 33.1. The Hotelling model satisfies the *bounded costs (BC)* condition, if

$$|c_A - c_B| < 3tl.$$

We note that under the BC condition the prices are higher than the production costs $\underline{p}_A > c_A$ and $\underline{p}_B > c_B$. Hence, there is a Nash price strategy if and only if the BC condition holds. Furthermore, under the BC condition, the pair of prices $(\underline{p}_A, \underline{p}_B)$ is the Nash price strategy.

By Eq. (33.4), the consumer surplus $\underline{CS}$ with respect to the Nash price strategy $(\underline{p}_A, \underline{p}_B)$ is given by

$$\underline{CS} = \int_0^l v_T - v(x)\mathrm{d}x = v_T l - \frac{3}{2}tl^2 - \frac{c_A + 2c_B}{3}l + \frac{(c_B - c_A + 3tl)^2}{36t}. \quad (33.10)$$

By Eq. (33.5), the welfare $\underline{W}$ is given by

$$\underline{W} = v_T l - \frac{1}{4}tl^2 - \frac{c_A + c_B}{2}l + \frac{5(c_A - c_B)^2}{36t}. \quad (33.11)$$

33.3 Uncertainty in the Production Costs

In this section, we introduce a simple notation that is fundamental for the elegance and understanding of the effects of uncertainty in the production costs of both firms.

Let the triples (I_A, Ω_A, q_A) and (I_B, Ω_B, q_B) represent (finite, countable or uncountable) sets of types I_A and I_B with σ-algebras Ω_A and Ω_B and probability measures q_A and q_B, over I_A and I_B, respectively.

We define the expected values $E_A(f)$, $E_B(f)$ and $E(f)$ with respect to the probability measures q_A and q_B as follows:

$$E_A(f) = \int_{I_A} f(z, w)\mathrm{d}q_A(z); \quad E_B(f) = \int_{I_B} f(z, w)\mathrm{d}q_B(w)$$

and

$$E(f) = \int_{I_A} \int_{I_B} f(z, w)\mathrm{d}q_B(w)\mathrm{d}q_A(z).$$

Let $c_A : I_A \to \mathbb{R}_0^+$ and $c_B : I_B \to \mathbb{R}_0^+$ be measurable functions where $c_A^z = c_A(z)$ denotes the production cost of firm A when the type of firm A is $z \in I_A$ and $c_B^w = c_B(w)$ denotes the production cost of firm B when the type of firm B is $w \in I_B$. Furthermore, we assume that the expected values of c_A and c_B are finite

$$E(c_A) = E_A(c_A) = \int_{I_A} c_A^z \mathrm{d}q_A(z) < \infty; \quad E(c_B) = E_B(c_B) = \int_{I_B} c_B^w \mathrm{d}q_B(w) < \infty.$$

We assume that $\mathrm{d}q_A(z)$ denotes the probability of the *belief* of the firm B on the production costs of the firm A to be c_A^z. Similarly, assume that $\mathrm{d}q_B(w)$ denotes the probability of the belief of the firm A on the production costs of the firm B to be c_B^w.

The simplicity of the following cost deviation formulas is crucial to express the main results of this article in a clear and understandable way. The *cost deviations* of firm A and firm B

$$\Delta_A : I_A \to \mathbb{R}_0^+ \quad \text{and} \quad \Delta_B : I_B \to \mathbb{R}_0^+$$

are given respectively by $\Delta_A(z) = c_A^z - E(c_A)$ and $\Delta_B(w) = c_B^w - E(c_B)$. The *cost deviation* between the firms

$$\Delta_C : I_A \times I_B \to \mathbb{R}_0^+$$

is given by $\Delta_C(z, w) = c_A^z - c_B^w$. Since the meaning is clear, we will use through the paper the following simplified notation:

$$\Delta_A = \Delta_A(z); \quad \Delta_B = \Delta_B(w) \text{ and } \Delta_C = \Delta_C(z, w).$$

The *expected cost deviation* Δ_E between the firms is given by $\Delta_E = E(c_A) - E(c_B)$. Hence,

$$\Delta_C - \Delta_E = \Delta_A - \Delta_B.$$

Let V_A and V_B be the variances of the production costs c_A and c_B, respectively. We observe that

$$E(\Delta_C) = \Delta_E; \quad E(\Delta_A^2) = E_A(\Delta_A^2) = V_A; \quad E(\Delta_B^2) = E_B(\Delta_B^2) = V_B.$$

Furthermore,

$$E_A(\Delta_C^2) = \Delta_B^2 + V_A + \Delta_E(\Delta_E - 2\Delta_B);$$
$$E_B(\Delta_C^2) = \Delta_A^2 + V_B + \Delta_E(\Delta_E + 2\Delta_A);$$
$$E(\Delta_C^2) = \Delta_E^2 + V_A + V_B. \tag{33.12}$$

33.4 Complete Information

Let us consider the case where the productions costs are revealed to both firms before they choose the prices. In this case, the competition between the firms is under complete information.

A *price strategy* (p_A^{CI}, p_B^{CI}) is given by a pair of functions $p_A^{CI} : I_A \times I_B \to \mathbb{R}_0^+$ and $p_B^{CI} : I_A \times I_B \to \mathbb{R}_0^+$ where $p_A^{CI}(z, w)$ denotes the price of firm A and $p_B^{CI}(z, w)$ denotes the price of firm B when when the type of firm A is $z \in I_A$ and the type of firm B is $w \in I_B$.

Definition 33.2. The Hotelling model satisfies the *bounded costs (BC)* condition, if

$$|\Delta_C| < 3tl.$$

Under the BC condition, by Eqs. (33.6) and (33.7), the Nash price strategy $(\underline{p}_A^{CI}, \underline{p}_B^{CI})$ is given by

$$\underline{p}_A^{CI}(z, w) = tl + c_B + \frac{2}{3}(\Delta_C)$$

and

$$\underline{p}_B^{CI}(z, w) = tl + c_A - \frac{2}{3}(\Delta_C).$$

By Eq. (33.8), the profit $\pi_A^{CI} : I_A \times I_B \to \mathbb{R}_0^+$ of firm A is given by

$$\underline{\pi}_A^{CI}(z, w) = \frac{(3tl - \Delta_C)^2}{18t}.$$

Similarly, by Eq. (33.9), the profit $\pi_B^{CI} : I_A \times I_B \to \mathbb{R}_0^+$ of firm B is given by

$$\underline{\pi}_B^{CI}(z, w) = \frac{(3tl + \Delta_C)^2}{18t}.$$

The expected profit $E_B(\underline{\pi}_A^{CI})$ for firm A is given by

$$E_B(\underline{\pi}_A^{CI}) = \frac{(6tl - 2\Delta_A - 2\Delta_E)^2 + 4V_B}{72t}$$

Similarly, the expected profit $E_A(\underline{\pi}_B^{CI})$ for firm B is given by

$$E_A(\underline{\pi}_B^{CI}) = \frac{(6tl - 2\Delta_B + 2\Delta_E)^2 + 4V_A}{72t}$$

The expected profit $E(\underline{\pi}_A^{CI})$ for firm A is given by

$$E(\underline{\pi}_A^{CI}) = \frac{(3tl - \Delta_E)^2 + V_A + V_B}{18t}.$$

Similarly, the expected profit $E(\underline{\pi}_B^{CI})$ for firm B is given by

$$E(\underline{\pi}_B^{CI}) = \frac{(3tl + \Delta_E)^2 + V_A + V_B}{18t}$$

By Eq. (33.10), the consumer surplus is given by

$$\underline{CS}^{CI}(z, w) = v_T l - \frac{3}{2}tl^2 - \frac{E(c_A) + 2E(c_B) + \Delta_A + 2\Delta_B}{3}l + \frac{(3tl - \Delta_C)^2}{36t}.$$

The expected value of the consumer surplus $E(\underline{CS}^{CI})$ is

$$E(\underline{CS}^{CI}(z,w)) = v_T l - \frac{3}{2}tl^2 - \frac{E(c_A) + 2E(c_B)}{3}l + \frac{(3tl - \Delta_E)^2 + V_A + V_B}{36t}.$$

By Eq. (33.11), the welfare is given by

$$\underline{W}^{CI}(z,w) = v_T l - \frac{1}{4}tl^2 - \frac{E(c_A) + E(c_B) + \Delta_A + \Delta_B}{2}l + \frac{5\Delta_C^2}{36t}.$$

Using equality (33.12), we obtain that the expected value of the welfare $E(\underline{W}^{CI})$ is given by

$$E(\underline{W}^{CI}(z,w)) = v_T l - \frac{1}{4}tl^2 - \frac{E(c_A) + E(c_B)}{2}l + \frac{5(V_A + V_B + \Delta_E^2)}{36t}.$$

33.5　Incomplete Information

Let us consider the case where the productions costs of each firm is not revealed to the other firm. In this case, the competition between the firms is under incomplete information.

A *price strategy* (p_A, p_B) is given by a pair of functions $p_A : I_A \to \mathbb{R}_0^+$ and $p_B : I_B \to \mathbb{R}_0^+$ where $p_A^z = p_A(z)$ denotes the price of firm A when the type of firm A is $z \in I_A$ and $p_B^w = p_B(w)$ denotes the price of firm B when the type of firm B is $w \in I_B$. We note that $E(p_A) = E_A(p_A)$ and $E(p_B) = E_B(p_B)$. The *indifferent consumer* $x : I_A \times I_B \to (0, l)$ is given by

$$x^{z,w} = \frac{p_B^w - p_A^z + tl}{2t}.$$

The *ex-post profits* $\pi_A^{EP} : I_A \times I_B \to \mathbb{R}_0^+$ and $\pi_B^{EP} : I_A \times I_B \to \mathbb{R}_0^+$ are given by

$$\pi_A^{EP}(z,w) = \pi_A(z,w) = (p_A^z - c_A^z)x^{z,w}$$

and

$$\pi_B^{EP}(z,w) = \pi_B(z,w) = (p_B^w - c_B^w)(l - x^{z,w}).$$

The *ex-ante profits* $\pi_A^{EA} : I_A \to \mathbb{R}_0^+$ and $\pi_B^{EA} : I_B \to \mathbb{R}_0^+$ are given by

$$\pi_A^{EA}(z) = E_B(\pi_A^{EP}) \quad and \quad \pi_B^{EA}(w) = E_A(\pi_B^{EP}).$$

We note that, the *expected profit* $E(\pi_A^{EP})$ of firm A is equal to $E_A(\pi_A^{EA})$ and the *expected profit* $E(\pi_B^{EP})$ of firm B is equal to $E_B(\pi_B^{EA})$.

A price strategy $(\underline{p}_A, \underline{p}_B)$ for both firms is a *Bayesian-Nash*, if for every $z \in I_A$ and for every deviation of the price $\underline{p}_A^z$ the ex-ante profit $\pi_A^{EA}(z)$ of firm A decreases, and for every $w \in I_B$ and for every deviation of the price $\underline{p}_B^w$ the ex-ante profit $\pi_B^{EA}(w)$ of firm B decreases.

For $i \in \{A, B\}$, we define

$$c_i^m = \min_{z \in I_i}\{c_i^z\} \quad \text{and} \quad c_i^M = \max_{z \in I_i}\{c_i^z\}.$$

Definition 33.3. The Hotelling model satisfies the *bounded uncertain costs (BUC)* condition, if

$$|3(c_A^z - c_B^w) + E(c_B) - E(c_A)| = |3\Delta_C - \Delta_E| < 6tl,$$

for all $z \in I_A$ and for all $w \in I_B$;

$$3\left(c_A^M + c_B^M - 2c_A^m\right) + E(c_A) - E(c_B) \leqslant 3tl + \frac{\left(E(c_A) + 2E(c_B) - 3c_A^M\right)^2}{12tl};$$

and

$$3\left(c_A^M + c_B^M - 2c_B^m\right) + E(c_B) - E(c_A) \leqslant 3tl + \frac{\left(2E(c_A) + E(c_B) - 3c_B^M\right)^2}{12tl}.$$

Let

$$\overline{\Delta} = \max_{i,j \in \{A,B\}}\{c_i^M - c_j^m\}$$

Thus, the bounded uncertain costs condition *BUC* is implied by the following stronger *SBUC* condition.

Definition 33.4. The Hotelling model satisfies the *bounded uncertain costs (SBUC)* condition, if

$$7\overline{\Delta} < 3tl$$

Theorem 33.1. *Under the BUC condition, there is a Bayesian-Nash price strategy $(\underline{p}_A, \underline{p}_B)$. Furthermore, the expected prices of the Bayesian-Nash price strategy are*

$$E(\underline{p}_A) = tl + E(c_A) - \frac{\Delta_E}{3};$$

$$E(\underline{p}_B) = tl + E(c_B) + \frac{\Delta_E}{3}.$$

and the Bayesian-Nash price strategy is

$$\underline{p}_A^z = E(\underline{p}_A) + \frac{\Delta_A}{2}; \quad \underline{p}_B^w = E(\underline{p}_B) + \frac{\Delta_B}{2}. \qquad (33.13)$$

The pair of prices $(\underline{p}_A, \underline{p}_B)$ satisfies $\underline{p}_A^z > c_A^z$ and $\underline{p}_B^w > c_B^w$. Furthermore, for different production costs, the differences between the optimal prices of a firm are proportional to the differences of the production costs

$$\underline{p}_A^{z_1} - \underline{p}_A^{z_2} = \frac{c_A^{z_1} - c_A^{z_2}}{2}.$$

and

$$\underline{p}_B^{w_1} - \underline{p}_B^{w_2} = \frac{c_B^{w_1} - c_B^{w_2}}{2}.$$

for all $z_1, z_2 \in I_A$ and $w_1, w_2 \in I_B$. Hence, half of the production costs value is incorporated in the price.

The ex-post profit of the firms is the effective profit of the firms given a realization of the production costs for both firm. Hence it is the main economic information for both firms. By Eq. (33.13), the ex-post profit of firm A is

$$\underline{\pi}_A^{EP}(z, w) = \frac{(6tl + \Delta_E - 3\Delta_C)(6tl + \Delta_E - 3\Delta_C - 3\Delta_B)}{72t}$$

and the ex-post profit of firm B is

$$\underline{\pi}_B^{EP}(z, w) = \frac{(6tl - \Delta_E + 3\Delta_C)(6tl - \Delta_E + 3\Delta_C - 3\Delta_A)}{72t}.$$

We observe that the differences between the ex-post profits of both firms has a very useful and clear economical interpretation in terms of the expected cost deviations.

$$\underline{\pi}_A^{EP}(z, w) - \underline{\pi}_B^{EP}(z, w) = \frac{6tl(\Delta_A - \Delta_B) + (\Delta_E - 3\Delta_C)(8tl - \Delta_A - \Delta_B)}{24t}$$

Furthermore, for different production costs, the differences between the ex-post profit of a firm is given by

$$\underline{\pi}_A^{EP}(z_1, w) - \underline{\pi}_A^{EP}(z_2, w) = \frac{(c_A^{z_2} - c_A^{z_1})(12tl - \Delta_E + 3(c_B^w + E(c_A) - c_A^{z_1} - c_A^{z_2}))}{24t}$$

and

$$\underline{\pi}_B^{EP}(z, w_1) - \underline{\pi}_B^{EP}(z, w_2)$$
$$= \frac{(c_B^{w_2} - c_B^{w_1})(12tl + \Delta_E + 3(c_A^z + E(c_B) - c_B^{w_1} - c_B^{w_2}))}{24t}$$

for all $z, z_1, z_2 \in I_A$ and $w, w_1, w_2 \in I_B$.

The ex-ante profit of a firm is the expected profit of the firm that knows its production cost but is uncertain about the production costs of the competitor firm. The ex-ante profit of firm A is

$$\underline{\pi}_A^{EA}(z) = \frac{(6tl - 3\Delta_A - 2\Delta_E)^2}{72t}.$$

Similarly, the ex-ante profit of firm B is

$$\underline{\pi}_B^{EA}(w) = \frac{(6tl - 3\Delta_B + 2\Delta_E)^2}{72t}.$$

We observe that the differences between the ex-ante profits of both firms has a very useful and clear economical interpretation in terms of the expected cost deviations.

$$\underline{\pi}_A^{EA}(z) - \underline{\pi}_B^{EA}(w) = \frac{(4tl - \Delta_A - \Delta_B)(3(\Delta_B - \Delta_A) - 4\Delta_E)}{24t}$$

Furthermore, for different production costs, the differences between the ex-ante profits of a firm are given by

$$\underline{\pi}_A^{EA}(z_1) - \underline{\pi}_A^{EA}(z_2) = \frac{(c_A^{z_2} - c_A^{z_1})(3(4tl + 2E(c_A) - c_A^{z_1} - c_A^{z_2}) - 4\Delta_E)}{24t}$$

and

$$\underline{\pi}_B^{EA}(w_1) - \underline{\pi}_B^{EA}(w_2) = \frac{(c_B^{w_2} - c_B^{w_1})(3(4tl + 2E(c_B) - c_B^{w_1} - c_B^{w_2}) + 4\Delta_E)}{24t}$$

for all $z, z_1, z_2 \in I_A$ and $w, w_1, w_2 \in I_B$.

We observe that the ex-ante and the ex-posts profits of both firms are strictly positive with respect to the Bayesian-Nash price strategy determined in Theorem 33.1. Hence, the expected profits of both firms are also strictly positive.

The expected profit of the firm is the expected gain of the firm. The expected profit of firm A is given by

$$E(\underline{\pi}_A^{EP}) = \frac{(3tl - \Delta_E)^2}{18t} + \frac{V_A}{8t}$$

Similarly, the expected profit of firm B is given by

$$E(\underline{\pi}_B^{EP}) = \frac{(3tl + \Delta_E)^2}{18t} + \frac{V_B}{8t}.$$

The ex-post consumer surplus is the realized gain of the consumers community for given outcomes of the production costs of both firms. Under incomplete information, by Eq. (33.4), the ex-post consumer surplus is

$$\underline{CS}^{EP} = v_T l - \frac{3}{2}tl^2 - \frac{l}{3}(2E(c_B) + E(c_A)) - \frac{\Delta_B l}{2} + \frac{(6tl - 3\Delta_C + \Delta_E)^2}{144t}.$$

The expected value of the consumer surplus is the expected gain of the consumers community for all possible outcomes of the production costs of both firms. The expected value of the consumer surplus $E(\underline{CS}^{EP})$ is given by

$$E(\underline{CS}^{EP}) = v_T l - \frac{3}{2}tl^2 - \frac{l}{3}(2E(c_B) + E(c_A)) + \frac{(6tl - 2\Delta_E)^2 + 9(V_A + V_B)}{144t}.$$

The ex-post welfare is the realized gain of the state that includes the gains of the consumers community and the gains of the firms for a given outcomes of the production costs of both firms. By Eq. (33.5), the ex-post welfare is

$$\underline{W}^{EP} = v_T l - \frac{1}{4}tl^2 - \frac{E(c_A) + E(c_B) + \Delta_A + \Delta_B}{2} - \frac{3\Delta_C(2\Delta_E - 9\Delta_C) + (\Delta_E)^2}{144t}.$$

The expected value of the welfare is the expected gain of the state for all possible outcomes of the production costs of both firms. The expected value of the welfare $E(\underline{W}^{EP})$ is given by

$$E(\underline{W}^{EP}) = v_T l - \frac{1}{4}tl^2 - \frac{E(c_A) + E(c_B)}{2} + \frac{27(V_A + V_B) + 20\Delta_E^2}{144t}.$$

33.6 Incomplete Versus Complete Information

Corollary 33.1. *The difference between the ex-post profit and the profit, under complete information, for firm A is*

$$\underline{\pi}_A^{EP}(z, w) - \underline{\pi}_A^{CI}(z, w) = \frac{(\Delta_A - \Delta_B)(\Delta_A + 2\Delta_B) - 2(3tl - \Delta_C)(2\Delta_A + \Delta_B)}{72t}.$$

The difference between the ex-post profit and the profit, under complete information, for firm B is

$$\underline{\pi}_B^{EP}(z,w) - \underline{\pi}_B^{CI}(z,w) = \frac{(\Delta_B - \Delta_A)(\Delta_B + 2\Delta_A) - 2(3tl + \Delta_C)(2\Delta_B + \Delta_A)}{72t}.$$

Corollary 33.2. *The difference between the ex-ante profit $E_B(\pi_A^{EP})$ and $E_B(\pi_A^{CI})$ for firm A is*

$$E_B(\underline{\pi}_A^{EP}) - E_B(\underline{\pi}_A^{CI}) = \frac{\Delta_A(5\Delta_A - 4(3tl - \Delta_E))}{72t} - \frac{V_B}{18t}.$$

The difference between the ex-ante profit $E_A(\pi_B^{EP})$ and $E_A(\pi_B^{CI})$ for firm B is

$$E_A(\underline{\pi}_B^{EP}) - E_A(\underline{\pi}_B^{CI}) = \frac{\Delta_B(5\Delta_B - 4(3tl + \Delta_E))}{72t} - \frac{V_A}{18t}.$$

The difference between the expected profits of firm A with complete and incomplete information is given by

$$E(\underline{\pi}_A^{EP}) - E(\underline{\pi}_A^{CI}) = \frac{5V_A - 4V_B}{72t}. \tag{33.14}$$

The difference between the expected profits of firm B with complete and incomplete information is given by

$$E(\underline{\pi}_B^{EP}) - E(\underline{\pi}_B^{CI}) = \frac{5V_B - 4V_A}{72t}. \tag{33.15}$$

Corollary 33.3. *The difference between the ex-post consumer surplus and the consumer surplus, under complete information, is*

$$\underline{CS}^{EP} - \underline{CS}^{CI} = \frac{(\Delta_A + \Delta_B)l}{4} + \frac{(\Delta_B - \Delta_A)(\Delta_B - \Delta_A - 4\Delta_C)}{144t}. \tag{33.16}$$

Therefore, Eq. (33.16) determines in which cases it is better to have uncertainty in the production costs instead of complete information in terms of consumer surplus $\underline{CS}^{EP} > \underline{CS}^{CI}$.

The difference between expected value of the consumer surplus and the expected value of the consumer surplus under complete information, is

$$E(\underline{CS}^{EP}) - E(\underline{CS}^{CI}) = \frac{5(V_A + V_B)}{144t}. \tag{33.17}$$

Therefore, in expected value the consumer surplus is greater with incomplete information than with complete information.

The difference between the ex-post welfare and the welfare, under complete information, is

$$\underline{W}^{EP} - \underline{W}^{CI} = \frac{7(\Delta_C)^2 - 6\Delta_C\Delta_E - (\Delta_E)^2}{144t}. \tag{33.18}$$

Therefore, Eq. (33.18) determines in which cases it is better to have uncertainty in the production costs instead of complete information in terms of welfare $\underline{W}^{EP} > \underline{W}^{CI}$.

The difference between expected value of the welfare and the expected value of the welfare under complete information, is

$$E(\underline{W}^{EP}) - E(\underline{W}^{CI}) = \frac{7(V_A + V_B)}{144t}. \tag{33.19}$$

Therefore, in expected value the welfare is greater with incomplete information than with complete information.

33.7 Symmetric Hotelling

A Hotelling game is *symmetric*, if $(I_A, \Omega_A, q_A) = (I_B, \Omega_B, q_B)$ and $c = c_A = c_B$. Hence, we observe that all the formulas of this paper hold with the following simplifications $\Delta_E = 0$, $E(c) = E(c_A) = E(c_B)$ and $V = V_A = V_B$. Let $c^M = c_A^M = c_B^M$ and $c^m = c_A^m = c_B^m$. The bounded uncertain costs in the symmetric case can be written in the following way

Definition 33.5. The symmetric Hotelling model satisfies the *bounded uncertain costs (BUC)* condition, if

$$|\Delta_C| < 2tl,$$

for all $z \in I_A$ and for all $w \in I_B$;

$$2\left(c^M - c^m\right) \leq tl + \frac{\left(c^M - E(c)\right)^2}{4tl};$$

Thus, the bounded uncertain costs condition *BUC* is implied by the following stronger condition.

Definition 33.6. The symmetric Hotelling model satisfies the *bounded uncertain costs (SBUC)* condition, if

$$2\overline{\Delta} < tl$$

Under the *BUC* condition, the expected prices of Bayesian-Nash price strategy are given by

$$E(\underline{p}_A) = E(\underline{p}_B) = tl + E(c).$$

We observe that the difference between the expected profits of firm A and B with complete and incomplete information is positive and it is given by

$$E(\underline{\pi}_A^{EP}) - E(\underline{\pi}_A^{CI}) = E(\underline{\pi}_B^{EP}) - E(\underline{\pi}_B^{CI}) = \frac{V}{72t}.$$

Furthermore, the difference between the welfare and the welfare with complete and incomplete information is positive and it is given by

$$\underline{W}^{EP} - \underline{W}^{CI} = \frac{7(\Delta_C)^2}{144t}.$$

33.8 Conclusion

Under complete and incomplete information, we studied the economic impact of the Nash price strategies in the profits of the firms, consumer surplus and welfare. Under incomplete information, we observed that the Bayesian-Nash price strategies do not depend upon the distributions of the production costs of the firms, except on their first moments. Furthermore, the prices of each firm, at equilibrium, are affine with respect to the expected costs of both firms and to their own costs. The corresponding expected profits are quadratic in the expected cost of both firms, in its own cost and in the transportation cost. We computed the ex-post consumer surplus and the ex-post welfare and we explicitly determined in which cases it is better to have uncertainty in the production costs instead of complete information in terms of profits, consumer surplus and in terms of welfare.

Acknowledgements We are thankful to the anonymous referees for their suggestions. We acknowledge the financial support of LIAAD-INESC TEC through 'Strategic Project—LA 14—2013–2014' with reference PEst-C/EEI/LA0014/2013, USP-UP project, IJUP, Faculty of Sciences, University of Porto, Calouste Gulbenkian Foundation, FEDER, POCI 2010 and COMPETE Programmes and Fundação para a Ciência e a Tecnologia (FCT) through Project "Dynamics and Applications", with reference PTDC/MAT/121107/2010. Telmo Parreira thanks FCT, for the PhD scholarship SFRH/BD/33762/2009.

References

1. Biscaia, R., Sarmento, P.: Spatial Competition and Firms' Location Decisions under Cost Uncertainty. FEP Working Papers No. 445 (2012)
2. Boyer, M., Laffont, J., Mahenc, P., Moreaux, M.: Location Distortions under incomplete information. Reg. Sci. Urban Econ. **24**(4), 409–440 (1994)
3. Boyer, M., Mahenc, P., Moreaux, M.: Asymmetric information and product differentiation. Reg. Sci. Urban Econ. **33**(1), 93–113 (2003)
4. D'Aspremont, C., Gabszewicz, J., Thisse, J.-F.: On hotelling's "stability in competition". Econometrica **47**(5), 1145–1150 (1979)
5. Ferreira, F.A., Pinto, A.A.: Uncertainty on a Bertrand duopoly with product differentiation. In: Machado, J.A., Silva, M.F., Barbosa, R.S., Figueiredo, L.B. (eds.) Nonlinear Science and Complexity, pp. 389–395. Springer, Netherlands (2011)
6. Ferreira, F., Ferreira, F.A., Ferreira, M., Pinto, A.A.: Flexibility in a Stackelberg leadership with differentiated goods. Optimization (accepted)
7. Ferreira, F., Ferreira, F.A., Pinto, A.A.: Bayesian price leadership. In: Tas, K., et al. (eds.) Mathematical Methods in Engineering, pp. 371–379. Springer, Netherlands (2007)
8. Ferreira, F.A., Ferreira, F., Pinto, A.A.: Unknown costs in a duopoly with differentiated products. In: Tas, K., et al. (eds.) Mathematical Methods in Engineering, pp. 359–369. Springer, Netherlands (2007)
9. Ferreira, F., Ferreira, F.A., Pinto, A.A.: Flexibility in stackelberg leadership. In: Machado, J.A., Patkai, B., Rudas, I.J. (eds.) Intelligent Engineering Systems and Computational Cybernetics, pp. 399–405. Springer, Netherlands (2009)
10. Ferreira, F., Ferreira, F.A., Pinto, A.A.: Price-setting dynamical duopoly with incomplete information. In: Machado, J.A., Silva, M.F., Barbosa, R.S., Figueiredo, L.B. (eds.) Nonlinear Science and Complexity, pp. 397–403. Springer, Netherlands (2011)
11. Ferreira, M., Figueiredo, I.P., Oliveira, B.M.P.M., Pinto, A.A.: Strategic optimization in R&D Investment. Optim. J. Math. Program. Oper. Res. **61**(8), 1013–1023 (2012)
12. Gibbons, R.: A Primer in Game Theory. Financial Times/Prentice Hall, New York (1992)
13. Graitson, D.: Spatial competition à la Hotelling: a selective survey. J. Ind. Econ. **31**, 11–25 (1982)
14. Hotelling, H.: Stability in competition. Econ. J. **39**, 41–57 (1929)
15. Lederer, P., Hurter, A.: Competition of Firms: discriminatory pricing and location. Econometrica **54**(3), 623–640 (1986)
16. Osborne, M.J., Pitchick, C.: Equilibrium in Hotelling's model of spatial competition. Econometrica **55**(4), 911–922 (1987)
17. Pinto, A.A., Parreira, T.: Price competition in the Hotelling model with uncertainty on costs. Optim. J. Math. Program. Oper. Res. (accepted)
18. Pinto, A.A., Parreira, T.: Bayesian-Nash prices in linear Hotelling model (submited)
19. Pinto, A.A., Parreira, T.: A hotelling-type network. In: Peixoto, M., Pinto, A.A., Rand, D. (eds.) Dynamics, Games and Science I. Springer Proceedings in Mathematics, vol. 1, pp. 709–720. Springer, Berlin (2011)
20. Pinto, A.A., Parreira, T.: Maximal differentiation in the Hotelling model with uncertainty. In: Pinto, A.A., Zilberman, D. (eds.) Modeling, Dynamics, Optimization and Bioeconomy. Springer Proceedings in Mathematics and Statistics Series. Springer, Heidelberg (2014)
21. Pinto, A.A., Parreira, T.: Optimal localization of firms in Hotelling networks. In: Pinto, A.A., Zilberman, D. (eds.) Modeling, Dynamics, Optimization and Bioeconomy. Springer Proceedings in Mathematics and Statistics Series. Springer, Heidelberg (2014)
22. Pinto, A.A., Ferreira, F.A., Ferreira, M., Oliveira, B.M.P.M.: Cournot duopoly with competition in the R&D expenditures. In: Proceedings of Symposia in Pure Mathematics, vol. 7. Wiley-VCH Verlag, Weinheim (2007)
23. Pinto, A.A., Oliveira, B. M. P. M., Ferreira, F.A., Ferreira, F.: Stochasticity favoring the effects of the R&D strategies of the firms. In: Machado, J.A., Patkai, B., Rudas, I.J. (eds.) Intelligent

Engineering Systems and Computational Cybernetics, pp. 415–423. Springer, Dordrecht (2008)
24. Salop, S.: Monopolistic competition with outside goods. Bell J. Econ. **10**, 141–156 (1979)
25. Tabuchi, T., Thisse, J.F.: Asymmetric equilibria in spatial competition. Int. J. Econ. Theory **13**(2), 213–227 (1995)
26. Ziss, S.: Entry deterrence, cost advantage and horizontal product differentiation. Reg. Sci. Urban Econ. **23**, 523–543 (1993)

Chapter 34
Maximal Differentiation in the Hotelling Model with Uncertainty

Alberto Adrego Pinto and Telmo Parreira

34.1 Introduction

Since the seminal work of Hotelling [14], the model of spatial competition has been seen by many researchers as an attractive framework for analyzing oligopoly markets (see [2,13,15–21,24,25]). In his model, Hotelling present a city represented by a line segment where a uniformly distributed continuum of consumers have to buy a homogeneous good. Consumers have to support linear transportation costs when buying the good in one of the two firms of the city. The firms compete in a two-staged location-price game, where simultaneously choose their location and afterwards set their prices in order to maximize their profits. Hotelling concluded that firms would agglomerate at the center of the line, an observation referred as the "Principle of Minimum Differentiation".

In 1979, D'Aspremont et al. [4] show that the "Principle of Minimum Differentiation" is invalid, since there was no price equilibrium solution for all possible locations of the firms, in particular when they are not far enough from each other. Moreover, in the same article, D'Aspremont et al. introduce a modification in the Hotelling model, considering quadratic transportation costs instead of linear. The introduction of this feature removed the discontinuities verified in the profit and

A.A. Pinto
LIAAD-INESC TEC, Porto, Portugal

Faculty of Sciences, Department of Mathematics, University of Porto, Rua do Campo Alegre, 687, 4169-007, Porto, Portugal
e-mail: aapinto@fc.up.pt

T. Parreira (✉)
Department of Mathematics, University of Minho, Campus de Gualtar, 4710-057 Braga, Portugal

LIAAD-INESC TEC, Porto, Portugal
e-mail: telmoparreira@hotmail.com

A.A. Pinto and D. Zilberman (eds.), *Modeling, Dynamics, Optimization and Bioeconomics I*, Springer Proceedings in Mathematics & Statistics 73, DOI 10.1007/978-3-319-04849-9_34,
© Springer International Publishing Switzerland 2014

demand functions, which was a problem in Hotelling model and they show that, under quadratic transportation costs, a price equilibrium exists for all locations and a location equilibrium exists and involves maximum product differentiation, i.e. the firms opt to locate at the extremes of the line.

Hotelling and D'Aspremont et al. consider that the production costs of both firms are equal to zero. Ziss [26] introduce a modification in the model of D'Aspremont et al. by allowing for different production costs between the two firms and examines the effect of heterogeneous production technologies on the location problem. Ziss shows that a price equilibrium exists for all locations and concludes that when the difference between the production costs is small, a price and location equilibrium exists in which the firms prefer to locate in different extremes of the line. However, if the difference between the production costs is sufficiently large, a location equilibrium does not exist.

Boyer et al. [3] and Biscaia and Sarmento [1] extended the work of Ziss by consider that the uncertainty on the productions costs exists only during the first subgame in location strategies. Then the production costs are revealed to the firms before the firms have to choose their optimal price strategies and so the second subgame has complete information.

In this work, we study the quadratic Hotelling model with incomplete information in the production costs of both firms. The incomplete information consists in each firm to know its production cost but to be uncertain about the competitor cost as usual in oligopoly theory (see [5–12, 22, 23]). However, in contrast with Boyer et al. [3], the production costs are not revealed to the firms before the firms have to choose their price strategy. Furthermore, our results are universal, in the incomplete information scenario, in the sense that they apply to all probability distributions in the production costs.

We say that the Bayesian-Nash price strategy has the duopoly property if both firms have non-empty market for every pair of production costs. We introduce the bounded uncertain costs and location $BUCL1$ condition that defines a bound for the production costs in terms only of the exogenous variables that are the transportation cost; the road length of the segment line; and the localization of both firms (see Sect. 34.6). We prove that there is a local optimum price strategy with the duopoly property if and only if the bounded uncertain costs and location $BUCL1$ condition holds. We compute explicitly the formula for the local optimum price strategy that is simple and leaves clear the influence of the relevance economic exogenous quantities in the price formation. In particular, we observe that the local optimum price strategy do not depend on the distributions of the production costs of the firms, except on their first moments.

We introduce two mild additional bounded uncertain costs and location $BUCL2$ and $BUCL3$ conditions. Under the $BUCL1$ and $BUCL2$ conditions, we prove that the local optimum price strategy is a Bayesian-Nash price strategy. Assuming that the firms choose the Bayesian-Nash price strategy, under the $BUCL3$ condition, we prove that the maximal differentiation is a local optimum for the localization strategy of both firms.

All the results presented in this chapter are proved in [21].

34.2   Local Optimum Price Strategy Under Complete Information

The buyers of a commodity will be supposed uniformly distributed along a line with length l, where two firms A and B located at respective distances a and b from the endpoints of the line sell the same commodity with unitary *production costs* c_A and c_B. We assume without loss of generality that $a \geqslant 0$, $b \geqslant 0$ and $l - a - b \geqslant 0$. No customer has any preference for either seller except on the ground of price plus *transportation cost* t. We will assume that each consumer buys a single unit of the commodity, in each unit of time and in each unit of length of the line. Denote A's *price* by p_A and B's *price* by p_B. The point of division $x = x(p_A, p_B) \in]0, l[$ between the regions served by the two entrepreneurs is determined by the condition that at this place it is a matter of indifference whether one buys from A or from B. The point x is the location of the *indifferent consumer* to buy from firm A or firm B, if

$$p_A + t (x - a)^2 = p_B + t (l - b - x)^2$$

Let

$$m = l - a - b; \quad \Delta_l = a - b \quad \text{and} \quad \Delta_C = c_A - c_B.$$

Solving for x, we obtain

$$x = \frac{p_B - p_A}{2 t m} + \frac{l + \Delta_l}{2}.$$

Both firms have a non-empty market share if, and only if, $x \in]0, l[$. Hence, the prices will have to satisfy

$$|p_A - p_B - t m \Delta_l| < t m l \tag{34.1}$$

Assuming inequality (34.1), both firms A and B have a non-empty demand (x and $l - x$) and the *profits* of the two firms are defined respectively by

$$\pi_A = (p_A - c_A) x = (p_A - c_A) \left(\frac{p_B - p_A}{2 t m} + \frac{l + \Delta_l}{2} \right) \tag{34.2}$$

and

$$\pi_B = (p_B - c_B) (l - x) = (p_B - c_B) \left(\frac{p_A - p_B}{2 t m} + \frac{l - \Delta_l}{2} \right). \tag{34.3}$$

Definition 34.1. A price strategy (p_A, p_B) for both firms is a *local optimum price strategy* if (i) for every small deviation of the price p_A the profit π_A of firm A

decreases, and for every small deviation of the price p_B the profit π_B of firm B decreases (local optimum property); and (ii) the indifferent consumer exists, i.e. $0 < x < l$ (duopoly property).

Let us compute the local optimum price strategy (p_A, p_B). Differentiating π_A with respect to p_A and π_B with respect to p_B and equalizing to zero, we obtain the first order conditions (FOC). The FOC implies that

$$p_A = t\,m\left(l + \frac{\Delta_l}{3}\right) + c_A - \frac{\Delta_C}{3} \tag{34.4}$$

and

$$p_B = t\,m\left(l - \frac{\Delta_l}{3}\right) + c_B + \frac{\Delta_C}{3}. \tag{34.5}$$

We note that the first order conditions refer to jointly optimizing the profit function (34.2) with respect to the price p_A and the profit function (34.3) with respect to the price p_B.

Since the profit functions (34.2) and (34.3) are concave, the second-order conditions for this maximization problem are satisfied and so the prices (34.4) and (34.5) are indeed maxima for the functions (34.2) and (34.3), respectively. The corresponding equilibrium profits are given by

$$\pi_A = \frac{(m\,(3\,l + \Delta_l)\,t - \Delta_C)^2}{18\,t\,m} \tag{34.6}$$

and

$$\pi_B = \frac{(m\,(3\,l - \Delta_l)\,t + \Delta_C)^2}{18\,t\,m}.$$

Furthermore, the consumer indifference location corresponding to the maximizers p_A and p_B of the profit functions π_A and π_B is

$$x = \frac{l}{2} + \frac{\Delta_l}{6} - \frac{\Delta_C}{6\,t\,m}.$$

Finally, for the pair of prices (p_A, p_B) to be a local optimum price strategy, we need assumption (34.1) to be satisfied with respect to these pair of prices. We observe that assumption (34.1) is satisfied with respect to the pair of prices (p_A, p_B) if and only if the following condition with respect to the production costs is satisfied.

Definition 34.2. The Hotelling model satisfies the *bounded costs and location* (*BCL*) condition, if

$$|\Delta_C - t\,m\,\Delta_l| < 3\,t\,m\,l.$$

We note that under the *BCL* condition the prices are higher than the production costs $p_A > c_A$ and $p_B > c_B$. Hence, there is a local optimum price strategy if and only if the *BCL* condition holds. Furthermore, under the *BCL* condition, the pair of prices (p_A, p_B) is the local optimum price strategy.

A strong restriction that the *BCL* condition imposes is that Δ_C converges to 0 when m tends to 0, i.e. when the differentiation in the localization tends to vanish.

34.3 Nash Price Strategy Under Complete Information

We note that, if a Nash price equilibrium satisfies the duopoly property then it is a local optimum price strategy. However, a local optimum price strategy is only a local strategic maximum. Hence, the local optimum price strategy to be a Nash equilibrium must also be global strategic maximum. In this section, we are going to show that this is the case.

Following D'Aspremont et al. [4], we note that the profits of the two firms, valued at local optimum price strategy are globally optimal if they are at least as great as the payoffs that firms would earn by undercutting the rivals's price and supplying the whole market.

Let (p_A, p_B) be the local optimum price strategy. Firm A may gain the whole market, undercutting its rival by setting

$$p_A^M = p_B - t\,m\,(l - \Delta_l).$$

In this case the profit amounts to

$$\pi_A^M = \frac{2}{3}\,(t\,m\,\Delta_l - \Delta_C)\,l.$$

A similar argument is valid for store B. Undercutting this rival, setting

$$p_B^M = p_A - t\,m\,(l + \Delta_l),$$

it would earn

$$\pi_B^M = \frac{2}{3}\,(\Delta_C - t\,m\,\Delta_l)\,l.$$

The conditions for such undercutting not to be profitable are $\pi_A \geqslant \pi_A^M$ and $\pi_B \geqslant \pi_B^M$. Hence, proving that

$$\frac{(m\,(3l + \Delta_l)\,t - \Delta_C)^2}{18\,t\,m} \geqslant \frac{2}{3}\,(t\,m\,\Delta_l - \Delta_C)\,l \tag{34.7}$$

is sufficient to prove that $\pi_A \geqslant \pi_A^M$. Similarly, proving that

$$\frac{(m\,(3\,l-\Delta_l)\,t+\Delta_C)^2}{18\,t\,m} \geqslant \frac{2}{3}\,(\Delta_C - t\,m\,\Delta_l)\,l \tag{34.8}$$

is sufficient to prove that $\pi_B \geqslant \pi_B^M$.

However, conditions (34.7) and (34.8) are satisfied because they are equivalent to

$$(m\,(3\,l-\Delta_l)\,t+\Delta_C)^2 \geqslant 0$$

and

$$(m\,(3\,l+\Delta_l)\,t-\Delta_C)^2 \geqslant 0.$$

Therefore, if (p_A, p_B) is a local optimum price strategy then (p_A, p_B) is a Nash price equilibrium.

34.4 Optimum Localization Equilibrium Under Complete Information

We are going to find when the maximal differentiation is a local optimum strategy assuming that the firms in second subgame choose the Nash price equilibrium strategy. For a complete discussion see Ziss [26].

We note that from (34.4) and (34.6), we can write the profit of firm A as

$$\pi_A = \frac{(p_A - c_A)^2}{2\,t\,(l-a-b)}.$$

Since

$$\frac{\partial p_A}{\partial a} = -\frac{2}{3}\,t\,(l+a),$$

we obtain that

$$\frac{\partial \pi_A}{\partial a} = -\frac{p_A - c_A}{6\,t\,(l-a-b)^2}\,(\Delta_C + t\,(l-a-b)\,(l+3\,a+b)).$$

Similarly, we obtain that

$$\frac{\partial \pi_B}{\partial b} = \frac{p_B - c_B}{6\,t\,(l-a-b)^2}\,(\Delta_C - t\,(l-a-b)\,(l+a+3\,b)).$$

Therefore, the maximal differentiation $(a, b) = (0, 0)$ is a local optimum strategy if and only if

$$\frac{\partial \pi_A}{\partial a}(0, 0) = -\frac{p_A - c_A}{6 t \, l^2} \left(\Delta_C + t \, l^2\right) < 0$$

and

$$\frac{\partial \pi_B}{\partial b}(0, 0) = \frac{p_B - c_B}{6 t \, l^2} \left(\Delta_C - t \, l^2\right) < 0$$

Since

$$\frac{p_A - c_A}{6 t \, l^2} > 0 \quad \text{and} \quad \frac{p_B - c_B}{6 t \, l^2} > 0$$

the maximal differentiation $(a, b) = (0, 0)$ is a local optimum strategy if and only if

$$|\Delta_C| < t \, l^2.$$

34.5 Incomplete Information on the Production Costs

The incomplete information consists in each firm to know its production cost but to be uncertain about the competitor cost. In this section, we introduce a simple notation that is fundamental for the elegance and understanding of the results presented in this paper.

Let the triples (I_A, Ω_A, q_A) and (I_B, Ω_B, q_B) represent (finite, countable or uncountable) sets of types I_A and I_B with σ-algebras Ω_A and Ω_B and probability measures q_A and q_B, over I_A and I_B, respectively.

We define the expected values $E_A(f)$, $E_B(f)$ and $E(f)$ with respect to the probability measures q_A and q_B as follows:

$$E_A(f) = \int_{I_A} f(z, w) \, dq_A(z); \quad E_B(f) = \int_{I_B} f(z, w) \, dq_B(w)$$

and

$$E(f) = \int_{I_A} \int_{I_B} f(z, w) \, dq_B(w) dq_A(z).$$

Let $c_A : I_A \to \mathbb{R}_0^+$ and $c_B : I_B \to \mathbb{R}_0^+$ be measurable functions where $c_A^z = c_A(z)$ denotes the production cost of firm A when the type of firm A is $z \in I_A$ and $c_B^w = c_B(w)$ denotes the production cost of firm B when the type of firm B is $w \in I_B$. Furthermore, we assume that the expected values of c_A and c_B are finite

$$E(c_A) = E_A(c_A) = \int_{I_A} c_A^{\bar{z}} \, dq_A(z) < \infty; \;\; E(c_B) = E_B(c_B) = \int_{I_B} c_B^w \, dq_B(w) < \infty.$$

We assume that $dq_A(z)$ denotes the probability of the *belief* of the firm B on the production costs of the firm A to be $c_A^{\bar{z}}$. Similarly, assume that $dq_B(w)$ denotes the probability of the belief of the firm A on the production costs of the firm B to be c_B^w.

The simplicity of the following cost deviation formulas is crucial to express the main results of this article in a clear and understandable way. The *cost deviations* of firm A and firm B

$$\Delta_A : I_A \to \mathbb{R}_0^+ \;\; \text{and} \;\; \Delta_B : I_B \to \mathbb{R}_0^+$$

are given respectively by $\Delta_A(z) = c_A^{\bar{z}} - E(c_A)$ and $\Delta_B(w) = c_B^w - E(c_B)$. The *cost deviation* between the firms

$$\Delta_C : I_A \times I_B \to \mathbb{R}_0^+$$

is given by $\Delta_C(z, w) = c_A^{\bar{z}} - c_B^w$. Since the meaning is clear, we will use through the paper the following simplified notation:

$$\Delta_A = \Delta_A(z); \;\;\; \Delta_B = \Delta_B(w) \;\; \text{and} \;\; \Delta_C = \Delta_C(z, w).$$

The *expected cost deviation* Δ_E between the firms is given by $\Delta_E = E(c_A) - E(c_B)$. Hence,

$$\Delta_C - \Delta_E = \Delta_A - \Delta_B.$$

34.6 Local Optimum Price Strategy Under Complete Information

In this section, we introduce incomplete information in the classical Hotelling game and we find the local optimal price strategy. We introduce the bounded uncertain costs condition that allows us to find the local optimum price strategy.

A *price strategy* (p_A, p_B) is given by a pair of functions $p_A : I_A \to \mathbb{R}_0^+$ and $p_B : I_B \to \mathbb{R}_0^+$ where $p_A^{\bar{z}} = p_A(z)$ denotes the price of firm A when the type of firm A is $z \in I_A$ and $p_B^w = p_B(w)$ denotes the price of firm B when the type of firm B is $w \in I_B$. We note that $E(p_A) = E_A(p_A)$ and $E(p_B) = E_B(p_B)$. The *indifferent consumer* $x : I_A \times I_B \to (0, l)$ is given by

$$x^{\bar{z}, w} = \frac{p_B^w - p_A^{\bar{z}} + t\,m\,(l + \Delta_l)}{2\,t\,m}. \tag{34.9}$$

The ex-post profit of the firms is the effective profit of the firms given a realization of the production costs for both firm. Hence, it is the main economic information for both firms. However, the incomplete information prevents the firms to have access to their ex-post profits except after the firms have already decided their price strategies. The *ex-post profits* $\pi_A^{EP} : I_A \times I_B \rightarrow \mathbb{R}_0^+$ and $\pi_B^{EP} : I_A \times I_B \rightarrow \mathbb{R}_0^+$ are given by

$$\pi_A^{EP}(z, w) = \pi_A(z, w) = (p_A^{\tilde{z}} - c_A^{\tilde{z}}) \, x^{z,w}$$

and

$$\pi_B^{EP}(z, w) = \pi_B(z, w) = (p_B^w - c_B^w)(l - x^{z,w}).$$

The ex-ante profit of the firms is the expected profit of the firm that know their production cost but are uncertain about the production cost of the competitor firm. The *ex-ante profits* $\pi_A^{EA} : I_A \rightarrow \mathbb{R}_0^+$ and $\pi_B^{EA} : I_B \rightarrow \mathbb{R}_0^+$ are given by

$$\pi_A^{EA}(z) = E_B(\pi_A^{EP}) \quad \text{and} \quad \pi_B^{EA}(w) = E_A(\pi_B^{EP}).$$

We note that, the *expected profit* $E(\pi_A^{EP})$ of firm A is equal to $E_A(\pi_A^{EA})$ and the *expected profit* $E(\pi_B^{EP})$ of firm B is equal to $E_B(\pi_B^{EA})$.

The incomplete information forces the firms to have to choose their price strategies using their knowledge of their ex-ante profits, to which they have access, instead of the ex-post profits, to which they do not have access except after the price strategies are decided.

Definition 34.3. A price strategy (p_A, p_B) for both firms is a *local optimum price strategy* if (i) for every $z \in I_A$ and for every small deviation of the price $p_A^{\tilde{z}}$ the ex-ante profit $\pi_A^{EA}(z)$ of firm A decreases, and for every $w \in I_B$ and for every small deviation of the price p_B^w the ex-ante profit $\pi_B^{EA}(w)$ of firm B decreases (local optimum property); and (ii) for every $z \in I_A$ and $w \in I_B$ the indifferent consumer exists, i.e. $0 < x^{z,w} < l$ (duopoly property).

We introduce the *BUCL1* condition that has the crucial economical information that can be extracted from the exogenous variables. The *BUCL1* condition allow us to know if there is, or not, a local optimum price strategy in the presence of uncertainty for the production costs of both firms.

Definition 34.4. The Hotelling model satisfies the *bounded uncertain costs and location (BUCL1)* condition 1, if

$$|\Delta_E - 3 \Delta_C + 2 \Delta_l \, t \, m| < 6 \, t \, m \, l.$$

for all $z \in I_A$ and for all $w \in I_B$.

A strong restriction that the *BUCL1* condition imposes is that Δ_C converges to 0 when m tends to 0, i.e. when the differentiation in the localization tends to vanish.

For $i \in \{A, B\}$, we define

$$c_i^m = \min_{z \in I_i}\{c_i^z\} \quad \text{and} \quad c_i^M = \max_{z \in I_i}\{c_i^z\}.$$

Let

$$\overline{\Delta} = \max_{i,j \in \{A,B\}} \{c_i^M - c_j^m\}$$

Thus, the bounded uncertain costs and location $BUCL1$ is implied by the following stronger $SBUCL1$ condition.

Definition 34.5. The Hotelling model satisfies the *bounded uncertain costs and location (SBUCL1)* condition, if

$$\overline{\Delta} < t\,l\,m.$$

The following theorem is a key economical result in oligopoly theory. First, it tell us about the existence, or not, of a local optimum price strategy only by accessing a simple inequality in the exogenous variables and so available to both firms. Secondly, give us explicit and simple formulas that allow the firms to know the relevance of the exogenous variables in their price strategies and corresponding profits.

Theorem 34.1. *There is a local optimum price strategy (p_A, p_B) if and only if the BUCL1 condition holds. Under the BUCL1 condition, the expected prices of the local optimum price strategy are given by*

$$E(p_A) = t\,m\left(l + \frac{\Delta_l}{3}\right) + E(c_A) - \frac{\Delta_E}{3}; \quad E(p_B) = t\,m\left(l - \frac{\Delta_l}{3}\right) + E(c_B) + \frac{\Delta_E}{3}.$$

Furthermore, the local optimum price strategy (p_A, p_B) is unique and it is given by

$$p_A^{\tilde{z}} = E(p_A) + \frac{\Delta_A}{2}; \quad p_B^w = E(p_B) + \frac{\Delta_B}{2}. \tag{34.10}$$

We observe that the differences between the expected prices of both firms has a very useful and clear economical interpretation in terms of the localization and expected cost deviations.

$$E(p_A) - E(p_B) = \frac{2\,t\,m\,\Delta_l + \Delta_E}{3}.$$

Furthermore, for different production costs, the differences between the optimal prices of a firm are proportional to the differences of the production costs

$$p_A^{z_1} - p_A^{z_2} = \frac{c_A^{z_1} - c_A^{z_2}}{2}.$$

and

$$p_B^{w_1} - p_B^{w_2} = \frac{c_B^{w_1} - c_B^{w_2}}{2}.$$

for all $z_1, z_2 \in I_A$ and $w_1, w_2 \in I_B$. Hence, half of the production costs value is incorporated in the price.

The following equation give us the information of the market size of both firms by giving the explicit localization of the indifferent consumer $x^{z,w}$ with respect to the local optimum price strategy

$$x^{z,w} = \frac{1}{2}\left(l + \frac{\Delta_l}{3}\right) + \frac{\Delta_E - 3\,\Delta_C}{12\,t\,m}.$$

The ex-ante profit of firm A is

$$\pi_A^{EA}(z) = \frac{(2\,t\,m\,(3\,l + \Delta_l) - 3\,\Delta_A - 2\,\Delta_E)^2}{72\,t\,m}. \tag{34.11}$$

Similarly, the ex-ante profit of firm B is

$$\pi_B^{EA}(w) = \frac{(2\,t\,m\,(3\,l - \Delta_l) - 3\,\Delta_B + 2\,\Delta_E)^2}{72\,t\,m}.$$

34.7   Bayesian-Nash Equilibrium

We note that, if a Bayesian-Nash price equilibrium satisfies the duopoly property then it is a local optimum price strategy. However, a local optimum price strategy is only a local strategic maximum. Hence, the local optimum price strategy to be a Bayesian-Nash equilibrium must also be global strategic maximum. In this section, we are going to show that this is the case.

Following D'Aspremont et al. [4], we note that the profits of the two firms, valued at local optimum price strategy are globally optimal if they are at least as great as the payoffs that firms would earn by undercutting the rivals's price and supplying the whole market for all admissible subsets of types I_A and I_B.

Let (p_A, p_B) be the local optimum price strategy. Given the type w_0 of firm B, firm A may gain the whole market, undercutting its rival by setting

$$p_A^M(w_0) = p_B^{w_0} - t\,m\,(l - \Delta_l).$$

Hence, by *BUCL1* condition $p_A^M(w_0) \leqslant p_A^z$ for all $z \in I_A$. We observe that if firm A chooses the price $p_A^M(w_0)$ then by equalities (34.9) and (34.10) the whole market belongs to Firm A for all types w of firm B with $c^w \geqslant c^{w_0}$. Let

$$x(w; w_0) = \min\left\{ l, \frac{p_B^w - p_A^M(w_0)}{2tm} + \frac{l + \Delta_l}{2} \right\}.$$

Thus, the *expected profit* with respect to the price $p_A^M(w_0)$ for firm A is

$$\pi_A^{EA,M}(w_0) = \int_{I_B} \left(p_A^M(w_0) - c_A^z \right) x(w; w_0) \, dq_B(w).$$

Let $w_M \in I_B$ such that $c^{w_M} = c_B^M$. Since $c^{w_M} \geqslant c_B^{w_0}$ for every $w_0 \in I_B$, we obtain

$$\pi_A^{EA,M}(w_0) \leqslant \left(p_A^M(w_0) - c_A^z \right) l \leqslant \left(p_A^M(w_M) - c_A^z \right) l$$

Given the type z_0 of firm A, firm B may gain the whole market, undercutting its rival by setting

$$p_B^M(z_0) = p_A^{z_0} - t\,m\,(l + \Delta_l).$$

Hence, by *BUCL1* condition $p_B^M(z_0) \leqslant p_B^w$ for all $w \in I_B$. We observe that if firm B chooses the price $p_B^M(z_0)$ then by equalities (34.9) and (34.10) the whole market belongs to Firm B for all types z of firm A with $c^z \geqslant c^{z_0}$. Let

$$x(z; z_0) = \max\left\{ 0, \frac{p_B^M(z_0) - p_A^z}{2tm} + \frac{l + \Delta_l}{2} \right\}.$$

Thus, the *expected profit* with respect to the price $p_B^M(z_0)$ of firm B is

$$\pi_B^{EA,M}(z_0) = \int_{I_A} \left(p_B^M(z_0) - c_B^w \right) (l - x(z; z_0)) \, dq_A(z).$$

Let $z_M \in I_A$ such that $c_A^{z_M} = c_A^M$. Since $c^{z_M} \geqslant c^{z_0}$ for every $z_0 \in I_A$, we obtain

$$\pi_B^{EA,M}(z_0) \leqslant \left(p_B^M(z_0) - c_B^w \right) l \leqslant \left(p_B^M(z_M) - c_B^w \right) l.$$

Remark 34.1. Under the *BUCL1* condition, the strategic equilibrium (p_A, p_B) is the unique pure Bayesian Nash equilibrium with the duopoly property if for every $z \in I_A$ and every $w \in I_B$,

$$\pi_A^{EA,M}(w) \leqslant \pi_A^{EA}(z) \quad \text{and} \quad \pi_B^{EA,M}(z) \leqslant \pi_B^{EA}(w).$$

Definition 34.6. The Hotelling model satisfies the *bounded uncertain costs and location (BUCL2)* condition 2, if

$$\Delta_E + 3\left(c_A^M + c_B^M - 2c_A^m\right) + \frac{\Delta_l\left(3c_A^M - E(c_A) - 2E(c_B)\right)}{3l} \leq$$

$$\leq \frac{(3l - \Delta_l)^2\, t\, m}{3l} + \frac{\left(3c_A^M - E(c_A) - 2E(c_B)\right)^2}{12\, t\, m\, l}$$

and

$$-\Delta_E + 3\left(c_A^M + c_B^M - 2c_B^m\right) - \frac{\Delta_l\left(3c_B^M - E(c_B) - 2E(c_A)\right)}{3l} \leq$$

$$\leq \frac{(3l + \Delta_l)^2\, t\, m}{3l} + \frac{\left(3c_B^M - E(c_B) - 2E(c_A)\right)^2}{12\, t\, m\, l}.$$

Thus, the bounded uncertain costs condition *BUCL2* is implied by the following stronger *SBUCL2* condition.

Definition 34.7. The Hotelling model satisfies the *strong bounded uncertain costs and location (SBUCL2)* condition 2, if

$$6\overline{\Delta} < l\, t\, m$$

We observe that the *SBUCL2* condition implies *SBUCL1* condition and so implies the *BUCL1* condition.

Theorem 34.2. *If the Hotelling model satisfies the BUCL1 and BUCL2 conditions the local optimum price strategy (p_A, p_B) is a Bayesian Nash equilibrium.*

Corollary 34.1. *If the Hotelling model satisfies SBUCL2 condition the local optimum price strategy (p_A, p_B) is a Bayesian Nash equilibrium.*

34.8 Optimum Localization Equilibrium Under Incomplete Information

We note that from (34.10) and (34.11), we can write the profit of firm A as

$$\pi_A^{EA}(z) = \frac{(p_A^{\tilde{z}} - c_A)^2}{2t\,(l - a - b)}.$$

Since

$$\frac{\partial p_A^{\tilde{z}}}{\partial a} = -\frac{2}{3}\, t\,(l + a)$$

we have

$$\frac{\partial \pi_A^{EA}}{\partial a} = \frac{p_A - c_A}{12\, t\, (l - a - b)^2}\, \left(-2\, t\, (l - a - b)\, (l + 3\, a + b) - 3\, \Delta_A - 2\, \Delta_E \right).$$

Similarly, we obtain that

$$\frac{\partial \pi_B^{EA}}{\partial b} = \frac{p_B - c_B}{12\, t\, (l - a - b)^2}\, \left(-2\, t\, (l - a - b)\, (l + 3\, b + a) - 3\, \Delta_B + 2\, \Delta_E \right).$$

Therefore, the maximal differentiation $(a, b) = (0, 0)$ is a local optimum strategy if and only if

$$\frac{\partial \pi_A^{EA}}{\partial a}(0, 0) = -\frac{p_A - c_A}{12\, t\, l^2}\, \left(2\, t\, l^2 + 3\, \Delta_A + 2\, \Delta_E \right) < 0$$

and

$$\frac{\partial \pi_B^{EA}}{\partial b}(0, 0) = -\frac{p_B - c_B}{12\, t\, l^2}\, \left(2\, t\, l^2 + 3\, \Delta_B - 2\, \Delta_E \right) < 0$$

Since

$$\frac{p_A - c_A}{6\, t\, l^2} > 0 \quad \text{and} \quad \frac{p_B - c_B}{6\, t\, l^2} > 0$$

the maximal differentiation $(a, b) = (0, 0)$ is a local optimum strategy if and only if the following condition holds.

Definition 34.8. The Hotelling model satisfies the *bounded uncertain costs and location (BUCL3)* condition, if

$$2\, t\, l^2 + 3\, \Delta_A + 2\, \Delta_E > 0$$

for all $z \in I_A$ and

$$2\, t\, l^2 + 3\, \Delta_B - 2\, \Delta_E > 0.$$

for all $w \in I_B$.

34.9 Conclusion

We proved that there is a local optimum price strategy with the duopoly property if and only if the bounded uncertain costs and location *BUCL1* condition holds. The explicit formulas of the local optimum price strategy determine prices for both firms that are affine with respect to the expected costs of both firms and to its own costs. Under the $BUCL1$ and $BUCL2$ conditions, we proved that the local optimum price strategy is a Bayesian-Nash price strategy. Assuming that the firms choose the Bayesian-Nash price strategy, under the $BUCL3$ condition, we proved that the maximal differentiation is a local optimum for the localization strategy of both firms.

Acknowledgements We are thankful to the anonymous referees for their suggestions. We acknowledge the financial support of LIAAD-INESC TEC through 'Strategic Project—LA 14—2013–2014' with reference PEst-C/EEI/LA0014/2013, USP-UP project, IJUP, Faculty of Sciences, University of Porto, Calouste Gulbenkian Foundation, FEDER, POCI 2010 and COMPETE Programmes and Fundação para a Ciência e a Tecnologia (FCT) through Project "Dynamics and Applications", with reference PTDC/MAT/121107/2010. Telmo Parreira thanks FCT, for the PhD scholarship SFRH/BD/33762/2009.

References

1. Biscaia, R., Sarmento, P.: Spatial Competition and Firms' Location Decisions under Cost Uncertainty. FEP Working Papers No. 445 (2012)
2. Boyer, M., Laffont, J., Mahenc, P., Moreaux, M.: Location Distortions under Incomplete information. Reg. Sci. Urban Econ. **24**(4), 409–440 (1994)
3. Boyer, M., Mahenc, P., Moreaux, M.: Asymmetric information and product differentiation. Reg. Sci. Urban Econ. **33**(1), 93–113 (2003)
4. D'Aspremont, C., Gabszewicz, J., Thisse, J.-F.: On hotelling's "stability in competition". Econometrica **47**(5), 1145–1150 (1979)
5. Ferreira, F.A., Pinto, A.A.: Uncertainty on a Bertrand duopoly with product differentiation. In: Machado, J.A., Silva, M.F., Barbosa, R.S., Figueiredo, L.B. (eds.) Nonlinear Science and Complexity, pp. 389–395. Springer, Netherlands (2011)
6. Ferreira, F., Ferreira, F.A., Ferreira, M., Pinto, A.A.: Flexibility in a Stackelberg leadership with differentiated goods. Optimization (accepted)
7. Ferreira, F., Ferreira, F.A., Pinto, A.A.: Bayesian price leadership. In: Tas, K., et al. (eds.) Mathematical Methods in Engineering, pp. 371–379. Springer, Netherlands (2007)
8. Ferreira, F.A., Ferreira, F., Pinto, A.A.: Unknown costs in a duopoly with differentiated products. In: Tas, K., et al. (eds.) Mathematical Methods in Engineering, pp. 359–369. Springer, Netherlands (2007)
9. Ferreira, F., Ferreira, F.A., Pinto, A.A.: Flexibility in stackelberg leadership. In: Machado, J.A., Patkai, B., Rudas, I.J. (eds.) Intelligent Engineering Systems and Computational Cybernetics, pp. 399–405. Springer, Netherlands (2009)
10. Ferreira, F., Ferreira, F.A., Pinto, A.A.: Price-setting dynamical duopoly with incomplete information. In: Machado, J.A., Silva, M.F., Barbosa, R.S., Figueiredo, L.B. (eds.) Nonlinear Science and Complexity, pp. 397–403. Springer, Netherlands (2011)
11. Ferreira, M., Figueiredo, I.P., Oliveira, B.M.P.M., Pinto, A.A.: Strategic optimization in R&D Investment. Optim. J. Math. Program. Oper. Res. **61**(8), 1013–1023 (2012)
12. Gibbons, R.: A Primer in Game Theory. Prentice Hall (1992)
13. Graitson, D.: Spatial competition à la Hotelling: a selective survey. J. Ind. Econ. **31**, 11–25 (1982)
14. Hotelling, H.: Stability in competition. Econ. J. **39**, 41–57 (1929)
15. Lederer, P., Hurter, A.: Competition of Firms: discriminatory pricing and location. Econometrica **54**(3), 623–640 (1986)

16. Osborne, M.J., Pitchick, C.: Equilibrium in Hotelling's model of spatial competition. Econometrica **55**(4), 911–922 (1987)
17. Pinto, A.A., Parreira, T.: A hotelling-type network. In: Peixoto, M., Pinto, A.A., Rand, D. (eds.) Dynamics, Games and Science I. Springer Proceedings in Mathematics, vol. 1, pp. 709–720. Springer, Berlin (2011)
18. Pinto, A.A., Parreira, T.: Complete versus incomplete information in the Hotelling model. In: Pinto, A.A., Zilberman, D. (eds.) Modeling, Dynamics, Optimization and Bioeconomy. Springer Proceedings in Mathematics and Statistics Series. Springer, Heidelberg (2014)
19. Pinto, A.A., Parreira, T.: Optimal localization of firms in Hotelling networks. In: Pinto, A.A., Zilberman, D. (eds.) Modeling, Dynamics, Optimization and Bioeconomy. Springer Proceedings in Mathematics and Statistics Series. Springer, Heidelberg (2014)
20. Pinto, A.A., Parreira, T.: Price competition in the Hotelling model with uncertainty on costs. Optim. J. Math. Program. Oper. Res. (accepted)
21. Pinto, A.A., Parreira, T.: Localization and prices in the quadratic Hotelling model with uncertainty (submited)
22. Pinto, A.A., Ferreira, F.A., Ferreira, M., Oliveira, B.M.P.M.: Cournot duopoly with competition in the R&D expenditures. In: Proceedings of Symposia in Pure Mathematics, vol. 7. Wiley-VCH Verlag, Weinheim, 2007
23. Pinto, A.A., Oliveira, B.M.P.M., Ferreira, F.A., Ferreira, F.: Stochasticity favoring the effects of the R&D strategies of the firms. In: Machado, J.A., Patkai, B., Rudas, I.J. (eds.) Intelligent Engineering Systems and Computational Cybernetics, pp. 415–423. Springer, Netherlands (2009)
24. Salop, S.: Monopolistic competition with outside goods. Bell J. Econ. **10**, 141–156 (1979)
25. Tabuchi, T., Thisse, J.F.: Asymmetric equilibria in spatial competition. Int. J. Econ. Theory **13**(2), 213–227 (1995)
26. Ziss, S.: Entry deterrence, cost advantage and horizontal product differentiation. Reg. Sci. Urban Econ. **23**, 523–543 (1993)

Chapter 35
A Cost Sharing Mechanism for Job Scheduling Problems

Joss Sánchez-Pérez

35.1 Introduction

Cost sharing among agents in a queue is a fundamental problem in many practical applications. For example, computer programs are regularly scheduled on servers, data are scheduled to be transmitted over networks, jobs are scheduled in shop-floor on machines, and queues appear in many public services (post offices, banks, etc...). Study of queueing problems has attracted economists for a long time.

In this chapter we study job scheduling situations, where agents have to process jobs. In particular, we consider the problem of a planner who has to provide a facility to a finite group of agents. Each agent has one job to process using this facility. The facility can be used by only one agent at a time; therefore, the planner will have to order the agents in a queue.

An stream of literature related to job scheduling problems is on *sequencing games*, first introduced by Curiel et al. [1], where they assume the presence of an initial ordering of jobs. In summary, the focus of this stream of research is how to share the savings in costs form the initial ordering to the optimal ordering amongst jobs (also see Hamers et al. [4] and Curiel et al. [2]).

In this work, we propose a cost sharing mechanism for the previous problem. We have two main assumptions: First, we suppose we know the costs for processing any set of jobs in any order. Second, we suppose that agents will process their jobs according to the order that generates the minimum cost to them, therefore they have no preferences in the order their jobs are processed.

We use the concept of potential of a cooperative game to establish this cost sharing mechanism. Hart and Mas-Colell [5] established a new form to generate

J. Sánchez-Pérez (✉)
Facultad de Economía, UASLP, San Luis Potosí, México
e-mail: joss.sanchez@uaslp.mx

A.A. Pinto and D. Zilberman (eds.), *Modeling, Dynamics, Optimization*
and Bioeconomics I, Springer Proceedings in Mathematics & Statistics 73,
DOI 10.1007/978-3-319-04849-9__35,
© Springer International Publishing Switzerland 2014

the value of a game. They start with the definition of a set of games in which a real function is defined, called the potential of the game. This function must be defined in such a way that the following two conditions are satisfied: (1) the gain or participation of each player is equal to the difference in potential between the game that is considered and the one that results when he abandons the game, and, (2) the sum of gains or participations add to what all the players get in that game. Surprisingly, these participations coincide with its corresponding Shapley value.

The chapter is organized as follows. We first recall the main basic features of job scheduling problems and the potential of a cooperative game. In Sect. 35.3 we show that the previous ideas could be applied, with appropriate modification, to determine a cost sharing mechanism for job scheduling problems. Finally, in Sect. 35.4, we present some properties of the mechanism proposed in this work.

35.2 The Model

Let us consider a finite set of participating agents, denoted by $N = \{1, 2, \ldots, n\}$. The group of permutations of a set of agents $S \subseteq N$, will be denoted by $\theta_S = \{\theta : S \to S \mid \theta \text{ is bijective}\}$. For each subset $S \subseteq N$, let us denote by S_π a list of jobs ordered according to permutation $\pi \in \theta_S$. Thus, the list $S_\pi = [2, 1, 4]$ represents the situation where agent 2 processes his job in first place, then agent 1 processes his job in second place and finally, agent 4 processes his job, for $S = \{1, 2, 4\}$. A list with no jobs is denoted by $[\varnothing]$.

On the other hand, $\Pi(S)$ will denote the set of lists of S with all possible orders for its process, i.e., $\Pi(S) = \{S_\pi \mid \pi \in \theta_S\}$. For example, if $N = \{1, 2, 3, 4\}$ and $S = \{1, 3, 4\}$, then $\Pi(S) = \{[1, 3, 4], [1, 4, 3], [3, 1, 4], [3, 4, 1], [4, 1, 3], [4, 3, 1]\}$. Additionally, we will denote the cardinality of a set by its corresponding lower-case letter, for instance $n = |N|$, $s = |S|$, $t = |T|$, and so on.

Also, let $E_N = \cup_{S \subseteq N} \Pi(S)$ be the set of all lists of jobs in any order.

Definition 35.1. A mapping

$$c : E_N \to \mathbb{R} \tag{35.1}$$

that assigns a real value, $c(S_\pi)$, to each list of jobs S_π is called a job scheduling problem. The set of job scheduling problems with agent set N is denoted by P^N, i.e.,

$$P^N = \{c : E_N \to \mathbb{R} \mid c[\varnothing] = 0\}$$

If $c \in P^N$ and $S_\pi \in E_N$, then the value $c(S_\pi)$ represents the cost for processing the jobs of the list S_π.

Given $c_1, c_2 \in P^N$ and $\lambda \in \mathbb{R}$, we define the sum $c_1 + c_2$ and the product λc_1, in P^N, in the usual form, i.e.,

$$(c_1 + c_2)(S_\pi) = c_1(S_\pi) + c_2(S_\pi) \quad \text{and} \quad (\lambda c_1)(S_\pi) = \lambda c(S_\pi)$$

respectively. It is easy to verify that P^N is a vector space with these operations.

A *solution* on P^N is a function $\varphi : P^N \to \mathbb{R}^n$. If φ is a solution and $c \in P^N$, then we can interpret $\varphi_i(c) \in \mathbb{R}$ as the cost which agent i should expect from the problem c.

Next, we present some basic ideas related to cooperative games and potential of a cooperative game. For a brief revision of the concepts of cooperative games that are mentioned here, such as the Shapley value, see Driessen [3].

Shapley [6] characterized a unique solution (denoted by Sh) for cooperative games[1]:

$$Sh_i(N, v) = \sum_{\{S \subseteq N : i \notin S\}} \frac{s!(n - s - 1)!}{n!} [v(S \cup \{i\}) - v(S)]$$

Now, let (N, v) a cooperative game and let us denote by $G = \{(S, v_S) \mid S \subseteq N\}$ the family of subgames of (N, v). The potential of a game is defined as a function

$$P : G \to \mathbb{R}$$

such that for every $S \subseteq N$,

$$\sum_{j \in S} [P(S, v_S) - P(S \setminus \{j\}, v_{S \setminus \{j\}})] = v(S)$$

with the condition that $P(\emptyset, v_\emptyset) = 0$. Furthermore, it is assumed that the loss of potential

$$P(S, v_S) - P(S \setminus \{j\}, v_{S \setminus \{j\}})$$

when the player j abandons the game, is equal to the amount that corresponds to this player in the game (S, v_S). Moreover, it turns out that this difference is equal to the Shapley value of player j in the game (S, v_S).

35.3 The Potential of Job Scheduling Problems

We need to define the loss of potential when one agent abandons a problem. A priori it could be from any list, so we propose the average loss of potential instead. Let us formalize this idea.

---

[1]A cooperative game is a pair (N, v), where $N = \{1, \dots, n\}$ is a finite set of players and v is a function $v : 2^N \to \mathbb{R}$ with the property that $v(\emptyset) = 0$.

For each $S \subseteq N$, S_o will denote the list in $\Pi(S)$ that generates the lowest cost for processing the jobs in S, i.e., S_o is such that $c(S_o) = \min \{c(S_\pi)\}_{\pi \in \theta_S}$. Let us denote

$$\overline{Pot}(S \backslash \{i\}) = \sum_{T_\pi \in \Pi(S \backslash \{i\})} \frac{Pot(T_\pi)}{(s-1)!} \tag{35.2}$$

Definition 35.2. We define the potential of the job scheduling problem $c \in P^N$ as a function

$$Pot : E_N \to \mathbb{R} \tag{35.3}$$

such that $Pot([\varnothing]) = 0$ and

$$\sum_{i \in S} \left[Pot(S_\pi) - \overline{Pot}(S \backslash \{i\}) \right] = c(S_\pi) \tag{35.4}$$

for every $S \subseteq N$ and every $\pi \in \theta_S$.

As in Hart and Mas-Colell [5], we will suppose that

$$Pot(N_o) - \overline{Pot}(N \backslash \{i\}) \tag{35.5}$$

is the amount that agent i should pay for processing his job. This amount is the average of the loss of potential when the agent i abandons the system. It is easy to see that (35.4) is equivalent to the following recursive expression,

$$Pot(S_\pi) = \frac{c(S_\pi)}{s} + \sum_{i \in S} \sum_{T_\pi \in \Pi(S \backslash \{i\})} \frac{Pot(T_\pi)}{s!} \tag{35.6}$$

for every $S \subseteq N$ and every $\pi \in \theta_S$, which determines the potential of the job scheduling problem. Now, $\varphi(c)$ denotes the vector of costs for processing the jobs of agents N, i.e., the vector with entry i is given by

$$\varphi_i(c) = Pot(N_o) - \overline{Pot}(N \backslash \{i\}) \tag{35.7}$$

and we will call it a cost sharing mechanism for the job scheduling problem c.

Example 35.1. In this example, we have three agents with set of jobs $N = \{1, 2, 3\}$, where the cost function c is given by

S_π	[1]	[2]	[3]	[1, 2]	[2, 1]	[1, 3]	[3, 1]	[2.3]	[3, 2]
$c(S_\pi)$	10	18	12	21	25	19	22	27	26

S_π	$[1, 2, 3]$	$[1, 3, 2]$	$[2, 1, 3]$	$[2, 3, 1]$	$[3, 1, 2]$	$[3, 2, 1]$
$c(S_\pi)$	35	37	31	30	40	35

Notice that $N_o = [2, 3, 1]$. In order to compute $Pot(N_o)$ using (35.6), we need to generate all subproblems as the result of removing one job. We repeat this process in each case until there are no jobs. Observe that for lists with one job, say $[a]$, $Pot([a]) = c([a])$. While if the list has two jobs, then $Pot([b, d]) = \frac{1}{2}[c([b, d]) + c([b]) + c([d])]$. Thus

S_π	$[1]$	$[2]$	$[3]$	$[1, 2]$	$[2, 1]$	$[1, 3]$	$[3, 1]$	$[2.3]$	$[3, 2]$	$[2, 3, 1]$
$Pot(S_\pi)$	10	18	12	$49/2$	$53/2$	$41/2$	22	$57/2$	28	35

Now, we can calculate the costs for the job scheduling problem using (35.7):

$$\overline{Pot}(N\setminus\{1\}) = \frac{1}{2}\left(\frac{57}{2} + 28\right) = \frac{113}{4}$$

$$\overline{Pot}(N\setminus\{2\}) = \frac{1}{2}\left(\frac{41}{2} + 22\right) = \frac{85}{4}$$

$$\overline{Pot}(N\setminus\{3\}) = \frac{1}{2}\left(\frac{49}{2} + \frac{53}{2}\right) = \frac{51}{2}$$

Thus, the vector of costs for processing the jobs is

$$\varphi(c) = \begin{bmatrix} 35 - 113/4 \\ 35 - 85/4 \\ 35 - 51/2 \end{bmatrix} = \begin{bmatrix} 27/4 \\ 55/4 \\ 19/2 \end{bmatrix} = \begin{bmatrix} 6.75 \\ 13.75 \\ 9.5 \end{bmatrix}$$

35.4 Properties of the Potential

The next theorem establishes an explicit expression for the potential of the job scheduling problem.

Theorem 35.1. *For each $S_\pi \in E_N$, we have that*

$$Pot(S_\pi) = \frac{c(S_\pi)}{s} + \sum_{T \subsetneq S L_\pi \in \Pi(T)} \sum \frac{(s-t)!}{s!t} \cdot c(L_\pi) \qquad (35.8)$$

Proof. The proof is by induction on the cardinality of S_π. For one job, i.e., $|S_\pi| = 1$, both (35.6) and (35.8) are equal to $c(S_\pi)$. Let us suppose that (35.6) and (35.8) are equal for $T_\pi \in \Pi(S\setminus\{i\})$ for every $\pi \in \theta_S$ and every $j \in N$. Then,

$$\sum_{i \in S}\sum_{T_\pi \in \Pi(S\setminus\{i\})} \frac{Pot(T_\pi)}{s!} = \sum_{i \in S}\sum_{T_\pi \in \Pi(S\setminus\{i\})} \left[\frac{c(T_\pi)}{s!t} + \sum_{R \subsetneq T}\sum_{L_\pi \in \Pi(R)} \frac{(t-r)!}{s!t!r} \cdot c(L_\pi) \right]$$

Now, since in the third sum $T = S\setminus\{i\}$, then $t = s - 1$ and we have that

$$= \sum_{i \in S}\sum_{T_\pi \in \Pi(S\setminus\{i\})} \frac{c(T_\pi)}{s!(s-1)} + \sum_{i \in S}\sum_{R \subsetneq S\setminus\{i\}}\sum_{L_\pi \in \Pi(R)} \frac{(s-r-1)!(s-1)!}{s!(s-1)!r} \cdot c(L_\pi)$$

$$= \sum_{i \in S}\sum_{T_\pi \in \Pi(S\setminus\{i\})} \frac{c(T_\pi)}{s!(s-1)} + \sum_{i \in S}\sum_{R \subsetneq S\setminus\{i\}}\sum_{L_\pi \in \Pi(R)} \frac{(s-r-1)!}{s!r} \cdot c(L_\pi)$$

$$= \sum_{i \in S}\sum_{R \subsetneq S\setminus\{i\}}\sum_{L_\pi \in \Pi(R)} \frac{(s-r-1)!}{s!r} \cdot c(L_\pi)$$

$$= \sum_{R \subsetneq S}\sum_{L_\pi \in \Pi(R)} \frac{(s-r-1)!(s-r)}{s!r} \cdot c(L_\pi) = \sum_{R \subsetneq S}\sum_{L_\pi \in \Pi(R)} \frac{(s-r)!}{s!r} \cdot c(L_\pi)$$

$$= Pot(S_\pi) - \frac{c(S_\pi)}{s}$$

Now, the next result provides us an alternative way to compute the cost sharing mechanism $\varphi(c)$ for the job scheduling problem c. For $S \subseteq N$, we will denote by

$$\mu(S) = \sum_{T_\pi \in \Pi(S)} c(T_\pi)$$

Theorem 35.2. *The cost for agent i in the job scheduling problem c is given by*

$$\varphi_i(c) = \frac{c(N_o)}{n} + \sum_{\{S \subsetneq N : i \in S\}} \left[\frac{(n-s)!}{n!s} \cdot \mu(S) - \frac{(s-1)!}{n!} \cdot \mu(N\setminus S) \right] \qquad (35.9)$$

Proof. By the previous theorem, we have that

$$\overline{Pot}(N\setminus\{i\}) = \sum_{S_\pi \in \Pi(N\setminus\{i\})} \frac{Pot(S_\pi)}{(n-1)!}$$

$$= \sum_{S_\pi \in \Pi(N\setminus\{i\})} \left[\frac{c(S_\pi)}{(n-1)!s} + \sum_{T \subsetneq S}\sum_{L_\pi \in \Pi(T)} \frac{(s-t)!}{(n-1)!s!t} \cdot c(L_\pi) \right]$$

Since $S = N\setminus\{i\}$, then $s = n - 1$ and so

$$= \frac{\mu(N\backslash\{i\})}{(n-1)!(n-1)} + \sum_{S_\pi \in \Pi(N\backslash\{i\})} \sum_{T \subsetneq S} \frac{(s-t)!}{(n-1)!s!t} \cdot \mu(T)$$

$$= \frac{\mu(N\backslash\{i\})}{(n-1)!(n-1)} + \sum_{T \subsetneq N\backslash\{i\}} \frac{(n-t-1)!(n-1)!}{(n-1)!(n-1)!t} \cdot \mu(T)$$

$$= \frac{\mu(N\backslash\{i\})}{(n-1)!(n-1)} + \sum_{T \subsetneq N\backslash\{i\}} \frac{(n-t-1)!}{(n-1)!t} \cdot \mu(T)$$

Notice that for $T = N\backslash\{i\}$, $t = n-1$ and $\frac{(n-t-1)!}{(n-1)!t} = \frac{1}{(n-1)!(n-1)}$. Then

$$= \sum_{T \subseteq N\backslash\{i\}} \frac{(n-t-1)!}{(n-1)!t} \cdot \mu(T) = \sum_{\{T \subsetneq N : i \in T\}} \frac{(t-1)!}{(n-1)!(n-t)} \cdot \mu(N\backslash T)$$

On the other hand,

$$Pot(N_o) = \frac{c(N_o)}{n} + \sum_{T \subsetneq N} \sum_{L_\pi \in \Pi(T)} \frac{(n-t)!}{n!t} \cdot c(L_\pi)$$

$$= \frac{c(N_o)}{n} + \sum_{T \subsetneq N} \frac{(n-t)!}{n!t} \cdot \mu(T)$$

$$= \frac{c(N_o)}{n} + \sum_{\{T \subsetneq N : i \in T\}} \frac{(n-t)!}{n!t} \cdot \mu(T) + \sum_{\{T \subsetneq N : i \notin T\}} \frac{(n-t)!}{n!t} \cdot \mu(T)$$

$$= \frac{c(N_o)}{n} + \sum_{\{T \subsetneq N : i \in T\}} \frac{(n-t)!}{n!t} \cdot \mu(T) + \sum_{\{T \subsetneq N : i \in T\}} \frac{t!}{n!(n-t)} \cdot \mu(N\backslash T)$$

Therefore,

$$\varphi_i(N, c) = Pot(N_o) - \overline{Pot}(N\backslash\{i\})$$

$$= \frac{c(N_o)}{n} + \sum_{\{T \subsetneq N : i \in T\}} \frac{(n-t)!}{n!t} \cdot \mu(T) + \sum_{\{T \subsetneq N : i \in T\}} \frac{t!}{n!(n-t)} \cdot \mu(N\backslash T)$$

$$- \sum_{\{T \subsetneq N : i \in T\}} \frac{(t-1)!}{(n-1)!(n-t)} \cdot \mu(N\backslash T)$$

$$= \frac{c(N_o)}{n} + \sum_{\{S \subsetneq N : i \in S\}} \left[\frac{(n-s)!}{n!s} \cdot \mu(S) - \frac{(s-1)!}{n!} \cdot \mu(N\backslash S) \right]$$

Example 35.2. We compute the cost sharing mechanism for the job scheduling problem in Example 35.3:

$$\mu(\{1\}) = 10, \mu(\{2\}) = 18, \mu(\{3\}) = 12$$

$$\mu(\{1, 2\}) = 46, \mu(\{1, 3\}) = 41, \mu(\{2, 3\}) = 53$$

and therefore,

$$\varphi_1(c) = \frac{30}{3} + \frac{1}{3}(10) - \frac{1}{6}(53) + \frac{1}{12}(46) - \frac{1}{6}(12) + \frac{1}{12}(41) - \frac{1}{6}(18) = \frac{27}{4}$$

$$\varphi_2(c) = \frac{30}{3} + \frac{1}{3}(18) - \frac{1}{6}(41) + \frac{1}{12}(46) - \frac{1}{6}(12) + \frac{1}{12}(53) - \frac{1}{6}(10) = \frac{55}{4}$$

$$\varphi_3(c) = \frac{30}{3} + \frac{1}{3}(12) - \frac{1}{6}(46) + \frac{1}{12}(41) - \frac{1}{6}(18) + \frac{1}{12}(53) - \frac{1}{6}(10) = \frac{19}{2}$$

The next result establishes a relation between the cost sharing mechanism given by (35.7) and the Shapley value. The process of bargaining generates a cooperative game in a natural way: when the coalition of agents $S \subsetneq N$ is formed they can generate an efficiency $\frac{\mu(S)}{s!}$, and if N is formed, they can generate $c(N_o)$. In other words, if $c \in P^N$, we can associate a cooperative game in characteristic function form w_c as follows:

$$w_c(S) = \begin{cases} c(N_o) & \text{if } S = N \\ \frac{\mu(S)}{s!} & \text{if } S \subsetneq N \end{cases} \tag{35.10}$$

Theorem 35.3. *For $c \in P^N$ we have that*

$$\varphi(c) = Sh(N, w_c)$$

Proof. By the proof of the previous theorem,

$$Pot(N_o) = \frac{c(N_o)}{n} + \sum_{T \subsetneq N} \frac{(n-t)!}{n!t} \cdot \mu(T)$$

$$= \frac{w(N)}{n} + \sum_{T \subsetneq N} \frac{(n-t)!(t-1)!}{n!} \cdot \frac{\mu(T)}{t!}$$

$$= \frac{w(N)}{n} + \sum_{T \subsetneq N} \frac{(n-t)!(t-1)!}{n!} \cdot w_c(T)$$

Observe that $\frac{(n-t)!(t-1)!}{n!} = \frac{1}{n}$ for $T = N$, then

$$= \sum_{T \subseteq N} \frac{(n-t)!(t-1)!}{n!} \cdot w_c(T)$$

$$= \sum_{\{T \subseteq N : i \in S\}} \frac{(n-t)!(t-1)!}{n!} \cdot w_c(T) + \sum_{\{T \subseteq N : i \notin S\}} \frac{(n-t)!(t-1)!}{n!} \cdot w_c(T)$$

$$= \sum_{\{T \subseteq N : i \notin S\}} \frac{(n-t-1)!t!}{n!} \cdot w_c(T \cup \{i\}) + \sum_{\{T \subseteq N : i \notin S\}} \frac{(n-t)!(t-1)!}{n!} \cdot w_c(T)$$

In a similar way,

$$\overline{Pot}(N \setminus \{i\}) = \sum_{T \subseteq N \setminus \{i\}} \frac{(n-t-1)!}{(n-1)!t} \cdot \mu(T)$$

$$= \sum_{T \subseteq N \setminus \{i\}} \frac{(n-t-1)!(t-1)!}{(n-1)!} \cdot \frac{\mu(T)}{t!}$$

$$= \sum_{\{T \subseteq N : i \notin S\}} \frac{(n-t-1)!(t-1)!}{(n-1)!} \cdot w_c(T)$$

Therefore,

$$\varphi_i(N, c) = Pot(N_o) - \overline{Pot}(N \setminus \{i\})$$

$$= \sum_{\{T \subseteq N : i \notin S\}} \frac{(n-t-1)!t!}{n!} \cdot w_c(T \cup \{i\})$$

$$+ \sum_{\{T \subseteq N : i \notin S\}} \frac{(n-t)!(t-1)!}{n!} \cdot w_c(T)$$

$$- \sum_{\{T \subseteq N : i \notin S\}} \frac{(n-t-1)!(t-1)!}{(n-1)!} \cdot w_c(T)$$

$$= \sum_{\{T \subseteq N : i \notin S\}} \frac{(n-t-1)!t!}{n!} [w_c(T \cup \{i\}) - w_c(T)]$$

$$= Sh_i(N, w_c)$$

Example 35.3. The cooperative game associated with the job scheduling problem in Example 35.3 is given by

S	$\{1\}$	$\{2\}$	$\{3\}$	$\{1, 2\}$	$\{1, 3\}$	$\{2, 3\}$	$\{1, 2, 3\}$
$w_c(S)$	10	18	12	23	41/2	53/2	30

Remark 35.1. The solution for the job scheduling problem c given by (35.9), is efficient with respect to N_o by construction. According to (35.4),

$$\sum_{i \in N} \varphi_i(c) = \sum_{i \in N} \left[Pot(N_o) - \overline{Pot}(S \backslash \{i\}) \right] = c(N_o)$$

Observe that the group of permutations of the set of agents, θ_N, acts on E_N in a natural way:

$$\theta\left([l_1, l_2, \ldots, l_s]\right) = \left[l_{\theta(1)}, l_{\theta(2)}, \ldots, l_{\theta(s)} \right]$$

for $\theta \in \theta_N$ and $[l_1, l_2, \ldots, l_s] \in E_N$. For every $\theta \in \theta_N$ and $c \in P^N$, we define $(\theta \cdot c) \in P^N$ as:

$$(\theta \cdot c)(S_\pi) = c(\theta^{-1}(S_\pi))$$

We can easily verify that $w_{\theta \cdot c} = \theta \cdot w_c$ for every $\theta \in \theta_N$ and every $c \in P^N$.

Remark 35.2. From the previous observation, the symmetry of the Shapley value and Theorem 7, the solution given by (35.9) satisfies a symmetry condition:

$$\varphi(\theta \cdot c) = \theta \cdot \varphi(c)$$

for every $\theta \in \theta_N$ and $c \in P^N$. This is true since

$$\varphi(\theta \cdot c) = Sh(N, w_{\theta \cdot c}) = Sh(N, \theta \cdot w_c) = \theta \cdot Sh(N, w_c) = \theta \cdot \varphi(c)$$

Such symmetry condition implies that the selected allocation only depends on the cost function c, and not, for instance, on the numbering of the agents.

35.5 Conclusion

We studied the problem of sharing costs for a job scheduling problem on a single server, when jobs are ordered in a queue. We took a cooperative game theory approach and propose a cost sharing mechanism for the previous problem, by using a modification of the concept of potential of a cooperative game. In fact, it was established a relation between such cost sharing mechanism and the Shapley value of a certain cooperative game.

This cost sharing mechanism was provided for the case where we suppose we know the costs for processing any set of jobs in any order and where agents will process their jobs according to the order that generates the minimum cost to them, therefore they have no preferences in the order their jobs are processed.

In future, we plan to further look at cost sharing mechanisms other than the Shapley value. Investigating the strategic power of jobs in such mechanisms is another line of future research.

Acknowledgements The author acknowledges support from CONACYT research grant 130515.

References

1. Curiel I., Pederzoli G., Tijs S.: Sequencing games. J. Oper. Res. **40**, 344–351 (1989)
2. Curiel I., Potters J., Prasad R., Tijs S., Veltman B.: Sequencing and cooperation. Oper. Res. **42**(3), 556–568 (1994)
3. Driessen T.: Cooperative games, solutions and applications. In: Theory and Decision Library. Kluwer Academic, Dordrecht (1988)
4. Hamers H., Suijs J., Tijs S., Borm P.: The split core for sequencing games. Games Econ. Behav. **15**, 165–176 (1996)
5. Hart S., Mas-Colell A.: Potential, value and consistency. Econometrica **57**(3), 589–614 (1989)
6. Shapley L.: A value for n-person games. Contrib. Theory Games **2**, 307–317 (1953)

Chapter 36
Welfare Assessment of the Renewable Fuel Standard: Economic Efficiency, Rebound Effect, and Policy Interactions in a General Equilibrium Framework

Farzad Taheripour and Wallace E. Tyner

36.1 Introduction

The USA Energy Independence and Security Act of 2007 (RFS2) defined mandatory annual targets to increase consumption of conventional, advanced, and cellulosic biofuels until 2022. The cellulosic component of the RFS has largely been waived up to date, but not the other categories. The annual volumetric targets for corn ethanol have been achieved, and the industry is expected to reach the target of 15 billion gallons of ethanol consumption prior to 2015. The economic and environmental consequences of this rapid expansion have been the focal point of many studies in recent years. These studies have used partial and general equilibrium economic models and analyzed the price impacts, welfare consequences, displacement between gasoline and ethanol consumption (rebound effect), induced land use changes, and reduction in emissions due to ethanol expansion. Most of partial equilibrium models developed in these analyses captured only the interactions between the energy and agricultural markets and ignored other economic activities and the fact that economic resources are limited. In addition, these studies failed to fully capture the interactions between the biofuel mandates and other existing distortionary tax policies. The general equilibrium analyses have taken into account the interactions between the agricultural, energy and other economic activities and considered the exiting resource constraints in their economic assessments of biofuel polices. However, like the partial equilibrium analyses, they ignored the fact that biofuel policies and other distortionary taxes interact, and these interactions could alter the economic implications of biofuel mandates.

F. Taheripour (✉) • W.E. Tyner
Department of Agricultural Economics, Purdue University, 403 West State St.
West Lafayette, IN 47907-2056, USA
e-mail: tfarzad@purdue.edu; wtyner@purdue.edu

A.A. Pinto and D. Zilberman (eds.), *Modeling, Dynamics, Optimization and Bioeconomics I*, Springer Proceedings in Mathematics & Statistics 73, DOI 10.1007/978-3-319-04849-9_36, © Springer International Publishing Switzerland 2014

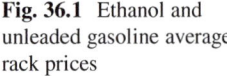
Fig. 36.1 Ethanol and
unleaded gasoline average
rack prices

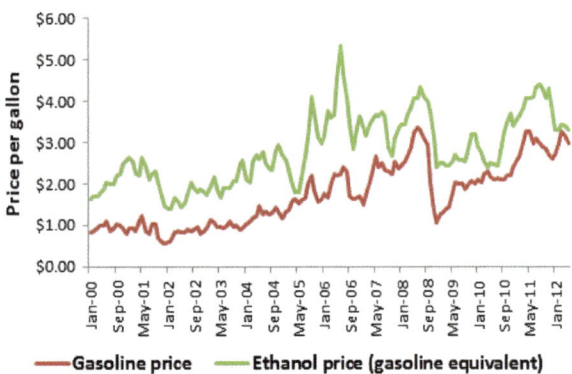

In general, there are two approaches to stimulate biofuel production: taxes/subsidies or mandates. The government could use economic incentives such as tax policies to encourage producers and/or consumers to produce and consume more biofuels. In this approach, the economy will pay the costs of policy through the tax system, and the burden of the policy depends on the efficiency of the tax system and the type of tax incentives. As an alternative the government can mandate biofuels and define penalties to force the economy to produce/consume the mandated level. In both cases the burden of the policy will be divided between consumers, blenders, and producers of gasoline, corn, and biofuel. However, how the cost ends up being absorbed can change from one option to another. To examine the economic consequences of a biofuel subsidy or mandate policy, we need to recognize how the policy is implemented in practice.

The US government has used a combination of these policies to boost ethanol production. The government has announced annual targets for corn ethanol with a maximum of 15 billion gallons for 2015 and stays at that level thereafter. To force the market to produce the mandated target the obligated parties (the blenders) are fined if the target is not achieved. In addition, the government was paying a tax credit to the blenders per gallon of ethanol, and an ethanol tariff was in place too. These two policy provisions expired at the end of 2011. While many studies have examined these policies, the fact that these policies interact with other exiting policies is ignored. The implemented biofuel policies have raised the price of agricultural commodities and hence reduced the need for agricultural subsidies. The commodity subsides paid in 2000 were more than $20 billion. This figure was about $6 billion in 2010. Hence, we can expect that the reductions in agricultural subsidies relieved a portion of the burden of the ethanol subsidy policy. On the other hand, the mandate has forced blenders to mix ethanol with gasoline. Over the past decade the wholesale price of ethanol (gasoline equivalent) was significantly higher than the wholesale price of pure gasoline (see Fig. 36.1). This means that the price of the blended fuel was higher in cost than pure gasoline, and the economy has paid the higher price due to the mandate. One can consider the difference between the prices of gasoline and ethanol (gasoline equivalent) as an implicit tax on the blended fuel. Of course consumers and producers shared the burden of this implicit tax. To correctly assess the economic impacts of the ethanol policy, the explicit and implicit component of this policy and the interplay between the mandate and other exiting tax policies should be recognized and taken into account.

This paper develops analytical and numerical general equilibrium models to examine the importance of the implicit and explicit portions of the US ethanol policy and their interactions with other pre-exiting distortionary policies (such as agricultural or income taxes) for the economic analyses of this policy. We first develop a stylized analytical general equilibrium framework which represents interactions between economic activities and government policies in a simple economy. The stylized analytical model is developed based on the work done by Goulder et al. [9] and Taheripour et al.[18]. The first paper examined the cost effectiveness of alternative air pollution reduction policies in a second-best setting, and the second paper analyzed welfare impacts of alternative polices for agricultural pollution control again in a second best setting. The analytical work decomposes the welfare impacts of a representative mandate policy into several components and shows how they affect welfare and interact with the implemented policy. The analytical work also indicates how the economy substitutes ethanol with gasoline and examines under what conditions the mandate could induce a rebound effect. Then we use a computable general equilibrium model to quantify the economic impacts of US ethanol policy. For the numerical analyses we rely on the GTAP-BIO-ADV model developed by Taheripour et al. [21].

36.2 Literature Review

The economic and environmental consequences of the US ethanol mandate have been examined from different angles using a wide range of economic modeling approaches. Many papers have used partial equilibrium models and highlighted the economic implications of this policy for the US economy. Early studies in this area examined the role of ethanol as an additive to gasoline and argued that using ethanol as an alternative for MTBE (another oxygenate) could increase economic welfare and reduce emissions (for example see [7]). Several papers examined the importance of government fixed ethanol subsidy for agricultural and energy markets and its economic consequences. In this context [8] examined the choice between crop and ethanol subsidies and claimed that a deficiency payment program (a direct subsidy to corn producers) that costs the same to taxpayers as an ethanol subsidy will induce an annual deadweight losses of $37 million in the long run. His corresponding estimate for the deadweight loss of ethanol subsidy was about $665 million. He missed the fact that an increase in ethanol subsidy could reduce the need for agricultural subsidies. Tyner and Quear [23] and Tyner and Taheripour [24] showed that replacing the fixed per gallon ethanol subsidy with a variable-rate subsidy could reduce the social costs of the government intervention and still protect the ethanol industry from the adverse consequences of down ward shifts in crude price. Taheripour and Tyner [17] and de Gorter and Just [2] studied the efficiency and distributive effects of a biofuel subsidy. Rajagopal et al. [15] argued that the ethanol tax credit reduces the price of gasoline by 3 % and could improve the welfare of the US economy by $11 billion. On the other hand, de Gorter and Just [4] showed that in the presence of farm subsidies the ethanol tax credit will reduce the welfare of the US economy by $1.3 billion.

In another line of research several papers examined the implications of ethanol subsidy for fuel demand. These papers usually employed partial equilibrium models and argued that the ethanol subsidy could reduce the price of E10 and increase the demand for this product. These papers claim that the increased demand for E10 due to the ethanol subsidy may mitigate environmental and security benefits of ethanol production because the subsidy generates a rebound effect which eventually leads to higher demand for gasoline and more imports of crude oil (for example: see [13,26]). Several papers also examined the effect of the ethanol mandate on gasoline demand. For example, de Gorter and Just [3] examine the effects of a tax credit in the presence of a blend mandate. They showed that a tax credit with a mandate results in a subsidy to fuel consumers and higher fuel consumption. Hochman et al. [11] also claimed that introducing biofuels into the energy market generates a rebound effect at a global scale.

Almost all of the analyses mentioned above are based on partial equilibrium models, which manly highlighted consequences of US ethanol policy for agricultural and/or fuel markets. These analyses usually ignore the rest of the economy and disregard the fact that resources are limited and that the biofuel mandates interact with other policies and pre-existing distortionary taxes. By the second half of the 2007, the importance of indirect land use emissions induced by biofuel production were introduced in to the literature. The early papers in this field suggested that biofuel production could have extraordinary land use implications [6, 12, 16, 22]. For example, Searchinger et al. [16] provided the first peer-reviewed estimate for the ILUC (about 0.73 hectares of new cropland area per 1,000 gallon of ethanol capacity). Those authors used a partial equilibrium modeling framework (FAPRI) to assess the ILUC due to the US ethanol program. However, the more recent studies find the early estimates have overstated the land use implications of US ethanol production [1, 5, 10, 14, 19, 25]. For example, Hertel et al. [10] using a general equilibrium model showed that full accounting for market mediated price responses to ethanol production, as well as the geography of world trade, contributed to significant reductions in estimated ILUC impacts. Those authors estimated that the ILUC for the US ethanol program is about 0.29 hectares per 1,000 gallons of ethanol.

Almost all research studies which examined the induced land use changes also ignored that US ethanol policy has reduced the need for agricultural subsidies, and their experiments poorly represent the way that the policy is implemented in the real world. In this paper we show that including reduction in agricultural production subsidies could significantly alter the induced land use changes due to ethanol production.

36.3   Why Simple Partial Equilibrium Models Could Be Misleading

In this section we employ a simple partial equilibrium model which has been frequently used to assess the welfare impacts of biofuel policies. In this analysis it is assumed that gasoline and ethanol are perfect substitutes and that they have identical energy contents. Consider the left panel of Fig. 36.2 which represents the market for ethanol. In this market the supply of and demand for ethanol with no government intervention are shown with S_{e0} and D_e. The demand curve for ethanol represents the derived demand of the blender blends ethanol with gasoline at an arbitrary rate and supplies the blend to the market for the blended fuel. This figure assumes that with no government support, ethanol production is zero. This means that the marginal cost of ethanol production is higher than the blender's willingness to pay for ethanol at any production level of this fuel. Now assume that the government subsidizes ethanol production with a fixed rate of t_e per gallon of ethanol. Ignore the fact that the government needs to finance the policy. In this partial equilibrium framework the supply of ethanol will shift to S_{e1}. With this subsidy the equilibrium price and quantity of ethanol will be p_{e1} and QS_{e1}. At this equilibrium the ethanol producer receives p_e per gallon of ethanol and the blender pays p_1 per gallon. With this set up the change in benefits received by the ethanol producer, the change in benefits received by the blender, and the amounts of subsidy paid by the government are equal to the areas of a, b, and $a + b + c$, respectively. The deadweight losses observed due to the ethanol subsidy would be equal to c.[1] Now for a moment assume that the policy has no other welfare impacts. With this assumption in mind now consider the right panel of Fig. 36.2 which depicts the market for the blended fuel. Given that the market for pure ethanol (say E85) is negligible we assumed that the consumers only consume the blended fuel and that the curve D_b represents their demand curve for this product. When ethanol is not subsidized the supply curve of $S_b = S_g + S_{e0}$ represents the market supply. With this supply curve the market equilibrium for the blend would be at E_0 with no ethanol blended with gasoline. When the government pays ethanol subsidy then the curve $S_b = S_g + S_{e1}$ represents the supply curve of the blend and the market equilibrium for the blended fuel moves to E_1. At this equilibrium supplies of gasoline, ethanol, and the blended fuel would be equal to $O_b Q_g$, $O_e Q_{Se1} = Q_g Q_1$, and $O_b Q_1$, respectively. In this situation one can conclude that the ethanol subsidy creates a rebound effect because it increases the consumption of the blended fuel by $Q_o Q_1$.

We now take into account the fact that the government needs to finance the ethanol subsidy. There are several ways to finance the policy. Reduction in existing subsidies on other products, an additional fuel tax on gasoline production, an income

---

[1]The deadweight losses mention here belongs to changes in ethanol market. Ethanol subsidy could affect consumers and producers surpluses in other markets as well. A partial equilibrium model can trace the changes in consumers and producers surpluses in few other markets such as markets for corn and E10, but they fail to capture the welfare impacts through the entire economy.

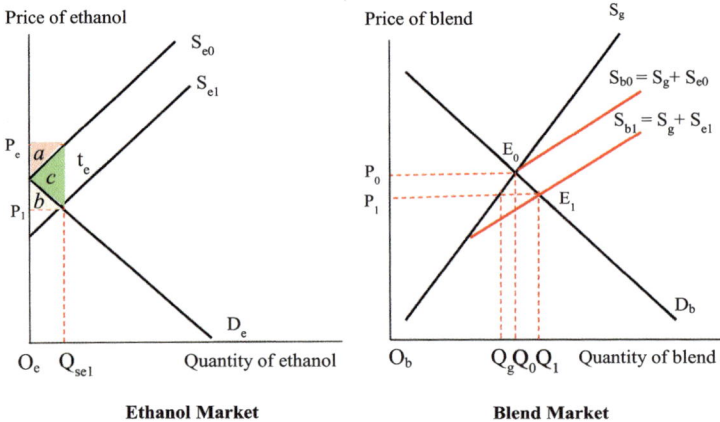

Fig. 36.2 Impacts of an ethanol subsidy on fuel market

tax, and/or changing in tax rates imposed on other goods and services are some options to finance the ethanol subsidy. Using either of these options or a combination of them will affect the above partial equilibrium analyses. To examine how the financing issue could alter the above rebound effect conclusion, consider a simple income tax. The income tax shrinks the households' disposable incomes which eventually reduces demands for goods and services including the demand for fuel. Consider now a case where the income elasticity of demand for fuel is high and the income tax hits the demand for this commodity significantly. The left panel of Fig. 36.3 represents this situation. This figure indicates that if the induced income tax effect of the ethanol subsidy is high, then the demand curve for ethanol shifts back significantly and no rebound effect is observed in the new equilibrium of E1 where $Q_1 < Q_0$. On the other hand, if the income elasticity of the demand for fuel is low, then the demand shifts back slightly and results in a minor rebound effect (see the right panel of Fig. 36.3). This simple example shows that including the possibility of an income tax for supporting the ethanol subsidy could alter the results of our partial equilibrium analyses. In general, studies which argued for rebound effect used partial equilibrium models which ignore the fact that the ethanol subsidy needs to be financed through the tax system.

Consider now a case where the government does not pay any subsidy but forces a fuel blend including a certain share of ethanol per gallon of the blend, say α. Given that the price of ethanol is more expensive than the price of gasoline we can assume that: $p_e = (1 + \beta)p_g$. Here p_e and p_g represent the prices of ethanol and gasoline, respectively, and $\beta > 0$. Following an average pricing rule the supply price of the blend will be: $p_b = (1 + \mu)p_g$ where $\mu = \beta(1 - \alpha) > 0$. Hence, in this case the blending mandate increases the supply price of the fuel with an equivalent ad valorem rate of $\mu = \beta(1 - \alpha) > 0$. The left panel of Fig. 36.4 represents the supply curve of the blended fuel for this case with S_b. In this panel the market equilibrium is at E1 with the equilibrium price of p_b (higher than the initial price of

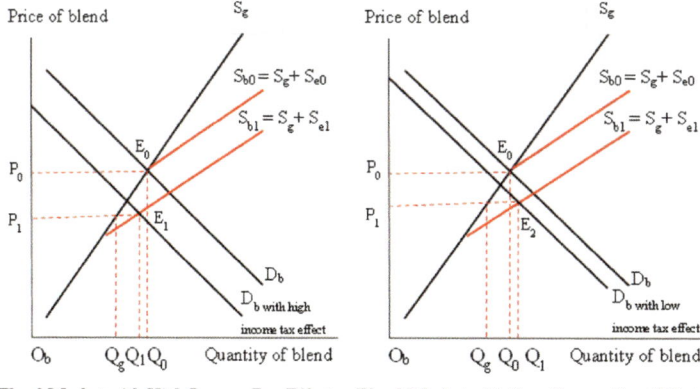

Fig. 36.3 Impacts of an ethanol subsidy on fuel market in the presence of income tax

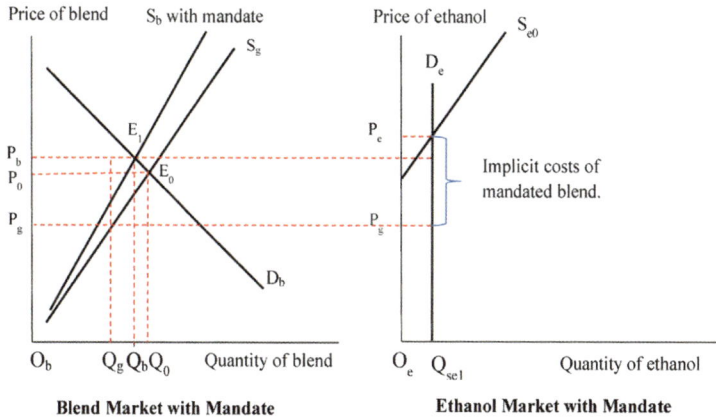

Fig. 36.4 Impacts of an ethanol mandate for fuel market with no explicit economic incentive

p_0) and the equilibrium quantity of Q_b (less than the initial quantity of Q_0). With a blending mandate in place the ethanol producer receives pe and the gasoline price received by gasoline producer is p_g. As shown in the right panel of Fig. 36.4 the difference between these two prices represents the social costs of the mandate per gallon of produced ethanol. In this case the partial equilibrium analysis does not produce a rebound effect because the price paid by the consumer of the blended fuel increases due to the mandate. However, it shows that the price received by the gasoline producer drops from p_0 to p_g and the quantity of gasoline supplied falls from Q_0 to Q_g. The reduction in gasoline price received by the gasoline producer with reduction in gasoline consumption in a country with mandate can open room for other countries to expand their gasoline consumption, and that could lead to a rebound effect at the global scale. The global numerical general equilibrium analysis provided in this paper indicates that the US ethanol mandate causes a week rebound effect at the global scale.

The above partial equilibrium analysis showed that imposing a mandate could not cause a rebound effect in the country which imposes the mandate. This could be a misleading conclusion. For example, in the US case the ethanol mandate could lead to increases in crop prices, raise farmers' incomes, increase land prices, generate higher income from trade of commodities, reduction in agricultural subsidies, and cause many other impacts. The compound effect of these changes could alter our conclusion from the above partial equilibrium analysis.

36.4 An Analytical General Equilibrium Framework

The analytical model developed in this section follows the work done by Goulder et al. [9]. These authors employed a stylized analytical general equilibrium framework and examined the cost effectiveness of alternative air emissions reduction policies in the presence of pre-existing distortionary labor tax. Taheripour et al. [18] extended their work and examined the economic efficiency of agricultural pollution reduction policies in the presence of labor and agricultural support policies. We revise this model by introducing ethanol into the modeling structure.

Consider an open economy with three commodities—gasoline (X), ethanol (E), and food (Y) with constant returns to scale production technologies. Gasoline consumption generates two externality costs. It increases emissions and reduces national security. The per gallon social costs of gasoline consumption are ω. The economy consists of three producers each producing only one commodity; a representative consumer who consumes good and services and owns endowments including labor ($\bar{L}$), land ($\bar{R}$), and capital ($\bar{K}$); and a government which determines income tax rates on labor (t_L), land (t_R), and capital (t_K), regulates externality costs of gasoline using a fixed tax rate (t_X) per gallon of produced gasoline, supports food production using a production subsidy (S_E), regulates imports of gasoline to match the world price of gasoline with its domestic market price using a tariff rate of t_m on imported gasoline, and pays transfer payments (G). The representative consumer derives utility from consumption of gasoline (C_X), Ethanol (C_E), Food (C_Y), and leisure (l) and disutility from gasoline externality costs (N) through the following utility function:

$$u = u(C_X, C_E, C_Y, l) - \phi(N). \tag{36.1}$$

Here $l = \bar{L} - L$, where L represents labor supply. The consumer receives disposable income from work ($(1-t_L)L$), return on capital ($(1-t_K)r_K \bar{K}$), land rent ($(1-t_R)r_R\bar{R}$), and government transfer payments (G). In these revenue components, r_K, and r_R represent capital and land rents and it is assumed that the wage is the numeraire and hence equals one. Henceforth, we represent the non-labor income with Q. The representative consumer allocates its labor and non-labor incomes to purchase gasoline (C_X), Ethanol (C_E), and (C_Y) with market prices of p_X, p_E, and p_Y, respectively. Hence the consumer budget constraint is:

$$p_X C_X + p_E C_E + p_Y C_Y = (1 - t_L)L + Q . \tag{36.2}$$

The economy is competitive, exports (y) some part of food production, and imports (x) some part of its gasoline consumption. The economy imports gasoline (the dirty good) at the world price (p_{XW}) and exports food at domestic market price (p_Y). The trade is in balance as shown in the following:

$$p_{XW}x = p_Y y(p_Y) . \tag{36.3}$$

Here $p_{XW} = p_X - t_m$, where t_m stands for any difference between the domestic and world prices of gasoline, which implies a tariff/subsidy per unit of imported gasoline.

In this economy producers use Constant Returns to Scale (CRS) technologies. This implies zero profits in production process of gasoline, ethanol, and food in a competitive market zero profit condition. Under these assumptions the marginal and average costs are equal to each other in the absence of regulation. These assumptions in combination with the existing regulations introduced above imply:

$$P_X = MC_X(r_R, r_K) + t_X , \tag{36.4}$$

$$P_E = MC_E(r_R, r_K) + t_E , \tag{36.5}$$

$$P_Y = MC_Y(r_R, r_K) + t_Y . \tag{36.6}$$

Finally, with the assumptions and the regulations defined above, the government budget constraint can be defined in as follow.

$$t_X O_X + x(t_m) + t_L L + t_R r_R \bar{R} + t_K r_K \bar{K} = S_E O_E + S_Y O_Y + G . \tag{36.7}$$

Here O_X, O_E, and O_Y represent outputs of gasoline, ethanol, and food. In this equation the first two elements of the left hand side show government tax revenues from production and imports of gasoline. Other components of the left hand side measure revenues from income taxes. The right hand side of the government constraint measures government's subsidies to support ethanol and food production plus the transfer payments.

In order to reduce emissions caused by gasoline consumption the government has several options to follow in this simple economy. Some important options are: an increase in gasoline tax, an increase in ethanol subsidy, a mandate on gasoline consumption or production, a mandate on ethanol consumption or production, introducing a tax on emissions, introducing an emissions reduction subsidy, and/or a combination of these policies. These policies could induce different welfare impacts and affect the economy in different ways. Given that the US has in the past used an ethanol tax credit, we examine the welfare impacts of an increase in ethanol subsidy to reduce total externality costs of gasoline consumption ($N = \omega C_Y$).

To achieve this goal consider a marginal increase in ethanol subsidy. To finance this policy the government can increase income tax rates, reduce food subsidies, change gasoline tariff rate, reduce transfer payments, or a combinations of these methods. To assess a general case assume that all of these options are all on the table.

To determine the welfare impacts of a marginal increase in ethanol subsidy in a general case, we differentiate the utility function with respect to S_E, enforce the household budget constraint, impose the trade balance, take into account the government budget constraint, apply the market clearing condition, and use Slutsky equation and Shepard's lemma and several other microeconomic theories.[2] The welfare impacts are classified into several components and are shown in the following equation:

$$\frac{du}{\lambda ds_E} = \underbrace{-S_Y \frac{dO_Y}{dS_E}}_{\text{Primary food effect}} + \underbrace{t_m(O_X + x)\varepsilon_X \theta_X - S_E O_E \varepsilon_E \theta_E}_{\text{Primary Rebound effect}}$$

$$= \underbrace{-(1 + \varepsilon_x)\frac{dp_X}{ds_E}x + p_X \frac{dx}{ds_E} + (1 - \varepsilon_y)\frac{dp_Y}{ds_E}y + p_Y \frac{dy}{ds_E} + x\frac{dt_m}{ds_E}}_{\text{Primary Trade effect}}$$

$$= \underbrace{+M\left(\frac{dN_{LTR}}{ds_E}\right)}_{\text{Revenue Recycling Effect}} - \underbrace{\sum_{J=X,E,Y}\left(\tau t_L\left(-\frac{\partial L}{\partial p_J}\right) + \tau'(C_J)S_G\right)\frac{dp_J}{ds_E}}_{\text{Tax Interaction Effect}}$$

$$\underbrace{+\tau t_L \varepsilon_{LQ}\left(\frac{L}{Q}\right)\left(\frac{dQ}{dt_N} - \frac{dG}{dt_N}\right)}_{\text{Non-Labor Income Effect}}. \tag{36.8}$$

The first three components measure the primary impacts of an increase in ethanol subsidy. The primary food effect is expected to be welfare improving. An increase in ethanol subsidy moves resources away from food to ethanol production and reduces the need for food subsidy. This item shows only the direct efficiency gain due to reduction in food subsidy. The second terms measures the impact of rebound effect on welfare. In this component θ_X and θ_E show percentage changes in consumption of gasoline and ethanol, and ε_X and ε_E stand for the demand price elasticities of these commodities. This component could increase or decrease welfare. In this component $-s_E O_E \varepsilon_E \theta_E$ is always welfare improving. However, $t_m(O_X + x)\varepsilon_X \theta_X$ could be positive or negative. The percentage change in gasoline consumption, θ_X,

---

[2]A similar approach is used in [9, 18]. The decomposition process used in this paper is available upon request from the authors.

determines the sign of this subcomponent. If $\theta_X < 0$, then this subcomponent is positive and hence the overall primary rebound effect is welfare improving. However, if $\theta_X > 0$, then the welfare impact of primary rebound effect could be positive, negative, or zero.

The next component is the primary trade effect. In general, when the demand for the exported food is inelastic, the world price of gasoline remains constant (or goes down), and the tariff (t_m) remains unchanged, then the trade effect will be welfare improving. Otherwise it could be either positive, negative, or zero.

The next component in Eq. (36.8) shows the revenue recycling effect. If the ethanol subsidy is financed using a labor tax, then the revenue recycling effect would be welfare decreasing. The next component is labeled tax interaction effect. The tax interaction effect measures efficiency costs due to interaction between the ethanol and labor tax. This secondary effect could be either positive or negative. Finally, the last component of the above equation measures efficiency costs due to interaction between labor supply and non-labor incomes. The policy will likely increase the price of land, which leads to an increase in leisure and reduces labor supply. This will reduce welfare. For more detail about the last three components of Eq. (36.8) see [18].

We now analyze the consequences of the ethanol subsidy for the substitution between gasoline and ethanol. Using the budget constraint defined in Eq. (36.2) in combination with some standard derivations it is straightforward to show that:

$$\frac{dC_X}{dC_E} = -\frac{1}{\varepsilon_{EX}} \frac{C_X}{C_E} - \left(\frac{\varepsilon_y + 1}{\varepsilon_{YE}}\right) \frac{C_Y}{C_E} \frac{p_Y}{p_X} + \left(\frac{\alpha_E}{\varepsilon_{IE}} - \frac{1 + \varepsilon_E}{\varepsilon_E}\right) \frac{p_E}{p_X}. \quad (36.9)$$

In the derivation process of this equation, it is assumed the government does not increase income tax rates to support the ethanol subsidy. This equation indicates that the displacement ratio between gasoline and ethanol (dC_X/dC_E) is a function of own and cross price elasticities, relative prices, and relative consumption of commodities. This ratio measures rebound effect according to the following criteria:

- If $dC_X/dC_E \geq 0$, then an increase in ethanol consumption due to an increase ethanol subsidy does not reduce consumption of gasoline. In this case total consumption of fuel $(C_X + C_E)$ goes up by $(dC_E + dC_X$: where $dC_X \geq 0)$. We refer to this as strong rebound effect.
- If $0 > dC_X/dC_E > -1$, then an increase in ethanol consumption due to an increase in ethanol subsidy decreases consumption of gasoline with an amount less than the increase in ethanol production. In this case total consumption of fuel $(C_X + C_E)$ goes up by $(dC_E + dC_X$: where $dC_X < 0)$. We refer to this as weak rebound effect.
- If $dC_X/dC_E \leq -1$, then an increase in ethanol consumption due to an increase in ethanol subsidy decreases consumption of gasoline with an amount equal to or larger than the increase in ethanol production. In this case total consumption of

fuel $(C_X + C_E)$ goes down or stays the same. We refer to this case as no rebound effect.

Since ethanol is a substitute for gasoline, then $\varepsilon_{EX} > 0$, and hence the first term in Eq. (36.8) is always negative. Consider the sign of the next two components of this equation. If food and fuel (ethanol) are compliments, then $\varepsilon_{YE} < 0$. Therefore, with an inelastic food demand the second term is positive. Finally, if the income elasticity of demand for ethanol is positive, and its own price elasticity is less than one then the third term is positive too. With these assumptions the displacement ratio can be either positive or negative. From the above analysis it clear that if ethanol and gasoline are compliments then the first component of Eq. (36.8) will become positive. The combination of this assumption and other assumptions on the income and price elasticities noted above implies a strong rebound effect. This means that if ethanol is an additive for gasoline, then it is likely to observe a strong rebound effect due to the ethanol subsidy.

36.5 Numerical Model

To evaluate the economic impacts of the US ethanol policy we modify the GTAP-BIO-ADV model developed in [21].[3] This model is designed and used to assess the land use impacts of alternative biofuel pathways. The GTAP-BIO-ADV is a CGE model which takes into account the interactions between a wide range of economic activities (including biofuels) and handles production, consumption, and trade of goods and services at a global scale, while it allocates scarce resources such as land, labor and capital among economic activities. This model covers production and consumption of the first and second generation of biofuels and links them with other industries and services.

This model includes the traditional fuels markets as well. The oil, gas, and coal industries supply materials to the processed petroleum, electricity, and other industries at the global scale. In general, the model considers liquid biofuels (ethanol, biodiesel, and bio-gasoline), as direct substitutes for gasoline. However, it assumes low degrees of substitutions among all energy commodities at the firm and household levels as well.

The model takes into account the competition for land among the land using industries such as forestry, livestock, and crops. Production of biofuel (except for corn stover) increases competition for land among the land use sectors. In this model cropland pasture is a part of cropland and is an input in the production processes of the livestock sector.

---

[3]The model is an advanced version of GTAP-BIO model which developed by Taheripour et al. [19], Hertel et al. [10] Taheripour et al. [20], Tyner et al. [21].

The model handles the production, consumption, and trade of a wide range of commodities at a global scale. It aggregates the world economy into 43 groups of commodities (including biofuels, DDGS, and oilseed meals) and 19 regions and represents the world economy in 2004.

The GTAP-BIO model and its successors substitute ethanol and gasoline volumetrically. Given that the energy content of ethanol is about 67 % of gasoline, the volumetric approach could generate misleading results. To fix this problem we made proper changes in the model to compare ethanol and gasoline based on their energy contents. We use the modified GTAP–BIO_AEZ model to examine the economic impacts of the US ethanol mandate.

The RFS2 is the core component of the US ethanol policy. This mandate implicitly forces the economy to consume 15 billion gallons of corn ethanol in 2015. At the same time the mandate has been supported by an ethanol tax credit and a trade tariff until the end of 2011. We can consider reduction in agricultural subsidy as a portion of ethanol policy as well. To introduce all components of the ethanol policy into the GTAP simulation process we have several options to follow. Consider the following three options.

36.5.1 Option 1

To implement the mandate and enforce the market to produce and consume 15 billion gallons of ethanol we need to introduce a market incentive into the GTAP-BIO-AEZ Model. The market incentive could be a revenue neutral tax credit for ethanol production financed by a gasoline tax. This method imposes the main burden of the policy on parties involved in the fuel market (ethanol producer, refineries, and fuel consumers). This method does not simulate the actual ethanol policy, but measures the economic impacts if we ignore other components of the policy. We refer to this simulation as experiment I.

36.5.2 Option 2

This option brings reduction in agricultural production subsidy into account. We know that in reality ethanol production has decreased the need for agricultural production subsidies because they are in part linked to commodity prices, which have partially increased due to ethanol demand for corn. The GTAP model uses ad valorem subsidies and thus cannot adjust them as commodity prices increase. However, many tax rates are flexible in the real world. For example, the US production subsidies go down if crop prices go up. We fixed this problem in option 2 by reducing US agricultural production output subsidies to zero, while other agricultural subsidies remain in effect according to their 2004 rates presented in the base data. In this option a portion of required ethanol subsidy comes from reduction in agricultural subsidy, and as a result, a lower gasoline tax is required to achieve the mandated level of 15 billion gallons of ethanol. We refer to this simulation as experiment *II*.

36.5.3 Option 3

Options 1 and 2 mainly impose the burden of the mandate policy on the fuel market. In option 3 we assume that the government cuts agricultural production subsidies, and then finances the rest of required ethanol subsidies to produce 15 billion gallons of ethanol using an income tax increase. This method spreads the burden of the mandate to all economic activities. We refer to this option as experiment *III*.

In conclusion the above experiments can be defined as:

Experiment I: An increase in US ethanol production from its 3.41 billion gallons in 2004 to 15 billion gallons mandated for 2015 using an incentive production subsidy per gallon of ethanol financed using a gasoline production tax.

Experiment II: An increase in US ethanol production from its 3.41 billion gallons in 2004 to 15 billion gallons mandated for 2015 using a production subsidy per gallon of ethanol financed using a gasoline production tax and by reduction of US agricultural output subsidies to zero.

Experiment III: An increase in US ethanol production from its 3.41 billion gallons in 2004 to 15 billion gallons mandated for 2015 using a production subsidy per gallon of ethanol financed using a an income tax and by reduction of US agricultural output subsidies to zero.

36.6   Numerical Results

The numerical analyses cover impacts of the ethanol mandate on commodity prices, crude oil and gasoline prices, fuel production, trade balance, and welfare.

36.6.1 Price Impacts

The ethanol mandate increases market prices of crop commodities. Among the alternative options, experiment *I* generates the lowest impact on crop prices. This is because this experiment assumes that agricultural activities will continue to receive production subsidies in the presence of the ethanol mandate. The crop price impacts obtained from experiments *II* and *III* are very similar and significantly higher than the price impacts of experiment *I*. For example, experiments *I*, *II*, and *III* predict that the ethanol mandate increases the price of coarse grains by about 7.2 %, 16.9 %, and 16.8 % respectively (Table 36.1).

Table 36.1 Price impacts of ethanol mandate under alternative experiments

Commodity	Experiment I	Experiment II	Experiment III
Paddy rice	2.4	7.2	6.9
Wheat	1.9	1.6	1.4
Coarse grains	7.2	16.9	16.8
Oilseeds	2.5	2.6	2.6
Sugar crops	3.5	0.9	0.9
Other crops	2.5	3.0	3.0
Crude oil	−2.9	−2.5	−1.2
Gasoline	6.2	4.3	−1.6

Figures are percentage changes due to ethanol shock

On the other hand experiment *I* predicts the highest price impact for gasoline and crude oil, because this experiment ignores the fact that a portion of required subsidy for ethanol is financed due to reduction in agricultural subsidies. On the other hand, experiment *III* predicts the lowest price impact for gasoline, because it spread the burden of the policy on all economic actives and finances a portion of the required subsidy by reduction in agricultural subsides. Indeed experiment *III* shows that if the mandate is supported by an income tax, then it could reduce the price of gasoline. The real world functioning of the ethanol mandate may be somewhere between experiments *II* and *III* (with no explicit ethanol subsidy in effect after 2011, most likely closer to experiment *II*). Thus the impact of the ethanol mandate on the price of gasoline will be somewhere between a reduction of 1.6% to an increase of 4.3%. The bottom line is that the impact of the ethanol subsidy on the gasoline price is small. Finally, all experiments show that the ethanol mandate barely reduces the crude oil price (by a number between −1.2% to −2.9%).

36.6.2   Impact on Fuel Production and Rebound Effect

An increase in US ethanol production from its 3.41 billion gallons in 2004 to 15 billion gallons increases the supply of ethanol by 11.59 billion gallons. This is identical to 7.77 billion gallons of gasoline and is not a large number compared to the global energy market. However, it is large enough to affect the US gasoline market. The impacts of adding 7.77 billion gallons of ethanol gasoline equivalent (EGA) on the US and world gasoline market are shown in Table 36.2. This table shows that the ethanol mandate reduces US gasoline consumption by about 11.44 billion gallons if we put the burden of the mandate on the fuel market and ignore reduction in agricultural subsidy as shown in experiment I. When we bring reduction in agricultural subsidy into account, the mandate reduces US gasoline consumption by 11.49 billion gallons in experiment *II*. Finally, in experiment *III* when we spread the burden of the mandate among all economic activities and take into account reduction in agricultural subsidies, then the mandate only reduces gasoline

Table 36.2 Impacts of ethanol mandate on gasoline consumption

Experiments	Increase in US Ethanol supply	Reduction in US gasoline consumption	Reduction in global gasoline consumption	US rebound effect[a]	Global rebound effect[a]
Experiment I	7.77	−12.44	−9.36	−1.60	−1.21
Experiment II	7.77	−11.49	−8.62	−1.48	−1.11
Experiment III	7.77	−8.06	−5.92	−1.04	−0.76

Figures are in billion gallon gasoline equivalent except otherwise noted
[a]Rebound effect is defined as: reduction in gasoline/increase in ethanol

consumption by 8.06 billion gallons. Indeed experiment *III* is the most efficient policy, among those which are examined in this paper, and has the lowest economic burden. In this case the reduction in gasoline consumption and the increase in EGA are close to each other. Table 36.2 shows that the mandate does not generate any rebound effect in the US.

The US biofuel mandate reduces global gasoline consumption less than the expected reduction in US under all policy settings. In experiment *III*, the mandate causes a rebound effect at the global scale. The global price of gasoline goes down and that encourages some countries to increase their gasoline consumption. In this case in response to an increase in EGA by 7.77 billion gallons, the global consumption of gasoline decreases only by 5.92 billion gallons.

36.6.3 Trade Impacts

Table 36.3 shows that the ethanol mandate generates a positive trade balance of $1.323 billion in experiment *I*, which induces a sharp reduction in US gasoline consumption (Table 36.3). In experiments *II* and *III*, the mandate causes negative trade balances of $−1.034 billion and $−5.018 billion. Table 36.3 indicates that US biofuel mandate can affect the trade balances of other counties as well. The regional impacts are not the same across the examined experiments.

36.6.4 Welfare Impacts

Finally, consider the welfare impacts in Table 36.4. The ethanol mandate reduces welfare under all experiments. The first experiment represents the worst case which causes about $16.8 billion in welfare losses. The second and third experiments generate about $15.3 billion and $14.8 billion welfare losses, respectively. Hence, experiment *III* which considers reduction in agricultural subsidies and implements a revenue neutral tax on gasoline financed using an income tax is the least cost policy. The overall global welfare impact of the US ethanol mandate does not very

Table 36.3 Impacts of ethanol mandate on trade balance by region

Region	Experiment I	Experiment II	Experiment III
USA	1,323	−1,034	−5,018
EU27	−413	505	2,343
Brazil	18	67	130
Can	10	56	141
Japan	−101	309	1,054
Chihkg	222	369	498
India	−65	−12	105
C_C_Amer	228	403	411
S_o_Amer	−157	−80	42
E_Asia	−12	24	78
Mala_Indo	1	30	46
R_SE_Asia	−40	23	111
R_S_Asia	−7	8	44
Russia	−329	−301	−245
Oth_CEE_CIS	−46	4	120
Oth_Europe	−30	6	55
MEAS_NAfr	−640	−515	−223
S_S_AFR	26	74	148
Oceania	14	62	160

Figures are in millions of 2004 dollars

significantly with the alternative policy set ups defined in experiments *I*, *II*, and *III*. However, regional impacts vary from one experiment to another one in each region.

36.6.5 Land Use Impacts

The induced land use impacts due to ethanol production have been the focal point several studies in recent years. Figure 36.5 represents impacts of ethanol production on expansion in cropland by region. In general, this figure shows that reduction in US agricultural production subsidies reduces the global cropland expansion due to ethanol production from about 2 million hectares in experiment *I* to about 1.9 million hectares in experiments *II* and *III*. This figure also shows that reduction in US agricultural production subsidies shifts the induced land use impacts of ethanol production from US to other regions. The expansion in US cropland is close to 1 million hectares in experiment *I*. This figure falls to about 0.3 million hectares in experiments *II* and *III*. This substantial difference means ignoring the fact that ethanol production reduces the need for agricultural subsidies leads to misleading estimates for induced land use changes due to ethanol policy.

Table 36.4 Welfare impacts of US ethanol mandate by region

Region	Experiment I	Experiment II	Experiment III
USA	−16,822	−15,339	−14,795
EU27	1,952	1,312	132
Brazil	173	194	151
Can	−624	−588	−316
Japan	318	−168	−528
Chihkg	134	67	−44
India	503	448	262
C_C_Amer	−1,597	−1,616	−833
S_o_Amer	−643	−557	−279
E_Asia	411	175	−63
Mala_Indo	−66	−53	−9
R_SE_Asia	221	215	145
R_S_Asia	63	51	25
Russia	−940	−917	−724
Oth_CEE_CIS	42	26	−3
Oth_Europe	−443	−427	−330
MEAS_NAfr	−3,611	−3,503	−2,501
S_S_AFR	−897	−808	−474
Oceania	92	123	134
World	−21,734	−21,365	−20,050

Figures are in millions of 2004 dollars

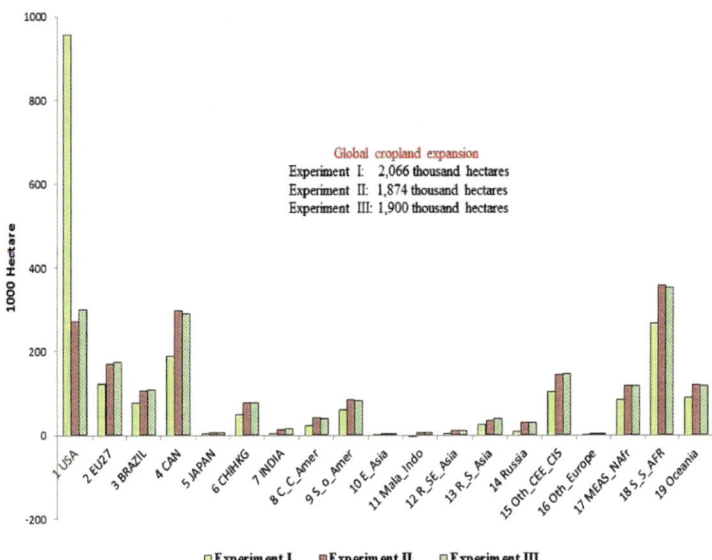

Fig. 36.5 Expansion in cropland due to expansion in US ethanol production

36.7 Conclusion

In this paper, we have shown that partial equilibrium evaluations of biofuels policies can lead to misleading results. We then develop a stylized theoretical model to show how a general equilibrium setup can improve the analysis of price, welfare, rebound, and other impacts. Finally, we implement an empirical analysis of the US corn ethanol mandate and show that inclusion of agricultural subsidies and income tax impacts are very important. For example, previous work (including our own) has seriously underestimated the price impacts on coarse grains because the financing of the implicit subsidy did not consider the reduction of agricultural subsidies. Also, other studies in the literature have estimated huge gasoline price decreases due to the US ethanol program. Here, we show that the gasoline price impact is essentially zero. These other studies did not include all the economy wide impacts. We also show the rebound, trade, and welfare impacts of the policy cases. The welfare impacts, interestingly, do not differ significantly across the cases.

We also show that ignoring the reduction of agricultural output subsidies due to higher coarse grain prices induced by biofuels demand leads to very misleading geographical distribution of land use changes. Taking into account the agricultural subsidy reduction diminishes land use change in the US by about 70 %, while reducing global land use change only about 5 %.

References

1. Al-Riffai, P., Dimaranan, B., Laborde, D.: Global Trade and Environmental Impact Study of the EU Biofuels Mandate. International Food Policy Research Institute, Washington, DC (2010)
2. de Gorter, H., Just, D.: The welfare economics of an excise-tax exemption for biofuels. Working Paper 13, Department of Applied Economics and Management, Cornell University, Ithaca (2007)
3. de Gorter, H., Just, D.: The law of unintended consequences: how the U.S. biofuel tax credit with a mandate subsidizes oil consumption and has no impact on ethanol consumption. Department of Applied Economics and Management, Cornell University, Ithaca (2008)
4. de Gorter, H., Just, D.: The welfare economics of a biofuel tax credit and the interaction effects with price-contingent farm subsidies. Am. J. Agr. Econ. **91**(2), 477–488 (2009)
5. EPA: Renewable Fuel Standard Program (RFS2) Regulatory Impact Analysis. United States Environmental Protection Agency, Washington, DC (2010)
6. Fargione, J., Hill, J., Tilman, D., Polasky, S., Hawthorne, P.: Land clearing and the biofuel carbon debt. Science **319**(5867), 1235–1238 (2008)
7. Gallagher, P., Shapouri, H., Price, J., Schamel, G., Brubaker, H.: Some long-run effects of growing markets and renewable fuel standards on additives markets and the US ethanol industry. J. Pol. Model. **25**(6–7), 585–608 (2003)
8. Gardner, B.: Fuel ethanol subsidies and farm price support. J. Agr. Food Ind. Organ. **5**(article 4), 1–20 (2007)
9. Goulder, L., Parry, I., Williams, R., III, Burtraw, D.: The cost-effectiveness of alternative instruments for environmental protection in a second best setting. J. Public Econ. **72**, 329–60 (1999)

10. Hertel, T., Golub, A., Jones, A., O'Hare, M., Pelvin, R., Kammen, D.: Effects of US maize ethanol on global land use and greenhouse gas emissions: estimating market-mediated responses. BioScience **60**(3), 223–231 (2010)
11. Hochman, G., Rajagopal, D., Zilberman, D.: The effect of biofuels on crude oil markets. AgBioForum **13**(2), 112–118 (2010)
12. Kammen, D., Farrell, A., Plevin, R., Jones, A., Delucchi, M., Nemet, G.: Energy and greenhouse impacts of biofuels: a framework for analysis. Discussion paper 2. Joint Transport Research Centre (2007)
13. Khanna, M., Ando, A., Taheripour, F.: Welfare effects and unintended consequences of ethanol subsidies. Rev. Agr. Econ. **30**(3), 411–421 (2008)
14. Laborde, D.: Assessing the Land Use Change Consequences of European Biofuels Policies. International Food Policy Research Institute, Washington, DC (2011)
15. Rajagopal, D., Sexton, S., Roland-Holst, D., Zilberman, D.: Challenge of biofuel: filling the tank without emptying the stomach? Environ. Res. Let.t **2**, 1–9 (2007)
16. Searchinger, T., Heimlich, R., Houghton, R., Dong, F., Elobeid, A., Fabiosa, J., Tokgoz, S., Hayes, D., Yu, T.: Use of U.S. Croplands for biofuels increases greenhouse gases through emissions from land-use change. Science **319**(5867), 1238–1240 (2008)
17. Taheripour, F., Tyner, W.: Ethanol subsidies, who gets the benefits? In: Outlaw, Duffield, Ernstes (eds.) Biofuel, Food & Feed Tradeoffs. Proceeding of a Conference Held by the Farm Foundation/USDA, at St. Louis, Missouri, 12–13 April 2007, pp. 91–98. Farm Foundation, Pak Brook (2008)
18. Taheripour, F., Khanna, M., Nelson, C.: Welfare impacts of alternative policies for environmental protection in agriculture in an open economy: a general equilibrium framework. Am. J. Agr. Econ. **90**(3), 701–718 (2008)
19. Taheripour, F., Hertel, T., Tyner, W., Bechman, J., Birur, D.: Biofuels and their by-products: global economic and environmental implications. Biomass Bioenergy **34**(3), 278–289 (2010)
20. Taheripour, F., Hertel, T., Tyner, W.: Implications of biofuels mandates for the global livestock industry: a computable general equilibrium analysis. Agric. Econ. **42**(3), 325–342 (2011)
21. Taheripour, F., Tyner, W., Wang, M.: Global Land Use Changes due to the U.S. Cellulosic Biofuel Program Simulated with the GTAP Model. Department of Agricultural Economics, Purdue University, Research Report Prepared for Argonne National Laboratory (2011)
22. Tokgoz, S., Elobeid, A., Fabiosa, J., Hayes, D., Babcock, B., Yu, T., Dong, F., Hart, C., Beghin, J.: Emerging biofuels: outlook of effects on U.S. grain, oilseed, and livestock markets. Staff Report 07-SR 101, Center for Agricultural and Rural Development, Iowa State University (2007)
23. Tyner, W., Quear, J.: Comparison of a fixed and variable corn ethanol subsidy. Choices **21** (3), 199–202 (2006)
24. Tyner, W., Taheripour, F.: Renewable energy policy alternatives for the future. Am. J. Agr. Econ. **89**(5), 1303–1310 (2007)
25. Tyner, W., Taheripour, F., Zhuang,Q., Birur, D., Baldos, U.: Land Use Changes and Consequent CO2 Emissions due to US Corn Ethanol Production: A Comprehensive Analysis. Department of Agricultural Economics, Purdue University (2010)
26. Vedenov, D., Wetzstein, M.: Toward an optimal U.S. ethanol fuel subsidy. Energy Econ. **30**(5), 2073–2090 (2008)

Chapter 37
New Uncertainties in Land Use Changes Caused by the Production of Biofuels

Wyatt Thompson, Nicholas Kalaitzandonakes, James Kaufman, and Seth Meyer

37.1 Introduction

Biofuels were initially heralded as a means to substitute a less polluting, renewable, and domestic fuel for imported fossil fuels. However, there is concern that as current practices of ethanol and biodiesel production continue to expand they may not reduce greenhouse gas (GHG) emissions. An important and contentious component of these GHG emission calculations involves land use changes. As higher feedstock demand raises crop prices it can draw more land into agricultural production, possibly emitting more GHGs, especially when forest is destroyed, burnt, and the soil carbon released [18,28]. According to this view, while biofuels made in the US, the EU and elsewhere from corn, wheat, or vegetable oil can emit fewer GHGs, the reduction in tailpipe emissions might take decades to offset the initial burst of GHGs caused by such land conversion [24,28]. Here, we reinforce the sensitivity of GHG estimates to assumptions, and we identify the critical questions relating to context that have not been addressed.

W. Thompson (✉)
Department of Agricultural and Applied Economics Department and Food and Agricultural
Policy Research Institute (FAPRI), University of Missouri, 200 Mumford Hall, Columbia, MO,
65211, USA
e-mail: thompsonw@missouri.edu

N. Kalaitzandonakes
University of Missouri, Columbia, MO, USA
e-mail: kalaitzandonakesn@missouri.edu

J. Kaufman
Economics and Management of Agrobiotechnology Center, University of Missouri, Columbia,
MO, USA

S. Meyer
Office of the Chief Economist, USDA, Washington, DC, USA

A.A. Pinto and D. Zilberman (eds.), *Modeling, Dynamics, Optimization
and Bioeconomics I*, Springer Proceedings in Mathematics & Statistics 73,
DOI 10.1007/978-3-319-04849-9__37,
© Springer International Publishing Switzerland 2014

Public policy recognizes the potential of land use changes to negate the GHG emission benefits from biofuels. For example, both US and EU policies extend support only to biofuels made from crops grown on land already in agricultural use.[1] However, indirect land use changes caused by market-wide price signals add quite a different level of measurement complexity. Increased production of biofuel feedstocks can be achieved through land conversion or higher yields. The last response expands supply without increasing the land base for feedstock production, although it can nevertheless affect greenhouse gas emissions if, for example, more fertilizer or machinery are used.

Additional land resources might be drawn into feedstock production from other crops, pastures or forests. Land conversion from certain crops (e.g. rice) to biofuel feedstock production can, in fact, lead to reductions in GHG emissions but increases in emissions are also possible. Conversion of forest to agricultural uses must similarly be qualified. Conversion of forest to agricultural land for feedstock production can be achieved through increased deforestation or decreased afforestation (forest growth) which have entirely different GHG implications. Deforestation would cause an initial burst in non-recurring emissions of GHGs while reduced afforestation would cause a decline in ongoing and gradual sequestration.

Changes in land use follow different paths in different contexts. If a biofuel demand shock is introduced when market conditions are characterized by tight markets with high and rising agricultural commodity prices, then the added pressure of biofuel feedstock purchases will tend to draw more land into agricultural use. If the biofuel demand shock takes place in a setting of falling agricultural commodity prices and idled crop area, then the greater output can be achieved without deforestation. No deforestation contradicts the common assumption of an initial burst of GHG emissions from deforestation. The context defined by the broader agricultural commodity markets is critical to determining at least the time path of GHG emissions and leads to persistent differences if there is discounting, but its role in calculating GHG emissions has not been addressed.

Furthermore, much like land use, crop yields respond to price changes [19]. Farm-level yields are price responsive since the optimal input levels depend on the marginal factor profitability and hence on product price. Yield changes through innovation and efficiency gains are also price responsive. Higher prices can lead to greater amount of industrial research and development (R&D) due to higher anticipated payoffs [16, 25]. R&D in seeds, chemistry, capital equipment and other inputs can increase average yields over time (e.g. [2, 9, 13, 23]). Higher prices can also accelerate adoption of the new technologies in agriculture [12].

---

[1]In the US, the Renewable Fuel Standard mandates that an auditor verifies feedstocks used by a producer or importer of renewable fuel meet the definition of renewable biomass in order to qualify for the biofuel mandates. In practice, the auditor makes an assessment by obtaining supporting documentation related to feedstock purchases. The definition of renewable biomass in US legislation excludes feedstocks from land that is new to agricultural production or managed forest [22]. US regulations to implement these biofuel use mandates demand that foreign-made biofuels also meet this requirement [29].

Land and yield responses thus have two roles in determining GHG emissions. First, as increasingly recognized, they substitute for one another in meeting a surge in demand caused by increasing biofuel feedstock purchases. The second as-yet unrecognized role is that they help to set the context. For example, along with other driving forces, such as income and population, trends in yields help to determine if commodity markets can meet a demand surge with or without drawing new lands into agricultural use.

The degree of variability of land use changes and ensuing GHG emissions is, ultimately, an empirical question, one which we take up in this study. We are interested in the overall variability of land use change measures and specifically in their sensitivity to context. We find that land use change measures can be quite sensitive to context, and particularly to crop yields. We also find that variations in the estimates of land use changes can imply rather different GHG emission paths.

37.2   Review of the Literature

Choosing an optimal biofuels policy that accounts for land use changes requires that such effects are measured with some degree of accuracy. This impact assessment must involve the development of a proper counterfactual case. Since the counterfactual is not actually observed, the construction of a model is often required. Ideally, the appropriate impact assessment models represent the global nature of agricultural commodity markets, their interactions, an array of land uses, and the potential for crop yields to adjust over time. These models are intended to provide information about policies whose affects go into the future, so the analysis should be based on a context that is relevant for forward-looking decisions, particularly in the case that results are sensitive to the context.

Given the demanding nature of these modeling features, researchers have resorted to modifying existing large-scale models of global agriculture in order to measure land use changes associated with the demand for biofuels. So far, there have been two broad modeling approaches. The first supplements computable general equilibrium (CGE) models with more detailed representations of land uses and measures how biofuel feedstock demand shifts supply and demand conditions for aggregate commodities (such as coarse grains and oilseeds) as well as land uses, relative to their values in some base year. The second extends partial equilibrium models of global agricultural commodity markets. In the first step, commodity market effects from an increase in biofuels are estimated over forward-looking (e.g. 10 year) projections. In the second step, estimates of land use changes are obtained with input from the first step estimates. For both model approaches, a final step is to add up GHG emission effects of biofuels including from land use changes.

37.2.1 Partial Equilibrium Analysis

Searchinger et al. [24] find that an increase in US ethanol production of 56 billion liters brings 10.8 million ha of additional land into cultivation and this change causes significant deforestation. These land use changes include diversions of 2.8 million ha in Brazil, 2.3 million ha in China and India, and 2.2 million ha in the United States. The authors adjust the FAPRI model of agricultural commodity markets and exogenously shift US ethanol demand to derive alternative scenarios of US ethanol production increases. Crop yields are assumed to grow at a constant rate but not to respond to price changes. The analysis yields a forward-looking baseline and the effects of ethanol production on commodity markets are measured by comparing the model results with the greater volume of ethanol to the baseline in each year.

Estimated changes in the commodity markets serve as input to the calculations for land use effects. The authors also use historical deforestation trends to estimate how much land may be drawn from forest or other uses to achieve the crop land changes caused by the increase in ethanol in the commodity model. In other words, all crop land use changes implied by the commodity model are summed up and this increase in land is assumed to be met by decreases in other land uses based on past trends. However, there is no feedback effect from land use changes to the commodity markets and back. The authors use GREET to calculate GHG emissions taking the calculated land use changes and implications for carbon sequestration into account. Given that the increases in cropland area are small, constituting no more than 1 % of forest area, even if all of the increase were associated with deforestation, their findings suggest that emissions from the conversion of small amounts of forest land to crop uses dominate GHG emissions from all other types of adjustments.

In Dumortier et al. [6], some of these authors return to the question of GHG emissions caused by changes in US ethanol production and further adjust the underlying commodity model assumptions and GHG calculations. Assumption changes include the reduction of U.S. deforestation trends to negligible levels. Model adjustments include the addition of more cross-price effects among crop areas, so that the own-price effects are partially offset in the event that an increase in biofuel feedstock demand drives crop prices higher. In the absence of any other representation of trade-off among land uses, the addition of cross-price effects tends to reduce the overall pull on land into agricultural production. Furthermore, in this second study the authors estimate GHG emissions based on the Intergovernmental Panel on Climate Change (IPCC) methods. Their calculations include considerations of cropland management, livestock manure, rice growing, and changes in land use (counting land moving out of agricultural production) but exclude some emissions associated with agricultural production, such as fuel use in machinery.

Dumortier et al. [6] test the sensitivity of their results against those derived in Searchinger et al. [24] and relative to some of their model assumptions (e.g. the projected petroleum price and yield trends). The authors find that their model is most sensitive to their assumptions about yields, though changes in yields are still introduced by adjusting the model output rather than allowing the model to come to a new simulated equilibrium solution. Irrespectively, they find that under higher rates of exogenous yield growth an additional billion gallons of ethanol is accompanied by a reduction in land used for agriculture.

In Fabiosa et al. [8] many of the same authors revisit the issue of land use effects from the expanding ethanol consumption in the United States and elsewhere using a version of the FAPRI model. In one scenario, the authors exogenously increase U.S. demand for ethanol by 10 % and in a second scenario they add a 5 % increase in ethanol demand of Brazil, China, the EU, and India. For these scenarios, these authors calculate prices, crop outputs, trade and area impact multipliers for selected countries. Of primary importance are the crop area multipliers, which serve as a convenient basis of comparison to [24]. These imply that the estimated land use changes in [8] are substantially lower for Brazil, China, India, while they are reversed in Argentina and Canada. The authors attribute the divergence of results to the underlying differences in certain assumptions (i.e. [24] impose a long-run equilibrium assumption for the U.S. ethanol market and do not constrain demand for E-85 fuel). Fabiosa et al. [8] nevertheless do not consider forest area simultaneously with crop area nor do they calculate land use change effects.

37.2.2 General Equilibrium Analysis

General equilibrium approaches estimating land use changes often employ GTAP data and the GTAP model itself. The extension of GTAP to represent land more carefully typically involves disaggregating country or regional total area into agro-ecological zones (AEZs), while assuming constant elasticities to govern reallocation of land among uses in each AEZ [1, 3, 4, 11, 17, 21, 30]. Other studies replace the constant elasticity framework with a representation that includes land conversion costs [14] or append to GTAP a different disaggregation of land based on a grid coupled with an upward-sloping land supply to agriculture [26].

In a particularly relevant CGE study, Keeney and Hertel [20] approach the problem based on a version of GTAP that includes some elaboration of the energy sector and AEZs. They represent the 2006 biofuel market by adjusting the 2001 base data through three specific changes: the rise in petroleum prices; the switch from one particular fuel additive to ethanol in the US; the addition of US and EU subsidies to biofuel markets (following [3]).

Keeney and Hertel [20] find that a one billion gallon increase in biofuel consumption in the US results in loss of forest and pasture land. The US, where the largest share of land use change occurs, experiences a 0.35 % decrease in forest land and a 0.53 % decrease in pasture. Land use changes are also experienced in other countries especially in Brazil and Canada which loose 0.16 and 0.10 % of forest cover respectively. The authors also find that an increase in the price elasticity of crop yields from zero to 0.25 implies a 30 % crop output response to an increase in biofuel demand, taking some of the pressure off land conversion. They further conclude that the estimated land use changes are sensitive to the model elasticities for yields, area, and trade. Although Keeney and Hertel usefully explore the sensitivity of land use changes to assumed model parameters, they do not carry such calculations to changes in GHG emissions. Hertel et al. [17] extend this work by testing the sensitivity to parameter values, and they conclude that the next logical step is to extend this work to calculate GHG emission effects.

Another relevant analysis based on GTAP is presented in a working paper by Tyner et al. [27]. Their starting GTAP variant is similar to Keeney and Hertel's, but several adjustments are made. Changes to land representation include *unused land* in the US, which is equal to the area committed to the Conservation Reserve Program and which competes with crop production. In the US and Brazil, additional competition between pasture and cropland is introduced through a new *cropland-pasture* variable. Tyner et al. [27] conduct scenario analysis for various changes in US ethanol production, some of which use the standard GTAP 2001 reference year. They also test the impacts starting from a database partly updated to 2006.[2] Tyner et al. [27] go on to project forward to the 2006–2015 period with GTAP by extending the 2001–2006 average population growth as a proxy for demand changes and assuming a 1 % yield growth in each year. They observe that one interesting consequence is that their assumed supply shifts outpace their demand shifts, leading to a growing forest area in the 2006–2015 period.

As for results, Tyner et al. [27] find that changes in US ethanol production of 13 billion gallons in the time period between 2001 and 2015 converts, globally, 1.7 million ha into cropland, 32.5 % from deforestation and 67.5 % from grassland. The authors also calculate GHG emission effects from land use changes and foregone sequestration based on the results of the GTAP scenarios. Because their results vary widely with their assumptions they state that *..one cannot escape the conclusion that modeling land use change is quite uncertain* (p. iii). Wang et al. [30] update their energy balances and GHG emissions estimates based on [27] and other work. The sensitivity of results to context is identified as an important remaining research question: *the future growth in the demand and supply of agricultural commodities— particularly coarse grains—is a critical determinant of the impacts of biofuel programs* (p. 1888). Authors base this judgment on the projections [27] produce using assumed population and yield growth with the corresponding 30-year GHG estimates generated using the standard assumption that more agricultural land and less forest must be explained by deforestation.

---

[2]The updating consists of using more recent data representing crop production, harvested area, forest areas, gross capital formation, labor force (skilled and unskilled), gross domestic product, and population for the whole world at the country level and adjusting simulated harvested area and forest land use to match observed data (p. 23).

37.2.3   Summary

The limited empirical literature that has estimated land use changes associated with increased demand for biofuels readily demonstrates the inherent difficulties in modeling such a complex phenomenon. It also presents a muddled picture of the likely land use changes from the demand of biofuel feedstocks. In some cases, feedstock demand is accompanied by significant land conversion into agricultural uses, deforestation and large non-recurring bursts of GHG emissions, while in others, agricultural land use is found to decrease despite such demand. It is currently unclear whether these differences in empirical findings are the result of differences in assumptions, shortcomings in modeling, diversity in the counterfactuals evaluated or other factors. Yet if land use changes are to be a useful policy instrument, the sources and degree of variability must be well understood.

The possible role of model assumptions in shaping the empirical results deserves close scrutiny. Existing studies allow for limited substitution among commodities and land uses (e.g. [6, 8, 24]) or trade (e.g. Armington structure in GTAP); limited price responsiveness of yields (e.g. [24]) and limited price feedbacks between prices and yields (e.g. [6, 8, 24]). These and other relevant restrictions, in turn, limit the ability of models to represent the intrinsic ability of commodity markets to respond to changing market conditions.

Even seemingly innocuous modeling assumptions can shape the results of the analytics. Keeney and Hertel [20] show that yield response is critical. However, the importance of cross-commodity effects, such as the impact of corn price increases on high-emitting rice and livestock production, has not been emphasized. The standard approach to represent increased crop area as a one-time surge deforestation and associated emissions has remained unquestioned, perhaps because the studies typically rely on base-period analysis or historical patterns in land use changes.

In this study we add to the emerging empirical evidence by developing a global partial equilibrium model emphasizing land use changes from increased feedstock demand of biofuels. Our empirical model allows for price responsive yields; endogenous and simultaneously determined yields, land uses and commodity prices; detailed disaggregation of land uses, and substitution of commodities within, and across, countries; and a forward-looking context.

37.3   Empirical Model

We are interested in the potential variability of GHG emissions in general and especially in their sensitivity to alternative paths of yield changes. We evaluate the impact of a hypothetical discontinuation in US and EU biofuel feedstock demands under different sets of conditions:

1. *Normal* yield price responsiveness and *normal* rate of annual yield growth;
2. *High* yield responsiveness achieved through intensification of production and *normal* annual rate of yield growth and
3. *Normal* yield price responsiveness and *high* annual rate of yield growth achieved through accelerated innovation.

The set of conditions we consider are rather stylized and we effectively assume that yields in any given year can grow in two ways: through ongoing innovation (which is assumed to be unresponsive to short term price changes in a 10-year period) and through intensification of production (which is price responsive).

To carry out the analysis we develop a global multi-country, multi-commodity partial equilibrium model that explicitly represents land uses which include forest, pasture and crops—with crop areas identified for the major crops grown in temperate zones. We assess how these commodity markets evolve over a 10-year period with US and EU biofuel feedstock use growing, then remove biofuel feedstock use and reassess the market situation. We present this model in an Appendix. The difference in land use, crop yields, and livestock output between the baseline and the scenario serve as the basis of our greenhouse gas calculations.

Since we focus on the sensitivity of land use measures, we consider the parallel implications for GHG emissions as land use and crop production patterns change without going so far as complete lifecycle analysis. Calculated GHG emissions in carbon dioxide equivalent (CO_2-e) are recursive to the commodity market area, yield, and production data based on country- and commodity-specific relationships, mostly as identified by the EPA [28]. Crop and palm oil production emissions calculations include nitrogen emitted due to synthetic fertilizers and crop residues, agricultural chemical production emissions, and methane emitted from flooded rice paddy fields. Livestock emissions, meaning those from beef, pork, and poultry production, are enteric methane and from manure. EPA estimates of emissions associated with agricultural energy use, including direct fuel use, heat and electricity, and energy inputs, are applied per acre of agricultural land. Forest sequestration is weighted averages of factors identified in EPA [28, p. 389]. We take any decrease in forest area over the period, allocate it proportionally among alternative land uses that increase, and apply fixed GHG emission factors for land conversion [15] to calculate non-recurring land use change emissions.

37.4 Empirical Results

The impact of biofuels on land use is determined in our analysis by examining how an elimination of US and EU biofuel feedstock demand for grains and vegetable oils would affect agricultural commodity markets and, consequently, land use in the represented countries and country groups. A complete elimination of biofuel production from these feedstocks in the next 10 years may be improbable, but the exercise allows us to estimate the total impact of biofuels on markets and land use. US feedstock purchases for biofuels are mainly corn and soybean oil, whereas a more diverse set of feedstocks are used in the EU with wheat playing the largest part among grains (see Fig. 37.1). Sugar is also important and, according to these projections, a growing source of biofuels. In our scenario analysis then, we effectively eliminate the production of 72 billion liters of ethanol and biodiesel in 2018.

37.4.1 Land Use Effects

Our base case is characterized by slight rises in forest area as the amount of land allocated to agricultural uses, including crops, declines. As discussed earlier, the initial conditions and the context of the analysis are important. The start of our analysis is characterized by high prices for many commodities. As supply responds and demand is weak due to a weak global economy, commodity prices recede from their initial peaks in the following years. Sugar is an exception, with the world sugar price rising somewhat from its initial level.

The removal of US and EU biofuel feedstock purchases reduces overall demand and hastens the decrease in prices. With normal yield growth and price responsiveness (base case), quickly eliminating US and EU biofuel feedstock purchases causes the simple average of grain prices in 2017 to be 25 % lower than if feedstock purchases continued to expand and the vegetable oil price to be 31 % lower. The price effect is almost as large even when the rates of yield growth are high. However, the price reductions are approximately halved if yields, and consequently total supplies, are more price responsive. All these scenarios are characterized by a 10-year period of generally falling crop prices from a high starting point.

In the base case, agricultural land use in these countries rises by 15.6 million hectares over the period, but the increase is 4.2 million hectares less if US and EU biofuel demands are eliminated (Table 37.1). With faster yield growth, there is a smaller agricultural land increase less in the base case over the period and 6.5 million hectares are diverted to other uses in the biofuel scenario. With faster growth rates in yields, the impact on forest area is lowest with or without biofuels.

It is worth emphasizing the market context once more. Given falling prices, there is less pressure to reallocate land to agricultural uses, tending to slow the pace of deforestation. Land use changes also vary by region. In the case of China, the long-standing trend towards afforestation at a national level is hastened if prices are falling (Table 37.1). In some instances, however, the lower prices of the projection period can reverse the trend and cause afforestation, in aggregate, particularly in the case of either faster rates of yield growth or no biofuel feedstock purchases—or both, in which case we find the sum of forest area in all countries actually increases. The calculated relative changes in land use should not be overstated. Although millions and tens of millions of hectares are important for GHG calculations discussed later, the calculated shifts constitute less than one percent of agricultural or forest land (see Table 37.2).

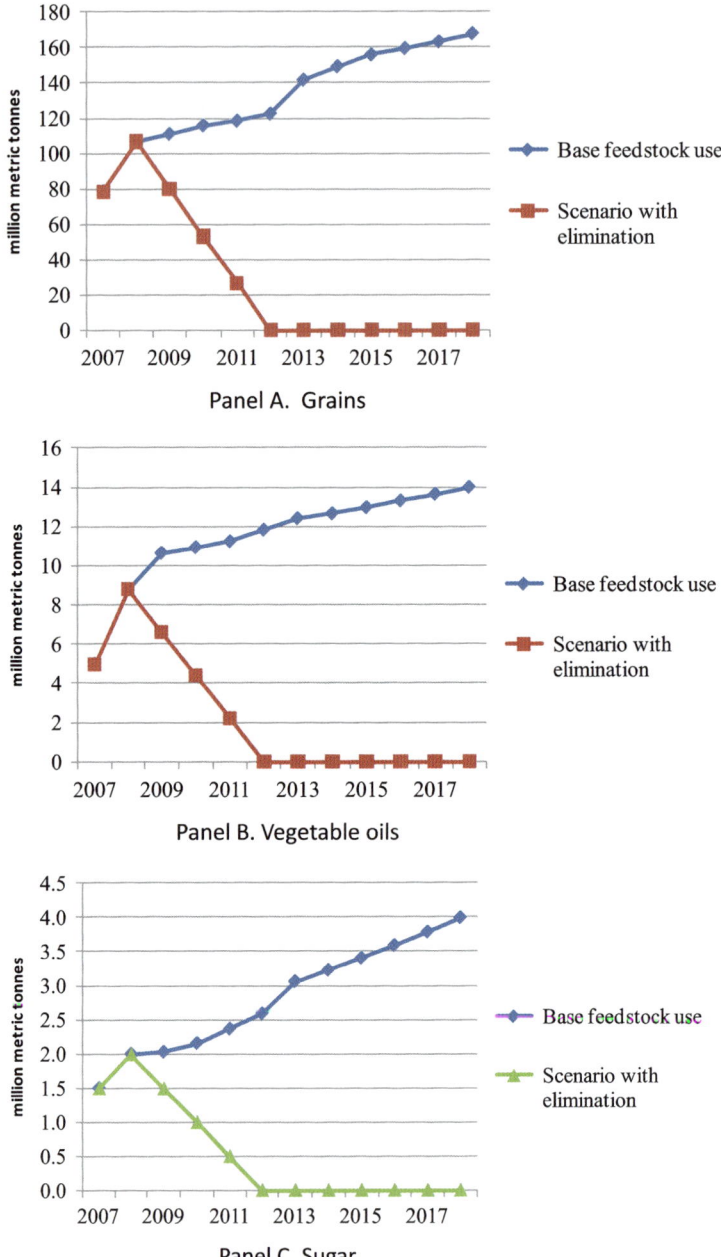

Fig. 37.1 Feedstock use in base case and scenario

Table 37.1 Changes in forest and agricultural land use when EU and US biofuel production is eliminated, 2008–2017

	Argentina and Brazil	Canada, Japan and EU	China and India	Mexico, Indonesia and Malaysia	Developing countries	Total of these countries
(millions of hectares)						
Agricultural land use						
Base						
Biofuels	6.0	2.6	−0.5	1.3	6.2	15.6
No biofuels	3.0	0.8	−1.9	0.3	2.0	4.2
High trend						
Biofuels	1.8	1.2	−2.1	0.4	1.8	3.0
No biofuels	−0.7	−0.3	−3.2	−0.5	−1.7	−6.5
High elasticity						
Biofuels	6.0	2.6	−0.5	1.3	6.2	15.6
No biofuels	4.4	1.2	−1.4	0.6	3.5	8.4
Forest area						
Base						
Biofuels	−5.6	−2.9	−0.3	−1.4	−5.0	−15.3
No biofuels	−3.3	−1.6	0.7	−0.5	−2.4	−7.1
High trend						
Biofuels	−2.4	−1.8	0.8	−0.6	−2.2	−6.3
No biofuels	−0.5	−0.7	1.6	0.2	0.0	0.7
High elasticity						
Biofuels	−5.6	−2.9	−0.3	−1.4	−5.0	−15.3
No biofuels	−4.3	−1.9	0.3	−0.8	−3.3	−10.0

There are reallocations within agricultural uses, as well, that reflect changing relative profitability of alternative uses. Removing US and EU biofuel feedstock purchases causes land reallocation away from annual crops and oil-producing palm trees (see Table 37.2). Cropland falls by six to nine million hectares in total. The change in palm oil area constitutes some ten thousand hectares almost all of which are in Indonesia and Malaysia. Sugar market effects tend to be indirect through interactions with other crops in area allocation itself and food use—both limited— since only small share of world sugar is used for ethanol production in the EU.[3] Pasture area affects are mixed as lower crop prices reduce competition for land but also increase competition for feeding animals. The net effect is typically lower pasture area even though livestock output rises slightly over the period, with total production rising about 1–2 % by the end of the period if US and EU biofuel feedstock purchases are eliminated. This net effect is small and contributes only slightly to the positive effect of eliminating US and EU biofuels on forest area, but it

---

[3]Brazil's sugar-ethanol production is not changed in this analysis.

Table 37.2 Biofuel feedstock demand elimination effects on land use, 2014/2015-17/18 averages

	Argentina and Brazil	Canada, Japan and EU	China and India	Mexico, Indonesia and Malaysia	Developing countries	Total of these countries
(percent change if US and EU biofuel feedstock purchases are eliminated)						
Forest area						
Base	0.42 %	0.24 %	0.35 %	0.48 %	0.36 %	0.35 %
Higher trend	0.34 %	0.21 %	0.29 %	0.42 %	0.30 %	0.30 %
Higher elasticity	0.30 %	0.20 %	0.26 %	0.38 %	0.27 %	0.27 %
Agricultural land use						
Base	−0.67 %	−0.56 %	−0.17 %	−0.59 %	−0.31 %	−0.37 %
Higher trend	−0.55 %	−0.48 %	−0.14 %	−0.50 %	−0.26 %	−0.31 %
Higher elasticity	−0.46 %	−0.47 %	−0.12 %	−0.47 %	−0.24 %	−0.28 %
Palm land use						
Base	−0.71 %	−0.55 %	−0.17 %	−1.94 %	−0.39 %	−1.76 %
Higher trend	−0.59 %	−0.47 %	−0.14 %	−1.81 %	−0.31 %	−1.64 %
Higher elasticity	−0.52 %	−0.47 %	−0.12 %	−1.93 %	−0.28 %	−1.74 %
Crop land use						
Base	−0.14 %	−1.14 %	−0.40 %	−0.95 %	−0.70 %	−0.63 %
Higher trend	−0.13 %	−0.96 %	−0.39 %	−0.86 %	−0.64 %	−0.57 %
Higher elasticity	−0.60 %	−1.11 %	−0.55 %	−1.15 %	−0.80 %	−0.77 %
Pasture area						
Base	−0.91 %	−0.33 %	−0.09 %	−0.61 %	−0.25 %	−0.35 %
Higher trend	−0.75 %	−0.34 %	−0.04 %	−0.49 %	−0.20 %	−0.27 %
Higher elasticity	−0.47 %	0.01 %	0.12 %	−0.24 %	−0.12 %	−0.13 %

highlights the importance of representing both supply and demand effects on pasture area in the analysis as either one alone could bias the land use change estimates.

37.4.2   Consequences for Greenhouse Gas Emissions

Our GHG emission calculations do not track all GHG emissions. We focus on the production of selected agricultural commodities to illustrate two important points. First, there is the sometimes over-looked potential for offsetting effects among commodities that spill over to livestock production. Second, there is the previously unrecognized role of the time path and context.

Eliminating US and EU biofuel feedstock purchases leads to the commodity market adjustments discussed above, which in turn cause changes in the pattern of GHG emissions (see Table 37.3). While crop emissions associated with synthetic fertilizers, crop residues, and agricultural chemicals fall by 7 mt CO_2-e as demand for such commodities as corn, wheat, and other feed grains falls, these reductions

Table 37.3 Greenhouse gas emissions of selected agricultural activities and land use in modeled countries

	Base yield scenarios		
	Biofuel	No biofuel	Change
	(CO_2 mt equivalents)		
Recurring annual emissions, 2014–2017 averages			
Crops total	962	964	3
Synth. fert. and crop resid.	273	267	−6
Chemical production	68	66	−1
Rice methane	621	631	10
Livestock (meat)	1,125	1,162	36
Energy use	617	615	−2
Forest sequestration	−6,803	−6,827	−24
Total recurring emissions	*−4,717*	*−4,701*	*15*
Non-recurring emissions			
Land use change	5,080	2,759	−2,320

are more than offset by 10 mt CO_2-e greater methane emissions as some land shifts into rice production, with the net effect being almost no change in crop emissions. Greater livestock production results in 36 mt CO_2-e more emissions.[4] The elimination of US and EU biofuel feedstock purchases means more land in forest, so more GHGs are sequestered in these areas.

As illustrated in Table 37.1, total deforestation calculated at the national level is reduced in the base yield case if US and EU biofuel feedstock demands are eliminated, which effectively halves emissions associated with deforestation based on our calculations. In the case of some countries like China, however, a trend of afforestation continues, and the falling agricultural commodity prices of the baseline only encourages this pattern of land use change, rather than change the rate of deforestation. In this case, the implication is that a changing rate of afforestation leads to more or less GHG sequestration, but no change in the emissions from deforestation over the 10-year projection period based on national forest area.

Context matters because the potential for afforestation is greater if prices are falling, and prices are more likely to be falling if yields are growing more quickly. If yield growth is high, non-recurring emissions caused by deforestation are lower (see Fig. 37.2). Even with US and EU biofuel feedstock purchases growing rapidly, non-recurring emissions are halved relative to the base yield case if yield growth rates are higher. Eliminating these biofuel demands leads to even smaller non-recurring emissions over the 10-year projection period. The elimination causes a decrease in non-recurring emissions of 1.6 billion tons CO_2-e, or about one-third less than the

---

[4]This increase is slightly greater than the relative increase in total production of beef, pork, or poultry, and reflects changes in composition and location of meat production as some regional practices generate more emissions than others.

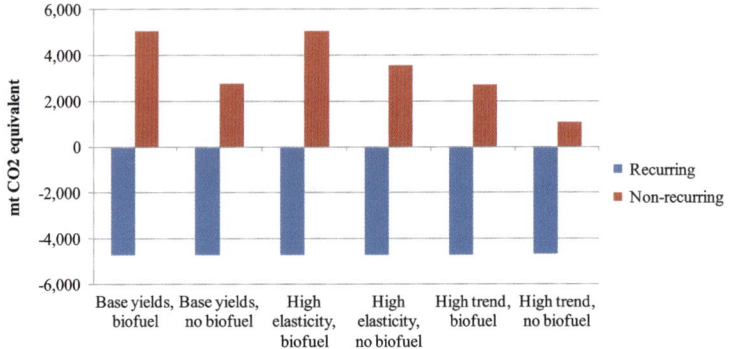

Fig. 37.2 Selected GHG emissions under alternative yield scenarios

reduction of 2.3 billion tons CO_2-e in the base yield case. The impact of eliminating US and EU biofuel feedstocks on GHG emissions is also about a third lower if crop yields are more sensitive to prices as less land conversion is necessary before markets have adjusted to the shock of lower demand.

The recurring emissions are largely stable in all the scenarios we explore, although the composition fluctuates. As discussed in the context of the base yield case, the elimination of US and EU biofuel feedstock purchases causes more methane from rice and livestock production and fewer emissions from other crop activities. In our experiments, these offsetting GHG effects tend to cancel one another.

37.5 Discussion and Conclusions

Our empirical results suggest that measures of land use changes are quite sensitive to model and context. Modest changes in the price responsiveness of yields, or their anticipated annual rates of growth from innovation, lead to drastic shifts of up to 50 % or more in GHG emission estimates. The sensitivity of these measures to assumptions about future yields is important as possible deviations from historical yield trends are not hypothetical (e.g. [5, 7]).

Sensitivity to other assumptions in the literature may also contribute to the variation of measured land use changes. Previous studies have often allowed only limited substitution among commodities, land uses and trade and have not fully represented the dynamic interplay between prices and yields. We relax some of these assumptions to explore the intrinsic ability of commodity markets to adjust and reduce land use changes. We draw some tentative conclusions about the potential influence of such assumptions by comparing our results with those of similar partial equilibrium models used in recent studies (see Table 37.4). The other studies summarized there are comparable partial equilibrium model approaches that make

Table 37.4 Estimated land use changes in selected partial equilibrium studies

Study	Base year	Period	Impact of	Estimated land use change
Searchinger et al. [24]	2007	2007–2016	56 billion litters	193
Dumortier et al. [6]	2008	2007–2016	30 billion litters	203
Fabiosa et al. [8]	2007	2007–2016	50 billion litters	127
This study				
Base Scenario	2008	2008–2017	72 billion litters	158
High Yield Scenario	2008	2008–2017	72 billion litters	132
High Elasticity Scenario	2008	2008–2017	72 billion litters	100

use of similar data sets, represent global agricultural commodity markets, and give forward-looking analysis over a projection period. Significant departures in our modeling assumptions for selected countries, however, include a complete array of cross-commodity effects in land allocation, feed demand, and food demand; price responsive yields that are determined at the same time as other market outcomes; and a complete set of land uses, including forest and pasture areas, that are also simultaneously determined with agricultural commodity areas, quantities, and prices. Despite the small sample and the inherently crude nature of the comparison, the results suggest that more restrictive modeling assumptions on demand and supply substitution possibilities may lead to higher estimated land use changes.

Finally, the sensitivity of land use measures to the initial conditions and the market context of the analysis has not been explored. Even if all impact assessment models yielded the same calculated supply responses to a biofuels demand shock, use of different comparators could lead to different land use measures and conclusions. Searchinger et al. [24] for instance, uses historical trends to infer how the land use change estimates from their commodity model might be achieved, and specifically how much comes from forest; and general equilibrium approaches (e.g. [17, 20, 27]) calculate reductions in forest area from their base data as deforested land. Whether calculated recursively or from base data, those studies assume forest area reductions amount to deforestation and that they occur at the outset of their analysis period, when biofuel feedstock purchases begin. We compare the baseline and the scenario as alternative forest area paths over time. Hence, one of our model results is forest contraction or expansion paths that are consistent with agricultural commodity market prices, so biofuel production impacts are more precisely expressed as quickening the pace of deforestation in some places and slowing or reversing afforestation in others.

The sensitivity to initial conditions and market context cannot be ignored, even though our estimates certainly can be improved. Our estimates could be updated to the latest data, land could be divided into sub national units, and GHG emission calculations could be refined. However, none of this would be likely to affect the following general conclusions that are omitted from the existing literature, as even [30], who note in passing that context matters, do not go so far as to identify these points. First, the frequent statement about the payback period of a biofuel, determined as the number of years of recurring GHG reductions from using the biofuel required to offset the non-recurring burst of GHG caused by initial deforestation, is unwarranted. As we show, there is no initial burst from deforestation if land use change amounts to slower afforestation instead of outright deforestation. With reduced afforestation, there is no payback period because emissions are lower from the outset. Unless estimates of land use change take into account the context, as determined by such factors as yield technology development and demand growth, the payback assumption is unsupported.

Second, context matters for decision making. The time path of GHG emissions can be important for policy decisions that seek to achieve societal objectives by implementing specific target GHG reductions by specific dates. Thus, arguments to assume deforestation or simply to ignore the time path and focus exclusively on the eventual end point are not strictly relevant to defining intermediate targets, nor measuring progress.

More critically still, if the policy is set over a defined period or if there is any discounting, then ignoring the time path of GHG emissions can bias estimated GHG emissions. If initial deforestation is assumed in biofuel analysis, then the accompanying burst of GHG emissions is less likely to be offset by lower recurring emissions in the future if the period of analysis is truncated or if future benefits are discounted. In contrast, reduced afforestation has no initial burst of GHG emissions so biofuel reduces GHG emissions from the first year, then calculations could have the opposite sign over a defined period or with discounting.

We add to the growing body of literature that finds variability and uncertainty in land use change measures. Our findings reinforce the importance of price responsiveness and identify context as a new source of uncertainty. Because deforestation is believed to cause a very large surge in non-recurring emissions, modestly different estimates of the scale or timing of deforestation can be equivalent to many years of recurring GHG emission reductions from biofuels, potentially changing the sign of total biofuel benefits if summed over a fixed time horizon or if future reductions are discounted. Ignoring the time path of land use change reduces the relevance to policy making and may reduce the accuracy of GHG emission calculations.

Acknowledgements This study was funded in part by the Office of Science (BER), U.S. Department of Energy under Grant No. DE-FG02-07ER64504. However, all findings, errors, and views expressed here are the authors' own.

Appendix

Model Documentation

This document summarizes the commodity models for Argentina, Brazil, Canada, China, European Union, Indonesia, India, Japan, Malaysia, Mexico, and the four developing country aggregates. These models are solved simultaneously, along with

Table 37.5 Countries modeled explicitly and four developing country aggregates

Argentina	Japan
Brazil	Malaysia
Canada	Mexico
China	Low food and low trade
EU-27	Low food and high trade
Indonesia	High food and low trade
India	High food and high trade

Table 37.6 Selected commodities

Corn (CR)	Rapeseed/canola meal (RM)
Distillers grains (DG)	Rapeseed/canola oil (RO)
Palm oil (ML)	Rapeseed/canola (RS)
Palm kernel meal (KM)	Soybean (SB)
Total oilseed meals (ME)	Soybean meal (SM)
Other annual crops (OC)	Soybean oil (SO)
Other coarse grains (OG)	Sugar (SU)
Total vegetable oils (OL)	Sunflower (UF)
Other oilseed meals (OM)	Sunflower meal (UM)
Other oilseed oils (OO)	Sunflower oil (UO)
Pork (PK)	Wheat (WH)
Petroleum (PT)	Macroeconomic: income (MAGDPR)
Rice (RC)	Macroeconomic: population (MAPOP)

Table 37.7 Land aggregates

Agriculture land (LA)	Other land (LO)
Agricultural land in annual crops (LC)	Palm land (LP)
Total land (LD)	Sugar land (LS)
Forest land (LF)	Agricultural land in pasture (LU)
Agricultural land in groves or orchards (LG)	

the FAPRI–MU.[5] model and a rest-of-world aggregate trade, for world market-clearing balances for the commodities covered. The following tables detail selected components of the model (Tables 37.5, 37.6, 37.7, and 37.8).

---

[5]Documentation for the FAPRI–MU model can be found at:
http://www.fapri.missouri.edu/outreach/publications/umc.asp?current_page=outreach
Specifically, the stochastic US crop model documentation
http://www.fapri.missouri.edu/outreach/publications/2011/FAPRI_MU_Report_09_11.pdf.

Table 37.8 Other selected quantities and prices

Crush of oilseeds (DCRU)	Expected producer price (PPRDE)
Total demand for domestic use (DDOM)	Expected net returns (ENRT)
Net exports (DEXN)	Index indicating land price (LDPS)
Feed and residual demand (DFED)	Domestic price (PDOM)
Food and edible product demand (DFOD)	Consumer price (PFOD)
Fuel demand (DFUL)	Area harvested (SHAR)
Industrial and inedible demand (DIND)	Area planted (SPLT)
Other demand (DOTH)	Production (SPRD)
Ending stock demand (DTES)	Yield (SYLD)
Oilseed yields of oil (OLSYLD)	Total area (TAR)
Meal (MESYLD)	World price (PWLD)
Producer price (PPRD)	

Summary of Equations of a Country or Developing Country Aggregate

Select equations for a representative country are summarized below. Cost indices and the link from prices to expected prices and returns are omitted here for brevity. The abbreviation "CC" denotes a relevant commodity or land use and "OC" indicates the other commodities or land uses in the same set. Here, soybean oil and meal prices (SO and SM) are used as key prices for vegetable oil and oilseed meal aggregates (Table 37.9).

Elasticities

Elasticities are required for all land use, yield, feed demand, stock, livestock product supply, and domestic use equation for all the countries. Due to the large number of elasticities in the model, only simple averages are reported here (Tables 37.10, 37.11, 37.12, and 37.13).

Yield Trend Estimates

Yield trend estimates, apart from price effects, are relevant to the projections reported in the paper. The table reports the simple average trend rates of yield growth by crop (see Table 37.14).

Table 37.9 Selected country equations

CC_TAR = f(CC_LDPS, OC_LDPS, LD_TAR), CC= LA, LF, LO
CC_TAR = f(CC_LDPS, OC_LDPS, LA_TAR), CC=LC, LU, LG, LP, SU
CC_TAR = f(CC_ENRT, OC_ENRT, LC_TAR), CC=WH, CR, OG, SB, RS, UF, RC
CC_HAR = f(CC_TAR), CC=WH, CR, OG, SB, RS, UF, SU, RC
CC_YLD = f(CC_PPRDE, TREND), CC=WH, CR, OG, SB, RS, UF, SU, RC, ML, KM
CC_SPRD=CC_HAR*CC_YLD, CC=WH, CR, OG, SB, RS, UF, SU, RC
CC_DCRU = CC_SPRD+CC_DTES$_{-1}$-CC_DTES, SB, RS, UF
CC_PDOM = CC_OLSYLD*SO_PDOM+CC_MESYLD*SM_PDOM, CC=SB, RS, UF
CC_SPRD = CC_HAR*CC_YLD, CC=ML, KM
OL_SPRD=SB_DCRU*SB_OLSYLD+RS_DCRU*RS_OLSYLD+UF_DCRU*UF_OLSYLD +ML_SYLD*LP_TAR+OO_SPRD
ME_SPRD=SB_DCRU*SB_MESYLD+RS_DCRU*RS_MESYLD+UF_DCRU*UF_MESYLD +KM_SYLD*LP_TAR+OM_SPRD
CC_DFED = f(CC_PDOM, OC_PDOM, BF_SPRD, PK_SPRD, PL_SPRD), CC =WH, CR, OG, ME, LU, RC
CC_DTES = f(CC_PDOM, CC_SPRD),CC=WH, CR, OG, ME, OL, SU, RC, BF, PK, PL
CC_SPRD = f(CC_PPRD, WH_PDOM, RC_PDOM, CR_PDOM, OG_PDOM, SM_PDOM, LU_LDPS), CC=BF, PK, PL
OL_DIND = f(OL_PDOM, OC_PDOM, MAGDPR/MAPOP)
CC_DFOD =f(CC_PFOD, OC_PFOD, MAGDPR/MAPOP),CC=WH, CR, OG, OL, SU, RC, BF, PK, PL
CC_DEXN = CC_SPRD+CC_DTES$_{-1}$-CC_DTES-CC_DFED-CC_DFOD-CC_DFUL-CC DOTH, CC=WH, CR, OG, ME, OL, SU, RC, BF, PK, PL
CC_PDOM=f(CC_PWLD), CC=WH, CR, OG, ME, OL, SU, RC, BF, PK, PL
CC_PFOD = f(CC_PDOM), CC=WH, CR, OG, ME, OL, SU, RC, BF, PK, PL
CC_PPRD = f(CC_PDOM), CC=WH, CR, OG, ME, OL, SU, RC, BF, PK, PL

Table 37.10 Average food demand elasticities

	Wheat	Corn	Other grains	Rice	Sugar	Oil	Beef	Pork	Poultry	Income
Wheat	−0.17	0.02	0.01	0.02	0.00	0.02	0.00	0.01	0.00	0.33
Corn	0.05	−0.45	0.02	0.06	0.02	0.01	0.03	0.02	0.02	0.42
Other grains	0.07	0.04	−0.41	0.07	0.02	0.01	0.02	0.01	0.02	0.38
Rice	0.10	0.02	0.01	−0.68	0.05	0.04	0.08	0.06	0.06	0.44
Sugar	0.00	0.01	0.01	0.00	−0.22	0.02	0.00	0.01	0.00	0.52
Oil	0.06	0.01	0.01	0.06	0.04	−0.66	0.08	0.05	0.06	0.55
Beef	0.01	0.01	0.01	0.00	0.01	0.02	−0.46	0.02	0.04	0.71
Pork	0.00	0.01	0.01	0.00	0.00	0.01	0.01	−0.34	0.01	0.73
Poultry	0.02	0.01	0.01	0.01	0.02	0.02	0.07	0.03	−0.56	0.67

Greenhouse Gas (GHG) Emission Calculations

Primary sources are the IPCC and the EPA. Unless otherwise specified, references to EPA are to Assessment and Standards Division Office of Transportation and Air Quality U.S. Environmental Protection Agency, *Draft Regulatory Impact Analysis: Changes to Renewable Fuel Standard Program* 2009

Table 37.11 Average feed demand elasticities

	Wheat	Corn	Other grains	Rice	Meal
Wheat	−1.95	0.99	0.62	0.53	0.02
Corn	0.36	−1.82	0.65	0.67	0.06
Other grains	0.43	0.90	−1.71	0.43	0.02
Rice	0.20	0.77	0.38	−1.66	0.15
Meal	0.02	0.05	0.02	0.12	−0.20

Table 37.12 Average crop area elasticities

	Wheat	Corn	Oth. gr.	Rice	Rapeseed	Soybean	Sunflower
Wheat	0.67	−0.26	−0.22	−0.09	−0.02	−0.05	−0.02
Corn	−0.13	0.61	−0.05	−0.18	−0.02	−0.17	−0.01
Other grains	−0.17	−0.06	0.47	−0.14	−0.01	−0.06	−0.01
Rice	−0.06	−0.22	−0.09	0.48	−0.01	−0.05	0.00
Rapeseed	−0.12	−0.07	−0.16	−0.10	0.39	−0.01	−0.01
Soybean	−0.32	−0.63	−0.53	−0.33	−0.06	1.21	−0.03
Sunflower	−0.25	−0.05	−0.17	−0.04	−0.11	−0.04	0.46

Table 37.13 Average aggregate area elasticities

Broad land use: agriculture, forest, and other	0.03
Agriculture versus forest	0.07
Among broad agricultural uses (e.g. palm, grove, annual crops, sugar, pasture)	0.17

Table 37.14 Average yield trends

Wheat	2.40 %
Corn	1.10 %
Other coarse grains	1.30 %
Rice	1.30 %
Rapeseed	2.60 %
Soybean	1.80 %
Sunflower	3.00 %
Palm	3.30 %
Sugar	1.10 %

http://www.epa.gov/OMS/renewablefuels/420d09001.pdf
with supplemental data files.

Emissions from Crop Production

Emissions from Agricultural Chemical Production

Emissions from agricultural chemical production were based on EPA methods. National fertilizer and agricultural chemical data was taken from the FAO's

FERISTAT database, which reports fertilizer use by crop and country for a single year between 1988 and 2004. GREET factors are applied to obtain the total per unit lifecycle emissions of the respective agricultural chemicals. The agricultural chemicals per hectare emissions are regressed on yield for country-and-commodity pairs with data. We calculate an elasticity of emissions to yield at mean values. The emission from each hectare of a crop in a country is equal to the base emission level for that country and that emission rate increases according to the yield change from the base value and the crop-specific elasticity. A lower limit is set at zero; we do not allow a negative relationship between yields and emissions.

Crop Land Uses

Crop- and country-specific GHG emission calculations and include emissions associated with synthetic fertilizers and crop residues for all crops and methane emissions in the case of rice. These emissions can rise with yields depending on cross-country comparison of emissions to yields.

Crop N_2O Emissions

For international N_2O emissions we considered both direct and indirect emissions from synthetic fertilizer application, and crop residue N in a method synonymous with EPA. Rate of N application was obtained from FAO's FERISTAT. From this direct N20 emissions were calculated assuming an appropriate rate of N_2O volatilization. Indirect emissions were adjusted for leaching and runoff and utilized a leaching and runoff emissions factor. We regress the synthetic fertilizer emissions on yield (using only country-and-commodity pairs with data, before inserting averages). We calculate an elasticity of emissions to yield at mean values. The emission from each hectare of a crop in a country is equal to the base emission level for that country and that crop increased according to the yield change from the base and the crop-specific elasticity. Emissions from crop residue are also included.

Rice CH_4 Emissions

Emissions from rice cultivation followed the EPA methods. The default IPCC emission rates were used and scaled for each cropping regime: irrigated, rainfed lowland, upland and deepwater by country by day. Rice cultivation season lengths were taken from the International Rice Research Institute (IRRI) to obtain annual emissions factors. Rice emissions of this type are not linked to yields.

Total Crop Land Emissions

Crop land N_2O (fertilizer and residue), agricultural chemical emissions, and rice CH_4 emissions are multiplied by the corresponding areas, converted into CO_2-e terms, and added. For developing country aggregates of the model, we take a weighted average of those countries within the aggregate for which emission data are available. The weight assigned each country for aggregating emissions is usually the harvested area, but production data for sugar and palm oil (PSD data). We use average values in cases where there are no data.

Agricultural Fuel Use

Lifecycle emissions are inclusive of both direct fuel use and heat and electricity consumption. Original data comes of agricultural fuel use emissions are from IEA [19] "CO_2 Emissions from Fuel Combustion" from which EPA Lifecycle GHG/Tailpipe CO_2 factors are applied. The resulting 2005 international agricultural sector emissions are then divided by 2005 national agricultural area to derive an average per area emission factor as found in EPA. Totals for developing countries are weighted averages of component countries based on FAOSTAT data.

Land Conversion

Calculated GHG emissions from land conversion include only emissions caused by deforestation. Emissions per hectare converted from forest use to select other uses are drawn from EPA. We use the average emissions of land conversion among developing countries represented on the table for developing country aggregates in the model. Our land uses do not match exactly the uses given in the original data. Land conversion from forest is calculated by taking the difference between initial and final forest area. If the net change is negative, then the former forest land is allocated among other broad land uses. The allocation is based on the weighted share of each land use with a positive net change. Thus, this measure is exclusively an estimate based on net effects, so deforestation in one place that is offset by an increase in forest area elsewhere would give a zero deforestation number in these calculations. This presumably overlooks some amount of normal turn-over in forest area as regards absolute numbers. However, when we consider the implications of a change from baseline paths to alternative scenario paths the focus moves to the changes in land use, this error might be less critical to scenario analysis.

Forest Sequestration

The main concern associated with sequestration is the annual difference between carbon sequestered in forestland compared to that of cropland. This is calculated in similar fashion to the EPA document using above- and below-ground sequestration data that decompose the long-term deforestation emission effects. These data give forest sequestration rates for some countries in the model. We calculate a ratio of the sequestration rate to the dry mass for those cases where there are data for both. This is applied to the dry mass on forest land for other countries to give an estimate of the sequestration rate of those countries. However, this assumes the same ratio would apply for above- and below-ground sequestration rates. For this, IPCC estimates of the dry mass on forest land are used. We aggregate these data by country according to FAO forest type data (Global Forest Resource Assessment [10]). For developing country aggregates of the commodity model, we use the countries with the most forest area and ignore some of the countries with much smaller forest area. We ignore any differences in sequestration rates of new or old forests.

Livestock Production

GHG emissions caused by enteric fermentation and manure management are country- and commodity-specific, but these are not tied to any measure of feed use changes relative to the base data. Rates of emissions are from EPA, but are on a per head per year basis. To convert to a per unit of livestock product basis, FAOSTAT data representing meat production and animal inventories are used. The inventories at one point of time in the year are taken as estimates of the average number of head in that year, although this may ignore seasonal differences. The ratio of output to animal inventory is used to convert the original per head emission rate into a per unit of output emission rate. Chicken is used to represent all poultry. For each of the four developing country aggregates, a weighted average rate is used. The weights are production of the animal output in question (e.g. beef, pork, and poultry).

Fuel Use

Transportation, storage, and tailpipe emissions of ethanol and biodiesel emissions are included. Tailpipe emissions of different GHGs are given by EPA and transportation and storage for ethanol are provided in a supplemental file. We apply the same transportation and storage emissions to biodiesel. We assume at present that ethanol and biodiesel displace gasoline and diesel at a rate of one BTU for one BTU.

References

1. Ahammad, H., Mi, R.: Land use change modeling in GTEM: accounting for forest sinks. Australian bureau of agricultural and resource economics. Paper presented at EMF22: Climate Change Control Scenarios, Stanford University, California. 25–27 May 2005
2. Alston, J., Wyatt, T.J., Pardey, P.G., Marra, M.C., Chan-Kang, C.: A meta-analysis of rates of return to agricultural R & D: ex pede Herculem?. Research reports 113, International Food Policy Research Institute (IFPRI) (2000)
3. Birur, D.K., Hertel, T.W., Tyner, W.E.: Impact of biofuel production on world agricultural markets: a computable general equilibrium analysis. GTAP Working Paper No 53, Center for Global Trade Analysis, Purdue University, West Lafayette (2008)
4. Birur, D.K., Hertel, T.W., Tyner, W.E.: The biofuels boom: implications for world food markets. In: Bunte, F., Dagevos, H. (ed.) Paper Presented at the Food Economy Conference, The Hague, Netherlands, October 2007, and Published as a Chapter in The Food Economy - Global Issues and Challenges, pp. 61–75. Wageningen Academic Publishers, The Hague, Netherlands (2009)
5. Cassman, K.G., Dobermann, A., Walters, D.T., Yang, H.: Meeting cereal demand while protecting natural resources and improving environmental quality. Annu. Rev. Environ. Resour. **28**, 315–358 (2003)
6. Dumortier, J., Hayes, D.J., Carriquiry, M., Dong, F., Du, X., Elobeid, A., Fabiosa, J.F., Tokgoz, S.: Sensitivity of carbon emission estimates from indirect land-use change. CARD Working Paper 09-WP 493 (2009)
7. Edgerton, M.D.: Increasing crop productivity to meet global needs for feed, food, and fuel. Plant Physiol. **149**, 7–13 (2009)
8. Fabiosa, J.F., Beghin, J.C., Dong, F., Elobeid, A., Tokgoz, S., Tun-Hsiang, Y.: Land allocation effects of the global ethanol surge: predictions from the international FAPRI model. Land Econ. **86**(4), 687–706 (2010)
9. Fernandez-Cornejo, J.: The Seed Industry in U.S. Agriculture: An Exploration of Data and Information on Crop Seed Markets, Regulation, Industry Structure, and Research and Development. Agriculture Information Bulletin Number, vol. 786. Economic Research Service, U.S. Department of Agriculture, Washington, DC (2004)
10. Food and Agricultural Organization (FAO) of the United Nations. Global Forest Resource Assessment 2000, Rome, Italy (2001)
11. Golub, A., Hertel, T., Sohngen, B.: Projected Land-Use Change in the Dynamic GTAP Framework. In: 10th Annual Conference on Global Economic Analysis, West Lafayette, 7–9 June 2007
12. Griliches, Z.: Hybrid corn: an exploration in the economics of technical change. Econometrica **25**(4), 501–522 (1957)
13. Griliches, Z.: Research costs and social returns: hybrid corn and related innovations. J. Polit. Econ. **66**(5), 419–431 (1958)
14. Gurgel, A., Reilly, J., Paltsev, S.: Potential land use implications of a global biofuels industry. J. Agr. Food Ind. Organ. **5**(2), 9 (2007)
15. Harris, N., Grimland, S., Brown, S.: GHG Emission Factors for Different Land-Use Transitions in Selected Countries of the World. Report to the EPA (2009)
16. Hayami, Y., Ruttan, V.M.: Agricultural Development: An International Perspective. John Hopkins University Press, Baltimore (1985)
17. Hertel, T., Tyner, W., Birur, D.: The global impacts of biofuel mandates. Energy J. **31**(1), 75–100 (2010)
18. Intergovernmental Panel on Climate Change (IPCC): 2006 IPCC Guidelines for National Greenhouse Gas Inventories. http://www.ipcc-nggip.iges.or.jp/public/2006gl/index.html
19. International Energy Agency (IEA): CO_2 Emissions from Fuel Combustion. IEA, Paris, France. http://www.oecdbookshop.org/oecd/display.asp?K=5L4TGCBJLGZX (22 November 2007)

20. Keeney, R., Hertel, T.W.: The indirect land use impacts of united states biofuel policies: the importance of acreage, yield, and bilateral trade responses. Am. J. Agr. Econ. **91**(4), 895–909 (2009)
21. Lee, H., Hertel, T.W., Rose, S., Avetisyan, M.: An integrated global land use data base for CGE analysis of climate policy options. In: Hertel, T., Rose, S., Tol, R. (ed.) Economic Analysis of Land Use in Global Climate Change Policy. Routledge, London (2008)
22. Public Law (P.L.) 110–140: Energy Independence and Security Act of 2007. http://frwebgate. access.gpo.gov. Accessed 10 Mar 2008
23. Ruttan, V.: Productivity growth in world agriculture: sources and constraints. J. Econ. Perspect. **16**(4), 161–184 (2002)
24. Searchinger, T., Heimlich, R., Houghton, R.A., Dong, F., Elobeid, A., Fabiosa, J., Tokgoz, S., Hayes, D., Tun-Hsiang, Y.: Use of U.S. croplands for biofuels increases greenhouse gases through emissions from land-use change. Science **319**(5867), 1238–1240 (2008)
25. Schmookler, J.: Invention and Economic Growth. Harvard University Press, Cambridge (1966)
26. Tabeau, A., Eickhout, B., van Meijl, H.: Endogenous agricultural land supply: estimation and implementation in the GTAP model. In: GTAP Conference, Addis Ababa, Ethiopia (2006)
27. Tyner, W., Taheripour, F., Zhuang, Q., Birur, D., Baldos, U.: Land Use Changes and Consequent CO_2 Emissions due to US Corn Ethanol Production: A Comprehensive Analysis. Report of the Department of Agricultural Economics, Purdue University (2010)
28. U.S. Environmental Protection Agency (EPA): Draft Regulatory Impact Analysis: Changes to Renewable Fuel Standard Program. EPA-420–D–09–001 (2009)
29. U.S. Environmental Protection Agency (EPA): Regulation of Fuels and Fuel Additives: Changes to Renewable Fuel Standard Program. 40 CFR, Part 80 (2009)
30. Wang, M., Han, J., Haq, Z., Tyner, W., Wu, M., Elgowainy, A: Energy and greenhouse gase emission effects of corn and cellulosic ethanol with technology improvements and land use changes. Biomass Bioenergy **35**, 1885–1896 (2011)

Chapter 38
Inequalities on the Parameters of a Strongly Regular Graph

Vasco Moço Mano, Enide Andrade Martins, and Luís Almeida Vieira

38.1 Introduction

In the 1963 paper *Strongly regular graphs, partial geometries and partially balanced designs* [1], Bose introduced a class of regular graphs with an additional property: the number of neighbors of every pair of vertices depends only from the fact that those vertices are adjacent or not. These graphs are called *strongly regular* and are defined by a set of four parameters. One problem on the study of these graphs is to find admissibility conditions that can rule out unrealistic parameter sets. In this work we apply the theory of Euclidean Jordan algebras to strongly regular graphs to deduce new inequalities over their parameter sets, that is necessary conditions for the existence of strongly regular graphs.

Euclidean Jordan algebras were introduced in 1934 in the paper *On an algebraic generalization of the quantum mechanical formalism* [10], by Jordan et al., and since then the concept has had a wide range of applications, for instance in statistics (see [13]), interior point methods (see [5, 6]) and combinatorics (see [2, 12, 15]).

This paper is organized as follows. In Sects. 38.2 and 38.3, we present some basic concepts concerning Euclidean Jordan algebras and strongly regular graphs, respectively. Next, in Sect. 38.4 we associate a three dimensional Euclidean Jordan algebra to the adjacency matrix of a strongly regular graph and in Sect. 38.5 we

L.A. Vieira (✉)
CMUP - Centro de Matemática da Universidade do Porto Faculty of Engineering, University of Porto, Porto, Portugal
e-mail: lvieira@fe.up.pt

E.A. Martins • V.M. Mano
CIDMA - Center for Research and Development in Mathematics and Applications, Department of Mathematics, University of Aveiro, Aveiro, Portugal
e-mail: enide@ua.pt; vascomocomano@gmail.com

A.A. Pinto and D. Zilberman (eds.), *Modeling, Dynamics, Optimization and Bioeconomics I*, Springer Proceedings in Mathematics & Statistics 73, DOI 10.1007/978-3-319-04849-9__38,
© Springer International Publishing Switzerland 2014

deduce new inequalities over the parameters and spectra of a strongly regular graph. Finally, in Sect. 38.6, we present some experimental results and conclusions.

38.2 Power Associative Algebras and Euclidean Jordan Algebras

In this section we present a brief review of concepts, properties and results from Euclidean Jordan algebras. Detailed literature can be found in the monograph by Faraut and Korányi, [4], and in Koecher's lecture notes, [11].

A real vector space $\mathcal{V}$ with a commutative bilinear map $(x, y) \mapsto x \cdot y$ in $\mathcal{V} \times \mathcal{V} \to \mathcal{V}$ is a real Jordan algebra if $x \cdot y = y \cdot x$ and $x \cdot (x^2 \cdot y) = x^2 \cdot (x \cdot y)$, where $x^2 = x \cdot x$. From now on we suppose that if $\mathcal{V}$ is a Jordan algebra, then $\mathcal{V}$ is a finite dimensional real algebra and has a unit element denoted by $\mathbf{e}$.

If $\mathcal{V}$ is a Jordan algebra then $\mathcal{V}$ is power associative, that is an algebra such that for any x in $\mathcal{V}$ the algebra spanned by x and $\mathbf{e}$ is associative.

Let $\mathcal{V}$ be a Jordan algebra. The rank of x in $\mathcal{V}$ is the least natural number k such that $\{\mathbf{e}, x, \ldots, x^k\}$ is linearly dependent and we write $\mathrm{rank}(x) = k$. Since $\mathrm{rank}(x) \leq n$ we define the rank of $\mathcal{V}$ as being the natural number $\mathrm{rank}(\mathcal{V}) = \max\{\mathrm{rank}(x) : x \in \mathcal{V}\}$. An element x in $\mathcal{V}$ is regular if $\mathrm{rank}(x) = \mathrm{rank}(\mathcal{V})$. Let x be a regular element of $\mathcal{V}$ and $r = \mathrm{rank}(x)$. Then, there exist real scalars $a_1(x), a_2(x), \ldots, a_{r-1}(x)$ and $a_r(x)$ such that

$$x^r - a_1(x)x^{r-1} + \cdots + (-1)^r a_r(x)\mathbf{e} = 0, \tag{38.1}$$

where 0 is the null vector of $\mathcal{V}$. Taking into account (38.1) we conclude that the polynomial

$$p(x, \lambda) = \lambda^r - a_1(x)\lambda^{r-1} + \cdots + (-1)^r a_r(x) \tag{38.2}$$

is the minimal polynomial of x. When x is not regular the minimal polynomial of x has a degree less than r. The roots of the minimal polynomial of x are the eigenvalues of x.

Example 38.1. The real vector space of real symmetric matrices of order n, $\mathrm{Sym}(n, \mathbb{R})$, equipped with the bilinear map $x \bullet y = (xy + yx)/2$ is a real power associative algebra whose unit is $\mathbf{e} = I_n$, the identity matrix of order n.

Remark 38.1. Let $\mathcal{V}$ be a finite dimensional associative real algebra with the bilinear map $(x, y) \mapsto x \cdot y$. We introduce on $\mathcal{V}$ a structure of Jordan algebra by considering a new product $\bullet$ defined by $x \bullet y = (x \cdot y + y \cdot x)/2$ for all x and y in $\mathcal{V}$. The product $\bullet$ is called the Jordan product.

Example 38.2. The real vector space $\mathcal{V} = \text{Sym}(n, \mathbb{R})$ is a real Jordan algebra when endowed with the bilinear map $\bullet$ given by $x \bullet y = (xy + yx)/2$ for all x and y in $\mathcal{V}$, where xy is the usual matrix multiplication of x and y.

An Euclidean Jordan algebra $\mathcal{V}$ is a Jordan algebra with an inner product $< \cdot, \cdot >$ such that

$$< x \cdot y, z > = < y, x \cdot z > \tag{38.3}$$

for all x, y and z in $\mathcal{V}$.

Example 38.3. The real vector space $\text{Sym}(n, \mathbb{R})$ is a real Euclidean Jordan algebra when endowed with the Jordan product and with the inner product $< x, y >= \text{tr}(xy)$, where tr denotes the usual trace of matrices.

Let $\mathcal{V}$ be a real Euclidean Jordan algebra with unit element $\mathbf{e}$. An element c in $\mathcal{V}$ is an idempotent if $c^2 = c$. Two idempotents c and d are orthogonal if $c \cdot d = 0$. The set $\{c_1, c_2, \ldots, c_l\}$ is a complete system of orthogonal idempotents if the following three conditions hold.

(*i*) $c_i^2 = c_i$, for $i = 1, \ldots, l$,
(*ii*) $c_i \cdot c_j = 0$, if $i \neq j$,
(*iii*) $\sum_{i=1}^{l} c_i = \mathbf{e}$.

An idempotent c is primitive if it is a nonzero idempotent of $\mathcal{V}$ and if it can't be written as a sum of two non-zero idempotents. We say that $\{c_1, c_2, \ldots, c_k\}$ is a Jordan frame if $\{c_1, c_2, \ldots, c_k\}$ is a complete system of orthogonal idempotents such that each idempotent is primitive.

Theorem 38.1 ([4], p. 43). *Let $\mathcal{V}$ be a real Euclidean Jordan algebra. Then for x in $\mathcal{V}$ there exist unique real numbers $\lambda_1, \lambda_2, \ldots, \lambda_k$, all distinct, and a unique complete system of orthogonal idempotents $\{c_1, c_2, \ldots, c_k\}$ such that*

$$x = \lambda_1 c_1 + \lambda_2 c_2 + \cdots + \lambda_k c_k. \tag{38.4}$$

The numbers λ_j's of (38.4) are the eigenvalues of x and the decomposition (38.4) is the first spectral decomposition of x. If x is an element of a real Euclidean Jordan algebra $\mathcal{V}$ and has the first spectral decomposition $x = \lambda_1 c_1 + \lambda_2 c_2 + \cdots + \lambda_k c_k$ then the minimal polynomial of x is the polynomial p such that $p(x, \lambda) = \prod_{i=1}^{k} (\lambda - \lambda_i)$.

Theorem 38.2 ([4, p. 44]). *Let $\mathcal{V}$ be a real Euclidean Jordan algebra with rank$(\mathcal{V}) = r$. Then, for each x in $\mathcal{V}$ there exist a Jordan frame $\{c_1, c_2, \ldots, c_r\}$ and real numbers $\lambda_1, \ldots, \lambda_{r-1}$ and λ_r such that*

$$x = \lambda_1 c_1 + \lambda_2 c_2 + \cdots + \lambda_r c_r. \tag{38.5}$$

The numbers λ_j's (with their multiplicities) are uniquely determined by x.

The decomposition (38.5) is called the second spectral decomposition of x. Regard that the second spectral decomposition of x is not unique.

38.3 Strongly Regular Graphs

In this section we introduce the basic definitions and properties of strongly regular graphs. Additional information can be found in [7].

Herein we only consider non-empty, simple graphs (graphs with no loops nor parallel edges) and not complete graphs (graphs that have some non-adjacent pair of vertices).

Considering a graph G, we denote its vertex set by $V(G)$ and its edge set by $E(G)$. An edge of G with endpoints x and y is denoted by xy. In this case the vertices are called adjacent or neighbors. The number of vertices of G, $|V(G)|$, is called the order of G. If all vertices of G have k neighbors, then G is a k-regular graph.

Let G be a graph of order n. Then G is a (n, k, a, c)-strongly regular graph if it is k-regular and any pair of adjacent vertices have a common neighbors and any pair of non-adjacent vertices have c common neighbors. The parameters of a (n, k, a, c)-strongly regular graph are not independent and are related by the equality

$$k(k - a - 1) = (n - k - 1)c. \tag{38.6}$$

The adjacency matrix of G, $A = [a_{ij}]$, is a binary matrix of order n such that $a_{ij} = 1$, if the vertex i is adjacent to j and 0 otherwise. The adjacency matrix of a strongly regular graph satisfies the equation

$$A^2 = kI_n + aA + c(J_n - A - I_n),$$

where J_n is the all one matrix of order n.

It is well known (see, for instance, [7]) that the eigenvalues of a (n, k, a, c)-strongly regular graph G are k, θ and τ, where θ and τ are given by

$$\theta = \frac{a - c + \sqrt{(a - c)^2 + 4(k - c)}}{2} \quad \text{and} \quad \tau = \frac{a - c - \sqrt{(a - c)^2 + 4(k - c)}}{2},$$

and it's multiplicities can also be expressed in terms of k, a and c.

Equation (38.6) is an example of a condition that must be satisfied by the parameters of any strongly regular graph. Among the most important feasibility conditions there are the Krein conditions obtained in 1973 by Scott, Jr. [14], and the Absolute Bounds by Seidel [3]. However, there are still many parameter sets for which we do not know if they correspond to a strongly regular graph. In this work we deduce some new inequalities on the parameters and on the spectra of a strongly regular graph.

38.4   The Three Dimensional Euclidean Jordan Algebra $\mathscr{A}$

Throughout this section we consider the Euclidean Jordan algebra $\mathrm{Sym}(n, \mathbb{R})$ defined as in Example 38.3. Let G be a (n, k, a, c)-strongly regular graph such that $0 < c < k < n - 1$ and A be its adjacency matrix with three distinct eigenvalues, namely k, θ and τ. Herein, k and θ are the positive eigenvalues and τ is the negative eigenvalue. Now we consider the Euclidean Jordan subalgebra of $\mathrm{Sym}(n, \mathbb{R})$, $\mathscr{A}$, spanned by I_n, and the natural powers of A. Since A has three distinct eigenvalues, then $\mathscr{A}$ is a three dimensional real Euclidean Jordan algebra with $\mathrm{rank}(\mathscr{A}) = 3$. Let $\mathscr{B} = \{E_1, E_2, E_3\}$ be the unique complete system of orthogonal idempotents of $\mathscr{A}$ associated to A, with

$$E_1 = \frac{1}{n}I_n + \frac{1}{n}A + \frac{1}{n}(J_n - A - I_n),$$

$$E_2 = \frac{-\tau n + \tau - k}{n(\theta - \tau)}I_n + \frac{n + \tau - k}{n(\theta - \tau)}A + \frac{\tau - k}{n(\theta - \tau)}(J_n - A - I_n),$$

$$E_3 = \frac{\theta n + k - \theta}{n(\theta - \tau)}I_n + \frac{-n + k - \theta}{n(\theta - \tau)}A + \frac{k - \theta}{n(\theta - \tau)}(J_n - A - I_n).$$

Let $M_n(\mathbb{R})$ be the set of square matrices of order n with real entries. For $B = [b_{ij}]$, $C = [c_{ij}]$ in $M_n(\mathbb{R})$, we denote by $B \circ C = [b_{ij}c_{ij}]$ the Hadamard product of matrices B and C and by $B \otimes C = [b_{ij}C]$, the Kronecker product of matrices B and C (see [9]).

For B in $M_n(\mathbb{R})$ and for l in $\mathbb{N}$ we denote by $B^{\circ l}$ and $B^{\otimes l}$ the Hadamard power and the Kronecker power of order l of B, respectively, with $B^{\circ 1} = B$ and $B^{\otimes 1} = B$.

38.5   Inequalities on the Parameters of a Strongly Regular Graph

Proceeding in a similar way like in [12], we establish inequalities over the parameters and the spectra of a strongly regular graph G with adjacency matrix A by the analysis of the spectra of particular Hadamard convergent series obtained from the unique complete system of orthogonal idempotents of the Euclidean Jordan algebra $\mathscr{A}$ associated to A.

Let G be a (n, k, a, c) strongly regular graph with adjacency matrix A. Let $\mathscr{B} = \{E_1, E_2, E_3\}$ be the unique complete system of orthogonal idempotents of the Euclidean Jordan algebra $\mathscr{A}$ associated to A defined in the previous section. Consider the following spectral decomposition of A, $A = kE_1 + \theta E_2 + \tau E_3$. For l in $\mathbb{N}$, let:

$$S_{3l}^{\otimes} = \epsilon E_3^{\otimes 2} \otimes J_n^{\otimes(4l-4)} + \frac{\epsilon^3}{3!}E_3^{\otimes 6} \otimes J_n^{\otimes(4l-8)} + \cdots + \frac{\epsilon^{2l-1}}{(2l-1)!}E_3^{\otimes(4l-2)},$$

$$(38.7)$$

where each summand is a Kronecker product with $4l - 2$ factors and ϵ is a real positive number less than one. The sum $S_{3l}^{\otimes}$ has a principal submatrix given by:

$$S_{3l}^{\circ} = \epsilon E_3^{\otimes 2} \circ J_n^{\circ(4l-4)} + \frac{\epsilon^3}{3!} E_3^{\circ 6} \circ J_n^{\circ(4l-8)} + \cdots + \frac{\epsilon^{2l-1}}{(2l-1)!} E_3^{\circ(4l-2)}.$$

(38.8)

Since J_n is the identity for the Hadamard product of matrices, which is associative, it follows that

$$S_{3l}^{\circ} = \sum_{i=1}^{l} \frac{\epsilon^{2i-1}}{(2i-1)!} E_3^{\circ(4i-2)}.$$

Let q_{3l}^1, q_{3l}^2 and q_{3l}^3 be the real numbers such that $S_{3l}^{\circ} = \sum_{i=1}^{3} q_{3l}^i E_i$. Since the set

$$\mathscr{C} = \{E_{i_1} \otimes E_{i_2} \otimes \cdots \otimes E_{i_{4l-2}} : i_1, i_2, \ldots, i_{4l-2} \in \{1, 2, 3\}\}$$

is a complete system of orthogonal idempotents that is a basis of the real Euclidean Jordan subalgebra $\mathscr{A}^{\otimes(4l-2)}$ of the real Euclidean Jordan algebra $\mathrm{Sym}(n^{4l-2}, \mathbb{R})$ spanned by $I_n^{\otimes(4l-2)}$ and the natural powers of $A^{\otimes(4l-2)}$, then the minimal polynomial of $S_{3l}^{\otimes}$ is

$$p(\lambda) = (\lambda - 0) \prod_{i=1}^{l} (\lambda - \frac{\epsilon^{2i-1}}{(2i-1)!} n^{4l-4i}).$$

Note that, to obtain the minimal polynomial p, we use the complete system of orthogonal idempotents, $\mathscr{C}$, in each summand of (38.7) (see [4, p. 44]). Now, since the matrix (38.8) is a principal submatrix of $S_{3l}^{\otimes}$ and p is the minimal polynomial of $S_{3l}^{\otimes}$ then by the interlacing theorem (see [8, Theorem 4.3.15]), the eigenvalues of S_{3l}° are all nonnegative. Regarding that

$$S_{3l}^{\circ} = \sum_{j=1}^{l} \frac{\epsilon^{2j-1}}{(2j-1)!} \left(\frac{\theta n + k - \theta}{n(\theta - \tau)} \right)^{4j-2} I_n +$$

$$+ \sum_{j=1}^{l} \frac{\epsilon^{2j-1}}{(2j-1)!} \left(\frac{-n + k - \theta}{n(\theta - \tau)} \right)^{4j-2} A +$$

$$+ \sum_{j=1}^{l} \frac{\epsilon^{2j-1}}{(2j-1)!} \left(\frac{k - \theta}{n(\theta - \tau)} \right)^{4j-2} (J_n - A - I_n),$$

we conclude that

$$
S_{3l}^{\circ} = \sum_{j=1}^{l} \frac{\epsilon^{2j-1}}{(2j-1)!} \left[\left(\frac{\theta n + k - \theta}{n(\theta - \tau)} \right)^2 \right]^{2j-1} I_n +
$$

$$
+ \sum_{j=1}^{l} \frac{\epsilon^{2j-1}}{(2j-1)!} \left[\left(\frac{-n + k - \theta}{n(\theta - \tau)} \right)^2 \right]^{2j-1} A +
$$

$$
+ \sum_{j=1}^{l} \frac{\epsilon^{2j-1}}{(2j-1)!} \left[\left(\frac{k - \theta}{n(\theta - \tau)} \right)^2 \right]^{2j-1} (J_n - A - I_n).
$$

The series

$$
\sum_{i=1}^{\infty} \frac{\epsilon^{2i-1}}{(2i-1)!} \left((E_3)^{\circ 2} \right)^{\circ (2i-1)}
$$

is convergent with sum S_3,

$$
S_3 = \sinh \left(\sqrt{\epsilon} \frac{-\theta n + k - \theta}{n(\theta - \tau)} \right)^2 I_n + \sinh \left(\sqrt{\epsilon} \frac{-n + k - \theta}{n(\theta - \tau)} \right)^2 A +
$$

$$
+ \sinh \left(\sqrt{\epsilon} \frac{k - \theta}{n(\theta - \tau)} \right)^2 (J_n - A - I_n). \tag{38.9}
$$

Consider the real numbers $q_{3\infty}^1$, $q_{3\infty}^2$ and $q_{3\infty}^3$ such that

$$
S_3 = q_{3\infty}^1 E_1 + q_{3\infty}^2 E_2 + q_{3\infty}^3 E_3.
$$

Since $q_{3\infty}^1 = \lim_{l \to \infty} q_{3l}^1$, $q_{3\infty}^2 = \lim_{l \to \infty} q_{3l}^2$ and $q_{3\infty}^3 = \lim_{l \to \infty} q_{3l}^3$, and the eigenvalues of S_{3l}° are nonnegative, it follows that $q_{3\infty}^1$, $q_{3\infty}^2$ and $q_{3\infty}^3$ are also nonnegative. Then from (38.9) and doing some algebraic manipulation we obtain:

$$
q_{3\infty}^1 = \sinh \left(\sqrt{\epsilon} \frac{-\theta n + k - \theta}{n(\theta - \tau)} \right)^2 + \sinh \left(\sqrt{\epsilon} \frac{-n + k - \theta}{n(\theta - \tau)} \right)^2 k +
$$

$$
+ \sinh \left(\sqrt{\epsilon} \frac{k - \theta}{n(\theta - \tau)} \right)^2 (n - k - 1);
$$

$$
q_{3\infty}^2 = \sinh \left(\sqrt{\epsilon} \frac{-\theta n + k - \theta}{n(\theta - \tau)} \right)^2 + \sinh \left(\sqrt{\epsilon} \frac{-n + k - \theta}{n(\theta - \tau)} \right)^2 \theta +
$$

$$
+ \sinh \left(\sqrt{\epsilon} \frac{k - \theta}{n(\theta - \tau)} \right)^2 (-\theta - 1);
$$

$$q_{3\infty}^3 = \sinh\left(\sqrt{\epsilon}\,\frac{\theta n + k - \theta}{n(\theta - \tau)}\right)^2 + \sinh\left(\sqrt{\epsilon}\,\frac{-n + k - \theta}{n(\theta - \tau)}\right)^2 \tau +$$

$$+ \sinh\left(\sqrt{\epsilon}\,\frac{k - \theta}{n(\theta - \tau)}\right)^2 (-\tau - 1).$$

Now, we consider the matrix $S_{33} = E_3 \circ S_3$. Then $S_{33} = q_3^1 E_1 + q_3^2 E_2 + q_3^3 E_3$. The nonnegativity of the eigenvalues of S_{33}, q_3^i, $i \in \{1, 2, 3\}$, follows from the property

$$\lambda_{min}(A \circ B) \geq \lambda_{min}(A)\lambda_{min}(B), \tag{38.10}$$

for any matrices $A, B \in \mathcal{M}_n(\mathbb{R})$, (see [9, p. 312]), the nonnegativity of the parameters $q_{3\infty}^i$, $i \in \{1, 2, 3\}$, beyond the fact that E_3 is an idempotent matrix. It follows that

$$q_3^1 = \frac{\theta n + k - \theta}{n(\theta - \tau)} \sinh\left(\sqrt{\epsilon}\,\frac{\theta n + k - \theta}{n(\theta - \tau)}\right)^2$$

$$+ \frac{-n + k - \theta}{n(\theta - \tau)} \sinh\left(\sqrt{\epsilon}\,\frac{-n + k - \theta}{n(\theta - \tau)}\right)^2 k +$$

$$+ \frac{k - \theta}{n(\theta - \tau)} \sinh\left(\sqrt{\epsilon}\,\frac{k - \theta}{n(\theta - \tau)}\right)^2 (n - k - 1); \tag{38.11}$$

$$q_3^2 = \frac{\theta n + k - \theta}{n(\theta - \tau)} \sinh\left(\sqrt{\epsilon}\,\frac{\theta n + k - \theta}{n(\theta - \tau)}\right)^2$$

$$+ \frac{-n + k - \theta}{n(\theta - \tau)} \sinh\left(\sqrt{\epsilon}\,\frac{-n + k - \theta}{n(\theta - \tau)}\right)^2 \theta +$$

$$+ \frac{k - \theta}{n(\theta - \tau)} \sinh\left(\sqrt{\epsilon}\,\frac{k - \theta}{n(\theta - \tau)}\right)^2 (-\theta - 1);$$

$$q_3^3 = \frac{\theta n + k - \theta}{n(\theta - \tau)} \sinh\left(\sqrt{\epsilon}\,\frac{\theta n + k - \theta}{n(\theta - \tau)}\right)^2$$

$$+ \frac{-n + k - \theta}{n(\theta - \tau)} \sinh\left(\sqrt{\epsilon}\,\frac{-n + k - \theta}{n(\theta - \tau)}\right)^2 \tau +$$

$$+ \frac{k - \theta}{n(\theta - \tau)} \sinh\left(\sqrt{\epsilon}\,\frac{k - \theta}{n(\theta - \tau)}\right)^2 (-\tau - 1).$$

Regard that $q_3^i \geq 0$, for $i \in \{1, 2, 3\}$, constitute new inequalities over the parameters of a strongly regular graph. Furthermore, for parameter sets (n, k, a, c) that satisfy $k < n/2$, we obtain the following result.

Theorem 38.3. *Let G be a (n, k, a, c)-strongly regular graph, such that $0 < c < k < n - 1$, whose adjacency matrix has the eigenvalues k, θ and τ. If $k < n/2$ then*

$$k < (\theta + 1)^2 \theta \frac{n^2}{(n - 2(k - \theta))(n - k + \theta)}. \tag{38.12}$$

Proof. Let ϵ be a real positive number less than one. Since q_3^1 is nonnegative and $(k - \theta)(n - k - 1) = -(\theta n + k - \theta) - (-n + k - \theta)k$ rewriting (38.11) we deduce

$$0 \leq \frac{\theta n + k - \theta}{n(\theta - \tau)} \left[\sinh\left(\sqrt{\epsilon} \frac{-\theta n + k - \theta}{n(\theta - \tau)}\right)^2 - \sinh\left(\sqrt{\epsilon} \frac{k - \theta}{n(\theta - \tau)}\right)^2 \right] -$$

$$- \frac{(n - k + \theta)k}{n(\theta - \tau)} \left[\sinh\left(\sqrt{\epsilon} \frac{-n - k + \theta}{n(\theta - \tau)}\right)^2 - \sinh\left(\sqrt{\epsilon} \frac{k - \theta}{n(\theta - \tau)}\right)^2 \right]. \tag{38.13}$$

Applying the Lagrange Theorem in the right member of inequality (38.13) to the function $f(x) = \sinh(\epsilon x)$ on the interval $](k-\theta)^2/(n(\theta-\tau))^2, (\theta n + k - \theta)^2/(n(\theta - \tau))^2[$ and on the interval $](k-\theta)^2/(n(\theta-\tau))^2, (n-k+\theta)^2/(n(\theta-\tau))^2[$ we conclude

$$0 \leq \epsilon \frac{\theta n + k - \theta}{n(\theta - \tau)} \cosh\left(\sqrt{\epsilon} \frac{-\theta n + k - \theta}{n(\theta - \tau)}\right)^2 \frac{\theta n + 2(k - \theta)}{n(\theta - \tau)} \frac{\theta}{\theta - \tau} -$$

$$- \epsilon \frac{(n - k + \theta)k}{n(\theta - \tau)} \cosh\left(\sqrt{\epsilon} \frac{k - \theta}{n(\theta - \tau)}\right)^2 \frac{n - 2(k - \theta)}{n(\theta - \tau)} \frac{1}{\theta - \tau}. \tag{38.14}$$

Multiplying both members of inequality (38.14) by $1/\epsilon$ and applying limits to both members of the remaining inequality when ϵ tends to zero, we obtain the inequality (38.15):

$$0 \leq \frac{(\theta n + k - \theta)(\theta n + 2(k - \theta))\theta}{n^2} - \frac{k(n - k + \theta)(n - 2(k - \theta))}{n^2}. \tag{38.15}$$

Finally because $(\theta n + k - \theta)/n < (\theta + 1)$, and since $k < n/2$ we conclude that $(\theta n + 2(k - \theta))/n < (\theta + 1)$, thus obtaining

$$0 < (\theta + 1)^2 \theta - \frac{k(n - k + \theta)(n - 2(k - \theta))}{n^2},$$

and therefore $k < (\theta + 1)^2 \theta n^2 / ((n - 2(k - \theta))(n - k + \theta))$. □

Remark 38.2. From inequality (38.12) we conclude that if G is a (n, k, a, c) strongly regular such that $0 < c < k < n - 1$ with the eigenvalues k, θ and τ then for a fixed value of $k < n/2$, θ cannot be too small relatively to the value k. This assertion is stronger for values $k \ll n/2$. For example, when $k < n/4$, from (38.12) we conclude that $k < \frac{8}{3}(\theta + 1)^2\theta$, and therefore for a sufficiently small value of θ this inequality will fail.

Using similar arguments as before to obtain the matrix S_{33}, we consider the matrix $S_{22} = E_2 \circ S_2$, where

$$
\begin{aligned}
S_2 &= \sum_{i=1}^{\infty} \frac{\epsilon^{2i-1}}{(2i-1)!} \left((E_2)^{\circ 2} \right)^{\circ(2i-1)} \\
&= \sinh\left(\sqrt{\epsilon} \, \frac{|\tau|n + \tau - k}{n(\theta - \tau)} \right)^2 I_n + \sinh\left(\sqrt{\epsilon} \frac{n + \tau - k}{n(\theta - \tau)} \right)^2 A + \\
&\quad + \sinh\left(\sqrt{\epsilon} \frac{\tau - k}{n(\theta - \tau)} \right)^2 (J_n - A - I_n).
\end{aligned}
\tag{38.16}
$$

Then $S_{22} = q_2^1 E_1 + q_2^2 E_2 + q_2^3 E_3$. By property (38.10) and since the parameters $q_{2\infty}^1, q_{2\infty}^2$, and $q_{2\infty}^3$ are nonnegative and E_2 is an idempotent matrix we conclude that the eigenvalues q_2^i of S_{22} for $i \in \{1, 2, 3\}$ are also nonnegative. It follows that the eigenvalues q_2^i for i in $\{1, 2, 3\}$ have the expressions presented below.

$$
\begin{aligned}
q_2^1 &= \frac{|\tau|n + \tau - k}{n(\theta - \tau)} \sinh\left(\sqrt{\epsilon} \frac{|\tau|n + \tau - k}{n(\theta - \tau)} \right)^2 \\
&\quad + \frac{n + \tau - k}{n(\theta - \tau)} \sinh\left(\sqrt{\epsilon} \frac{n + \tau - k}{n(\theta - \tau)} \right)^2 k + \\
&\quad + \frac{\tau - k}{n(\theta - \tau)} \sinh\left(\sqrt{\epsilon} \frac{\tau - k}{n(\theta - \tau)} \right)^2 (n - k - 1);
\end{aligned}
\tag{38.17}
$$

$$
\begin{aligned}
q_2^2 &= \frac{|\tau|n + \tau - k}{n(\theta - \tau)} \sinh\left(\sqrt{\epsilon} \frac{|\tau|n + \tau - k}{n(\theta - \tau)} \right)^2 \\
&\quad + \frac{n + \tau - k}{n(\theta - \tau)} \sinh\left(\sqrt{\epsilon} \frac{n + \tau - k}{n(\theta - \tau)} \right)^2 \theta + \\
&\quad + \frac{\tau - k}{n(\theta - \tau)} \sinh\left(\sqrt{\epsilon} \frac{\tau - k}{n(\theta - \tau)} \right)^2 (-\theta - 1);
\end{aligned}
$$

$$
q_2^3 = \frac{|\tau|n + \tau - k}{n(\theta - \tau)} \sinh\left(\sqrt{\epsilon} \frac{|\tau|n + \tau - k}{n(\theta - \tau)} \right)^2
$$

$$+ \frac{n+\tau-k}{n(\theta-\tau)} \sinh\left(\sqrt{\epsilon}\frac{-n+\tau-k}{n(\theta-\tau)}\right)^2 \tau +$$

$$+ \frac{\tau-k}{n(\theta-\tau)} \sinh\left(\sqrt{\epsilon}\frac{\tau-k}{n(\theta-\tau)}\right)^2 (-\tau-1).$$

Again, note that $q_2^i \geq 0$, for $i \in \{1,2,3\}$, gives us new inequalities over the parameters of a strongly regular graph. In Theorem 38.3 we presented an inequality for parameter sets (n,k,a,c) with $k < n/2$. Next we have a complementary result for $k > n/2$.

Theorem 38.4. *Let G be a (n,k,a,c)-strongly regular graph, such that $0 < c < k < n-1$, whose adjacency matrix has the eigenvalues k, θ and τ. If $k > n/2$ then*

$$k < (|\tau|-1)\tau^2 \frac{n^2}{(n+\tau-k)(-n+2(k-\tau))}. \tag{38.18}$$

Proof. Since q_2^1 is nonnegative and since $(\tau-k)(n-k-1) = -(|\tau|n+\tau-k) - (n+\tau-k)k$, rewriting (38.17) we deduce that

$$0 \leq \frac{|\tau|n+\tau-k}{n(\theta-\tau)}\left[\sinh\left(\sqrt{\epsilon}\frac{|\tau|n+\tau-k}{n(\theta-\tau)}\right)^2 - \sinh\left(\sqrt{\epsilon}\frac{\tau-k}{n(\theta-\tau)}\right)^2\right] -$$

$$- \frac{(n+\tau-k)k}{n(\theta-\tau)}\left[\sinh\left(\sqrt{\epsilon}\frac{\tau-k}{n(\theta-\tau)}\right)^2 - \sinh\left(\sqrt{\epsilon}\frac{-n+\tau-k}{n(\theta-\tau)}\right)^2\right].$$

Applying the Lagrange Theorem to the function $f(x) = \sinh(\epsilon x)$ on the interval $](\tau-k)^2/(n(\theta-\tau))^2, (|\tau|n+\tau-k)^2/(n(\theta-\tau))^2[$ and on the interval $](\tau-k)^2/(n(\theta-\tau))^2, (n+\tau-k)^2/(n(\theta-\tau))^2[$ we obtain the inequality (38.19).

$$0 \leq \epsilon\frac{|\tau|n+\tau-k}{n(\theta-\tau)}\cosh\left(\sqrt{\epsilon}\frac{|\tau|n+\tau-k}{n(\theta-\tau)}\right)^2 \frac{|\tau|n+2(\tau-k)}{n(\theta-\tau)}\frac{|\tau|}{\theta-\tau} -$$

$$- \epsilon\frac{(n+\tau-k)k}{n(\theta-\tau)}\cosh\left(\sqrt{\epsilon}\frac{-n+\tau-k}{n(\theta-\tau)}\right)^2 \frac{-n+2(k-\tau))}{n(\theta-\tau)}\frac{1}{\theta-\tau}.$$

$$\tag{38.19}$$

Multiplying by $1/\epsilon$ both members of inequality (38.19) and applying limits to both members of the remaining inequality when ϵ tends to zero, we obtain the inequality (38.20):

$$0 \leq \frac{(|\tau|n+\tau-k)(|\tau|n+2(\tau-k))|\tau|}{n^2} - \frac{(n+\tau-k)k(-n+2(k-\tau))}{n^2}.$$

$$\tag{38.20}$$

Table 38.1 Numerical
results when $k < n/2$

	P_1	P_2	P_3	P_4	P_5
θ	1	4	4	1	3
τ	-221	-56	-61	-11	-57
$q^1_{\theta\tau kn}$	-456.5	-60.1	-43.9	-4.0	-112.7

Table 38.2 Numerical
results when $k > n/2$

	P_6	P_7	P_8	P_9	P_{10}
θ	60	123	86	80	143
τ	-5	-3	-4	-4	-2
$q^2_{\theta\tau kn}$	-96.4	-699.0	-428.0	-499.7	-476.3

Now since $(|\tau|n + \tau - k)/(n(\theta - \tau)) < (|\tau| - 1)/(\theta - \tau)$ and $(|\tau|n + 2(\tau - k))/(n(\theta - \tau)) < |\tau|/(\theta - \tau)$ then

$$0 < (|\tau| - 1)\tau^2 - \frac{k(n + \tau - k))(-n + 2(k - \tau))}{n^2}$$

and therefore $k < (|\tau| - 1)\tau^2 n^2/((n + \tau - k)(-n + 2(k - \tau)))$.   □

Remark 38.3. From inequality (38.18) we conclude that if G is a (n, k, a, c) strongly regular such that $0 < c < k < n - 1$ with the eigenvalues k, θ and τ then for a fixed value of $k > n/2$, the value of $|\tau|$ cannot be too small relatively to the value of k. This assertion is stronger for values of $k \gg n/2$.

38.6 Experimental Results and Conclusions

In this section we present in Table 38.1 some examples of parameter sets (n, k, a, c) that do not verify the inequality (38.12) of Theorem 38.3. We consider the parameter sets $P_1 = (1{,}296, 481, 40, 260)$, $P_2 = (1{,}288, 312, 36, 88)$, $P_3 = (1{,}275, 364, 63, 120)$, $P_4 = (63, 22, 1, 11)$ and $P_5 = (936, 255, 30, 84)$. For each example we have $k < n/2$ and we present the respective eigenvalues θ, τ and the value of

$$q^1_{\theta\tau kn} = (\theta + 1)^2\theta n^2/((n - 2(k - \theta))(n - k + \theta)) - k,$$

defined from (38.12) in Theorem 38.3. Next, in Table 38.2, we present some examples of parameter sets (n, k, a, c) that do not verify the inequality (38.18) of Theorem 38.4. We consider the parameter sets $P_6 = (1{,}275, 910, 665, 610)$, $P_7 = (1{,}296, 861, 612, 492)$, $P_8 = (1{,}296, 860, 598, 516)$, $P_9 = (1{,}275, 896, 652, 576)$ and $P_{10} = (841, 520, 375, 234)$. For each example we have $k > n/2$ and we present the respective data as in Table 38.1 but in the last line we compute the value of

$$q^2_{\theta\tau kn} = (|\tau| - 1)\tau^2 \frac{n^2}{(n + \tau - k)(-n + 2(k - \tau))} - k,$$

defined from (38.18) in Theorem 38.4.

From the data in Table 38.1 we confirm the result of Theorem 38.12 that when $k < n/2$ then θ cannot be too small relatively to the fixed value of k and from the data in Table 38.2 we confirm the result of Theorem 38.18 that when $k > n/2$ then $|\tau|$ cannot be to small relatively to the value of k.

Acknowledgements Vasco Mano and Enide A. Martins were supported by Portuguese funds through the CIDMA—Center for Research and Development in Mathematics and Applications, and the Portuguese Foundation for Science and Technology ("FCT—Fundação para a Ciência e a Tecnologia"), within project PEst-OE/MAT/UI4106/2014. Enide Martins was also supported by Project PTDC/MAT/112276/2009.

Luís Vieira was partially funded by the European Regional Development Fund Through the program COMPETE and by the Portuguese Government through the FCT—Fundação para a Ciência e a Tecnologia under the project PEst C/MAT/UI0144/2013.

References

1. Bose, R.C.: Strongly regular graphs, partial geometries and partially balanced designs. Pac. J. Math. **13**, 389–419 (1963)
2. Cardoso, D.M., Vieira, L.A.: Euclidean Jordan algebras with strongly regular graphs. J. Math. Sci. **120**, 881–894 (2004)
3. Delsarte, Ph., Goethals, J.M., Seidel, J.J.: Bounds for system of lines and Jacobi polynomials. Philips Res. Rep. **30**, 91–105 (1975)
4. Faraut, J., Korányi, A.: Analysis on Symmetric Cones. Oxford Mathematical Monographs. Clarendon Press, Oxford (1994)
5. Faybusovich, L.: Euclidean Jordan algebras and interior-point algorithms. Positivity **1**, 331–357 (1997)
6. Faybusovich, L.: Linear systems in Jordan algebras and primal-dual interior-point algorithms. J. Comput. Appl. Math. **86**, 149–175 (1997)
7. Godsil, C., Royle, G.: Algebraic Graph Theory. Chapman & Hall, New York (1993)
8. Horn, R.A., Johnson, C.R.: Matrix Analysis. Cambridge University Press, Cambridge (1985)
9. Horn, R.A., Johnson, C.R.: Topics in Matrix Analysis. Cambridge University Press, Cambridge (1991)
10. Jordan, P., Neuman, J.V., Wigner, E.: On an algebraic generalization of the quantum mechanical formalism. Ann. Math. **35**, 29–64 (1934)
11. Koecher, M.: The Minnesota Notes on Jordan Algebras and Their Applications. Springer, Berlin (1999)
12. Mano, V.M., Vieira, L.A.: Admissibility conditions and asymptotic behavior of strongly regular graph. Int. J. Math. Models Methods Appl. Sci. Methods **5**(6), 1027–1033 (2011)
13. Massam, H., Neher, E.: Estimation and testing for lattice condicional independence models on Euclidean Jordan algebras. Ann. Stat. **26**, 1051–1082 (1998)
14. Scott, L.L., Jr.: A condition on Higman parameters. Notices Am. Math. Soc. **20**, A-97 (1973)
15. Vieira, L.A.: Euclidean Jordan algebras and inequalities on the parameters of a strongly regular graph. AIP Conf. Proc. **1168**, 995–998 (2009)

Chapter 39
Financial, Real, and Quasi Options: Similarities and Differences

Justus Wesseler

39.1 Introduction

"Es sei hier nur noch erwähnt, dass die Bachelierschen Betrachtungen jeder mathematischen Strenge gänzlich entbehren" [17, p. 417]. This quote refers to a comment of Andrei Nikolajewitsch Kolmogoroff on the works of Louis Bachelier. What is remarkable about this quote is that it has been published in 1930s but it took more than 60 years to recover the contribution of Bachelier [9] to the evaluation of financial and real assets and to appreciate the contributions of among others Albert Einstein, Adriaan Fokker, Andrei N. Kolmogoroff, Max Planck, and Norbert Wiener for evaluating financial and real options. They laid the foundations for evaluating the movement of particles under uncertainty. The interest of Albert Einstein was not to describe the precise place of a molecule but the probability that a molecule would be at a certain place at a certain time considering its initial position [12]. The mathematics have been further developed by Max Planck and Adriaan Fokker. The Fokker-Planck equation describes the evolution of a probability distribution over time. A similar result has been obtained by Kolmogoroff and known as the Kolmogoroff forward or backward equation. These equations have become a central tool for deriving analytical as well as for developing numerical solutions for investments under uncertainty (e.g. [10, 38]). An important building block of models has been the Wiener Process, named after Norbert Wiener, who formalized

J. Wesseler (✉)
Technische Universität München, München, Germany
e-mail: justus.wesseler@wzw.tum.de

A.A. Pinto and D. Zilberman (eds.), *Modeling, Dynamics, Optimization and Bioeconomics I*, Springer Proceedings in Mathematics & Statistics 73, DOI 10.1007/978-3-319-04849-9_39,

random walks more rigorous than Einstein did,[1] while first known applications of real options at least date back to the ancient Greeks [6].

While all these developments did happen in the field of mathematics and physics it took until the late 1960s that these methods had been picked up by economists to first evaluate the prices of financial assets under uncertainty where the price of an asset can be seen as being equivalent to a particle in physics. It took again about 10 more years before a number of papers did appear to use the same mathematical tools to evaluate real instead of financial assets. In the early 1970s Kenneth Arrow and Anthony Fisher did publish their seminal paper on valuing environmental preservation under uncertainty and irreversibility [2]. In the same year Claude Henry published his paper on investment and uncertainty and the irreversibility effect [13]. Both, the Arrow and Fisher as well as Henry contribution point out that irreversibility effects create a bias towards delayed investment in comparison to assessments that do not consider uncertainty, irreversibility, and flexibility in decision making explicitly. Arrow and Fisher call the size of the bias the quasi option value while Henry calls it the irreversibility effect.

The main message it that even so the expected value of an investment under uncertainty is positive, the value of the investment considering postponement might be even larger—implying that the profit maximizing strategy is to postpone the investment. This is similar to the evaluation of a financial call option. Exercising a call option might be profitable, the option is "in the money", but further waiting to exercise the option can increase profits.

In the following the three approaches, the financial, real, and quasi option approach will be presented in a discrete time discrete state model. In Sect. 39.3 the three approaches will be compared. The differences and similarities will be illustrated using a numerical example. Section 39.4 discusses applications and challenges for modeling in particular with respect to the bioeconomy as well as an outlook for future research while Sect. 39.5 concludes.

39.2 The Three Approaches

39.2.1 The Financial Call Option

A financial call option gives the holder of the call option the right but not the obligation to buy a financial instrument S, $S : [0, T] \rightarrow \Re$ at time t_0, expiring at time T, $T \in \Re^+$ with an exercise price K, $K \in \Re$, T at a given price, C, today, t_0 [21]. The call can only be exercised at maturity date T (European Call Option), the price movement can either be up, u, or down, d, with probabilities q

---

[1]The Wiener process is a Markov process with a normal distributed variance that increases linear in time.

and $1 - q$. Therefore,

$$C_u = \max[0, uS - K] \quad \text{with probability} \quad q$$

$$\nearrow$$

$$C$$

$$\searrow$$

$$C_d = \max[0, dS - K] \quad \text{with probability} \quad 1 - q.$$

The question to be answered in the context of this paper is how much a potential holder of the call option should pay for the call option today.

Following Cox et al. [8] the "fair price" of the call option will be:

$$C = [\rho C_u + (1 - \rho)C_d]/(1 + r), \tag{39.1}$$

with $\rho \equiv \frac{(1+r)-d}{u-d}$ and $1 - \rho \equiv \frac{u-(1+r)}{u-d}$, r the riskless interest rate, $r \in \Re^+$ over the period $0 \to T$, $u - 1 > 0$ the upward move of the stock price and $d - 1 < 0$ the downward movement of the stock price, $C_u = \max[0, uS - K]$ and $C_d = \max[0, dS - K]$, while the probability of an up-ward move is q and the probability of a downward move is $1 - q$. The result of Eq. 39.1 is obtained by assuming that risks in the movement of the financial instrument can be hedged using a portfolio of riskless bonds and n shares of the financial instrument S. Since the value of the portfolio depends on S, it matches the risk of the call. ρ and (1- ρ) change if a dividend equivalent to $nS(r - 1)$ will be paid. In that case $\hat{\rho} \equiv \frac{1-d}{u-d}$ and $1 - \hat{\rho} \equiv \frac{u-1}{u-d}$. The remarkable result of Eq. 39.1 is that the "fair price" of the call option is independent of the probabilities of the upward, q, or downward $(1 - q)$ movement of the price of the financial instrument.[2] If investors agree on the size of the upward and downward movement, and, the riskless interest rate is the same everyone, then all investors would price the call the same, independently of their attitudes towards risk. This is a noteworthy property which will be relevant when the real option and quasi option value approach will be discussed.

39.2.2   The Real Option Value

The valuing of call options on financial instruments has been translated to the valuation of investments under uncertainty and flexibility. An investment opportunity has properties similar to those of a call option. An investor has the right but not the obligation to invest. The question is whether or not to exercise the option immediately, or to postpone and decide at a later point in time whether or

---

[2]This not necessarily applies to dividend paying financial instruments.

not to invest. To introduce the problem, consider a simple investment where the value of the investment option depends on the movement of the net product price, p, $p \in \mathfrak{R}^+$, of the product to be produced. Investment option I, $I : [0,1] \rightarrow \mathfrak{R}$ at time t_0, expiring at time t_1, $T \in \mathfrak{R}^+$ and $T \rightarrow \infty$. The investment option can only be exercised today, t_0, or at maturity t_1. The product price p_0, can either move up with size u, $u - 1 > 0$ and probability q, or move down with size d, $d - 1 < 0$ and probability $1 - q$. The value of the immediate investment is $V_0 = p_0 + q\frac{up_0}{r} + (1 - q)\frac{dp_0}{r}$. The value of the investment opportunity if one has to invest at t_0 is $\Omega_0 = \max[V_0 - I, 0]$. The value of postponed investment to t_1 is $V_1^u = \frac{up_0}{r}(1 + r)$ or $V_1^d = \frac{dp_0}{r}(1 + r)$ and the value of investment at t_1 is $F_1 = \max[V_1 - I, 0]$. V_1 and F_1 are random variables form the perspective at t_0 and the value at t_0 is $E[F_1] = \{(q \max[V_1^u - I, 0] + (1 - q) \max[V_1^d - 1, 0])/(1 + r)\}$ and the optimal decision to be taken at t_0 is $F_0 = \max[V_0 - I, E(F_1)]$. Assuming $I < \frac{dp_0}{r}(1 + r) + p_0$ it pays to invest immediately. In case $\frac{dp_0}{r}(1 + r) + p_0 \leq I < \frac{up_0}{r}(1 + r)$ it pays to delay and decide after uncertainty has been resolved whether or not to invest. Gains from waiting arise as long as $\frac{dp_0}{r}(1 + r) + p_0 \leq I < \frac{up_0}{r}(1 + r)$ as in this case by Jensen's Inequality $-I + p_0 + q\frac{up_0}{r} + (1 - q)\frac{dp_0}{r} < q(-\frac{I}{1+r} + \frac{up_0}{r})$.

39.2.3 The Quasi Option Value

The quasi-option value approach originates from the paper by Arrow and Fisher [2]. The basic question being asked if whether or not converting a piece of land with amenity values in a different form of use such as e.g. housing when future benefits from preservation as well as development are uncertain but uncertainty be resolved over time generates opportunity costs that are not captured by standard cost-benefit-analysis using expected values of uncertain future benefits from preservation as well as development. They show a bias towards development exists, if the assessment will be based on expected values. The bias reduces the opportunity costs of development and Arrow and Fisher name the bias *quasi option value*. The bias is a result of ignoring that as time passes new information arrives and uncertainty about future states of nature will be reduced. The model is a bit more complex than the real option model presented as benefits from development say investment as well as benefits from preservation say non-investment are considered while in the basic real option model presented benefits from non-investment are zero.

Two future states of nature are considered, A_1 and A_2 with the future denoted as t_1. If A_1 occurs development is the better option while if A_2 occurs preservation is the better one. If A_2 occurs but development has been chosen at t_0 the decision cannot be reversed to preservation. The benefits and costs from development include onetime development costs either c_0 or c_1 and present and future benefits b_{d0} and b_{d1}. The preservation option includes only present and future benefits b_{p0} and b_{p1}. The net-present-value of the opportunity to either preserve or develop at t_0, NPV_0 includes the following mutually exclusive payment streams where all values are expressed in present values:

$$\max NPV = \begin{cases} b_{d0} - c_{d0} + b_{d1} \\ b_{p0} + b_{d1} - c_{d1} \\ b_{p0} + b_{p1}. \end{cases} \qquad (39.2)$$

Immediate development would take place if $b_{d0} - c_{d0} - b_{p0} + b_{d1} - \max\{(b_{d1} - c_{d1}), b_{p1}\} > 0$. As the future will be uncertain, benefits and costs at t_1 can be replaced by their expected value resulting in $b_{d0} - c_{d0} - b_{p0} + E[b_{d1}] - \max\{E[(b_{d1} - c_{d1})], E[b_{p1}]\} > 0$. This approach would be appropriate as the standard method applied in cost-benefit analysis, if new information is unavailable. If the arrival of new information can be used, the maximum of the following two alternatives will be the optimal decision:

$$\max NPV = \begin{cases} b_{d0} - c_{d0} + E[b_{d1}] & \text{develop immediately} \\ b_{p0} + \max\{E[(b_{d1} - c_{d1})], E[b_{p1}]\} & \text{postpone decision.} \end{cases} \qquad (39.3)$$

The development option will be chosen if $b_{d0} - c_{d0} + E[b_{d1}] - b_{p0} - \max\{E[(b_{d1} - c_{d1})], E[b_{p1}]\} > 0$. Now the decision under uncertainty and irreversibility including and excluding future information can be compared. The difference yields: $\max\{E[(b_{d1} - c_{d1})], E[b_{p1}]\} - \max\{E[(b_{d1} - c_{d1})], E[b_{p1}]\}$. Again, by Jensen's Inequality this difference is larger than or equal to zero. This difference is the quasi option value that needs to be considered for an appropriate assessment of an investment that includes irreversibilities and uncertainties.

39.3 A Comparison

For comparing the three models a numerical example illustrating similarities and differences will be used. The numerical examples use the parameter values of Chap. 2 in [10] used as well in [11] and [25]. For the purposes of comparison, the *nomenclatura* by Dixit and Pindyck will be used. The equivalent *nomenclatura* used either within financial economics or the quasi option literature has been listed in Table 39.1. The irreversible investment costs I are 1,600. The current price p at $t = 0$, p_0 is 200. The future price after one period at $t = 1$, p_1, is either in the case of an upward jump, u, $p_1^u = 300$ or in the case of a downward jump, d, $p_1^d = 100$. For simplicity, it is assumed that p reflects the net-price, revenues minus reversible costs and that the prices will stay constant until infinity after the end of period one. The probability of an upward jump q is equivalent to the probability of a downward jump $1 - q$ with $q = 1 - q = 0.5$. The discount rate r will be constant with $r = 0.1$. Hence, the value of the investment, if exercised today will be

Table 39.1 Numerical example comparing the three approaches

Numerical example	FOV	ROV	QOV
1,600	Exercise price	Irreversible investment	Development costs
2,200	Current stock price	Current project value, V	Benefits from immediate development
3,300	Future stock price (underlying) high	Future Price of underlying asset high	Future benefits from development high
1,100	Future stock price (underlying) low	Future Price of underlying asset low	Future benefits from development low
773	Value of the call	Value of the option to invest	Value of the development opportunity considering irreversibility and arrival of additional information
600	Intrinsic option value	Value of immediate investment, $V - I$	Benefits from immediate development
173	Option time value	Value of waiting/flexibility	
227			QOV

$$V_0 = p_0 + q \sum_{t=1}^{\infty} \frac{p_1^u}{(1+r)^t} + (1-q) \sum_{t=1}^{\infty} \frac{p_1^d}{(1+r)^t}$$

$$= p_0 + q \frac{p_1^u}{r} + (1-q) \frac{p_1^d}{r} = 200 + 1500 + 500 = 2,200.$$

The value of the investment at $t = 1$ in the case of a price increase will be $V_1^u = \sum_{\tau=0}^{\infty} \frac{p_1^u}{(1+r)^\tau} = 3,300$ and in the case of a price decrease $V_1^d = \sum_{\tau=0}^{\infty} \frac{p_1^d}{(1+r)^\tau} = 1,100$. The expected value of an immediate investment is $NPV_i = V_0 - I = 600$. The expected value of a postponed investment valued at $t = 0$ is $NPV_{p,i} = q(V_1^u - I)/(1+r) = 733$). The difference $NPV_{p,i} - NPV_i = 173$ is the value of waiting or in the terminology of financial options the option time value.

In the quasi-option approach of Arrow and Fisher the quasi option value is $E[\max\{0, NPV_{p,i}^u\}] - \max\{0, E[q(NPV_{p,i}^u) + (1-q)(NPV_{p,i}^d)]\}$ yielding $0.5 \cdot 1,545.45 - 0.5(1,545.45 - 454.54) = 227.23$.[3] The quasi option value is higher than the value of waiting. The difference between the quasi option value and the value of waiting yields: $(1-q)NPV_{p,i}^d - (qNPV_{p,i}^u - NPV_i)$. Collecting terms and simplifying yields $QOV - VOW = \frac{I}{1+r} - I + p_0$. Hence, the difference is the difference in foregone benefits and costs by a postponed investment as pointed out by Mensink and Requate [20]. This includes two components, the benefits arising when immediate investments being made p_0 but reduced by the savings on

---

[3]Note, the opportunity costs in this example are considered to be zero.

investment costs that can be made by investing later, $\frac{rI}{1+r}$. The foregone benefits and costs have been considered within the real option value approach but not explicitly in the quasi option value approach. In case there are no immediate benefits, i.e. $p_0 = 0$ and decisions will be made continuously, $t \to 0$, ROV and VOW will be equivalent. In case decisions will be considered at incremental time steps the QOV needs to be reduced by foregone benefits and costs to yield the same result for the irreversibility effect within the real option value and quasi option value approach. The foregone benefits and costs can be negative in case $p_0 < \frac{rI}{1+r}$. In this case, the VOW will be larger than the QOV. For the numerical example, setting $p_0 = 0$ results in a value of waiting of about 418 while the quasi option value remains the same.[4] The foregone benefits as well as the savings in investment costs would be captured by standard benefit-cost-analysis comparing delayed with immediate investment and what matters are the gains from additional information.

Applying the financial option pricing, ρ and $(1 - \rho)$ change if a dividend equivalent to nSr will be paid. In that case $\hat{\rho} = \frac{1-d}{u-d}$ and $1 - \hat{\rho} = \frac{u-1}{u-d}$. Using the evaluation of financial options with $u = 0.5$, $d = 1.5$, $i = 0.1$, $\hat{\rho} = (1 - \hat{\rho}) = 0.5$, $C_u = 3,300 - 1,600 = 1,700$, and $C_d = 0$, yields a value of $C = 1,700 \cdot 0.5/1.1 = 773$. If dividend payments will be included, then the value of the call as well as the real option value will depend on the size of q as q has an effect on the expected rate of price changes. A higher q in this case will increase the probability of immediate investment.

39.4 Modifications, Applications, and Outlook

The three approaches discussed have been presented in discrete time discrete state. Most applications deviate from discrete time, discrete state by analyzing investments in a continuous time, continuous state framework. Model applications include not only irreversible costs but also irreversible benefits, optimal abandonment, entry and exit, uncertainty over several variables such as reversible and irreversible costs and benefits, discount rates, and many more as discussed in more detail in recent reviews by Mezey and Conrad [22] and Perrings and Brock [23]. In the following, the discrete time discrete state model discussed in Sect. 39.3 will be presented in continuous time continuous state by choosing as an example the introduction of a new technology.

39.4.1 An Illustrative Case: Introducing Transgenic Crops

Consider a decision maker or a decision making body, similar to an EU Agency, or, the United States Environmental Protection Agency (USEPA) that has the authority

---

[4]Please not in this case $V_p^u = 3,000$ and $V_p^d = 1,000$.

to decide whether or not a particular transgenic crop, e.g. a toxin producing crop like Bt-corn,[5] should be released for commercial planting. The agency can approve an application for release or postpone the decision. The objective of the agency is to maximize the welfare of consumers living in the economy and ignore positive and negative transboundary effects. The supply for all transgenic crops is perfectly elastic and demand perfectly inelastic per unit of time. Ex-ante effects of the decision to release transgenic crops on the up-stream sector of the economy are ignored by the agency. The welfare effect of releasing a specific transgenic crop can be described as the net-present-value from T until infinity of the additional net benefits at the farm level, V, which will be further defined below, minus the difference between irreversible costs, I, and irreversible benefits, R. R and I are assumed to be known and constant.

This is a useful simplification for two reasons. Firstly, not much is known about the magnitude of irreversible costs I. As will be shown later, the model can be solved for the irreversible costs and provide information about an acceptable level, which can then be compared with available information. Secondly, information about the irreversible damages from pesticide use on a per hectare level, which are the irreversible benefits of planting transgenic crops, R, are available and can easily be included into the model.

As the agency has the possibility to postpone the decision on whether or not to release the transgenic crop, the agency has to maximize the value resulting from this decision, $F(V)$, to maximize the welfare. This objective can be described as maximizing the expected value from releasing the transgenic crop:

$$\max F(V) = \max E[(V_T - (I - R)) \, e^{-\rho T}] \tag{39.4}$$

where E is the expectation operator, T the unknown future point in time when the transgenic crop is released into the environment and ρ the discount rate.

As the release of a transgenic crop has almost no effect on the fixed costs, the net-benefits from a transgenic crop at farm level for a specific region are the total sum of gross margins over all farms. The welfare effect at farm level, hence, is the difference between the sums of gross margins from transgenic crops minus the total sum of gross margins from the alternative non-transgenic crop (further called conventional crop). From now on this difference will be called the additional net-benefit from transgenic crops B. The instantaneous additional net-benefit, B, at time T, B_T, is then the difference in gross margin between the transgenic and traditional crops. The gross margin for each crop type is defined as the difference between the revenues and variable costs at T. Other additional benefits arising from the application of the new technology, such as, e.g., through "peace of mind", are assumed to be balanced by concerns about the new technology, on average, and, therefore are ignored.

---

[5]Modified corn that produces the δ-endotoxins of the soil bacterium Bacillus thuringiens which control the European Corn Boxer.

The benefits and costs used to calculate B_T are those that are reversible. The instantaneous additional net-benefits under the given level of information are known with certainty. Future additional net-benefits are uncertain as new information about prices, costs and yields arrives continuously. The uncertainty can be modelled by choosing a stochastic process that describes the future development path of the additional net-benefits B_T.

The geometric Brownian motion has frequently been used to model uncertain returns from agricultural crops [28], returns from pig-raising [24] and on-farm investments [16, 29, 39]. Richards and Green [30] suggest decomposing returns from agricultural crops. They model crop prices as a geometric Brownian motion and crop yields as a geometric Brownian motion combined with a Poisson process, where the geometric Brownian motion represents "normal" years and the Poisson process years with extreme yields. Because additional net-benefits are chosen as the stochastic variable, we assume that extreme yields are smoothed, and, hence, a decomposition of prices and yields is not necessary.

A mean-reverting process could also model additional net-benefits where it is assumed that additional net-benefits decrease over time. The decrease could be explained e.g. by the observation that pests are becoming resistant to plant produced pesticides and weeds to broadband herbicides. Wesseler [34] compares the results of modelling additional net-benefits with a geometric Brownian motion and a mean-reverting process and shows that the different processes could result in different decisions.[6] This leads to the problem of identifying the relevant process. The identification of the relevant process based on time series data is difficult, as the results are ambiguous [27]. Dixit and Pindyck [11] therefore recommend identifying the process based on theoretical arguments.

This case uses the geometric Brownian motion to model the uncertain future additional net-benefits, for the following two reasons: firstly, it is reasonable to assume that technical change for a transgenic crop will be continuous and secondly, the process is analytically tractable.

The geometric Brownian motion is a non-stationary continuous time stochastic process with Markov properties where α is the constant drift rate, σ the constant variance rate and dz the Wiener process, with $E(dz) = 0$ and $E(dz)^2 = dt$:

$$dB = \alpha B dt + \sigma B dz. \tag{39.5}$$

The geometric Brownian motion is the limit of a random walk [7]; hence it is consistent with assuming log-normality of the stochastic variable with zero drift. The expected value of this process grows at the rate α. The use of the geometric Brownian motion also assumes that B_t will not turn negative, which is similar for continuous differentiation of the process at the boundary of zero (see [11], Eq. (39.17), p. 191). This assumes that growers will immediately stop (start) planting without having to bear additional costs as soon as they realize the gross

---

[6]See also [31] for similar results.

margin will turn out to be negative (positive). Which is a reasonable assumption, as farmers can and do easily move from one crop variety to another.

When today's additional net-benefits, B_T, are known, follow a geometric Brownian motion until infinity and are discounted at μ (the risk adjusted rate of return derived from the capital asset pricing model (CAPM), then the expected present value of additional net-benefits from transgenic crops, V_T at time $t = T$ is:

$$E[V_T] = B_T \int_{t=0}^{\infty} e^{(\alpha-\mu)t} dt = \frac{B_T}{\mu - \alpha}. \tag{39.6}$$

As V_T is a constant multiple of B_T, also V_T follows a geometric Brownian motion with the same drift parameter α and variance parameter σ. If speculative bubbles are ruled out and as $V(0) = 0$, Eq. (39.3) will also be the value of releasing transgenic crops into the environment. Equation (39.4) can then be rewritten:

$$\max F(B) = \max E\left[\left(\frac{B_T}{\mu - \alpha} - (I - R)\right) e^{-\rho T}\right].$$

As the irreversible costs I and the irreversible benefits R are assumed to be constant, the option pricing approach using contingent claim analysis as described by Dixit and Pindyck (1994, Chap. 5)[11] can be applied. This results in the standard second order differential or Fokker-Planck equation, which has to be solved:

$$\frac{1}{2}\sigma^2 B^2 F''(B) + (r - \delta)BF'(B) - rF(B) = 0. \tag{39.7}$$

A solution to this homogenous second order differential equation is:

$$F(B) = A_1 B^{\beta_1} + A_2 B^{\beta_2}. \tag{39.8}$$

Solving Eq. (39.8) according to the boundary conditions (Dixit and Pindyck 1994) provides the following solutions:

$$B^* = \frac{\beta_1}{\beta_1 - 1}\delta(I - R) \tag{39.9}$$

$$A_2 = 0 \tag{39.10}$$

$$A_1 = \frac{(\beta_1 - 1)^{\beta_1 - 1}}{(I - R)^{\beta_1 - 1}(\delta\beta_1)^{\beta_1}} \quad \text{with} \tag{39.11}$$

$$\beta_1 = \frac{1}{2} - \frac{r - \delta}{\sigma^2} + \sqrt{\left(\frac{r - \delta}{\sigma^2} - \frac{1}{2}\right)^2 + \frac{2r}{\sigma^2}} > 1, \tag{39.12}$$

where B^* is the optimal level of additional net-benefits B, r the risk-free rate of return, δ the convenience yield, which is the difference between the risk adjusted discount rate μ and the growth rate α, σ the variance parameter of the geometric Brownian motion of equation (39.2), and β_1 the positive root of the quadratic equation (39.7), in the following called β for short.

Equation (39.9) says it is optimal to release a transgenic crop into the environment immediately, if the additional net-benefits B_T are equal to the with δ annualised difference between irreversible cost and irreversible benefits multiplied by the so called "hurdle rate" or option multiplier $\beta/(\beta - 1)$.

Equation (39.9) can be rearranged to:

$$I^* = R + \frac{B_T}{\delta} / \frac{\beta}{\beta - 1} = R + \frac{B_T}{\delta} - \frac{B_T/\delta}{\beta}, \qquad (39.13)$$

where I^* are the maximum incremental social tolerable irreversible costs of releasing a transgenic crop into the environment. This shifts the attention from the additional net-benefits to the irreversible cost. The irreversible costs are now the critical variable, whereas the additional net-benefits are assumed to be known. This is a more reasonable expression, as far more information is available about the additional net-benefits from field trails, releases of similar crops or from other countries. Equation (39.13) can be formulated as a rule that the agency should follow when it has to decide whether or not a transgenic crop should be released:

> Postpone the release of a transgenic crop into the environment, if the irreversible costs are higher than the irreversible benefits plus the present value of an infinite stream of instantaneous additional net-benefits, using the convenience yield as the relevant discount rate, divided by the hurdle rate.

This rule has two important properties, which result out of the use of the contingent claim analysis (see Appendix). Firstly, future costs and benefits have been discounted using rates provided by the market. No individual discount rates have been used. Secondly, uncertainty about the additional net-benefits has been included by using a riskless hedge portfolio and, hence, the evaluation of the benefits is independent of attitudes towards risk, which reduces the impact of risk-preferences on decision-making.

The last formulation of the maximum incremental social tolerable irreversible costs in Eq. (39.13) illustrates the effect of waiting due to uncertainty and irreversibility. The first two terms, R and B/δ, illustrate the results of the orthodox approach. Without recognizing explicitly irreversibility and uncertainty, the benefits are the sum of the irreversible benefits plus the present value of infinite additional net-benefits. By including irreversibility and uncertainty, a proportion of the present value of infinite additional net-benefits, $\frac{B_T}{\delta}/\beta$, has to be deducted. This proportion in this context can be interpreted as the economic value of uncertainty and irreversibility of releasing transgenic crops.

The maximum incremental social tolerable irreversible costs as explained in Eq. (39.13) will change over time with new information about additional net-benefits. These changes will not only consist of changes in yields but also of changes in product prices and variable costs due to regulatory and other polices.

These policies will have either an increasing or decreasing effect on I^*. An increase (decrease) in I^* can be seen as an increase (decrease) in the likelihood to release transgenic crops earlier, as the higher (lower) the maximum incremental social tolerable irreversible costs are the lower (higher) the chances that they will be crossed. The impact of changes in the growth rate α and the standard deviation σ on the annualised hurdle rate are illustrated in Table 39.1. An increase in the growth rate α increases the maximum incremental social tolerable irreversible costs as the first derivative of I^* with respect to α is positive (proof in Appendix 2):

$$\frac{\partial I^*}{\partial \alpha} = \frac{\partial (B/\delta)}{\partial \alpha} \frac{\beta}{\beta - 1} + \frac{B}{\delta} \frac{\partial ((\beta - 1)/\beta)}{\partial \alpha} > 0 \tag{39.14}$$

$$= \frac{\partial (B/\delta)}{\partial \alpha} \frac{\beta}{\beta - 1} + \frac{B}{\delta} \beta^{-2} \frac{\partial \beta}{\partial \alpha} > 0.$$

The overall effect can be decomposed into two effects. The first term on the right-hand-side of Eq. (39.14) shows the impact on current additional net-benefits B, which is positive. An increase in α reduces the discounting effect, increases total benefits, increases the maximum incremental social tolerable irreversible costs and hence, increases the probability of an earlier release. The second term on the right-hand-side reduces the effect, as the partial derivative of β with respect to α is negative. This is the effect of a higher growth rate on the option value. An increase in the growth rate increases the value of releases in the future, which increases the value of the option to release at a later point in time and hence increases the probability of a later release. As the effect on the present value is greater then the effect on the future value the overall effect is positive as mentioned earlier.

An increase in the uncertainty of additional net-benefits has the opposite effect, as the impact of an increase in the variance parameter σ on I^* is negative as the partial derivative of β with respect to σ is negative (proof in Appendix 2):

$$\frac{\partial I^*}{\partial \sigma} = \frac{B}{\delta} \frac{\partial ((\beta - 1)/\beta)}{\partial \sigma} < 0 \tag{39.15}$$

$$= \frac{B}{\delta} \beta^{-2} \frac{\partial \beta}{\partial \sigma} < 0.$$

An increase in uncertainty decreases the likelihood of an early release, as the future benefits increase while future losses can be avoided by waiting. This is the standard result from the literature on financial economics.

A change in the risk-free rate of return, r, also has a negative impact on the maximum incremental social tolerable irreversible costs I^* (proof in Appendix 2):

$$\frac{\partial I^*}{\partial r} = \frac{B}{\delta} \frac{\partial ((\beta - 1)/\beta)}{\partial r} < 0 \tag{39.16}$$

$$= \frac{B}{\delta} \beta^{-2} \frac{\partial \beta}{\partial r} < 0.$$

The decreasing effect of an increase in the risk-free rate of return can be explained by the decrease of the opportunity costs of the option to release transgenic crops (see Appendix 1).

Also of interest is a simultaneous change in the growth and the variance rate. Considering Young's theorem this can be modelled by getting the derivative of I^* with respect to α and σ (proof in Appendix 2):

$$\frac{\partial^2 I^*}{\partial\sigma\,\partial\alpha} = \frac{\partial(B/\delta)}{\partial\alpha}\beta^{-2}\frac{\partial\beta}{\partial\sigma} + \frac{B}{\delta}\frac{\partial\left(\beta^{-2}\frac{\partial\beta}{\partial\sigma}\right)}{\partial\alpha} < 0. \tag{39.17}$$

The first term of Eq. (39.17) shows the change the growth rate α has on the current additional net benefits, which is positive and multiplied by the negative effect of σ on β. Hence, the total effect of the first term on I^* is negative. This negative effect is augmented by the second term also being negative. The overall impact of a simultaneous marginal change is a decrease in the maximal tolerable irreversible costs I^*. The positive effect of an increase in the growth rate on the likelihood of an earlier release is surpassed by the negative effect of an increase in uncertainty on the likelihood of an earlier release.

The continuous time continuous state result of the simple investment problem presented above provides well-known results (see e.g. [11, 19]). While considering uncertainty over one or more variables as long as they follow the same stochastic process can often still be solved analytically, most models have to be solved numerically. One of the major building blocks has been the Wiener process. Other processes include jump processes to consider drastic environmental changes or ex-post liabilities, mean-reverting processes for deviations from long-term equilibria, Brownian bridges, and more as discussed in detail in a number of text books such as [1, 14, 32] or [11].

Problems can be solved either by using a dynamic programming or a contingent claim approach. The major problem within the dynamic programming approach is the right choice of the discount rate or discount rates. This is not a trivial issue as the debate about climate change policies illustrates. Within the contingent claim analysis this less of a problem as market are used, but the problem will be the identification of the appropriate matching portfolio that replicates the uncertainty of the investment under consideration, i.e. the quasi option value. Nevertheless within the debate about environmental problems such as the conservation of biological diversity this would be a promising approach as it would allow to identify the "fair" market price of biological diversity.

In general, for applications related to the bioeconomy whether at the micro or at the macro level benefits and costs have to be differentiated between reversible and irreversible benefits and costs. Also, for many assessments a differentiation between private and external benefits and costs is useful and in particular if also sustainability issues are of concern [37].

In particular the postponement of investments has in recent years become an issue of importance. The costs of delays caused by regulations can be substantially undervalued if forgone benefits are irreversible [36]. Economists in general agree that the optimal level of regulations is where marginal benefits equal the marginal costs of regulations [3]. The calculation of marginal benefits and marginal costs will be complicated if uncertainties and irreversibilities need to be considered, which holds for almost all regulations in food production. At the micro level in particular regulatory issues such as labeling requirements and different production standards become increasingly important resulting in new forms of contractual arrangements [33]. These arrangements generally include ex-ante regulations as well as ex-post liability rules in case non-compliance happens. The optimal design of those arrangements often includes irreversible ex-ante compliance costs while ex-post liability follows a jump process. Both can be combined to model the economics of contractual arrangements allowing for more detailed insights about incentives to participate in new contractual arrangements as well as the incentives to comply with the arrangements (see e.g. [5]). One of the major insights from that literature is that irreversible ex-ante regulatory costs provide incentives to delay adoption of more stringent regulations as irreversible investment costs can be delayed and costs avoided if future benefits will be low but that also the size of the firm will be important if irreversible regulatory costs increase nonlinear with firm size.

At the macro level during the last decade the concept of genuine investment as an indicator for sustainable development has been proposed [4]. Yet, the concept does not consider possible irreversible benefits and costs and uncertainty of genuine investments explicitly. This will be another fruitful area of research. First attempts in that direction can be found in [37] and [35]. Within a real option framework not the value of the economy as measured by genuine investment but the value of an economy as measured by the option value and hence changes in option values would be the relevant indicator for sustainable development. In this context future opportunities become important and for policy makers the major question will be if their policies increase or decrease the option value.

Some criticism has been raised against stressing the importance of irreversibilities as in the end all costs are irreversible. This will be correct if decisions are made within continuous time but this is hardly the case. Take agriculture as an example. Decisions about the quantity of hog production are made on about a 6 month basis. These decisions are reversible while the specific investment in the pig barn cannot be reversed after a 6 month period. There will be substantial losses involved if the barn has been constructed to last for several years bit production closes after 1 year. In crop production on annual basis seed expenditures can be considered reversible as the crops have been harvested within a year and production choice can be adjusted if economic circumstances change in favor corn instead of wheat. Investments in seeds and pesticides can be considered irreversible within the cropping season. Decisions on pesticide use are made under uncertainty as future pest and disease problems are not know with certainty. They depend on future whether conditions, what neighboring farmers are doing and more. The expenditures for pesticides within a cropping season can be considered as irreversible. Analyzing optimal pesticide use within a real option framework helps to explain why farmers rationally use less pesticides than standard benefit-costs-analysis would suggest [18].

39.5 Conclusion

Three approaches have evolved to model investments under uncertainty, irreversibility, and flexibility. The methods allow considering irreversibilities and uncertainties of a decision explicitly and enable researchers to recognize the risk associated with the investment at the theoretical level. Not only irreversible costs but also irreversible benefits matter as discussed by e.g. [37] and [26]. Including irreversible benefits and costs into the benefit-cost framework results in a different decision rule in comparison to the standard deterministic neoclassical framework. This is now well known in the literature on real options.

A comparison between the real and quasi option approach shows the quasi option value does not include foregone benefits and costs of a delayed investment, but is a measure of the economic value of uncertain information.

The decision rule for investments using contingent claim analysis allows solutions to be derived that are independent of risk and time preference. Individuals that are highly concerned about a new technology and those who are not, but both want to maximize their income, would come to the same conclusion about the timing of introduction. The risk-adjusted rate of return μ derived from the CAPM in the example depends on the risk free interest rate r and the market price of risk; hence the optimal decision to release transgenic crops is not independent of changes in interest rates.

The effects of policies on the timing of investments were analysed in a two-step procedure. First, the impacts on model parameters were identified and then the effect of the parameter changes on the maximum incremental social tolerable irreversible costs. The most counterintuitive result was the increase in the likelihood of an earlier investment with a decrease in additional net-benefits. This is explained by the opposite impact a simultaneous change in the growth rate and the variance rate has on the maximum incremental social tolerable irreversible costs.

Future applications within the evaluation of genuine investments and analysing the effect of changes in government regulations on compliance incentives as well as the optimal design of regulation are fruitful areas for future research in this domain.

Appendix 39.1: Solving for F(B) Using Contingent Claim Analysis as Explained by Dixit and Pindyck (Chap. 5, pp. 150–152)

Assuming that an asset or a portfolio of assets exists that allows the risk of the additional net-benefits to be tracked, the arbitrage pricing principle can be applied to value the portfolio that includes the additional benefits from transgenic crops [25]. In this case, a portfolio can be constructed consisting of the option to release transgenic crops into the environment, $F(B)$, and a short position of $n = F'(B)$ units of the additional benefits of transgenic crops. The value of this portfolio is

$\Phi = F(B) - F'(B)B$. A short position will require a payment to the holder of the corresponding long position of $\delta F'(B)Bdt$. The total return from holding this portfolio over a short time interval $(t, t + dt)$ holding $F'(B)$ constant will be:

$$d\Phi = dF(B) - F'(B)dB - \delta BF'(B)dt. \tag{39.18}$$

Applying Itô's Lemma[7] to $dF(B)$, equating the return of the risk less portfolio to the risk free rate of return $r[F(B) - F'(B)B]dt$ and rearranging terms results in the following differential equation:

$$\frac{1}{2}\sigma^2 B^2 F''(B) + (r - \delta)BF'(B) - rF(B) = 0. \tag{39.19}$$

A solution to this homogenous second order differential equation is:

$$F(B) = A_1 B^{\beta_1} + A_2 B^{\beta_2}. \tag{39.20}$$

As the value of the option to release transgenic crops into the environment is worthless, if there are no additional net-benefits, A_2 has to be 0. The other boundary conditions are the 'value matching', Eq. (39.8), and 'smooth pasting', Eq. (39.9), conditions[8]:

$$F(B^*) = V(B^*) - I + R \tag{39.21}$$

$$F'(B^*) = V'(B^*). \tag{39.22}$$

Solving Eq. (39.7) according to the boundary conditions provides the following solutions:

$$B^* = \frac{\beta_1}{\beta_1 - 1}\delta(I - R) \tag{39.23}$$

$$A_1 = \frac{(\beta_1 - 1)^{\beta_1 - 1}}{(I - R)^{\beta_1 - 1}(\delta\beta_1)^{\beta_1}} \quad \text{with} \tag{39.24}$$

$$\beta_1 = \frac{1}{2} - \frac{r - \delta}{\sigma^2} + \sqrt{\left(\frac{r - \delta}{\sigma^2} - \frac{1}{2}\right)^2 + \frac{2r}{\sigma^2}} > 1. \tag{39.25}$$

---

[7]See, e.g., Sect. 22 [15] for an introduction of Itô-stochastic processes.

[8]The value matching condition sustains that the value of the option to release the transgenic crop is equivalent to the value of immediate release. The smooth pasting condition says that at the point of value matching a marginal change in the value of the option to release the transgenic crop has to be equal to a marginal change in the value of immediate release [11, pp. 130–132].

Appendix 39.2: Proof of the Results of the Partial Derivatives

To improve the readability of the equations the following notation will be introduced:

$$v = \sqrt{\left(\frac{r - \delta}{\sigma^2} - \frac{1}{2}\right)^2 + \frac{2r}{\sigma^2}} \qquad (39.26)$$

$$\chi = \left(\frac{r - \delta}{\sigma^2} - \frac{1}{2}\right) \qquad (39.27)$$

$$\theta = \frac{r - \delta}{\sigma^2}. \qquad (39.28)$$

The following assumptions will be made for the proofs:

1. $B, r, \mu, \sigma > 0$.
2. $\mu - \alpha = \delta > 0$.
3. $\beta = -\chi + v > 1$.

Proof. 1: $\frac{\partial I^*}{\partial \alpha} > 0$:

$$\frac{\partial I^*}{\partial \alpha} = \frac{B}{\delta^2} - \frac{B}{\delta^2}\beta^{-1} - \frac{B}{\delta}\beta^{-2}\frac{\partial \beta}{\partial \alpha} > 0 \qquad (39.29)$$

$$\frac{B}{\delta^2} - \frac{B}{\delta^2}\beta^{-1} - \frac{B}{\delta}\beta^{-2}\frac{\partial \beta}{\partial \alpha} > 0$$

$$\Rightarrow \frac{1}{\delta} - \frac{1}{\delta\beta} > \frac{1}{\beta^2}\frac{\partial \beta}{\partial \alpha}. \qquad (39.30)$$

The left-hand-side of Eq. (39.30) is positive as δ is positive and $\beta > 1$. Equation (39.29) would be correct if the right-hand-side of Eq. (39.30) is negative, that is if $\partial \beta / \partial \alpha < 0$.

$$\frac{\partial \beta}{\partial \alpha} = -\frac{1}{\sigma^2} + \frac{\chi}{v\sigma^2} < 0 \qquad (39.31)$$

$$-\frac{1}{\sigma^2} + \frac{\chi}{v\sigma^2} < 0$$

$$\Rightarrow -\chi + v > 0$$

$$\Rightarrow \beta > 0.$$

□

Proof. 2: $\frac{\partial I^*}{\partial \sigma} < 0$:

$$\frac{\partial I^*}{\partial \sigma} = \frac{B}{\delta} \beta^{-2} \frac{\partial \beta}{\partial \sigma} < 0. \qquad (39.32)$$

Equation (39.32) will be correct if $\partial \beta / \partial \sigma < 0$:

$$\frac{\partial \beta}{\partial \sigma} = \frac{2(r - \delta)}{\sigma^3} - \frac{2\chi \left(\frac{r-\delta}{\sigma^3} \right) + \frac{2r}{\sigma^3}}{\upsilon} < 0. \qquad (39.33)$$

Equation (39.33) can be rearranged to:

$$\frac{(r - \delta)\upsilon - \chi(r - \delta) - r}{\upsilon \sigma^3} < 0. \qquad (39.34)$$

Equation (39.34) will be correct if the nominator is negative as the denominator is always positive:

$$(r - \delta)\upsilon - \chi(r - \delta) - r < 0 \qquad (39.35)$$

$$\Rightarrow \qquad (r - \delta)(-\chi + \upsilon) < r$$

$$\text{or} \qquad \beta(r - \delta) < r.$$

This has to hold for the case $(r - \delta) < 0$ and $(r - \delta) > 0$.
 For the case $(r - \delta) < 0$ follows that $\beta > \frac{r}{r-\delta}$ is correct as $\beta > 1$.
 For the case $(r - \delta) > 0$ Eq. (39.35) can be rearranged to:

$$\upsilon < \frac{r}{(r - \delta)} + \chi. \qquad (39.36)$$

Equation (39.36) can be rearranged to:

$$\left(\frac{r}{r - \delta} \right)^2 + \frac{r}{\sigma^2} - \frac{r}{r - \delta} > 0. \qquad (39.37)$$

Equation (39.37) is correct if $\frac{r}{r-\delta} > 1$. This holds if $(r - \delta) > 0$. □

Proof. 3:

$$\frac{\partial^2 I^*}{\partial \alpha \partial \sigma} < 0:$$

Using the result of Eq. (39.32) and differentiation according to σ provides:

$$\frac{\partial^2 I^*}{\partial \alpha \partial \sigma} = \frac{B}{\delta^2} \beta^{-2} \frac{\partial \beta}{\partial \sigma} + \frac{B}{\delta} \left(\frac{\partial^2 \beta}{\partial \alpha \partial \sigma} \beta^{-2} - 2\beta^{-3} \frac{\partial \beta}{\partial \alpha} \frac{\partial \beta}{\partial \sigma} \right) < 0. \qquad (39.38)$$

Equation (39.38) can easily be rearranged to

$$\frac{B}{\delta} > 2\beta^{-3}\frac{\partial\beta}{\partial\alpha}. \tag{39.39}$$

This is correct if $\partial\beta/\partial\alpha < 0$ which has already been proven. □

Proof. 4: $\frac{\partial I^*}{\partial r} < 0$:

$$\frac{\partial I^*}{\partial r} = \frac{B}{\delta}\beta^{-2}\frac{\partial\beta}{\partial r} < 0. \tag{39.40}$$

Equation (39.40) is correct if $\frac{\partial\beta}{\partial r} < 0$.

$$\frac{\partial\beta}{\partial r} = -\frac{1}{\sigma^2} + \frac{\partial\upsilon}{\partial r} < 0$$

$$-\frac{1}{\sigma^2} + \frac{\partial\upsilon}{\partial r} < 0$$

$$\Rightarrow \quad -\frac{1}{\sigma^2} + \frac{\chi+1}{\upsilon} < 0$$

$$\Rightarrow \quad \chi + 1 < \upsilon$$

$$\Rightarrow \quad \beta > 1.$$

□

References

1. Amram, M., Kulatilaka, N.: Real Options. Harvard Business School Press, Boston (1999)
2. Arrow, K., Fisher, A.: Environmental preservation, uncertainty, and irreversibility. Q. J. Econ. **88**, 312–319 (1974)
3. Arrow, K.J., Cropper, M.L., Eads, G.C., Hahn, R.W., Lave, L.B., Noll, R.G., Portney, P.R., Russell, M., Schmalensee, R., Smith, V.K., Stavins, R.N.: Is there a role for benefit-cost analysis in environmental, health, and safety regulation? Science **272**, 221–222 (1996)
4. Arrow, K.J., Dasgupta, P., Goulder, L.H., Mumford, K.J., Oleson, K.: Sustainability and the measurement of wealth. Environ. Dev. Econ. **17**, 317–353 (2012)
5. Beckmann, V., Soregaroli, C., Wesseler, J.: Ex-ante regulation and ex-post liability under uncertainty and irreversibility: governing the coexistence of gm crops. economics: the open-access. Open-Assessment E-J. (2010). http://dx.doi.org/10.5018/economics-ejournal.ja. 2010-9
6. Bernstein, P.L.: Capital Ideas: The Improbable Origins of Modern Wall Street. Wiley, Hoboken (2005)
7. Cox, D., Miller, H.: The Theory of Stochastic Processes. Chapman and Hall, London (1965)
8. Cox, J.C., Ross, S.A., Rubinstein, M.: Option pricing: a simplified approach. J. Fin. Econ. **7**, 229–263 (1979)

9. Davis, M., Etheridge, A.: Louis Bachelier's Theory of Speculation: The Origins of Modern Finance. Princeton University Press, Princeton (2005)
10. Dixit, A., Pindyck, R.S.: Investment Under Uncertainty. Princeton University Press, Princeton (1994)
11. Dixit, A., Pindyck, R.S.: The options approach to capital investment. Harv. Bus. Rev. (May–June) 105–115 (1995)
12. Einstein, A.: Concerning the motion, as required by the molecular-kinetic theory of heat, of particles suspended in liquids at rest. Annalen der Physik **17**, 549–560 (1905)
13. Henry, C.: Investment decision under uncertainty: the irreversibility effect. Am. Econ. Rev. **64**, 1006–1012 (1974)
14. Hull, J.C.: Options, Futures and Other Derivatives: Global Edition, 8th edn. Pearson, Essex (2011)
15. Kamien, M.I., Schwartz, N.L.: Dynamic Optimization. Amsterdam, North-Holland (1991)
16. Khanna, M., Isik, M., Winter-Nelson, A.: Investment in site-specific crop management under uncertainty: implications for nitrogen pollution control and environmental policy. Agr. Econ. **24**, 9–21 (2000)
17. Kolmogoroff, A.N.: Über die analytischen Methoden in der Wahrscheinlichkeitsrechnung. Mathematische Annalen **104**(1), 415–458 (1931)
18. Ndeffo Mbah, M.L., Forster, G., Wesseler, J., Gilligan, C.: Economically optimal timing of crop disease control under uncertainty: an options approach. J. R. Soc. Interface **7**(51), 1421–1428 (2010)
19. McDonald, R., Siegel, D.: The value of waiting to invest. Q. J. Econ. **101**, 707–728 (1986)
20. Mensink, P., Requate, T.: The Dixit-Pindyck and the Arrow-Fisher-Hanemann-Henry option values are not equivalent: a note on Fisher. Resource Energy Econ. **27**, 83–88 (2000)
21. Merton, R.C.: Application of option pricing theory: twenty-five years later. Am. Econ. Rev. **88**, 323–349 (1998)
22. Mezey, E.W., Conrad, J.M.: Real options in resource economics. Annu. Rev. Resour. Econ. **2**, 33—52 (2010)
23. Perrings, C., Brock, W.: Irreversibility in economics. Annu. Rev. Resour. Econ. **1**, 219–238 (2009)
24. Pietola, K.S., Wang, H.H.: The value of price- and quantity-fixing contracts for piglets in Finland. Eur. Rev. Agr. Econ. **27**(4), 431–447 (2000)
25. Pindyck, R.S.: Irreversibility, uncertainty, and investment. J. Econ. Lit. **29**, 1340–1351 (1991)
26. Pindyck, R.S.: Irreversibilities and the timing of environmental policy. Resource Energy Econ. **22**, 233–259 (2000)
27. Pindyck, R.S., Rubinfeld, D.: Econometric Models and Economic Forecasts, 3rd edn. McGraw-Hill, New York (1991)
28. Price, J., Wetzstein, M.: Irreversible investment decisions in perennial crops with yield and price uncertainty. J. Agr. Resource Econ. **24**(1), 173–185 (1999)
29. Purvis, A., Boggess, W.G., Moss, C.B., Holt, J.: Adoption of emerging technologies under output uncertainty: an ex-ante approach. Am. J. Agr. Econ. **77**(3), 541–551 (1995)
30. Richards, T.J., Green, G.P.: Economic hysteresis in variety selection. J. Agr. Appl. Econ. **35**, 1–14 (2003)
31. Slade, M.E.: Valuing managerial flexibility: an application of real-option theory to mining investments. J. Environ. Econ. Manag. **41**, 193–233 (2001)
32. Trigeorgis, L.: Real Options. MIT, Cambridge (1996)
33. Weaver, R., Wesseler, J.: Monopolistic pricing power for transgenic crops when technology adopters face irreversible benefits and costs. Appl. Econ. Lett. **11**(15), 969–973 (2005)
34. Wesseler, J.: Assessing the risk of transgenic crops - the role of scientific belief systems. In: Matthies, M., Malchow, H., Kriz, J. (eds.) Integrative Systems Approaches to Natural and Social Sciences - Systems Science, pp. 319–327. Springer, Berlin (2001)
35. Wesseler, J. The Santaniello theorem of irreversible benefits. AgBioForum **12**(1), 8–13 (2009)
36. Wesseler, J., Zilberman, D.: The economic power of the Golden Rice opposition. Environ. Dev. Econ. (2014). doi:10.1017/S1355770X1300065X

37. Wesseler, J., Scatasta, S., Nillesen, E.: The Maximum Incremental Social Tolerable Irreversible Costs (MISTICs) and other benefits and costs of introducing transgenic maize in the EU-15. Pedobiologia **51**(3), 261–269 (2007)
38. Wilmott, Paul.: The Mathematics of Financial Derivatives. Cambridge University Press, Cambridge (1995)
39. Winter-Nelson, A., Amegbeto, K.: Option values to conservation and agricultural price policy: application to terrace construction in Kenya. Am. J. Agr. Econ. **80**(2), 409–418 (1998)

Chapter 40
A Review and New Contribution on Conic Multivariate Adaptive Regression Splines (CMARS): A Powerful Tool for Predictive Data Mining

Fatma Yerlikaya-Özkurt, İnci Batmaz, and Gerhard-Wilhelm Weber

Abbreviations

ACCR	Average Correct Classification Rate
ANNs	Artificial Neural Networks
AUC	Area under ROC
BB1;BB2;BB3	Box-Behnken
BCMARS	Bootstrapping CMARS
BFs	Basis Functions
CART	Classification and Regression Trees
C&C	Communities and Crime
CCS	Concrete Compressive Strength
CIs	Confidence Intervals
CMARS	Conic (Continuous, Convex) Multivariate Adaptive Regression Splines
CQP	Conic Quadratic Problem
CST	Concrete Slump Test
CV	Cross-Validation
CVR	Congressional Voting Records
DM	Data Mining
D-Opt	D optimal
DRM	Dynamic Regression Model

F. Yerlikaya-Özkurt • G.-W. Weber
Institute of Applied Mathematics, Middle East Technical University, 06800 Ankara, Turkey
e-mail: fatmayerlikaya@gmail.com; gweber@metu.edu.tr

İ. Batmaz
Department of Statistics, Middle East Technical University, 06800 Ankara, Turkey
e-mail: ibatmaz@metu.edu.tr

A.A. Pinto and D. Zilberman (eds.), *Modeling, Dynamics, Optimization and Bioeconomics I*, Springer Proceedings in Mathematics & Statistics 73, DOI 10.1007/978-3-319-04849-9__40,
© Springer International Publishing Switzerland 2014

EPT	Electricity prices of Turkey
ER	Error Rate
FCC	Face Center Cube
FF	Forest Fires
FPR	False Positive Rate
GAMs	Generalized Additive Models
GAMs & CQP	Generalized Additive Models with Conic Quadratic Programming
GCV	Generalized Cross-Validation
HEPA	Hepatitis
ICD	International Credit Default
IKL	Infinite Kernel Learning
IP	Infinite Programming
IPMs	Interior Points Methods
LSD	Least Significant Difference
MAE	Mean Absolute Error
MARS	Multivariate Adaptive Regression Splines
MAUC	Mean AUC
MC	Metal Casting
MD	Minitab Data
MER	Mean ER
MKL	Multiple Kernels Learning
MLR	Multiple Linear Regression
MSE	Mean Squared Error
NSY	Noisy Data
PID	Pima Indians Diabetes
PM10	PM10
PRESS	Predicted Error Sum of Squares
PRSS	Penalized Residual Sum of Squares
PT	Parkinsons Telemonitoring
PWI	Proportion of Residuals within Three Sigma
R^2	Multiple Coefficient of Determination
RANOVA	Repeated Analysis of Variance
RCMARS	Robust CMARS
RMSE	Root Mean Squared Error
RO	Robust Optimization
ROC	Receiver Operating Characteristic Curve
RSS	Residual Sum of Square
RWQ	Red Wine Quality
Std Dev Error	Standard Deviation of Error
SVMs	Support Vector Machines
TPR	True Positive Rate
TR	Tikhonov Regularization
US	Uniform Sampling

40.1 Introduction

Recent developments in computer hardware and software provide environments to automatically collect gross amount of data from various sources. Data Mining (DM) methods enable us to analyze such data stored in various data repositories gathered for different purposes such as quality improvement in industry [17] or developing early warning systems [3] in various fields such as finance, environment, energy and so on. There is a number of predictive DM tools useful for establishing mathematical or statistical relationships between several factors of interest. Some widely used ones are Classification and Regression Trees (CART), Artificial Neural Networks (ANNs), Multiple Linear Regression (MLR), Support Vector Machines (SVMs), and so on. Among them *Multivariate Adaptive Regression Splines (MARS)* is a well-known predictive DM method capable of modeling high-dimensional data with nonlinear structure [11]. The flexible nature of MARS modeling leads to successful implementation of the method in various application areas [11]. Reported success of the method attracted the attention of many researchers to tackle with the problems of MARS method or alternatively extend it further to improve its capability. In one of these studies, *Conic Multivariate Adaptive Regression Splines (CMARS)* is developed as an alternative to the backward stepwise part of the MARS algorithm. Later on, to assess the power of the method, several real-life as well as simulation data applications are implemented. Also, CMARS is rigorously evaluated and compared with some other predictive DM methods such as CART, IKL and GAMs for classification and MLR and MARS for prediction. On the other hand, in other studies, some attempts are made to reduce the complexity of CMARS method by using bootstrapping technique. This leads to a new method named as Bootstrapping CMARS (BCMARS) [41, 42]. In addition, Robust Optimization (RO) is employed, and Robust CMARS (RCMARS) is developed [28, 40] to enhance the capability of CMARS method to be able to handle not only fixed but also random input and output variables. Hence, the main purpose of this study is to sum up the findings and conclusions of these studies to critically report the current achievements as well as pitfalls of the CMARS method. The review reveals that CMARS seems to be a very powerful DM tool, and deserves further interest to make it more powerful than ever.

The chapter is organized as follows. In Sect. 40.2, the development of CMARS method is presented. Background on evaluating model performances is given in Sect. 40.3. In Sect. 40.4, data sets used in comparative studies of CMARS are explained. Various applications of CMARS and its comparison with the other methods are summarized in Sect. 40.5. In Sect. 40.6, improvements on CMARS are mentioned. Conclusions and future research are stated in the last section.

40.2 The CMARS Method

MARS is a nonparametric regression method which considers the following general relationship between variables [11]:

$$y = f(\boldsymbol{x}) + \epsilon, \tag{40.1}$$

where y is a response variable, $\boldsymbol{x} = (x_1, x_2, \ldots, x_p)^T$ is a vector of predictors, and ϵ is an error term having zero mean and finite variance. It aims to construct reflected pairs for each input component x_j $(j = 1, 2, \ldots, p)$ with p-dimensional knots $\boldsymbol{\tau}_i = (\tau_{i,1}, \tau_{i,2}, \ldots, \tau_{i,p})^T$ $(i = 1, 2, \ldots, N)$ at each input vectors $\tilde{\boldsymbol{x}}_i = (\tilde{x}_{i,1}, \tilde{x}_{i,2}, \ldots, \tilde{x}_{i,p})^T$. (Actually, we may, without loss of generality, assume that the knot values are very close to data points to prevent from non-differentiability in our optimization problem later on.)

Here, the reflected pairs are piecewise linear functions whose forms are given by

$$c^+(x, \tau) = [+(x - \tau)]_+, \quad c^-(x, \tau) = [-(x - \tau)]_+, \tag{40.2}$$

where $[q]_+ := \max\{0, q\}$ and τ is a univariate knot. As a result, the set of basis functions (BFs) becomes

$$\wp := \left\{ (x_j - \tau)_+, (\tau - x_j)_+ \mid \tau \in \{x_{1,j}, x_{2,j}, \ldots, x_{N,j}\}, \ j \in \{1, 2, \ldots, p\} \right\}. \tag{40.3}$$

Thus, we can represent $f(\boldsymbol{x})$ in (40.1) by a general model involving an intercept and a linear combination of BFs contained in (40.3), as follows:

$$y = \theta_0 + \sum_{m=1}^{M} \theta_m \psi_m(\boldsymbol{x}^m) + \epsilon. \tag{40.4}$$

Here, ψ_m $(m = 1, 2, \ldots, M)$ are BFs or their products from (40.3), and θ_m, are the parameters representing the mth BF $(m = 1, 2, \ldots, M)$ or the constant one $(m = 0)$. *Interaction BFs*, on the other hand, are created by multiplying an existing BF with a truncated linear function involving a new predictor. For given data, $(\tilde{\boldsymbol{x}}_i, \tilde{y}_i)$ $(i = 1, 2, \ldots, N)$, the mth BF is

$$\psi_m(\boldsymbol{x}^m) := \prod_{j=1}^{K_m} [s_{\kappa_j^m} \cdot (x_{\kappa_j^m} - \tau_{\kappa_j^m})]_+, \tag{40.5}$$

where K_m is the number of truncated linear functions multiplied in the mth BF; $x_{\kappa_j^m}$ is the input variable corresponding to the jth truncated linear function in the mth BF; $\tau_{\kappa_j^m}$ is the knot value of the variable $x_{\kappa_j^m}$; and $s_{\kappa_j^m}$ is the $+/-$ sign.

The MARS algorithm consists of two main parts [11]. In the first part (forward algorithm), the BFs are selected to minimize the *lack-of-fit* until a user-specified maximum number of BFs, M_{max}, is attained. This process results in a large model which overfits the data. In the second part (backward algorithm), to overcome this disadvantage, the BFs contributing least to the residual squared errors are removed, thus causing less complex model. In both parts, MARS algorithm uses Generalized Cross-Validation (GCV) as a variable selection criterion [7].

CMARS is developed as an alternative to the backward part of the MARS algorithm. CMARS uses up to all BFs generated by the forward algorithm of MARS, and minimizes *Penalized Residual Sum of Squares (PRSS)* as described below [39, 43]:

$$PRSS := \sum_{i=1}^{N} (y_i - f(\tilde{x}_i))^2 + \sum_{m=1}^{M_{max}} \lambda_m \sum_{\substack{|\alpha|=1 \\ \alpha=(\alpha_1,\alpha_2)^T}}^{2} \sum_{\substack{r<s \\ r,s \in V_m}} \int_{Q^m} \theta_m^2 \left[D_{r,s}^\alpha \psi_m(t^m) \right]^2 dt^m,$$
(40.6)

where $V_m := \left\{ \kappa_j^m \mid j = 1, 2, \ldots, K_m \right\}$ are the predictors associated with the mth BF, ψ_m, and $t^m = \left(t_{m_1}, t_{m_2}, \ldots, t_{m_{K_m}} \right)^T$ represents the vector of predictors contributing to the mth BF. Besides,

$$D_{r,s}^\alpha \psi_m(t^m) := \frac{\partial^{|\alpha|} \psi_m}{\partial^{\alpha_1} t_r^m \, \partial^{\alpha_2} t_s^m}(t^m),$$
(40.7)

for $\alpha = (\alpha_1, \alpha_2)^T$, $|\alpha| := \alpha_1 + \alpha_2$, where $\alpha_1, \alpha_2 \in \{0, 1\}$. More information on the implementation of derivatives can be found in [43]. As can be seen from (40.6), PRSS is composed of two parts representing accuracy and complexity, respectively [13]. They are tried to be compromised by using the penalty parameters λ_m [39].

In Eq. (40.6), the integral symbol, "$\int$", is used as dummy in the sense of $\int_{Q^m}$, where Q^m is some appropriately large K_m-dimensional parallel-pipe where the integration takes place. Since the multi-dimensional integrals in (40.6) are difficult to evaluate, we discretize these integrals and rearrange PRSS in the following form [36]:

$$PRSS \approx \sum_{i=1}^{N} \left(y_i - \theta^T \psi(\tilde{d}_i) \right)^2$$

$$+ \sum_{m=1}^{M_{max}} \lambda_m \theta_m^2 \sum_{i=1}^{(N+1)^{K_m}} \left(\sum_{\substack{|\alpha|=1 \\ \alpha=(\alpha_1,\alpha_2)^T}}^{2} \sum_{\substack{r<s \\ r,s \in V_m}} \left[D_{r,s}^\alpha \psi_m(\hat{x}_i^m) \right]^2 \right) \Delta \hat{x}_i^m, \quad (40.8)$$

where $\boldsymbol{\psi}(\tilde{\boldsymbol{d}}_i) := \left(1, \psi_1(\tilde{\boldsymbol{x}}_i^1), \ldots, \psi_M(\tilde{\boldsymbol{x}}_i^M), \psi_{M+1}(\tilde{\boldsymbol{x}}_i^{M+1}), \ldots, \psi_{M_{max}}(\tilde{\boldsymbol{x}}_i^{M_{max}})\right)^T$,

and $\boldsymbol{\theta} := (\theta_0, \theta_1, \ldots, \theta_{M_{max}})^T$ with the point $\tilde{\boldsymbol{d}}_i := \left(\tilde{\boldsymbol{x}}_i^1, \tilde{\boldsymbol{x}}_i^2, \ldots, \tilde{\boldsymbol{x}}_i^M, \tilde{\boldsymbol{x}}_i^{M+1}, \ldots,\right.$

$\left.\tilde{\boldsymbol{x}}_i^{M_{max}}\right)^T$ in the argument with $(\sigma^{\kappa_j})_{j \in \{1,2,\ldots,p\}} \in \{0, 1, 2, \ldots, N+1\}^{K_m}$ and

$$\hat{\boldsymbol{x}}_i^m = \left(\tilde{x}_{l_1^{\kappa_1^m}, \kappa_1^m}^m, \ldots, \tilde{x}_{l_{K_m}^{\kappa_{K_m}^m}, \kappa_{K_m}^m}^m\right), \quad \Delta\hat{\boldsymbol{x}}_i^m := \prod_{j=1}^{K_m}\left(\tilde{x}_{l_{\kappa_j}^{\kappa_j}+1, \kappa_j^m}^m - \tilde{x}_{l_{\kappa_j}^{\kappa_j}, \kappa_j^m}^m\right).$$

(40.9)

To put the PSSS in (40.8) a simpler form in order to be able to handle the problem easily, we use a uniform penalization, λ, for each derivative term. Then, PRSS turns into a *Tikhonov Regularization (TR) problem* as can be seen below [1]:

$$PRSS \approx \left\|\boldsymbol{y} - \boldsymbol{\psi}(\tilde{\boldsymbol{d}})\boldsymbol{\theta}\right\|_2^2 + \lambda \left\|\boldsymbol{L\theta}\right\|_2^2,$$

(40.10)

which can be reformulated as a *Conic Quadratic Problem (CQP)* as follows [24,35]:

$$\min_{t,\boldsymbol{\theta}} \quad t,$$

$$\text{subject to} \quad \left\|\boldsymbol{y} - \boldsymbol{\psi}(\tilde{\boldsymbol{d}})\boldsymbol{\theta}\right\|_2 \leq t,$$

$$\left\|\boldsymbol{L\theta}\right\|_2 \leq \sqrt{\tilde{M}}.$$

(40.11)

Here, the optimization problem given in (40.11) is solved by using Interior Points Methods (IPMs) via the optimization software MOSEK [23,25]. There can be many solutions to this problem for different $\tilde{M}$ values which are determined by trial and error [13]. As a representative solution, however, we have selected the one that is closest to the corner of an efficiency (or L) curve, in which $\left\|\boldsymbol{L\theta}\right\|_2$ plotted versus $\left\|\boldsymbol{y} - \boldsymbol{\psi}(\tilde{\boldsymbol{d}})\boldsymbol{\theta}\right\|_2$ [39,43].

40.3 Background on Evaluating Model Performances

The performance evaluation techniques used in the comparison studies reviewed are briefly summarized below.

40.3.1 Validation Techniques

In order to compare different model performances with that of CMARS, different validation techniques are used. In case of data availability, the models developed are evaluated by using *new (collected or generated) data*. However, in most cases, the data used for benchmarking is obtained from either well-known data repositories or previous researches in literature where data is studied in various respects. In these data sets, usually, *Cross-Validation (CV)* is used for model validation [19]. This technique, called *k-fold CV*, basically, subdivides data randomly into some (say k) approximately equal size folds (subsamples). In this technique, each time, one of the subsamples is assumed to be the test sample, and the other $(k - 1)$ samples are combined to form a training sample. Initially, the training sample is used to construct the model, and then, its performance is evaluated by using the test sample. In the end, the average of each performance measure considered (see Sect. 40.3.2 below for different measures used in comparisons) over k-folds is computed. Here, k is usually determined depending on the sample size. When size is large enough, tenfold CV is preferred; for the smaller sample sizes, five- or three-fold CV is usually used. Note also that two-fold CV is called as the *hold-out* method.

An extended form of k-fold CV is known as *r-times replicated k-fold CV*. In this approach, the k-fold process is replicated r times with new partitions. Overall, a total of $(r.k)$ results are obtained for each measure studied. To obtain better estimates, the k results obtained from k-folds of a replication are averaged, as well as all $(r.k)$ results.

40.3.2   Performance Criteria and Measures

In the studies reviewed, the prediction performances of the models developed by the predictive data mining tools, MLR and MARS, are compared with respect to several criteria such as accuracy, complexity, stability, efficiency, and robustness, where applicable. To evaluate the model accuracies for prediction models, various measures are considered. These include *Multiple Coefficient of Determination* (R^2), *Mean Absolute Error (MAE)*, *Predicted Error Sum of Squares (PRESS)*, *Proportion of Residuals within Three Sigma (PWI)*, and *Root Mean Squared Error (RMSE)*. As another performance criterion, complexity, in the comparison studies, is evaluated by means of *Mean Squared Error (MSE)*. Stability of a measure indicates how much the model performance is consistent in training and test samples. In comparison studies, two different formulations are used for this purpose [27, 43]. Definitions of both accuracy and stability measures are presented in Appendix 1. Robustness criterion, on the other hand, helps us evaluate model performances under different data sets with different characteristics. Standard deviations of the measures are used for evaluating robustness of the methods. To measure the method efficiency, computational run times of the data sets are recorded, and compared.

Classification models, on the other hand, are evaluated only with respect to accuracies. These include, *Type I* and *Type II Errors*, *Specificity*, *Sensitivity* and *Error Rate (ER)*. These measures are the misclassification/correct classification probabilities taken into consideration for justifying the performance of classification models. They are also compared by *Receiver Operating Characteristic Curve*

(ROC), which is a curve of *True Positive Rate (TPR)* versus *False Positive Rate (FPR)*. Because performance comparisons are difficult with ROC curves, a related statistic, named *Area under ROC (AUC)*, is preferably used. While cross validating, *Mean AUC (MAUC), Mean ER (MER)* as well as *Standard Deviation of Error (Std Dev Error)* are also computed. Note that definitions of these measures are given in Appendix 2.

40.3.3   Statistical Tests for Performance Comparisons

In some comparison studies, to decide whether there are statistically significant differences among m prediction models developed, the following statistical hypotheses are tested

$$H_0 : \ M_{Method\ 1} = M_{Method\ 2} = \ldots = M_{Method\ m}$$

versus

$$H_1 : \text{At least one of them is different.}$$

In the null hypothesis, M denotes the expected value of a performance measure such as MAE, MSE, R^2 used in the comparisons. When two methods are compared, that is $m = 2$, to conduct the above test, a *paired-t test* is used; when more than two methods are compared, *Repeated Analysis of Variance (RANOVA)* is performed [9]. Following a significant RANOVA test, Fisher's *Least Significant Difference (LSD)* test is also applied to identify statistically different models. The above testing procedure is applied for training and test data sets as well as the stabilities of the measures.

40.4   Data Sets Used in Comparisons

In this section, all data sets used in CMARS applications and comparisons with the other classification/prediction techniques are summarized in Table 40.1. In addition, characteristics of data sets and modeling methods used are given in Table 40.2. Note that before developing any models, all data sets are preprocessed to handle missing values, inconsistencies and outliers if there are any. Besides, all variables are standardized to make models comparable. An exploratory data analysis is also performed to search for possible relations between variables where necessary.

As can be seen on the tables, 21 data sets having different features are used in the studies reviewed. The size of data sets range from 20 to 1599, and the number of predictors varies from 2 to 35. These data sets are gathered from various study areas such as life, finance, industry, social, business and so on. They are mostly real-life data (14 data sets); only seven of them are simulated.

Table 40.1 Data sets used in applications and comparisons

Name	Label	Type	Study Area	Original Source	Study Used
Pima Indians Diabetes	PID	Real Life	Life	[38]	[37]
International Credit Default	ICD	Real Life	Finance	[33]	[33]
Metal Casting	MC	Real Life	Industry	[2]	[39, 44]
Concrete Slump Test	CST	Real Life	Computer	[38]	[39, 41]
PM10	PM10	Real Life	Environment	[34]	[39, 41]
Forest Fires	FF	Real Life	Physical	[38]	[39, 41]
Parkinsons Telemonitoring	PT	Real Life	Life	[38]	[39]
Concrete Compressive Strength	CCS	Real Life	Physical	[38]	[39]
Red Wine Quality	RWQ	Real Life	Business	[38]	[39]
Communities and Crime	C&C	Real Life	Social	[38]	[39]
Uniform Sampling	US	Simulation	Aeronautics	[15]	[5, 39, 41]
Face Center Cube	FCC	Simulation	Aeronautics	[15]	[5]
D optimal	D-Opt	Simulation	Aeronautics	[15]	[5]
Box-Behnken	BB1;BB2;BB3	Simulation	Aeronautics	[15]	[5]
Noisy Data	NSY	Simulation	Math	[14]	[39]
Electricity prices of Turkey	EPT	Real Life	Energy	[29]	[29]
Minitab Data	MD	Real Life	Industry	[22]	[40]
Congressional Voting Records	CVR	Real Life	Social	[38]	[8]
Hepatitis	HEPA	Real Life	Life	[38]	[8]

40.5   Various Applications of CMARS and Comparison with Other Methods

In reviewed studies, CMARS has been applied to various domains, and its performance is empirically compared with those of three well-known classification and two prediction data mining methods. Classification methods are CART, IKL and GAM & CQP, and prediction methods are MLR and MARS. In this section, first, these methods are introduced briefly, and then, the results of performance comparisons are presented.

40.5.1   Comparison with Classification and Regression Trees (CART)

CART is a nonparametric technique that can be used for variable selection as well as model construction [6]. Generally, it can model both continuous and categorical dependent variables. For categorical/continuous variables, it produces a binary classification/regression tree starting from the root. The basic idea of CART is to

Table 40.2 Characteristics of data sets and modeling methods used

Data Label	n	p	Response(s)	Response Type	Predictor Type	Model Type	Predictive Data Mining Tool
PID	768	8	Whether the patient shows signs of diabetes or not	Binary	Integer, Real	Classification	MARS
ICD	1019	12	If the country is defaulted or not	Binary	Real	Classification	CART, MARS, GAMs&CQP
MC	92	35	Percentage of defective items	Continuous	Real	Prediction	MLR, MARS
CST	103	7	28-day Compressive Strength (Mpa)	Continuous	Real	Prediction	MARS, BCMARS
PM10	500	7	Hourly values of the concentration of PM10 (particles)	Continuous	Real	Prediction	MARS, BCMARS
FF	517	11	Area—the burned area of the forest (in ha); 0.00 to 1090.84	Continuous	Real	Prediction	MARS, BCMARS
PT	578	21	Motor UPDRS—Clinician's motor UPDRS score, linearly interpolated	Continuous	Integer, Real	Prediction	MARS
CCS	1030	8	Concrete compressive strength	Continuous	Real	Prediction	MARS
RWQ	1599	11	Quality (score between 0 and 10)	Discrete	Real	Prediction	MARS
C&C	879	24	ViolentCrimesPerPop: total number of violent crimes per population	Continuous	Real	Prediction	MARS
US	160	10	Total impulse	Continuous	Real	Prediction	MARS, BCMARS
FCC	149	10	Maximum acceleration rate	Continuous	Real	Prediction	MARS
D-Opt	66	10	Total impulse	Continuous	Real	Prediction	MARS
BB	161	10	Total impulse; Maximum pressure; Launch acceleration	Continuous	Real	Prediction	MARS
NSY	100	2	Artificially created function	Continuous	Real	Prediction	MARS
EPT	30	–	A day-ahead electricity prices of Turkey	Continuous	Real	Prediction	MARS, RCMARS, DRM
MD	20	3	–	Continuous	Real	Prediction	RCMARS, MARS
CVR	52	16	The voting result: democrat or republican	Binary	Categorical	Classification	IKL
HEPA	155	19	A person die or live due to hepatitis	Binary	Categorical, Integer, Real	Classification	IKL

construct a tree that will separate the data in the "best" way by finding binary splits on variables. The best splitting decision attains the homogeneity of each child node called the impurity; it is a special function defined as

$$i(t) = \phi(p(1\,|\,t), p(2\,|\,t), \ldots, p(c\,|\,t)),$$

where

$$p(c\,|\,t) = \frac{p(c,t)}{p(t)}, \quad p(t) = \sum_{c=1}^{C} p(c,t), \quad p(c,t) = \pi(c)\frac{N_c(t)}{N_c}, \quad \pi(c) = \frac{N_c}{N}.$$

Here, N_c is the number of observations belongs to the class c; N is the number of observations; $\pi(c)$ is the prior probability of the class c; $N_c(t)$ is the number of observations in the node t belonging to the class c; $p(c,t)$ is the probability of having an observation both in class c and in node t, and $p(c\,|\,t)$ is the probability of class c in node t.

The best splitting rule is the one with the maximum change of impurity. The best known criteria are the Gini Index and the Twoing rule. The Gini impurity function can be linear or quadratic; its change of impurity function is given by

$$i(t) = \sum_{\substack{k \neq l \\ k,l=1,2,\ldots,C}} p(k\,|\,t)p(l\,|\,t).$$

In the Twoing rule, for each node, the impurity is estimated as a binary problem. At a node t, with s splitting into t_L and t_R, which are left and right child nodes, s is chosen by maximizing the following formula (its change of impurity function):

$$P_L P_R \left[\sum_{c=1}^{C} |p(c\,|\,t_L) - p(c\,|\,t_R)| \right]^2 /4,$$

where P_L and P_R are the probabilities split into the left or the right. The $p(c\,|\,t_L)$ and $p(c\,|\,t_R)$ are conditional probabilities split into the left or the right at a node t.

The tree building process is stopped when a maximum tree is reached, or there is a limit on the number of levels in the tree. Then, CART starts the pruning process. A child is pruned away if the resulting change in the predicted misclassification cost is less than the complexity parameter times the change in tree complexity. Such a process continues until the optimal tree is achieved.

In one of the CMARS related studies [33], CMARS and CART are applied to the area of finance by the data set labeled as ICD (see Tables 40.1 and 40.2 for the data description) to model sovereign default. In this application, since the dependent variable is of type categorical, particularly binary, classification trees are utilized to construct the CART model by using MATLAB 7 Statistics Toolbox [20]. Note that in this study Gini Index is preferred as a splitting criterion, because it requires relatively less computation time and is more robust to outliers.

In the study, tenfold CV is implemented, and the classification capabilities of the models developed are determined by the scoring results of training and test samples. Results indicate that, in the training sample, CART has a better classification capability than CMARS with respect to the Average Correct Classification Rate (ACCR). In the test sample, however, CMARS gives more accurate results than CART with respect to the same measure. In addition, as the classification accuracy of the test samples of CMARS is only slightly lower than those of the corresponding training samples.

On the other hand, CART and CMARS have similar small Type II errors. In terms of this error, CART detects default records slightly better than CMARS. However, in terms of Type I error, CMARS has a better estimate. And it clearly detects non-default records better than CART.

If we look at discriminative power of the methods, there is a big difference in the test and training sample results of CART; it does not sustain its discrimination success in training sample, and becomes the worst model in the test sample results.

40.5.2 Comparison with Infinite Kernel Learning (IKL)

Multiple Kernels Learning (MKL) is a classification method constructed by selection of finite combinations of kernels. The main idea of MKL is to combine finitely many pre-chosen kernels, k_κ ($\kappa = 1, 2, \ldots, K$), in a convex combination [32]

$$k_\gamma(\boldsymbol{x}_i, \boldsymbol{x}_{i*}) := \sum_{\kappa=1}^{K} \gamma_\kappa k_\kappa(\boldsymbol{x}_i, \boldsymbol{x}_{i*}) \ (i, i^* = 1, 2, \ldots, N),$$

where $\gamma_\kappa \geq 0$ ($\kappa = 1, 2, \ldots, K$), $\sum_{\kappa=1}^{K} \gamma_\kappa = 1$ with the input vectors $\boldsymbol{x}_i$ and $\boldsymbol{x}_{i*}$. Nevertheless, MKL may not be successful for classifying large and inhomogeneous (heterogeneous) data sets (i.e. data from a number of sources, largely unknown and unlimited, and in many varying formats), usually, due to finite number of choices of kernels. To overcome this problem, IKL method is developed for learning problems with the help of infinite and semi-infinite programming by considering all elements in kernel space [32]. Thus, the problem becomes infinite in both its dimension and number of constraints, and called as Infinite Programming (IP) problem. An infinite combination has the following form

$$k_\gamma(\boldsymbol{x}_i, \boldsymbol{x}_{i*}) := \int_\Omega k(\boldsymbol{x}_i, \boldsymbol{x}_{i*}, w) d\gamma(w),$$

where γ is a monotonically increasing function of integral one (1), or is a probability measure on Ω, and $w \in \Omega$ is a kernel parameter. The kernel function, $k(\boldsymbol{x}_i, \boldsymbol{x}_{i*}, w)$, is assumed to be a twice continuously differentiable function with respect to w, i.e., $k(\boldsymbol{x}_i, \boldsymbol{x}_{i*}, \cdot) \in C^2$.

Infinitely many kernels mean infinitely many coefficients expressed with an increasing monotonic function via positive measures. The formulation of IKL is given as [32]

$$\max_{b, \gamma} \quad b \quad (b \in \mathbb{R}, \ \gamma : \text{a positive measure on } \Omega),$$

$$\text{such that } b - \int_\Omega T(w, \boldsymbol{\alpha}) d\gamma(w) \leq 0 \quad (\boldsymbol{\alpha} \in A),$$

$$\int_\Omega d\gamma(w) = 1,$$

where $T(w, \boldsymbol{\alpha}) := S(w, \boldsymbol{\alpha}) - \sum_{i=1}^N \alpha_i$, $S(w, \boldsymbol{\alpha}) := \frac{1}{2} \sum_{i,i*=1}^N \alpha_i \alpha_{i*} \chi_i \chi_{i*} k$ $(\boldsymbol{x}_i, \boldsymbol{x}_{i*}, w)$ and $\Omega := [0, 1]$ and $A := \left\{ \boldsymbol{\alpha} \in \mathbb{R}^N \mid \boldsymbol{0} \leq \boldsymbol{\alpha} \leq C\boldsymbol{1} \text{ and } \sum_{i=1}^N \alpha_i y_i = 0 \right\}$ are index sets. In this problem, the objective function is continuous and the index set is compact, where $C \in \mathbb{R}$, and $(\boldsymbol{1} = (1, 1, \ldots, 1)^T \in \mathbb{R}^N$.

In one of the studies reviewed [8], CMARS and IKL are applied to the data sets CVR and HEPA (see Tables 40.1 and 40.2 for the data descriptions). Here, the data sets CVR and HEPA are homogeneous and heterogeneous data, respectively. To develop IKL models, a toolbox written in JAVA is used. To evaluate the model performances, five-fold CV is applied, and the classification measures MER, Std Dev Error and AUC are calculated. According to the results, CMARS performs better than IKL in both data sets with respect to most of the (different) measures. On the other hand, when AUC is considered, CMARS perform even better on the heterogeneous data than IKL.

40.5.3 Comparison with Generalized Additive Models (GAMs) and CQP

GAM is a flexible modeling technique that can be used to estimate parametric, non-parametric and semi-parametric additive models [12]. It is constructed by smoothed functions (like splines), and the optimization problem involved is solved by the CQP. Besides, it can handle high-dimensional data with discrete and nonnormal dependent variable.

The GAMs are defined as follows

$$G(\mu(\boldsymbol{x}_i)) = y_i = \beta_0 + \sum_{j=1}^p f_j(x_{ij}) + \epsilon \quad (i = 1, 2, \ldots, N),$$

where N is the number of observations; p is the number of predictor variables; μ denotes the expected value; G is the link function. Here, it is assumed that $E\left(f_j(x_{ij})\right) = 0$ $(j = 1, 2, \ldots, p)$ [12], and f_j, estimated from data, are smoothed functions containing a convex combination of d_j base-splines, of the following form

$$f_j(x_j) = \sum_{k=1}^{d_j} \theta_k^j h_k^j(x_j) \quad (k = 1, 2, \ldots, d_j),$$

where $h_k^j : \mathbb{R} \to \mathbb{R}$ presents the kth base-spline of variable x_j and θ_k^j is the associated coefficient. To estimate parameters of GAMs, CQP is used, and the approximated version of the problem, called GAMs & CQP is given in the form below:

$$\min_{t, \beta, f} \quad t,$$

$$t \geq 0,$$

$$\text{subject to} \quad \sum_{i=1}^{N} \left\{ y_i - \beta_0 - \sum_{j=1}^{p} f_j(x_{ij}) \right\}^2 \leq t^2,$$

$$\int \left[f_j''(t_j) \right] dt_j \leq \tilde{M}_j \quad (j = 1, 2, \ldots, p).$$

In the comparison study, the data sets used for CART (see Sect. 40.5.1) are also applied to GAMs & CQP [33]. Results indicate that the classification capabilities of GAMs & CQP and CMARS in the test sample are accurate. CMARS performs better than GAM & CQP but the latter one discovers the main structure of data better. The prediction performance of the models is also compared according to the statistical error definitions (i.e. Type I and Type II Errors). GAM & CQP has a better estimate compared to CMARS with regard to Type I error. In test sample, however, error definitions show that CMARS and GAM & CQP have the same capability in determining the non-default country assignation, i.e. the response in classification. Moreover, ROC plots reveal that CMARS have more accurate results in test samples.

40.5.4 Comparison with Multiple Linear Regression (MLR)

MLR is an extensively used powerful modeling technique [26]. An MLR model can be written as

$$y = \beta_0 + \beta_1 x_1 + \cdots + \beta_p x_p + \epsilon,$$

where y is the response variable (or the dependent variable); x_j $(j = 1, 2, \ldots, p)$ are the regressor variables (predictor or independent variables) and ϵ is a random error term. Here, the intercept β_0, the regression coefficients β_j $(j = 1, 2, \ldots, p)$ and the error variance σ^2 are to be estimated. For this purpose, the least squares estimation (LSE) method which depends on the minimization of sum of the squared error terms is usually used. If $\boldsymbol{X}^T \boldsymbol{X}$ is nonsingular, the unique solution is obtained by the following equation

$$\hat{\boldsymbol{\beta}} = (\boldsymbol{X}^T \boldsymbol{X})^{-1} \boldsymbol{X}^T \boldsymbol{y}.$$

Here, $\boldsymbol{y}$ is the $N \times 1$ response vector; $\boldsymbol{X}$ is the $N \times (p + 1)$ predictor variable matrix and $\hat{\boldsymbol{\beta}}$ is the $(p + 1) \times 1$ regression coefficients vector including the intercept. Note that the MLR method requires satisfaction of some (so-called white-noise) assumptions to obtain statistically valid inferences.

In the study where CMARS and MLR are compared, the models are developed for two data sets [44]: MC and US (see Tables 40.1 and 40.2 for the data descriptions) by using the statistical software MINITAB 15 [22]. While developing MLR models, white noise assumptions are all tested. If any of the assumptions are not validated, corrective measures such as transformation of the response or predictor(s) are taken.

In the study, three-times replicated three-fold CV is applied, and MAE, MSE and R^2 measures and their stabilities are computed. RANOVA and LSD tests indicate that MLR method performs worse than CMARS in the MC data. Besides, MLR models are not statistically valid. CMARS models provide a good fit to data with respect to R^2 measure; it does not perform, on the other hand, as well as MLR does with respect to the stabilities of measures MAE and MSE. All these findings are attributed to the complex structure of the data set studied.

For the US data, no differences are detected among the performances of the models built according to the measures of interest, namely R^2, MAE, MSE, PRESS, PWI, and their stabilities. This may also be due to the structure of the data; it reveals rather linear relationships between its response and predictor variables. As a result, all models studied provide a good fit to this data set.

40.5.5   Comparison with Multivariate Adaptive Regression Splines (MARS)

As stated in Sect. 40.2, MARS is a powerful nonparametric regression technique for handling nonlinear and high-dimensional data [11]; it has no specific assumptions regarding the form of relationship between variables, and it estimates the general model function by using piecewise linear functions, called BFs. In this section, comparison results of CMARS with MARS are summarized. In these studies, the

MARS models are constructed by using Salford Systems MARS version 2 and 3 [18].

40.5.5.1   Comparison of MARS and CMARS for Classification

Two studies have been conducted for comparison of MARS and CMARS for classification. In one of them, both methods are applied to PID data to determine if a person has diabetes or not [37, 38]. For this purpose, the hold-out validation technique (75 % of observations used to train data) is used, and ACCR measure is obtained. According to the results, both methods perform almost the same. However, in term of the true diagnose of the disease, CMARS performs better than MARS. In both training and test samples, MARS and CMARS scoring models are almost the same; on the other side, CMARS discovers the main structure of data better. Besides, in terms of Type II error, CMARS detects diabetes better. Depending on these findings, the study in [37] concludes that CMARS gives a superior estimate compared to MARS.

In the other study, comparison of MARS and CMARS models are made for ICD data to predict the countries' default possibilities show variations in training and tests samples [33]. CMARS outperforms MARS in terms of the ACCR measure. Also, it is more robust than MARS, and its performance measures reveal higher stabilities. According to the default country assignation, MARS and CMARS have close Type II errors. However, CMARS seriously reduces the chance of committing Type I error compared to that of MARS.

40.5.5.2   Comparison of MARS and CMARS for Prediction

The prediction performances of CMARS and MARS are evaluated both on real-life and simulated data sets. In one of the simulation applications, both MARS and CMARS models are developed involving lower- as well as higher-order interactions [5]. According to the RANOVA results, since no statistically significant differences are detected between MARS and CMARS models developed, all training and test data sets are combined, and a paired t-test is performed to compare model performances with respect to each measure. According to the results, CMARS performs better than MARS with respect to MAE, R^2 and PRESS. And also CMARS is more stable than MARS with respect to the stability of MAE.

Yet, in another study, an extensive comparison is conducted by using nine real-life data sets obtained from well-known data repositories [39]. These data sets are classified according to the level (small, medium and large) of their size and scale (i.e. number of predictors). In these comparisons, three-times replicated three-fold CV is applied. According to results, for all training data sets, CMARS performs better and it is more robust than MARS with respect to all measures considered. In addition, although CMARS is more stable than MARS, MARS is more robust than

CMARS in stability with respect to most of the measures. Furthermore, an effect of sample size on the performance of methods is also detected. For example, CMARS performs better than MARS mostly on medium to large training samples. If we consider the performance with respect to scale, almost for all measures, MARS and CMARS perform better (or the same) as the scale changes from small to large. To compare the efficiencies of methods, computational run times (in seconds) for each training sample is recorded on the same computer. Run times seem to be related to the sample size but not the problem scale. MARS may provide solutions very fast compared to CMARS since its applications are run on professional software, Salford MARS. The run times of CMARS method increase almost up to three to five times as much to that of MARS as sample size increases. Moreover, in the same paper, a simulation study conducted shows that CMARS method has a better performance than MARS on the noisy data [39].

40.6 Improvements on CMARS

CMARS provides a new contribution to well-known predictive data mining methods, which have been applied in many areas. Studies in these areas necessitate further researches/extensions on the CMARS method to make it less complex and more robust. These are explained in the following subsections.

40.6.1 Bootstrapping CMARS (BCMARS)

As we have mentioned in Sect. 40.2, CMARS generates models with maximum number of BFs. However, some of these coefficients may not contribute significantly to the model. By detecting and removing these coefficients, the model size (i.e. complexity) can be reduced. For this purpose, bootstrapping methods are applied on CMARS, and this new technique is called as Bootstrapping CMARS (BCMARS) [41].

40.6.1.1 Methodology of the BCMARS

Bootstrapping is a data-based simulation method useful for making inferences [4, 10, 41]. The bootstrap applications include estimation of standard errors and bias, constructing *Confidence Intervals (CIs)*, testing hypothesis and so on. The bootstrap is an iterative procedure consisting of the following few steps. First, a random sample of size (the same as the original data) from the empirical distribution is generated with replacement. Next, the statistic of interest is computed for the generated sample. Then, the first two steps are repeated k times. In the study mentioned, different bootstrapping regression methods (Random-X, Fixed-X and Wild Bootstrap) are applied to the CMARS method.

40.6.1.2 Applications and Comparisons of BCMARS

BCMARS models using different bootstrapping methods are developed on four data sets, namely CS, US, PM10 and FF (see Tables 40.1 and 40.2 for data descriptions). In the comparison of models, three-fold CV technique is applied. Then, performances of the models are compared according to complexity, stability, accuracy, precision, robustness and efficiency criteria, whose definitions are given in Sect. 40.3.2. For the significance level selected, the statistical significance of model parameters are tested by using the CI approach. In the comparisons, effects of certain data characteristics such as the size and scale on the methods' performances are also evaluated. Note here that to fit a MARS model to a data set, the R package version "*Earth*" [21] is preferred. For all other computations, including nonparametric bootstrap, the code written in MATLAB is run.

Comparison results show that, in general, BCMARS methods perform better than MARS and CMARS methods with respect to most of the measures considered. Either one of the BCMARS methods yield less complex models than MARS or CMARS. It is apparent from the results that by reducing the complexity via bootstrapping, the CIs of model parameters become narrower than those of CMARS. Moreover, standard errors of the model parameters decrease after bootstrapping, leading more precise estimates [42].

40.6.2 Robust CMARS (RCMARS)

The importance and benefit of CQP in manufacturing is mentioned in [43]. As a further study, the robustification of CMARS with Robust Optimization (RO) is done via CQP [28]. RO developed by Aharon Ben-Tal, and used by Laurent El Ghaoui in the area of data mining. This new technique is called as a Robust CMARS (RCMARS). In the following sections, the theory, applications and comparisons of RCMARS are summarized.

40.6.2.1 Methodology of the RCMARS

The CMARS model depends on parameters. Small perturbations in data can result in different model parameters causing unstable solutions. In RCMARS, the purpose is to decrease the estimation error while keeping efficiency as high as possible. To achieve this purpose, we apply some approaches such as usage of more robust estimators, scenario optimization and robust counterpart. RCMARS is a kind of regularization in the input and output domain. Therefore, some changes are done in both input and output variables by including uncertainty with the help of RO in the part of Residual Sum of Square (RSS).

To robustify CMARS, RO on the BFs provided by the MARS model is employed, and it is assumed that input and output variables of CMARS model are all random variables. They lead us to *uncertainty sets*, which are assumed to contain CIs. MARS method employs expansions of piecewise linear BFs based on the new data set that have uncertainties. All expressions for MARS and CMARS are reformulated by changing input and output symbols with new ones containing uncertainty [28, 31].

The following general model is considered to represent the relation between input variables and the response

$$y = f(\underbrace{\check{x}}_{noisy\ data}) + \epsilon,$$

where y is the response variable; $\check{x} = (\check{x}_1, \check{x}_2, \ldots, \check{x}_p)^T$ is a vector of predictors, and ϵ is an error term. Here, each $\check{x}_j$ is assumed to be a normally distributed random variable defined by

$$\check{x}_j = \bar{x} + \xi_j \quad (j = 1, 2, \ldots, p).$$

The mth BFs ($m = 1, 2, \ldots, M_{max}$), based on the new data set $(\check{\check{x}}_i, \check{\check{y}}_i)$ ($i = 1, 2, \ldots, N$) that contain uncertainties are expressed as given below by incorporating uncertainty sets $U_1 \subseteq \mathbb{R}^{N \times M_{max}}$ and $U_2 \subseteq \mathbb{R}^N$ into the data $(\check{\check{x}}_i, \check{\check{y}}_i)$ ($i = 1, 2, \ldots, N$):

$$\psi_m(\check{\check{x}}_i^m) := \prod_{j=1}^{K_m} [(\check{\check{x}}_{i\kappa_j^m} - \tau_{\kappa_j^m})]_\pm.$$

Then, the PRSS of the CMARS model in (40.6) with uncertainty will have the following form:

$$PRSS := \sum_{i=1}^{N} (\check{\check{y}}_i - f(\check{\check{x}}_i))^2 + \sum_{m=1}^{M_{max}} \lambda_m \sum_{\substack{|\alpha|=1 \\ \alpha=(\alpha_1,\alpha_2)^T}} \sum_{\substack{r<s \\ r,s \in V_m}}^{2} \int_{\Omega_m} \theta_m^2 \left[D_{r,s}^\alpha \psi_m(t^m)\right]^2 dt^m.$$

After discretization is applied to approximate the multi-dimensional integrals, the λ_m ($m = 1, 2, \ldots, M_{max}$) in PRSS with uncertainty is unified as λ as it is done in CMARS (see Sect. 40.2). Then, PRSS takes the following form

$$PRSS \approx \left\| \check{\check{y}} - \psi(\check{\check{d}})\theta \right\|_2^2 + \lambda \left\| L\theta \right\|_2^2,$$

and it resembles a classical TR problem with $\lambda \geq 0$, i.e., $\lambda = \phi^2$ for some $\phi \in \mathbb{R}$. Here, the TR problem can be tackled with the CQP. Note that regularization from CMARS is already some kind of robustification. Additionally, robustifying CMARS by RO approach needs changes to be carried out in the $\left\| \overset{\smile}{y} - \psi(\overset{\smile}{d})\theta \right\|_2^2$ part. However, no change is required in the integration function of complexity part of the PRSS model; as a result, the part of $\|L\theta\|_2^2$ remains the same as it is in CMARS.

The RO approach makes the optimization models robust against constraint violations by solving *robust counterparts* in some prespecified *uncertainty sets* for the uncertain parameters. When the uncertainty set employed has a special shape such as *ellipsoidal* or *polyhedral*, the RO problem can be solved efficiently. Using ellipsoidal uncertainty sets provides more successful results than using polyhedral uncertainty sets. Unfortunately, they increase the complexity of optimization problems. Therefore, while robustifying the CMARS, *polyhedral uncertainty* with different uncertain scenarios is preferred.

When the uncertainty sets U_1 and U_2 are assumed to be *polyhedral* for input and output data, the robust counterpart can be defined as follows

$$\min_{\theta} \quad \max_{W \in U_1, \, z \in U_2} \quad \|z - W\theta\|_2^2 + \lambda \|L\theta\|_2^2.$$

Here, U_1 and U_2 are polytopes with $2^{N \cdot M_{max}}$ vertices $W^1, W^2, \ldots, W^{2^{N \cdot M_{max}}}$ and 2^N vertices $z^1, z^2, \ldots, z^{2^N}$, respectively. They are not exactly known, but belong to convex bounded uncertain domains given by

$$U_1 = \left\{ \sum_{j=1}^{2^{N \cdot M_{max}}} \delta_j W^j \mid \delta_j \geq 0 \; (j \in \{1, 2, \ldots, 2^{N \cdot M_{max}}\}), \; \sum_{j=1}^{2^{N \cdot M_{max}}} \delta_j = 1 \right\},$$

$$U_2 = \left\{ \sum_{i=1}^{2^N} \varphi_i z^i \mid \varphi_i \geq 0 \; (i \in \{1, 2, \ldots, 2^N\}), \; \sum_{i=1}^{2^N} \varphi_i = 1 \right\},$$

where $U_1 = \text{conv}\left\{ W^1, W^2, \ldots, W^{2^{N \cdot M_{max}}} \right\}$ and $U_2 = \text{conv}\left\{ z^1, z^2, \ldots, z^{2^N} \right\}$ are the corresponding convex hulls. Then, the robust CQP takes the form

$$\min_{t, \alpha} \quad t,$$

$$\text{subject to} \quad \|W\alpha - z\|_2 \leq t, \quad \forall \; W \in U_1, \; z \in U_2,$$

$$\|L\alpha\|_2 \leq \sqrt{\overset{\smile}{M}}.$$

Here $W = \sum_{j=1}^{2^{N \cdot M_{max}}} \delta_j W^j$ and $z = \sum_{i=1}^{2^N} \varphi_i z^i$. The robust CQP can be equivalently stated as a standard CQP by using polytopes U_1 and U_2 as follows:

$$\min_{t,\alpha} \; t,$$

subject to $\left\| W^j \alpha - z^i \right\|_2 \le t \; (i = 1, 2, \dots, 2^N; \; j = 1, 2, \dots, 2^{N \cdot M_{max}}),$

$$\|L\alpha\|_2 \le \sqrt{\breve{M}}.$$

Then, the CQP problem including convex combinations of the vertices is constructed and solved by using MOSEK [28, 29, 40].

40.6.2.2 Applications and Comparisons of RCMARS

RCMAR method is applied to a simple data set containing three predictors and 20 observations [28, 30]. To develop RCMARS model, uncertainty is incorporated into the data by following the methodology presented above. For this purpose, the uncertainty matrices and vectors based on polyhedral uncertainty sets are obtained. To overcome the computational problems faced due to large size of the uncertainty matrix for input data, the PRSS as a CQP problem are formulated for each observation by using the combinatorial approach, which is named as *weak robustification*. The 20 weak RCMARS models are then solved by MOSEK optimization software.

The RCMARS method is also applied on a real-life data to forecast electricity prices of Turkey. Then, its performance is compared with that of CMARS and *Dynamic Regression Model (DRM)*. It a powerful forecasting technique having the following form

$$y_t = \beta_0 x_{t-1} + \beta_1 x_{t-2} + \cdots + \beta_n x_{t-n} + \vartheta_0 y_{t-1} + \vartheta_1 y_{t-2} + \cdots + \vartheta_n y_{t-n} + \epsilon_t,$$

where $\beta_l, \; \vartheta_l \; (l = 1, 2, \dots, r)$ are the coefficients and ϵ_t is the error term assumed to be independent and normally distributed with mean zero and constant variance. Here, $(t = 1, 2, \dots, T)$ denotes a time series. In this model, the $(y_t)_{t=1,2,\dots,T}$ state of the process is linked to past processes, $y_{t-1}, y_{t-2}, \dots, y_{t-r}$, and the other factors $x_{t-1}, x_{t-2}, \dots, x_{t-r}$. Then, the models can be estimated by using the standard ordinary least-squares regression.

The comparison results show that CMARS and RCMARS perform better than DRM [29].

40.7 Conclusion and Outlook

There are many researches in literature that reveals the power of MARS method, particularly, for modeling high-dimensional and nonlinear so-called complex data relations. These studies encouraged us to improve the MARS algorithm further.

As a result, a group of studies has been carried out, and CMARS is developed as an alternative to the backward algorithm of the MARS method. Following, several other studies are applied the CMARS method to various application areas with different data characteristics. Its performance is also compared to that of very well-know modeling techniques such as CART, GAMs, etc. to evaluate its achievements. These studies show that CMARS is a powerful predictive data mining tool and a strong alternative to the MARS backward algorithm.

In this chapter, we aimed at gathering all these studies and their findings together to summarize what has been done up to now on the CMARS method, and what else can be done as a future study to improve or extend it further. The followings are main conclusions that can be drawn from the studies reviewed:

- Modeling an additive structure with CART is not easy, and the estimation variance of a CART model increases with continuous explanatory variables. On the contrary, CMARS captures such a structure easily, and thus, seems more suitable for future predictions than CART.
- CMARS performs better than IKL with respect to (most of) the comparison measures considered. CMARS is even good at for classifying the heterogeneous data.
- The performances of CMARS and GAMs & CQP are competing with each other; they seem to have the same capability in classification.
- When the structure of data is linear, CMARS performs as good as MLR does; when there is non-linearity, however, CMARS outperforms MLR. There are two main disadvantages of MLR over CMARS. First, human expertise is usually needed, and as a result, it may take a long time to develop MLR models. Secondly, the white-noise assumptions should be satisfied to obtain statistically valid inferences but these assumptions may not be realistic in most real-life applications.
- CMARS performs better than MARS according to most of the measures considered, or their performances are competing with each other. In addition, CMARS is more robust and stable than MARS overall. Furthermore, better performances of MARS method are obtained for small size-small scale, and that of CMARS method for small size-medium scale data sets. Run times seem to be related to the sample size but not the problem scale. As a result, the least amount of run times for both methods is obtained for small size samples regardless of the scale. Moreover, CMARS method has a better performance than MARS on noisy data.

In addition to the performance comparisons, CMARS method is further improved to overcome some of its problems such as reducing complexity. BCMARS method developed for this purpose enables us to obtain more precise parameter estimates than CMARS. Nevertheless, it takes significantly more time than MARS and CMARS methods. On the other hand, RCMARS can be thought as an extension of CMARS, in which random variables can be modeled more realistically. Results reveal that RCMARS provides more precise estimates than CMARS in such cases.

Table 40.3 Prediction performance measures

Criterion	Name of Measure	Symbol	Explanation	Interpretation	Formula
Accuracy	Multiple Coefficient of Determination	R^2	Percentage of variation in response explained by the model	Higher values are the better	$R^2 := 1 - \left(\frac{\sum_{i=1}^N (y_i - \hat{y}_i)^2}{\sum_{i=1}^N (y_i - \bar{y})^2}\right)$
	Mean Absolute Error	MAE	Measures the average magnitude of errors	Smaller values are the better	$MAE := \frac{1}{N}\sum_{i=1}^N |y_i - \hat{y}_i|$
	Mean Squared Error	MSE	Measures average of the squared errors	Smaller values are the better	$MSE := \frac{1}{N}\sum_{i=1}^N (y_i - \hat{y}_i)^2$
	Root Mean Square Error	RMSE	Square root of the MSE	Smaller values are the better	$RMSE := \sqrt{\frac{1}{N}\sum_{i=1}^N (y_i - \hat{y}_i)^2}$
	Prediction Error Sum of Squares	PRESS	Reveals predictive capability of the model	Smaller values are the better	$PRESS := \sum_{i=1}^N \left(\frac{e_i}{1-h_i}\right)^2$
	Proportion of Residuals within Three Sigma	PWI	Percentage of residuals within some user-specified thresholds (3 sigma is taken)	Higher values are the better	$PWI = \frac{a}{N}$
Complexity	Mean Squared Error	MSE	Measures average of the squared errors by regarding the number of terms in the model	Smaller values are the better	$MSE := \frac{1}{N-p}\sum_{i=1}^N (y_i - \hat{y}_i)^2$
Stability	Stability Measure 1	–	Compares the performance of the method on both training and test sample	Values closer to zero (0) indicate better stability	$\frac{CR_{TR}-CR_{TE}}{CR_{TR}+CR_{TE}}$ where CR_{TR} and CR_{TE} are the performance measure values for the training and the test, respectively
	Stability Measure 2	–	Compares the performance of the method on both training and test sample	Values closer to zero (1) indicate better stability	$\min\left\{\frac{CR_{TR}}{CR_{TE}}, \frac{CR_{TE}}{CR_{TR}}\right\}$

Table 40.4 General notation

		Condition	
		Positive	Negative
Model Outcome	Positive	True Positive (TP)	False Positive (FP)
	Negative	False Negative (FN)	True Negative (TN)

Studies are still being carried out on MARS and CMARS. In a recent study, the forward algorithm of MARS is tackled with a mapping algorithm to reduce the complexity of the model generated by MARS [16]. Based on the success of this new approach, it is going to be adapted to CMARS to reduce the complexity of CMARS models [16]. Last but not least, the CMARS code will be made available soon to all interested researchers to let them conduct their comparative studies, and provide us more feedback on the performance of the CMARS algorithm.

Acknowledgements One of the authors, Fatma Yerlikaya-Özkurt, has been supported by the TUBITAK Domestic Doctoral Scholarship Program.

Appendix 1: Prediction Performance Measures

General Notation:

y_i: ith observed response,
$\hat{y}_i$: ith fitted response,
$\bar{y}$: mean response,
N: number of observations,
p: number of terms in the model,
$e_i = y_i - \hat{y}_i$: ith ordinary residual,
$\bar{e}$: mean of ordinary residual,
$\sigma = \sqrt{\frac{\sum_{i=1}^{N}(e_i - \bar{e})^2}{N}}$: standard deviation of ordinary residual,
$std = 3 \cdot \sigma$: 3σ as a threshold,
a: the sum of indicator variables where the condition $|e_i| < std$ is satisfied,
h_i: leverage value for the ith observation (i.e., ith diagonal element of the hat matrix, $H = X(X^T X)^{-1}X^T$.
X: an $N \times p$ observation matrix with $rank(X) = p$ $(p \leq n)$.

See Table 40.3 for description and explanation of the prediction performance measures.

Table 40.5 Classification performance measures

Criterion	Name of Measure	Symbol/ Abbreviation	Explanation	Interpretation	Formula
Accuracy	Sensitivity	$1-\beta$	Measures the actual proportion of positives correctly identified (i.e. the power of decision)	Higher values are the better	$Sensitivity = \frac{TP}{(TP+FN)}$
	Specificity		Measures the actual proportion of negatives correctly identified	Higher values are the better	$Specificity = \frac{TN}{(TN+FP)}$
	Type I Error	α	Wrong decision made when a negative condition is classified as positive (i.e. false positive rate)	Smaller values are the better	$Type\ I\ Error = 1 - Specificity = \frac{FP}{(TP+FN)}$
	Type II Error	β	Wrong decision made when a positive condition is classified as negative (i.e. false negative rate)	Smaller values are the better	$Type\ II\ Error = 1 - Sensitivity = \frac{FN}{(TN+FP)}$
	Average Correct Classification Rate	ACCR	Average rate of correct classification	Higher values are the better	$ACCR = \frac{(TN+TP)}{(TP+TN+FP+FN)}$
	Area under the Curve	AUC	Area under Receiving Operating Characteristic (ROC) Curve	An area of one (1) represents perfect classification; an area of 0.5 represents a worthless classification	$AUC = \frac{G+1}{2}$
	Mean Area under the Curve	MAUC	Average of the AUCs over all folds of the CV process	Smaller values are the better	$MAUC = \frac{\sum_{i=1}^{k} AUC_i}{k}$ where k is the number of folds
	Error Rate	ER	Rate of incorrect classification	Smaller values are the better	$ER = \frac{(FN+FP)}{(TP+TN+FP+FN)}$
	Mean Error Rate	MER	Average of error rate over all folds of the CV process	Smaller values are the better	$MER = \frac{\sum_{i=1}^{k} ER_i}{k}$ where k is the number of folds
	Standard Deviation of Error	Std Dev Error	Standard deviation of the error rate	Smaller values are the better	$Std\ Dev\ Error = \sqrt{\frac{\sum_{i=1}^{k}(ER_i - MER_i)^2}{k}}$ where k is the number of folds

Appendix 2: Classification Performance Measures

General Notation:

X_i: The cumulated proportion of the population variable, for $i = (1, 2, \ldots, N)$ with
$X_0 = 0$, $X_N = 1$,
Y_i: The cumulated proportion of the income variable, for $i = (1, 2, \ldots, N)$ with
$Y_0 = 0$, $Y_N = 1$, and $Y_{i+1} > Y_i$,
G: The *Gini coefficient*, where $G = 1 - \sum_{i=1}^{N} (X_i - X_{i-1})(Y_i + Y_{i-1})$.

See Tables 40.4 and 40.5 for description and explanation of the classification performance measures.

References

1. Aster, R.-C., Borchers, B., Thurber, C.: Parameter Estimation and Inverse Problems. Academic Press, Burlington (2004)
2. Bakır, B.: Defect Cause Modelling with Decision Tree and Regression Analysis: A Case Study in Casting Industry. MSc., Middle East Technical University (2006)
3. Batmaz, İ., Köksal, G.: Overview of knowledge discovery in databases process and data mining for surveillance technologies and EWS. In: Koyuncugil, A.S., Özgülbaϑ, N. (eds.) Surveillance Technologies and Early Warning Systems: Data Mining Applications for Risk Detection. IGI Global Publisher (Idea Group Publisher), Hershey, PA (2011)
4. Batmaz, İ., Yazıcı, C., Yerlikaya-Özkurt, F.: Bootstrapping conic multivariate adaptive regression splines. Technical Report, 4, (2012)
5. Batmaz, İ., Yerlikaya-Özkurt, F., Kartal-Koç, E., Köksal, G., Weber, G.-W.: Evaluating the CMARS performance for modeling nonlinearities. In: Proceedings of the 3rd Global Conference on Power Control and Optimization, Gold Coast (Australia), 1239, pp. 351–357 (2010)
6. Breiman, L., Friedman, J.-H., Olshen, R., Stone, C.: Classification and regression trees. Wadswort International Group, Belmont, California (1984)
7. Craven, P., Wahba, G.: Smoothing noisy data with spline functions: estimating the correct degree of smoothing by the method of generalized cross-validation. Numerische Mathematik 31, 377–403 (1979)
8. Çelik, G.: Parameter Estimation in Generalized Partial Linear Models with Conic Quadratic Programming. MSc., Middle East Technical University (2010)
9. Davis, C.-S.: Statistical Methods for the Analysis of Repeated Measures. Springer, New York (2003)
10. Efron, B., Tibshirani, R.-J.: An Introduction to the Bootstrap. Chapman & Hall, New York (1993)
11. Friedman, J.-H.: Multivariate adaptive regression splines. Ann. Stat. 19, 1–141 (1991)
12. Hastie, T., Tibshirani, R.: Generalized additive models: some applications. J. Am. Stat. Assoc. 82, 371–386 (1987)
13. Hastie, T., Tibshirani, R., Friedman, J.-H.: The Elements of Statistical Learning, Data Mining, Inference and Prediction. Springer, New York (2001)
14. Hock, W., Schittkowski, K.: Test Examples for Nonlinear Programming Codes. Springer, New York (1981)

15. Kartal, E.: Metamodelling Complex Systems Using Liner and Nonlinear Regression Methods. MSc., Middle East Technical University (2007)
16. Kartal-Koç, E., İyigün, C.: Restructuring forward stepwise algorithm of multivariate adaptive regression splines (MARS) by using self organizing maps (SOM). In: INFORMS 2011 Annual Meeting. Charlotte, North Carolina, USA, November 13–16 (2011)
17. Köksal, G., Batmaz, İ., Testik, M.-C.: A review of data mining applications for quality improvement in manufacturing industry. Expert Syst. Appl. (2011). Doi.10.1016/j.eswa.2011.04.063
18. MARS from Salford Systems. Available at http://www.salfordsystems.com/mars/phb
19. Martinez, W.-L., Martinez, A.-R.: Computational Statistics Handbook with MATLAB. Chapman and Hall, CRC, London (2002)
20. MATLAB Version 7.5 (R2007b)
21. Milborrow, S.: Earth: Multivariate Adaptive Regression Spline Models, R Software Package (2009). Available at http://cran.r-project.org/web/packages/earth/index.html
22. MINITAB 15 Statistical Software. Available at www.minitab.com
23. MOSEK, A very powerful commercial software for CQP. Available at http://www.mosek.com
24. Nemirovski, A.: A Lectures on Modern Convex Optimisation, Israel Institute of Technology (2002). Available at http://iew3.technion.ac.il/Labs/Opt/opt/LN/Final.pdf
25. Nesterov, Y.-E., Nemirovski, A.-S.: Interior Point Polynomial Algorithms in Convex Programming. SIAM, Philadelphia (1994)
26. Neter, J., Kutner, M., Wasserman, W., Nachtsheim, C.: Applied Liear Statistical Models. WCB/McGrawHill, Boston (1996)
27. Osei-Bryson, K.-M.: Evaluation of decision trees: A multi-criteria approach. Comput. Oper. Res. **31**, 1933–1945 (2004)
28. Özmen, A.: Robust Conic Quadratic Programming Applied to Quality Improvement - A Robustification of CMARS. MSc., Middle East Technical University (2010)
29. Özmen, A., Yıldırım, M.-H., Türker-Bayrak, Ö., Weber, G.-W.: Electricity Price Modelling for Turkey, OR 2011 Zurich, August 30 September 2, 2011, Zurich, Switzerland
30. Özmen, A., Weber, G.-W., Batmaz, İ.: The new robust CMARS (RCMARS) method. In: 24th Mini EURO Conference - On Continuous Optimization and Information-Based Technologies in the Financial Sector, MEC EurOPT Selected Papers, ISI Proceedings, pp. 362–368 (2010)
31. Özmen, A., Weber, G.-W., Batmaz, İ., Kropat, E.: RCMARS: Robustification of CMARS with different scenarios under polyhedral uncertainty set. Communications in Nonlinear Science and Numerical Simulation (CNSNS): Nonlinear, Fractional and Complex. DOI: 10.1016/j.cnsns.2011.04.001
32. Özöğür-Akyüz, S., Weber, G.-W.: Infinite kernel learning via infinite and semi-infinite programming. Optim. Method Software **6**, 937–970 (2010)
33. Sezgin-Alp, Ö., Büyükbebeci, E., Işcanoğlu-Çekiç, A., Yerlikaya-Özkurt, F., Taylan, P., Weber, G.-W.: -CMARS and GAM and CQP- modern optimization methods applied to international credit default prediction. J. Comput. Appl. Math. **235**, 4639–4651 (2011)
34. StatLib: Datasets Archive. Available at http://lib.stat.cmu.edu/datasets/
35. Taylan, P., Weber, G.-W., Beck, A.: New approaches to regression by generalized additive models and continuoues optimisation for modern applications in finance, science and technology. J. Optim. **56**, 675–698 (2007)
36. Taylan, P., Weber, G.-W., Yerlikaya, F.: Continuous optimisation applied in MARS for modern applications in finance, science and technology. In: ISI Proceedings of 20th Mini-EURO Conference Continuous Optimisation and Knowledge-Based Technologies, pp. 317–322 (2008)
37. Taylan, P., Weber, G.-W., Yerlikaya-Özkurt, F.: A new approach to multivariate adaptive regression spline by using Tikhonov regularizati on and continuous optimization. TOP (Oper. Res. J. SEIO (Spanish Stat. Oper. Res. Soc.) **18**, 377–395 (2010)
38. UCI: Machine Learning Repository. Available at http://archive.ics.uci.edu/ml/
39. Weber, G.-W., Batmaz, İ., Köksal, G., Taylan, P., Yerlikaya-Özkurt, F.: CMARS: A new contribution to nonparametric regression with multivariate adaptive regression splines supported by continuous optimisation. Inverse Probl. Sci. Eng. **20**(3), 371–400 (2011)

40. Weber, G.-W., Özmen, A., Batmaz, İ.: RCMARS: The new robust CMARS method. In: VI Moscow International Conference on Operations Research (ORM 2010), Moscow, Russia, October 20–25 (2010)
41. Yazıcı, C.: A Computational Approach to Nonparametric Regression: Bootstrapping the CMARS Method. MSc., Middle East Technical University (2011)
42. Yazıcı, C., Yerlikaya-Özkurt, F., Batmaz, İ.: A computational approach to nonparametric regression: bootstrapping the CMARS method. In: CFE-ERCIM 2011: 4th International Conference of the ERCIM (European Research Consortium for Informatics and Mathematics) Working Group on Computing and Statistics. London, UK. December 17–19 (2011). Book of Abstracts, 115
43. Yerlikaya, F.: A New Contribution to Nonlinear Robust Regression and Classification with MARS and Its Application to Data Mining for Quality Control in Manufacturing. MSc., Middle East Technical University (2008)
44. Yerlikaya Özkurt, F., Taylan, P., Batmaz, İ., Köksal, G., Weber, G.-W.: A modification of MARS by Tikhonov regularization and conic quadratic programming for modeling quality data. In: EURO XXIII 2009, Bonn, Germany, July 5–8 (2009)

Chapter 41
Survey and Evaluation on Modelling of Next-Day Electricity Prices

Miray Hanım Yıldırım, Özlem Türker Bayrak, and Gerhard-Wilhelm Weber

41.1 Introduction

Economic and industrial growths, technological improvements and rapid increases in population and urbanization cause an over reliance on electricity in all over the world. A sustainable and reliable electricity supply, which can be achieved by the actions not only on the strategic or tactical level (i.e., long or medium term actions), but also on the operational level (i.e., very short term), is one of the major problem of the world. Thus, many countries have been making a radical paradigm change in their policies, strategies, actions and market structures to guarantee a sustainable and reliable electricity supply.

Success in operational level actions is very critical, especially, for electricity market. In the last two decades, market structures have been changed from a regulated form to a deregulated form in many developed or developing country. These reforms aim at competition in the electricity supply, which triggers to minimize the cost while improving the quality. Besides, there exists an environmental benefit of the deregulated form. In general, competitive markets attract innovation and new investments which may be an important opportunity for producing green electricity.

M.H. Yıldırım (✉)
Institute of Applied Mathamatics, Middle East Technical University, Ankara, Turkey

Department of Industrial Engineering, Cankaya University, Ankara, Turkey
e-mail: maslan@cankaya.edu.tr

Ö.T. Bayrak
Department of Industrial Engineering, Cankaya University, Ankara, Turkey
e-mail: ozlemt@cankaya.edu.tr

G.-W. Weber
Institute of Applied Mathamatics, Middle East Technical University, Ankara, Turkey
e-mail: gweber@metu.edu.tr

A.A. Pinto and D. Zilberman (eds.), *Modeling, Dynamics, Optimization and Bioeconomics I*, Springer Proceedings in Mathematics & Statistics 73, DOI 10.1007/978-3-319-04849-9_41,
© Springer International Publishing Switzerland 2014

Fig. 41.1 Categorization
of the methods used in next
day's electricity price
forecasting

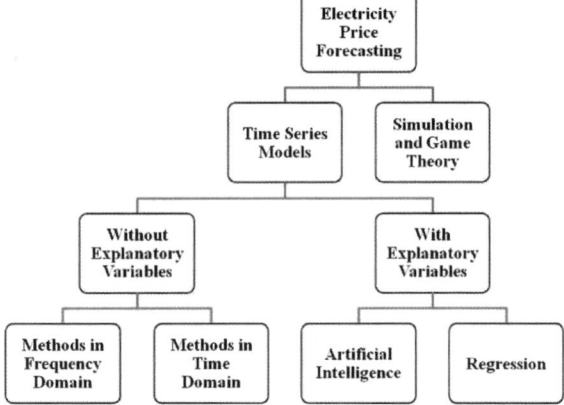

One of the main advantages of deregulated markets is the transparency in electricity prices. Transparent electricity prices are the results of competitive markets. Operational level actions taken in these markets directly affect the electricity prices in a very short term. There are reviews on electricity price forecasting methods. For instance, [1, 2, 33] present a review on electricity price forecasting under deregulated market conditions. Among these studies, [2, 33] focus on short-term forecasting. However, there is no substantial review work specifically on next day's electricity price forecasting. Here, an in-depth survey on forecasting methods of next day's electricity prices is presented.

Forecasting methods examined here are classified mainly in two categories: time-series based methods and simulation and game theory based methods. Time-series based methods are also divided into two sub-categories regarding the existence of explanatory variables in the model. Figure 41.1 illustrates this taxonomy. Regarding this categorization, a detailed review of time-series methods with sub-categories is given in Sect. 41.2. Simulation and game theory based methods are outlined in Sect. 41.3. Different approaches like hybrid methods that are used in next day's price forecasting are elaborated in Sect. 41.4. The survey is concluded in Sect. 41.5 with an evaluation of the methods used in forecasting of next-day electricity prices.

41.2   Time Series Models

Next day's electricity price data generally exhibit seasonality and high volatility, while involving outliers, high rate of recurrence, and non-constant mean and variance [16]. Time-series models can handle all these structural issues of the data. Therefore, time-series methods are one among of the mostly used forecasting methods. In this section, the models are divided into two categories by considering the existence of explanatory variables. Neural networks (NN), support

Table 41.1 Mathematical equations for time-series based models

Model	Mathematical Equation
AR	$(1 - \sum_{i=1}^{p} \phi_i \cdot L^i) \cdot y_t = c + \epsilon_t$
ARMA	$(1 - \sum_{i=1}^{p} \phi_i \cdot L^i) \cdot y_t = c + (1 + \sum_{i=1}^{q} \theta_i \cdot L^i) \cdot \epsilon_t$
ARIMA	$(1 - \sum_{i=1}^{p} \phi_i \cdot L^i) \cdot (1 - L)^d \cdot y_t = c + (1 + \sum_{i=1}^{q} \theta_i \cdot L^i) \cdot \epsilon_t$
GARCH	$(1 - \sum_{i=1}^{p} \phi_i \cdot L^i) \cdot y_t = c + (1 + \sum_{i=1}^{q} \theta_i \cdot L^i) \cdot \epsilon_t$
DR	$(1 - \sum_{i=1}^{p} \phi_i \cdot L^i) \cdot y_t = c + \sum_{i=1}^{n} \sum_{j=1}^{r_i} a_{i,j} \cdot L^j \cdot x_{i,t} \cdot \epsilon_t$
ARMAX	$(1 - \sum_{i=1}^{p} \phi_i \cdot L^i) \cdot y_t = c + (1 + \sum_{i=1}^{q} \theta_i \cdot L^i) \cdot \epsilon_t + \sum_{i=1}^{n} a_i \cdot x_{i,t}$

vector machines (SVM), data mining, generalized autoregressive conditional heteroskedasticity (GARCH), and dynamic regression (DR) are reviewed under the methods with explanatory variables. On the other hand, wavelet transforms (WT) in frequency domain, autoregressive (AR) models, integrated (I) models, moving average (MA) models and their combinations (ARIMA, ARMA, ARMAX, etc.) in time domain are considered when the model is constructed without explanatory variables. Table 41.1 gives mathematical formulations of some of these models [2].

In these formulas, y_t and ϵ_t represent price and error at time t, respectively, L represents back shift operator and $L^p y_t$ is a difference operator. Coefficients of polynomial function are shown by ϕ and θ_i, and c represents a constant number. Furthermore, p, q, d, n, and r_i are the model order parameters. In DR and ARMAX, explanatory variables are shown by $x_{i,t}$, whereas $a_{i,j}$ and a_i represent regression coefficients.

41.2.1 Models with Explanatory Variables

Electricity prices are affected by several factors such as demand and temperature. Models with explanatory variables are used to identify these factors and the value of their effects. These methods can be categorized as artificial intelligence and regression-based methods.

Artificial intelligence (AI) methods simulate human brain in order to train future prices by using historical electricity price and the factors that affect it. Both Fig. 41.2 and the following equation show the basic form of a neural network:

$$\hat{y} = f(x_1, x_2, \ldots, x_n).$$

Fig. 41.2 Basic
representation of NN

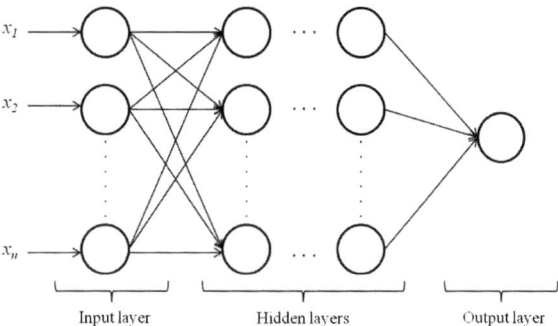

Input layer   Hidden layers   Output layer

Here, $x_1, x_2, \ldots, x_n$ represent n independent variables and $\hat{y}$ is the forecasted
output. The model takes the following form when the electricity prices forecast:

$$\hat{y}_{t+1} = f(y_t, y_{t-1}, \ldots, y_{t-n}),$$

where y_t means the price at time t.

Neural network and its derivations are the mostly used methods in this category.
The paper [55] represents a model that forecasts next day's electricity price,
especially for the weekends and public holidays in United Kingdom whereas [25]
proposes a model for all days with an acceptable error. Three multilayer feed-
forward neural networks are constructed to predict the price. Mean absolute
percentage errors of these models show that the networks can be used to develop
bidding strategies for the traders in the market. The paper [53] forecasts electricity
prices of the Pennsylvania-New Jersey-Maryland (PJM) market using a three-step
approach: preparation of the inputs, artificial neural network (ANN) training and
simulation, and accuracy assessment. Comparing the results in terms of errors shows
that ANN performs better than ARIMA and Wavelet Transform. ANN is used in [38]
to forecast next day's electricity price by considering both mean and hourly prices.
Cascaded NN developed with a feature selection technique is proposed in [7]. The
new approach involves a chain of NN based forecast engines, which is compared
with classical techniques by using Spanish and Australian national electricity
markets' data, shows a better performance than the classical form. Fuzzy NN is
another method derived from NNs. A new approach that uses fuzzy logic, which
is applied for Spanish market, is proposed in [3]. NN has a feed-forward structure
with three layers: input, hidden and output. Input vector of node j is defined as
$I_j = (W_{1j}^{(1)} X_1, W_{2j}^{(1)} X_2, \ldots, W_{Nj}^{(1)} X_N)$, where $W_{ij}^{(1)}$ is the weight between input
and hidden nodes. Smaller error is obtained compared to DR, ARIMA, and transfer
functions. Fuzzy NN follows the process given below in order to obtain forecasted
outputs:

$$O_j = \frac{HV_j}{HV'_j},$$

$$HV_j = \prod_{k=1}^{N}(UP_{jk} - LOW_{jk}),$$

$$HV'_j = \prod_{k=1}^{N} \max\{(UP_{jk} - LOW_{jk}), (UP_{jk} - I_{jk})(I_{jk} - LOW_{jk})\},$$

$$out = \sum_{j=1}^{N_H} O_j W_j^{(2)}$$

for $j = 1, 2, \ldots, N_H$. Here, O_j is output of node j, N_H is the number of hidden nodes, HV_j represents hypervolume of hidden node j, LOW_{jk} and UP_{jk} are its lower and upper bounds. $W_j^{(2)}$ is the weight between hidden node and output node.

The ANN model given in [42] uses Spanish market's data in a different training method in which the input vectors are classified according to their similarities. In order to show the strong dependency between the prices and the demand, Ranjbar et al. [43] uses ANN. This method gives better results when the prices have no spikes. In case of spikes, an enhanced probability neural network (EPNN) yields accurate results [36]. EPNN add a new layer, summation, and a new process, orthogonal experimental design, to the network in order to decrease the forecast error. Hidden nodes H_k and output nodes O_j are defined as follows:

$$H_k = \exp\left\{-\sum_{i=1}^{n} \frac{(x_i - w_{ki}^{IH})^2}{2\sigma_k^2}\right\},$$

$$O_j = \frac{\sum_{k=1}^{K} w_{jk}^{HS} H_k}{\sum_{k=1}^{K} H_k} = \frac{S_j}{\sum_{k=1}^{K} H_k},$$

where $x_i(k)$ is the ith variable of the kth predicted output, σ_k is the predetermined parameter, w_{jk}^{HS} represents weights between hidden node H_k and the summation node S_j.

Input-Output Hidden Markov Model (IOHMM) is another ANN based approach used to forecast next day's electricity price. IOHMM for the Spanish electricity market is presented in [28]. The model provides comprehending the market and predicting the prices with a reasonable accuracy. NN is applied for the Iranian electricity market with pay-as-bid pricing mechanism [13] and for PJM electricity market in order to decrease prediction error for peak prices [51].

There are hybrid approaches that use AI philosophy. SVM and evolutionary algorithms (EA) are also used to forecast next day's electricity market. The article [22] uses a SVM algorithm with particle swarm optimization (PSO) for several electricity markets. A flexible support vector regression (SVR) model is defined as follows:

$$f(x) = \sum_{i=1}^{n} w_i \phi_i(x_i) + b,$$

where x_i is the ith input, w_i and b are the unknown parameters. The model takes the following form in order to minimize model complexity:

$$\text{Min } \frac{1}{2} \|w\|^2$$

subject to

$$y_i - \phi(w, x_i) - b \leq \epsilon$$

$$\phi(w, x_i) + b_i - y \leq \epsilon$$

for $i = 1, 2, \ldots, n$. The performance of the method is measured by using Italy, New England, and New York markets' data. Results show a better performance compared to classical ARIMA models.

Other combinations are made by Saini et al. [44] and Fan et al. [20]. In [44], the parameters of SVM are optimized by a genetic algorithm and results are obtained with acceptable accuracy. On the other hand, Fan et al. [20] uses a self-organized map network (SOM) to group the input data set as an unsupervised learning mechanism and SVM to fit the training data as a supervised learning mechanism. This hybrid method is applied for New England electricity market. NN and evolutionary algorithms with an iterative parameter search process are combined for PJM and Spanish electricity market [6]. SVM, PSO and SOM are used together to forecast electricity prices of PJM in [39].

The forecasting methods based on regression, model the relationship between electricity prices and the factors that affect the prices. Thus, electricity price can be estimated by using exogenous variables like demand. Dynamic regression and generalized autoregressive conditionally heteroskedastic (GARCH) method are the most common techniques in this category. However, in most studies, dynamic regression is used to compare the prediction performances. For instance, in [40], dynamic regression and transfer function methods are applied for the Spanish and the Californian markets. The relationship between fundamental factors (such as demand, demand slope and curvature, demand volatility, excess of generation capacity, scarcity, price volatility, etc.) and their effects over time via several versions of regression are modelled by Karakatsani and Bunn [34]. These models give the best results, especially for British market's data. In [26], multivariate regression is used to analyze the effect of renewable energy to electricity prices in the Spanish market. Forecasting for the Spanish and the Californian electricity market is made by GARCH method in [23]. The model's performance is better than the one by ARIMA, in particular, when volatility and price spikes exist. GARCH seasonal dynamic factor analysis (GARCH-SeaDFA) is a specific approach for the structure of electricity price data. This approach is proposed in [24] and applied for the Spanish market. Forecasting performance of GARCH models is especially better when a volatility is included. Because of this fact, [30, 31] use GARCH models for their analyses.

Generalized additive models (GAM) are used to maximize quality of prediction via involving nonlinear effects [32]. They have the following form:

$$E(Y|X_1, X_2, \ldots, X_p) = \alpha + f_1(X_1) + f_2(X_2) + \ldots + f_p(X_p).$$

Here, $X_1, X_2, \ldots, X_p$ represent predictors, Y represents the outcome, and f_i's are unspecified smooth functions. The paper [45] uses generalized additive models via location, scale and shape estimation of specific time instances. The estimated parameters are used as an input for dynamically chancing prices. Another approach is proposed by Özmen et al. [41] by using the logic given in [49, 50]. In [41], GAM generates an initial model of next day's price which is improved by robust optimization technique.

41.2.1.1 Models Without Explanatory Variables

Electricity prices can be predicted by the models that have no explanatory variables. A classification of these models can be made based on whether time domain or frequency domain is used. AR, ARIMA, ARMA, etc., are the most frequently used time-based models. On the other hand, wavelet transform (WT) enlarges the time space to the time-frequency space. For instance, Amjady and Keynia [5] applies WT as preprocessor in order to make this expansion and then to forecast prices with a better performance via combination of NN and EA. One of the main advantages of WT is that the method can decompose time series in time and frequency. By considering this advantage, Benaouda and Murtagh [11] proposes a WT with multi-resolution decomposition, implemented for Australia's data. The method performs better than single resolution forms.

In liberal markets, electricity price data generally have a high frequency, non-constant mean and variance, and multiple seasonality. Thus, AR, ARIMA and ARMA are very suitable methods for this kind of data. The ARIMA model proposed by Contreras et al. [18] gives reasonable prediction errors for the Spanish and the Californian markets. In some studies, models without explanatory variables are combined with models with explanatory variables. For instance, Swider and Weber [47] suggests a model by using ARMA extended by GARCH and applies the model for the German market. Similarly, Conejo et al. [17] proposes a WT-based model combined with ARIMA, and Tan et al. [48] presents a WT-based model combined with ARIMA and GARCH. ARIMA and its variations are used to forecast

Midwest Independent System Operator in [15]. WT decomposes and restructures the data set and, then, ARMA and GARCH predict next day's prices for PJM and Spanish markets in spite of the data set reveals nonstationarity, nonlinearity and high volatility. AR models with nonparametric extensions proposed in [56] are applied for Nordic and Californian markets. In [34], regression and AR is combined with time-varying parameter effects. AR and regression models give better results when they are combined with time-varying coefficients.

41.3 Simulation and Game Theory

Game-theory and simulation-based methods are generally devoted to improve strategies for market participants. These methods are developed for predicting market operators' buying and selling bids. Moreover, simulation models try to imitate the real market and its conditions directly. For instance, Bernal-Agustin et al. [12] develops a market simulator for the Spanish market. The algorithm includes the following steps: (i) calculation of intersection of supply, demand and market clearing price for each hour in a day, (ii) assignment of selling bid, (iii) assignment of buying bid, (iv) acceptance of the bids if the maximum variation of the unit output between two consecutive hours is between the required limits, (v) verification of bids for non-divisible quantity rule, (vi) verification of minimum revenue rule. The algorithm used in [12] guarantees a feasible result.

Game theory is generally used to determine bidding strategies. A generation of companies' bidding strategies under operational constraints is investigated in [14]. A static game theory and a cost-minimization unit-commitment algorithm are developed for generating companies. Thus, the companies can analyze bidding strategies in the market.

There are also combined versions of game theory and simulation in order to use the advantages of both methods. For example, Guerci et al. [29] proposes a multi-agent based simulation model for physical power exchange markets such as Spanish and Italian markets. Stochastic control theory is used by Giabardo et al. [27], where a Cournot competition model is considered for bidding strategies. In addition, simulations are made to see long-term optimal strategies.

41.4 Other Approaches Used in Next Day's Price Forecasting

Forecasting next day's price is a challenging problem since the prices can be affected by many factors. As a consequence, different approaches are developed to tackle these issues and to suppress the disadvantages of classical methods. For instance, in [57], market price is investigated for New York Independent System Operator day-ahead market with different demand values. Satisfactory predictions can be made for the Spanish electricity market by using weighted nearest neighbors; that is presented in [37]. A forecasting system with multi-component that consists of a fuzzy inference system, an intelligent system, and least-squares estimation is developed by Li et al. [35]. Designing the input vectors of electricity prices is an important issue that affects the forecasting performance. A hybrid NN model is proposed by Amjady et al. [4] with a relief algorithm. This algorithm is used to select the features of the input vector. In order to handle daily seasonality, Vilar et al. [54] uses a functional nonparametric model. Here, electricity price is considered as discrete-time realization of a continuous time stochastic process, $\{X(t)\}_{t \in \mathbb{R}}$ observed for $t \in [a, b)$ and $X(t)$ is a seasonal process whose interval $[a, b)$ includes $n + 1$ seasonal periods of length τ and $b = a + (n + 1)\tau$. Functional data and its corresponding functional nonparametric model are given below:

$$X_i(t) = X(a + (i - 1)\tau + t),$$

$$X_{i+1}(r) = m(X_i) + \varepsilon_{i+1},$$

for $t \in C = [0, \tau)$ and $i = 1, \ldots, n$.

Forecasting of price spikes is another issue in day-ahead electricity markets. In [9], a specific method that consists of a probabilistic NN and a hybrid neuro-evolutionary system is developed. Another hybrid neuro-evolutionary system is developed by Amjady and Keynia [8] to improve forecasting accuracy by using PJM and Spanish markets' data. A hybrid nonlinear chaotic dynamic and evolutionary strategy-based approach is developed by Unsihuay-Vila et al. [52]. This method applies a training procedure for increasing performance of forecasting. The paper [10] uses autoregressive-moving-average model with exogenous inputs (ARMAX), where fuzzy logic is employed. A hybrid wavelet-ARIMA and radial basis function neural model is proposed by Shafie-khah et al. [46] to obtain an improved accuracy with less input data. Stochastic programming is also used in price forecasting especially for bidding strategies. For instance, a quadratic mixed-integer stochastic programming model is proposed by Corchero and Heredia [19] for optimal-4qbid strategies of the Spanish market. In [21], a stochastic mixed-integer linear programming model is used for the bidding problem of the Nordic power market.

41.5 Evaluations

There is a variety of methods developed to forecast next day electricity prices of many different market. Among these, NN-based methods are the most common. In NN based methods, the level of accuracy ranges between 0.5 and 47 % (Table 41.2). Among these methods, ANN-based hybrid methods give the promising results. However, superiority of ANN could not be proved yet. In order to assess NN-based methods, traditional methods such as AR, ARIMA, GARCH, linear regression (LR), and multiple regression (MR) are used. Most of these traditional methods do not involve explanatory variables. Although these methods are applied to assess NN-based approaches, a wider error range is observed (Table 41.3). In some electricity markets, explanatory variables (e.g., electricity demand, fuel price,

Table 41.2 Papers used NN-based methodology

Ref. No	Method	Error Type	Accuracy
[16]	ARIMA, DR, TF, NN and WT	FMSE, DE	1–23 %
[55]	NN	DMAPE	8.93–12.19 %
[25]	NN	PE	Error less than 1c€ in 85 % of the cases
[53]	ANN	MAE, MAPE	1–9 %
[38]	ANN	APE	1–25 %
[7]	Cascaded NN	MAPE, WMAPE	4–7 %
[3]	Fuzzy NN	WMAPE	7.5 %
[37]	Weighted NN	MRE, MAE, MMRE	5–14 %
[42]	ANN	MAPE	3–10 %
[43]	ANN	MAE, RMSE	0.5–9 %
[36]	Enhanced Probability NN	RMSE, MAPE, MAE	1–8 %
[13]	ANN	RMSE, MAPE, MAE	0.7–11 %
[51]	NN	MAPE	11–33 %
[5]	WT-Hybrid forecast method (NN and EA)	WME, WPE	4–26 %
[4]	Modified relief algorithm and hybrid NN	WMAPE, WMAE	4–9 %
[9]	FST-probabilistic NN-HNES	MAE, MAPE	5–47 %
[46]	WT-ARIMA-RBFNN	WFE	4–7 %

MAE: mean absolute error, *APE*: absolute percentage error, *MAPE*: mean APE, *MPE*: mean percentage error, *MSE*: mean square error, *RMSE*: root mean square error, *FMSE*: forecast MSE, *RMSFE*: root mean square forecasting error, *PE*: prediction error, *DE*: daily error, *DMAPE*: daily MAPE, *WMAPE*: weekly MAPE, *WMAE*: weekly MAE, *MRE*: mean relative error, *MMRE*: mean error relative to mean price, *WME*: weekly mean error, *WPE*: weekly peak error, *WFE*: weekly forecast error

Table 41.3 Papers based on time series methods without explanatory variables

Ref. No	Method	Error Type	Accuracy
[40]	DR-TFM	FMSE	2–8 %
[34]	AR, LR	RMSE, MAE, MAPE, Max APE	0–13 %
[26]	MR	–	–
[23]	GARCH	FMSE	1–43 %
[24]	GARCH-SeaDFA	MAPE	5–9 %
[30]	GARCH	MFE, MAFE, RMSFE	1–53 %
[31]	GARCH	–	–
[45]	GAM	MWE	6–23 %
[5]	WT-Hybrid NN and EA	WME, WPE	4–26 %
[11]	Hybrid WT	APE, MAPE, RMSE	2–44 %
[18]	ARIMA	FMSE, WME	4–21 %
[47]	ARMA-GARCH	MAE, MAPE	4–13 %
[17]	WT-ARIMA	MAPE	5–11 %
[48]	WT-ARIMA and GARCH	MAPE	0–2 %
[15]	ARIMA, ARIMA-EGARCH, and ARIMA-EGARCH-M	RMSE, MAPE, MAE	10–96 %
[56]	AR and its extensions	WME	2–50 %

Table 41.4 Factors affecting electricity price

Ref. No	Factors Affecting Price
[3, 9, 15, 17, 31, 37, 46–48, 52–54]	Historical prices
[4, 6–8, 11, 16, 20, 23, 24, 35, 38–42, 45]	Historical prices, demand
[26]	Demand, composition of electricity production by each energy resource (renewables, cogeneration, hydro, nuclear, combined cycle, fuel and natural gas), net electricity exports, pumping and distribution losses
[22]	Fuel price, market concentration index, reserve margin
[18]	Historical prices (demand and available daily production of hydro units in with explanatory variable case)
[25]	Historical prices, day and month type
[55]	Historical prices, demand, settlement period
[5]	Historical prices, demand, available generation
[43]	Historical prices, demand, change in demand, time slot of the day, day of week
[13]	Historical prices, demand, change in demand, time slot of the day, day of week
[34]	Historical prices, demand, demand volatility, demand slope and curvature, scarcity, spread, diurnal and weekly effects, seasonality, trend, excess in generation capacity
[28]	Historical prices, demand, hydro generation, nuclear generation, thermal generation
[10]	Historical prices, demand, imports
[56]	Historical prices, demand, temperature
[44]	Historical prices, demand, temperature, humidity, crude oil prices, wind speed
[30]	Historical prices, natural gas prices
[36]	Historical prices, system load and temperature
[51]	Weekly variation data of electricity price and demand

available generator capacities, temperature, humidity) are highly affective. Especially in such systems, traditional methods yield significant errors.

The most critical factors affecting these systems are price history and electricity demand (Table 41.4). Other common factors include resource prices, generator capacities, climate effects, and time slot. These factors variate with respect to the market of concern. Among the electricity markets, mainly European markets—especially, the Spanish—are studied (Table 41.5). This is triggered by early revolution into a competitive market structure. In addition to Europe, PJM is also commonly investigated.

Table 41.5 Paper categorization of electricity markets

Market	Number of Papers	Reference Number(s)
Spanish electricity market	21	[3,6–8,12,17–19,22–24,26,28, 29,37,40,42,46,48,52,54]
PJM	11	[5,6,8,9,16,35,36,39,48,51,53]
Italian electricity market	4	[4,22,38,45]
New England	4	[4,20,22,52]
New York	4	[22,30,31,57]
National Electricity Market-Victoria-New South Wales	3	[7,11,44]
Nord Pool	3	[21,27,56]
United Kingdom	2	[34,55]
European Energy Exchange	2	[25,47]
Ontario	2	[4,10]
Iran electricity market	1	[13]
Turkish electricity market	1	[41]
Five hubs of the Midwest Independent System Operator	1	[15]

41.6 Conclusion and Outlook

The electricity markets' structures in the world have been changing especially in the last decades. These reforms aim at a higher competition between service providers to supply electricity with a low price but high quality. In competitive markets, it is very critical to predict actions in a very short period of time for both demand and supply side. Therefore, next-day electricity price forecasting is a very promising field for the researchers.

In this study, day-ahead markets and their electricity price forecasting models are reviewed. A new taxonomy is given to categorize the papers in an efficient way. It classifies the methods according to the most simple form whereas at the same time it covers all approaches used for day-ahead markets. Thus, the main categorization is made according to the model or approach used. In addition, some sub-categories, such as influencing factors of electricity price, market types whose data are used, error types and accuracy levels, are listed and compared. Moreover, the taxonomy can be up-to-date all the time because of its general hierarchial structure.

This study may guide the researchers through determination of the approach and its infrastructure as well as through evaluation and comparison of results. For instance, existing classifications of short-term price forecasting methods do not consider the effects of explanatory variables. On the other hand, the taxonomy presented in this study classifies the methods considering the existence of explanatory variables. As a result, a researcher may select a proper method to analyze by investigating the explanatory variables in the system. In addition, accuracies of the methods are presented in the study. Therefore, both market participants and researchers may benefit from these data in order to assess their approaches and bidding strategies.

References

1. Aggarwal, S.K., Saini, L., Kumar, A.: Electricity price forecasting in deregulated markets: A review and evaluation. Int. J. Electr. Power Energy Syst. **31**, 13–22 (2009a)
2. Aggarwal, S.K., Saini, L.M., Kumar, A.: Short term price forecasting in deregulated electricity markets: A review of statistical models and key issues. Int. J. Energy Sect. Manag. **3**, 333–358 (2009b)
3. Amjady, N.: Day-ahead price forecasting of electricity markets by a new fuzzy neural network. IEEE Trans. Power Syst. **21**, 887–896 (2006)
4. Amjady, N., Daraeepour, A., Keynia, F.: Day-ahead electricity price forecasting by modified relief algorithm and hybrid neural network. IET Gen. Trans. Distrib. **4**, 432–444 (2010)
5. Amjady, N., Keynia, F.: Day ahead price forecasting of electricity markets by a mixed data model and hybrid forecast method. Int. J. Electr. Power Energy Syst. **30**, 533–546 (2008)
6. Amjady, N., Keynia, F.: Day-ahead price forecasting of electricity markets by mutual information technique and cascaded neuro-evolutionary algorithm. Power **24**, 306–318 (2009a)
7. Amjady, N., Keynia, F.: Day-ahead price forecasting of electricity markets by a new feature selection algorithm and cascaded neural network technique. Energy Convers. Manag. **50**, 2976–2982 (2009b)
8. Amjady, N., Keynia, F.: Application of a new hybrid neuro-evolutionary system for day-ahead price forecasting of electricity markets. Appl. Soft Comput. **10**, 784–792 (2010)
9. Amjady, N., Keynia, F.: A new prediction strategy for price spike forecasting of day-ahead electricity markets. Appl. Soft Comput. **11**, 4246–4256 (2011)
10. Arciniegas, A., Arciniegasrueda, I.: Forecasting short-term power prices in the Ontario Electricity Market (OEM) with a fuzzy logic based inference system. Util. Pol. **16**, 39–48 (2008)
11. Benaouda, D., Murtagh, F.: Hybrid Wavelet model for electricity pool-price forecasting in a deregulated electricity market. IEEE Int. Conf. Eng. Intell. Syst. 1–6 (2006)
12. Bernal-Agustin, J.L., Contreras, J., Martin-Flores, R., Conejo, A.J.: Realistic electricity market simulator for energy and economic studies. Electr. Power Syst. Res. **77**, 46–54 (2007)
13. Bigdeli, N., Afshar, K., Amjady, N.: Market data analysis and short-term price forecasting in the Iran electricity market with pay-as-bid payment mechanism. Electr. Power Syst. Res. **79**, 888–898 (2009)
14. Borghetti, A., Massucco, S., Silvestro, F.: Influence of feasibility constrains on the bidding strategy selection in a day-ahead electricity market session. Electr. Power Syst. Res. **79**, 1727–1737 (2009)
15. Bowden, N., Payne, J.E.: Short term forecasting of electricity prices for MISO hubs: Evidence from ARIMA-EGARCH models. Energy Econ. **30**, 3186–3197 (2008)
16. Conejo, A.J., Contreras, J., Espy, R., Plazas, M.A.: Forecasting electricity prices for a day-ahead pool-based electric energy market. Int. J. Forecast. **21**, 435–462 (2005)
17. Conejo, A.J., Plazas, M.A., Espinola, R., Member, S., Molina, A.B.: Day-ahead electricity price forecasting using the wavelet transform and ARIMA models. Power **20**, 1035–1042 (2005)
18. Contreras, J., Espinola, R., Member, S., Nogales, F.J.: ARIMA models to predict next-day electricity prices. Power **18**, 1014–1020 (2003)
19. Corchero, C., Heredia, F.J.: A stochastic programming model for the thermal optimal day-ahead bid problem with physical futures contracts. Comput. Oper. Res. **38**, 1501–1512 (2011)
20. Fan, S., Mao, C., Chen, L.: Next-day electricity-price forecasting using a hybrid network. IET Gener. Transm. Distrib. **1**, 176–182 (2007)
21. Fleten, S.E., Kristoffersen, T.K.: Stochastic programming for optimizing bidding strategies of a Nordic hydropower producer. Eur. J. Oper. Res. **181**, 916–928 (2007)
22. Gao, C., Bompard, E., Napoli, R., Cheng, H.: Price forecast in the competitive electricity market by support vector machine. Phys. A **382**, 98–113 (2007)

23. Garcia, R.C., Contreras, J., Member, S., Akkeren, M.V., Garcia, J.B.C.: A GARCH forecasting model to predict day-ahead electricity prices. Power **20**, 867–874 (2005)
24. Garcia-Martos, C., Rodriguez, J., Sanchez, M.J.: Forecasting electricity prices and their volatilities using unobserved components. Energy Econ. **33**, 1227–1239 (2011)
25. Gareta, R., Romeo, L.M., Gil, A.: Forecasting of electricity prices with neural networks. Energy Convers. Manag. **47**, 1770–1778 (2006)
26. Gelabert, L., Labandeira, X., Linares, P.: An ex-post analysis of the effect of renewables and cogeneration on Spanish electricity prices. Energy Econ. **33**, S59–S65 (2011)
27. Giabardo, P., Zugno, M., Pinson, P., Madsen, H.: Feedback, competition and stochasticity in a day ahead electricity market. Energy Econ. **32**, 292–301 (2010)
28. Gonzalez, A.M., Munoz, A., Roque, S., Garcia-Gonzalez, J.: Modeling and forecasting electricity prices with input/output hidden Markov models. Power **20**, 13–24 (2005)
29. Guerci, E., Ivaldi, S., Pastore, S., Cincotti, S.: Modeling and implementation of an artificial electricity market using agent-based technology. Phys. A **355**, 69–76 (2005)
30. Guirguis, H.S., Felder, F.A.: Further advances in forecasting day-ahead electricity prices using time series models. New York **4**, 159–166 (2004)
31. Hadsell, L.: The impact of virtual bidding on price volatility in New York's wholesale electricity market. Econ. Lett. **95**, 66–72 (2007)
32. Hastie, T., Tibshirani, R., Friedman, J.: The Elements of Statistical Learning: Data Mining, Inference, and Prediction. Springer, New York (2001)
33. Hu, L., Taylor, G., Wan, H.B., Irving, M.: A review of short-term electricity price forecasting techniques in deregulated. In: Univ. Power Eng. Conf. (UPEC), Proc. 44th Int. pp. 1–5, (2009)
34. Karakatsani, N., Bunn, D.: Forecasting electricity prices: The impact of fundamentals and time-varying coefficients. Int. J. Forecast. **24**, 764–785 (2008)
35. Li, G., Member, S., Liu, C.C., Mattson, C., Lawarree, J.: Day-ahead electricity price forecasting in a grid environment. Power **22**, 266–274 (2007)
36. Lin, W.M., Gow, H.J., Tsai, M.T.: Electricity price forecasting using enhanced probability neural network. Energy Convers. Manag. **51**, 2707–2714 (2010)
37. Lora, A.T., Santos, J.M.R., Exposito, A.G., Luis, J., Ramos, M., Member, S., Santos, J.C.R.: Electricity market price forecasting based on weighted nearest neighbors techniques. Power **22**, 1294–1301 (2007)
38. Menniti, D., Scordino, N., Sorrentino, N.: Forecasting next-day electricity prices by a neural network approach. In: 8th Int. Conf. European Energy Market (EEM), pp. 209–215 (2011)
39. Niu, D., Liu, D., Wu, D.D.: A soft computing system for day-ahead electricity price forecasting. Appl. Soft Comput. **10**, 868–875 (2010)
40. Nogales, F.J., Contreras, J., Conejo, A.J., Member, S.: Forecasting next-day electricity prices by time series models. Power **17**, 342–348 (2002)
41. Özmen, A., Yıldırım, M.H., Bayrak, Ö.T., Weber, G.-W.: Electricity price modelling for Turkey. In: Klatte, D., Schmedders, K., Luethi, H.-J. (eds.) OR 2011 Proceedings
42. Pino, R., Parreno, J., Gomez, A., Priore, P.: Forecasting next-day price of electricity in the Spanish energy market using artificial neural networks. Eng. Appl. Artif. Intell. **21**, 53–62 (2008)
43. Ranjbar, M., Soleymani, S., Sadati, N., Ranjbar, A.M.: Electricity price forecasting using artificial neural network. In: 2006 Int. Conf. Power Electron. Drives Energy Syst., pp. 1–5 (2006)
44. Saini, L.M., Aggarwal, S.K., Kumar, A.: Parameter optimisation using genetic algorithm for support vector machine-based price-forecasting model in National electricity market. IET Gener. Transm. Distrib. **4**, 36–49 (2010)
45. Serinaldi, F.: Distributional modeling and short-term forecasting of electricity prices by generalized additive models for location, scale and shape. Energy Econ. **33**, 1216–1226 (2011)
46. Shafie-khah, M., Moghaddam, M.P., Sheikh-El-Eslami, M.K.: Price forecasting of day-ahead electricity markets using a hybrid forecast method. Energy Convers. Manag. **52**, 2165–2169 (2011)

47. Swider, D.J., Weber, C.: Extended ARMA models for estimating price developments on day-ahead electricity markets. Electr. Power Syst. Res. **77**, 583–593 (2007)
48. Tan, Z., Zhang, J., Wang, J., Xu, J.: Day-ahead electricity price forecasting using wavelet transform combined with ARIMA and GARCH models. Appl. Energy **87**, 3606–3610 (2010)
49. Taylan, P., Weber, G.-W.: New approaches to regression in financial mathematics by additive models. J. Comput. Technol. **12**, 3–22 (2007)
50. Taylan, P., Weber, G.-W., Beck, A.: New approaches to regression by generalized additive models and continuous optimization for modern applications in finance, science and technology. Optimization **56**, 675–698 (2007)
51. Toyama, H., Senjyu, T., Areekul, P., Chakraborty, S., Yona, A., Funabashi, T.: Next-day electricity price forecasting on deregulated power market. In: Transm. Distrib. Conf. Expo.: Asia and Pacific. pp. 1–4 (2009)
52. Unsihuay-Vila, C., Zambroni de Souza, A.C., Marangon-Lima, J.W., Balestrassi, P.P.: Electricity demand and spot price forecasting using evolutionary computation combined with chaotic nonlinear dynamic model. Int. J. Electr. Power Energy Syst. **32**, 108–116 (2010)
53. Vahidinasab, V., Jadid, S., Kazemi, A.: Day-ahead price forecasting in restructured power systems using artificial neural networks. Elect. Power Syst. Res. **78**, 1332–1342 (2008)
54. Vilar, J.M., Cao, R., Aneiros, G.: Forecasting next-day electricity demand and price using nonparametric functional methods. Int. J. Electr. Power Energy Syst. **39**, 48–55 (2012)
55. Wang, A.J., Ramsay, B.: A neural network based estimator for electricity spot-pricing with particular reference to weekend and public holidays. Neurocomputing **23**, 47–57 (1998)
56. Weron, R., Misiorek, A.: Forecasting spot electricity prices: A comparison of parametric and semiparametric time series models. Int. J. Forecast. **24**, 744–763 (2008)
57. Zhang, N.: Generators' bidding behavior in the NYISO day-ahead wholesale electricity market. Energy Econ. **31**, 897–913 (2009)

Chapter 42
Calculus and "Digitalization" in Finance: Change of Time Method and Stochastic Taylor Expansion with Computation of Expectation

Fikriye Yılmaz, Hacer Öz, and Gerhard-Wilhelm Weber

42.1 Introduction

The close relations between the various dimensions in the analysis, forecasting and decision making, particularly, *state variables* (stochastic processes) and *time*, has been a crucial issue in stochastic calculus since many decades. In fact, it was decisive in definitions (normalization and standardizations) and computational methods, including the needs of *rescaling*. We recall that such normalizations and scalings with respect to volatility, time intervals and jump hight, are in the center of the axioms which determine what a Brownian motion and a Lévy process have to fulfill, respectively. Already here, intervals [0,1] are addressed, in other words, values 0 and 1 are aimed at and used for core and elegant representations and evaluations. Moreover, 0 and 1 can be used to encode a "yes" or a "no", regarding whether or not an integration of deterministic (Riemann or Stieltjes or Lebesgue) or stochastic (Itô or Stratonovich) kind should be performed. It this way, strings of values 0 and 1 come into play; let us call it a "digitalization"; with those strings we can really work and execute continuous-analytic operations. This contributes very much to the "algebraization" or "automization" of our activities of modeling, computation and decision making in the financial sector.

Itô formula establishes many fundamental results in stochastic calculus. The product rule is one of the implications of the Itô formula. If X_t and Y_t are semi-martingales, it holds [16,24]

F. Yılmaz (✉)
Gazi University, Ankara, Turkey
e-mail: yfikriye@gmail.com

H. Öz • G.-W. Weber
Institute of Applied Mathematics, Middle East Technical University, Ankara, Turkey
e-mail: haceroz.oz@gmail.com; gweber@metu.edu.tr

A.A. Pinto and D. Zilberman (eds.), *Modeling, Dynamics, Optimization and Bioeconomics I*, Springer Proceedings in Mathematics & Statistics 73, DOI 10.1007/978-3-319-04849-9_42,
© Springer International Publishing Switzerland 2014

$$d(X_t, Y_t) = X_t dY_t + Y_t dX_t + d[X, Y]_t, \tag{42.1}$$

where $d[X, Y]_t$ denotes the *quadratic covariation* of X_t and Y_t.

The change of time method of the stochastic differential equations (SDEs) is employed in one of the areas where the product rule is naturally applied. The following stochastic differential equation

$$dX_t = \alpha(t, X_t) dW_t \tag{42.2}$$

can be solved by the random time change. This problem was studied by Ikeda and Watanabe [26]. The more general SDE

$$dX_t = \beta(t, X_t) dt + \alpha(t, X_t) dW_t \tag{42.3}$$

can be obtained in the form of (42.2) by Girsanov transformation.

The time change is based on the idea of changing time from t to a nonnegative process with nondecreasing sample paths. Given the stochastic process $X = (X_t)_{t \geq 0}$ on the filtered probability space $(\Omega, \mathscr{F}, \mathbb{P})$ with $\mathscr{F} = (\mathscr{F}_t)_{t \geq 0}$, to construct the time change, a process $\phi = (\phi_t)_{t \geq 0}$ is defined such that

- $\phi_0 = 0$,
- ϕ is continuous and strictly increasing,
- $\phi_t \to \infty$ as $t \to \infty$.

The change of time method is applied in many financial mathematical problems. There have been a great deal of utilizations in the stochastic calculus of financial applications recently [2, 14, 17, 22, 29].

Johnson and Shanno [15] studied pricing of options using time changed *stochastic volatility model (SVM)* [9, 11, 13]. German and Co-workers [9] used subordinated process to construct SVM for Lévy process. One of the important mathematical finance problems is represented by *Heston model* (1993):

$$\begin{cases} dS_{\S_t} = r_t dt + \sigma_t dW_t^1, \\ d\sigma_t^2 = \kappa(\theta^2 - \sigma_t^2) dt + \gamma \sigma_t dW_t^2. \end{cases} \tag{42.4}$$

The second equation of the system is well-known *Cox, Ingersoll and Ross process* [10]. The change of time method can be applied to such an equation [30] resulting in the pricing of variance swaps and volatility swaps.

There has been a great interest in the simulation methods of SDEs [18, 21, 23, 25, 28]. Stochastic Taylor expansion provides a source for the discrete-time approximation methods. One of the simplest ways to discretize the process is *Euler* method, which approximates the integrals by using the left-point rule. The *Milstein* scheme (1974), which has the order 1.0 of strong convergence, is stronger than *Euler* method. By adding more stochastic integrals using the stochastic Taylor expansion, more accurate schemes can be obtained.

Kloeden and Platen [18] have given a methodical means of deriving the Taylor series for both Stratonovich and Itô form of a SDE. By recursively using of Itô formula, obtained Taylor series can be related to a tree theory. The tree expansion is given for the true solution in [4]. Runge-Kutta method has been constructed, which has the order 1.5, in [3–5].

In the deterministic case, the accuracy of the numerical scheme can be obtained by comparing the obtained result with the exact solution. In the stochastic differential equations, there are two ways to measure the accuracy of the solution: strong and the weak convergence. A time discrete approximation X^h is said to *converge strongly* with order $p > 0$ at time T if there is a positive constant c, independent of h, and $h_0 > 0$ such that

$$E(|X_T - X^h(T)|) \le ch^p, \quad h \in (0, h_0).$$

On the other hand, X_h is said to *converge weakly* with order $p > 0$ at time T if there is a positive constant c, independent of h, and $h_0 > 0$ such that

$$|E(X_T) - E(X^h(T))| \le ch^p, \quad h \in (0, h_0).$$

As a second part of this chapter, an explicit expansion of the stochastic Taylor expansion is given. The expectation of the multiple Itô integrals is calculated briefly.

Through our research and computational-operational agenda of *digitalization*, *algebraization* and *automization*, we benefit from the close interplay between (financial) states (processes), time and Brownian motions and, in future, Lévy processes, via scientific methods of transformation, scaling, "change" and refined enumeration, to more elegantly work out intrinsic structures and to more efficiently solve challenging and coupled problems of finance. When, in daily life, we sometimes say *"Time will tell"*, by our study, we give a new and enriched interpretation to it. What is more, the "elastic" and "flexible" use of time and in all the coordinates does also allow a more focussed study around of times of "knockouts" and various kinds of crash, crisis and catastrophe.

42.2 Change of Time Method

Random time change is one of the probabilistic methods to solve SDEs. It is central to the work of Doeblin [12]. Dambis and Dubins-Schwarz have developed a theory of random time changes for semimartingales in the 1960s [16, 27]. In [1, 26, 30], the class of time changes are formulated.

42.2.1 Change of Time for Martingales

Theorem 42.1 (Dambis, Dubins-Schwartz Theorem, [12, 16, 27]). *Let* $M_t \in M_2^{c,loc}$ *be a continuous local* $(\mathscr{F}_t)$*–martingale with existing* $[M]_t$*, being the* quadratic variation of M_t, such that $\lim_{t\to\infty}[M]_t = \infty$ a.s.. Then, if we set $\tau_t := \inf\{u : [M]_u > t\}$ and $\tilde{\mathscr{F}}_t := \mathscr{F}_{\tau_t}$, the time change process $W_t := M_{\tau_t}$ is an $(\mathscr{F}_{\tau_t})$-Brownian motion, and

$$M_t = W_{[M]_t}.$$

We note that the local martingale M_t can be expressed by W_t and $(\tilde{\mathscr{F}}_t)$-stopping time since $\{[M]_t \le u\} = \{\tau_u \ge t\}$.

42.2.2 Change of Time for Itô Integral

Definition 42.1. Let $(\Omega, \mathscr{F}, \mathbb{P})$ be a probability space with a filtration $\mathscr{F} = (\mathscr{F}_t)_{t\ge 0}$ and let I be the class of functions

$$\phi : [0, \infty) \longrightarrow [0, \infty),$$

$$t \mapsto \phi_t,$$

which satisfies

- $\phi_0 = 0$,
- ϕ is continuous and strictly increasing,
- $\phi_t \to \infty$ as $t \to \infty$.

Clearly, if ϕ^{-1} is the inverse function of $\phi \in I$ then $\phi^{-1} \in I$. Each $\phi \in I$ defines a transformation T^ϕ of $C := C([0, \infty))$ (continuous functions w defined on $[0, \infty)$ with values in $\mathbb{R}$) into itself by

$$T^\phi : C \longrightarrow C,$$

$$w \mapsto (T^\phi w),$$

where

$$(T^\phi w)(t) := w(\phi_t^{-1}) \quad (t \in [0, \infty)).$$

Here, T^ϕ is called the *time change* defined by $\phi \in I$ and $\phi = \phi_t(\omega)$ is called a *process of the time change* for $\omega \in \Omega$. It is clear that $\phi = \phi_t(\omega) \in \Omega$ is an $(\mathscr{F}_t)$-adapted increasing process, so that the inverse function ϕ_t^{-1} of ϕ_t is an $(\mathscr{F}_t)$-stopping time for each fixed $t \in [0, \infty)$. We note that often in literature, ω is in the role of a continuous w, indeed.

Let $\tilde{M}_t := \int_0^t \sigma(s)dW(s)$ be an Itô integral with

$$\lim_{t\to\infty}[\tilde{M}]_t = \int_0^t \sigma^2(s)ds = +\infty \quad \text{and} \quad \phi_t := \inf\{u : [\tilde{M}]_u > t\}.$$

Then, $B_t = \tilde{M}_{\phi_t}$ is a Brownian motion. Here, the change of time is

$$\phi_t^{-1} = [\tilde{M}]_t = (\int_0^t \sigma^2(s)ds).$$

Thus, an SDE in $\mathbb{R}^1$ of type

$$X_t = X_0 + \int_0^t \sigma(s, X_s)dW_s + \int_0^t b(s, X_s)ds$$

can be rephrased in the form

$$X_t - X_0 = B_{[\tilde{M}]_t} + \int_0^t b(s, X_s)ds$$

$$= B_{\int_0^t \sigma^2(s,X_s)ds} + \int_0^t b(s, X_s)ds,$$

with a Brownian motion $(B_t)_{t\geq 0}$. Then, one-dimensional Itô-Doeblin formula takes the form

$$f(t, X_t) - f(0, X_0) = B_{\int_0^t \sigma^2(s,X_s)f'(s,X_s)^2 ds}$$

$$+ \int_0^t (\frac{\partial f}{\partial s} + \mathscr{L}f)(s, X_s)ds,$$

with $\mathscr{L}f := \frac{1}{2}\sigma^2 f'' + bf'$. Now, we verify this result.

If $X_t = X_0 + \int_0^t \sigma(s, X_s)dW_s + \int_0^t b(s, X_s)ds$, then by applying Itô formula we get

$$f(t, X_t) - f(0, X_0) = \int_0^t \frac{\partial f}{\partial s}ds + \int_0^t (\frac{\partial f}{\partial x}(b(s, X_s)ds + \sigma(s, X_s)dW_s)$$

$$+ \frac{1}{2}\int_0^t (\frac{\partial^2 f}{\partial x^2}\sigma^2(s, X_s)ds$$

$$= \int_0^t \frac{\partial f}{\partial s}ds + \frac{1}{2}\int_0^t \frac{\partial^2 f}{\partial x^2}\sigma^2(s, X_s)ds + \int_0^t \frac{\partial f}{\partial x}b(s, X_s)ds$$

$$+ \int_0^t \frac{\partial f}{\partial x}(\sigma(s, X_s)dW_s$$

$$= B_{\int_0^t \sigma^2(s,X_s)f'(s,X_s)^2 ds} + \int_0^t (\frac{\partial f}{\partial s} + \mathscr{L}f)(s, X_s)ds.$$

42.2.3 Change of Time for SDEs

We consider the SDE given in the following form (without drift)

$$dX_t = \alpha(t, X_t)dW_t, \tag{42.5}$$

where $(W_t)_{t \geq 0}$ is a Brownian motion and $\alpha(t, X)$ is a continuous and measurable function and $(X_t)_{t \geq 0}$ is a continuous process on $[0, \infty)$. If we can solve Eq. (42.5), then we can also resolve the equations having drift term $\beta(t, X_t)dt$ by the method of transformation of drift or Girsanov transformation. The following theorem provides us to solve Eq. (42.5).

Theorem 42.2. [26] *Let* $\tilde{W} = (\tilde{W}_t)_{t \geq 0}$ *be a one-dimensional* $((\mathscr{F}_t)_{t \geq 0})$*-Brownian motion with* $\tilde{W}_0 = 0$ *for a given probability space* $(\Omega, \mathscr{F}, \mathbb{P})$ *with a filtration* $(\mathscr{F}_t)_{t \geq 0}$, *and let* X_0 *be an* $(\mathscr{F}_0)$*-adapted random variable. We define a continuous process* $V := (V_t)_{t \geq 0}$ *by*

$$V_t = X_0 + \tilde{W}_t.$$

Let ϕ_t *be the change of time process such that*

$$\phi_t = \int_0^t \alpha^{-2}(\phi_s, X_0 + \tilde{W}_s)ds.$$

If $X_t := V_{\phi_t^{-1}} = X_0 + \tilde{W}_{\phi_t^{-1}}$ *and* $\tilde{\mathscr{F}}_t := \mathscr{F}_{\phi_t^{-1}}$, *then there exists an* $(\tilde{\mathscr{F}}_t)$*-adapted Brownian motion* $W = (W_t)_{t \geq 0}$ *such that* (X, W) *is a solution of Eq. (42.5) on the probability space* $(\Omega, \mathscr{F}, \mathbb{P})$

Remark. The converse of this theorem also holds.

Proof. By definition of time change,
$M_t := \tilde{W}_{\phi_t^{-1}}$ is a martingale with quadratic variation $[M]_t = \phi_t^{-1}$, where

$$\phi_t = \int_0^t \alpha^{-2}(\phi_s, V_s)ds \Rightarrow d\phi_t = \alpha^{-2}(\phi_t, V_t)dt$$

$$\Rightarrow dt = \alpha^2(\phi_t, V_t)d\phi_t$$

$$\Rightarrow \int_0^t ds = \int_0^t \alpha^2(\phi_s, V_s)d\phi_s$$

$$\Rightarrow t = \int_0^t \alpha^2(\phi_s, V_s)d\phi_s$$

$$\Rightarrow \phi_t^{-1} = \int_0^{\phi_t^{-1}} \alpha^2(\phi_s, V_s)d\phi_s$$

$$\Rightarrow \phi_t^{-1} = \int_0^t \alpha^2(s, V_s)ds.$$

Hence, ϕ_t^{-1} satisfies the equation

$$\phi_t^{-1} = \int_0^t \alpha^2(s, X_0 + \tilde{W}_{\phi_s^{-1}})ds.$$

We set $W_t = \int_0^t \alpha^{-1}(s, X_s)dM_s$. Then,

$$[M]_t = \int_0^t \alpha^{-2}(s, X_s)d[W]_s$$

$$= \int_0^t \alpha^{-2}(s, X_s)\alpha^2(s, X_s)ds.$$

$$= \int_0^t ds = t.$$

This result implies that W_t is an $(\mathscr{F}_t)$-Brownian motion. Since $M_t := \tilde{W}_{\phi_t^{-1}} = X_t - X_0 = \int_0^t \alpha(s, X_s)dW_s$, (X, W) is a solution of Eq. (42.5). □

42.2.4   Application of Time Change

42.2.4.1   Heston Model

Let $(\Omega, \mathscr{F}, \mathbb{P})$ be a probability space. We assume that underlying asset S_t in the risk neutral world and the variance follow the model (42.4):

$$\begin{cases} dS_{\S_t} = r_t dt + \sigma_t d W_t^1, \\ d\sigma_t^2 = k(\theta^2 - \sigma_t^2)dt + \gamma\sigma_t d W_t^2, \end{cases}$$

where r_t is the deterministic interest rate, σ_0 and θ are short and long volatility, respectively. Furthermore, k is a reversion speed, $\gamma > 0$ is a volatility parameter, W_t^1 and W_t^2 are independent Brownian motions. The Heston asset process has a variance σ_t^2 that follows *Cox, Ingersoll and Ross process* defined in the second equation.

Lemma 42.1 ([30]). *A solution of the following SDE*

$$d\sigma_t^2 = k(\theta^2 - \sigma_t^2)dt + \gamma\sigma_t dW_t^2$$

has the following look:

$$\sigma_t^2 = e^{-kt}\left(\sigma_0^2 - \theta^2 + \tilde{W}^2(\phi_t^{-1})\right) + \theta^2,$$

where $\tilde{W}^2(\phi_t^{-1})$ is an $\mathscr{F}_{\phi_t^{-1}}$-measurable one-dimensional Brownian motion. Here, ϕ_t^{-1} is the inverse of ϕ_t defined as:

$$\phi_t = \gamma^{-2} \int_0^t \left\{ e^{ks}\left(\sigma_0^2 - \theta^2 + \tilde{W}^2(\phi_t^{-1})\right) + \theta^2 e^{2k\phi_s} \right\}^{-1} ds.$$

Proof. We define the following process:

$$V(t) = e^{kt}\left(\sigma_t^2 - \theta^2\right).$$

Then, using the Itô product rule

$$\begin{aligned}
dV(t) &= ke^{kt}\left(\sigma_t^2 - \theta^2\right)dt + e^{kt}d\sigma_t^2 \\
&= ke^{kt}\left(\sigma_t^2 - \theta^2\right)dt + e^{kt}\left(k(\theta^2 - \sigma_t^2) + \gamma\sigma_t dW_t^2\right) \\
&= \gamma e^{kt}\sigma_t dW_t^2 \\
&= \gamma e^{kt}\sqrt{e^{-kt}V_t + \theta^2}dW_t^2.
\end{aligned}$$

Using the change of time method to the general equation, we get

$$dX_t = \alpha(t, X_t)dW_t^2,$$

where $\alpha(t, X_t) = e^{kt}\sqrt{e^{-kt}V_t + \theta^2}$. Since $X(t) = V(t)$, $X(0) = \sigma_0^2 - \theta^2$, $V_t = \sigma_0^2 - \theta^2 + \tilde{W}^2(\phi_t^{-1})$. Then,

$$\begin{aligned}
e^{kt}\left(\sigma_t^2 - \theta^2\right) &= \sigma_0^2 - \theta^2 + \tilde{W}^2(\phi_t^{-1} \\
&\Rightarrow \sigma_t^2 = e^{-kt}\left(\sigma_0^2 - \theta^2 + \tilde{W}^2(\phi_t^{-1})\right) + \theta^2.
\end{aligned}$$

Note that $\tilde{W}^2_{\phi_t^{-1}}$ is an $\mathscr{F}_{\phi_t^{-1}}$-measurable one-dimensional Brownian motion and ϕ_t^{-1} is the inverse of ϕ_t:

$$\phi_t = \gamma^{-2} \int_0^t \left\{ e^{ks}\left(\sigma_0^2 - \theta^2 + \tilde{W}^2(\phi_t^{-1})\right) + \theta^2 e^{2k\phi_s} \right\}^{-1} ds.$$

$\square$

42.2.4.2 Variance and Volatility Swap

A *variance swap* is a forward contract on annualized variance, the square of realized volatility. Its payoff at expiration is given by [30]

$$N(\sigma_R^2(S) - K_{\text{var}}),$$

where $V = \sigma_R^2(S)$ is the *realized stock variance* over the life of the contract,

$$\sigma_R^2(S) := \frac{1}{T} \int_0^T \sigma_s^2 ds.$$

Moreover, the price of a forward contract P on the future realized variance with strike price K_{var} is the expected present value of the future payoff in the risk-neutral world:

$$P_{var} = E(e^{-rT}(\sigma_R^2(S) - K_{var})).$$

We remark that

$$E(\sigma_t^2) = E(e^{-kt}(\sigma_0^2 - \theta^2 + \tilde{W}^2(\phi_t^{-1})) + \theta^2)$$
$$= e^{-kt}(\sigma_0^2 - \theta^2) + e^{-kt}E(\tilde{W}^2(\phi_t^{-1})) + \theta^2$$
$$= e^{-kt}(\sigma_0^2 - \theta^2) + \theta^2.$$

In fact,

$$E(V) = \frac{1}{T} \int_0^T E(\sigma_s^2)ds$$
$$= \frac{1}{T} \int_0^t (e^{-ks}(\sigma_0^2 - \theta^2) + \theta^2)ds$$
$$= \frac{1}{T} \left(\frac{e^{-ks}}{k}(\sigma_0^2 - \theta^2) + \theta^2 s \right)\Big|_0^T$$
$$= \frac{1}{T} \left(\frac{e^{-kT}}{-k}(\sigma_0^2 - \theta^2) + \theta^2 T + \frac{1}{k}(\sigma_0^2 - \theta^2) \right)$$
$$= \frac{1 - e^{kT}}{kT}(\sigma_0^2 - \theta^2) + \theta^2,$$

so that $P_{var} = e^{-rT}\left(\frac{1-e^{kT}}{kT}(\sigma_0^2 - \theta^2) + \theta^2 - K_{var} \right)$.

In a similar way, a *volatility swap* can be studied [30]; its value (price) can be represented by

$$P_{vol} = e^{-rT}\left((\frac{1 - e^{kT}}{kT}(\sigma_0^2 - \theta^2) + \theta^2)^{1/2} \right.$$

$$- \frac{\gamma^2 e^{-2kT}}{2k^3 T^2}\left[(2e^{2kT} - 4e^{kT}kT - 2)(\sigma_0^2 - \theta^2) \right.$$

$$\left. + (2e^{2kT} - 3e^{2kT} + 4e^{kT} - 1)\theta^2 \right] / \left[8\left(\frac{1 - e^{-kT}}{kT}(\sigma_0^2 - \theta^2) + \theta^2 \right)^{3/2} \right] \left. - K_{vol} \right).$$

42.3   Stochastic Taylor Expansion

We consider Eq. (42.3) with just the notations α and β exchanged, which can be written in the integral form as [18]

$$X(t) = X(0) + \int_0^t \alpha(X(s))ds + \int_0^t \beta(X(s))dW_s. \qquad (42.6)$$

Itô formula implies that

$$f(X(t)) = f(X(0)) + \int_0^t L^0(X(s))ds + \int_0^t L^1(X(s))dW_s,$$

where $\mathscr{L}^0 = \alpha\frac{\partial}{\partial x} + \frac{1}{2}\beta^2\frac{\partial^2}{\partial x^2}$ and $\mathscr{L}^1 = \beta\frac{\partial}{\partial x}$.

Applying Itô formula for $f(X) = X$ gives Eq. (42.6). We use Itô formula for the terms in the integrals. We apply it for $f(X) = \alpha(X)$ and $f(X) = \beta(X)$, respectively. Then, we obtain

$$\alpha(X(t)) = \alpha(X(0)) + \int_0^t \mathscr{L}^0\alpha(X(s))ds + \int_0^t \mathscr{L}^1\alpha(X(s))dW_s, \quad (42.7)$$

$$\beta(X(t)) = \beta(X(0)) + \int_0^t \mathscr{L}^0\beta(X(s))ds + \int_0^t \mathscr{L}^1\beta(X(s))dW_s. \quad (42.8)$$

Substituting (42.7) and (42.8) in (42.6) implies that

$$X(t) = X(0) + \int_0^t \left[\alpha(X(0)) + \int_0^s \mathscr{L}^0\alpha(X(\tau))d\tau + \int_0^s \mathscr{L}^1\alpha(X(\tau))dW_\tau\right]ds$$

$$+ \int_0^t \left[\beta(X(0)) + \int_0^s \mathscr{L}^0\beta(X(\tau))d\tau + \int_0^s \mathscr{L}^1\beta(X(\tau))dW_\tau\right]dW_s.$$

Similarly we will apply Itô formula for $f(X) = \mathscr{L}^0\alpha(X)$, $f(X) = \mathscr{L}^1\alpha(X)$, $f(X) = \mathscr{L}^0\beta(X)$ and $f(X) = \mathscr{L}^1\beta(X)$. Then,

$$X(t) = X(0) + \alpha(X(0))\int_0^t ds + \beta(X(0))\int_0^t dW_s + \mathscr{L}^0\alpha(X(0))\int_0^t\int_0^s d\tau ds$$

$$+ \mathscr{L}^1\alpha(X(0))\int_0^t\int_0^s dW_\tau ds + \mathscr{L}^0\beta(X(0))\int_0^t\int_0^s d\tau dW_s$$

$$+ \mathscr{L}^1\beta(X(0))\int_0^t\int_0^s dW_\tau dW_s$$

$$+ \int_0^t\int_0^s \left[\int_0^z \mathscr{L}^0\mathscr{L}^0\alpha(X(z))dz + \int_0^z \mathscr{L}^1\mathscr{L}^0\alpha(X(z))dW_z\right]d\tau ds$$

$$
+ \int_0^t \int_0^s \left[\int_0^z \mathscr{L}^0 \mathscr{L}^1 \alpha(X(z)) dz + \int_0^z \mathscr{L}^1 \mathscr{L}^1 \alpha(X(z)) dW_z \right] dW_\tau ds
$$

$$
+ \int_0^t \int_0^s \left[\int_0^z \mathscr{L}^0 \mathscr{L}^0 \beta(X(z)) dz + \int_0^z \mathscr{L}^1 \mathscr{L}^0 \beta(X(z)) dW_z \right] d\tau dW_s
$$

$$
+ \int_0^t \int_0^s \left[\int_0^z \mathscr{L}^0 \mathscr{L}^1 \beta(X(z)) dz + \int_0^z \mathscr{L}^1 \mathscr{L}^1 \beta(X(z)) dW_z \right] dW_\tau ds.
$$

We now consider $\mathscr{L}^0 \mathscr{L}^0 \alpha(X)$, $\mathscr{L}^1 \mathscr{L}^0 \alpha(X)$, $\mathscr{L}^0 \mathscr{L}^1 \alpha(X)$, $\mathscr{L}^1 \mathscr{L}^1 \alpha(X)$, $\mathscr{L}^0 \mathscr{L}^0 \beta(X)$, $\mathscr{L}^1 \mathscr{L}^0 \beta(X)$, $\mathscr{L}^0 \mathscr{L}^1 \beta(X)$ and $\mathscr{L}^1 \mathscr{L}^1 \beta(X)$, and we apply Itô formula again:

$$
\begin{aligned}
X(t) = {} & X(0) + \alpha(X(0)) I_0 + \beta(X(0)) I_1 + \mathscr{L}^0 \alpha(X(0)) I_{00} + \mathscr{L}^1 \alpha(X(0)) I_{10} \\
& + \mathscr{L}^0 \beta(X(0)) I_{01} + \mathscr{L}^1 \beta(X(0)) I_{11} + \mathscr{L}^0 \mathscr{L}^0 \alpha(X(0)) I_{000} \\
& + \mathscr{L}^1 \mathscr{L}^0 \alpha(X(0)) I_{100} + \mathscr{L}^0 \mathscr{L}^1 \alpha(X(0)) I_{010} + \mathscr{L}^1 \mathscr{L}^1 \alpha(X(0)) I_{110} \\
& + \mathscr{L}^0 \mathscr{L}^0 \beta(X(0)) I_{001} + \mathscr{L}^1 \mathscr{L}^0 \beta(X(0)) I_{101} + \mathscr{L}^0 \mathscr{L}^1 \beta(X(0)) I_{011} \\
& + \mathscr{L}^1 \mathscr{L}^1 \beta(X(0)) I_{111} + R,
\end{aligned}
$$

where R denotes the remainder term and $I_{i_1, i_2, \ldots, i_k}$ represent the Itô integral, where integration is with respect to ds if $i_k = 0$, or dW_s if $i_k = 1$. For example,

$$
I_{010} = \int_0^t \int_0^{s_1} \int_0^{s_2} ds_3 dW_{s_2} ds_1.
$$

We note that there is a rooted tree expansion in terms of elementary differentials for the Itô-Taylor series expansions of the solution of Eq. (42.3). The rooted tree theory was introduced by Butcher [8] for the deterministic expansions. In [8], an elementary differential tree expansions are covered. Since the integration is with respect to ds and $dW_k(s)$ for $k = 1, 2, \ldots, d$, where d denotes the number of iterated integrals, each node of a tree can be colored with any one of $d + 1$ colorings. A node colored with the 0 label corresponds to an integration with ds and will be called a *deterministic node*. Nodes with colored with any other label from 1 to d will be called *stochastic*. In [3–5, 7, 20], a detailed coverage of rooted-tree theory and its application with *Runge-Kutta* type methods can be found.

42.3.1 Expectations in SDEs

In order to compute the order of a numerical scheme, we have to deal with expectations [26]. It is needed to calculate the expected values of the product of iterated Itô integrals.

42.3.2 Itô Product Rule

Let X and Y be two semi-martingales starting at 0. *Itô's Product Rule* is the analogue of the *Leibnitz Product Rule* for standard calculus:

$$X_t Y_t = \int_0^t X_s dY_s + \int_0^t Y_s dX_s + [X, Y]_t,$$

where $[X, Y]_t$ denotes the *quadratic covariation* of X_t and Y_t. If X_t and Y_t satisfy the system

$$\begin{cases} dX_t = \mu_x(X_t, Y_t, t)X_t dt + X_t \sigma_x(X_t, Y_t, t)dW_t, \\ dY_t = \mu_y(X_t, Y_t, t)X_t dt + Y_t \sigma_y(X_t, Y_t, t)dW_t, \end{cases}$$

then,

$$d[X, Y]_t = Cov[dX_t, dY_t | \mathscr{F}_t] = \sigma_x \sigma_y X_t Y_t dt.$$

42.3.3 Itô Integration by Parts

In a similar manner, the product of stochastic integrals can be considered. Let I_a and I_b be two stochastic integrals. Then,

$$I_a I_b = \int_0^t I_a d(I_b)_s + \int_0^t I_b d(I_a)_s + [I_a, I_b]_t,$$

with $a = (a_1, a_2, \ldots, a_k)$ and $b = (b_1, b_2, \ldots, b_l)$; moreover,

$$d(I_a) = \int_0^t \int_0^{s_{k-1}} \ldots \int_0^{s_2} dW_{a_1}(s_1)dW_{a_2}(s_2) \ldots dW_{a_{k-1}}(s_{k-1}) := I_{a^-}.$$

Let $l := l(a)$ denotes the length of a. For instance, $l((0, 1, 1)) = 3$. We write a^- for the multi-index obtained from deleting the last component of a. The multiple Itô integral I_a is defined as

$$I_a = \begin{cases} 1 & \text{if } k = 0, \\ \int_0^t I_{a^-} ds & \text{if } k \geq 1 \text{ and } a_k = 0, \\ \int_0^t I_{a^-} dW_s^{a_k} & \text{if } k \geq 1 \text{ and } a_k \geq 1. \end{cases}$$

For instance,

$$I_{101} = \int_o^t \int_0^{s_3} \int_0^{s_2} dW_{s_1}^1 ds_2 dW_{s_3}^1.$$

Thus,

$$I_a I_b = \int_0^t I_a I_{b-} dW_s^{b_k} + \int_0^t I_b I_{a-} dW_s^{a_k} + [I_a, I_b]_t.$$

Now, we will verify that $[I_a, I_b]_t = \int_0^t I_{a-} I_{b-} \mathrm{I}_{\{a_k = b_l \neq 0\}} ds$.
Where $\mathrm{I}_{\{a_k = b_k \neq 0\}}$ is the indicator function with the given condition.
Since $[\int H_- dX, Y] = \int H_- d[X, Y]$ [14, 19], therefore

$$[I_a, I_b]_t = \left[\int_0^t I_{a-} dW_s^{a_k}, \int_0^t I_{b-} dW_s^{b_k} \right]$$

$$= \int_0^t I_{a-} d\left[dW_s^{a_k}, \int_0^t I_{b-} dW_s^{b_k} \right]$$

$$= \int_0^t I_{a-} I_{b-} d\left[dW_s^{a_k}, dW_s^{b_k} \right]$$

$$= \int_0^t I_{a-} I_{b-} \mathrm{I}_{\{a_k = b_l \neq 0\}} ds,$$

Thus, we have the following formula [6]:

$$I_a I_b = \int_0^t I_a I_{b-} dW_s^{b_k} + \int_0^t I_b I_{a-} dW_s^{a_k} + \int_0^t I_{a-} I_{b-} \mathrm{I}_{\{a_k = b_l \neq 0\}} ds.$$

As an example we consider $a = (0, 1)$ and $b = (1, 0)$. Here, since $a_2 \neq b_2$ the indicator function becomes 0. Then,

$$I_{01} I_{10} = \int_0^t I_{01} I_1 dW_0 + \int_0^t I_0 I_{10} dW_1 + \int_0^t I_1 I_0 (\mathrm{I} - \text{indicator}) ds$$

$$= \int_0^t \left\{ \int_0^t I_{01} dW_1 + \int_0^t I_0 I_1 dW_1 \right\} dW_0 + \int_0^t I_{10} dW_1 + 0$$

$$= \int_0^t \left(I_{011} + I_{11} \right) dW_0 + I_{101}$$

$$+ I_{0110} + I_{110} + I_{101}.$$

Kloeden and Platen [18] give the following formula for the computation of expected values of multiple stochastic integrals:

$$E(I_a I_b) = \begin{cases} 0, & \text{if } a^+ \neq b^+, \\ \frac{h^w(a,b)}{w(a,b)!} \prod_{i=0}^{l(a^+)} \frac{(k_i(a) + k_i(b))!}{k_i(a)! k_i(b)!}, & \text{otherwise.} \end{cases} \tag{42.9}$$

As an example, we consider $E(I_{01}I_{10})$. Now, $a = (0,1)$ and $b = (1,0)$. Then, $a^+ = 1$, $b^+ = 1$, $l(a) = 2$, $l(b) = 2$ and $l(a^+) = 1$, $l(b^+) = 1$. Let $k(\cdot)$ denotes the number of 0's before 1's. Then, consecutively, $k_0(a) = 1$, $k_1(a) = 0$, $k_0(b) = 0$ and $k_1(b) = 1$. We compute $w(a,b)$, given in the above formula, by

$$w(a,b) = l(a^+) + \sum_{i=0}^{l(a^+)} (k_i(a) + k_i(b))$$

$$= 1 + (1 + 0 + 0 + 1) = 3.$$

Thus, $E(I_{01}I_{10}) = \frac{h^3}{3!}$.

42.4   Conclusion

In this chapter, we tried to display a part of the inner beauty of stochastic dynamics and, herewith, of various kinds of optimization and optimal control problems subject to those dynamical constraints. Here, this beauty expresses itself in terms of *digitalization*, *algebraization* and *automization* which are not only very aesthetic indeed, but also very practical.

Future research in this field can fruitfully address various extensions and many real-world applications in the areas of finance, but also actuarial sciences, microbiology, communication and social sciences, and the wide field of modern economics.

Moreover, *digitalization* of expectations can be considered for the order issues of higher-order weak schemes. As a mathematical application, control problems of SDEs may be worked out, further along the lines of this chapter.

Acknowledgements The authors of this chapter would like to express their gratitude to the editors of that book, Prof. Dr. D. Zilberman, Prof. Dr. A. Pinto, for giving them the opportunity to introduce their representations and reflections to the readers.

References

1. Andreas, E.: Stochastic Analysis. Lecture Notes, University of Bonn (2012). http://wt.iam.uni-bonn.de/fileadmin/WT/Inhalt/people/Andreas_Eberle/StoAn1213/StochasticAnalysisNotes1213.pdf
2. Applebaum, D.: Lévy Processes and Stochastic Calculus. Cambridge University Press, Cambridge (2004)
3. Burrage, P.M.: Runge-Kutta methods for stochastic differential equations. Ph.D. thesis, Department of Mathematics, University of Queensland, Australia (1999)
4. Burrage, K., Burrage, P.M.: High strong order methods for non-commutative stochastic ordinary differential equation system and the Magnus formula. Phys. D **133**, 34–48 (1999)

5. Burrage, K., Burrage, P.M.: Order conditions of stochastic Runge-Kutta methods by B-series. SIAM J. Numer. Anal. **38**, 1626–1646 (2000)
6. Burrage, K., Burrage, P.M.: Numerical Methods for Stochastic Differential Equations. Fields Institute, Toronto (2001)
7. Burrage, K., Burrage, P.M., Tian, T.H.: Numerical methods for solving stochastic differential equations on parallel computers. In: Proceedings of the 5th International Conference on High-Performance Computing in the Asia-Pacific Region (2001)
8. Butcher, J.C.: The Numerical Analysis of Ordinary Differential Equations. Wiley, Chichester (1987)
9. Carr, P., Geman, H., Madan, D.B., Yor, M.: Stochastic volatility for Lévy processes. Math. Finance **13**(3), 345–382 (2003)
10. Cox, J.C., Ingersoll, J.E., Ross, S.A.: An intertemporal general equilibrium model of asset prices. Econometrica **53**, 363–384 (1985)
11. Demeterfi, K., Derman, E., Kamal, M., Zou, J.: A Guide to volatility and variance swaps. J. Derivatives Summer **6**(4), 9–32 (1999)
12. Doeblin, W., Yor, M.: Sur l'équation de Kolmogoroff, vol. 331. Pli Cachetè à l'Académie des Sciences, Paris (2000)
13. Elliott, R., Swishchuk, A.: A pricing options and volatility swaps in Markov-modulated Brownian and fractional brownian markets. In: RJE Conference, Calgary (2005)
14. Jacod, J.: Calcul Stochastique et Problemés de Martingales. Lecture Notes in Mathematics, vol. 714. Springer, Berlin (1979)
15. Johnson, H.E., Shanno, D.: Option pricing when the variance is changing. J. Financ. Quant. Anal. **22**, 143–151 (1987)
16. Karatzas, I., Shreve, S.E.: Brownian Motion and Stochastic Calculus. Springer, New York (2010)
17. Kilic, E., Karimov, A. Weber, G.-W.: Applications of stochastic hybrid systems in portfolio optimization. In: Thomaidis, N., Dash, G.H., Jr. (eds.) Recent Advances in Computational Finance, Chap. 5, pp. 75–98. Nova Science Publishers, Inc., New York (2013)
18. Kloeden, P.E., Platen, E.: Numerical Solution of Stochastic Differential Equations. Springer, Berlin (1992)
19. Kobayashi, K.: Stochastic calculus for a time-changed semimartingale and the associated stochastic differential equations. J. Theor. Probab. **24**(3), 789–820 (2010)
20. Komori, Y., Sugiura, H., Mitsui, T.: Rooted Tree Analysis of the Order Conditions of ROW-Type Schemes for Stochastic Differential Equations, Report No. 7, Mathematical Sciences, Nagoya University, Nagoya (1994)
21. Kunita, H.: Stochastic Flows and Stochastic Differential Equations. Cambridge Studies in Advanced Mathematics, vol 24. Cambridge University Press, Cambridge (1990)
22. Mele A.: Lectures on Financial Mathematics. Lecture Notes, University of Lugano, Swiss-Finance-Institute (2012). http://www.antoniomele.org/files/fin_eco.pdf
23. Milstein, G.N.: Numerical Integration of Stochastic Differential Equations. Kluwer, Dordrecht/Boston/London (1995)
24. Movellan, J.R.: Tutorial on Stochastic Differential Equations. MPLab Tutorials Version 06.1 (2011)
25. Platen, E.: An introduction to numerical methods for stochastic differential equations. Acta Numerica **8**, 197–246 (1999)
26. Platen, E., Ikeda, N., Watanabe, S.: Stochastic Differential Equations and Diffusion Processes, vol. 24. North Holland Publ. Co., Amsterdam, 480 S., Dfl. 175 (1982)
27. Revuz, D., Yor, M.: Continuous Martingale and Brownian Motion. Springer, Berlin (2005)
28. Rümelin, W.: Numerical treatment of stochastic differential equations. SIAM J. Numer. Anal. **19**, 604–613 (1982)
29. Stele, J.M.: Stochastic Calculus and Financial Applications. Application of Mathematics, vol. 45. Springer, New York (2001)
30. Swishchuk, A.: Change of Time Method: Application to Mathematical Finance I, Math & Comp. Finance Lab, Department of Mathematics and Statistic, U of C "Lunch at the Lab" Talk (2005)

Printed by Printforce, the Netherlands